Lecture Notes in Networks and Systems

Volume 289

The series "Lecture Notes in Networks and Systems" publishes the latest developments in Networks and Systems—quickly, informally and with high quality. Original research reported in proceedings and post-proceedings represents the core of LNNS.

Volumes published in LNNS embrace all aspects and subfields of, as well as new challenges in, Networks and Systems.

The series contains proceedings and edited volumes in systems and networks, spanning the areas of Cyber-Physical Systems, Autonomous Systems, Sensor Networks, Control Systems, Energy Systems, Automotive Systems, Biological Systems, Vehicular Networking and Connected Vehicles, Aerospace Systems, Automation, Manufacturing, Smart Grids, Nonlinear Systems, Power Systems, Robotics, Social Systems, Economic Systems and other. Of particular value to both the contributors and the readership are the short publication timeframe and the world-wide distribution and exposure which enable both a wide and rapid dissemination of research output.

The series covers the theory, applications, and perspectives on the state of the art and future developments relevant to systems and networks, decision making, control, complex processes and related areas, as embedded in the fields of interdisciplinary and applied sciences, engineering, computer science, physics, economics, social, and life sciences, as well as the paradigms and methodologies behind them.

Indexed by SCOPUS, INSPEC, WTI Frankfurt eG, zbMATH, SCImago.

All books published in the series are submitted for consideration in Web of Science.

For proposals from Asia please contact Aninda Bose (aninda.bose@springer.com).

More information about this series at https://link.springer.com/bookseries/15179

Petros Nicopolitidis · Sudip Misra ·
Laurence T. Yang · Bernard Zeigler · Zhaolng Ning
Editors

Advances in Computing, Informatics, Networking and Cybersecurity

A Book Honoring Professor Mohammad S. Obaidat's Significant Scientific Contributions

 Springer

Editors
Petros Nicopolitidis
Department of Informatics
Aristotle University of Thessaloniki
Thessaloniki, Greece

Laurence T. Yang
Department of Computer Science
St. Francis Xavier University
Antigonish, NS, Canada

Zhaolng Ning
Dalian University of Technology
Dalian, China

Sudip Misra
Department of Computer Science
and Engineering
Indian Institute of Technology
Kharagpur, West Bengal, India

Bernard Zeigler
Department of Electrical and Computer
Engineering
University of Arizona
Tucson, AZ, USA

ISSN 2367-3370   ISSN 2367-3389 (electronic)
Lecture Notes in Networks and Systems
ISBN 978-3-030-87048-5   ISBN 978-3-030-87049-2 (eBook)
https://doi.org/10.1007/978-3-030-87049-2

This Springer imprint is published by the registered company Springer Nature Switzerland AG
The registered company address is: Gewerbestrasse 11, 6330 Cham, Switzerland

Preface

Information and Communication Technologies (ICTs) play an integral role in today's society. Four major driving pillars in the field are computing, which nowadays enables data processing in unprecedented speeds, informatics, which derives information stemming for processed data to feed relevant applications, networking which interconnects the various computing infrastructures and cybersecurity for addressing the growing concern for secure, private and lawful use of the ICT infrastructure and services.

These fields have proven to be essential for any nation to have a place in today's digital and knowledge economy era. COVID-19 has proven that these technologies and systems will be needed for the coming several decades and more. They are the air that other systems and fields breathe.

This book is written to honor Prof. Mohammad Salameh Obaidat's worldwide abundant, lasting and significant recognized technical contributions to the areas of computing, informatics, networking and cybersecurity. He is not only a worldwide researcher, academic, teacher, scholar, engineer and scientist, but also an international technical leader who also has served the international professional community in various aspects such as serving as President and Chair of Board of Directors of the Society for Modeling and Simulation International, SCS, chaired numerous international conferences, founded or Co-founded five international conferences, and given numerous keynote speeches all over the world. He received numerous awards for his worldwide technical contributions and leadership services from IEEE, SCS and other professional entities.

The present book aims to survey as well as introduce recent worldwide research and development efforts in the above-mentioned hot areas, present related case studies together with trends and challenges. These areas—computing, informatics, networking and cybersecurity—have received numerous technical contributions from the renowned worldwide scholar, Prof. Mohammad S. Obaidat, to whom the book is dedicated.

The chapters of the book have been written by worldwide experts in the respective fields. Despite the fact that there are different authors of these chapters, we managed book materials to be coherent so as to be easy to follow by readers.

The content organization is based on the four field categories reflected in the book title. The first part of the book is related to the area of computing. Chapter "Workload Scheduling in Fog and Cloud Environments: Emerging Concepts and Research Directions" addresses fog and cloud environments. It provides the necessary background in this field and presents an overview of the emerging concepts and techniques as well as exploring future research directions.

Chapter "A Comprehensive Survey of Estimator Learning Automata and Their Recent Convergence Results" addresses estimator learning automata algorithms. For almost three decades, the reported proofs of pursuit algorithms possessed a flaw, which is referred to as the claim of the "monotonicity" property. The chapter first records all the reported estimator algorithms and then provides a comprehensive survey of the proofs from a different perspective.

Chapter "Multimodal Data Fusion" studies multimodal data fusion algorithms from the aspects of incomplete modal analysis fusion, incremental modal clustering fusion, heterogeneous modal migration fusion and low-dimensional modal sharing fusion. A series of multimodal data fusion models and algorithms are designed.

Chapter "Efficient Parallel Implementation of Cellular Automata and Stencil Computations in Current Processors" discusses optimization of cellular automata (CA) implementations in current high-performance computational systems, focusing on those especially aimed at current processors, including hardware alternatives like GPUs, BTBs, different forms of parallelism such as instruction-level parallelism, thread-level parallelism, data-level parallelism, as well as software approaches like "if-else" statement elimination, loop unrolling, data pipelining and blocking.

Chapter "Smart Healthcare: Rough Set Theory in Predicting Heart Disease" discusses the use of discrete event system specification (DEVS) as the basic modeling and simulation framework for model-based system engineering methodology that supports the critical stages in top-down design of complex networks. Focusing on design of communication networks for emergency response, it shows how such networks pose challenges to current technologies that current simulators cannot address.

The second part covers research contributions in informatics and emerging applications. Chapter "Healthcare Patient Flow Modeling and Analysis with Timed Petri Nets" discusses rough-set-based methodologies for developing a predictive model for the occurrence of heart disease. Classification accuracy and explanatory power of the rough-set-based methods are compared with these for other machine learning languages and non-rule-based methods.

Chapter "Enabling and Enforcing Social Distancing Measures at Smart Parking Infrastructures Using Blockchain Technology in COVID-19" continues on applications in health care and proposes a method for patient flow modeling and analysis with timed Petri nets. Details of patient flow performance analysis based on stochastic timed Petri nets, such as the average patient waiting time, resource constraints modeling, task durations modeling and patient flow hierarchical modeling that handles complexity, are discussed.

Chapter "Using DEVS for Full Life Cycle Model-Based System Engineering in Complex Network Design", motivated by the COVID-19 pandemic, proposes a method for enabling and enforcing social distancing measures at smart parking infrastructures using blockchain technology. It provides a practical model where blockchain technology is enforced to find solutions and to deal with the problems in maintaining social distancing at smart parking in pandemic outbreaks situations.

Chapter "Touchless Palmprint and Fingerprint Recognition" presents an overview of the various methods reported in the literature for touchless palmprint and fingerprint recognition; describing the corresponding acquisition methodologies and processing methods.

Chapter "A Survey of IoT Software Platforms" provides an in-depth survey of IoT software platforms. It proposes a broad evaluation scheme for IoT software platforms based on an analysis of the characteristics found in contemporary platforms that are surveyed.

The third part of the book is concerned with networking.

Chapter "FLER: Fuzzy Logic-Based Energy Efficient Routing for Wireless Sensor Networks" presents a fuzzy logic-based energy-efficient routing scheme for wireless sensor networks. The proposed approach chooses the optimal forwarder node, which is predicted with low energy consumption, high stability, low retransmission rate and low average transmission delay.

Chapter "Application of Device-to-Device Communication in Video Streaming for 5G Wireless Networks" presents one of the key-enabling technologies in 5G wireless networks, device-to-device (D2D) communication. It reviews work in the literature on D2D communication and its applications in video streaming and discusses an architecture that provides dynamic adaptive streaming over HTTP-based peer-to-peer (P2P) video streaming in cellular networks.

Chapter "5G Green Network" proposes a multi-user multi-antenna random cellular network model targeted to energy efficiency. A spectrum efficiency model and an energy efficiency model are presented based on the random cellular network model and the maximum achievable energy efficiency of the considered multi-user multi-antenna HCPP random cellular networks.

Chapter "Geocommunity Based Data Forwarding in Social Delay Tolerant Networks" investigates the problem of data forwarding in social-based delay-tolerant networks (DTNs). It also studies the existing location-based routing schemes and highlights the current challenges and future research direction for geocommunity-based routing in social delay-tolerant networks.

Chapter "Resource Allocation Challenges in the Cloud and Edge Continuum" considers bandwidth allocation and suggests modeling and architectural paradigms for integration of satellite communications and networking into the highly virtualized 5G wireless networks and the forthcoming 6G systems.

Chapter "Collaborative Caching Strategy in Content-Centric Networking" studies caching strategies in the content-centric networking (CCN). In order to efficiently and reasonably place contents in routers and reduce the transmission delay, it proposes a caching scheme based on on-path caching and off-path caching. In order to further reduce the energy consumption, it also proposes a caching scheme to minimize energy consumption.

Chapter "Blockchain-Based Software-Defined Vehicular Networks for Intelligent Transportation System Beyond 5G" discusses the integration of software-defined networking (SDN) and cloud computing over 5G mobile communication in vehicular networks. A programmable, efficient and controllable network architecture is introduced to achieve sustainable network development.

Chapter "Application Layer Protocols for Internet of Things" surveys application layer protocols for Internet of Things (IoT) covering also light-way protocols. A performance evaluation of the analyzed IoT application layer protocol is also provided.

Chapter "Resource Allocation in Satellite Networks—From Physical to Virtualized Network Functions" discusses resource allocation challenges in the cloud and edge continuum. It presents key computing technologies of the past and the present, along with related networking technologies. It continues by describing the important resource allocation challenges and concludes by formulating and evaluating a basic resource allocation problem for assigning application's workload in an edge–fog–cloud hierarchical infrastructure.

The fourth part of the book contains research contributions in cybersecurity.

Chapter "Secure D2D in 5G Cellular Networks: Architecture, Requirements and Solutions" reviews the security architecture, security requirements and existing solutions for the 5G D2D networks. It classifies various security challenges and requirements of a secure 5G D2D network and major research works according to their application scenarios. Finally, it discusses open related research issues.

Chapter "New Waves of Cyber Attacks in the Time of COVID19" surveys cyberattacks in relation to the COVID-19 pandemic crisis. The analysis covers different technical and socioeconomical aspects, and the needed countermeasures in response to such attacks are also evaluated.

Chapter "A Comparison of Performance of Rough Set Theory with Machine Learning Techniques in Detecting Phishing Attack" applies classical rough set analysis (CRSA) for detecting phishing attacks. It focuses primarily on comparing the classification performance of CRSA with the traditional machine learning (ML) tools.

Chapter "Towards Owner-Controlled Data Sharing" discusses some of the main problems related to allowing data owners to share their data with interested consumers in a controlled way in the context of digital data market. It illustrates some recent proposals for the specification and enforcement of the owners' requirements. It also reviews proposals addressing the issue of ensuring that data owners remain in control of their data.

Chapter "Emerging Role of Block Chain Technology in Maintaining the Privacy of Covid-19 Public Health Record is also influenced by the COVID-19 pandemic and in the context of blockchain technology. It introduces a related distributed protocol for handling the amount of potential data and the usage of distributed data provisions for maintaining the privacy of COVID-19 public health records in an effort to guarantee secure decentralized data record protection.

Chapter "Malware Forensics: Legacy Solutions, Recent Advances, and Future Challenges" covers the background of malware across different platforms, existing solutions and malware analysis techniques for malware prevention and detection, as well as recent advances in the malware domain using cutting-edge technologies and future directions.

Chapter "Security of Cyber-Physical-Social Systems: Impact of Simulation-Based Systems Engineering, Artificial Intelligence, Human Involvement, and Ethics" deals with cyber-physical-social systems. Sources of failures in cyber-physical-social systems and the fact that with increased connectedness, the systems become more vulnerable. Some possibilities to get prepared for unexpected conditions are discussed. The impact of human involvement is explained as sources of additional and severe problems.

Chapter "Machine Learning Methods for Enhanced Cyber Security Intrusion Detection System" discusses the origin and evolution of an intrusion detection system (IDS), followed by the classification of IDSs and the contribution of machine learning (ML) in the field. It briefs the prominent current works. An outline of the datasets frequently used for evaluation purposes is presented. Moreover, it describes the collaborative IDS that enhances the Big Data security and presents IDS research issues and challenges.

Chapter "5G Network Slicing Security" is concerned of 5G network slicing security. It introduces the technical background and architecture of 5G and explains the architecture of SDN/NFV and the threats to information security and privacy in 5G systems. It discusses network slicing security, including inter-slice and intra-slice security threats and also reviews security risks related to 5G network slicing and corresponding solutions identified by 3GP.

Chapter "A Security-Driven Scheduling Model for Delay-Sensitive Tasks in Fog Networks" proposes a security-driven scheduling model for delay-sensitive tasks in fog networks. The contribution of the proposed method is twofold: First is to provide the robust security service to the delay-sensitive tasks, and second is to enhance the performance of the system without violating the scheduling constraints of delay-sensitive tasks.

We are very thankful to all the authors of the chapters of this book, who have worked very hard to bring forward excellent-quality chapters to the aid of students, instructors, researchers and community practitioners. As the individual chapters of this book were written by different authors, the responsibility for the contents of each chapter lies with the respective authors.

We would also like to thank solicited chapter reviewers for their constructive suggestions. Many thanks to Springer Nature editors and editorial assistants for their outstanding work. We hope that this book will be a token of appreciation to the abundant, lasting and original numerous contributions of our distinguished world-wide colleague, Prof. Mohammad S. Obaidat, who has been a first-class dedicated scholar, academic, researcher and teacher. Finally, we hope that the book will be a valuable reference for students, instructors, researchers, scientists, engineers and developers.

Thessaloniki, Greece   Petros Nicopolitidis
Kharagpur, India   Sudip Misra
Antigonish, Canada   Laurence T. Yang
Tucson, USA   Bernard Zeigler
Dalian, China   Zhaolng Ning

About Professor Mohammad S. Obaidat

Prof. Mohammad S. Obaidat (Fellow of IEEE 2005, Life Fellow of IEEE 2018) is an internationally known academic/researcher/scientist/ scholar. He received his Ph.D. degree in computer engineering with a minor in computer science from The Ohio State University, Columbus, USA. He has received extensive research funding and published to date (2019) about one thousand (1000) refereed technical articles. About half of them are journal articles, over 95 books and over 70 book chapters. He is Editor-in-Chief of three scholarly journals and an editor of many other international journals. He is the founding Editor-in-Chief of *Wiley Security and Privacy Journal*. Moreover, he is founder or co-founder of five international conferences.

Among his previous positions are Advisor to the President of Philadelphia University for Research, Development and Information Technology; President and Chair of Board of Directors of the Society for Molding and Simulation International, SCS; Senior Vice President of SCS; Dean of the College of Engineering at Prince Sultan University; Chair and tenured Professor at the Department of Computer and Information Science and Director of the MS Graduate Program in Data Analytics at Fordham university; Chair and tenured Professor of the Department of Computer Science and Director of the Graduate Program at Monmouth University; Tenured Full Professor at King Abdullah II School of Information Technology, University of Jordan; The PR of China Ministry of Education Distinguished Overseas Professor at the University of Science and Technology Beijing, China; and an Honorary Distinguished Professor at the Amity University-A Global University. He is now the Founding Dean and Professor, College of Computing and Informatics at The University of Sharjah, UAE.

He has chaired numerous (over 175) international conferences and has given numerous (over 175) keynote speeches worldwide. He has served as ABET/CSAB evaluator and on IEEE CS Fellow Evaluation Committee. He has served as IEEE CS Distinguished Speaker/Lecturer and an ACM Distinguished Lecturer. Since 2004, he has been serving as an SCS Distinguished Lecturer. He received many best paper awards for his papers including ones from IEEE ICC, IEEE Globecom, AICSA, CITS, SPECTS and DCNET International conferences. He also received best paper awards from IEEE Systems Journal in 2018 and in 2019 (two best paper awards). In 2020, he received four best paper awards from IEEE Systems Journal. In 2021, he also received the IEEE Systems best paper award.

He also received many other worldwide awards for his technical contributions including: The 2018 IEEE ComSoc-Technical Committee on Communications Software Technical Achievement Award for contribution to Cybersecurity, Wireless Networks Computer Networks and Modeling and Simulation, SCS prestigious McLeod Founder's Award, Presidential Service Award, SCS Hall of Fame—Lifetime Achievement Award for his technical contribution to modeling and simulation and for his outstanding visionary leadership and dedication to increasing the effectiveness and broadening the applications of modeling and simulation worldwide. He also received the SCS Outstanding Service Award. He was awarded the IEEE CITS Hall of Fame Distinguished and Eminent Award.

He is a Life Fellow of IEEE and a Fellow of SCS.

Contents

Networking

Computing

Workload Scheduling in Fog and Cloud Environments: Emerging Concepts and Research Directions

Georgios L. Stavrinides and Helen D. Karatza

Abstract In recent years, we have been witnessing the growing adoption of infrastructure virtualization technologies and cloud computing. A wide range of applications has been migrated from traditional computing environments to the cloud. On the other hand, organizations with existing on-premises infrastructure investments are making the shift to hybrid cloud, in order to leverage the security provided by the private cloud and the virtually unlimited resources of the public cloud. With the rapid expansion of the Internet of Things, fog computing emerged as a new paradigm, extending the cloud to the network edge, closer to where the data are generated. The workloads on such platforms tend to be complex, featuring various degrees of parallelism. Consequently, one of the major challenges involved with fog and cloud computing, is the effective and efficient scheduling of the workload. In this chapter, we provide the necessary background in this field and present an overview of the emerging concepts and techniques, exploring future research directions.

Keywords Cloud computing · Fog computing · Workload classification · Scheduling objectives · Scheduling techniques

1 Introduction

Virtualization is the cornerstone of modern technological trends, such as fog and cloud computing. It introduces a thin software abstraction layer between the hardware and the operating system (OS) of a physical host. This abstraction layer is called *virtual machine monitor (VMM)* or *hypervisor* and it directly controls the physical resources of the host. It enables multiple *virtual machines (VMs)* to run on the same physical host. Examples of hypervisors are VMware vSphere, Microsoft Hyper-V

G. L. Stavrinides (✉) · H. D. Karatza
Department of Informatics, Aristotle University of Thessaloniki, 54124 Thessaloniki, Greece
e-mail: gstavrin@csd.auth.gr

H. D. Karatza
e-mail: karatza@csd.auth.gr

and Citrix Hypervisor. A VM is an isolated and self-contained software environment that emulates a physical computer, with its own CPU, memory, storage, network interface, OS and applications. In essence, a VM is a computer file, called an image, that behaves like a physical computer. This type of virtualization is typically called *hypervisor-based virtualization* and it enables running multiple, different operating systems on the same hardware [4, 50].

Another type of virtualization that is gaining momentum in recent years, is *container-based virtualization* or *containerization*. A *container* is an isolated, lightweight unit of software that contains all the necessary code, runtime environment, system tools, system libraries and settings required to run an application. Multiple containers can run atop the same operating system of a single physical host or VM. Containers are created and managed by a containerization engine, such as Docker. Containers use fewer system resources than VMs. However, as they provide process-level isolation, they are less secure than VMs, which provide complete isolation from the host operating system and other VMs. Containers in cloud and fog environments are usually used in conjunction with VMs [41].

Virtualization also enables the aggregation and pooling of physical resources such as storage and networking equipment and services (e.g., logical ports, switches, routers, firewalls etc.). Deploying virtualized environments provides greater flexibility, scalability, security and operational cost reduction, as well as efficient load balancing and better adaptability to workload variations [53].

The rest of this chapter is organized as follows: Sect. 2 presents the main characteristics and the established service and deployment models of cloud computing. Section 3 describes related concepts, characteristics, service and deployment models of the emerged new paradigm, fog computing. A classification and the characteristics of the workloads typically processed in fog and cloud environments are presented in Sect. 4. The scheduling problem is defined in Sect. 5, where the main scheduling objectives in these platforms are presented. Established scheduling techniques for each workload type are described in Sect. 6. Section 7 gives an overview of recent scheduling trends and challenges in fog and cloud environments. Section 8 concludes this chapter, illuminating future research directions.

2   Cloud Computing

The ongoing advances in virtualization technologies have led to the widespread adoption of *cloud computing*. Cloud computing emerged as a new paradigm that offers computational services to consumers as utilities [33].

2.1   *Characteristics*

The main characteristics of cloud computing are the following [42, 78]:

1. *Rapid elasticity*: Cloud computing provides virtually unlimited resources that can be provisioned and released (i.e., scaled up and down) dynamically, depending on demand.
2. *Resource Pooling*: Cloud computing resources can be pooled to serve multiple end-users, using a multi-tenant model (e.g., the same physical resource may be shared through the virtualization layer between different end-users).
3. *On-demand self-service*: End-users can request, customize, pay and use cloud computing services on-demand, nearly instantly and without the intervention of human operators.
4. *Broad network access*: Cloud computing services can be accessed over the network via a broad range of platforms, such as workstations, laptops, mobile phones and tablets.
5. *Measured service*: The usage of cloud computing services can be measured, monitored, controlled, reported and billed as a utility, providing transparency for both the service provider and the end-user.

As virtualization is at the heart of cloud computing, all of its advantages, such as greater flexibility, scalability and more efficient load balancing, also hold for cloud computing. From the end-users perspective, the major advantage of cloud computing is that computing resources, platforms and applications are all hosted into the cloud provider's data centers, relieving the consumers from the burden of acquiring, maintaining and monitoring expensive hardware and software infrastructure.

2.2 Service Models

Cloud computing services are typically classified into the following three categories [72]:

1. *Infrastructure as a Service (IaaS)*: End-users can lease and provision virtualized computing resources, typically in the form of VMs. End-users cannot manage or control the underlying cloud infrastructure, but have control over the leased VMs. Examples of IaaS cloud services are: Amazon Elastic Compute Cloud (Amazon EC2), Google Cloud Compute Engine and Microsoft Azure IaaS.
2. *Platform as a Service (PaaS)*: End-users are provided with a development framework (programming environment and tools) and a platform to develop and deploy their applications. End-users cannot manage or control the underlying cloud infrastructure, such as network, servers, operating systems and storage, but have control over the configuration and deployment of developed applications. Examples of PaaS cloud services are: Google App Engine, Microsoft Azure App Service and Amazon Web Services (AWS).
3. *Software as a Service (SaaS)*: End-users can use the provider's applications running on cloud infrastructure. End users cannot manage or control the underlying cloud infrastructure, such as network, servers, operating systems, storage or even individual application capabilities. However, they may be able to configure limited

user-specific application settings. Examples of SaaS cloud services are: Microsoft 365, Salesforce, Google Workspace and Adobe Creative Cloud.

A cloud service may be billed:

- on a *pay-as-you-go basis*, according to which end-users pay only for the period of time they use the service,
- on a *subscription basis*, according to which end-users pay a fixed amount for a fixed period of time during which they can use the service (e.g., a monthly or annual subscription fee), or
- on a *quotas basis*, according to which the cloud provider may offer a fixed amount of computing resources for free, defined by a set of quotas. In case more resources are required beyond the free usage tiers, the end-user is charged for the additional resources consumed.

2.3 Deployment Models

The typical deployment models of cloud computing infrastructure are as follows [71]:

1. *Private cloud*: The cloud infrastructure resides on the premises of a single organization and is exclusively used by its users (Fig. 1a).
2. *Public cloud*: The cloud infrastructure is provisioned for open use by the general public. It exists on the premises of the cloud provider, commonly spanning several geographically distributed data centers (Fig. 1b).
3. *Community cloud*: The cloud infrastructure encompasses two or more private clouds of a respective number of different organizations that have common privacy, security and regulatory considerations or other shared concerns. It is exclusively used by the community of users of the participating organizations (Fig. 2).
4. *Hybrid cloud*: The cloud infrastructure is a composition of a private cloud and one or more public clouds. The participating infrastructures remain unique entities, but are connected together by standardized or proprietary technologies that enable data and application portability between them (Fig. 3).

A private cloud typically offers greater security than public clouds, as its resources cannot be accessed from users outside the organization that owns the cloud infrastructure. However, its resources are usually limited. On the other hand, the resources in a public cloud are virtually unlimited, offering greater elasticity and scalability [9]. However, due to the utilization of resource multi-tenancy, public clouds are often considered less secure than private clouds. Recent legislation, such as the General Data Protection Regulation (GDPR) of the European Union (EU), makes organizations more reluctant to transfer sensitive, personal data to public clouds, since a data leak would result in substantial fines. Consequently, more and more organizations with existing on-premises infrastructure investments choose to shift to a hybrid cloud computing model, as it offers the best of both worlds [80]. This paradigm

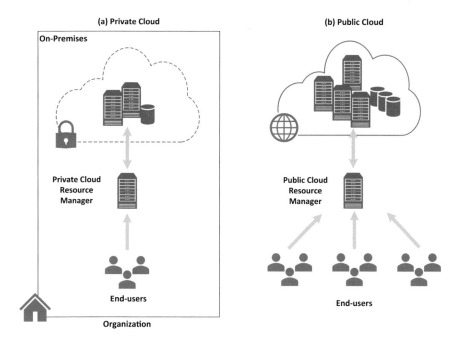

Fig. 1 A private cloud environment (**a**) and a public cloud environment (**b**)

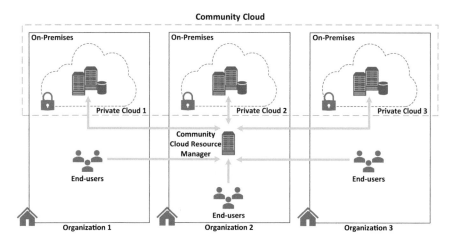

Fig. 2 A community cloud environment

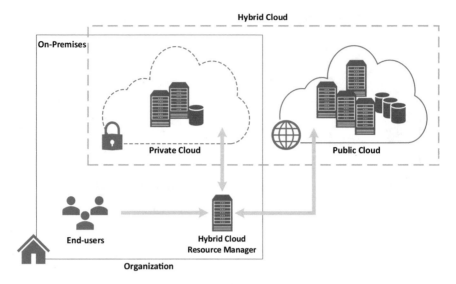

Fig. 3 A hybrid cloud environment

shift is usually facilitated by off-the-shelf solutions, such as VMware Cloud Foundation. In a hybrid cloud setting, supplementary public cloud resources can be utilized for offloading eligible jobs, when the private cloud resources cannot cope with the workload—this technique is known as *cloud bursting*. For instance, the on-premises resources can be utilized for sensitive, business-critical operations, such as financial reporting. On the other hand, public cloud resources can be used for high-volume, non-critical applications, such as data analytics operations on non-sensitive data [3].

3 The Internet of Things and Fog Computing

A wide spectrum of everyday objects, such as sensors, actuators and mobile devices are connected to the Internet, storing, processing and exchanging data collected from their surrounding environment. All of these objects form the *Internet of Things (IoT)* [14]. The growing number of IoT sensors and devices generate an unprecedented volume and variety of data at staggering speeds—due to these characteristics the data are typically referred to as *big data*—often requiring processing within strict time constraints, in a real-time manner. The advent of IoT facilitates the realization of concepts such as [27, 45, 74]:

- *Smart homes*: Automation of traditional home appliances can enable them to interact with each other, send real-time notifications to users and allow remote control via the Internet.

- *Smart cities*: A variety of sensors can enable more efficient waste management, water supply and road traffic control, quicker response to critical and life-threatening incidents, as well as safer autonomous transportation.
- *Smart grid*: Through sensors and actuators wind farms and solar parks can be more efficient and the electrical grid can be upgraded, allowing real-time pricing and consumer participation in power supply.
- *Smart healthcare*: Wearable and mobile devices can provide real-time fall detection and automatically call for help, as well as offer real-time exposure notifications and contact tracing in case of a pandemic of an infectious disease, such as COVID-19.

3.1 Fog Computing

Typically, IoT sensors and devices have limited processing capabilities, storage capacity and battery life. Therefore, it is often required to send the generated data to the cloud for further processing and analysis. This leads to heavy data traffic and significant service delays, since it is difficult to meet the delay-sensitive and context-aware service requirements of IoT applications by using cloud computing alone. Facing these restrictions and challenges, *fog computing* emerged as a new paradigm, supplementing cloud computing. The fog extends the cloud to the network edge, close to where the IoT data are generated. Instead of sending vast amounts of data to the cloud, data are processed by nearby fog resources, as physical proximity significantly affects latency [10, 17, 46].

3.1.1 Characteristics

The core component of a fog computing infrastructure is the *fog node* [52]. A fog node can be any device, physical or virtual, with computing capabilities, storage capacity and network connectivity. Examples of physical components that can be fog nodes are gateways, switches, routers and servers. On the other hand, virtualized switches, VMs and cloudlets (i.e., micro cloud data centers) are examples of virtual components that can be fog nodes. Several fog nodes can form clusters, where each participating fog node is aware of its geographical and logical location. Fog nodes are tightly coupled with the IoT devices or access networks, providing computing resources to these devices. Connectivity is typically facilitated by wireless technologies, such as Bluetooth, Wi-Fi and cellular networks (e.g., 5G networks). As fog resources are typically limited compared to those in a cloud infrastructure, fog nodes are typically cloud-aware and provide communication to centralized cloud resources. Supplementary cloud resources can be used to offload a part of the IoT workload when needed, taking into account its strict latency requirements.

The main characteristics of fog computing, catering to the requirements of an IoT ecosystem, are the following [25, 29]:

1. *Low latency*: as fog computing infrastructures are physically located in close proximity to the IoT layer, analysis and response to data generated by IoT actuators, sensors and devices take significantly less time compared to the case of a centralized cloud.
2. *Context awareness*: as fog nodes are located close to IoT devices, context awareness can be exploited, enabling improved services. For instance, if IoT sensors in a highway detect a car accident, other vehicles in the vicinity of the accident can be immediately notified. This would not be possible if the IoT data were analyzed by a central cloud, due to lack of local context.
3. *Real-time processing*: fog computing applications typically have firm deadlines within which they must finish execution.
4. *Heterogeneity*: fog computing supports the processing of a wide spectrum of data, collected via multiple types of network communication capabilities. Fog nodes themselves may be heterogeneous.
5. *Geographical distribution and mobility support*: fog computing supports wider geographically distributed deployments, as well as mobility techniques and protocols for direct communication with mobile IoT devices, compared to the more centralized cloud. For example, fog computing infrastructure deployed along highways can facilitate the realization of fully autonomous vehicles, by providing continuous real-time and low-latency processing of data collected by numerous moving autonomous vehicles.
6. *Scalability and interoperability*: as in the case of cloud computing, fog computing supports elasticity and resource pooling. Furthermore, fog resources by different fog providers can cooperate in order to provide a particular service. When required, fog resources can be supplemented with cloud resources from various cloud providers.

3.1.2   Service and Deployment Models

A fog infrastructure acts as a middle layer between the IoT devices and the cloud, resulting in a three-tier architecture, as shown in Fig. 4. As the fog computing paradigm is an extension of traditional cloud computing, the same service models (i.e., IaaS, PaaS and SaaS) and deployment models (i.e., private, public, community and hybrid) are also supported, as described in Sects. 2.2 and 2.3, respectively. Consequently, through fog computing, cloud-like services can be moved closer to the IoT devices in order to reduce latency, network traffic, power consumption and operational cost [49].

3.1.3   Mist Computing and Edge Computing

As fog computing continues to evolve, *mist computing* has emerged as a lightweight form of fog computing. Mist computing forms a layer between the fog computing layer and the IoT layer [24]. The mist computing layer comprises *mist nodes*, which

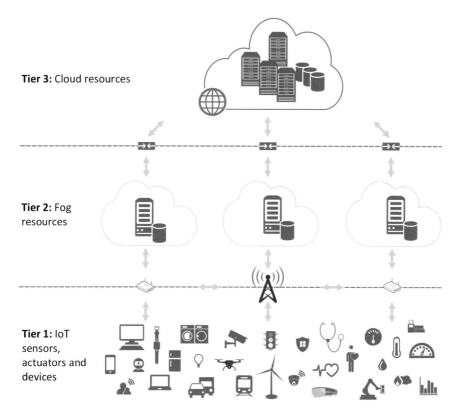

Fig. 4 Fog computing three-tier architecture

are low-latency microcomputers and microcontrollers. Mist nodes are placed even closer to the IoT sensors and devices they serve, at the very edge of the network, in an attempt to provide even lower latency services. Mist nodes collaborate with fog nodes, which are typically more powerful, essentially bringing the fog computing layer closer to the IoT layer.

Fog computing is often confused with *edge computing* [30]. However, there are key differences between the two paradigms [11]:

- Edge computing is typically provided in the IoT layer, by local computing capabilities embedded on IoT sensors and devices (i.e., it is end-device dependent). On the other hand, fog computing is end-device independent and supports a hierarchical and scalable architecture placed between the IoT and cloud layers.
- Edge computing is typically cloud unaware, whereas fog computing is cloud-aware, extending the cloud closer to the IoT layer.
- Edge computing is typically limited to specific applications and services, whereas fog computing is versatile. A fog computing infrastructure can run a variety of applications and provide a wide spectrum of services.

- Edge computing does not typically utilize virtualization and resource pooling, in contrast to fog computing.

4   Workloads in Fog and Cloud Environments

A wide range of applications are processed on fog and cloud environments, catering to every aspect of our daily life. Some prominent examples are the following [15]:

- *Productivity applications*: email management, word processing, spreadsheet creation and other office productivity tools, web conferencing and collaboration tools, virtual assistants etc.
- *Multimedia applications*: photo and video editing suites, video and game streaming services etc.
- *Critical infrastructures monitoring*: road traffic monitoring, real-time tsunami alerts, power grid and water supply network monitoring etc.
- *Healthcare monitoring*: telemedicine, remote monitoring of patients' vital signs, real-time contact tracing during pandemics etc.
- *Financial applications*: web banking, online stock and foreign exchange trading, customer relationship management, credit card payments processing etc.
- *Scientific applications*: drug discovery and molecular simulation, detection and measurement of gravitational waves, climate modeling and simulation etc.
- *Social media applications*: photo and video sharing, microblogging, social networking etc.

Furthermore, machine learning-driven big data analytics leverage these platforms in order to provide deeper insights to many industry and research sectors.

The workload in fog and cloud environments typically comprises complex jobs, each consisting of several component tasks. At one end of the spectrum, is the case where the component tasks require frequent communication with each other during their execution. At the other end of the spectrum, is the case where the component tasks are completely independent, not requiring any communication with each other. In between, is the case where the component tasks require communication with each other, but only before or after their execution. Consequently, the workload on such platforms can be classified into the following categories [64, 65, 69]:

1. *fine-grained parallel applications*,
2. *coarse-grained parallel applications* and
3. *embarrassingly parallel applications*.

Schematic examples of jobs of each workload type are illustrated in Fig. 5. The black arrows between the component tasks represent the communication between them. The orange arrows represent the data input required for each component task. Hereafter, for simplicity purposes, the terms application and job are used interchangeably.

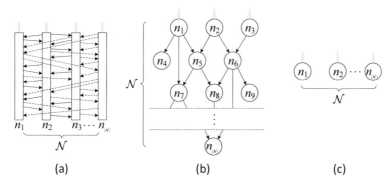

Fig. 5 Workload types in fog and cloud environments: **a** a fine-grained parallel job comprising a set $\mathcal{N} = \{n_1, n_2, \ldots, n_{|\mathcal{N}|}\}$ of frequently communicating component tasks, **b** a coarse-grained parallel job comprising a set $\mathcal{N} = \{n_1, n_2, \ldots, n_{|\mathcal{N}|}\}$ of component tasks with precedent constraints among them and **c** an embarrassingly parallel job comprising a set $\mathcal{N} = \{n_1, n_2, \ldots, n_{|\mathcal{N}|}\}$ of independent component tasks

4.1 Fine-Grained Parallel Applications

A fine-grained parallel application consists of a large number of frequently communicating parallel component tasks. The tasks in a fine-grained parallel application are usually smaller compared to those in coarse-grained and embarrassingly parallel applications. The size of a task is typically expressed as the number of instructions required to be executed by the particular task. The component tasks of a fine-grained parallel application can be assigned to different computational resources. However, the frequent communication between the parallel tasks during their execution requires effective synchronization techniques, in order to avoid a great number of context switches.

The communication and synchronization between the parallel tasks of a fine-grained application, especially in a high performance computing context, can be achieved using a message-passing programming paradigm, such as the *message passing interface (MPI)* [55]. MPI is a standardized message-passing library interface specification that is practical, portable, efficient and flexible. It defines the syntax and semantics of a core set of library routines that can be utilized in a message-passing program [28].

4.2 Coarse-Grained Parallel Applications

A coarse-grained parallel application consists of parallel component tasks that do not require any communication with each other during processing, but only before or after their execution. Typically, the number of tasks in a coarse-grained parallel application is smaller than in a fine-grained parallel application, but larger than the number of tasks in an embarrassingly parallel application. On the contrary, the size of the component tasks in a coarse-grained parallel application is usually larger compared to those in fine-grained parallel applications, but smaller compared to the size of tasks in embarrassingly parallel applications. The communication dependencies of the component tasks in a coarse-grained parallel application form precedence constraints among the tasks, in such a way that the output data of a task are used as input by other tasks. A component task can only start execution (i.e., it is *ready*) when its predecessor tasks have completed processing. A task without any parent tasks is called an *entry task*, whereas a task without any child tasks is called an *exit task*.

A coarse-grained parallel application is usually called a *workflow application*. It is typically represented by a *directed acyclic graph (DAG)* or *task graph*, $\mathcal{G} = (\mathcal{N}, \mathcal{E})$, where $\mathcal{N}$ is the set of the nodes of the graph, whereas $\mathcal{E}$ is the set of the directed edges between the nodes. Each node in the graph represents a component task of the application, whereas a directed edge between two nodes represents the data that must be transmitted from the first task to the other. Each node has a weight that represents the computational cost of its corresponding task, typically expressed as the amount of time required to execute the number of instructions of the particular task. Each edge between two tasks has a weight that denotes the communication cost that is incurred when transferring data from the first task to the other, typically expressed as the amount of time required to transfer the bytes of data between the two tasks [68].

The *level* of a task in a DAG is equal to the length of the longest path from the particular task to an exit task in the graph. The length of a path is the sum of the computational and communication costs of all of the nodes and edges, respectively, along the path. The *critical path* of a DAG is the longest path from an entry task to an exit task in the graph. The *communication to computation ratio (CCR)* of a DAG is the ratio of its average communication cost to its average computational cost on the target system. It is given by:

$$CCR = \frac{\sum_{e_{ij} \in \mathcal{E}} \overline{Comm(e_{ij})}}{\sum_{n_i \in \mathcal{N}} \overline{Comp(n_i)}} \tag{1}$$

where $\mathcal{N}$ and $\mathcal{E}$ are the sets of the nodes and edges of the DAG, respectively. $\overline{Comm(e_{ij})}$ is the average communication cost of the edge e_{ij} over all of the communication links in the system, whereas $\overline{Comp(n_i)}$ is the average computational cost of the component task n_i over all of the computational resources in the system.

4.3 Embarrassingly Parallel Applications

An embarrassingly parallel application consists of parallel component tasks that are independent, do not communicate with each other and can be executed in any order. The number of tasks in embarrassingly parallel applications is usually smaller than in fine-grained and coarse-grained parallel applications. On the other hand, the size of the component tasks in an embarrassingly parallel application is typically larger than those in fine-grained and coarse-grained parallel applications. Embarrassingly parallel applications are often called *bag-of-tasks (BoT)* applications. Due to the independence between their component tasks, BoT applications are well suited for execution on widely distributed resources where communication can become a bottleneck for more tightly-coupled parallel applications, such as fine-grained and coarse-grained parallel applications [58].

A programming paradigm that enables the processing of massively parallel applications is *MapReduce* [18]. A MapReduce application typically consists of two types of tasks: (a) *map tasks* and (b) *reduce tasks*. A map task takes a set of data and converts it into another set of data, where individual elements are broken down into tuples of key/value pairs. Parallel map tasks can process different chunks of data. A reduce task takes as input the output from map tasks and combines those data tuples into a smaller set of tuples, in order to produce the final result. A reduce task is always performed after the map tasks. In case a MapReduce application has only map tasks, it can be considered an embarrassingly parallel application. In case it has one or more reduce tasks, it can be considered a coarse-grained parallel application [21].

5 Scheduling Problem

5.1 Definition

In a distributed environment, such as a cloud or fog platform, the scheduling problem concerns the assignment of a set of tasks $\mathcal{N} = \{n_1, n_2, \ldots, n_{|\mathcal{N}|}\}$ to a set of computational resources $\mathcal{V} = \{vm_1, vm_2, \ldots, vm_{|\mathcal{V}|}\}$, which are connected by a set of communication links $\mathcal{L} = \{l_1, l_2, \ldots, l_{|\mathcal{L}|}\}$, where:

$$|\mathcal{V}| - 1 \leq |\mathcal{L}| \leq \frac{|\mathcal{V}| \, (|\mathcal{V}| - 1)}{2} \tag{2}$$

The goal is to process all tasks in order to optimize one or more objectives, taking into account any imposed constraints. For instance, the objective may be to minimize the average response time or the average makespan of the tasks. The imposed constraints may for example concern the execution order of the tasks or a deadline before which all tasks must be completed [12, 34]. In its general form, the scheduling problem has been shown to be NP-hard [20].

5.2 Objectives in Cloud and Fog Platforms

In cloud and fog platforms, scheduling objectives must satisfy two different perspectives: the cloud/fog provider's perspective and the end-user's perspective [7], as
shown in Fig. 6. Cloud and fog providers seek to maximize profit and minimize *service level agreement (SLA)* violations. An SLA is a contract between the cloud/fog
provider and the end-user and defines the provided service, the availability and *quality of service (QoS)* requirements, as well as the adopted pricing scheme. An SLA
violation occurs when the provider does not meet the agreed commitments, typically
those related to the availability and QoS of the provided service. On the other hand,
the end-users demand an acceptable QoS and *quality of experience (QoE)*, while minimizing the monetary cost involved with executing their jobs on cloud/fog resources.
QoS is defined from the system's perspective, using objective metrics related to the
characteristics of the target system and workload. On the contrary, QoE is defined
entirely from the end-user's perspective and is a subjective measure of the delight or
annoyance of the user's experience with the provided service. Clearly, the perceived
QoE depends heavily on the provided QoS [77].

Cloud and fog providers can maximize their profit by minimizing their carbon
footprint. This can be achieved by utilizing energy-aware scheduling strategies. By
minimizing data center running costs, cloud and fog services can be provided to end-
users at lower prices. Furthermore, efficient load balancing and consolidating techniques can be used to minimize workload offloading to resources of other providers
and thus minimize additional costs imposed on end-users. From the SLA perspective, availability can be enhanced by employing fault-tolerant scheduling methods.
On the other hand, QoS—and thus QoE—can be maximized by utilizing efficient and
effective scheduling policies that leverage data locality, as well as the characteristics
of the workload, in order to provide high quality results in a timely manner.

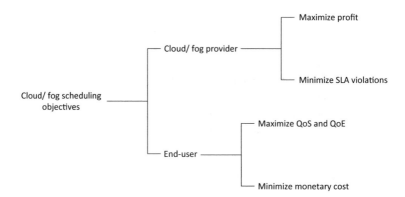

Fig. 6 Scheduling objectives in cloud and fog computing platforms

5.2.1 Typical QoS Objectives

A component task n_i of a job comprising a set of tasks $\mathcal{N} = \{n_1, n_2, \ldots, n_{|\mathcal{N}|}\}$ is typically characterized by the following time-related parameters [61]:

- *arrival time* $a(n_i)$: it is the time at which the task arrives at the target system.
- *start time* $s(n_i)$: it is the time at which the task starts its execution.
- *finish time* $f(n_i)$: it is the time at which the task finishes its execution.
- *deadline* $d(n_i)$: it is the time before which the task should finish its execution.

These time-related parameters are illustrated schematically in Fig. 7.

In cloud and fog platforms, the scheduling objectives typically associated with the provided QoS are as follows:

1. Minimization of the *average response time* $\overline{R}$ of the tasks $n_i \in \mathcal{N}$, where $\overline{R}$ is given by:

$$\overline{R} = \frac{1}{|\mathcal{N}|} \sum_{n_i \in \mathcal{N}} R(n_i) \tag{3}$$

   where $R(n_i) = f(n_i) - a(n_i)$ and $|\mathcal{N}|$ is the number of tasks in $\mathcal{N}$.
2. Minimization of the *makespan* (i.e., total execution time) M of the tasks $n_i \in \mathcal{N}$, where M is defined as:

$$M = \max_{n_i \in \mathcal{N}} \{f(n_i)\} - \min_{n_i \in \mathcal{N}} \{s(n_i)\} \tag{4}$$

3. Maximization of the *task guarantee ratio* G of the tasks $n_i \in \mathcal{N}$, where G is given by:

$$G = \frac{1}{|\mathcal{N}|} \sum_{n_i \in \mathcal{N}} g(n_i) \tag{5}$$

   where

$$g(n_i) = \begin{cases} 1, & \text{if } f(n_i) \leq d(n_i) \\ 0, & \text{otherwise} \end{cases} \tag{6}$$

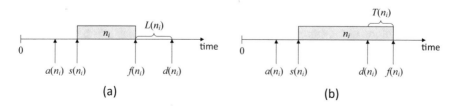

(a)   (b)

Fig. 7 Typical time-related parameters characterizing a component task n_i of a parallel application. Figure **a** depicts the case where the task finishes before its deadline $d(n_i)$ and thus its laxity is $L(n_i) = d(n_i) - f(n_i) > 0$. Figure **b** depicts the case where the task finishes after its deadline $d(n_i)$ and thus its tardiness is $T(n_i) = f(n_i) - d(n_i) > 0$

4. Maximization of the *average laxity* $\overline{L}$ of the tasks $n_i \in \mathcal{N}$, where $\overline{L}$ is given by:

$$\overline{L} = \frac{1}{|\mathcal{N}|} \sum_{n_i \in \mathcal{N}} L(n_i) \tag{7}$$

where

$$L(n_i) = \begin{cases} d(n_i) - f(n_i), & \text{if } d(n_i) > f(n_i) \\ 0, & \text{otherwise} \end{cases} \tag{8}$$

5. Minimization of the *average tardiness* $\overline{T}$ of the tasks $n_i \in \mathcal{N}$, where $\overline{T}$ is defined as:

$$\overline{T} = \frac{1}{|\mathcal{N}|} \sum_{n_i \in \mathcal{N}} T(n_i) \tag{9}$$

where

$$T(n_i) = \begin{cases} f(n_i) - d(n_i), & \text{if } f(n_i) > d(n_i) \\ 0, & \text{otherwise} \end{cases} \tag{10}$$

6 Established Scheduling Techniques

Some of the most widely-used scheduling techniques in cloud and fog environments are presented below. The scheduling methods are categorized according to the workload type that they are applicable to (i.e., fine-grained, coarse-grained or embarrassingly parallel applications).

6.1 Gang Scheduling

An effective way to schedule fine-grained parallel applications is *gang scheduling*. According to this approach, the parallel tasks of an application form a gang and are scheduled and processed simultaneously on different computational resources, so that all of the tasks start execution at the same time. In this way, the efficient synchronization of the component tasks is achieved and the risk of a task waiting to communicate with another task that is currently not running is avoided. In this case, the component task with the largest execution time determines the execution time of the whole gang. The size of a gang is defined as the number of its component tasks.

Without this technique, the synchronization of the component tasks would require more context switches and therefore additional overhead. However, this approach requires that the number of available computational resources is greater than or equal to the number of parallel tasks of the application. Furthermore, due to the fact that all of the tasks of a gang must start execution at the same time, there may be cases where some of the computational resources are idle, even when there are tasks of other applications waiting to be processed. This may happen when a task at the head of the queue of an idle resource waits for the other tasks of its gang in order to start execution. The other tasks of the gang may not be able to start execution at the same time, due to other workload processing priorities on their assigned resource [56].

Two of the most widely used gang scheduling strategies are the following [70]:

1. *Adapted First Come First Served (AFCFS)*: According to this policy, the gang that arrived first at the system has the highest priority for execution. A gang starts execution when its tasks are at the head of their assigned queues and the respective resources are idle. When there are not enough idle resources for a gang with a large number of parallel tasks, a smaller gang with tasks waiting behind those of the larger gang can start execution. This technique is also referred to as *backfilling*. The main disadvantage of this scheduling strategy is that it tends to favor smaller gangs, resulting in greater response times for larger gangs.
2. *Largest Gang First Served (LGFS)*: According to this policy, the largest gang that arrived at the system has the highest priority for execution. This gang scheduling policy tends to favor larger gangs, at the expense of smaller gangs. However, this is sometimes desirable and it may lead to a better overall performance, compared to the AFCFS strategy.

6.2 Workflow Scheduling

Workflow scheduling policies are divided into the following four categories:

1. *List scheduling*: This approach consists of two steps: (a) a *task prioritization step* and (b) a *resource selection step*. In the first step, the ready tasks are arranged in a list according to their priority, based on specific criteria. The ready task with the highest priority is selected first for scheduling. In the second step, the task that was selected in the first step is assigned to the resource that minimizes a specific cost function. One of the most widely used list scheduling approaches, suitable for heterogeneous resources, is the *Heterogeneous Earliest Finish Time (HEFT)* policy [75]. In the first step, it selects the ready task with the highest level. For the calculation of the level of each task, the average computational and communication costs of the tasks and edges, respectively, are used. In the second step, it selects the resource that can provide the selected task with the earliest finish time, utilizing idle time slots in the schedule of the particular resource. List scheduling heuristics are generally simpler, more practical, easier to implement and they usually outperform other workflow scheduling approaches.
2. *Clustering*: This approach is an iterative process. At first, each task is an independent cluster. At each iteration, clusters are merged, according to specific criteria. At the end of the process, a cluster merging step is performed, so that the number of clusters is equal to the number of available resources. Subsequently, a cluster mapping step is performed, in order to map each cluster of tasks to a resource.

Finally, a task ordering step is performed, which determines the execution order of the tasks on each resource, taking into account the precedence constraints of the tasks [31]. One of the most well-known clustering heuristics is the *Dominant Sequence Clustering (DSC)* policy [84]. According to this approach, the tasks in a DAG are clustered in such a way, so that the dominant sequence (i.e., longest path) of the graph is minimized.

3. *Task duplication*: This approach utilizes idle resource time by duplicating predecessor tasks in a DAG, so that the makespan of the particular DAG is minimized. One of the most established task duplication policies is the *Duplication Scheduling Heuristic (DSH)* [35]. In each scheduling step, it selects the task with the highest level and allocates it to the resource that can provide it with the earliest start time. In order to calculate the earliest possible start time of the selected task on each resource, first its start time is calculated without duplication of any predecessor tasks. Subsequently, it determines the idle time period between the finish time of the last scheduled task on the particular resource and the start time of the currently examined task. It then tries to duplicate the predecessors of the task into the idle time slot in a recursive manner, starting from the parent task from which the data arrives the latest, until either the idle time slot cannot accommodate other predecessor tasks or the start time of the examined task is not improved.

4. *Guided random search*: This approach seeks to find in an iterative manner the best schedule for a DAG, based on specific criteria and taking into account the precedence constraints of the tasks. In each step, the previously generated schedule is improved, by utilizing random parameters for the generation of the new schedule. The iterations continue until a terminating condition is satisfied. The most commonly used algorithms of this category are *genetic algorithms*, according to which each new schedule is generated by applying evolutionary techniques from nature, known as fitness functions [2].

6.3   Bag-of-Tasks Scheduling

Three of the most well-known techniques for scheduling bag-of-tasks applications are the following:

1. *Min-Min*: This strategy consists of two iterative steps. In the first step, the minimum completion time (MCT) of each unassigned task is calculated, over all of the available resources. In the second step, the task with the minimum MCT is assigned to the corresponding resource. In each iteration of the scheduling process, the MCT of each unassigned task is determined taking into account the current load of the resources, as resulted by the scheduling decision of the previous iteration [81].

2. *Max-Min*: This iterative strategy differs from the Min-Min technique, in that the task with the maximum (instead of the minimum) MCT is assigned to the corresponding resource in the second step of the scheduling process. Consequently, in

cases where the application consists of a large number of tasks with small execution times and a few tasks with large execution times, the Max-Min strategy is likely to give a smaller makespan than the Min-Min policy, since it schedules the tasks with larger execution times at earlier iterations [73].

3. *Sufferage*: This strategy is a two-step iterative process, like the Min-Min and Max-Min policies. However, in this case in addition to the MCT of each task, its second MCT is also calculated during the first step of the scheduling process. Subsequently, the sufferage value of each task is determined, by subtracting its MCT from its second MCT. In the second step, the task with the largest sufferage value is assigned to the resource that can provide it with the MCT. Hence, this technique is based on the idea that the highest priority for scheduling should be given to the task that would suffer the most (in terms of completion time) if it is not assigned to the resource that can provide it with the MCT [40].

7 Scheduling Trends and Challenges

Due to its vital importance in achieving the goals of both the cloud/fog providers and end-users perspectives, scheduling in cloud and fog environments continues to attract extensive attention in industry and academia. Novel heuristics have been proposed in an attempt to tackle major challenges that are still open, such as hybrid deployment models, security, monetary cost, real-time constraints, energy efficiency and dynamic scaling of resources.

7.1 Hybrid Clouds

More and more organizations with existing on-premises infrastructure investments and severe security concerns choose to transition to a hybrid cloud computing model. Consequently, the adoption of effective scheduling techniques in such environments is crucial. Bittencourt and Madeira [8] proposed a heuristic for the scheduling of real-time workflow applications in a hybrid cloud. The objective of the proposed method was to minimize the monetary cost of the leased public cloud resources, while meeting the deadlines of the workflows. According to their approach, an initial schedule was first created, using only resources in the private cloud. Subsequently, while the makespan of the workflow was larger than its deadline, it selected: (a) which tasks to reschedule so that their execution time would be reduced by using more powerful resources in the public cloud and (b) which public cloud resources to use for the rescheduled tasks so that the total monetary cost would be minimized. They showed that their algorithm reduced monetary cost while meeting the deadlines of the workflows.

Calheiros and Buyya [13] proposed a cost-aware approach for the scheduling of deadline-constrained bag-of-tasks jobs in a hybrid cloud. Their approach involved a dedicated architecture for the dynamic provisioning and scheduling of the jobs, as well as an auxiliary accounting mechanism responsible for taking budget-related decisions. Zhang et al. [86] proposed a task dispatching strategy in which real-time tasks were dispatched to a hybrid cloud, taking into consideration the total monetary cost required for the resources of the public cloud. However, their method did not take into account the substantial impact on the monetary cost incurred by transferring data to the public cloud. In order to overcome this problem, Zhang et al. [85] proposed a class of heuristics based on a task rescheduling policy, which considered the cost of transferring data to public cloud resources.

Based on the fact that linear programming is a general technique suitable for tackling optimization problems, Van den Bossche et al. [76] proposed a binary integer program formulation of the problem of scheduling preemptible deadline-constrained bags-of-tasks in a hybrid cloud with multiple public cloud providers. Their objective was to maximize the utilization of the on-premises data center and minimize the monetary cost of running the outsourced tasks on public cloud resources. The experimental evaluation showed that the proposed approach provided good results, requiring however very high solve time variances. Wang et al. [79] proposed a heuristic in order to tackle the problem of scheduling real-time bag-of-tasks jobs in a hybrid cloud. In order to utilize public cloud resources, their technique selected the resource with the highest cost-performance ratio.

Chunlin et al. [16] proposed a heuristic for scheduling heterogeneous workloads in a hybrid cloud. Their method first assigned tasks to the appropriate private cloud resources, based on the type of each task and the utilization of each resource. Subsequently, the execution time of the tasks was predicted, using a back propagation neural network algorithm. If according to the prediction results some tasks would not meet their deadline, they were executed on the public cloud resources requiring the lowest monetary cost. In order to harness the security provided by the private cloud and the virtually unlimited resources of the public cloud, Stavrinides and Karatza [67] proposed cost-aware heuristics for the scheduling of real-time bags-of-tasks in a hybrid cloud. The proposed heuristics were based on the idea of trading off timeliness for monetary cost savings, but in a restrictive manner. Furthermore, they took into consideration that some of the component tasks of the jobs required input data that were sensitive and thus should not be transferred to the public cloud. The experimental evaluation revealed that the proposed heuristics outperformed other scheduling policies in the particular framework.

7.2 Fog and Cloud Collaboration

Despite the fact that cloud bursting is a concept originating from the hybrid cloud computing paradigm, it can also be applied in a fog environment, in order to handle workload fluctuations by offloading tasks to the cloud. This technique has been the focus of several recent research efforts [47, 54, 62]. Aburukba et al. [1] proposed an approach to minimize the overall latency of IoT workload in a fog-cloud environment.

They used integer linear programming in order to model the minimum service time for IoT requests. A heuristic approach using a genetic algorithm was developed so as to obtain a feasible solution with a good quality in a reasonable computational time. The genetic algorithm was customized for the proposed model in order to minimize latency.

Enguehard et al. [23] utilized queueing theory in order to build an analytical framework for evaluating the performance of a cloud offloading strategy in a fog environment. Load-balancing was achieved using an approach based on request popularity. The experimental evaluation showed that the proposed method performed significantly better than an optimized blind load-balancer. The collaboration between fog and cloud platforms was investigated by Deng et al. [19]. They examined the tradeoff between power consumption and transmission latency, using an approximate approach. The proposed technique decomposed the primary problem into three subproblems of corresponding subsystems, which could be respectively solved. Simulations and numerical results revealed that by sacrificing modest computational resources in order to save communication bandwidth and reduce latency, fog resources could improve the performance of a cloud environment.

Pham et al. [48] proposed a real-time workflow scheduling strategy based on the collaboration between fog and cloud resources. The objective of the proposed approach was to achieve a tradeoff between the application makespan and the monetary cost of the cloud resources, under user-defined deadlines. The proposed strategy also utilized idle time slots during the scheduling process. However, it considered that both the performance and monetary cost factors had equal weight during the decision making process. Furthermore, it did not take into account the communication cost incurred by the transfer of data from the IoT layer to the fog and cloud layers. On the other hand, Stavrinides and Karatza [66] proposed a fog and cloud aware heuristic, suitable for the dynamic scheduling of multiple real-time workflows, utilizing possible schedule gaps. The proposed approach took into account the communication cost incurred by the transfer of data from the IoT devices to the resources in the fog and cloud layers. It also took into consideration both the communication latency and monetary cost involved with the cloud resources, in addition to the real-time constraints of the workload. The proposed scheduling approach was based on the tradeoff between performance and monetary cost, utilizing different contribution factors of these two parameters during resource selection, as opposed to [48].

7.3 Real-Time Constraints

One of the most challenging QoS requirements that cloud and fog providers have to meet, is the real-time processing of the workload, within strict deadlines. Naha et al. [44] addressed the problem of satisfying deadline-based dynamic user requirements in a fog-cloud environment, through resource allocation and scheduling algorithms. Their approached used resource ranking and provision of resources in a hybrid and hierarchical fashion. The experimental results indicated that the performance of the proposed heuristics was better compared with existing algorithms, in terms of overall data processing time, monetary cost and network delay. Ghose et al. [26] presented a scheduling method for real-time tasks in a cloud environment. The objective of the proposed approach was to schedule the real-time workload in an energy-efficient way, by dividing the problem into several subproblems and solving them individually. The proposed method, utilizing different task division techniques, was compared to a baseline policy, providing promising results.

In addition to the real-time constraints of the workload in a fog-cloud environment, Auluck et al. [5] proposed a scheduling technique that was also security-aware. In the examined framework, jobs submitted by the end-users were tagged as private, semi-private or public. Furthermore, fog and cloud resources were classified as trusted, semi-trusted or untrusted. Their approach chose suitable fog and cloud resources for the scheduling of the workload, by taking into account the network delay and the security tags of the jobs. The scheduling algorithm assigned private jobs to user's local fog resources or pre-trusted cloud resources. On the other hand, it assigned semi-private and public jobs to remote semi-trusted and untrusted fog and cloud resources, respectively. The experimental evaluation revealed that in the particular framework, the proposed algorithm offered a higher guarantee ratio compared to other scheduling algorithms. Xu et al. [83] proposed a real-time task scheduling algorithm that took into account the laxity of the tasks, as well as the energy consumption of the fog and cloud resources. The proposed approach formulated the scheduling problem as a constrained optimization problem and utilized an ant colony system algorithm for its solution. The experimental results showed that the proposed method could effectively reduce energy consumption, while minimizing the deadline miss ratio of the tasks.

Based on the observation that it is often more desirable for a real-time application to produce an imprecise result by its deadline, than to produce a precise result late, Lin et al. [38] proposed the *imprecise computations* technique. According to this method, a real-time application is allowed to return intermediate, approximate results of poorer, but still acceptable quality, when the deadline of the application cannot be met. Imprecise computations can be utilized especially in the case of applications with *monotone* component tasks, where the quality of a task's results is improved as more time is spent to produce them (e.g., statistical estimation and video processing tasks). Each monotone task typically consists of a *mandatory part*, followed by an *optional part*. In order for a task to return an acceptable result, its mandatory part must be completed. The optional part refines the result produced by the mandatory part. This technique provides scheduling flexibility, by trading off precision for timeliness, since it allows the scheduler to terminate a task that has completed its mandatory part at any time, depending on the workload conditions [57]. Stavrinides and Karatza [60] utilized imprecise computations in order to schedule real-time fine-grained parallel applications in a cloud environment with transient software failures. Their approach provided fault-tolerance through the use of application-directed checkpointing. They examined the impact of checkpointing interval selection on the system performance, under various failure probabilities. The simulation results showed that for higher failure probabilities and smaller service times (i.e., due to imprecise computations), checkpointing should be more frequent, in order to achieve good performance. However, the checkpointing interval should be above a particular threshold, as unnecessary frequent checkpointing may lead to performance degradation.

7.4 Energy Efficiency

As cloud and fog providers strive to reduce their carbon footprint, a large body of research has been focused on energy-efficient scheduling heuristics in such platforms. In this context, a widely used power management method is the *dynamic voltage and frequency scaling (DVFS)* technique. DVFS allows the dynamic adjustment of the supply voltage and operating frequency (i.e., speed) of a processor, based on the workload conditions [59]. Mhedheb et al. [43] proposed a load and thermal-aware VM scheduling mechanism in a cloud platform, capable of preventing the occurrence of over-loaded or over-heated physical servers. DVFS was utilized for the power management of the physical machines. Stavrinides and Karatza [63] proposed an energy-efficient, QoS-aware and cost-effective scheduling strategy for real-time workflow applications in a cloud environment. The proposed approach utilized per-core DVFS on the underlying heterogeneous multi-core processors, as well as imprecise computations, in order to fill in schedule gaps. The heuristic also took into account the effects of input error on the processing time of the component tasks. Their objective was to provide timeliness and energy efficiency by trading off result precision, while keeping the result quality of the completed jobs at an acceptable standard and the monetary cost required for the execution of the jobs at a reasonable level. The simulation experiments revealed that the proposed approach outperformed other baseline scheduling policies, providing promising results.

Wu and Lee [82] proposed a scheduling heuristic based on an integer linear programming model, in order to minimize the energy consumption in a heterogeneous fog architecture with IoT workflows. They showed that the proposed energy minimization scheduling technique could achieve near-optimal energy efficiency while maintaining fast execution times. The joint optimization of monetary cost and makespan of scheduling workflows in IaaS clouds was examined by Zhou et al. [87]. The authors proposed a novel workflow scheduling scheme, which integrated the fuzzy dominance sort mechanism with the HEFT list scheduling heuristic. They conducted extensive experiments using both real-world and synthetic workflows and demonstrated the efficacy of the proposed scheduling approach.

Based on the observation that in a fog-cloud setting a low task execution delay usually means high energy consumption, Li et al. [36] proposed a tradeoff strategy in order to achieve energy savings utilizing a delay threshold. For this purpose, they modeled a mathematical framework of a fog-cloud cooperation system with queueing theory. They also modeled the energy consumption and delay functions of the three system layers (IoT, fog and cloud). They formulated a joint energy and delay optimization problem, utilizing nonlinear programming to calculate the optimal workload of each layer for energy consumption optimization. Furthermore, they designed a fog-cloud cooperation algorithm for the scheduling and offloading of IoT tasks. The experimental results showed that in the proposed framework the energy consumption was reduced. On the other hand, Luo et al. [39] employed containers in order to improve the resource utilization of fog nodes and reduce service delay in a multi-cloud and multi-fog architecture. Task scheduling was performed using a novel energy-aware algorithm. The heuristic was based on the transmission energy consumption of IoT devices and used a dynamic threshold strategy to schedule requests in real-time. The proposed method guaranteed the energy balancing of IoT devices, without increasing the transmission latency.

Server consolidation is another approach that is predominantly used in cloud/fog environments in order to enhance load balancing and energy efficiency. It is typically achieved through the *live migration* of VMs/containers. According to this technique, running VMs/containers are moved from one physical host to another, without downtime—i.e., with no impact on the availability of the VM/container and without interrupting the applications currently running on the VM/container. Towards this direction, Beloglazov et al. [6] defined an architectural framework and principles for energy-efficient cloud computing. They investigated energy-aware resource provisioning and allocation algorithms that provisioned data center resources to client applications in a way that improved the energy efficiency of the data center, without violating the negotiated SLAs. Furthermore, they developed algorithms for energy-efficient mapping of VMs to suitable cloud resources, utilizing dynamic server consolidation through VM live migrations. Khan et al. [32] proposed a consolidation algorithm in a cloud environment that favored the most effective live migration among VMs and/or containers that ran on VMs. They investigated how migration decisions should be made in order to save energy, without any negative impact on the system performance. Through simulation experiments they demonstrated that it was more performance-efficient to migrate VMs than containers. However, migrating containers was more energy-efficient than VMs.

7.5 Dynamic Scaling

Cloud and fog platforms are characterized by the elasticity of their resources. Consequently, one of the main aspects of workload scheduling in such environments is the dynamic scaling of the resources according to the system load. With regard to this issue, Ramirez et al. [51] proposed predictive autoscaling policies for the scheduling of microservice applications in a cloud environment with VMs and containers. The proposed policies utilized horizontal, vertical or hybrid scaling for both VMs and containers. The type of scaling is called horizontal when containers or VMs are added or removed. On the other hand, it is called vertical when resources allocated to a container or a VM are either increased or decreased. The combination of these two scaling types is called hybrid scaling. According to the proposed approach, the long-term forecast of the workload and the capacity model of microservices were taken into account in order to produce the sequence of scaling actions scheduled for execution in the future, aiming to meet the service level objectives and minimize the monetary cost. The proposed autoscaling policies were evaluated for several workload patterns. The experimental results demonstrated that hybrid scaling provided more flexibility and reduced costs.

Li and Xia [37] designed a platform that could autoscale containerized web applications in a hybrid cloud. They developed a hybrid scheduling controller using a combination of prediction and reaction algorithms for determining when to scale up or down. They showed that containers could be effectively deployed in both the private and the public cloud. Furthermore, they demonstrated that the proposed approach could dynamically adjust the number of containers in a few seconds in order to meet the resource requirements. In order to address the problem of determining dynamically the minimum number of fog nodes required for the processing of IoT workloads, Kafhali and Salah [22] developed a queueing analytical model that captured the dynamics and behavior of fog nodes and IoT workloads. They derived mathematical formulas from the analytical model for key performance metrics. Numerical examples were given in order to show how the proposed analytical model could be used to predict the performance of the fog resources, and also to dynamically determine the required number of fog nodes under variable IoT workload conditions. The analytical model was validated using discrete-event simulations.

8 Conclusions

In this chapter, the main characteristics, service and deployment models of cloud and fog computing were presented. We gave a classification of the workload typically processed on such platforms, along with the characteristics of each workload type. The definition and the main objectives of the scheduling problem in fog and cloud environments were presented. Established scheduling techniques for each workload type were described. Furthermore, we presented an overview of recent scheduling trends and challenges in fog and cloud environments, such as hybrid deployment models, security, monetary cost, real-time constraints, energy efficiency and dynamic scaling of resources.

Workload scheduling in fog and cloud environments still remains an active research area, attracting significant attention from both academia and industry. With the explosive growth of the Internet of Things and big data, workloads requiring processing on fog and cloud platforms are expected to get even more complex and computationally demanding. The research community has been dealing with the challenges that derive from the distinctive characteristics of these platforms. Nevertheless, there are still several open issues and challenges that need to be addressed. Especially with the emergence of concepts such as smart homes, smart cities and smart healthcare, security and privacy in the context of workload scheduling is an important domain where research efforts are expected to focus on in the near future.

References

1. Aburukba, R.O., AliKarrar, M., Landolsi, T., El-Fakih, K.: Scheduling Internet of Things requests to minimize latency in hybrid fog-cloud computing. Future Gener. Comput. Syst. **111**, 539–551 (2020). https://doi.org/10.1016/j.future.2019.09.039
2. Ali, I.M., Sallam, K.M., Moustafa, N., Chakraborty, R., Ryan, M.J., Choo, K.K.R.: An automated task scheduling model using non-dominated sorting genetic algorithm II for fog-cloud systems. IEEE Trans. Cloud Comput. 1–15 (2020). https://doi.org/10.1109/TCC.2020.3032386
3. Amiri, M.J., Maiyya, S., Agrawal, D., Abbadi, A.E.: SeeMoRe: a fault-tolerant protocol for hybrid cloud environments. In: Proceedings of the IEEE 36th International Conference on Data Engineering (ICDE'20), pp. 1345–1356 (2020). https://doi.org/10.1109/ICDE48307.2020.00120
4. Anand, A., Chaudhary, A., Arvindhan, M.: The need for virtualization: when and why virtualization took over physical servers. In: Proceedings of the First International Conference on Advanced Communication & Computational Technology (ICACCT'19), pp. 1351–1359 (2019). https://doi.org/10.1007/978-981-15-5341-7_102
5. Auluck, N., Rana, O., Nepal, S., Jones, A., Singh, A.: Scheduling real time security aware tasks in fog networks. IEEE Trans. Serv. Comput. 1–14 (2019). https://doi.org/10.1109/TSC.2019.2914649
6. Beloglazov, A., Abawajy, J., Buyya, R.: Energy-aware resource allocation heuristics for efficient management of data centers for cloud computing. Future Gener. Comput. Syst. **28**(5), 755–768 (2012). https://doi.org/10.1016/j.future.2011.04.017
7. Bittencourt, L.F., Goldman, A., Madeira, E.R.M., da Fonseca, N.L.S., Sakellariou, R.: Scheduling in distributed systems: a cloud computing perspective. Comput. Sci. Rev. **30**, 31–54 (2018). https://doi.org/10.1016/j.cosrev.2018.08.002
8. Bittencourt, L.F., Madeira, E.R.M.: HCOC: a cost optimization algorithm for workflow scheduling in hybrid clouds. J. Internet Serv. Appl. **2**, 207–227 (2011). https://doi.org/10.1007/s13174-011-0032-0
9. Bittencourt, L.F., Madeira, E.R.M., da Fonseca, N.L.S.: Scheduling in hybrid clouds. IEEE Commun. Mag. **50**(9), 42–47 (2012). https://doi.org/10.1109/MCOM.2012.6295710
10. Bonomi, F., Milito, R., Zhu, J., Addepalli, S.: Fog computing and its role in the Internet of Things. In: Proceedings of the First Edition of the MCC Workshop on Mobile Cloud Computing (MCC'12), pp. 13–16 (2012). https://doi.org/10.1145/2342509.2342513
11. Bonomi, L.: Fog vs edge computing. Tech. Rep. 1.1.01, Nebbiolo Technologies Inc. (2019)
12. Buttazzo, G.C.: Hard Real-Time Computing Systems: Predictable Scheduling Algorithms and Applications, 3rd edn. Springer (2011). https://doi.org/10.1007/978-1-4614-0676-1
13. Calheiros, R.N., Buyya, R.: Cost-effective provisioning and scheduling of deadline-constrained applications in hybrid clouds. In: Proceedings of the 13th International Conference on Web Information Systems Engineering (WISE'12), pp. 171–184 (2012). https://doi.org/10.1007/978-3-642-35063-4_13
14. Chen, Y.: Service-Oriented Computing and System Integration: Software, IoT, Big Data, and AI as Services, 6th edn. Kendall Hunt Publishing (2018)
15. Chen, Y., Tsai, W.T.: Service-Oriented Computing and Web Software Integration: From Principles to Development, 5th edn. Kendall Hunt Publishing (2015)
16. Chunlin, L., Jianhang, T., Youlong, L.: Hybrid cloud adaptive scheduling strategy for heterogeneous workloads. J. Grid Comput. **17**, 419–446 (2019). https://doi.org/10.1007/s10723-019-09481-3
17. Cisco: Fog computing and the Internet of Things: extend the cloud to where the things are. Tech. Rep. C11-734435-00, Cisco Systems, Inc. (2015)
18. Dean, J., Ghemawat, S.: MapReduce: simplified data processing on large clusters. Commun. ACM **51**(1), 107–113 (2008). https://doi.org/10.1145/1327452.1327492

19. Deng, R., Lu, R., Lai, C., Luan, T.H., Liang, H.: Optimal workload allocation in fog-cloud computing toward balanced delay and power consumption. IEEE Internet Things J. **3**(6), 1171–1181 (2016). https://doi.org/10.1109/JIOT.2016.2565516

20. Drozdowski, M.: Scheduling for Parallel Processing, 1st edn. Springer (2009). https://doi.org/10.1007/978-1-84882-310-5

21. Ekanayake, J., Fox, G.: High performance parallel computing with clouds and cloud technologies. In: Proceedings of the First International Conference on Cloud Computing (Cloud-Comp'09), pp. 20–38 (2009)

22. El Kafhali, S., Salah, K.: Efficient and dynamic scaling of fog nodes for IoT devices. J. Supercomput. **73**, 5261–5284 (2017). https://doi.org/10.1007/s11227-017-2083-x

23. Enguehard, M., Carofiglio, G., Rossi, D.: A popularity-based approach for effective cloud offload in fog deployments. In: Proceedings of the 30th International Teletraffic Congress (ITC'18), pp. 55–63 (2018). https://doi.org/10.1109/ITC30.2018.00016

24. Galambos, P.: Cloud, fog, and mist computing: advanced robot applications. IEEE Syst. Man Cybern. Mag. **6**(1), 41–45 (2020). https://doi.org/10.1109/MSMC.2018.2881233

25. Ghobaei-Arani, M., Souri, A., Rahmanian, A.A.: Resource management approaches in fog computing: a comprehensive review. J. Grid Comput. **18**, 1–42 (2020). https://doi.org/10.1007/s10723-019-09491-1

26. Ghose, M., Kaur, S., Sahu, A.: Scheduling real time tasks in an energy-efficient way using VMs with discrete compute capacities. Computing **102**, 263–294 (2020). https://doi.org/10.1007/s00607-019-00738-z

27. Gouda, O.M., Hejji, D.J., Obaidat, M.S.: Privacy assessment of fitness tracker devices. In: Proceedings of the 2020 International Conference on Computer, Information and Telecommunication Systems (CITS'20), pp. 1–8 (2020). https://doi.org/10.1109/CITS49457.2020.9232503

28. Grama, A., Gupta, A., Karypis, G., Kumar, V.: Introduction to Parallel Computing, 2nd edn. Addison-Wesley (2003)

29. Iorga, M., Feldman, L., Barton, R., Martin, M.J., Goren, N., Mahmoudi, C.: Fog computing conceptual model. Tech. Rep. 500-325, National Institute of Standards and Technology, U.S. Department of Commerce (2018). https://doi.org/10.6028/NIST.SP.500-325

30. Jararweh, Y., Doulat, A., AlQudah, O., Ahmed, E., Al-Ayyoub, M., Benkhelifa, E.: The future of mobile cloud computing: integrating cloudlets and mobile edge computing. In: Proceedings of the 23rd International Conference on Telecommunications (ICT'16), pp. 1–5 (2016). https://doi.org/10.1109/ICT.2016.7500486

31. Jiang, H.J., Huang, K.C., Chang, H.Y., Gu, D.S., Shih, P.J.: Scheduling concurrent workflows in HPC cloud through exploiting schedule gaps. In: Proceedings of the 11th International Conference on Algorithms and Architectures for Parallel Processing (ICA3PP'11), pp. 282–293 (2011). https://doi.org/10.1007/978-3-642-24650-0_24

32. Khan, A.A., Zakarya, M., Khan, R., Rahman, I.U., Khan, M., Khan, A.U.R.: An energy, performance efficient resource consolidation scheme for heterogeneous cloud datacenters. J. Netw. Comput. Appl. **150**, 102497 (2020). https://doi.org/10.1016/j.jnca.2019.102497

33. Khayer, A., Talukder, M.S., Bao, Y., Hossain, M.N.: Cloud computing adoption and its impact on SMEs' performance for cloud supported operations: a dual-stage analytical approach. Technol. Soc. **60**, 101225 (2020). https://doi.org/10.1016/j.techsoc.2019.101225

34. Kołodziej, J.: Evolutionary Hierarchical Multi-Criteria Metaheuristics for Scheduling in Large-Scale Grid Systems, 1st edn. Springer (2012). https://doi.org/10.1007/978-3-642-28971-2

35. Kruatrachue, B., Lewis, T.G.: Duplication scheduling heuristic, a new precedence task scheduler for parallel systems. Tech. Rep. 87-60-3, Oregon State University (1987)

36. Li, G., Yan, J., Chen, L., Wu, J., Lin, Q., Zhang, Y.: Energy consumption optimization with a delay threshold in cloud-fog cooperation computing. IEEE Access **7**, 159688–159697 (2019). https://doi.org/10.1109/ACCESS.2019.2950443

37. Li, Y., Xia, Y.: Auto-scaling web applications in hybrid cloud based on docker. In: Proceedings of the 5th International Conference on Computer Science and Network Technology (ICCSNT'16), pp. 75–79 (2016). https://doi.org/10.1109/ICCSNT.2016.8070122

38. Lin, K.J., Natarajan, S., Liu, J.W.S.: Imprecise results: utilizing partial computations in real-time systems. In: Proceedings of the 8th IEEE Real-Time Systems Symposium (RTSS'87), pp. 210–217 (1987)
39. Luo, J., Yin, L., Hu, J., Wang, C., Liu, X., Fan, X., Luo, H.: Container-based fog computing architecture and energy-balancing scheduling algorithm for energy IoT. Future Gener. Comput. Syst. **97**, 50–60 (2019). https://doi.org/10.1016/j.future.2018.12.063
40. Maheswaran, M., Ali, S., Siegel, H.J., Hensgen, D., Freund, R.F.: Dynamic mapping of a class of independent tasks onto heterogeneous computing systems. J. Parallel Distrib. Comput. **59**(2), 107–131 (1999). https://doi.org/10.1006/jpdc.1999.1581
41. Mavridis, I., Karatza, H.: Combining containers and virtual machines to enhance isolation and extend functionality on cloud computing. Future Gener. Comput. Syst. **94**, 674–696 (2019). https://doi.org/10.1016/j.future.2018.12.035
42. Mell, P., Grance, T.: The NIST definition of cloud computing. Tech. Rep. 800-145, National Institute of Standards and Technology, U.S. Department of Commerce (2011). https://doi.org/10.6028/NIST.SP.800-145
43. Mhedheb, Y., Jrad, F., Tao, J., Zhao, J., Kołodziej, J., Streit, A.: Load and thermal-aware VM scheduling on the cloud. In: Proceedings of the 13th International Conference on Algorithms and Architectures for Parallel Processing (ICA3PP'13), pp. 101–114 (2013). https://doi.org/10.1007/978-3-319-03859-9_8
44. Naha, R.K., Garg, S., Chan, A., Battula, S.K.: Deadline-based dynamic resource allocation and provisioning algorithms in fog-cloud environment. Future Gener. Comput. Syst. **104**, 131–141 (2020). https://doi.org/10.1016/j.future.2019.10.018
45. Obaidat, M.S., Nicopolitidis, P.: Smart Cities and Homes: Key Enabling Technologies, 1st edn. Morgan Kaufmann Publishers Inc. (2016)
46. OpenFog: OpenFog Architecture Overview. Tech. Rep. OPFWP001.0216, OpenFog Consortium Architecture Working Group (2016)
47. Pham, X.Q., Huh, E.N.: Towards task scheduling in a cloud-fog computing system. In: Proceedings of the 18th Asia-Pacific Network Operations and Management Symposium (APNOMS'16), pp. 1–4 (2016). https://doi.org/10.1109/APNOMS.2016.7737240
48. Pham, X.Q., Man, N.D., Tri, N.D.T., Thai, N.Q., Huh, E.N.: A cost- and performance-effective approach for task scheduling based on collaboration between cloud and fog computing. Int. J. Distrib. Sens. Netw. **13**(11), 1–16 (2017). https://doi.org/10.1177/1550147717742073
49. Puliafito, C., Mingozzi, E., Longo, F., Puliafito, A., Rana, O.: Fog computing for the Internet of Things: a survey. ACM Trans. Internet Technol. **19**(2), 18:1–18:41 (2019). https://doi.org/10.1145/3301443
50. Ramakrishnan, J., Shabbir, M.S., Kassim, N.M., Nguyen, P.T., Mavaluru, D.: A comprehensive and systematic review of the network virtualization techniques in the IoT. Int. J. Commun. Syst. **33**(7), e4331 (2020). https://doi.org/10.1002/dac.4331
51. Ramirez, Y.M., Podolskiy, V., Gerndt, M.: Capacity-driven scaling schedules derivation for coordinated elasticity of containers and virtual machines. In: Proceedings of the 2019 IEEE International Conference on Autonomic Computing (ICAC'19), pp. 177–186 (2019). https://doi.org/10.1109/ICAC.2019.00029
52. Ren, J., Zhang, D., He, S., Zhang, Y., Li, T.: A survey on end-edge-cloud orchestrated network computing paradigms: transparent computing, mobile edge computing, fog computing, and cloudlet. ACM Comput. Surv. **52**(6), 125:1–125:36 (2019). https://doi.org/10.1145/3362031
53. Sahoo, J., Mohapatra, S., Lath, R.: Virtualization: a survey on concepts, taxonomy and associated security issues. In: Proceedings of the Second International Conference on Computer and Network Technology (ICCNT'10), pp. 222–226 (2010). https://doi.org/10.1109/ICCNT.2010.49
54. Shah-Mansouri, H., Wong, V.W.S.: Hierarchical fog-cloud computing for IoT systems: a computation offloading game. IEEE Internet Things J. **5**(4), 3246–3257 (2018). https://doi.org/10.1109/JIOT.2018.2838022
55. Stamatakis, A., Ott, M.: Exploiting fine-grained parallelism in the phylogenetic likelihood function with MPI, Pthreads, and OpenMP: a performance study. In: Proceedings of the Third

IAPR International Conference on Pattern Recognition in Bioinformatics (PRIB'08), pp. 424–435 (2008). https://doi.org/10.1007/978-3-540-88436-1_3

56. Stavrinides, G.L., Karatza, H.D.: Fault-tolerant gang scheduling in distributed real-time systems utilizing imprecise computations. Simul.: Trans. Soc. Model. Simul. Int. **85**(8), 525–536 (2009). https://doi.org/10.1177/0037549709340729

57. Stavrinides, G.L., Karatza, H.D.: The impact of input error on the scheduling of task graphs with imprecise computations in heterogeneous distributed real-time systems. In: Proceedings of the 18th International Conference on Analytical and Stochastic Modelling Techniques and Applications (ASMTA'11), pp. 273–287 (2011). https://doi.org/10.1007/978-3-642-21713-5_20

58. Stavrinides, G.L., Karatza, H.D.: The impact of data locality on the performance of a SaaS cloud with real-time data-intensive applications. In: Proceedings of the 21st IEEE/ACM International Symposium on Distributed Simulation and Real Time Applications (DS-RT'17), pp. 1–8 (2017). https://doi.org/10.1109/DISTRA.2017.8167683

59. Stavrinides, G.L., Karatza, H.D.: Energy-aware scheduling of real-time workflow applications in clouds utilizing DVFS and approximate computations. In: Proceedings of the IEEE 6th International Conference on Future Internet of Things and Cloud (FiCloud'18), pp. 33–40 (2018). https://doi.org/10.1109/FiCloud.2018.00013

60. Stavrinides, G.L., Karatza, H.D.: The impact of checkpointing interval selection on the scheduling performance of real-time fine-grained parallel applications in SaaS clouds under various failure probabilities. Concurr. Comput. Pract. Exp. **30**(12), e4288 (2018). https://doi.org/10.1002/cpe.4288

61. Stavrinides, G.L., Karatza, H.D.: Scheduling data-intensive workloads in large-scale distributed systems: trends and challenges. *Studies in Big Data*, vol. 36, 1st edn., chap. 2, pp. 19–43. Springer (2018). https://doi.org/10.1007/978-3-319-73767-6_2

62. Stavrinides, G.L., Karatza, H.D.: Cost-effective utilization of complementary cloud resources for the scheduling of real-time workflow applications in a fog environment. In: Proceedings of the 7th International Conference on Future Internet of Things and Cloud (FiCloud'19), pp. 1–8 (2019). https://doi.org/10.1109/FiCloud.2019.00009

63. Stavrinides, G.L., Karatza, H.D.: An energy-efficient, QoS-aware and cost-effective scheduling approach for real-time workflow applications in cloud computing systems utilizing DVFS and approximate computations. Future Gener. Comput. Syst. **96**, 216–226 (2019). https://doi.org/10.1016/j.future.2019.02.019

64. Stavrinides, G.L., Karatza, H.D.: A hybrid approach to scheduling real-time IoT workflows in fog and cloud environments. Multimed. Tools Appl. **78**(17), 24639–24655 (2019). https://doi.org/10.1007/s11042-018-7051-9

65. Stavrinides, G.L., Karatza, H.D.: Scheduling different types of gang jobs in distributed systems. In: Proceedings of the 2019 International Conference on Computer, Information and Telecommunication Systems (CITS'19), pp. 1–5 (2019). https://doi.org/10.1109/CITS.2019.8862091

66. Stavrinides, G.L., Karatza, H.D.: Cost-aware cloud bursting in a fog-cloud environment with real-time workflow applications. Concurr. Comput. Pract. Exp. e5850 (2020). https://doi.org/10.1002/cpe.5850

67. Stavrinides, G.L., Karatza, H.D.: Dynamic scheduling of bags-of-tasks with sensitive input data and end-to-end deadlines in a hybrid cloud. Multimed. Tools Appl. 1–23 (2020). https://doi.org/10.1007/s11042-020-08974-8

68. Stavrinides, G.L., Karatza, H.D.: Orchestration of real-time workflows with varying input data locality in a heterogeneous fog environment. In: Proceedings of the Fifth International Conference on Fog and Mobile Edge Computing (FMEC'20), pp. 202–209 (2020). https://doi.org/10.1109/FMEC49853.2020.9144824

69. Stavrinides, G.L., Karatza, H.D.: Scheduling real-time bag-of-tasks applications with approximate computations in SaaS clouds. Concurr. Comput. Pract. Exp. **32**(1), e4208 (2020). https://doi.org/10.1002/cpe.4208

70. Stavrinides, G.L., Karatza, H.D.: Weighted scheduling of mixed gang jobs on distributed resources. In: Proceedings of the 2020 International Conference on Communications, Computing, Cybersecurity and Informatics (CCCI'20), pp. 1–6 (2020). https://doi.org/10.1109/CCCI49893.2020.9256505
71. Sun, L., Dong, H., Hussain, O.K., Hussain, F.K., Liu, A.X.: A framework of cloud service selection with criteria interactions. Future Gener. Comput. Syst. **94**, 749–764 (2019). https://doi.org/10.1016/j.future.2018.12.005
72. Surbiryala, J., Rong, C.: Cloud computing: history and overview. In: Proceedings of the 2019 IEEE Cloud Summit (CS'19), pp. 1–7 (2019). https://doi.org/10.1109/CloudSummit47114.2019.00007
73. Tabak, E.K., Cambazoglu, B.B., Aykanat, C.: Improving the performance of independent task assignment heuristics MinMin, MaxMin and Sufferage. IEEE Trans. Parallel Distrib. Syst. **25**(5), 1244–1256 (2014). https://doi.org/10.1109/TPDS.2013.107
74. Talaat, M., Alsayyari, A.S., Alblawi, A., Hatata, A.Y.: Hybrid-cloud-based data processing for power system monitoring in smart grids. Sustain. Cities Soc. **55**, 102049 (2020). https://doi.org/10.1016/j.scs.2020.102049
75. Topcuoglu, H., Hariri, S., Wu, M.Y.: Performance-effective and low-complexity task scheduling for heterogeneous computing. IEEE Trans. Parallel Distrib. Syst. **13**(3), 260–274 (2002). https://doi.org/10.1109/71.993206
76. Van den Bossche, R., Vanmechelen, K., Broeckhove, J.: Cost-optimal scheduling in hybrid IaaS clouds for deadline constrained workloads. In: Proceedings of the IEEE 3rd International Conference on Cloud Computing (CLOUD'10), pp. 228–235 (2010). https://doi.org/10.1109/CLOUD.2010.58
77. Varela, M., Skorin-Kapov, L., Ebrahimi, T.: Quality of service versus quality of experience, 1st edn., chap. 6. T-Labs Series in Telecommunication Services, pp. 85–96. Springer (2014). https://doi.org/10.1007/978-3-319-02681-7_6
78. Voorsluys, W., Broberg, J., Buyya, R.: Introduction to cloud computing, 1st edn., chap. 1, pp. 1–41. Wiley (2011). https://doi.org/10.1002/9780470940105.ch1
79. Wang, B., Song, Y., Sun, Y., Liu, J.: Managing deadline-constrained bag-of-tasks jobs on hybrid clouds. In: Proceedings of the 24th High Performance Computing Symposium (HPC'16), pp. 1–8 (2016). https://doi.org/10.22360/SpringSim.2016.HPC.039
80. Wang, W.J., Chang, Y.S., Lo, W.T., Lee, Y.K.: Adaptive scheduling for parallel tasks with QoS satisfaction for hybrid cloud environments. J. Supercomput. **66**(2), 783–811 (2013). https://doi.org/10.1007/s11227-013-0890-2
81. Weng, C., Lu, X.: Heuristic scheduling for bag-of-tasks applications in combination with QoS in the computational grid. Future Gener. Comput. Syst. **21**(2), 271–280 (2005). https://doi.org/10.1016/j.future.2003.10.004
82. Wu, H.Y., Lee, C.R.: Energy efficient scheduling for heterogeneous fog computing architectures. In: Proceedings of the 42nd IEEE Annual Computer Software and Applications Conference (COMPSAC'18), pp. 555–560 (2018). https://doi.org/10.1109/COMPSAC.2018.00085
83. Xu, J., Hao, Z., Zhang, R., Sun, X.: A method based on the combination of laxity and ant colony system for cloud-fog task scheduling. IEEE Access **7**, 116218–116226 (2019). https://doi.org/10.1109/ACCESS.2019.2936116
84. Yang, T., Gerasoulis, A.: DSC: scheduling parallel tasks on an unbounded number of processors. IEEE Trans. Parallel Distrib. Syst. **5**(9), 951–967 (1994). https://doi.org/10.1109/71.308533
85. Zhang, Y., Zhou, J., Sun, J.: Scheduling bag-of-tasks applications on hybrid clouds under due date constraints. J. Syst. Archit. **101**, 101654 (2019). https://doi.org/10.1016/j.sysarc.2019.101654
86. Zhang, Y., Zhou, J., Sun, L., Mao, J., Sun, J.: A novel firefly algorithm for scheduling bag-of-tasks applications under budget constraints on hybrid clouds. IEEE Access **7**, 151888–151901 (2019). https://doi.org/10.1109/ACCESS.2019.2948468
87. Zhou, X., Zhang, G., Sun, J., Zhou, J., Wei, T., Hu, S.: Minimizing cost and makespan for workflow scheduling in cloud using fuzzy dominance sort based HEFT. Future Gener. Comput. Syst. **93**, 278–289 (2019). https://doi.org/10.1016/j.future.2018.10.046

A Comprehensive Survey of Estimator Learning Automata and Their Recent Convergence Results

B. John Oommen, Xuan Zhang, and Lei Jiao

Abstract The pre-cursor field to Reinforcement Learning is that of Learning Automata (LA). Within this field, Estimator Algorithms (EAs) can be said to be the state-of-the-art. Further, the subset of *Pursuit* Algorithms (PAs), discovered by Thathachar and Sastry [34, 39], were the pioneering schemes. This chapter contains a comprehensive survey of the various EAs, and the most recent convergence results for PAs. Unlike the prior LA, EAs are based on a fundamentally distinct phenomenon. They are also the most accurate LA, converging in the least time. EAs operate on two vectors, namely, the action probability vector which is updated using responses from the Environment, and quickly-computed estimates of the reward probabilities of the various actions. The proofs that they are ε-optimal is thus very complex. They have to incorporate two rather snon-orthogonal phenomena, which are the convergence of these estimates *and* the convergence of the probabilities of selecting the various actions. For almost three decades, the reported proofs of PAs possessed an infirmity (or flaw), which we refer to as the claim of the "monotonicity" property. This flaw was discovered by the authors of [37], who also provided an alternate proof for a specific PA where the scheme's parameter decreased with time. This paper first records all the reported EAs. It then reports a comprehensive survey of the proofs from a different perspective. These proofs have not required that the sequence of action probabilities of selecting the optimal action satisfies the property of *monotonicity*. On the other hand, whenever any action probability is close enough to unity, we require that the process jumps to an absorbing barrier at the next time instant, i.e., in a single step. By requiring such a constraint, these proofs invoke the *weaker* property, i.e., the sub-martinagale property of $p_m(t)$, to demonstrate the ε-optimality. We have thus proven

B. John Oommen (✉)
School of Computer Science, Carleton University, Ottawa, Canada
e-mail: oommen@scs.carleton.ca

B. John Oommen · L. Jiao
Centre for Artificial Intelligence Research, University of Agder, Grimstad, Norway
e-mail: lei.jiao@uia.no

X. Zhang
NORCE, Norwegian Research Center, Jon Lilletuns vei 9, 4879 Grimstad, Norway

© The Author(s), under exclusive license to Springer Nature Switzerland AG 2022
P. Nicopolitidis et al. (eds.), *Advances in Computing, Informatics, Networking and Cybersecurity*, Lecture Notes in Networks and Systems 289,
https://doi.org/10.1007/978-3-030-87049-2_2

the ε-optimality for the Absorbing CPA [49, 50], the Discretized PA [51, 52], and for the family of Bayesian PA [53], where the estimates are obtained by a Bayesian (rather than a Maximum Likelihood (ML)) process.

Keywords Pursuit learning automata (LA) · Martingale properties of LA · Convergence proofs of LA

1 Introduction

1.1 Learning Automata

An *automaton* is a self-operating machine, attempting to achieve a certain goal. As part of its mandate, it is supposed to respond to a sequence of instructions to attain to this goal. These instructions are usually pre-determined, and could be deterministic or stochastic. The automaton is designed so as to work with a random Environment, and it does this in such a way that it achieves "Learning". From a psychological perspective, the latter implies that the automaton can acquire knowledge and consequently, modify its behavior based on the experience that it has gleaned. Thus, we refer to the automata studied in this Chapter as *adaptive automata* or *learning automata*, since they can adapt to the dynamics of the Environment that they interact with. More precisely, the adaptive automata that we study in this Chapter, glean information from the responses that they receive from the Environment by virtue of interacting with it. Further, the automata attempt to learn the most desirable action from the set of actions offered to them by the random Environment, which could be stationary or non-stationary. The Automaton, is thus, a model for a stochastic "decision maker", whose goal is to adaptively learn the best action.

Learning Automata (LA) are skeletal models for reinforcement learning. A LA interacts with a Random Environment and seeks to learn the best (i.e., optimal) action that the latter offers. The model of computation is as follows: At every single time instant, the LA chooses one action. This, in turn, warrants a response from the Environment, modelled as being either a "Reward" or a "Penalty". Based on this response and the historical knowledge that the LA has gleaned, it chooses a subsequent action, in a well-planned algorithmically strategic manner, in order to make "wiser" decisions in the future. Thus, the LA, even though it does not have a complete knowledge about the Environment, is able to attain to the optimal decision through repeated interactions.

LA have boasted scores of applications. These include theoretical problems like the graph partitioning problem [30]. They have been used in controlling intelligent vehicles [42]. When it concerns neural networks and hidden Markov models, Meybodi *et al.* [5, 20] have used them in adapting the former, and the authors of [15] have applied them in training the latter. Network call admission, traffic control and quality of service routing have been resolved using LA in [3, 4, 44], while the authors of [36] have studied the problem of achieving distributed scheduling using LA. They have also found applications in tackling problems involving network and communications issues [21, 24, 31, 32], and have been utilized in the path control of vehicles [42].

Game playing has been resolved using LA [9, 26], as has been parameter optimization [5]. Knapsack-like problems were solved using the fundamental tools of LA and then these solutions have been used in web polling and sampling [12]. Granmo and his co-authors have used them to allocate limited resources in a stochastically optimally manner [8, 10–12]. LA have also been applied in processing natural language data, and in string taxonomy [25]. One can achieve map learning [6], and service selection in stochastic environments [45] using LA. Finally, web crawling [41], microassembly path planning [19], numerical optimization [43], just-in-time manufacturing systems [14] and multiagent learning [7], have been achieved using the tools available in this field.

Apart from these, scores of applications have invoked the fertile fields of LA and stochastic learning, as one can see from the reference books [16, 22, 23, 33, 40].

In their simplest forms, LA have a Fixed Structure, and machines of such a design are classified as being *Fixed Structure* Stochastic Automata (FSSA). FSSA are characterized by having time-invariant state update and decision (or output) functions. The pioneering examples of these were due to Tsetlin, Krylov and Krinsky [23]. Subsequently-developed *Variable* Structure Stochastic Automata (VSSA) work with a different mechanism. They possess state update and output functions that modify the action selection probabilities. Examples of VSSA are the Linear Reward-Penalty (L_{R-P}) rule (also termed as a "scheme"), the Linear Reward-Inaction (L_{R-I}) scheme, the Linear Inaction-Penalty (L_{I-P}) scheme and the Linear Reward-εPenalty ($L_{R-\varepsilon P}$) scheme [23]. By definition, the L_{R-I} and $L_{R-\varepsilon P}$ algorithms apportion a greater significance or importance to the reward responses from the Environment than to penalties. With regard to their accuracies, the latter are also ε-optimal in all stationary Environments. This philosophy is, indeed, pertinent for FSSA, because unlike the Tsetlin automaton (which is only ε-optimal when the largest reward probability is greater than 0.5 [23]), in both the Krinsky and Krylov automata, the machines treat rewards noticeably "more seriously" than penalties, and are thus ε-optimal in all stationary Environments.

LA can also be characterized by their Markovian representations. They thus fall into one of two families, being either ergodic or those that possess absorbing barriers. Absorbing automata have underlying Markov Chains that get absorbed or locked into a barrier state. Sometimes this can occur even after a relatively small, *finite* number of iterations. The classic book [23] reports numerous LA families that contain such absorbing barriers. On the other hand, the literature has also reported scores of ergodic automata [23, 29], which converge in distribution. In these cases, the asymptotic *distribution* of the action probability vector converges to a value that is independent of its initial vector. Absorbing LA are usually designed to operate in stationary Environments. As opposed to these, ergodic LA are preferred for non-stationary Environments, namely those that possess time-dependent reward probabilities.

VSSA can also be characterized as being Continuous or Discretized. This depends on the values that the action probabilities can take. Continuous LA allow the action probabilities to assume any value in the interval [0, 1]. Such algorithms have a relatively slow rate of convergence. One can increase their speeds of convergence, by incorporating the concept of discretization [17, 18, 27]. This phenomenon is implemented by constraining the action selection probability to be one of a finite number of values in the interval [0, 1]. The discretization is said to be *linear* if the values allowed are equally spaced in this interval, otherwise, it is called *non-linear*. By incorporating discretization, almost all of the reported VSSA of the continuous type have been also discretized [27, 29, 47].

Within the field of LA, Estimator Algorithms (EAs) can be said to be the state-of-the-art. Further, within this family, the sub-family of *Pursuit* Algorithms (PAs), discovered by Thathachar and Sastry [34, 39], were the pioneering schemes.

With the above as a backdrop, we now concentrate on the main goal of our chapter, namely, to focus on the family of Estimator Algorithms (EAs). These algorithms, like the previous VSSA, maintain, and update a vector of action selection probabilities. However, unlike the previously-studied VSSA, these algorithms also keep running estimates of each action's reward probability. This is done using a *reward-estimate vector*. The convergence results (proofs) of all the EAs have an infirmity and should not, strictly speaking, be referred to as "proofs". The flaw was discovered by the authors of [37]. The details of this flaw and how it has been rectified in subsequent papers, has also been explained here. The algebraic and finer details of the proofs are, however, omitted.

1.2   Contributions of This Paper

The contributions of this chapter are listed below.

1. The first goal of this work is to give a comprehensive overview of various families of EAs and PAs.
2. We then mention the error referred to above, that was used to "prove" the convergence of the various families of PAs. This error was first reported by the authors of [37].
3. We then report a new method which is able to correctly prove the ε-optimality of all artificially-enforced absorbing EAs. These are EAs which are specifically augmented with artificially-enforced absorbing states [28]. This proof is, in particular, valid for the ACPA, which is the CPA with artificially-enforced absorbing states.
4. The method proposed in [37] is, of course, valid for the proofs of absorbing EAs. It resorts to the so-called monotonicity property. We have shown that it is *sufficient* for convergence, but it is not really *necessary* for obtaining a formal proof that the ACPA is ε-optimal.
5. We thus briefly highlight the main components (without the fine details) of a completely new proof methodology which is based on the convergence theory of submartingales and the theory of Regular functions [23]. This proof is, thus,

distinct in principle and argument from the proof reported in [37], and it adds insights into the convergence of different absorbing EAs.

6. Thereafter, we utilize the convergence theory of submartingales and the theory of Regular functions to allude to the ε-optimality of the DPA, although the formal proof is omitted in the interest of space.

7. Finally, we mention how we can invoke the same theories (of submartingales and of Regular functions) to prove the ε-optimality of the families of Bayesian PAs, where the estimates of the reward probabilities are not computed using a ML methodology, but rather due to a Bayesian paradigm.

8. The most interesting and significant contribution of this short article is that we have given a single comprehensive proof for the ACPA, DPA, and the corresponding continuous and discretized Bayesian PAs, *all within one single derivation*. We are not aware of any similar proof in the entire body of LA-related literature!

2   Estimator Algorithms

2.1   *Rationale and Motivation*

The speed of convergence of LA is one of the most important considerations in designing them. This motivated the invention of the family of "discretized" algorithms. Extending this need, Thathachar and Sastry designed the class referred to as *Estimator Algorithms* (EAs) [34, 38, 39]. These converge much faster than all the previous families. EAs maintain and update an action probability vector, but also keep running estimates for each action that is rewarded, using a *reward-estimate vector*. They then also use those estimates in the probability updating equations. The reward estimates vector is denoted in the literature by $\hat{D}(t) = [\hat{d}_1(t), \ldots, \hat{d}_r(t)]^T$. The corresponding state vector is denoted by $Q(t) = \langle P(t), \hat{D}(t) \rangle$.

EAs rely on the *confidence* in the reward capabilities of the different actions. For example, these algorithms initially process each action a number of times, and then (in one version) could increase the probability of the action with the highest reward estimate [1]. This leads to a scheme with better accuracy in choosing the correct action. The previous non-estimator VSSA algorithms update the probability vector directly and solely on the basis of the current response of the Environment, where, the action selection probabilities are increased or decreased based on the reward or penalty received, and on the type of vector updating scheme being used. However, EAs update the probability vector based on both the estimate vector, and the Environment's current feedback response. It thus influences the probability vector both directly and indirectly, the latter as being a result of the estimation of the reward estimates. This could lead to increases in action selection probabilities of actions that are not currently rewarded.

The added computational cost, involved in maintaining the reward estimates, are well compensated for. EAs have an order of magnitude superior performance than the non-estimator algorithms previously introduced. Lanctôt and Oommen [18] further introduced even faster discretized versions of these continuous EAs.

2.2 Continuous Estimator Algorithms

The family of Pursuit Algorithms (PAs) is a sub-class of EAs that *pursue* an action that the automaton "currently" perceives to be the optimal one.

The CP_{RP} algorithm, as it is referred to in the literature, was the first-reported PA. It was introduced by Thathachar and Sastry [34, 38, 39]. At every time instant, the algorithm chooses an action $\alpha(t)$ based on the action probability vector, $P(t)$. Whether the response is a reward or a penalty, it increases that component of $P(t)$ which has the maximal current reward estimate, and it decreases the probability corresponding to the rest of the actions. Finally, the algorithm updates the running estimates of the reward probability of the action chosen. The estimate vector $\hat{D}(t)$ can be computed using a ML paradigm:

$$\hat{d}_i(t) = \frac{W_i(t)}{Z_i(t)}, \quad \forall i = 1, 2, \ldots, r, \tag{1}$$

where $W_i(t)$ is the number of times the action α_i has been rewarded until the current time t, and $Z_i(t)$ is the number of times α_i has been chosen until the current time t. Thus, since the *currently perceived* "best action" is rewarded, *its* action probability value is increased with a value proportional to its distance to unity, namely $1 - p_m(t)$. To render the consistent nature of the updating, the "less optimal actions" are penalized by decreasing their probabilities proportionally.

Based on the above concepts, the CP_{RP} algorithm is formally given in [1, 18, 29].

The algorithm is similar in principle to the L_{RP} algorithm, because both the CP_{RP}, and the L_{RP} algorithms increase/decrease the action probabilities of the vector independent of whether the Environment responds to the automaton with a reward or a penalty. The major difference lies in the way the reward estimates are maintained, used, and are updated on both reward/penalty. It should be emphasized that whereas the non-pursuit algorithm moves the probability vector in the direction of the most recently rewarded action, the pursuit algorithm moves the probability vector in the direction of the action with the highest reward estimate. Thathachar and Sastry [34, 38, 39] have theoretically "proven" their ε-optimality, and experimentally shown that these PAs are more accurate, and several orders of magnitude faster than the non-pursuit algorithms.

The reward-inaction version of this pursuit algorithm is also similar in design, and is described in [1, 18, 29]. Also, other pursuit-like estimator schemes have also been devised, and they can be found in [1, 29].

2.3   The TSE Algorithm

A more advanced EA, which we refer to as the TSE algorithm to maintain consistency with the existing literature [1, 18, 29], was designed by Thathachar and Sastry [38, 39].

Like the other EAs, the TSE algorithm maintains the running reward estimates vector $\hat{D}(t)$, and uses it to calculate the action probability vector $P(t)$. When an action $\alpha_i(t)$ is rewarded, according to the TSE algorithm, the probability components with a reward estimate greater than $\hat{d}_i(t)$ are treated differently from those components with a value lower than $\hat{d}_i(t)$. The algorithm does so by increasing the probabilities for all the actions that have a higher estimate than the estimate of the chosen action, and decreasing the probabilities of all the actions with a lower estimate. This is done with the help of an indicator function $S_{ij}(t) = 1$ when $\hat{d}_i < \hat{d}_j$, and $S_{ij}(t) = 0$ when $\hat{d}_i >= \hat{d}_j$. In this way, the TSE algorithm uses both the probability vector $P(t)$ and the reward estimates vector $\hat{D}(t)$ to update the action probabilities. The algorithm is formally described in [1]. On careful inspection of the algorithm, it can be observed that $P(t + 1)$ depends indirectly on the response of the Environment to the LA. The feedback from the Environment changes the values of the components of $\hat{D}(t)$, which, in turn, affects the values of the functions $f(.)$ and $S_{ij}(t)$ [1, 18, 38, 39]. Analyzing the algorithm carefully, we obtain three cases. If the ith action is rewarded, the probability values of the actions with reward estimates higher than the reward estimate of the currently selected action are updated using the following equation [38]:

$$p_j(t + 1) = p_j(t) - \lambda \left[f(\hat{d}_i(t) - \hat{d}_j(t)) \frac{\{p_i(t) - p_j(t)p_i(t)\}}{r - 1} \right] \qquad (2)$$

When $\hat{d}_i(t) < \hat{d}_j(t)$, since the function $f(\hat{d}_i(t) - \hat{d}_j(t))$ is monotonic and increasing, $f(\hat{d}_i(t) - \hat{d}_j(t))$ is seen to be negative. This leads to a higher value of $p_j(t + 1)$ than that of $p_j(t)$, which indicates that the probability of choosing actions, that have estimates greater than that of the estimates of the currently chosen action, will increase.

For all the actions with reward estimates smaller than the estimate of the currently selected action, the probabilities are updated based on the following equation:

$$p_j(t + 1) = p_j(t) - \lambda f\left(\hat{d}_i(t) - \hat{d}_j(t)\right) p_j(t). \qquad (3)$$

The sign of the function $f\left(\hat{d}_i(t) - \hat{d}_j(t)\right)$ is positive, which indicates that the probability of choosing actions, that have estimates less than that of the estimate of the currently chosen action, will decrease.

Thathachar and Sastry have "proven" that the TSE algorithm is ε-optimal [38]. They have also experimentally shown that the TSE algorithm often converges several orders of magnitude faster than the L_{RI} scheme.

2.4   The Generalized Pursuit Algorithm

Agache and Oommen [2] proposed a generalized version of the Pursuit algorithm (CP_{RP}) proposed by Thathachar and Sastry [34, 38, 39]. Their algorithm, called the *Generalized Pursuit Algorithm* (GPA), generalizes Thathachar and Sastry's Pursuit algorithm by pursuing all those actions that possess higher reward estimates than the chosen action. In this way the probability of choosing a wrong action is minimized. Agache and Oommen experimentally compared their pursuit algorithm with the existing algorithms, and found that their algorithm is the best in terms of the rate of convergence [2].

In the CP_{RP} algorithm, the probability of the best estimated action is maximized by first decreasing the probability of all the actions in the following manner [2]:

$$p_j(t+1) = (1 - \lambda)p_j(t), \quad j = 1, 2, \ldots, r. \tag{4}$$

The sum of the action probabilities is made unity, by the help of the probability mass Δ, which is given by [2]:

$$\Delta = 1 - \sum_{j=1}^{r} p_j(t+1) = 1 - \sum_{j=1}^{r}(1-\lambda)p_j(t) = 1 - \sum_{j=1}^{r} p_j(t) + \lambda \sum_{j=1}^{r} p_j(t) = \lambda. \tag{5}$$

Thereafter, the probability mass Δ is added to the probability of the best-estimated action. The GPA algorithm, thus, equi-distributes the probability mass Δ to the action estimated to be superior to the chosen action. This gives us [2]:

$$p_m(t+1) = (1-\lambda)p_m(t) + \Delta = (1-\lambda)p_m(t) + \lambda, \tag{6}$$

where $\hat{d}_m = \max_{j=1,2,\ldots,r}(\hat{d}_j(t))$. Thus, the updating scheme is given by [2]:

$$p_j(t+1) = (1-\lambda)p_j(t) + \frac{\lambda}{K(t)}, \quad \text{if } \hat{d}_j(t) > \hat{d}_i(t), j \neq i$$

$$p_j(t+1) = (1-\lambda)p_j(t), \quad \text{if } \hat{d}_j(t) \leq \hat{d}_i(t), j \neq i$$

$$p_i(t+1) = 1 - \sum_{j \neq i} p_j(t+1), \tag{7}$$

where $K(t)$ denotes the number of actions that have estimates greater than the estimate of the reward probability of the action currently chosen. The formal algorithm is omitted, but can be found in [2].

2.5 Discretized Estimator Algorithms

As we have seen so far, discretized LA are superior to their continuous counterparts, and the estimator algorithms are superior to the non-estimator algorithms in terms of their rates of convergence. Utilizing the previously proven capabilities of discretization in improving the speed of convergence of the learning algorithms, Lanctôt and Oommen [18] enhanced the Pursuit and the TSE algorithms. This led to the designing of classes of learning algorithms, referred to in the literature as the *Discrete Estimator Algorithms* (DEA) [18]. To this end, as done in the previous discrete algorithms, the components of the action probability vector are allowed to assume a finite set of discrete values in the closed interval [0, 1], which is, in turn, divided into the number of sub-intervals that is proportional to the resolution parameter, N. Along with this, a reward estimate vector is maintained to keep an estimate of the reward probability of each action [18].

Lanctôt and Oommen showed that for each member algorithm belonging to the class of DEAs to be ε-optimal, it must possess a pair of properties known as the *Property of Moderation* and the *Monotone Property*. Together these properties help "prove" the ε-optimality of any DEA algorithm.

Moderation Property: A DEA with r actions and a resolution parameter N is said to possess the *property of moderation*, if the maximum magnitude by which an action probability can decrease per iteration is bounded by $\frac{1}{rN}$.

Monotone Property: Suppose there exists an index m and a time instant $t_0 < \infty$, such that $\hat{d}_m(t) > \hat{d}_j(t)$, $\forall j$ s.t. $j \neq m$ and $\forall t$ s.t. $t \geq t_0$, where $\hat{d}_m(t)$ is the maximal component of $\hat{D}(t)$. A DEA is said to possess the *Monotone Property*, if there exists an integer N_0 such that for all resolution parameters $N > N_0$, $p_m(t) \mapsto 1$ with probability one as $t \mapsto \infty$, where $p_m(t)$ is the maximal component of $P(t)$.

The discretized versions of the Pursuit Algorithm, and the TSE Algorithm possessing the moderation, and the monotone properties are presented below.

The Discretized Pursuit Algorithm: The Discretized Pursuit Algorithm (formally described in [18]), is referred to as the DPA in the literature, and is similar to a great extent, to its continuous pursuit counterpart, i.e., the CP_{RI} algorithm, except that the updates to the action probabilities for the DPA algorithm are made in discrete steps. Therefore, the equations in the CP_{RP} algorithm that involve multiplication by the learning parameter λ are substituted by the addition or subtraction by quantities proportional to the smallest step size.

As in the CP_{RI} algorithm, the DPA algorithm operates in three steps. If $\Delta = \frac{1}{rN}$ (where N denotes the resolution, and r the number of actions) denotes the smallest step size, the integral multiples of Δ denote the step sizes in which the action probabilities are updated. Like the continuous Reward-Inaction algorithm, when the chosen action $\alpha(t) = \alpha_i$, is penalized, the action probabilities remain unchanged. However, when the chosen action $\alpha(t) = \alpha_i$ is rewarded, and the algorithm has not converged, the algorithm decreases, by the integral multiples of Δ, the action probabilities which do not correspond to the highest reward estimate.

Lanctôt and Oommen have shown that the DPA algorithm possesses the properties of moderation and monotonicity, and that it is thus ε-optimal [18]. They have also experimentally proved that in different ranges of Environments from simple to complex, the DPA algorithm is at least 60% faster than the CP_{RP} algorithm [18].

The Discretized TSE Algorithm: Lanctôt and Oommen also discretized the TSE algorithm, and have referred to it as the *Discrete TSE algorithm* (DTSE) [18]. Since the algorithm is based on the continuous version of the TSE algorithm, it obviously has the same level of intricacies, if not more. Lanctôt and Oommen theoretically "proved" that like the DPA estimator algorithm, this algorithm also possesses the *Moderation* and the *Monotone* properties, while maintaining many of the qualities of the continuous TSE algorithm. They also provided the "proof" of convergence of this algorithm.

There are two notable parameters in the DTSE algorithm:

1. $\Delta = \frac{1}{rN\theta}$, where N is the resolution parameter as before, and
2. θ is an integer representing the largest value any of the action probabilities can change by in a single iteration.

A formal description of the DTSE algorithm is omitted here, but is in [18].

The Discretized Generalized Pursuit Algorithm: Agache and Oommen [2] provided a discretized version of their GPA algorithm presented earlier. Their algorithm, called the *Discretized Generalized Pursuit Algorithm* (DGPA) also essentially generalizes Thathachar and Sastry's Pursuit algorithm [38, 39]. But unlike the TSE, it pursues all those actions that possess higher reward estimates than the chosen action.

In essence, in any single iteration, the algorithm computes the number of actions that have higher reward estimates than the current chosen action, denoted by $K(t)$, whence the probability of all the actions that have estimates higher than the chosen action is increased by an amount $\frac{\Delta}{K(t)}$, and the probabilities for all the other actions are decreased by an amount $\frac{\Delta}{(r-K(t))}$, where $\Delta = \frac{1}{rN}$ denotes the resolution step, and N the resolution parameter. The DGPA algorithm has been "proven" to possess the *Moderation* and *Monotone* properties, and to thus "be" ε-optimal [2]. The detailed steps of the DGPA algorithm are omitted here.

2.6   The Use of Bayesian Estimates in PAs

There are many formal methods to achieve estimation, the most fundamental of them being the ML and Bayesian paradigms. These are age-old, and have been used for centuries in various application domains. As mentioned in the previous sections, PAs have been designed, since the 1980s, by using ML estimates of the Reward probabilities. However, as opposed these estimates, more recently, the present authors (and others) have shown that the family of Bayesian estimates can also be incorporated into the design and analysis of LA. Incorporating the Bayesian philosophy has led to various Bayesian Pursuit Algorithms (BPAs).

The first of these BPAs was the Continuous Bayesian Pursuit Algorithm (CBPA) (which unless otherwise stated, is referred to here as the BPA) [46]. This then made the way for the development of the Discretized Bayesian Pursuit Algorithm (DBPA) [48]. The BPAs obey the same "pursuit" paradigm for achieving the learning. However, since one can invoke the properties of their *a priori* and *a posteriori* distributions, the Bayesian estimates can provide more meaningful estimates when it concerns the field of LA, and are superior to their ML counterparts. The families of BPAs are, probably, the fastest and most accurate non-hierarchical LA reported in the literature.

3   Previously-Reported Inaccurate "Proofs"

It is relatively easy to design a new LA. However, it is extremely difficult to analyze a new machine. Indeed, the formal proofs of the convergence accuracies of LA is, probably, the most difficult part in the fascinating field involving the design and analysis of LA. Since this is a comprehensive chapter, it would be educative to record the distinct mathematical techniques used for the various families, namely for the FSSA, VSSA, Discretized and PAs.

- The proof methodology for the family of FSSA simply involves formulating the Markov chain for the LA. Thereafter, one computes its equilibrium (or steady state) probabilities, and then derives the asymptotic action selection probabilities.
- The proofs of the asymptotic convergence for VSSA involve the theory of small-step Markov processes and distance diminishing operators. In more complex cases, they resort to the theory of Regular functions.
- The proofs for Discretized LA also consider the asymptotic analysis of the LA's Markov chain in the discretized space. Thereafter, one derives the *total* convergence probability to the various actions.
- The convergence proofs become much more complex when we have to deal with two rather non-orthogonal inter-related concepts, as in the case of the families of EAs. These two concepts are the convergence of the reward estimates *and* the convergence of the action probabilities themselves. Ironically, the combination of these phenomena to achieve the updating rule is what grants the EAs their enhanced speed. But this leads to a surprising dilemma: If the accuracy of the estimates computed is not dependable because of an inferior estimation phase (i.e., if the non-optimal actions are not examined "enough number of times"), the accuracy of the EA converging to the optimal action can be forfeited. Thus the proofs for EAs have to consider criteria which the other types of LA do not have to.

The ε-optimality of the EAs, and the families of PAs including the CPA, GPA, DPA, TSE and the DTSE (explained in the previous sections) have been studied and presented in [1, 2, 18, 34, 38, 39]. The methodology for all these derivations was the same, and was erroneous. The estimate used was a MLE, and by the law of large numbers, when all the actions are chosen infinitely often, they claimed that the estimates are ordered in terms of their Reward probabilities. They also claimed that these estimates are properly ordered *forever* after some time instant, t_0 and thus the action probability of selecting the optimal action would increase monotonically, resulting in the convergence to the optimal action. We refer to this as *the monotonicity property*. If now the learning parameter is small/large enough, they will consequently converge to the optimal action with surely.

The fault in the argument is the following: While such an ordering is, indeed, true by the law of large numbers, it is only valid if all the actions are chosen infinitely often. This, in turn, forces the time instant, t_0, to also be infinite. If such an "infinite" selection does not occur, the ordering cannot be guaranteed for *all* time instants after t_0. In other words, the authors of all these papers misinterpreted the concept of ordering "forever" with the ordering "most of the time" after t_0, rendering the "proofs" of the ε-optimality invalid.

The authors of [37] discovered the error in the proofs of the above-mentioned papers, and the detailed explanation of this discovery is found in [37]. These authors also presented a correct proof for the ε-optimality of the CPA based on the monotonicity property, except that their proof required an external condition that the learning parameter, λ, decreased with time. This, apparently, is the only strategy by which one can prove the ε-optimality of the CPA. However, by designing the LA to jump to an absorbing barrier in a single step when any $p_j(t) \geq T$, where T is a user-defined threshold close to unity, the updated proofs have shown ε-optimality because of a *weaker* property, i.e., the submartinagale property of $p_m(t)$, where t is greater than a finite time index, t_0.

It is not an elementary exercise to generalize the arguments of [37] for the submartinagale property if the value of λ is kept fixed. However, the submartinagale arguments can avoid the previously-flawed methods of reasoning. This is achieved by them not requiring that the process $\{p_m(t)_{(t \geq 0)}\}$ satisfies the *monotonicity* property, but the *submartinagale* property. This phenomenon is valid for the corrected proofs of the CPA, DPA, and the family of Bayesian PAs.

4   The Proofs for the Pursuit Learning Paradigm

Since the proofs that we consider in greater detail all concern the so-called "Pursuit Learning Paradigm" we shall briefly formalize it first. As explained in the previous sections, all the algorithms follow the same pursuit leaning paradigm, in which, LAs maintain both action probabilities and reward estimates so as to make the best choice. In every single interaction between the LA and its Environment, the algorithm selects an action by sampling from the action probability vector, then the Environment stochastically provides feedback for the selected action with a reward or a penalty. Based on the response from the Environment, the LA updates its reward estimates, and pursues the current best action by increasing its action probability upon receiving a reward. Thus, in the interest of simplicity, we formalize the learning process in the following pursuit paradigm, which is valid for ML and Bayesian estimates:

Pursuit Learning Paradigm

Parameters:

α_j: The jth action of the r actions that can be selected by the LA.
p_j: The jth element of the action probability vector P.
λ: The learning parameter in continuous algorithms, where $0 < \lambda < 1$.
Δ: The learning parameter in discretized algorithms, where $\Delta = \frac{1}{rN}$.
a_j: The number of times α_j has been rewarded plus 1.
b_j: The number of times α_j has been penalized plus 1.
d_j: The jth element of the reward probability vector D.
$\hat{d}_j$: The jth element of the reward estimates vector $\hat{D}$.
m: The index of the optimal action.
h: The index of the greatest element of $\hat{D}$.
R: The response from the Environment, where $R = 1$ corresponds to a Reward, and $R = 0$ to a Penalty.
T: A Threshold, where $T \geq 1 - \varepsilon$.

Method:

1. Initialize the LA: $p_j(0) = \frac{1}{r}$, $a_j(0) = b_j(0) = 1$, and t:=1.
2. The LA selects an action based on sampling from $P(t)$. Suppose $\alpha(t) = \alpha_i$.
3. Increase $a_i(t)$ by 1 if the response from the Environment is a reward, otherwise, increase $b_i(t)$ by 1:

$$a_i(t) = a_i(t-1) + R(t),$$
$$b_i(t) = b_i(t-1) + R(t).$$

4. Update $\hat{d}_i(t)$ according to the estimating method:

   (a) Maximum Likelihood estimates:

   $$\hat{d}_i(t) = \frac{a_i(t)}{(a_i(t) + b_i(t))}.$$

   (b) The upper 95% reward probability bound of Bayesian estimates:

   $$\int\limits_0^{\hat{d}_j(t)} f_j(v; a_j(t), b_j(t))dv = \frac{\int_0^{\hat{d}_j(t)} v^{(a_j(t)-1)}(1-v)^{(b_j(t)-1)}dv}{\int_0^1 u^{(a_j(t)-1)}(1-u)^{(b_j(t)-1)}du} = 0.95,$$

   where $f_j(v; a_j(t), b_j(t)) = \frac{v^{(a_j(t)-1)}(1-v)^{(b_j(t)-1)}}{\int_0^1 u^{(a_j(t)-1)}(1-u)^{(b_j(t)-1)}du}$, is the probability density function of the *Beta* distribution of the jth action at time t.

5. Suppose $\hat{d}_h(t)$ is the largest element of $\hat{D}(t)$, then α_h is the best action in current iteration. The LA increases the probability of the current best action (i.e., pursues it) by increasing its action probability $p_h(t)$ upon receiving a reward:

   – For continuous algorithms, they are updated based on continuous L_{RI} rules:

$$\textbf{If } R(t) = 1 \textbf{ Then}$$
$$p_j(t+1) = (1-\lambda)p_j(t), j \neq h$$
$$p_h(t+1) = 1 - \sum_{j \neq h} p_j(t+1).$$
$$\textbf{Else}$$
$$P(t+1) = P(t)$$
$$\textbf{EndIf}$$
$$t = t+1$$

   – For discretized algorithms, the discretized L_{RI} rules are applied to update the action probabilities:

$$\textbf{If } R(t) = 1 \textbf{ Then}$$
$$p_j(t+1) = \max\{(p_j(t) - \Delta), 0\}, j \neq h$$
$$p_h(t+1) = 1 - \sum_{j \neq h} p_j(t+1).$$
$$\textbf{Else}$$
$$P(t+1) = P(t)$$
$$\textbf{EndIf}$$
$$t = t+1$$

6. Repeat from Step 2 until a stopping criterion (one of the following) is met.

   – For continuous algorithms, when an action probability surpasses the threshold T, *that specific* action probability will jump to 1 and break the loop. This operation renders continuous algorithms absorbing, and thus we call them absorbing continuous algorithms.

$$\textbf{If } p_j(t+1) \geq T, \ \forall j \in (1, 2, \ldots, r)$$
$$p_j(t+1) = 1$$
$$\textit{Break}$$
$$\textbf{EndIf}$$

   – in the case of discretized algorithms, the loop will be broken when an action probability reaches 1:

$$\textbf{If } p_j(t+1) = 1, \ \forall j \in (1, 2, \ldots, r)$$

Break

EndIf

End Pursuit Learning Paradigm

5 Proofs of Convergence of the Pursuit-Paradigm-Based LA

5.1 The Previous Erroneous "Proof"

The previous flawed proofs for the convergence of the Pursuit Algorithms followed a four-step process, which will be introduced in the next subsection. The flaw exists in Step 2, where the reward estimates are proven to converge to their true values in probability, but rather stated to converge to the true values for sure. This led to a mistaken "monotonicity" property of the probability sequence $\{p_m(t)_{t>t_0}\}$, instead of the submartingale property. The flawed proof was elaborated and rectified in [37, 50].

5.2 The New State-of-The-Art Proofs

A pursuit-paradigm-based algorithm is considered as ε-optimal if it follows the statement of Theorem 1.

Theorem 1 *Given any $\varepsilon > 0$ and $\delta > 0$, there exist a $\lambda_0 > 0$ for continuous algorithms ($\Delta_0 > 0$ for discretized algorithms),[1] and a $t_0 < \infty$ such that for all time $t \geq t_0$ and for any positive learning parameter $\lambda < \lambda_0$ ($\Delta < \Delta_0$),*

$$Pr\{p_m(t) > 1 - \varepsilon\} > 1 - \delta.$$

The standard proof for Theorem 1 contains the following four steps:

1. The Moderation Property of the LA. This is to guarantee that by using a sufficiently small learning parameter ($\lambda < \lambda_0$ or $\Delta < \Delta_0$), each action can be selected an arbitrarily large number of times before a certain time instant.
2. Based on the Moderation Property, there exists a time instant, t_0, after which, the reward estimates will be accurate enough.
3. When the reward estimates are sufficiently accurate, the action probability sequence of the optimal action, $\{p_m(t)_{t>t_0}\}$ is a submartingale.
4. Based on the submartingale convergence property and the theory of Regular functions, the probability that $p_m(\infty) = 1$ converges to 1.

---

[1] In the rest of the paper, we put the counterparts for discretized algorithms in parentheses.

Indeed, the proofs for the convergence of all the algorithms involved in this paper followed the same four-step method, with distinct proving details.

Proof for Moderation property:

The moderation property of continuous algorithms has been proven in [35], and for discretized algorithms, the property is elaborated in [27]. Both proofs applied sufficiently small values for the learning parameters, i.e., λ in the continuous algorithms, and Δ in the discretized algorithms, to ensure that before the LA reaches any absorbing state, each of its actions would be selected an arbitrarily large number of times.

Proof for accuracy of the Reward Estimates:

The estimates that are used to estimate action reward probabilities generally involve two classic estimating methods. They are the ML Estimates being used by the ACPA and the DPA, and the Bayesian Estimates that are used in the BPA family. The proof for their accuracy is based on the weak law of large numbers. In the proof for the accuracy of the MLE reward estimates [50, 52], the Hoeffding's inequality was applied to prove that each estimate will remain in a sufficiently small neighbourhood of its true value if the number of samples are sufficiently large. Besides, by virtue of the Hoeffding's inequality, the number of times each action needs to be selected can be lower-bounded in order to achieve a certain degree of estimating accuracy.

The proof for Bayesian reward estimates [53] is a bit different from that for MLE, as the former utilizes the 95% percentile instead of the mean as the quantity for estimation. The proof indeed consists of two parts. The first part is similar to the proof for MLE, where it shows the mean of the posterior *Beta* distribution can get arbitrarily close to the real reward probability if all actions get enough number of trials. The second part is then to prove that the 95% percentile of the posterior probability distribution will be arbitrarily close to the mean given the number of trials for each action is large enough.

Once the reward estimates are accurate enough, it means that the probability that the optimal action is ranked as the best action will be arbitrarily high, and this is the prerequisite for the rest of the convergence proof.

Proof for $\{p_m(t)_{t>t_0}\}$ being a submartingale:

This part of proof follows the definition of submartingales. Firstly, $p_m(t)$ is a probability, thus $E[p_m(t)] \leq 1 < \infty$. Secondly, to prove $E[p_m(t + 1)|p_m(t)] \geq p_m(t)$, we calculate $E[p_m(t + 1)|P(t)]$ explicitly so that we can compare the results with $p_m(t)$ afterwards. Calculating $E[p_m(t + 1)|P(t)]$ depends only on the action probability updating rules. If we define $q(t)$ as the probability that the optimal action is ranked as the best action at time instant t, then in the absorbing continuous algorithms:

$$E[p_m(t + 1)|P(t)] = p_m(t) + \lambda(q(t) - p_m(t)) \sum_{j=1...r} p_j(t)d_j,$$

and in the discretized algorithms:

$$E[p_m(t+1)|P(t)] = p_m(t) + (q(t)(c_t\Delta + \Delta) - \Delta) \sum_{j=1...r} p_j(t)d_j,$$

where $c_t = 1, 2, \ldots, r - 1$. In order for $E[p_m(t+1)|P(t)] \geq p_m(t)$, for absorbing continuous algorithms, $q(t)$ needs to be greater than $p_m(t)$. Therefore, if in the first two steps of the proof, we define the learning parameter λ_0 and the time instant t_0 to be such that for all $\lambda < \lambda_0$ and for all $t > t_0$, the reward estimates can achieve an accuracy that satisfies $q(t)_{t>t_0} > T$, then $q(t)_{t>t_0} > p_m(t)_{t>t_0}$, and $\{p_m(t)_{t>t_0}\}$ is a submartingale.

Similarly, in the discretized algorithms, when we define Δ_0 and t_0 such that for all $\Delta < \Delta_0$, and for all $t > t_0$, $q(t) > \frac{1}{2}$, then for the same Δ and t, $q(t) > \frac{1}{c_t+1}$, and then $E[p_m(t+1)|P(t)] \geq p_m(t)$, whence $\{p_m(t)_{t>t_0}\}$ is a submartingale.

Proof for $Pr\{p_m(\infty) = 1\} \to 1$:
Based on the submartingale convergence property, in the infinite time horizon, $p_m(t)$ converges either to 0 or to 1. Then the theory of Regular functions are utilized to prove $p_m(\infty)$ converges to 1 with an arbitrarily large probability. In brief, although it is difficult to observe directly the quantity of $Pr\{p_m(\infty)\}$, it can be studied indirectly by investigating a Regular function of P, which is, indeed, an equivalence to $Pr\{p_m(\infty)\}$. This Regular function is further lower-bounded by a subregular function of P, which itself converges to 1 when the learning rate λ (Δ) goes to 0. As the lower-bound subregular function converges to 1, the Regular function will also converge to 1, hence $Pr\{p_m(\infty) = 1\} \to 1$.

The key (and indeed, the most difficult aspect) in this part of the proof is to correlate $Pr\{p_m(\infty)\}$ with the Regular function of P, and to find also the subregular function of P to lower-bound the Regular function. All these need quite a bit of algebraic manipulations, and one can refer to [50, 52, 53] for additional details.

In summary, the Moderation property of the pursuit-paradigm-based algorithms guarantees enough trails for each action, and then according to the weak law of large numbers, both MLE reward estimates and the 95% percentile Bayesian estimates are able to achieve arbitrarily high estimating accuracy. High accuracy of reward estimates renders the probability sequence $\{p_m(t)_{t>t_0}\}$ to be a submartingale, and thereafter, based on the submartingale convergence property and the theory of Regular functions, $p_m(\infty)$ convergence to 1 with an arbitrarily high probability. The convergence of all these algorithms is this correctly proven.

6 Conclusions

Estimator Algorithms (EAs) have been recorded to be the fastest Learning Automata (LA) reported in the literature. Within the family of EAs, the sub-family of *Pursuit* algorithms are the ones which pioneered the field. This chapter first presents a complete survey of the various EAs and PAs reported, including those using a continuous or discretized philosophy.

We have then moved on to the formal proofs of their convergence accuracies, i.e., their ε-optimality, which is well-known to be the most intricate part in the design and analysis of LA. The convergence proofs for all the existing PAs in all the reported papers since the 1980s possessed a common error. This was discovered by the authors of [37]. The latter paper rectified the error and provided a new proof for the Continuous Pursuit Algorithm (CPA), and this had its foundations in the so-called monotonicity property of the $\{p_m(t)_{(t>t_0)}\}$ sequence. Their proof, however, required the resolution/parameter to decrease/increase gradually.

The authors of this chapter have earlier proven the convergence of the CPA which contains an artificial absorbing barrier, the ACPA. However, *this* current proof examined the *submartingale* property of $\{p_m(t)_{(t>t_0)}\}$. Thereafter, by virtue of the submartingale property and the weaker condition, the new proof invoked the theory of Regular functions, and used a constant resolution/learning parameter. The strategy of the proof has also been used to prove the convergence of the Discretized Pursuit Algorithm, and also the continuous and discretized version of the PAs that utilize Bayesian estimates instead of Maximum Likelihood estimates.

References

1. Agache, M.: Estimator Based Learning Algorithms. M.C.S. Thesis, School of Computer Science, Carleton University, Ottawa, Ontario, Canada (2000)
2. Agache, M., Oommen, B.J.: Generalized pursuit learning schemes: new families of continuous and discretized learning Automata. IEEE Trans. Syst. Man Cybern. Part B **32**(6), 738–749 (2002)
3. Atlassis, A.F., Loukas, N.H., Vasilakos, A.V.: The use of learning algorithms in ATM networks call admission control problem: a methodology. Comput. Netw. **34**, 341–353 (2000)
4. Atlassis, A.F., Vasilakos, A.V.: The use of reinforcement learning algorithms in traffic control of high speed networks. Advances in Computational Intelligence and Learning, pp. 353–369 (2002)
5. Beigy, H., Meybodi, M.R.: Adaptation of parameters of BP algorithm using learning automata. In: Proceedings of Sixth Brazilian Symposium on Neural Networks. JR, Brazil, pp. 24–31 (2000)
6. Dean, T., Angluin, D., Basye, K., Engelson, S., Aelbling, L., Maron, O.: Inferring finite automata with stochastic output functions and an application to map learning. Mach. Learn. **18**, 81–108 (1995)
7. Erus, G., Polat, F.: A layered approach to learning coordination knowledge in multiagent environments. Appl. Intell. **27**, 249–267 (2007)
8. Granmo, O.C.: Solving stochastic nonlinear resource allocation problems using a hierarchy of twofold resource allocation automata. IEEE Trans. Comput. **59**(4), 545–560 (2010)
9. Granmo, O.C., Glimsdal, S.: Accelerated Bayesian learning for decentralized two-armed bandit based decision making with applications to the Goore game. Appl. Intell. **38**, 479–488 (2013)
10. Granmo, O.C., Oommen, B.J.: On allocating limited sampling resources using a learning automata-based solution to the fractional knapsack problem. In: Proceedings of the 2006 International Intelligent Information Processing and Web Mining Conference, Advances in Soft Computing, vol. 35, Ustron, Poland, pp. 263–272 (2006)
11. Granmo, O.C., Oommen, B.J.: Optimal sampling for estimation with constrained resources using a learning automaton-based solution for the nonlinear fractional knapsack problem. Appl. Intell. **33**(1), 3–20 (2010)

12. Granmo, O.C., Oommen, B.J., Myrer, S.A., Olsen, M.G.: Learning automata-based solutions to the nonlinear fractional knapsack problem with applications to optimal resource allocation. IEEE Trans. Syst. Man Cybern. Part B **37**(1), 166–175 (2007)

13. Hoeffding, W.: Probability inequalities for sums of bounded random variables. J. Am. Stat. Assoc. **58**, 13–30 (1963)

14. Hong, J., Prabhu, V.V.: Distributed reinforcement learning control for batch sequencing and sizing in just-in-time manufacturing systems. Appl. Intell. **20**, 71–87 (2004)

15. Kabudian, J., Meybodi, M.R., Homayounpour, M.M.: Applying continuous action reinforcement learning automata (CARLA) to global training of hidden markov models. In: Proceedings of ITCC'04, the International Conference on Information Technology: Coding and Computing, Las Vegas, Nevada, 2004, pp. 638–642

16. Lakshmivarahan, S.: Learning Algorithms Theory and Applications. Springer (1981)

17. Lanctot, J.K., Oommen, B.J.: On discretizing estimator-based learning algorithms. IEEE Trans. on Systems, Man, and Cybernetics, Part B: Cybernetics **2**, 1417–1422 (1991)

18. Lanctot, J.K., Oommen, B.J.: Discretized estimator learning automata. IEEE Trans. Syst. Man Cybern. Part B: Cybern. **22**(6), 1473–1483 (1992)

19. Li, J., Li, Z., Chen, J.: Microassembly path planning using reinforcement learning for improving positioning accuracy of a 1 cm^3 omni-directional mobile microrobot. Appl. Intell. **34**, 211–225 (2011)

20. Meybodi, M.R., Beigy, H.: New learning automata based algorithms for adaptation of back-propagation algorithm parameters. Int. J. Neural Syst. **12**, 45–67 (2002)

21. Misra, S., Oommen, B.J.: GPSPA?: A new adaptive algorithm for maintaining shortest path routing trees in stochastic networks. Int. J. Commun. Syst. **17**, 963–984 (2004)

22. Najim, K., Poznyak, A.S.: Learning Automata: Theory and Applications. Pergamon Press, Oxford (1994)

23. Narendra, K.S., Thathachar, M.A.L.: Learning Automata: An Introduction. Prentice Hall (1989)

24. Obaidat, M.S., Papadimitriou, G.I., Pomportsis, A.S., Laskaridis, H.S.: Learning automata-based bus arbitration for shared-medium ATM switches. IEEE Trans. Syst. Man Cybern. Part B **32**, 815–820 (2002)

25. Oommen, B.J.: Stochastic searching on the line and its applications to parameter learning in nonlinear optimization. IEEE Trans. Syst. Man Cybern. Part B: Cybern. **27**(4), 733–739 (1997)

26. Oommen, B.J., Granmo, O.C., Pedersen, A.: Using stochastic AI techniques to achieve unbounded resolution in finite player Goore Games and its applications. In: Proceedings of IEEE Symposium on Computational Intelligence and Games, Honolulu, HI, pp. 161–167 (2007)

27. Oommen, B.J., Lanctot, J.K.: Discretized pursuit learning automata. IEEE Trans. Syst. Man Cybern. **20**, 931–938 (1990)

28. Oommen, B.J.: Absorbing and ergodic discretized two-action learning automata. IEEE Trans. Syst. Man Cybern. **16**, 282–296 (1986)

29. Oommen, B.J., Agache, M.: Continuous and discretized pursuit learning schemes: various algorithms and their comparison. IEEE Trans. Syst. Man Cybern. **31**(3), 277–287 (2001)

30. Oommen, B.J., Croix, T.D.S.: Graph partitioning using learning automata. IEEE Trans. Comput. **45**, 195–208 (1996)

31. Oommen, B.J., Roberts, T.D.: Continuous learning automata solutions to the capacity assignment problem. IEEE Trans. Comput. **49**, 608–620 (2000)

32. Papadimitriou, G.I., Pomportsis, A.S.: Learning-automata-based TDMA protocols for broadcast communication systems with bursty traffic. IEEE Commun. Lett. 107–109 (2000)

33. Poznyak, A.S., Najim, K.: Learning Automata and Stochastic Optimization. Springer, Berlin (1997)

34. Sastry, P.S.: Systems of Learning Automata: Estimator Algorithms Applications. Ph.D. Thesis, Department of Electrical Engineering, Indian Institute of Science, Bangalore, India (1985)

35. Rajaraman, K., Sastry, P.S.: Finite time analysis of the pursuit algorithm for learning automata. IEEE Trans. Syst. Man Cybern. Part B: Cybern. **26**, 590–598 (1996)

36. Seredynski, F.: Distributed scheduling using simple learning machines. Eur. J. Oper. Res. **107**, 401–413 (1998)
37. Ryan, M., Omkar, T.: On ε-optimality of the pursuit learning algorithm. J. Appl. Probab. **49**(3), 795–805 (2012)
38. Thathachar, M.A.L., Sastry, P.S.: A class of rapidly converging algorithms for learning automata. IEEE Trans. Syst. Man Cybern. **SMC-15**, 168–175 (1985)
39. Thathachar, M.A.L., Sastry, P.S.: Estimator algorithms for learning automata. In: Proceedings of the Platinum Jubilee Conference on Systems and Signal Processing, Department of Electrical Engineering, Indian Institute of Science, Bangalore, India, Dec 1986, pp. 29–32
40. Thathachar, M.A.L.T., Sastry, P.S.: Networks of Learning Automata?: Techniques for Online Stochastic Optimization. Kluwer Academic, Boston (2003)
41. Torkestani, J.A.: An adaptive focused web crawling algorithm based on learning automata. Appl. Intell. **37**, 586–601 (2012)
42. Unsal, C., Kachroo, P., Bay, J.S.: Multiple stochastic learning automata for vehicle path control in an automated highway system. IEEE Trans. Syst. Man Cybern. Part A **29**, 120–128 (1999)
43. Vafashoar, R., Meybodi, M.R., Momeni, A.A.H.: CLA-DE: a hybrid model based on cellular learning automata for numerical optimization. Appl. Intell. **36**, 735–748 (2012)
44. Vasilakos, A., Saltouros, M.P., Atlassis, A.F., Pedrycz, W.: Optimizing QoS routing in hierarchical ATM networks using computational intelligence techniques. IEEE Trans. Syst. Sci. Cybern. Part C **33**, 297–312 (2003)
45. Yazidi, A., Granmo, O.C., Oommen, B.J.: Service selection in stochastic environments: a learning-automaton based solution. Appl. Intell. **36**, 617–637 (2012)
46. Zhang, X., Granmo, O.C., Oommen, B.J.: The Bayesian pursuit algorithm: A new family of estimator learning automata. In: Proceedings of IEAAIE2011. pp. 608–620. Springer, New York, USA (2011)
47. Zhang, X., Granmo, O.C., Oommen, B.J.: Discretized Bayesian pursuit—a new scheme for reinforcement learning. In: Proceedings of IEAAIE2012. Dalian, China, pp. 784–793 (2012)
48. Zhang, X., Granmo, O.C., Oommen, B.J.: On incorporating the paradigms of discretization and Bayesian estimation to create a new family of pursuit learning automata. Appl. Intell. **39**, 782–792 (2013)
49. Zhang, X., Granmo, O.C., Oommen, B.J., Jiao, L.: On using the theory of regular functions to prove the ε-optimality of the continuous pursuit learning automaton. In: Proceedings of IEAAIE2013, pp. 262–271. Springer, Amsterdam, Holland (2013)
50. Zhang, X., Granmo, O.C., Oommen, B.J., Jiao, L.: A formal proof of the ε-optimality of absorbing continuous pursuit algorithms using the theory of regular functions. Appl. Intell. **41**, 974–985 (2014)
51. Zhang, X., Oommen, B.J., Granmo, O.C., Jiao, L.: Using the theory of regular functions to formally prove the ε-optimality of discretized pursuit learning algorithms. In: Proceedings of IEAAIE2014, pp. 379–388. Springer, Kaohsiung, Taiwan (2014)
52. Zhang, X., Oommen, B.J., Granmo, O.C., Jiao, L.: A formal proof of the ε-optimality of discretized pursuit algorithms. Appl. Intell. (2015). https://doi.org/10.1007/s10489-015-0670-1
53. Zhang, X., Oommen, B.J., Granmo, O.C.: The design of absorbing bayesian pursuit algorithms and the formal analyses of their ε-Optimality. Pattern Anal. Appl. **20**(3) (2015)

Multimodal Data Fusion

Zhikui Chen, Liang Zhao, Qiucen Li, Xin Song, and Jianing Zhang

Abstract In the era of big data, a large number of multimodal data are constantly generated and accumulated. In the practical application analysis of multimodal big data, the modal incompleteness, real-time processing, modal imbalance and high-dimensional attributes pose severe challenges to the multimodal fusion method. In view of the above characteristics of multimodal data and the shortcomings of existing multimodal fusion algorithms, this chapter studies the multimodal data fusion algorithms from four aspects: incomplete modal analysis fusion, incremental modal clustering fusion, heterogeneous modal migration fusion and low dimensional modal sharing fusion. A series of multimodal data fusion models and algorithms are designed.

Keywords Multimodal data · Data fusion · Unsupervised learning

1 Introduction

With the rapid development of Internet of Things, social network and electronic commerce, data are growing and accumulating at an unprecedented speed. For example, according to the research report released by IDC, a famous consulting company, the total amount of network data generated in the world in 2011 is 1.8 ZB, and it is expected to reach 35 ZB by 2020 [1]. On Facebook, 30 billion pieces of content are shared every month. Assuming that the size of each content is 2 KB, 60 TB of data will be generated every month [2]. Walmart, a famous global supermarket chain, needs to process more than 1 million user requests per hour and maintains the database with the size of more than 2.5 PB, while it is still expanding [3]. Thus, the era of big data has arrived [4].

Nowadays, big data are gradually integrated into all walks of life of social development. It is highly valued by governments at all levels, field experts and business

Z. Chen (✉) · L. Zhao · Q. Li · X. Song · J. Zhang
School of Software, Dalian University of Technology, No. 321 Tuqiang Street, Dalian Economic and Technological Development Zone, Dalian 116620, Liaoning, P.R. China
e-mail: zkchen@dlut.edu.cn

© The Author(s), under exclusive license to Springer Nature Switzerland AG 2022
P. Nicopolitidis et al. (eds.), *Advances in Computing, Informatics, Networking and Cybersecurity*, Lecture Notes in Networks and Systems 289,
https://doi.org/10.1007/978-3-030-87049-2_3

people, and has become a research hotspot in the academic and industrial circles [5, 6]. The rapid growth of data has brought valuable resources to many industries, but also brought great challenges. Effective feature learning of massive and complex big data, discovering the hidden knowledge and rules in big data, and then mining the potential value of big data will greatly promote the development of various fields of society [7–9].

In the era of big data, with the wide application of multimedia technology and the richness of data description means, multimodal data widely exist. Multimodal data refer to the data obtained through different domains or perspectives for the same description object, and each domain or perspective describing these data is called a modality [10, 11]. For example, in face recognition (as shown in Fig. 1a), multimodal data may be composed of images from different angles of the face; in video analysis, each video can be decomposed into audio, image, subtitle and other modal information; as shown in Fig. 1b, in news dissemination, each news can be described in multiple languages to obtain different modal representations of the same news; in the social network platform, a social information can be expressed as text, picture, link and other modal data; in the image classification, each image can be expressed as different modal features such as pixel intensity or gray, texture (as shown in Fig. 1c). In multimodal data, each modality can provide certain information for the rest modalities, that is, there is a certain correlation between the modalities [12, 13]. Different from traditional data analysis, when big data are used to solve practical problems, all relevant modal information should be considered at the same time. However, in multimodal data mining, the same processing of different modal data or simple connection and integration of all modal features cannot guarantee the effectiveness of the mining task. Therefore, it is the main difference between big data research and traditional data research to study the multimodal data fusion analysis method, learn more accurate complex data features through the complementary information between modalities, and support the subsequent big data decision-making and prediction. It is also the main problem and technical difficulty of big data research at this stage.

The key to big data analysis and mining is multimodal data fusion, however, the modal incompleteness, real-time processing, modal imbalance and high-dimensional

   (a) multi-angle features   (b) multi-language features   (c) multi-visual features

Fig. 1 Examples of multimodal data

attributes of multimodal data pose severe challenges to the design of fusion methods. Modal incompleteness refers to the loss of some modal information of some data instances in multimodal data. For example, in multi-language text classification, most multimodal text instances contain all language descriptions, while some text instances contain only one or several language descriptions. Real-time processing refers to the rapid generation and the need for real-time analysis and processing of multimodal data. Modal imbalance means that in multimodal data analysis, the number of some modal data instances is large, while the number of other modal data instances is relatively small, so it is necessary to use the modal data with more instances to assist the analysis and learning of the modal data with fewer instances. The high-dimensional property of attributes means that the modal feature description of multimodal data has high-dimensional attributes. For example, in the social network platform, the image and text of data instances are represented as high-dimensional vectors.

Therefore, in view of the characteristics of multimodal data such as modal incompleteness, real-time processing, modal imbalance and high-dimensional attributes, this chapter studies the multimodal data fusion algorithm in four aspects: incomplete modal analysis fusion, incremental modal clustering fusion, heterogeneous modal migration fusion and low-dimensional modal sharing fusion. The details are as follows:

(1) In order to solve the problem of modal incompleteness of multimodal data, an incomplete multimodal data fusion algorithm based on deep semantic matching is proposed. Based on the correlation of multimodal high-level semantics, a unified deep learning model is designed which combines the modal-private deep network and the modal-shared features. In addition, based on the spatial geometric characteristics of the modalities, the modal local invariant graph regularized factors are designed for coupling multimodal deep shared features and the original modal features. The model is updated by using the optimization algorithms such as coordinate descent and back propagation, and the semantic fusion features of multimodal data are obtained.

(2) Aiming at the problem of real-time processing of multimodal data, an incremental co-clustering fusion algorithm for parameter-free multimodal data is proposed. A new similarity measurement for multimodal data is defined, and three incremental clustering strategies are designed, which are cluster creation, cluster merging and instance division, for the incremental fusion of multimodal data. At the same time, an adaptive modal weight mechanism is designed to dynamically adjust the modal weight in the co-clustering fusion process.

(3) For the problem of modal imbalance of multimodal data, a heterogeneous modal data migration and fusion algorithm based on multi-layer semantic matching is proposed. A multi-layer semantic matching unified deep network architecture is designed by coupling modal deep network and modal correlation analysis model. At each level, the feature correlation fusion of multimodal data is carried out, and the modal output features of the top level are used to correlate the modal network as a whole. A new deep fusion objective function is defined, the heterogeneous modal-private deep matching network and modal

high-level semantic common space are optimized and learned, and the transfer and fusion from modal knowledge in source domain to task in target domain is completed.

(4)   As for the problem of high-dimensional attributes of multimodal data, a unsupervised multimodal data non-negative correlation feature sharing fusion algorithm is proposed. A co-learning model of modal-private (uncorrelated or negatively correlated) features and cross modal shared (correlated) features is designed. The coupling of shared features is used to establish the joint optimization objective function of each modality. Modal invariant graph regularization and projection matrix sparseness are used to assist the model optimization process. Finally, the robust cross modal data fusion features in low dimensional subspace are obtained by iterative co-learning of correlated and uncorrelated features between modalities.

To sum up, different data fusion algorithms are designed according to the characteristics of multimodal data to effectively fuse and mine big data features, and extract the potential semantics and hidden knowledge of big data, which has certain research significance and application value.

2   Incomplete Multimodal Data Fusion Algorithm Based on Deep Semantic Matching

This section proposes an incomplete multimodal data fusion algorithm based on deep semantic matching. First, a deep semantic matching model is builded, which combines a deep neural network to fuse modal and matrix decomposition to deal with incomplete multimodal. And this section gives its detailed design process, Then elaborates on the model update and optimization process.

2.1   Problem Description

Deep multimodal analysis can effectively compensate for the semantic deviation between modal through multi-layer nonlinear conversion. Inspired by it, this section proposes a deep incomplete multimodal data fusion method, utilizing deep learning networks and incomplete multimodal data fusion methods to jointly learn the deep semantic matching network from each modal to the shared space and the deep shared subspace between modal.

Assuming incomplete multimodal data set $X = \{X^{(v)}\}_{v=1}^{V} = \{X_C^{(v)}, X_I^{(v)}\}_{v=1}^{V}$ contains V data model and n data instances belonging to k clusters. $X_C^{(v)} \in R^{d_v \times c}$ and $X_I^{(v)} \in R^{d_v \times n_v}$ represent the v-th modal feature matrix of c complete modal data instances and the v-th modal of other n_v data instances respectively. Correspondingly, each modal contains $c + n_v$ data instances, every data instances contains

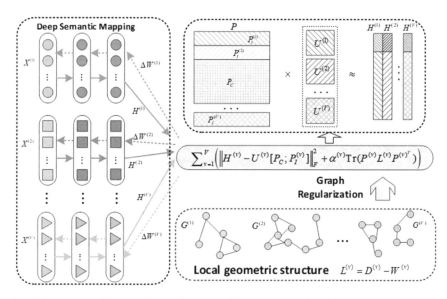

Fig. 2 Illustration of the work flow of the proposed incomplete multimodal deep semantic mapping method

d_v dimensional attribute feature. Since multimodal data usually contains different feature representations of the same data instance, therefore, different modal data instance descriptions generally have the same semantic representation. Based on it, this section proposes a high-level semantic matching method that couples deep multimodal matrix decomposition and local invariant graph regularization, which can mine deep semantic sharing features among incomplete multimodal data. As shown in Fig. 2, first, the modal private deep neural network converts each modal data instance $X^{(v)} = \{X_C^{(v)}, X_I^{(v)}\}$ into a deep feature representation $\{H^{(v)}\} = \{H_C^{(v)}, H_I^{(v)}\}$. Then, each modal representation $H^{(v)}$ is Decomposed into base matrix $U^{(v)}$ and uniform coding matrix $P^{(v)} = [P_C; P_I^{(v)}]$ by a non-negative matrix factorization model fused with local invariant graph regularization. Through joint training, optimizing the modal private deep network and base matrix, and uniform coding matrix, Multimodal deep semantic sharing feature in subspace will be acquired. In the following sections, incomplete multimodal non-negative matrix factorization, local invariant graph regularization, deep incomplete multimodal semantic matching algorithm and optimization algorithm process will be explained in detail.

2.2 Incomplete Multimodal Non-negative Matrix Factorization

In the existing multimodal non-negative matrix factorization algorithm, consistent coding is the basic principle [14, 15]. The incomplete multimodal nonnegative matrix factorization method proposed in this section is also based on coupling the consistent coding matrix between complete modal instances to learn the shared features of different modal in the subspace. Given the characteristic dimension k of the potential shared space, $\{P_C^{(v)}\}_{v=1}^V \in R^{k \times c}$ is the latent representation of modal data $\{X_C^{(v)}\}_{v=1}^V$. Because $\{X_C^{(v)}\}_{v=1}^V$ is complete modal data, the same data instance should have the same semantics in different modalities, therefore, the data characteristics of each modal in the latent subspace should be as similar as possible, So the latent representation of each modal is approximately equal to the uniform coding matrix P_C. Therefore, the basic incomplete multimodal data fusion learning model can be expressed as follows:

$$\min \sum_{v=1}^V \left\| [X_C^{(v)}; X_I^{(v)}] - U^{(v)}[P_C; P_I^{(v)}] \right\|_F^2,$$

$$s.t. \, U^{(v)} \geq 0, \quad P^{(v)} = [P_C; P_I^{(v)}] \geq 0, \tag{2.1}$$

where $\{U^{(v)}\}_{v=1}^V \in R^{d_v \times k}$ is the base matrix of the matrix factorization, $\{P_I^{(v)}\}_{v=1}^V \in R^{k \times n_v}$ is the latent representation of incomplete modal data instances $\{X_I^{(v)}\}_{v=1}^V$ in subspace. In this way, in Eq. (2.1), each modal can only have the same complete modal data coding matrix P_C, we can couple the basis matrix and the modal private coding matrix to minimize the objective function. However, the optimization model in Eq. (2.1) cannot guarantee the inherent geometry and discriminative structure of each modal data in the learned shared subspace [16], In order to solve this problem, The next section will introduce the local invariant graph rule division factor of modal data to optimize and improve it.

2.3 Local Invariant Graph Regularization

This section uses the invariant graph model to represent regularly the learned shared coding matrix P_C, to ensure that each modal data is consistent with its geometric structure in the latent subspace. Assuming two instances $x_i^{(v)}$ and $x_j^{(v)}$ in modal v are closer in the original data space, then their low-dimensional representations $p_i^{(v)}$ and $p_j^{(v)}$ are also as close as possible in the learned subspace. As shown in Fig. 2, the local geometric structure between data points can be described by constructing the nearest neighbor graph $G^{(v)}$ of each modal. Every data instance $x_i^{(v)}$ in modal v is represented as one point in $G^{(v)}$, By searching p points closest to $x_i^{(v)}$, we can construct the weight adjacency matrix $W^{(v)}$ of $G^{(v)}.W_{ij}^{(v)}$ represents the similarity of data instances $x_i^{(v)}$ and $x_j^{(v)}$ in $W^{(v)}$. In this section, the following heat kernel weighting method [17] is used to measure it:

$$W_{ij}^{(v)} = \begin{cases} \exp(-d_{ij}^{(v)}/\sigma), & x_i^{(v)} \in N_p(x_j^{(v)})||x_j^{(v)} \in N_p(x_i^{(v)}), \\ 0, & \text{otherwise}, \end{cases} \tag{2.2}$$

where $d_{ij}^{(v)}$ is Euclidean distance between data instances $x_i^{(v)}$ and $x_j^{(v)}$, $N_p(x_i^{(v)})$ and $N_p(x_j^{(v)})$ represent p nearest data instances of $x_i^{(v)}$ and $x_j^{(v)}$ respectively. Similarly, Euclidean distance can also be used to measure the similarity of two data instances $p_i^{(v)}$ and $p_j^{(v)}$ in the shared subspace. By fusing the similarity matrix $W^{(v)}$ in the original space of the modal and similarity measurement in shared space, The local invariant graph embedding function can be obtained for each modal, as follows:

$$\begin{aligned} \Re &= \frac{1}{2} \sum_{i,j=1}^{c+n_v} \left\| P_i^{(v)} - P_j^{(v)} \right\|^2 W_{ij}^{(v)} \\ &= \sum_{i=1}^{c+n_v} P_i^{(v)^T} P_i^{(v)} D_{ii}^{(v)} - \sum_{i,j=1}^{c+n_v} P_i^{(v)^T} P_j^{(v)} W_{ij}^{(v)} \\ &= Tr(P^{(v)} D^{(v)} P^{(v)^T}) - Tr(P^{(v)} W^{(v)} P^{(v)^T}) \\ &= Tr(P^{(v)} L^{(v)} P^{(v)^T}), \end{aligned} \tag{2.3}$$

where $Tr()$ is matrix trace, $D^{(v)}$ is diagonal matrix, Each data on the diagonal is the sum of each row or column of $W^{(v)}$, $L^{(v)} = D^{(v)} - W^{(v)}$ is Graph Laplacian matrix of $G^{(v)}$.

By minimizing $\Re$, Two similar data instances $x_i^{(v)}$ and $x_j^{(v)}$ ($W_{ij}^{(v)}$ is greater) in the modal original space are also similar ($\| P_i^{(v)} - P_j^{(v)} \|^2$ is smaller) in the shared subspace. By integrating local invariant graph embedding function for each modal and incomplete multimodal non-negative matrix factorization, Equation (2.1) can be rewritten as:

$$\min \sum_{v=1}^{V} \left(\left\| [X_C^{(v)}; X_I^{(v)}] - U^{(v)}[P_C; P_I^{(v)}] \right\|_F^2 + \alpha^{(v)} Tr(P^{(v)} L^{(v)} P^{(v)^T}) \right),$$

$$s.t.\, U^{(v)} \geq 0, \quad P^{(v)} = [P_C; P_I^{(v)}] \geq 0, \tag{2.4}$$

where $\alpha^{(v)} > 0$ represents regularization parameters.

2.4 Incomplete Multimodal Deep Semantic Matching Algorithm

Most of the existing incomplete multimodal data fusion algorithms use linear or nonlinear transformations to learn the shared characteristics of two modal data, and achieve good results. However, when arbitrary modal data and the feature distribution between the modalities is quite different, simple single-layer linear or nonlinear conversion may cause degradation of the performance of the fusion result. Deep neural networks can learn the deep semantic information of data through multilayer nonlinear transformation, Applying deep neural networks to learn multimodal data fusion features has attracted wide attention from scholars [18]. Therefore, this section combines deep learning networks and incomplete multimodal data analysis algorithms, joint mines deep semantic matching features of arbitrary modal data. Correspondingly, the incomplete multimodal deep semantic matching model can be expressed as:

$$\min \sum_{v=1}^{V} \left(\left\| [H_C^{(v)}; H_I^{(v)}] - U^{(v)}[P_C; P_I^{(v)}] \right\|_F^2 + \alpha^{(v)} Tr(P^{(v)} L^{(v)} P^{(v)^T}) \right),$$

$$s.t.\, U^{(v)} \geq 0, \quad P^{(v)} = [P_C; P_I^{(v)}] \geq 0, \tag{2.5}$$

where $H^{(v)} = [H_C^{(v)}; H_I^{(v)}] = f(W_v X^{(v)} + b_v)$ is feature output of modal private deep network, f is nonlinear activation function (the Sigmoid function is used in this section), W_v, b_v Are the weight matrix and bias vector respectively. Through joint optimization of modal private deep learning network, basis matrix and consistent coding matrix, multimodal deep semantic shared subspace can be obtained, fusion analysis of multimodal data features can be performed in the obtained subspace. The next section will describe in detail the optimization update process of the proposed incomplete multimodal deep semantic matching model.

2.5 Model Optimization

Equation (2.5) is a non-convex function of coupled variables $U^{(v)}$, $P^{(v)}$ and $\theta^{(v)} = \{W_v, b_v\}$, its optimization needs to use the block coordinate descent method to iteratively optimize each variable, and then reaches the local optimum [19]. In the specific optimization process, when optimizing a variable, other variables need to be kept unchanged. The detailed optimization process is as follows:

(1)   Given $U^{(v)}$ and $\theta^{(v)}$, update $P^{(v)}$:

Since different modalities contain the part original data of deep semantic spatial features, the introduction of local invariant graph regularization factors makes the optimization of Eq. (2.5) more difficult. As Fig. 2 shows, shared characteristics

$[P_C; P_I^{(1)}]$, $[P_C; P_I^{(2)}]$ and $[P_C; P_I^{(v)}]$ only contain part of P, therefore, in the specific optimization process, we update shared feature P_C and $P_I^{(v)}$ respectively, and the invariant graph restriction is relaxed as:

$$\alpha^{(v)} Tr(P^{(v)} L^{(v)} P^{(v)^T}) \approx \alpha^{(v)} Tr(P_C L_C^{(v)} P_C^T). \tag{2.6}$$

Given the base matrix $U^{(v)}$ and the modal deep network $\theta^{(v)}$ of each modal, By minimizing shared feature P_C and $P_I^{(v)}$, the objective function Eq. (2.5) can be written as:

$$\min_{P_C \geq 0} \sum_{v=1}^{V} \left(\left\| H_C^{(v)} - U^{(v)} P_C \right\|_F^2 + \alpha^{(v)} tr(P_C L^{(v)} P_C^T) \right), \tag{2.7}$$

$$\min_{P_I^{(v)} \geq 0} \left\| H_I^{(v)} - U^{(v)} P_I^{(v)} \right\|_F^2. \tag{2.8}$$

Lagrangian function optimization is used to obtain Eq. (2.7) and (2.8) respectively:

$$L(P_C) = \sum_{v=1}^{V} (Tr(H_C^{(v)} H_C^{(v)^T}) - 2Tr(H_C^{(v)} P_C^T U^{(v)^T})$$
$$+ Tr(U^{(v)} P_C P_C^T U^{(v)^T}) + \alpha^{(v)} Tr(P_C L_C^{(v)} P_C^T) + Tr(\Psi^{(v)} P_C)), \tag{2.9}$$

$$L(P_I^{(v)}) = Tr(H_I^{(v)} H_I^{(v)^T}) - 2Tr(H_I^{(v)} P_I^{(v)^T} U^{(v)^T})$$
$$+ Tr(U^{(v)} P_I^{(v)} P_I^{(v)^T} U^{(v)^T}) + Tr(\Phi^{(v)} P_I^{(v)}), \tag{2.10}$$

where Lagrange Multiplier $\Psi^{(v)}$, $\Phi^{(v)}$ restrict $P_C \geq 0$ and $P_I^{(v)} \geq 0$. Using KKT (Karush–Kuhn–Tucker) conditions $(\Psi^{(v)})_{ij}(P_C)_{ij} = 0$ and $(\Phi^{(v)})_{ij}(P_I^{(v)})_{ij} = 0$. The following equation about $(P_C)_{ij}$ and $(P_I^{(v)})_{ij}$ can be obtained:

$$\sum_{v=1}^{V} (-(U^{(v)^T} H_C^{(v)})_{ij}(P_C)_{ij} + (U^{(v)^T} U^{(v)} P_C)_{ij}(P_C)_{ij}$$
$$+ \alpha^{(v)} (P_C L_C^{(v)})_{ij}(P_C)_{ij}) = 0, \tag{2.11}$$

$$-(U^{(v)^T} H_I^{(v)})_{ij}(P_I^{(v)})_{ij} + (U^{(v)^T} U^{(v)} P_I^{(v)})_{ij}(P_I^{(v)})_{ij} = 0. \tag{2.12}$$

So the optimized update rules for shared features $(P_C)_{ij}$ and $(P_I^{(v)})_{ij}$ can be obtained:

$$(P_C)_{ij} \leftarrow \frac{\sum_{v=1}^{V} (U^{(v)^T} H_C^{(v)} + \alpha^{(v)} P_C W_C^{(v)})_{ij}}{\sum_{v=1}^{V} (U^{(v)^T} U^{(v)} P_C + \alpha^{(v)} P_C D_C^{(v)})_{ij}} (P_C)_{ij}, \tag{2.13}$$

$$(P_I^{(v)})_{ij} \leftarrow \frac{(U^{(v)^T} H_I^{(v)})_{ij}}{(U^{(v)^T} U^{(v)} P_I^{(v)})_{ij}} (P_I^{(v)})_{ij}. \tag{2.14}$$

(2)   Given $P^{(v)}$ and $\theta^{(v)}$, update $U^{(v)}$:

Given the consistent coding matrix $P^{(v)}$ and the modal depth network $\theta^{(v)}$ of each modal, by minimizing base matrix $U^{(v)}$, the objective function Eq. (2.5) can be further written as:

$$\min_{U^{(v)} \geq 0} \left\| H^{(v)} - U^{(v)} P^{(v)} \right\|_F^2. \tag{2.15}$$

Similar to the minimization of Eq. (2.8), we can get the update process of the basis matrix $U^{(v)}$ as follows. First the Lagrangian multiplier optimization of Eq. (2.15) is used:

$$L(U^{(v)}) = Tr(H^{(v)} H^{(v)^T}) - 2Tr(H^{(v)} P^{(v)^T} U^{(v)^T})$$
$$+ Tr(U^{(v)} P^{(v)} P^{(v)^T} U^{(v)^T}) + Tr(\Theta^{(v)} U^{(v)}). \tag{2.16}$$

The KKT condition $(\Theta^{(v)})_{ij}(U^{(v)})_{ij} = 0$ is used to get the equation about $(U^{(v)})_{ij}$:

$$-(H^{(v)} P^{(v)^T})_{ij}(U^{(v)})_{ij} + (U^{(v)} P^{(v)} P^{(v)^T})_{ij}(U^{(v)})_{ij} = 0, \tag{2.17}$$

the update rule of the base matrix $(U^{(v)})_{ij}$ is thus obtained:

$$(U^{(v)})_{ij} \leftarrow \frac{(H^{(v)} P^{(v)^T})_{ij}}{(U^{(v)} P^{(v)} P^{(v)^T})_{ij}} (U^{(v)})_{ij}. \tag{2.18}$$

(3)   Given $P^{(v)}$ and $U^{(v)}$, update $\theta^{(v)}$:

When the basis matrix $U^{(v)}$ and uniform coding matrix $P^{(v)}$ of each modal are fixed, the update of the modal network $\theta^{(v)}$ can be described as a sub-problem of minimizing the objective function Eq. (2.5), as follows:

$$\min_{W_v, b_v} O(\theta^{(v)}) = \left\| f(W_v X^{(v)} + b_v) - U^{(v)} P^{(v)} \right\|_F^2. \tag{2.19}$$

In order to optimize Eq. (2.19), we use the gradient descent method to sequentially update the network parameters of each layer in the back propagation process of the deep neural network [20]. For example, the weights W_v^l and bias vectors b_v^l of the l-th layer network are updated as follows:

$$W_v^l = W_v^l - \mu^{(v)} \frac{\partial O(\theta^{(v)})}{\partial W_v^l}, \tag{2.20}$$

$$b_v^l = b_v^l - \mu^{(v)} \frac{\partial O(\theta^{(v)})}{\partial b_v^l}, \tag{2.21}$$

where $\mu^{(v)}$ is learning rate of deep network, the gradient descent $\frac{\partial O(\theta^{(v)})}{\partial W_v^l}$ and $\frac{\partial O(\theta^{(v)})}{\partial b_v^l}$ of W_v^l and b_v^l can further be written as:

$$\frac{\partial O(\theta^{(v)})}{\partial W_v^l} = \delta_v^{l+1}(A_v^l)^T, \tag{2.22}$$

$$\frac{\partial O(\theta^{(v)})}{\partial b_v^l} = \delta_v^{l+1}, \tag{2.23}$$

where $A_v^l = f(z_v^l)$ is the l-th output of deep network for v-th modal. $H^{(v)} = A_v^{n_l^v}$ is the final layer output result of the deep network, n_l^v is the number of layers of the deep network. About δ_v^l,

(1) When $l = n_l^v$,

$$\delta_v^l = 2(A_v^l - U^{(v)} P^{(v)}) \cdot f'(z_v^l). \tag{2.24}$$

(2) When $l = 2, 3, \ldots, n_l^v - 1$,

$$\delta_v^l = (W_v^l)^T \delta_v^{l+1} \cdot f'(z_v^l), \tag{2.25}$$

where $\cdot$ is dot product of vector or matrix. The optimization update process of the entire incomplete multimodal deep semantic matching model is shown in Algorithm 2.1.

Algorithm 2.1: Deep incomplete multimodal semantic matching model optimization

---

Input: Incomplete multimodal data set $X = \{X^{(v)}\}_{v=1}^V$, Regularization parameters $\{\alpha^{(v)}\}_{v=1}^V$, Deep learning network parameters for each modal

Output: Deep matching network for each modal $\{\theta^{(v)}\}_{v=1}^V$, base matrix $\{U^{(v)}\}_{v=1}^V$ and uniform coding matrix P

1: Randomly initialize the depth matching network parameters $\theta^{(v)}$ for each modal, and use modal data $X^{(v)}$ and stacked autoencoder to pre-train modal deep network layer by layer;

2: By minimizing the objective function Eq. (2.7), initialize the modal base matrix $U^{(v)}$ and the shared characteristic matrix P_C;

3: Randomly initialize the private feature matrix $P_I^{(v)} \geq 0$ in the shared space of each modal;

4: **repeat**

5: Through the forward propagation of each modal depth network $\{\theta^{(v)}\}_{v=1}^V$, the deep conversion output characteristics $\{H^{(v)}\}_{v=1}^V$ of each modal input data $\{X^{(v)}\}_{v=1}^V$ are obtained;

6: According to Eqs. (2.13) and (2.14), update the inter-modal consistent coding matrix P_C and the private feature matrix $P_I^{(v)}$ in the shared space of each modal;

7: Update the basis matrix $\{U^{(v)}\}_{v=1}^V$ of each modal according to Eq. (2.18);

8: According to Eqs. (2.20) ... (2.25), in the back propagation process, the deep matching network of each modal $\{\theta^{(v)}\}_{v=1}^V$ is jointly fine-tuned;

9: **until** The objective function converges;

10: return $P = [P_C; P_I^{(1)}; P_I^{(2)}; \ldots; P_I^{(V)}]$;

In Algorithm 2.1, firstly, deep neural network of each modal is used to generate the deep conversion features $\{H^{(v)}\}_{v=1}^V$ of incomplete modal input data during the forward propagation process. Based on the generated transformation characteristics of each modal, using the incomplete multimodal non-negative matrix decomposition of the fusion invariant graph regularization factor can update the basis matrix and inter-modal consistent coding matrix of each modal. Then in the back propagation process of the modal deep network, the updated base matrix and the consistent coding matrix are used to fine-tune and optimize modal deep network. Through repeated joint optimization and adjustment, when the objective function converges, the deep matching network of each modal and the shared feature space between modalities can be obtained. In the obtained modal shared subspace, incomplete multimodal data can be effectively semantically fused to complete the analysis and mining of multimodal data.

2.6 Algorithm Complexity Analysis

In the algorithm implementation, it is necessary to save the deep learning network parameters $\theta^{(v)} = \{W_v, b_v\}$ of each modal, input data $X^{(v)}$, network output A_v^l, projection matrix $U^{(v)}$, private feature $P_I^{(v)}$, similarity matrix $W^{(v)}$, regularization parameters $\alpha^{(v)}$ and modal shared feature P_C, Assuming the number of nodes in each layer of the modal network is d_v, the corresponding space cost is $O((n_l^v - 1)d_v(d_v + 1))$, $O(d_v n)$, $O((n_l^v - 1)d_v n)$, $O(kd_v)$, $O(kn_v)$, $O(d_v c^2)$, $O(1)$ and $O(kc)$, where $n = c + n_v$. Therefore, when performing fusion analysis on V incomplete modal, the overall space cost is approximately $O(V(n_l^v d_v^2 + n_l^v d_v n + kd_v + kn_v + d_v c^2 + 1) + kc)$, Usually modal attribute dimensions d_v are less than the number n of data instances, therefore, the algorithm space complexity is approximately $O(d_v(n + c^2))$.

Algorithm 2.1 mainly includes two main components: deep modal network matching optimization and cross-modal shared feature correlation learning. In the deep modal network matching optimization, it is necessary to learn the modal deep features through forward propagation, and update and adjust the modal network parameters through back propagation. Assuming the number of nodes in each layer of the modal network is d_v, then the time cost of once matching optimization for each modal network is approximately $O(nn_l^v d_v^2)$. In cross-modal shared feature-related learning, it is necessary to calculate the similarity matrix $W^{(v)}$ of each modal, and iteratively update the modal projection matrix $U^{(v)}$, private features $P_I^{(v)}$ and modal shared features P_C, the corresponding time costs are approximately is $O(c^2 d_v)$, $O(nkd_v)$, $O(kd_v n_v)$ and $O(Vkd_v c)$. In summary, completing V modal t times iteration optimization, the overall time cost of the algorithm is $O(tV(nn_l^v d_v^2 + nkd_v) + Vc^2 d_v) \approx O(nd_v^2 + c^2 d_v)$.

3   Incremental Co-clustering Fusion Algorithm for Parameter-Free Multimodal Data

This section describes the research of incremental clustering fusion method for multimodal data. First, it analyzes and summarizes the scientific challenges brought by the real-time data to multimodal data fusion, and briefly summarizes the solutions proposed by existing related researches to face these challenges. In view of the problems of existing methods, this section proposes a parameter-free multimodal data incremental co-clustering fusion algorithm. First, the strategies of multimodal data similarity measurement, incremental cluster update and adaptive modal weight adjustment are designed and then the implementation details of the incremental clustering fusion algorithm for multimodal data without parameters is given.

3.1   Problem Description

Given Continuously collected multimodal data set $X = \{X^{(v)}\}_{v=1}^{V}$, which contains n data instances and V modal feature data set. $X^{(v)} \in R^{d_v \times n}$ is used to represent n data instances containing d_v dimensional feature vectors in the v-th modal feature set.

The goal of incremental co-clustering is to divide n multimodal data instances in the data set into K clusters of clustering results through fast and effective multimodal data feature fusion and measurement, which can be expressed as $R = \{R_k\}_{k=1}^{K}$ [21]. For example, in a picture sharing website (Flickr), the incremental multimodal co-clustering algorithm determines the similarity between pictures based on the visual and textual modal characteristics of the picture, obtains the latent semantic description of the image content by the effective fusion of the visual characteristics of the picture and the short text around the picture and improves the efficiency of the feature fusion task through efficient incremental algorithm design.

According to the analysis in the previous section, multimodal incremental co-clustering is a very difficult research task, especially for multimodal data with rapid growth, continuous evolution of class structure and low time-dependent correlation, its efficient incremental analysis More challenging. Specifically, one of the challenges is how to integrate multimodal features to define similarity measures between instances in multimodal incremental co-clustering. For example, when using multiple visual features (color histogram, boundary extraction, texture features, et al.) to cluster pictures, each visual feature cannot well express the latent semantics of the picture, as shown in Fig. 3. Different appearance entities (Fig. 3c) or different

(a) different background entities

(b) different background entities

(c) different appearance entities

Fig. 3 Examples of multimodal data

background entities (Fig. 3a, b) have different visual characteristics. It is particularly important to effectively fuse multimodal visual features, explore the latent semantics of image content, and then perform unified clustering of similar images. Another major challenge of incremental co-clustering of multimodal data is how to dynamically adjust the cluster structure that changes over time. For example, an existing data instance can be divided into kt classes at a certain moment, As data instances are added dynamically, certain classes may merge with each other, and the addition of new instances may also produce classes that have never appeared before, therefore, the dynamic adjustment of class structure is also necessary for multimodal incremental clustering. In the following subsections, we will give the detailed implementation process of the multimodal data incremental clustering fusion algorithm proposed in this section for the above two problems.

3.2 Multimodal Feature Similarity Measurement and Learning Strategy

In order to better measure the similarity between multimodal data instances, the PFICC algorithm extends the single-modal data measurement to the multimodal data space, and designs a new weighted similarity measurement method that incorporates multimodal data features. Given two multimodal data instances $\{X_i^{(v)}\}_{v=1}^V$ and $\{X_j^{(v)}\}_{v=1}^V$ containing V feature set, the weighted similarity between them is defined as follows:

$$Dis(X_i, X_j) = \frac{1}{2} \sum_{v=1}^V \alpha^{(v)} \left(\frac{||X_i^{(v)} - X_j^{(v)}||}{||X_i^{(v)}||} + \frac{||X_i^{(v)} - X_j^{(v)}||}{||X_j^{(v)}||} \right), \quad (3.1)$$

where modal contribution parameters $\alpha^{(v)}$ specify the weight of the v-th modal feature in the data similarity measurement process, two norm $\ell_2 = || \cdot ||$ is represented as $||y|| \equiv \sqrt{\sum_i (y_i)^2}$. Similarity measurement $Dis(X_i, X_j)$ is the mutual distance between data instances that is not affected by the modal feature dimension, the smaller the mutual distance, the higher the similarity between the two instances. In the PFICC algorithm proposed in this section, it is necessary to calculate the similarity between each newly acquired multimodal data instance and the existing cluster clusters to incrementally adjust the clustering results, so the dynamic description and update of the cluster structure of the multimodal data clustering results are particularly important. We use the multimodal center point to describe the cluster structure. When a new data instance $\{x^{(v)}\}_{v=1}^V$ is divided into the clustering result cluster R_k represented by the center point C_k, the corresponding update strategy of C_k is as follows:

$$\widehat{C}_k^{(v)} = (n_k C_k^{(v)} + x^{(v)})/(n_k + 1), \quad (3.2)$$

where n_k is the number of data instances in existing cluster R_k.

In multimodal data sets, some data feature sets only contain [0, 1] sparse representation instances, for example, the instance feature representation of different label text meta information exists in the web image data set. Based on this, we use the probability distribution of the label to construct a similar modal cluster structure description, that is the center point representation [21] of the modal clustering result cluster. Assuming clusters R_k contain n_k data instances v of $R_k = \{x_1, x_2, \ldots, x_{n_k}\}$, we define meta information feature vector in sparse modal v of $x_j (j = 1, 2, \ldots, n_k)$ is $x_j^{(v)} = [x_j^{(v),1}, x_j^{(v),2}, \ldots, x_j^{(v),d_v}]$, where $x_j^{(v),m}$ corresponds m-th label t_m in text label table, the values of $x_j^{(v),m}$ are as follows:

$$x_j^{(v),m} = \begin{cases} 1, & if\ t_m \in x_j, \\ 0, & otherwise. \end{cases} \quad (3.3)$$

Therefore, the v-th feature modal structure of cluster R_k can be represented as $C_k^{(v)} = [C_k^{(v),1}, C_k^{(v),2}, \ldots, C_k^{(v),d_v}]$ by its center point, where $C_k^{(v),m}$ is probability of the m-th label in the cluster R_k, the specific definition is as follows:

$$C_k^{(v),m} = p(t_m | C_k^{(v)}) = \frac{\sum_{j=1}^{n_k} x_j^{(v),m}}{n_k}. \quad (3.4)$$

When the newly added data instance $\{x^{(v)}\}_{v=1}^{V}$ is divided into cluster R_k, the probability of the m-th label in the cluster R_k is updated as follows:

$$p(t_m | \hat{C}_k^{(v)}) = \frac{\sum_{j=1}^{n_k+1} x_j^{(v),m}}{n_k+1} = \frac{n_k}{n_k+1} p(t_m | C_k^{(v)}) + \frac{x_{n_k+1}^{(v),m}}{n_k+1}, \quad (3.5)$$

thus general representation of learning strategy with the m-th label existing probability $C_k^{(v),m}$ in cluster is obtained, that is:

$$\hat{C}_k^{(v),m} = \frac{n_k}{n_k+1} C_k^{(v),m} + \frac{x_{n_k+1}^{(v),m}}{n_k+1}. \quad (3.6)$$

Because $x_{n_k+1}^{(v),m}$ is equal to 0 or 1, Eq. (3.6) can be further simplified:

$$\hat{C}_k^{(v),m} = \begin{cases} \eta C_k^{(v),m}, & if\ x_{n_k+1}^{(v),m} = 0, \\ \eta \left(C_k^{(v),m} + \frac{1}{n_k} \right), & otherwise, \end{cases} \quad (3.7)$$

where $\eta = \frac{n_k}{n_k+1}$.

3.3 *Incremental Cluster Update*

In this section, we will design three cluster adjustment strategies including cluster creation, cluster merging and instantiation, incrementally fuse the newly acquired data instances and dynamically update the corresponding cluster structure. First, based on the multimodal similarity measurement Eq. (3.1), construct the similarity distance matrix DM of the cluster center points of the existing clustering results, then select the maximum and minimum distance values ($max(DM)$ and $min(DM)$) between the center points in the distance matrix as the basic parameters of the cluster adjustment strategy. If the current clustering result is empty, several data instances obtained first are randomly selected as the initial cluster, which itself serves as the center point of the cluster. When a new data instance $\{x^{(v)}\}_{v=1}^{V}$ is obtained, First calculate the multimodal weighted distance between it and each center point to obtain the minimum value $minD(x)$. Based on minimum distance $minD(x)$ between $\{x^{(v)}\}_{v=1}^{V}$ and each center point and the maximum distance $max(DM)$ and minimum distance $min(DM)$ between the center points, the specific implementation of the incremental adjustment process of the clustering result cluster is as follows:

(1) if $minD(x)$ is greater than $max(DM)$, then newly added data instance $\{x^{(v)}\}_{v=1}^{V}$ is relatively far from the existing clusters (as shown in Fig. 4), so a cluster containing only data instances x is created, and the center point of the cluster is x itself.

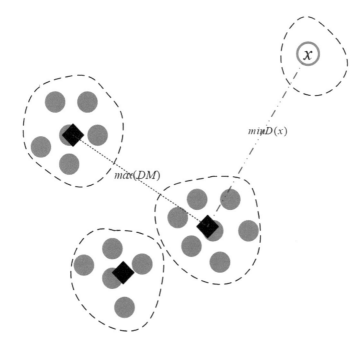

Fig. 4 Cluster creating when $minD(x) > max(DM)$

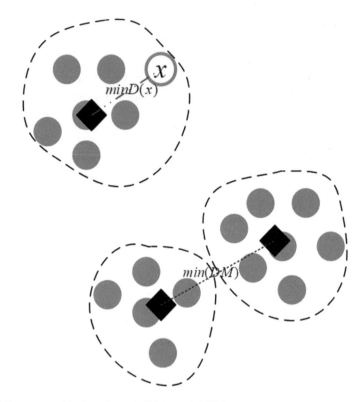

Fig. 5 Instance partitioning when $minD(x) < min(DM)$

(2)   if $minD(x)$ is less than $min(DM)$, as shown in Fig. 5, x will be divided into clusters closest to it, and the cluster center points of the corresponding clusters will be updated accordingly according to Eqs. (3.2) and (3.7).

(3)   if $minD(x)$ is less than $max(DM)$ and $minD(x)$ is greater than $min(DM)$, there will be the following two cluster adjustment strategies.

①   When the distance (the corresponding distances are $minD(x)$ and $min(x)$) between the nearest and the next two center points, C_i and C_j, to x is exactly $min(DM)$, first temporarily add x to the clusters corresponding to the center points C_i and C_j, and update the corresponding center points C_i and C_j. if the distance between x and the two updated center points is less than the minimum distance between the original center points $min(DM)$, Combine the two clusters with x (as shown in Fig. 6a), and use Eqs. (3.2) and (3.8) to update the center point of the merged cluster. Otherwise, x will be divided into the nearest cluster (as shown in Fig. 6b), and the corresponding cluster center point will be updated.

$$\hat{C}_i^{(v)}(\hat{C}_j^{(v)}) = (n_i C_i^{(v)} + n_j C_j^{(v)})/(n_i + n_j).   \tag{3.8}$$

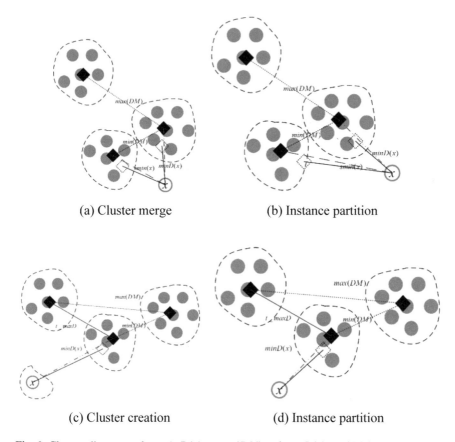

(a) Cluster merge   (b) Instance partition

(c) Cluster creation   (d) Instance partition

Fig. 6 Cluster adjustment when $minD(x) < max(DM)$ and $minD(x)min(DM)$

② When the distance between x and the two nearest and next center points,C_i and C_j, is not equal to $min(DM)$, first find the maximum distance t_maxD corresponding to the center point closest to x in the center point distance matrix, and update the center point of the cluster closest to x by adding x. If the distance between the updated center point and x is greater than t_maxD (as shown in Fig. 6c), a new cluster that only contains x is created, and the center point of the cluster is represented by x. Otherwise, x will be divided into the nearest cluster (as shown in Fig. 6d), and the cluster center point of the corresponding cluster will be updated according to Eqs. (3.2) and (3.7).

3.4 Adaptive Modal Weight Adjustment

In the process of incremental co-clustering and fusion of multimodal data, the parameters $\alpha^{(v)}$ specify the weight of each modal in the data similarity measurement. As shown in Eq. (3.1), the more effective it is to distinguish the characteristic modalities of the clustering mode, the higher the weight value should be. In the multimodal incremental co-clustering algorithm, with the continuous addition of data instances, the clustering structure is constantly updated, and the corresponding modal weights are constantly changing. Therefore, it is more important to design an adaptive update mechanism of modal weight parameters than to set a fixed value of modal weight through experiments. Usually the cluster corresponding to a robust characteristic modal has a compact distribution of data instances, The distribution compactness of this cluster can be measured by the difference between the data instances and the center point in the cluster [21]. Given cluster R_k containing V modal and the center point C_k and data instances $\{x_1, x_2, \ldots, x_{n_k}\}$ in cluster, the difference between data instances in the v-th modal of cluster R_k and the center point is defined as:

$$D_k^{(v)} = \frac{\frac{1}{n_k} \sum_{i=1}^{n_k} |C_k^{(v)} - x_i^{(v)}|}{|C_k^{(v)}|},$$ (3.9)

where $\ell_1 = |\cdot|$ is defined as $|y| = \sum_i |y_i|$. Thus, the difference measurement between the data instance containing all clusters and the center point in the v-th modal can be obtained as:

$$D^{(v)} = \frac{1}{K} \sum_{k=1}^{K} D_k^{(v)},$$ (3.10)

Therefore, the robustness of the modal v can be defined as:

$$B^{(v)} = \exp(-D^{(v)}).$$ (3.11)

When $D^{(v)}$ approaches 0, $B^{(v)}$ is 1. On the contrary, when $D^{(v)}$ approaches infinity, $B^{(v)}$ is 0. This means that the smaller the data difference $D^{(v)}$ in modal v, The higher the robustness of the modal $B^{(v)}$, therefore, modal v has higher weight and credibility in the clustering process, ensuring that similar data instances are divided into the same cluster. Above, the normalized modal weight parameter can be expressed as:

$$\alpha^{(v)} = \frac{B^{(v)}}{\sum_{v=1}^{V} B^{(v)}}.$$ (3.12)

In the clustering process, since the initial modalities have the same discriminative characteristics, the same value is set for each modal weight. As new data instances continue to increase, modal weights $\alpha^{(v)}$ need to be adjusted accordingly. Next, three dynamic adjustment strategies for modal weights corresponding to design data instance division, cluster merging and cluster creation will be adapted to update the weight value of each modal.

(1) Instance division

When a data instance x is divided into cluster R_k represented by C_k, We only need to consider the change of the data difference parameter $\{D_k^{(v)}\}_{v=1}^V$ within the cluster corresponding to the k-th cluster in each modal to update the weights of different modalities. First, update the cluster center points $C_k^{(v)}$ of each modal to $\widehat{C}_k^{(v)}$ according to Eqs. (3.2) and (3.7), then update $D_k^{(v)}$ of each modal based on $\widehat{C}_k^{(v)}$, details as follows:

$$
\begin{aligned}
\hat{D}_k^{(v)} &= \frac{\frac{1}{n_k+1} \sum_{i=1}^{n_k+1} |\widehat{C}_k^{(v)} - x_i^{(v)}|}{|\widehat{C}_k^{(v)}|} \\
&= \frac{\frac{n_k}{n_k+1} \times \left(\frac{1}{n_k} \sum_{i=1}^{n_k} |\widehat{C}_k^{(v)} - x_i^{(v)}| + \frac{1}{n_k} |\widehat{C}_k^{(v)} - x^{(v)}| \right)}{|\widehat{C}_k^{(v)}|} \\
&= \frac{\eta \times \left(\frac{1}{n_k} \sum_{i=1}^{n_k} (|\widehat{C}_k^{(v)} - C_k^{(v)} + C_k^{(v)} - x_i^{(v)}|) + \frac{1}{n_k} |\widehat{C}_k^{(v)} - x^{(v)}| \right)}{|\widehat{C}_k^{(v)}|} \\
&= \frac{\eta \times \left(\frac{1}{n_k} \sum_{i=1}^{n_k} |C_k^{(v)} - x_i^{(v)}| + |\widehat{C}_k^{(v)} - C_k^{(v)}| + \frac{1}{n_k} |\widehat{C}_k^{(v)} - x^{(v)}| \right)}{|\widehat{C}_k^{(v)}|} \\
&= \frac{\eta}{|\widehat{C}_k^{(v)}|} \left(|C_k^{(v)}| D_k^{(v)} + |C_k^{(v)} - \hat{C}_k^{(v)}| + \frac{1}{n_k} |\hat{C}_k^{(v)} - x^{(v)}| \right),
\end{aligned} \tag{3.13}
$$

Therefore, the weight parameters of all modalities $\alpha^{(v)}$ are incrementally updated according to Eqs. (3.10) … (3.12).

(2) Cluster merging

When two cluster R_k and R_j represented by two center points C_k and C_j in the existing clustering results are merged, first $\widehat{C}_k^{(v)}$ can be obtained by updating the center point of the merged cluster according to Eq. (3.8), and then update the $D_k^{(v)}$ value of each modal cluster, details as follows:

$$
\begin{aligned}
\hat{D}_k^{(v)} &= \frac{\frac{1}{n_k+n_j} \sum_{i=1}^{n_k+n_j} |\hat{C}_k^{(v)} - x_i^{(v)}|}{|\hat{C}_k^{(v)}|} \\
&= \frac{\sum_{i=1}^{n_k} |\hat{C}_k^{(v)} - x_i^{(v)}| + \sum_{t=1}^{n_j} |\hat{C}_k^{(v)} - x_t^{(v)}|}{(n_k + n_j)|\hat{C}_k^{(v)}|} \\
&= \frac{\sum_{i=1}^{n_k} |\hat{C}_k^{(v)} - C_k^{(v)} + C_k^{(v)} - x_i^{(v)}| + \sum_{t=1}^{n_j} |\hat{C}_k^{(v)} - C_j^{(v)} + C_j^{(v)} - x_t^{(v)}|}{(n_k + n_j)|\hat{C}_k^{(v)}|} \\
&= \frac{\frac{n_k}{n_k} \sum_{i=1}^{n_k} |C_k^{(v)} - x_i^{(v)}| + n_k |\hat{C}_k^{(v)} - C_k^{(v)}| + \frac{n_j}{n_j} \sum_{t=1}^{n_j} |C_j^{(v)} - x_t^{(v)}| + n_j |\hat{C}_k^{(v)} - C_j^{(v)}|}{(n_k + n_j)|\hat{C}_k^{(v)}|}
\end{aligned}
$$

$$= \frac{\left(n_k|C_k^{(v)}|D_k^{(v)} + n_k|\hat{C}_k^{(v)} - C_k^{(v)}| + n_j|\hat{C}_k^{(v)} - C_j^{(v)}| + n_j|C_j^{(v)}|D_j^{(v)}\right)}{(n_k + n_j)|\hat{C}_k^{(v)}|}, \qquad (3.14)$$

Correspondingly, the weight parameters $\alpha^{(v)}$ of all modalities can be incrementally updated according to Eqs. (3.10) ... (3.12).

(3) Cluster creation

When a new cluster R_k generated in the clustering result, all its corresponding values of $\{D_k^{(v)}\}_{v=1}^{V}$, $k = 1, 2, \ldots, K$ are zero. Therefore, the addition of new clusters only causes the proportional adjustment of the existing modal weight parameters, as follows:

$$\hat{\alpha}^{(v)} = \frac{\hat{B}^{(v)}}{\sum_{v=1}^{V} \hat{B}^{(v)}} = \frac{(B^{(v)})^{\frac{K}{K+1}}}{\sum_{v=1}^{V} (B^{(v)})^{\frac{K}{K+1}}}. \qquad (3.15)$$

In summary, through updating $D_k^{(v)}$, the incremental adjustment of the modal weight parameters can be realized.

3.5 Multimodal Incremental Co-clustering Fusion Algorithm

As shown in Fig. 7 and Algorithm 3.1. First, based on the labeled data, pre-define the clustering patterns existing in the data set, and calculate the multimodal clustering center point and the distance matrix DM between the center points. Then, initialize the modal weight $\alpha^{(v)}$, and calculate all intermediate variables needed for modal weights. If there is no labeled data, randomly select several multimodal data instances as the initial clustering mode, and set it as the center point of the corresponding cluster.

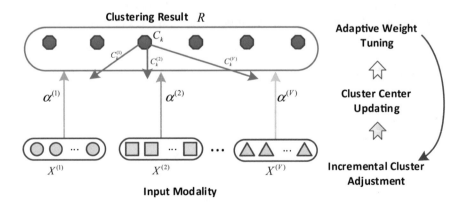

Fig. 7 The structure of our proposed PFICC method

When each new data instance x arrivals, calculate the multimodal weighted distance between it and the existing cluster center point, And select the minimum distance $minD(x)$. At the same time, select the maximum distance $max(DM)$ and minimum distance $min(DM)$ between the center points in the distance matrix DM. Based on $minD(x)$, $max(DM)$ and $min(DM)$, through three cluster adjustment strategies and corresponding modal weight update strategies, the incremental clustering fusion of multimodal data is completed. See Algorithm 3.2 for detailed cluster adjustment and weight update.

Algorithm 3.1 PFICC algorithm

Input: Multimodal data set $X = \{X^{(v)}\}_{v=1}^V$

Output: Clustering result $R = \{R_k\}_{k=1}^K$

1: Based on labeled data, predefine cluster mode R and calculate the distance between center points $\{C_k\}_{k=1}^K$ of each cluster mode to obtain DM, If there is no labeled data, randomly select several data as the initial clustering mode;

2: Calculate the difference $D_k^{(v)}$ between the center point and the modal data instance in each cluster;

3: Randomly initialize the private feature matrix $P_l^{(v)} \geq 0$ in the shared space of each modal, obtain modal weight α;

4: **for** (each data instance x) **do**:

5: Calculate the minimum distance $minD(x)$ between x and each center point according to Eq. (3.1);

6: Select the maximum and minimum distance, $max(DM)$ and $min(DM)$, between center points in DM;

7: Complete the incremental update of clusters according to Algorithm 3.2;

8: **end**

9: **return** clustering result $R = \{R_k\}_{k=1}^K$;

It can be seen from the Algorithms 3.1 and 3.2 that the PFICC algorithm can complete the incremental clustering fusion of multimodal data without introducing manual or experimental setting variables.

(2) Algorithm complexity analysis

The detailed complexity analysis of the PFICC algorithm is given below. In the clustering process of the algorithm, the distance matrix DM between the center points, the center point C, the cluster R, the difference parameter D and the modal weight α need to be stored. DM is a $K \times K$-dimensional matrix, C contains V $K \times d_v$-dimensional matrix, R is a n-dimensional vector, α is a V-dimensional vector. In incremental clustering, the number of data instances n is usually much larger than other variables, therefore, the space complexity of the PFICC algorithm is approximately equal to $O(n)$.

Algorithm 3.2: Incremental cluster update

Input: $x, minD(x), DM, max(DM), min(DM), R, C, D$ and α

Output: R, C, DM, D and α

1: **if** $minD(x) > max(DM)$ **then**

2: generate new cluster only containing x, x is the center point of cluster; update parameters α according to Eq. (3.15);

3: **else**

4: **if** $minD(x) < min(DM)$ **then**

5: x is divided into the nearest cluster; Update the center point C of the cluster according to Eqs. (3.2) and (3.7); Update D and modal weight α according to Eqs. (3.13) and (3.10) … (3.12);

6: **else**

7: **if** $min(DM)$ corresponding center point C_i and C_j are two center points nearest to x then

8: Temporarily add x to the cluster represented by the two center points and update C_i and C_j to $\widehat{C}_i$ and $\widehat{C}_j$;

9: if the distance between x and $\widehat{C}_i$ and the distance between x and $\widehat{C}_j$ are all less than $min(DM)$ then

10: merge R_i, R_j and x; And update the corresponding center point according to Eqs. (3.2), (3.7) and (3.8); update D and α according to Eqs. (3.13), (3.14) and (3.10) … (3.12);

11: **else**

12: x Is divided into the nearest cluster; update center points C, D and α;

13: **else**

14: Select the maximum distance t_maxD corresponding to the center point C_i closest to x in DM;

15: add x to the cluster corresponding to C_i and update C_i to $\widehat{C}_i$;

16: if the distance x and $\widehat{C}_i$ is greater than t_maxD then

17: generate new cluster only containing x, x is center point of new cluster; Update α according to Eq. (3.15);

18: **else**

19: x Is divided into the nearest cluster; update center points C, D and α;

20: Update the center point distance matrix DM;

In PFICC, the main steps of its realization include: cluster mode initialization, incremental cluster adjustment and modal parameter update. For clustering mode initialization, the weighted distance between any two cluster centers needs to be calculated according to Eq. (3.1), and the time cost is $O(Vd_vK^2)$. When a new data instance is added to an existing cluster, the cluster center point, center point distance matrix and modal weight parameters need to be dynamically updated according to the three strategies of cluster establishment, cluster merging, and instance division to complete the incremental clustering division of data instances. In the cluster creation process, the weight distance between the center point of the newly created cluster and the center point of each cluster needs to be calculated to update the cluster center point matrix DM, time cost is $O(Kd_vV)$, update the modal weight parameter according to Eq. (3.15), the time cost is $O(V)$. In the cluster merging process, it is necessary to update the center point of the cluster according to Eq. (3.8), calculate the distance between the updated center point and other center points according to Eq. (3.1), adjust the modal weight parameter according to Eqs. (3.14) and (3.10)... (3.12), the corresponding time cost is $O(Vd_v) + O(Kd_v V) + O(V) \approx O(Kd_v V)$.

In the process of dividing the data instance, the new data instance is divided into the cluster closest to it, so the center points [as shown in Eqs. (3.2) and (3.7)] and the distance matrix between center points [as shown in Eq. (3.1)] and modal weight parameters [as shown in Eqs. (3.10) ... (3.13)] need to be updated, the sum of the corresponding time costs is approximately equal to $O(Kd_v V)$. Suppose there are n data instances (normally n is much larger than d_v, V and K) participate in clustering, The overall time complexity of the algorithm PFICC is approximately equal to $O(nKd_v V) \approx O(n)$.

In summary, the parameter-free incremental multimodal co-clustering fusion algorithm proposed in this section has good time and space complexity and is suitable for real-time fusion analysis of massive multimodal data.

4 Deep Heterogeneous Transfer Learning Model

4.1 *Problem Formulation*

For feature-based deep heterogeneous transfer learning, one crucial problem is to learn the common feature representation subspace in which the labeled source data can be transferred for annotating target instances. Partially motivated by multi-view learning (MVL) [19], we exploit the co-occurrence data to learn the shared semantic subspace across domains. However, the problem setting is different in this chapter. MVL requires fully co-occurrence instances between two views of data, and assumes the labels of the co-occurrence instances to be available, in general, for task learning. However, Deep heterogeneous transfer learning model (DHTL) requires no label information of the co-occurrence instances and only exploits the labeled data in the source domain to train classifiers for unlabeled instances classification in the target domain.

Here, the unlabeled dataset in the target domain is denoted as $D^T = \{X_i^T\}_{i=1}^{n_t}$ the labeled dataset in the source domain is $D^S = \{X_i^S, Y_i^S\}_{i=1}^{n_s}$, and a co-occurrence unlabeled dataset of pairs of the source and target domains is $D^C = \{C_i^S, C_i^T\}_{i=1}^{n_c}$, where the superscripts T and S denote the target and source domains, X and C denote the features of instances represented by vectors, and Y and n denote the label vectors and the number of instances in domains, respectively. The goal of this article is to jointly learn two deep networks $\theta^T = \{W^T, b^T\}$ and $\theta^S = \{W^S, b^S\}$ based on D^C to achieve the high-level semantic mapping subspace Ω for feature representations in the source and target domains, such that the gap between domains in the new feature space is bridged favorably. With the learned networks θ^T, θ^S and the feature subspace

Ω, one can train a classifier Ψ from the source domain data D^S to make a prediction for the target domain data D^T by applying $\Psi(\Omega(D^T))$.

4.2 Deep Semantic Mapping Mechanism

The CCA [22] can maximize correlations between domains by deriving the projection subspace as a joint representation. Therefore, integrate CCA with deep neural networks to construct a multi-layer correlation matching model (see Fig. 8), which can identify a deep semantic mapping representation and learn the domain-specific networks for source and target domains.

As presented in Fig. 8, the proposed DHTL model consists of the source and target domain specific deep learning networks coupled by CCA for the deep semantic mapping subspace learning, and the prediction model on the learned feature subspace for transfer learning from the source to the target domain. To learn the correlating semantic subspace, CCA is employed to guide the domain network matching layer by layer. For each layer, the source and target domain networks learn the hidden features of co-occurrence data C^S, C^T in the forward propagation simultaneously. Then, CCA is conducted on the achieved hidden features to identify the correlation

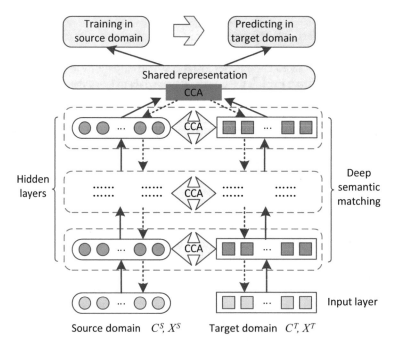

Fig. 8 Major components of the proposed deep heterogeneous transfer learning model

coefficients between heterogeneous features, which can be used for refining the network parameters in back propagation.

After the multi-layer networks are constructed, the additional top layer CCA is employed to fine-tune the whole correction matching network, from which the high-level semantic mapping subspace and domain-specific networks are obtained. When transforming the labeled source domain data X^S and unlabeled target data X^T to the common semantic subspace by the trained domain networks, the source labeled features can be used for the target data prediction.

This section employs the stacked auto-encoders (SAE) as the deep neural networks, thus given the co-occurrence pairs of two domains C^S, C^T, the domain-specific auto-encoders training and the cross domain matching with one hidden layer is presented in Fig. 8. First, the domain pairs C^S, C^T are encoded by the source and target auto-encoders to generate the hidden representations $A^{S(2)} = f(W^{S(1)}C^S + b^{S(1)})$, $A^{T(2)} = f(W^{T(1)}C^T + b^{T(1)})$, of these two domains, in which $W^{S(1)}$ and $W^{T(1)}$ are the weight matrices, $b^{S(1)}$ and $b^{T(1)}$ are the bias vectors, and f is a nonlinear activation function (the Sigmoid function is adopted in this article). After that, the hidden features $A^{S(2)}$ and $A^{T(2)}$ are utilized for canonical correlation analysis to learn the correlation coefficients $V^{S(2)}, V^{T(2)}$, which can project the source and target domains to a correlating subspace Ω, meanwhile promoting the domain parameters (W^S, b^S, W^T, b^T) refining to reconstruct more matching domain auto-encoders. Therefore, the objective of DHTL model is to minimize the reconstruction error for source and target domains, and maximize the correlation between them simultaneously, which can be defined as:

$$\min J = J_S(W^S, b^S) + J_T(W^T, b^T) + \alpha \Gamma(V^S, V^T), \tag{4.1}$$

The first two terms of the objective J are the reconstruction errors for the source and target domains, respectively, including the cost function and the regularization item, as shown in Eqs. (4.2) and (4.3):

$$J_S(W^S, b^S) = \left[\frac{1}{m} \sum_{i=1}^{m} \left(\frac{1}{2} \| h_{W^S, b^S}(C_i^S) - C_i^S \|^2 \right) \right] + \frac{\lambda^S}{2} \sum_{l=1}^{n^S - 1} \sum_{j=1}^{n_l^S} \sum_{k=1}^{n_{l+1}^S} (W_{kj}^{S(l)})^2, \tag{4.2}$$

$$J_T(W^T, b^T) = \left[\frac{1}{m} \sum_{i=1}^{m} \left(\frac{1}{2} \| h_{W^T, b^T}(C_i^T) - C_i^T \|^2 \right) \right] + \frac{\lambda^T}{2} \sum_{l=1}^{n^T - 1} \sum_{j=1}^{n_l^T} \sum_{k=1}^{n_{l+1}^T} (W_{kj}^{T(l)})^2, \tag{4.3}$$

where $h_{W^S, b^S}(C_i^S)$ and $h_{W^T, b^T}(C_i^T)$ are the output results for each co-occurrence pair, n^S and n^T are the number of layers of the source and target networks, n_l^S and n_l^T are the corresponding number of neurons in layer l, and λ^S and λ^T are the tradeoff parameters.

The third item in Eq. (4.1) is the correlation matching between domains, which can learn the projection vectors V^S and V^T to transform two domain features to maximize their correlations. Specifically, this item can be formalized as follows,

$$\Gamma(V^S, V^T) = \sum_{l=2}^{n^S-1} \frac{V^{S(l)^T} \Sigma_{ST} V^{T(l)}}{\sqrt{V^{S(l)^T} \Sigma_{SS} V^{S(l)}} \sqrt{V^{T(l)^T} \Sigma_{TT} V^{T(l)}}}, \tag{4.4}$$

in which $\Sigma_{ST} = A^{S(l)} A^{T(l)^T}$, $\Sigma_{SS} = A^{S(l)} A^{S(l)^T}$, $\Sigma_{TT} = A^{T(l)} A^{T(l)^T}$.

By minimizing J in Eq. (4.1), we can collectively train the domain matching networks $\theta^T = \{W^T, b^T\}$ and $\theta^S = \{W^S, b^S\}$. When multiple layers of the networks are constructed, an additional CCA on the top layer is employed to fine-tune the whole network in the back-propagation process, from which the high-level semantic mapping subspace between domains is obtained.

4.3   Training the DHTL Model

In Eq. (4.1), the variables $\theta^T = \{W^T, b^T\}$, $\theta^S = \{W^S, b^S\}$, V^S and V^T are coupled together, so it is generally difficult to optimize them at the same time. To solve this problem, we adopt the Lagrangian multiplier and stochastic gradient descent methods to update $[V^S, V^T]$ and $[\theta^T, \theta^S]$ in an alternating way. Thus, there are two subproblems in DHTL.

1.   Updating $[V^S, V^T]$ with $[\theta^T, \theta^S]$ fixed:

In Eq. (4.1), there is no direct relation to V^S and V^T in the first and second terms, so the updating of V^S and V^T only depends on the third term, namely Eq. (4.4). As proven in [22], the solution of Eq. (4.4) is not affected by rescaling V^S or V^T either together or independently, so that the optimization problem for each layer of $V^{S(l)}$ and $V^{T(l)}$ is equivalent to:

$$\max_{V^{S(l)}, V^{T(l)}} \frac{V^{S(l)^T} \Sigma_{ST} V^{T(l)}}{\sqrt{V^{S(l)^T} \Sigma_{SS} V^{S(l)}} \sqrt{V^{T(l)^T} \Sigma_{TT} V^{T(l)}}},$$
$$s.t.\ V^{S(l)^T} \Sigma_{SS} V^{S(l)} = 1 \text{ and } V^{T(l)^T} \Sigma_{TT} V^{T(l)} = 1, \tag{4.5}$$

Therefore, the corresponding Lagrangian is

$$L(\omega_l, V^{S(l)}, V^{T(l)}) = -V^{S(l)^T} \Sigma_{ST} V^{T(l)}$$
$$+ \frac{\omega_l^S}{2} (V^{S(l)^T} \Sigma_{SS} V^{S(l)} - 1)$$
$$+ \frac{\omega_l^T}{2} (V^{T(l)^T} \Sigma_{TT} V^{T(l)} - 1), \tag{4.6}$$

Taking derivatives with respect to $V^{S(l)}$ and $V^{T(l)}$, we get

$$\frac{\partial L}{\partial V^{S(l)}} = \Sigma_{ST} V^{T(l)} - \omega_l^S \Sigma_{SS} V^{S(l)} = 0, \tag{4.7}$$

$$\frac{\partial L}{\partial V^{T(l)}} = \Sigma_{ST}^T V^{S(l)} - \omega_l^T \Sigma_{TT} V^{T(l)} = 0, \tag{4.8}$$

Through the reduction, we have

$$V^{T(l)} = \frac{\Sigma_{TT}^{-1} \Sigma_{ST}^T V^{S(l)}}{\omega_l}, \tag{4.9}$$

$$\Sigma_{ST} \Sigma_{TT}^{-1} \Sigma_{ST}^T V^{S(l)} = \omega_l^2 \Sigma_{SS} V^{S(l)}, \tag{4.10}$$

where $\omega_l = \omega_l^S = \omega_l^T$. So $V^{S(l)}$ and ω_l in Eq. (4.10) can be solved by the generalized eigenvalue decomposition [23] and the corresponding $V^{T(l)}$ can be calculated by Eq. (4.9).

2.   Updating θ^T (θ^S) with $\theta^S(\theta^T)$ and $[V^S, V^T]$ fixed:

It can be seen from Eq. (4.1) that, θ^T and θ^S are independent of each other. When V^S, V^T and ω are fixed, the updating of θ^T and θ^S is similar. Hence, we only describe the parameter updating processes of the source domain network, which can be written as the following subproblem involving θ^S,

$$\min_{\theta^S} \phi(\theta^S) = J_S(W^S, b^S) + \alpha \Gamma(V^S, V^T). \tag{4.11}$$

In order to get the optimized parameters of source domain deep network, gradient descent method is used to adjust the network weight W^S and bias vector b^S in the process of back propagation [24], as following:

$$W^{S(l)} = W^{S(l)} - \mu^S \frac{\partial \phi}{\partial W^{S(l)}}, \tag{4.12}$$

$$b^{S(l)} = b^{S(l)} - \mu^S \frac{\partial \phi}{\partial b^{S(l)}}, \tag{4.13}$$

where μ^S is the learning rate. And the gradient of the objective function ϕ with respect to $\{W^{S(l)}, b^{S(l)}\}_{l=1}^{n^S-1}$ of the generalized layers can be calculated as

$$\begin{aligned}
\frac{\partial \phi}{\partial W^{S(l)}} &= \frac{\partial J_S(W^S, b^S)}{\partial W^{S(l)}} + \alpha \frac{\partial \Gamma(V^S, V^T)}{\partial W^{S(l)}} \\
&= \left(\delta^{S(l+1)} + \alpha \gamma^{S(l+1)} - \alpha \omega_l \beta^{S(l+1)}\right) * A^{S(l)} / n_c + \lambda^S W^{S(l)},
\end{aligned} \tag{4.14}$$

$$\frac{\partial \phi}{\partial b^{S(l)}} = \frac{\partial J_S(W^S, b^S)}{\partial b^{S(l)}} + \alpha \frac{\partial \Gamma(V^S, V^T)}{\partial b^{S(l)}}$$
$$= \left(\delta^{S(l+1)} + \alpha \gamma^{S(l+1)} - \alpha \omega_l \beta^{S(l+1)}\right)/n_c, \tag{4.15}$$

Herein, (i) for $l = n^S$, we have

$$\delta^{S(l)} = -(C^S - A^{S(l)}) \cdot A^{S(l)} \cdot (1 - A^{S(l)}), \tag{4.16}$$

$$\gamma^{S(l)} = \beta^{S(l)} = 0. \tag{4.17}$$

(ii) for $l = 2, \ldots, n^S - 1$, we have

$$\delta^{S(l)} = W^{S(l)^T} \delta^{S(l+1)} \cdot A^{S(l)} \cdot (1 - A^{S(l)}), \tag{4.18}$$

$$\gamma^{S(l)} = A^{T(l)} V^{T(l)} V^{S(l)^T} \cdot A^{S(l)} \cdot (1 - A^{S(l)}), \tag{4.19}$$

$$\beta^{S(l)} = A^{S(l)} V^{S(l)} V^{S(l)^T} \cdot A^{S(l)} \cdot (1 - A^{S(l)}). \tag{4.20}$$

The operator $\cdot$ stands for dot product. After each layer of the joint domain network is pre-trained by Eqs. (4.5) and (4.11). The CCA on the top hidden layer is employed to fine-tune all the parameters through the whole network in the back-propagation process. Because there are no labeled instances in the co-occurrence pairs for fine-tuning, we exploit the correlation of two domain networks and rewrite Eq. (4.1) as

$$\max_{\theta^S, \theta^T, V^S, V^T} J = \Gamma(V^S, V^T), \tag{4.21}$$

The procedures of updating $[V^S, V^T]$ are the same as that in Eq. (4.5). However, for updating $[\theta^T, \theta^S]$, some differences from Eq. (4.11) should be addressed. Specifically, we set $\delta^{S(l)} = 0$, $\alpha = -1$, and

(i) for $l = n^S$, we have

$$\gamma^{S(l)} = A^{T(l)} V^{T(l)} V^{S(l)^T} \cdot A^{S(l)} \cdot (1 - A^{S(l)}), \tag{4.22}$$

$$\beta^{S(l)} = A^{S(l)} V^{S(l)} V^{S(l)^T} \cdot A^{S(l)} \cdot (1 - A^{S(l)}). \tag{4.23}$$

(ii) for $l = 2, \ldots, n^S - 1$, we have

$$\gamma^{S(l)} = W^{S(l)^T} \gamma^{S(l+1)} \cdot A^{S(l)} \cdot (1 - A^{S(l)}), \tag{4.24}$$

$$\beta^{S(l)} = W^{S(l)^T} \beta^{S(l+1)} \cdot A^{S(l)} \cdot (1 - A^{S(l)}). \qquad (4.25)$$

The details of the proposed model training are summarized in Algorithm 4.1.

Algorithm 4.1 Deep heterogeneous transfer model training

Input: The co-occurrence domain data pairs $D^C = \{C_i^S, C_i^T\}_{i=1}^{n_c}$, the trade-off parameters λ^S,
λ^T, the learning rates μ^S, μ^T, and the number of nodes in hidden layers

Output: The domain-specific networks $\theta^S = \{W^S, b^S\}, \theta^T = \{W^T, b^T\}$, and the correlation
coefficients V^S, V^T between domains

1: Initialize θ^S and θ^T for the source and target domain SAEs randomly;

2: Pre-train the joint auto-encoders layer by layer for two domains with inputs $\{C_i^S, C_i^T\}_{i=1}^{n_c}$
   according to Eqs. (4.1), (4.5) and (4.11);

3: Randomly initialize the private feature matrix $P_I^{(v)} \geq 0$ in the shared space of each modal,
   obtain modal weight α;

4: **repeat**

5: Set $\delta^{S(l)} = 0, \delta^{T(l)} = 0$; //Fine-tune the whole model by the CCA on the top hidden layer;

6: Updating V^S, V^T by minimizing Eq. (4.5);

7: Updating θ^S, θ^T by minimizing Eq. (4.21);

8: **until** Convergence;

In Algorithm 4.1, the joint auto-encoders by CCA for each layer are pre-trained to learn the encoding and decoding weights of the proposed model, which is favorable for the multi-layer correlation mapping construction. After that, the whole model is fine-tuned by the CCA on the top hidden layer in the backpropagation process. When it is converged, the domain-specific networks and the correlation coefficients between domains are achieved for deep heterogeneous transfer learning. It is noteworthy that to guarantee the performance of the proposed DHTL model, the parameters are tuned with the training data by adopting the cross-validation, and empirically selected to yield the best results in a greedy fashion.

4.4 Prediction by Deep Transfer Model

In order to achieve promising prediction (classification) performances for a target domain with no labeled instances, we focus on how to learn the deep semantic subspace across domains with the aid of co-occurrence data sets that contain pairs of source and target domain instances. By applying the proposed multi-layer correlation mapping model on the co-occurrence pairs, the deep semantic subspace can be identified, where the classifiers for the label prediction of target instances can be constructed. Specifically, the labeled data X^S in the source domain and the unlabeled data X^T in the target domain are mapped by the proposed joint deep networks to generate high-level hidden features $A^{S(n^S)}$ and $A^{T(n^T)}$ at first. Then, the features $A^{S(n^S)}$, $A^{T(n^T)}$ are transformed by the correlation coefficients $V^{S(n^S)}$, $V^{T(n^T)}$ to $H^S = V^{S(n^S)}A^{S(n^S)}$ and $H^T = V^{T(n^T)}A^{T(n^T)}$ in the common semantic subspace Ω. After that, we apply a standard classification algorithm on $\{H_i^S, Y_i^S\}_{i=1}^{n_s}$ to train a target classifier Ψ (in this section, we choose the popular Support Vector Machine, SVM),which can be used for predicting the target instances X^T by $\Psi(H^T)$. Algorithm 4.2 presents the deep heterogeneous transfer model for target data prediction in detail.

The DHTL model has a major advantage when compared with the existing deep heterogeneous transfer learning methods—no labeled target data is required to build the classifier for target data prediction. Because the source and target domain data can be transformed to the common semantic space, the labeled source domain instances are sufficient to build the classifier for prediction in the target domain.

Algorithm 4.2 Deep heterogeneous transfer for target data prediction

---

Input: The labeled source domain data $D^S = \{X_i^S, Y_i^S\}_{i=1}^{n_s}$ and unlabeled target domain data $D^T = \{X_i^T\}_{i=1}^{n_t}$; The domain-specific networks $\theta^S = \{W^S, b^S\}, \theta^T = \{W^T, b^T\}$ and the correlation coefficients V^S, V^T between domains

Output: The label $\{Y_j^T\}_{j=1}^{n_t}$ for target domain instances

1: **for** $i = \{1, 2, \ldots, n_s\}$:

2: The instance X_i^S is transformed to $A_i^{S(n^S)}$ based on $\theta^S = \{W^S, b^S\}$ trained by Eqs. (4.1) and (4.11) in forward learning;

3: $A_i^{S(n^S)}$ is projected to $H_i^S = V^{S(n^S)}A_i^{S(n^S)}$ based on $V^{S(n^S)}$ learned by Eq. (4.5) in common semantic space Ω;

4: The classifier Ψ (SVM) is trained on the labeled source semantic features $\{H_i^S, Y_i^S\}_{i=1}^{n_s}$;

5: **for** $j = \{1, 2, \ldots, n_t\}$:

6: The instance X_j^T is transformed to $A_j^{T(n^T)}$ based on $\theta^T = \{W^T, b^T\}$ trained by Eqs. (4.1) and (4.11) in forward learning;

7: $A_j^{T(n^T)}$ is projected to $H_j^T = V^{T(n^T)}A_j^{T(n^T)}$ based on $V^{T(n^T)}$ learned by Eq. (4.5) in common semantic space Ω;

8: The label Y_j^T for X_j^T is predicted by $\Psi(H_j^T)$;

---

5 Unsupervised Multi-View Non-Negative Correlated Feature Learning

5.1 Problem Formulation

In practical applications, unlabeled multimodal data is very common. Generally, multimodal data has high-dimensional attribute features. Through effective multi-modal feature fusion, multi-modal high-dimensional features are mapped to low-dimensional unified feature sharing space, which can promote the mutual supplement and sharing of knowledge among modes, and then mine the hidden semantic information of multimodal data. Through the separation of private features, we can better learn the shared features between modes, and then get more accurate multimodal data fusion results.

Assuming a given dataset $X = \{X^{(v)}\}_{v=1}^{V}$ with V views and N instances, where $X^{(v)} \in R^{d_v \times n}$ represents the feature matrix of the v-th view for N instances with d_v-dimensional features, the goal of multi-view clustering, as defined in this section, is to partition the set of N instances into K clusters based on the learned common feature representation through correlated feature learning on multi-view data. In existing correlated feature learning methods, they usually assume that a latent subspace is shared by all views, and transform each view to the subspace to learn the common semantics. This may ignore the effect of view-specific information in different views, which would cause semantic gaps for multi-view learning. To address this problem, in this section, we learn view-specific features and capture inter-view feature correlations in the latent common subspace by exploring both correlated and uncorrelated features simultaneously. As shown in Fig. 9, the feature matrix $X^{(v)}$ for each view is transformed by structured sparse matrices $U_I^{(v)}$ and $U_C^{(v)}$ to view-specific feature matrix $V_I^{(v)}$ and view-sharing feature matrix V_C, respectively. By jointly optimizing these matrices with local invariance graph regularization and sparse projection constraint, the common feature representation VC can be achieved. This section will describe the design choices and optimization processes of UMCFL in details.

5.2 Correlated and Uncorrelated Feature Learning

In UMCFL, we aim to find a common feature representation V_C for multi-view clustering. Given the latent dimensions of the shared subspace m_c and view-specific feature representation m_v for each view, we can define the basic multi-view learning model as:

$$\min \sum_{v=1}^{V} \left\| \begin{bmatrix} U_C^{(v)} \\ U_I^{(v)} \end{bmatrix} X^{(v)} - \begin{bmatrix} V_C \\ V_I^{(v)} \end{bmatrix} \right\|_F^2,$$
$$s.t.\ U_C^{(v)}, U_I^{(v)}, V_C, V_I^{(v)} \geq 0, \tag{5.1}$$

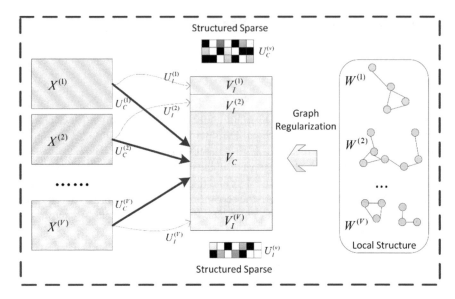

Fig. 9 The overview of unsupervised multi-view non-negative correlated feature learning (UMCFL)

where $U_C^{(v)} \in R^{m_c \times d_v}$, $U_I^{(v)} \in R^{m_v \times d_v}$, $V_C \in R^{m_c \times n}$ and $V_I^{(v)} \in R^{m_v \times n}$. In this way, each view is forced to have the same subspace feature matrix V_C, which couples the view-specific projection matrix and uncorrelated feature matrix for each view to minimize the objective function.

Since the global property in crucial for multi-view subspace learning [17], we propose to regularize the shared feature matrix V_C, by an affinity graph model. As shown in Fig. 9, we construct a local structure invariance graph as [17] represented by distance-weighted matrix $W^{(v)}$ between instances for each view. By integrating $W^{(v)}$ with the similarity measurement in low dimensional sharing subspace, the affinity graph model for the learned shared features V_C can be formulated as

$$
\begin{aligned}
\Re &= \frac{1}{2} \sum_{i,j=1}^{N} ||V_{Ci} - V_{Cj}||^2 W_{ij}^{(v)} \\
&= \sum_{i=1}^{N} (V_{Ci})^T V_{Ci} D_{ii}^{(v)} - \sum_{i,j=1}^{N} (V_{Ci})^T V_{Cj} D_{ij}^{(v)} \\
&= Tr(V_C D^{(v)} (V_C)^T) - Tr(V_C W^{(v)} (V_C)^T) \\
&= Tr(V_C L^{(v)} (V_C)^T),
\end{aligned}
\tag{5.2}
$$

in which $Tr(\cdot)$ is the trace of a matrix, $D^{(v)}$ and $L^{(v)} = D^{(v)} - W^{(v)}$ are the diagonal matrix of $W^{(v)}$ and the graph Laplacian matrix for each view v. By minimizing $\Re$, the two data instances $X_i^{(v)}$ and $X_j^{(v)}$ that are similar in the v-th feature set (i.e., $W_{ij}^{(v)}$

is big), should be similar in the learned latent subspace as well (i.e., $\left\| V_{Ci} - V_{Cj} \right\|^2$ is small).

Moreover, for projection matrices $U_C^{(v)}$ and $U_I^{(v)}$, the structured sparse regularizer should be added to make view v independent of the latent dimensions [19]. Thus, by combining the graph regularization and sparseness constraint to correlated feature learning, the model in Eq. (5.1) can be rewritten as

$$\min \left(\sum_{v=1}^{V} \left\| \begin{bmatrix} U_C^{(v)} \\ U_I^{(v)} \end{bmatrix} X^{(v)} - \begin{bmatrix} V_C \\ V_I^{(v)} \end{bmatrix} \right\|_F^2 + \alpha Tr(V_C L^{(v)} (V_C)^T) + \beta(\|U_C^{(v)}\|_{21} + \|U_I^{(v)}\|_{21}), \right)$$

$$s.t.\ U_C^{(v)}, U_I^{(v)}, V_C, V_I^{(v)} \geq 0, \tag{5.3}$$

Herein, the sparseness constraint is given by the l_{21}-norm, defined as $\|U^{(v)}\|_{21} = \sum_i \|U_i^{(v)}\|^2$. $\alpha \geq 0$ and $\beta \geq 0$ are the parameters that control the regularization of new representations and the sparsity of projection matrices, respectively.

5.3 Optimization

Since the optimization problem in Eq. (5.3) is not convex with respect to $U_C^{(v)}$, $U_I^{(v)}$, V_C and $V_I^{(v)}$ coupled together, it is unpractical to find a global minima. Hence, in this section, we propose to alternatively optimize the variables, which keeps other variables fixed when updating one variable, to achieve a local minima [17, 25].

Optimizing V_C and $V_I^{(v)}$:

Given the projection matrices $U_C^{(v)}$ and $U_I^{(v)}$, the minimizing objective over V_C and $V_I^{(v)}$ can be simplified as

$$\min_{V_C \geq 0} \sum_{v=1}^{V} \left(\left\| U_C^{(v)} X^{(v)} - V_C \right\|_F^2 + \alpha Tr(V_C L^{(v)} (V_C)^T) \right), \tag{5.4}$$

$$\min_{V_I^{(v)} \geq 0} \left\| U_I^{(v)} X^{(v)} - V_I^{(v)} \right\|_F^2. \tag{5.5}$$

To optimize Eqs. (5.4) and (5.5), the Lagrangian function is employed as follows:

$$L(V_C) = \sum_{v=1}^{V} (Tr(U_C^{(v)} X^{(v)} X^{(v)^T} U_C^{(v)^T}) - 2Tr(U_C^{(v)} X^{(v)} (V_C)^T)$$

$$+ Tr(V_C (V_C)^T) + \alpha Tr(V_C L^{(v)} (V_C)^T) + Tr(\varphi^{(v)} V_C)), \tag{5.6}$$

$$L(V_I^{(v)}) = Tr(U_I^{(v)} X^{(v)} X^{(v)^T} U_I^{(v)^T}) - 2Tr(U_I^{(v)} X^{(v)} (V_I^{(v)})^T)$$

$$+ Tr(V_I^{(v)}(V_I^{(v)})^T) + Tr(\phi^{(v)} V_I^{(v)}), \tag{5.7}$$

where $\varphi^{(v)}$ and $\phi^{(v)}$ are Lagrangian multipliers for constraints $V_C \geq 0$ and $V_I^{(v)} \geq 0$, respectively. The partial derivatives of objective function to V_C and $V_I^{(v)}$ are

$$\frac{\partial L(V_C)}{\partial V_C} = \sum_{v=1}^{V} (-2U_C^{(v)} X^{(v)} + 2V_C + 2\alpha V_C L^{(v)} + \varphi^{(v)}), \tag{5.8}$$

$$\frac{\partial L(V_I^{(v)})}{\partial V_I^{(v)}} = -2U_I^{(v)} X^{(v)} + 2V_I^{(v)} + \phi^{(v)}, \tag{5.9}$$

When applying the KKT conditions $(\varphi^{(v)})_{ij}(V_C)_{ij} = 0$ and $(\phi^{(v)})_{ij}(V_I^{(v)})_{ij} = 0$, the following updating rules for V_C and $V_I^{(v)}$ can be obtained,

$$(V_C)_{ij} \leftarrow \frac{\left(\sum_{v=1}^{V} (U_C^{(v)} X^{(v)} + \alpha V_C W^{(v)})\right)_{ij}}{\left(\sum_{v=1}^{V} (V_C + \alpha V_C D^{(v)})\right)_{ij}} (V_C)_{ij}, \tag{5.10}$$

$$(V_I^{(v)})_{ij} \leftarrow \frac{\left(U_I^{(v)} X^{(v)}\right)_{ij}}{\left(V_I^{(v)}\right)_{ij}} (V_I^{(v)})_{ij}. \tag{5.11}$$

Optimizing $U_C^{(v)}$ and $U_I^{(v)}$:

When V_C and $V_I^{(v)}$ are fixed, the optimizations of $U_C^{(v)}$ and $U_I^{(v)}$ are independent with different view v, which can be formulated as

$$\min_{U_C^{(v)} \geq 0} \left\| U_C^{(v)} X^{(v)} - V_C \right\|_F^2 + \beta \| U_C^{(v)} \|_{21}, \tag{5.12}$$

$$\min_{U_I^{(v)} \geq 0} \left\| U_I^{(v)} X^{(v)} - V_I^{(v)} \right\|_F^2 + \beta \| U_I^{(v)} \|_{21}. \tag{5.13}$$

Similar to the optimizing rules for Eqs. (5.4) and (5.5), the following updating rules for $U_C^{(v)}$ and $U_I^{(v)}$ can be achieved,

$$(U_C^{(v)})_{ij} \leftarrow \frac{\left(V_C^{(v)} X^{(v)T}\right)_{ij}}{\left(U_C^{(v)} X^{(v)} X^{(v)T} + \frac{1}{2}\beta M_C^{(v)} U_C^{(v)}\right)_{ij}} (U_C^{(v)})_{ij}, \tag{5.14}$$

$$(U_I^{(v)})_{ij} \leftarrow \frac{\left(V_I^{(v)} X^{(v)^T}\right)_{ij}}{\left(U_I^{(v)} X^{(v)} X^{(v)^T} + \frac{1}{2}\beta M_I^{(v)} U_I^{(v)}\right)_{ij}} (U_I^{(v)})_{ij}, \qquad (5.15)$$

where $M_C^{(v)}$ and $M_I^{(v)}$ are the diagonal matrices with the k-th diagonal element given by $(M_C^{(v)})_{kk} = \frac{1}{||(U_C^{(v)})_k||^2}$ and $(M_I^{(v)})_{kk} = \frac{1}{||(U_I^{(v)})_k||^2}$. $|| \cdot ||^2$ is the l_2-norm. $(U_C^{(v)})_k$ and $(U_I^{(v)})_k$ are the k-th row of $U_C^{(v)}$ and $U_I^{(v)}$, respectively.

Algorithm 5.1 gives the overall optimization processes for our proposed model. In Step 1, the distance-weighted matrix $W^{(v)}$ and corresponding $D^{(v)}$ in each view are calculated and the projection matrices $U_C^{(v)}$ and $U_I^{(v)}$ and new feature matrices V_C and $V_I^{(v)}$ are initialized randomly. After that, V_C, $V_I^{(v)}$, $U_C^{(v)}$ and $U_I^{(v)}$ are updated repeatedly in Step 3–5 until convergence. Based on the learned shared features V_C across multiple views, the final clustering result can be obtained by existing single-view clustering algorithms, e.g. K-means in this section.

Since the optimization of the proposed model Eq. (5.3) can be divided into four subproblems [see Eqs. (5.4), (5.5), (5.12) and (5.13)] and each of them is a convex problem with respect to one variable, we can solve the subproblems alternatively to find the optimal solution, and thus this model can be converge to a local minima [17, 25].

Algorithm 5.1 UMCFL algorithm

Input: Multi-view dataset $D = \{X^{(v)}\}_{v=1}^V$, parameter α and β, dimensions of the shared and view-specific features

Output: Shared features V_C and view-specific features $V_I^{(v)}$

1: Initialize V_C, $V_I^{(v)}$, $U_C^{(v)}$ and $U_I^{(v)}$ randomly. Calculate $W^{(v)}$ and $D^{(v)}$;

2: **repeat**

3: Update V_C and $V_I^{(v)}$ according to Eqs. (5.10) and (5.11);

4: Calculate $M_C^{(v)}$ and $M_I^{(v)}$;

5: Update $U_C^{(v)}$ and $U_I^{(v)}$ according to Eqs. (5.14) and (5.15);

6: **until** Convergence;

7: Output V_C and $V_I^{(v)}$;

---

6 Conclusion

In the era of big data, a large number of multimodal data were constantly generated and accumulated. How to effectively integrate different modal data and mine the hidden value of data through complementary learning of information between modalities was the main concern of big data research at this stage. However, in the practical application analysis of multimodal big data, the modal incompleteness, real-time processing, modal imbalance and high-dimensional attributes pose severe challenges to the multimodal fusion method. In view of the above characteristics of multimodal data and the shortcomings of existing multimodal fusion algorithms, this chapter studies the multimodal data fusion algorithms from four aspects: incomplete modal analysis fusion, incremental modal clustering fusion, heterogeneous modal migration fusion and low dimensional modal sharing fusion. A series of multimodal data fusion models and algorithms are designed.

References

1. Wang, Y., Jin, X., Cheng, X.: Network big data: present and future. Chin. J. Comput. **36**(6), 1125–1138 (2013)
2. Manyika, J., Chui, M., Brown, B., et al.: Big Data: The Next Frontier for Innovation, Competition, and Productivity, vol. 5(33), pp. 1–137 (2011)
3. Li, J., Liu, X.: An important aspect of big data: data usability. J. Comput. Res. Dev. **50**(6), 1147–1162 (2013)
4. Zhang, Q.: Research on deep computation model for big data feature learning. Dalian University of Technology, Dalian (2015)
5. Gao, Q., Zhang, F., Wang, R., et al.: Trajectory big data: a review of key technologies in data processing. J. Softw. **28**(4), 959–992 (2017)
6. Li, H., Wang, Y., Jia, Y., et al.: Network big data oriented knowledge fusion methods: a survey. Chin. J. Comput. **2017**(1), 1–27 (2017)
7. Wu, X., Zhu, X., Wu, G.Q., et al.: Data mining with big data. IEEE Trans. Knowl. Data Eng. **26**(1), 97–107 (2014)
8. Zhao, L., Chen, Z., Hu, Y., et al.: Distributed feature selection for efficient economic big data analysis. IEEE Trans. Big Data. (2016). https://doi.org/10.1109/TBDATA.2016.2601934
9. Du, X., Chen, Y.: Approaches for value extraction on big data. Big Data Res. **3**(2), 19–25 (2017)
10. Bengio, Y., Courville, A., Vincent, P.: Representation learning: a review and new perspectives. IEEE Trans. Pattern Anal. Mach. Intell. **35**(8), 1798–1828 (2013)
11. Lahat, D., Adali, T., Jutten, C.: Multimodal data fusion: an overview of methods, challenges, and prospects. Proc. IEEE **103**(9), 1449–1477 (2015)
12. Xing, J., Niu, Z., Huang, J., et al.: Towards robust and accurate multi-view and partially-occluded face alignment. IEEE Trans. Pattern Anal. Mach. Intell. (2017). https://doi.org/10.1109/TPAMI.2017.2697958
13. Zhang, C., Hu, Q., Fu, H., et al.: Latent multi-view subspace clustering. In: Proceedings of the 2017 IEEE Conference on Computer Vision and Pattern Recognition, pp. 4333–4341. IEEE, Honolulu (2017)
14. Liu, J., Wang, C., Gao, J., et al.: Multi-view clustering via joint nonnegative matrix factorization. In: Proceedings of the 2013 SIAM International Conference on Data Mining. Society for Industrial and Applied Mathematics, pp. 252–260. Philadelphia (2013)
15. Li, S.Y., Jiang, Y., Zhou, Z.H.: Partial multi-view clustering. In: Proceedings of 28th AAAI Conference on Artificial Intelligence, pp. 1968–1974. AI Access Foundation, Quebec City (2014)
16. Yin, Q., Wu, S., Wang, L.: Unified subspace learning for incomplete and unlabeled multi-view data. Pattern Recogn. **67**(67), 313–327 (2017)
17. Cai, D., He, X., Han, J., et al.: Graph regularized nonnegative matrix factorization for data representation. IEEE Trans. Pattern Anal. Mach. Intell. **33**(8), 1548–1560 (2011)

18. Wang, W., Arora, R., Livescu, K., et al.: On deep multi-view representation learning: objectives and optimization. arXiv preprint. arXiv: 1602.01024 (2016)
19. Guan, Z., Zhang, L., Peng, J., et al.: Multi-view concept learning for data representation. IEEE Trans. Knowl. Data Eng. **27**(11), 3016–3028 (2015)
20. Zhao, L., Chen, Z., Yang, Z., et al.: Local similarity imputation based on fast clustering for incomplete data in cyber-physical systems. IEEE Syst. J. https://doi.org/10.1109/JSYST.2016.2576026 (2016)
21. Meng, L., Tan, A.H., Xu, D.: Semi-supervised heterogeneous fusion for multimedia data co-clustering. IEEE Trans. Knowl. Data Eng. **26**(9), 2293–2306 (2014)
22. Hardoon, D.R., Szedmak, S., Shawe-Taylor, J.: Canonical correlation analysis: an overview with application to learning methods. Neural Comput. **16**(12), 2639–2664 (2004)
23. Pereira, J.C., Coviello, E., Doyle, G., et al.: On the role of correlation and abstraction in cross-modal multimedia retrieval. IEEE Trans. Pattern Anal. Mach. Intell. **36**(3), 521–535 (2014)
24. Shu, X., Qi, G.J., Tang, J., et al.: Weakly-shared deep transfer networks for heterogeneous-domain knowledge propagation. In: Proceedings of the 23rd ACM International Conference on Multimedia, pp. 35–44. ACM, Brisbane (2015)
25. Shao, W., He, L., Lu, C.T., et al.: Online unsupervised multi-view feature selection. In: Proceedings of 16th IEEE International Conference on Data Mining, pp. 1203–1208. IEEE, Barcelona (2016)

Efficient Parallel Implementation of Cellular Automata and Stencil Computations in Current Processors

Fernando Diaz-del-Rio, Daniel Cagigas-Muñiz, Jose Luis Guisado-Lizar, and Jose Luis Sevillano-Ramos

Abstract A Cellular Automaton is a bio-inspired discrete model of computation with multiple applications, consisting of a regular grid of cells that have different states along time. Our research group at the University of Seville (Spain) collaborated in the past with Prof. Mohammad S. Obaidat in the fusion of Cellular Automata (CA) with another bio-inspired approach, the Address-Event-Representation (AER) neuromorphic communication protocol, for implementing a vision filter based on 3×3 convolutions [22]. In the last years, our group has continued working on the optimization of CA implementations in current high-performance computational systems. In this chapter, several of these optimizations will be described, focusing on those especially aimed at current processors, including hardware alternatives (e.g. GPUs, BTBs, etc.), different forms of parallelism such as instruction-level parallelism, thread-level parallelism, data-level parallelism, as well as software approaches (such as 'if-else' statement elimination, loop unrolling, data pipelining and blocking, etc.). The effect of these optimizations will be qualitatively illustrated by means of the Roofline model, considering simple CAs such as the well-known Game-of-Life (GOL), which has been extensively used to explore CA characteristics. This CA was invented by John Horton Conway, an English mathematician that recently died of complications from COVID-19, so we want this case-study to serve also as a tribute to him.

F. Diaz-del-Rio · D. Cagigas-Muñiz · J. L. Guisado-Lizar · J. L. Sevillano-Ramos (✉)
Department of Computer Architecture and Technology, Universidad de Sevilla, Seville, Spain
e-mail: jlsevillano@us.es

F. Diaz-del-Rio
e-mail: fdiaz@us.es

D. Cagigas-Muñiz
e-mail: dcagigas@us.es

J. L. Guisado-Lizar
e-mail: jlguisado@us.es

© The Author(s), under exclusive license to Springer Nature Switzerland AG 2022
P. Nicopolitidis et al. (eds.), *Advances in Computing, Informatics, Networking and Cybersecurity*, Lecture Notes in Networks and Systems 289,
https://doi.org/10.1007/978-3-030-87049-2_4

93

1 Introduction

Cellular automata (CA) [15, 26] are mathematical models made up of "cells" whose state evolve in discrete time steps. They change their state at each time step according to certain rules, usually taking into account the state of neighboring cells. CA are commonly used in the modeling and simulation of complex systems, in which the overall behavior results from collective action of many simple components that interact locally. They have been used to build models in many different areas of science and technology, including biomedicine (viral infections, spread of epidemics), physics (dynamics magnetization in solids, reaction-diffusion processes), engineering (networks telecommunications, cryptography), ecology (population dynamics), and the economy (markets stock market) [1, 7]. They have also been widely used in theoretical computer science, mainly to study emergent properties of complex systems, self-reproduction and formal properties such as computational universality or reversibility [18].

The Game of Life (GoL) [9] is the best known CA model. It was introduced in the late 1960s by John Horton Conway and gained widespread popular interest when appeared in a 1970 article by the mathematician Martin Gardner in the Scientific American magazine [10]. It is based on a square cell grid, where each cell can take two states (alive or dead), and it evolves depending on the state of its 8 neighboring cells and a set of rules. Thus, this evolution is deterministic: each state at each time step depends exclusively on the initial state of its cells.

CA have an inherent parallel nature because each cell can be processed independently at each time step, only taking into account the states of its neighbors. In other words, the temporal evolution of its components takes place in a relatively independent way and communication between them is restricted to a local environment. Thanks to this, CA are very suitable for their implementation on parallel architecture computers taking full advantage of their efficiency, because the communication flow between processors can be maintained low. In the last three decades this potential has been used to build simulations using CA models on high-performance parallel systems, such as clusters [12], parallel computers with shared memory [4], or highly parallel GPUs [11]. Our research group at the University of Seville (Spain) collaborated with Prof. Mohammad S. Obaidat (among many other things) in the fusion of CA with another bio-inspired approach, the Address-Event-Representation (AER) neuromorphic communication protocol [22]. The aim of that work was to use the Cellular Automation approach to implement an AER neuro-inspired filter for vision processing, based on 3×3 convolutions. This was one of the first collaborations with Prof. Obaidat, who visited our University several times and gave a number of conferences and workshops on multiple topics: wireless sensor networks, simulation of computer and communication systems, etc. This long collaboration has inspired our research group, and helped us to improve our international projection and consolidate several of these lines of research. In the last years, our group has continued working on the optimization of CA implementations as well as stencil computations in current high-performance computational systems.

Stencil computations [21] are a wide class of applications in which a regular grid is swept iteratively, updating its cells using a fixed nearest-neighbors pattern. Algorithms and codes that are organised in this way are called Iterative Stencil Loops (ISLs) or simply stencil codes. They are used in many important application domains such as CA, numerical methods to solve partial differential equations for weather modeling, seismic simulations, fluid dynamics, heat diffusion, quantum dynamics simulations, image processing, etc. [6]. Usual CA employ stencil codes to sweep their cell grid, although most CA are more complex, from an algorithmic point of view, than other typical stencil codes for two reasons. First, stencils generally involve obtaining a numerical result by applying a matrix of coefficients (pattern) to each cell and its neighbors employing simple arithmetic operations. On the other hand, in CA the objective is not to obtain a numerical result, but a new state, that is computed by applying a transition rule which usually has a series of if-else structures, more complex to compute than arithmetic operations. Second, CA neighborhoods can be wider and more complex than usual stencil patterns. In the rest of this work we will refer specifically to CA, but many of the optimizations described can also be applied to most stencil algorithms.

We present here an exhaustive analysis of the whole range of techniques, algorithms and optimizations that can be used to implement cellular automata and also many stencil computations in the most efficient possible way on current CPUs. In order to analyse and discuss the different optimizations for a CA implementation, we center mainly in the well-known GoL. However our discussion is generic enough to be extended to most CA, with the possible exception of some particular CA variants, such as Random Boolean Networks (RBNs), in which each cell can interact with any other cell of the system, which involves a random memory access pattern that make many of the optimizations presented not applicable.

2 Related Works

Oxman et al. [28] presented efficient implementations of the GoL CA for single-core and multi-core CPUs, representing the state of a cell in memory by just a single bit, with value '0' meaning 'dead' and value '1' meaning alive. In this way they obtained the best reported performance up to this day for the GoL CA, but with the drawback that this technique is not generic. It can only be applied to very simple toy-model CA such as GoL and is not useful for many CA dealing with real-world applications, in which each cell can have more (sometimes many more) than 2 states.

Different works study the execution of CA on CPUs and GPUs, mainly focusing on highlighting the benefits of GPUs for this kind of applications. Significant examples are [2, 5, 23, 25, 30]. Rybacki et al. [30] found that the performance was strongly dependent on the CA model, running on a single core processor, a multi core processor and a GPU. None of these studies analyse in general the performance of algorithms or optimizations applicable to CAs.

Gibson et al. [11] present a thorough study of the performance of the GoL CA in two dimensions and multi-state generalizations of it, implemented on GPUs and multi-core CPUs using an OpenCL framework, with respect to different standard CA parameters like lattice and neighborhood sizes, number of states, complexity of the state transition rules, number of generations, etc. Their study is useful to help predict the performance impact of different CA parameters elections. They only consider a simple toy model CA (GoL and some generalizations of it) and not other types of CA used in real applications, whose higher complexity can affect performance. They do not take into account detailed computer architecture details to model CA performance, such as for instance [32] does for stencil computations.

Balasalle et al. [3] found that careful optimizations in the memory access patterns can produce a 65% improvement in run-time from a baseline implementation of the simplest two-dimensional CAs—the game of life. However, they did not study other more realistic CA models.

Besides, in the last decade noteworthy models have been presented to quantify the performance of iterative algorithms in modern processors. They are the basis to understand the performance of CAs, to detect their execution bottlenecks and to find out which additional optimizations are still applicable.

Stengel et al. [32] apply the Execution-Cache-Memory (ECM) model, which is an extension and refinement of the Roofline model [33], to quantify the performance bottlenecks of stencil algorithms on a contemporary Intel CPU. They don't consider explicitly CA, but the presented performance model for stencil computations can be applied to many CA. Detailed estimations of the data traffic through the different cache levels and to/from DRAM are made and the impact of some typical optimizations such as spatial and temporal blocking is studied. The model does not apply to GPUs.

Yang et al. [34] applies the Roofline performance model to analyze different applications, including stencil computations, on GPU. Their model takes into account L1, L2, device memory, and system memory bandwidths. Their work helps to understand performance bottlenecks on NVIDIA GPUs and can suggest different code optimizations.

3   Cellular Automata Model

The typical workflow of a Cellular Automata (CA) consists of declaring two grids arrays for the current and the next CA state at each time iteration. Let pcurr and pnext be these two arrays. Figure 1 shows a pseudocode that summarizes a typical CA implementation. After allocating memory for both arrays, the content of pcurr is pre-filled with the initial data of the CA.

Then, at each time iteration or computation step, the grid that is going to hold the next state is calculated based on the content of the CA current state (pcurr). At the end of the computation step there is an exchange of roles between these two grids for the next time iteration. In programming languages like C/C++ this implies

```
allocate_memory (pcurr, pnext);
fill_with_initial_values (pcurr);
step = 0;
while (step<MAX_STEPS) {
      compute_step (pcurr, pnext);
      swap (pcurr, pnext);
      step++;
}
print_or_save_results (pcurr);
```

Fig. 1 Pseudo code description of a generic CA program

swapping the pointers pcurr and pnext. Note that this scheme is the same for most of stencil algorithms, like those of Jacobi type. The compute_step function from Fig. 1 implies the evolution of one step for the CA.

Figure 2 describes the standard and simplest code of a generic computation step for a CA. We suppose that the column is the leading dimension because matrices are stored by rows (as in most languages like C/C++). Hence, for each column iteration (loop with iterator c) a single cell update to be stored in pnext[r][c] is computed according to the CA rules.

In the most generic implementation, firstly a set of conditional structures specifies which current state the cell contains (from the set CURR_STATE_1, CURR_STATE_2, …, CURR_STATE_M).

Secondly and for each possible cell state pcurr[r][c], the cell neighborhood value is obtained by the function f_neigh( ) applied to the selected set of neighbors (which is represented by variable NEIGHBORS in Fig. 2).

These neighborhood values are denoted as CURR_STATE_K_COND_N_L, and for each of them a new value, called NEXT_STATE_K_COND_N_L, K = 1, 2, …, M, L = 1, 2, …, N_K in the figure, is stored in pnext[r][c].

Note that a perfect synchronization is required in order not to mix the CA state of a certain step with that of the next step. This barrier at the end of each computation step supposes a source of running time waste, which can be important for massive parallel computers such as GPUs. Nevertheless, there are options that increase performance by holding a mixture of several temporal CA states at each same matrix (pcurr, pnext). These are discussed in Sect. 5.9.

In Sect. 5, the different optimizations for this CA code are discussed and their theoretical impact in performance is analyzed, based on the architecture and execution time model described in the next section

4 Current CPU Architecture

Since our main aim is the optimization of CA implementations in current high performance processors, in this section we will discuss the main characteristics of these

```
for (int r = 0; r < SIDE; r++)  // row loop
  {
  for (int c = 0; c < SIDE; c++) // column loop
    {
    if (pcurr[r][c] == CURR_STATE_1) {
      if ( (f_neigh(NEIGHBORS) == CURR_STATE_1_COND_1)
        pnext[r][c] = NEXT_STATE_1_COND_1;
      if ( (f_neigh(NEIGHBORS) == CURR_STATE_1_COND_2)
        pnext[r][c] = NEXT_STATE_1_COND_2;

      ...

      if ( (f_neigh(NEIGHBORS) == CURR_STATE_1_COND_N_1)
        pnext[r][c] = NEXT_STATE_1_COND_N_1;
      }
    if (pcurr[r][c] == CURR_STATE_2) {
      if ( (f_neigh(NEIGHBORS) == CURR_STATE_2_COND_1)
        pnext[r][c] = NEXT_STATE_2_COND_1;
      if ( (f_neigh(NEIGHBORS) == CURR_STATE_2_COND_2)
        pnext[r][c] = NEXT_STATE_2_COND_2;

      ...

      if ( (f_neigh(NEIGHBORS) == CURR_STATE_2_COND_N_2)
        pnext[r][c] = NEXT_STATE_2_COND_N_2;
      }
    ...
    if (pcurr[r][c] == CURR_STATE_M) {
      if ( (f_neigh(NEIGHBORS) == CURR_STATE_M_COND_1)
        pnext[r][c] = NEXT_STATE_M_COND_1;
      if ( (f_neigh(NEIGHBORS) == CURR_STATE_M_COND_2)
        pnext[r][c] = NEXT_STATE_M_COND_2;

      ...

      if ( (f_neigh(NEIGHBORS) == CURR_STATE_M_COND_N_M)
        pnext[r][c] = NEXT_STATE_M_COND_N_M;
      }
    } // end of column loop
  } // end of row loop
```

Fig. 2 Pseudo code of a generic CA compute_step()

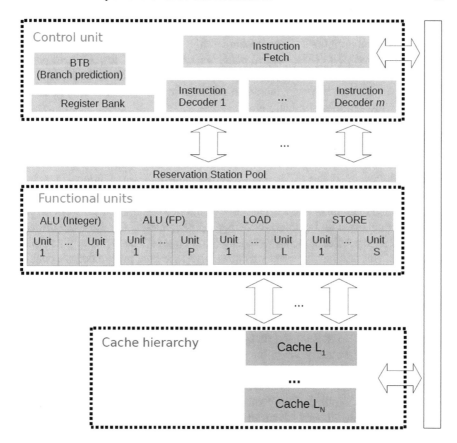

Fig. 3 Current CPU scheme

machines. This discussion may seem too general given the wide variety of processors available today. However, some key features of modern processors are common to the vast majority of them, allowing us to draw conclusions with sufficient generality [13]. For instance, while theoretically machine instructions are executed one at a time, all current microprocessors implement techniques to exploit the available instruction-level parallelism (ILP). So-called "superscalar" processors can issue 4 to 6 instructions per clock cycle that are distributed among the different functional units, where they wait for the operands to be available in the so-called "Reservation Stations". These instructions can therefore be executed in parallel, even out-of-order, as soon as the operands are available, while preserving the original logical program order. A simplified scheme of a current processor is shown in Fig. 3.

Since a huge number of transistors can be integrated onto a single chip, structural hazards due to lack of hardware resources are relatively rare. Therefore, the program execution in these modern microprocessors is essentially limited by the data dependencies in the code. Data-flow graphs become a very useful tool to represent the program execution, allowing to identify "critical paths" representing the data dependences that limit the ILP by imposing a strict order in which some instructions must be executed [29]. These critical paths are a fundamental lower bound on the execution time, and theoretically in current processors they should be close to the actual execution time. However, there are still some limitations to this ideal behavior. The main causes of these limitations are:

- Branch mispredictions: machine code includes conditional branches that limit ILP as the processor cannot execute instructions until the outcome of the branch is known. Current processors incorporate advanced mechanisms (such as Branch Target Buffers or BTBs [13]) to predict the outcome of these conditional branches. When the BTB correctly predicts the outcome of a branch, instructions at the branch target can be issued for parallel execution. This mechanism works really well particularly with branches following regular patterns. For the more difficult cases, hard to predict branches can many times be substituted by conditional data transfers. But despite all these techniques, given the high mispredictions penalty in modern processors (tens of cycles for current microprocessors [29]) there are still cases where this penalization cannot be ignored.
- Structural hazards: although as we mentioned current processors are full of hardware resources, there are always limiting situations where the hardware resources are not enough to sustain the data-flow, such as: insufficient number of functional units, particularly when the code involves too many concurrent similar arithmetic operations; non-pipelined long-latency functional units; not enough entries in input/output buffers of functional units; and so on.
- Memory wall: Despite the introduction of multiple levels of caches, improvements in DRAM bandwidth, and advanced prefetch and refill schemes, the memory system is the limiting factor that has the most influence on the performance of today's processors. This is particularly important with multicore superscalar processors, with aggressive mechanisms to take advantage of the available data level parallelism: vector instructions, Single Instruction Multiple Data (SIMD) instruction extensions, Graphics Processing Units (GPUs), etc. All these techniques further increase the memory system requirements of current processors, and many times memory bandwidth and (to a lesser extent) latency become insufficient to meet these requirements.

For the execution of CA these three limitations may be of importance. For instance, as shown in Fig. 2 conditional structures (if-then-else) are an essential part of most CA codes so branch mispredictions may be significant. However, as we will discuss later, most of these conditional structures can usually be substituted by predicated operations (for instance, conditional movs). So probably the two primary involved factors are the other two: structural hazards (mainly in terms of the number of functional units) and memory bandwidth. In other words, the key question is to what extent the CA execution time is limited by the memory bandwidth or by the number of computational units, given the parallel nature of CAs.

4.1 Roofline Model

For this kind of trade-offs (computing throughput versus memory bandwidth) a popular tool is the so-called Roofline model [33], that allows us to understand which factor is limiting our code, and how close to the limit we are. It introduces the concept of Arithmetic Intensity (A_I) that can be defined as the number of floating-point operations per byte read, which depends only on the application and not on the specific hardware. The Roofline model plots floating-point throughput as a function of A_I in a log-log scale. Two bounds can be represented in this model: peak floating-point (a horizontal roof) and peak memory bandwidth (a slope 1 roof). These bounds are fundamental limits of the machine, and the performance of a given application can be represented as a coordinate in this space according to its A_I. If the A_I of the application is low then its coordinate would be under the slope-1 roof meaning that its execution is bound by the memory system; if on the contrary A_I is high the execution is bound by the peak floating-point performance. Several variations of the Roofline-model exist such as the Cache-Aware Roofline Model [16] or the Integrated Roofline Model [20]. These variants make the model more flexible, even using it for other types of operations (not necessarily arithmetic) allowing for the representation of the effect of different application optimizations. For instance, as seen in Fig. 4 (blue arrow), consider a that a better caching technique is used that permits the working data set of an application to fit into a higher level of cache. Since this higher cache level offers a higher peak bandwidth the roof moves up indicating that the upper bound for the attainable performance also increases (the effect of "upgrading the loft" in [16]). Another possible optimization (black arrow in Fig. 4) would be to maximize in-core performance through a better use of the available data level parallelism using for instance vectorization. Vector-like instructions apply the same operation to more data elements and therefore the achievable performance can increase. The opposite effect can also occur: for instance, if the application includes too many operations using the same functional unit, the attainable performance would be reduced as other units remain unused. In general, if a given application is optimized so that it uses an appropriate mix of instructions that allows all available functional units to be in use, we would be closer to the upper horizontal ceiling. Finally, some applications can be optimized to reduce data movements, for instance by reusing some elements from previous iterations. In these cases, it can be considered as if the A_I increased (orange arrow in Fig. 4), as the net effect is that less memory accesses are needed to run the algorithm. This is a common case in stencil computations and CAs as we will see shortly. In this chapter we will use this more flexible roofline representation because it allows for a convenient graphical representation of the qualitative effect of optimizations and improvements on different implementations of CAs.

The execution time per cell may be more accurately estimated by using the Execution-Cache-Model (ECM) model [32], which takes into account all the executed instructions as well as the cycles spent by the load/store instructions in the CPU, plus the penalties suffered by the different memory levels along the life of a cache line. However, on the one hand, the Roofline model is enough to understand

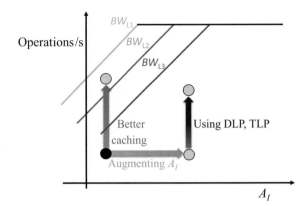

Fig. 4 Roofline model for GoL with single-precision floating-point cell types

the main timing involved in the CA execution, and on the other hand, ECM model is not accurate enough when analyzing CAs. The main reason is that conditional sentences are not modelled by ECM (it was conceived for Stencil codes), which are common in CAs. The Roofline model does not actually model conditional sentences either, but instead it is much simpler and its graphical character allows for a more intuitive visualization of optimizations. For these reasons, ECM model is out of the scope of this work. The interested reader is referred to [32] for a deeper analysis.

5 Optimization

In this section we will focus on how to write CA implementations than can be efficiently executed in current microprocessors, following the ideas outlined in the previous sections. As mentioned before, the main limitations to an ideal execution are: penalizations due to branch mispredictions, structural hazards (such as limited number of functional units or buffers) and insufficient memory bandwidth. We will discuss several techniques to overcome these limitations in the particular case of CA code. Since we are interested in general optimization techniques, we will not consider detailed CA implementations. The variety of CAs is enormous, so this work is not centered in studying specific optimizations that are valid only for a particular CA. Instead, we consider broader optimizations, which will in some cases be applied to a particular CA for a better understanding. We will take the pseudocode outlined in Figs. 2 and 3 as a representation of what would be a realistic implementation, and our discussion of the different optimizations will be based on them. Besides, when appropriate we will also discuss or indicate briefly if the described optimizations can be incorporated to multithread and GPU programming style.

5.1 Reducing Memory Size

The primary aspects to consider for an efficient CA parallel implementation are those accounting for the memory accesses. In this sense, the most direct way to reduce the memory size is that of coding the cell state with less bits. For the sake of compatibility, code reusing and extending a CA to future variations, programmers tend to use standard variable types. For example, in the case of GoL, data type 'int' is an option that allows up to 2^{32} possible states to be coded. This is fine if we want the cell state to be extended in future CA versions. However, this implies spending 32 bits per cell in memory when the classical GoL needs only one bit per cell (for its two possible states: live or dead).

Whereas some studies such as [28] encode the cell in a single bit, thus reaching an enormous speedup, this special coding is not extensible to other CAs, implies a considerable technical complexity, and its compatibility for different platforms is difficult to ensure. This complexity is high for CPUs, but at least once coded it can be extended for multithreading; however, it is far more difficult for GPUs. As a compromise between these two extreme options, and for the case of GoL, an 8-bit data type 'char' would be enough (it allows up to $2^8 = 256$ different states), while making memory alignment easier and allowing more compatibility and a smooth incorporation of future alternatives.

Another aspect that has some influence on the memory consumption is the matrix boundaries. Boundary conditions are usually managed differently in CAs and Stencil algorithms. Instead of fixing the boundary values, which is common when solving differential equations, CAs simulates boundaries as a periodic boundary conditions, that is, each side of the square grid has the opposite side as its neighbor. Hence, the CA grid forms a torus (an n-dimensional torus for other dimensions).

From the computation perspective, a first solution is to add ghost rows/columns to compute cells in the borders (sometimes called halo regions). Thus, it needs more memory. In the case of an $N \times N$ cell automaton, $2 * N + 2 * (N - 1)$ new cells are needed. In addition, these new cells are to be updated, by copying these 4 borders at the end of each computation step.

An alternative to the previous technique that prevents the use of extra memory is to mathematically calculate the adjacent cell coordinates for each border cell. In this case, insertion of conditional sentences to compute the border neighbor indexes should be prevented (see Sect. 5.4), by using for instance modular arithmetic functions.

5.2 Reusing CA Elements

As shown in Fig. 1, a typical CA implementation includes two matrices: one to hold the current CA state and another one to hold the next state. In what follows, we will not consider the possibility of using only one matrix that is updated as the CA evolves, because it introduces considerable complications (like additional buffers to hold the rows that are to be written, several difficulties when parallelizing the CA among several threads, etc.), and it does not provide a significant speedup. At this point, it is interesting to study in more detail the memory accesses that are needed to execute a typical CA code.

Since the next state of a given cell depends on the state of its neighbors, every iteration requires accessing all the cell's neighbors. These memory accesses must be carefully reorganized in order to promote caching, accesses that are very intricate and diverse depending on the neighboring definition. Different neighbors imply diverse strategies for reducing accesses to the lower levels of the memory hierarchy. In order to fix ideas, let us consider the usual 8-adjacency neighboring, that is, to process the central (C) cell we must read the values of (in clockwise direction) the East, North East, North, North West, West, South West, South and South East (E, NE, N, NW, W, SW, S, SE) cells (see Fig. 5). The basic, naive code would require 9 read accesses (1 central cell and 8 neighbor cells) and 1 write access (update of the central cell) per iteration. Analyzing this code using the Roofline model described in previous section, we can say that if the CA involves D arithmetic operations per element, the Arithmetic Intensity would be $D/10$ operations per byte (assuming that each cell occupies 1 byte, as discussed before).

However, this code can be easily optimized. If we assume row-major order as in C/C++ language (that is, consecutive elements of a row are contiguous in memory) then we know that it is desirable to go along the matrix rows when updating the CA states, that is, for each new cell updating we run an increment in the column index c (inner loop in Fig. 3). This would mean that for a given cell some neighbors have been previously read: neighbors on the left of the current cell pcurr[r][c] (that is, North, North West, West, South West and South cells) that have been recently read (by the pcurr[r][c-1] cell); and also East and North East neighbors that were previously read when updating the upper cell (the North one, that is pcurr[r-1][c]). This means that all the neighbor cells except the South East one have been previously accessed.

The different memory hierarchy levels can automatically cache these previously read neighbors, but the exact behavior requires a deeper analysis. In order to study the access to new cache lines, let us study the case when the central columns (containing the N, C and S cells) are hold in the last byte of different cache lines (all of them should fulfill simultaneously this condition, because matrix rows are desired to be aligned with line caches). This means that all the neighbor cells except the South East one must have been cached by the processor. More precisely, neighbors on the left of the current cell must reside in the first level cache, whereas the allocation of the East and North East neighbors depends on cache sizes. To get through this example, let us suppose that the two rows of these last cells (rows $r - 1$ and r) also fit into a certain intermediate level L_k. Finally, according to the row processing of the CA matrix (see Fig. 5), the South East (SE) cell must be freshly captured from that hierarchy level that can hold the CA matrix (namely L_m, which can be RAM memory). A capable and experienced programmer, being conscious of the reuse of the different cells, may have stored the most recently used elements (N, NW, W, SW,

Fig. 5 Caching neighbor cells: light blue cells are allocated in the L1 cache (the number of cells depends on the L1 line size); orange ones in some level L_k, green cells can reside in registers and the South East cell (in red) is the farthest one, residing in a further level L_m

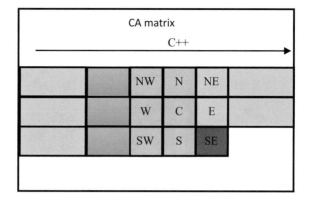

S and C) in CPU registers along the previous iterations of the column loop, thus saving some L1 accesses. In the end, the general picture of the memory state when processing a current cell C is the one expressed by the colors in Fig. 5. Figure 6 shows how the code of Fig. 3 would be transformed.

Note that with this optimization the Arithmetic Intensity would increase to $D/2$ operations per byte, since only two net accesses to L_m are needed: one read for the SE cell and one write to update the central cell. In the Roofline model, this optimization corresponds to the orange arrow in Fig. 4, which moves the application to a coordinate of the model where the ceiling (i.e. the performance bound) is higher thus leaving more room for an advanced processor to improve performance.

This latter reasoning about the Arithmetic Intensity is simple and useful to illustrate the effect of this optimization. However, it can also be refined. Not only does the SE cell introduce an access to a non-cached element but it must take part in the computation of the operations that determine the evolution of the current central cell C. Although there is no general rule to determine the required new operations that SE must be involved in, in many cases a smart reorganization of the `f_neigh(NEIGHBORS)` in Fig. 2 can lead to some recurrence procedure like the following one:

```
prev_neigh = f_recurrent_neigh (E, NE, N, NW, W, SW, S);
neigh = f_recurrent_neigh (prev_neigh, SE);
```

where the variables here represent the state of these cells.

Although simple, the most well-known case is when `f_neigh(NEIGHBORS)` computes the sum of the 8 neighbor states. In this case, if in previous cell computations the sums of NW, W, SW, of N, S, C and of E, NE were previously stored in several registers, they would be reused for the current cell, and the neighbor computation yields simply: f_{neigh} (NEIGHBORS) $= E + R1 + R2 + R3$, where R1, R2, R3 contains sum(NW, W, SW), sum(N, S), sum(E, NE); respectively. In addition, after this iteration, R2 must be set with the value of sum (NE, SE) to be ready for the next iteration. Likewise for R1 and R3.

Whereas all the above considerations are easy to understand, their practical implementation are far from being simple; and what is more, their implementation may be usually very different according to the CA rules and the neighboring set to be included in the computations. Moreover, it is a hard task for the programmer to include all these recommendations without committing mistakes, i.e. debugging can be also a lengthy task.

The code in Fig. 6, which includes all the previously discussed optimizations, should be now easy to understand. Also, a simplified computation of the execution time is possible. Note that all accesses can be supposed to be done to different cache levels except those of the new state updating and:

```
SE = pcurr[SE_row][SE_col]; // unique new CA matrix reading
```

Hence, if a cell state has b_c bytes and the CA matrix resides in RAM memory, the Roofline model tells us that if we are under the slope-1 roof (meaning that the execution is bound by the memory system) then the timing per cell and step yields to the time consumed by these two accesses to main memory, that is (being f the clock frequency, and BW the RAM bandwidth):

$$BW \times \frac{1cell}{2 \times b_c} \times \frac{1}{f}$$

A final remark must be done with respect to those RAM writes that are not to be reused anymore. In case of using no-write-allocate store instructions (also called "streaming stores" or non-temporal stores), some cycles would be saved. The exact number of cycles depends on great manner on the architecture.

5.3   Matrices to Hold the Neighbor Information and the CA Input and Output

With respect to the computation time, the actual behaviour depends on the cache policies. For instance, consider a write-allocate cache, that is, a cache that loads the missed-write location before the successful write-hit operation. In this case, using two matrices requires two accesses to level L_m for updating next state (one for allocating the line in L_1; a second one to flushing it to L_m to update it).

A very important advantage of using two matrices is that writing in the output matrix is only necessary when it is detected that the cell state has changed along the current step. Hence, this can contribute to reduce the number of writes (in the CA next state matrix).

There is another option used in some CPU implementations that incorporates an additional "neighbor matrix", holding the neighbor information for each cell. If this information were obtained by some simple operation (usually the sum of living neighbors like in GoL), the additional matrix would contain for each cell, the amount

```
for (int r = 0; r < SIDE; r++)  { // row loop
      NE_row=f_row(r-1, SIDE); // compute previous row index
      SE_row=f_row(r+1, SIDE); // compute the next row index
      // COLUMN PROLOG: prepare first values of N,NW,W,SW,S,C
      prev_neigh = f_recurrent_neigh (N, NW, W, SW, S);
      for (int c = 0; c < SIDE; c++) { // column loop
          SE_col = f_col(c+1, SIDE); // compute next column index
          SE = pcurr[SE_row][SE_col];
          // unique CA matrix reading from last memory level
          E = pcurr[r][SE_col];
          // matrix reading from certain intermediate cache level
          NE = pcurr[NE_row][SE_col];
          // matrix reading from certain intermediate cache level
          neigh = f_recurrent_neigh ( prev_neigh, NE, E, SE );
          if (C == CURR_STATE_1) {
                  // pnext[c] is written only once for the
                  //     true case of this list:
                  if (neigh == CURR_STATE_1_COND_1)
                      pnext[c] = NEXT_STATE_1_COND_1;
                  if (neigh == CURR_STATE_1_COND_2)
                      pnext[c] = NEXT_STATE_1_COND_2;
                  ...
                  if (neigh == CURR_STATE_1_COND_N_1)
                      pnext[c] = NEXT_STATE_1_COND_N_1;
            }
          if (C == CURR_STATE_2) {
                  if (neigh == CURR_STATE_2_COND_1)
                      pnext[c] = NEXT_STATE_2_COND_1;
                  ...
                  if (neigh == CURR_STATE_2_COND_N_2)
                      pnext[c] = NEXT_STATE_2_COND_N_2;
          }
          ...
          if (C == CURR_STATE_M) {
                  if (neigh == CURR_STATE_M_COND_1)
                      pnext[c] = NEXT_STATE_M_COND_1;
                  ...
                  if (neigh == CURR_STATE_M_COND_N_M)
                      pnext[c] = NEXT_STATE_M_COND_N_M;
          }
      // refresh the values of N, NW, W, SW, S, C by calling:
      {N, NW, W, SW, S, C} = refresh_neigh (E, NE, SE) ;
      prev_neigh = f_recurrent_neigh (N, NW, W, SW, S);
      }  //  end of column loop
  } // end of row loop
```

Fig. 6 Ideal CA code according to Roofline and ECM models. Only the vectors and matrices reside in memory. All the variables are allocated in registers (according to the scheme of Fig. 5)

of living neighbors around it [11]. Then, anytime a cell changes its state, those cells that are neighbor to it must update their corresponding cells in the "neighbor matrix". This introduces additional writes to a full-size matrix, whose cells must have the sufficient size to hold the number of neighbors.

This option may simplify the coding in some cases; note that recurrence function `f_recurrent_neigh ( )` is avoided, but a new updating function for the "neighbor matrix" is needed, which must be called by checking when the cell state changes. Nevertheless, the question is: how often is this new matrix written? If writes are rare this option may save some amount of time; otherwise it would introduce some penalty timing.

Note that the "neighbor matrix" does not save memory bandwidth with regard to the implementation of Fig. 6, because ideally the different neighbors are hold in registers. Again memory bandwidth has a determinant effect in performance.

Note that the "neighbor matrix" can neither be easily implemented in GPU nor would produce good timing: it requires atomic accesses to this written matrix and additional conditional structures, which is known that are inefficient in these platforms. Writing in multithreaded programs must also be carefully addressed in the boundaries between matrix fragments, because two threads can write simultaneously to these boundary elements.

5.4 Eliminating Conditional Structures

As mentioned in Sect. 4, another limitation of current processors are branch mispredictions, which become particularly important here as conditional structures ($if - then - else$) are usually present in CA naive codes when coding evolution rules (see for example Fig. 6). From the CPU point of view, if the state evolution presents an almost random behavior (usual in many CAs), the Branch Target Buffer (BTB) that tries to predict the branch outcome is going to commit many mispredictions. Being misprediction penalty very high for current processors (more than 10 cycles, which supposes the abort of around 50 instructions), avoiding this problem is very appreciated. On the contrary, if a CA evolves in a more predictable way, elimination of conditional branches is not so crucial. Things are different for GPUs, because no conditional sentence is convenient [19], and the programmer (or compiler) should always try to convert the $if - then - else$ structures into non-branched codes according to the following possibilities.

If a compiler (either CPU or, more often, GPU) predicts that a number of branches may have an irregular behavior, it will convert these sentences by means of predicative instructions. Each of these predicative instructions contains an operation that is executed (more exactly, its result is written) if and only if an additional condition register is true. That is, if the condition is true, the instruction is executed normally; on the contrary, if the condition is false, the execution continues as if the instruction were a no-operation. Such instructions can be used to eliminate branches. For instance, the "Conditional MOV if Zero" instruction:

	(C1, C2) = (0, 0)	(C1, C2) = (0, 1)	(C1, C2) = (1, 1)	(C1, C2) = (1, 0)
Current state = 0				
Current state = 1				

Fig. 7 An 8 element Karnaugh map for any set of rules based on the values of two conditions C1, C2 plus a binary current cell state. The 8 empty cells must contain the next state for each case

```
CMOVZ Rd,Rf,Rcond;
```

is equivalent to:

```
if (Rcond==0) Rd=Rf;
```

This way, any operation of the *if* body would be predicated by the condition of the (eliminated) branch, while the operations of the 'else' body are predicated by the complementary condition. A more sophisticated predicative execution can be found when compiling for SIMD kernels (see [13]).

If the compiler cannot eliminate the conditional structures of a CA, the programmer has several alternatives to do it; however, none of them is generic due to the extensive diversities of existing CA evolution rules. For simple rules, converting the condition into a mask to be applied (using a binary AND operator; like '&' in C++ language) is an option. For instance, if the code is simply:

```
if (curr_state == 0) next_state = A;
else next_state = B;
```

it can be transformed into:

```
int condition_mask = (curr_state == 0) - 1;
```

```
// it returns a mask having 0 × 0 (true) or 0xFF....F (false) values
```

```
next_state = A &  (condition_mask) + B & condition_
mask;
```

If the group of conditions is bigger, mask operations can be cumbersome, thus a Karnaugh map can solve this codification. For example, if the rules are based on the values of two conditions C1, C2 plus a binary current cell state, an 8 element Karnaugh map is sufficient to hold all the next possible states (see Fig. 7). Conditions C1 and C2 in classical GoL are those that compare the number of live neighbors, for instance for a live cell, `f_neigh(NEIGHBORS) == 2 || f_neigh(NEIGHBORS) == 3` according to Fig. 3. Hence, a set of AND/OR operations can condense the 8 cases into a fast and short formula.

Another alternative is to use some formula that compresses all the evolution cases. This is conceivable when the number of cases is short. Computing a number (index) for all the rule cases, then indexing a LUT (Look-Up-Table) with this number is a fast and generic enough way to carry out this alternative. A LUT is here an array indexed by the possible rule cases. In the case of the GoL, the index can be computed via the current cell state (values 0 or 1) and the sum of the state of its 8 neighbors

$\sum_{i=0}^{8} neigh_i =$	0	1	2	3	4	5	6	7	8
Current state = 0	0	0	0	1	0	0	0	0	0
Current state = 1	0	0	1	1	0	0	0	0	0

Next state

Fig. 8 A 18-element LUT for the classic GoL to avoid using conditional sentences. Each column gives the next state for the different number of neighboring live cells

(values from 0 to 9); that is, an array of 2 * 9 binary elements that indicates the state at which the cell should change in the next computation step (Fig. 8). For example:

```
int index = (curr_state * 9) + sum_neighbors; // it returns
```
an integer in the interval [0,17]
```
next_state = LUT[index]; // here LUT can be initialized with the GoL
```
rules or with another evolution rules.

Although CAs are usually memory bounded (i.e. they are located under the slope-1 roof of the Roofline model) note that a small LUT is going to reside always in the L1 cache, thus supposing only one additional Load instruction, which contributes with a very little time to the execution time. Therefore, this alternative is probably much faster than the naive code composed of conditional sentences.

Full LUT can also be defined for simple CAs, thus preventing any computation with the neighbors. For the case of GoL, instead of computing the sum of the neighbors, a full LUT can contain $2^9 = 512$ elements, because the current cell and its neighbors count for 9 binary values.

Nonetheless, the tricky coding appears when running along the CA matrix if one wants to reuse previous elements, i.e. registers E, NE, N, NW, W, SW, S and C in Fig. 6. As stated in Sect. 5.1, the computed sums of a previous cell can be reused if cells are stored in a 'char' type. Compressing even more the cell, for example, in only one bit, would require a set of intricate shift and binary operations with the previous 8 neighbors plus the central cell when moving to the next cell.

The concept of LUT applied to CAs is similar (but not equal) to that of the Hash Tables described in [14].

5.5 Loop Unrolling

Loop unrolling is a common technique to speed up big loops, which is frequently used by compilers. In the case of CAs, it introduces a clear benefit in relation to the total amount of data (cells) to be read. In addition, understanding this technique allows the programmer to introduce additional optimizations that are described in the next subsections (mainly for the data pipelining and temporal blocking techniques).

The general loop unrolling procedure is a useful technique in loops without loop-carried dependencies, that is, loops where iterations are independent of each other.

Current state reads without unrolling

```
for (int c = 0; c < SIDE; c++) { // column loop
        if (pcurr[r][c] == ...) {
                if ( (f_neigh(pcurr[r][c-1],pcurr[r][c+1])==...)...
                }
    ...
        }
} // end of column loop
```

Current state reads using unrolling

```
for (int c = 0; c < SIDE; c+=3) { // column loop
        if (pcurr[r][c] == ...) {
                if ( (f_neigh(pcurr[r][c-1],pcurr[r][c+1])==...)...
                }
        if (pcurr[r][c+1] == ...) {
                if ( (f_neigh(pcurr[r][c-0],pcurr[r][c+2])==...)...
                }
        if (pcurr[r][c+2] == ...) {
                if ( (f_neigh(pcurr[r][c+1],pcurr[r][c+3])==...)...
                }
    ...
        }
} // end of column loop
```

Fig. 9 The effect of unrolling in the number of state reads. The not unrolled version needs three loads for each cell. Bottom: the unrolled code requires 5 loads per each unrolled iteration, which supposes 5/3 loads per cell in the mean (see also Fig. 10)

In the case of standard CAs (Fig. 3) there are no loop-carried dependencies among iterations, because the CA output is written into a matrix which is different from the input matrix, so loop-unrolling is highly convenient. Hence, the reads across matrix pcurr[][] (directly for the central cell pcurr[][], or through the function f_neigh(NEIGHBORS) according to Fig. 3) are to be repeated when unrolling, as discussed below.

In order to fix ideas, let us suppose a CA that updates its state using only neighbors from the same row: more specifically the left and right cells. In Fig. 9 the necessary changes to unroll three times the inner loops are described, and the consequences with respect to the data reusing are schematized in Fig. 10, left. Note that this procedure is applicable for other neighboring cases; in particular for the well-known GoL the resulting scheme is depicted on the right.

The most beneficial consequence can be foreseen through the parameter A_I of the Roofline model. Supposing that each b_c-bit cell must compute D operations, and the column loop is unrolled U times, the A_I for an 8 neighbor CA changes from $\frac{D}{9 \times b_c}$ for the standard version to $\frac{U*D}{(9+(U-1)\times 3)\times b_c}$ for the unrolled one.

In the described example, A_I is increased from $\frac{D}{9 \times b_c}$ to $\frac{3*D}{15 \times b_c}$, that is, a ratio of $\frac{27}{15}$, which means an 80% increase that can have important consequences on execution time as Fig. 4 outlined.

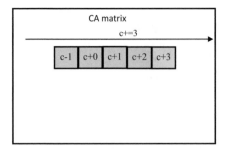 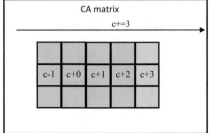

Fig. 10 Left: a scheme of unrolling 3 times a linear CA that uses left and right neighbors. Green cells are to be reused, but orange ones are not. Right: a similar scheme for the GoL case, which uses 8 neighbors

5.6 Vectorization and Packet Coding

The utilization of vector registers or SIMD (Single Instruction Multiple Data) computation can speed up greatly iterative algorithms that use massive vectors or matrices [13]. Speed up of a loop comes close to the size of the vector register (measured in number of elements) if bandwidth does not suppose a bottleneck and all the elements in an iteration use the same index (e.g. [i] for vectors). Hence, CPU vendors are progressively incorporating bigger registers and more flexible vector instructions in their kernels. Besides, whereas GPUs must be programmed using a different paradigm, which is sometimes called SIMT (Single Instruction Multiple Thread), their coding deeply resembles that of SIMD kernels.

However, when a loop contains accesses to different indexes, like combining the state of a central cell with these of its neighbors (the case of CAs), vectorization is much more complicated, and compilers do usually not translate high level code into its assembler vectorized version.

On one hand, in order to ease this issue and according to the success of SIMD kernels, vendors are incorporating many operations to shuffle, shift, permute the elements of a vector register plus simpler ways to insert assembler code into a high level program [17]. On the other hand, modern compilers can do a good vectorization job, grouping several iterations (and their memory accesses to vectors/matrixes) into a set of vector instructions.

Programmers must be aware that automatic vectorization of the CA inner loop can be done by compilers. This can be detected by: (a) looking if vector kernel instructions have been inserted in the assembler code (e.g. instructions with suffixes ps, pd, pb or prefixes VEX, v, p, etc. in x64/x86 machines; 'p' means packet); (b) checking that the base registers that access the CA matrices in the inner loop are incremented by the size of the vector register (usually 16, 32 or 64 bytes for 128, 256 or 512-bit registers, respectively). Furthermore, they should help the compiler do this task by simplifying their code. In general, programmers should convert complex and nested conditional structures into formulas as much as possible (see Sect. 5.4).

Conversely, some easy and compact algorithms, like accessing a Look-Up-Table (LUT), are not fully vectorizable because once the set of LUT indexes is computed (probably in a nice vector form), the accesses to the LUT must be done sequentially. This is inherent to a generic indexation, because the values to be extracted from the LUT are not consecutive but dispersed.

Overall, a CA whose rules depend on K neighbors are translated by the compiler into K vector memory accesses, which means a redundant consumption of accessed bytes. To illustrate this idea, let us consider a simple linear (unidimensional) Cellular Automata, where (in the simplest case) every cell has only two neighbors: left and right. Figure 11 shows how such a linear CA would be processed using 8-element vector registers.

Vectorizing a loop in this manner does not modify neither its theoretical A_I nor the consumed memory bandwidth, but it usually supposes a certain speed up, mainly because the number of executed instructions is reduced and because memory accesses are much more aligned. In terms of the Roofline model, this optimization supposes a movement of the application up to a coordinate of the model closer to the ceiling (i.e. the performance bound), just like the black arrow in Fig. 4.

Special vectorization reducing the number of accesses to the minimum, that is, a cell read (its east neighbor) per cell in the case of the linear CA, can be tricky, and it involves a set of shifting, masking and AND/OR logical operations of two vector registers (containing consecutive bytes), to extract the required neighbor vectors from these vector registers.

Vector registers are nowadays big enough (currently up to 512 bits) to hold a high number of cell states. The extreme case of comprising the cell state into one bit would yield to a striking vectorization of 512 cells per register. Nevertheless, as commented in the previous subsections, this would be only valid for very specific implementations. A more generic CA coding is that of a byte per cell, which yields also to a considerable 64-cell packing ($512 = 64 \times 8$).

Another interesting packet coding, which does not depend on the machine architecture (here the set of vector instructions), is that of using standard variables to hold several cells. For instance, as an int64 variable can hold 8 cells of 1 byte, one can reproduce the vectorization scheme but with pure high level code for these int64 variables This style is fully portable and does not depend on the machine architecture as SIMD instructions do.

5.7 Data Pipelining

Section 5.5 shows that unrolling a CA column loop means that several elements are reused in the unrolled iterations. Hence, CPU registering of those elements that are to be reused is a classical technique in assembler compilers that can be adapted to high level CA codes to favor a better compilation, while maintaining the bandwidth consumption to a minimum.

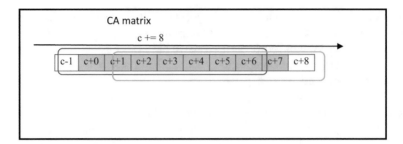

Fig. 11 Vector accesses required to compute the next sate of a linear CA that uses left and right neighbors through 8-element vector registers. Apart from reading 8 current states (light blue cells), the red rectangle (8 left neighbors of these cells) and the green rectangle (8 right neighbors) are needed. As a consequence, each cell is redundantly accessed three times

The compiler method called software pipelining is mainly related to separate real dependences [13] while reusing registers. With respect to data reusing, and as far as we know, its data version was introduced in [8]; there called "Register Pipelining". Because of the sequential nature of this iteration instruction reordering, it is only valid for sequential computing, that is, for a thread that processes several iterations. This implies that this beneficial method cannot be used in GPUs, because the common approach of GPUs is SIMT-like, that is, a thread is in charge of only one cell.

Once again let us explain this method with the simple case of the linear, unidimensional CA introduced in the previous section. Consider Fig. 12, where the data used by the iterations are represented in the above part, as the iterations go by towards the bottom. Below this representation, we rename the CA cells by registers `le`, `ce`, `ri` (left, central and right cells, respectively) from above to below. At the same time, the necessary renaming is represented with arrows. Accordingly, the resultant pseudocode is shown in the below part of Fig. 12. As a result and for each iteration, only one new element (in red) must be captured from the corresponding cache level, while the rest can be reused thanks to be CPU-registered. A similar role is played by the variables E, NE, N, NW, W, SW, S, C and prev_neigh in Fig. 6 before the end of the column loop. As a result, the net effect of this technique is similar to that of the one described in Sect. 5.2.

Finally note that the fusion of this technique and vectorization is also possible. Even though it gets to a complicated code programming for a CA using many neighbors, we must only be cautious with the renaming at the end of an iteration (like that of prev_neigh in Fig. 6).

5.8 Spatial Blocking

Spatial blocking (or "loop tiling") is a well-known optimization to improve data access locality. It is valid for those programs where some data is used several times.

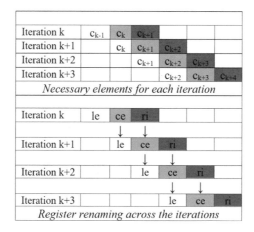

Iteration k	c_{k-1}	c_k	c_{k+1}			
Iteration k+1		c_k	c_{k+1}	c_{k+2}		
Iteration k+2			c_{k+1}	c_{k+2}	c_{k+3}	
Iteration k+3				c_{k+2}	c_{k+3}	c_{k+4}

Necessary elements for each iteration

Iteration k	le	ce	ri			
Iteration k+1		le	ce	ri		
Iteration k+2			le	ce	ri	
Iteration k+3				le	ce	ri

Register renaming across the iterations

Pseudocode of the data pipelining for the column loop

```
processing column 0;
//COLUMN PROLOG:
le = pcurr[r][0]; ce = pcurr[r][1];
for (int c = 1; c < SIDE-1; c++) { // column loop
      ri = pcurr[r][c+1];
      pnext[r][c] = func(le, ce, ri);
      // renaming to prepare the next iteration
      le=ce; ce=ri;
      }   // end of column loop
processing column SIDE-1;
//final actions at the end of a column
```

Fig. 12 Up: schematic representation of the data pipelining for a linear CA that uses left and right neighbors (variables `le`, `ri`, resp.). Red cells represent the new elements to be captured from memory and green cells the central cell (variable `ce`) whose state is to be updated. Down: pseudocode of the data pipelining for the column loop

In this method, the grid is divided into blocks of the same size that are processed as indivisible chunks (e.g. by a thread). Spatial blocking changes the element processing order so that those data that have been cached are promoted to be accessed in a short time window. Hence the size of the block is related to the amount of data that can be held in a certain cache level. In the case of CAs, this method is beneficial because the neighbor cells of a central cell would be accessed again in future iterations (as a different neighbor or as a central cell). Cells that are in the block border do not completely fulfill this reusing (because they need neighbors that are out of the block), whereas those that are in the block interior promote it. Thus, the best block to promote this reusing is the most compact one, that is the one with less perimeter in relation to its area. A square block would have the biggest compaction for simple codes that covers a matrix (a circle is the most compact planar figure, but a set of equal circles cannot cover a matrix without lots of intersections).

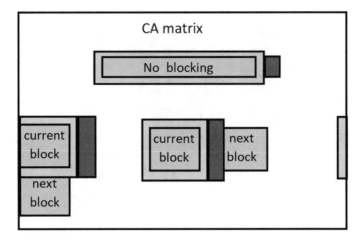

Fig. 13 Different loop organizations: left: vertical tiling (including the perimeter at the right of the matrix); center: blocking using two outer loops like in Fig. 14, and, up: classical scheme following the leading (column) dimension. Green boxes are the cell blocks that hold in a certain level. Orange cells are this block neighbors (perimeter of the blocks). Red cells denote the possible set of cells prefetched by the hardware

Besides, an important factor, which is hidden to the code but automatically managed by modern caches, is the hardware prefetching, that is, the additional cache uploading of those lines whose memory addresses are predicted by the CPU. Obviously, this can be beneficial if the CPU hits in its prediction, but it may waste a precious bandwidth otherwise. While prefetchers can vary from a processor to another, the most plausible prefetched lines are depicted in red in Fig. 13.

Figure 13 represents schematically three possible manners to organize the loops for a 2-D CA. In the upper part of this figure, the usual nesting of two loops (that of Fig. 6) generates blocks of small compaction degrees. Its advantage is that recently prefetched cells are probably used by the next iterations. Vertical tiling is drawn in the left of this figure. Its main advantages are that only one additional external loop is needed to manage row jumps and that its theoretical code reusing is as good as the classical tiling scheme (Figs. 13, right, and 14). However, most of the prefetched lines for vertical tiling are not to be reused by the next block; a drawback not present in the classical tiling scheme.

The speedup that these techniques may introduce for a CA is not easy to be modelled. Firstly, Roofline model does not analyze the caching behavior, it does not give different performance values for different blocking strategies. Secondly, the access to the output matrix (pnext) cannot take advantage of this technique because its elements are not reused. Nonetheless, the general rule of thumb is that the block should have a size around half of a certain cache level [13]. Moreover, the code given by Fig. 14 is usually the fastest, mainly for those big matrixes in which three rows (for a 3 × 3-neighbor CA) cannot be held in a cache level. And finally, the more neighbors the CA cells manage, the more favorable blocking is.

```
void compute_step( ... );
for (int i = 1; i < SIDE-1; i += n_rows_block)
    {
    for (int j = 1; j < SIDE-1; j += n_cols_block)
        {
        for (int ii = i; ii < min(i+n_rows_block,SIDE-1); ii++)
          {
          for (int jj=1; jj < min(j+n_cols_block, SIDE-1); jj++)
            {
            ...
            pnext[ii][jj] = ...;
            }
          }
        }
    }
```

Fig. 14 Loops reorganization for applying the classical spatial blocking method in a CA. Here, n_rows_block and n_cols_block hold for the number of rows and columns of the block, resp.

5.9 Temporal Blocking

Spatial blocking favors the spatial locality for those cells within a current temporal step. However, as cells of a block are to be used in subsequent steps, temporal blocking goes beyond spatial blocking by performing multiple updates on a block before it is evicted, thus reducing main memory bandwidth consumption. As far as we know, the first apparition of this technique was in [31], and it can be even applied to three dimensional grids (called 3.5D blocking in [27]). Since then it has not been profusely used, maybe because (a) the intermediate CA states are not directly available; (b) the code is somewhat tricky, and difficult to debug.

The challenge is that every cell needs current neighbor values to compute its evolution rules. Thus, if we wanted a block of a core to work independently of another block (to prevent remote access to another cores), its border neighbors (its ghost zones) would not be available. For instance, border cells could not be computed for the first temporal step. Indeed, the more radius the neighborhood would have, the more border cells would remain obsolete. And this will keep on for the subsequent steps; consequently, after a number of steps equal to the half side of a block, this block cannot compute valid evolution rules anymore.

This issue is better understood for the case of a linear CA if we follow the exact temporal evolution of any cell in a block (for each cell of Fig. 15, numbers represent the temporal step in which they are). We need the two buffers, previously called `pcurr` and `pnext`, which are going to hold a mix of cells, each living in a different temporal step. In order to fix ideas, Fig. 15 represents two blocks (of two independent cores) holding 8 cells each, for a total of 4 different states (3 computation steps). Temporal blocking needs two stages depicted with blue and green letters at the bottom of the figure. From left to right the CA evolves by writing exactly in those cells (of `pcurr` or `pnext`) that can be computed using only a block. For example, at the end

Fig. 15 Temporal blocking example for an unidimensional CA. Numbers represent the temporal step of each cell. Blue numbers at the stage 1 (left blocks) are the cells to be updated in the next step. Likewise for green numbers at stage 2

of the first step (stage 1), the pair of extreme cells could not be updated in the `pnext` buffer, likewise the two pairs of most external cells cannot in the second step. That is, at stage 1 the updating proceeds always with central cells: two less at each step. Note that cells that can be updated at each step are drawn in blue for the first stage, and in green for the second stage. At the end of the first stage, block synchronization is needed as each one needs the neighbors from its adjacent blocks. Stage 2 proceed from the extreme cells progressively to the central ones; that is, updating proceeds always with edge cells: a pair more at each step.

The speedup that temporal blocking can get for a CA is difficult to be modelled. Again, Roofline model does not contemplate the caching behavior; in addition, wasted bandwidth consumed by hardware prefetchers can be considerable as the pattern of memory accesses is not repetitive along the different steps. Finally, intermediate synchronization at the end of each stage would spoil an unpredictable time. Nevertheless, it is being progressively incorporated to CAs and Stencil loops even for GPU platforms [24].

6 Conclusions and Future Work

The regular grid of cells and the regular time iterative process in which Cellular Automaton are based allows to introduce a bunch of programming optimizations for their implementation in current high-performance computers. A thoughtfully revision of these optimizations has not been reviewed in depth until now. In this chapter, firstly, these optimizations are illustrated so that they can be adapted to any CA. Secondly, their impact in performance is analyzed and discussed under the umbrella of the Roofline model, which plots the operations throughput as a function of the Arithmetic Intensity A_I (the number of operations per byte read) in a log-log scale. At the same time that these measures are related to the modern hardware architecture, its various alternatives (e.g. GPUs, BTBs, etc.), and the different forms of parallelism such as instruction-level parallelism, thread-level parallelism and data-level parallelism.

The analytical performance measures allow to predict the impact of the different approaches in the execution time. In this respect and considering that the program execution in modern microprocessors is essentially limited by the availability of the input data, the limits to the ideal behavior are given by branch mispredictions, availability of hardware resources and limitations in the memory bandwidth and latency. According to this model, moving the coordinates of a system towards a position closer to higher ceilings (i.e. the performance bounds) optimizes execution. Thus, the different optimizations of Cellular Automata implementations should be focused on trying to overcome these limitations.

Acknowledgements This research was funded by the following research project of Ministerio de Economía, Industria y Competitividad, Gobierno de España (MINECO) and the Agencia Estatal de Investigación (AEI) of Spain, cofinanced by FEDER funds (EU): MABICAP (Bio-inspired machines on High Performance Computing platforms: a multidisciplinary approach, TIN2017-89842P), Par-HoT (Parallel Data Processing based on Homotopy Connectivity: Applications to Stereoscopic Vision and Biomedical Data, PID2019-110455GB-I00).

References

1. Hoekstra, A.G., Kroc, J., Sloot, P. (eds.): Simulating Complex Systems by Cellular Automata. Springer, Berlin, Heidelberg (2010)
2. Bajzát, T., Hajnal, E.: Cell automaton modelling algorithms: implementation and testing in GPU systems. In: INES 2011, 15th International Conference on Intelligent Engineering Systems (2011)
3. Balasalle, J., Lopez, M., Rutherford, M.: Optimizing Memory Access Patterns for Cellular Automata on GPUs, pp. 67–75. Elsevier–Morgan Kaufmann–NVIDIA (2011)
4. Bandman, O.: Using multi core computers for implementing cellular automata systems. Lect.ure Notes Comput. Sci. **6873**(1), 140–151 (2011)
5. Cagigas-Muñiz, D., Diaz-del Rio, F., López-Torres, M., Jiménez-Morales, F., Guisado, J.L.: Developing efficient discrete simulations on multicore and GPU architectures. Electronics **9**, 189 (2020). https://doi.org/10.3390/electronics9010189
6. Cattaneo, R., Natale, G., Sicignano, C., Sciuto, D., Santambrogio, M.D.: On how to accelerate iterative stencil loops: a scalable streaming-based approach. ACM Trans. Archit. Code Optim. **12**(4), 1–26 (2015)
7. Chopard, B., Droz, M.: Cellular Automata Modeling of Physical Systems. Cambridge University Press, Cambridge, MA, USA (1998)
8. Duesterwald, E., Gupta, R., Soffa, M.L.: Register pipelining: an integrated approach to register allocation for scalar and subscripted variables. In: Kastens, U., Pfahler, P. (eds.) Compiler Construction, pp. 192–206. Springer, Berlin, Heidelberg (1992)
9. Berlekamp, E.R., Conway, J.H., Guy, R.K.: Winning Ways for your Mathematical Plays, 2nd edn. A K Peters/CRC Press, New York, USA (2001)
10. Gardner, M.: Mathematical games: the fantastic combinations of John Conway's new solitaire game & "Life". Sci. Am. **223**(4), 120–123 (1970)
11. Gibson, M.J., Keedwell, E.C., Savić, D.A.: An investigation of the efficient implementation of cellular automata on multi-core CPU and GPU hardware. J. Parallel Distrib. Comput. **77**, 11–25 (2015)
12. Guisado, J., Jiménez-Morales, F., Fernández-de Vega, F.: Cellular automata and cluster computing: an application to the simulation of laser dynamics. Adv. Complex Syst. **10**(Suppl. 1), 167–190 (2007)
13. Hennessy, J.L., Patterson, D.A.: Computer Architecture, Sixth Edition: A Quantitative Approach, 6th edn. Morgan Kaufmann Publishers Inc., San Francisco, CA, USA (2017)
14. Hwu, W.m.: GPU Computing Gems Jade Edition, 1st edn. Morgan Kaufmann Publishers Inc., San Francisco, CA, USA (2011)

15. Ilachinski, A.: Cellular Automata: A Discrete Universe. World Scientific, Singapore (2001)
16. Ilic, A., Pratas, F., Sousa, L.: Cache-aware roofline model: upgrading the loft. IEEE Comput. Archit. Lett. **13**(1), 21–24 (2014). https://doi.org/10.1109/L-CA.2013.6
17. Intel: Intel intrinsics guide. https://software.intel.com/sites/landingpage/IntrinsicsGuide/
18. Kari, J.: Theory of cellular automata: a survey. Theor. Comput. Sci. **334**(1–3), 3–33 (2005)
19. Kirk, D.B., Hwu, W.m.W.: Programming Massively Parallel Processors: A Hands-on Approach. Morgan Kaufmann Publishers, Burlington, MA (2010)
20. Koskela, T., Matveev, Z., Yang, C., Adedoyin, A., Belenov, R., Thierry, P., Zhao, Z., Gayatri, R., Shan, H., Oliker, L., Deslippe, J., Green, R., Williams, S.: A novel multi-level integrated roofline model approach for performance characterization. In: Yokota, R., Weiland, M., Keyes, D., Trinitis, C. (eds.) High Performance Computing, pp. 226–245. Springer, Cham (2018)
21. Li, Z., Song, Y.: Automatic tiling of iterative stencil loops. ACM Trans. Progr. Lang. Syst. **26**(6), 975–1028 (2004)
22. Linares-Barranco, A., Sevillano, J., Obaidat, M.S.: AER filtering using glider: VHDL cellular automata description. In: 15th IEEE International Conference on Electronics, Circuits and Systems, pp. 614–617 (2008)
23. Lopez-Torres, M., Guisado, J., Jimenez-Morales, F., Diaz-del Rio, F.: GPU-based cellular automata simulations of laser dynamics. In: Proceedings of the XXIII Jornadas de Paralelismo: Jornadas SARTECO 2012, pp. 261–266. SARTECO, Elche (2012). http://www.jornadassarteco.org/js2012/papers/paper_151.pdf
24. Matsumura, K., Zohouri, H., Wahib, M., Endo, T., Matsuoka, S.: AN5D: automated stencil framework for high-degree temporal blocking on GPUS. In: International Symposium on Code Generation and Optimization, pp. 199–211 (2020). https://doi.org/10.1145/3368826.3377904
25. Millñin, E., Martínez, P., Gil Costa, G., Piccoli, M., Printista, A., Bederian, C., García Garino, C., Bringa, E.: Parallel implementation of a cellular automata in a hybrid CPU/GPU environment. In: XVIII Congreso Argentino de Ciencias de la Computación, pp. 184–193 (2013)
26. von Neumann, J.: Theory of Self-reproducing Automata. University of Illinois Press, Urbana (1966)
27. Nguyen, A.D., Satish, N., Chhugani, J., Kim, C., Dubey, P.: 3.5-D blocking optimization for stencil computations on modern CPUS and GPUS . In: SC, pp. 1–13. IEEE (2010)
28. Oxman, G., Weiss, S., Be'ery, Y.: Computational methods for Conway's Game of Life cellular automaton. J. Comput. Sci. **5**(1), 24–31 (2014)
29. Bryant, R.E., O'Hallaron, D.R.: Computer Systems: A Programmer's Perspective, 3rd edn. Pearson, London, UK (2016)
30. Rybacki, S., Himmelspach, J., Uhrmacher, A.: CPU and GPU based simulation of cellular automata—a performance comparison. In: Proceedings of the 1st SIMUL, pp. 62–67 (2009)
31. Song, Y., Li, Z.: New tiling techniques to improve cache temporal locality. In: Proceedings of the ACM SIGPLAN 1999 Conference on Programming Language Design and Implementation, PLDI '99, pp. 215–228. Association for Computing Machinery, New York, NY, USA (1999). https://doi.org/10.1145/301618.301668
32. Stengel, H., Treibig, J., Hager, G., Wellein, G.: Quantifying performance bottlenecks of stencil computations using the execution-cache-memory model. In: Proceedings of the 29th ACM on International Conference on Supercomputing, ICS '15, pp. 207–216. Association for Computing Machinery, New York, NY, USA (2015). https://doi.org/10.1145/2751205.2751240
33. Williams, S., Waterman, A., Patterson, D.: Roofline: an insightful visual performance model for multicore architectures. Commun. ACM **52**(4), 65–76 (2009)
34. Yang, C., Kurth, T., Williams, S.: Hierarchical Roofline analysis for GPUS: accelerating performance optimization for the NERSC-9 Perlmutter system. Concurr. Comput. **32**(20), 1–12 (2020)

A Comprehensive Review on Edge Computing: Focusing on Mobile Users

A. Dimou, C. Iliopoulos, E. Polytidou, S. K. Dhurandher, G. Papadimitriou, and P. Nicopolitidis

Abstract During the last decade, evolutions in Information and Communication technologies (ICT) such as cloud computing, wireless networks and mobile computing, have accelerated the proliferation of mobile users and consequently the development of sophisticated, complex and resource-intensive mobile applications, such as speech recognizers, image processors and multimedia services. Running such applications on mobile devices was challenging since mobile computing has inherent problems such as resource scarcity, frequent disconnections and mainly mobility. To address the shortcomings of mobile computing, researchers have used the concept of Cloud Computing (CC) to perform resource-intensive tasks outside mobile devices. This led to mobile and cloud computing convergence and the emergence of a new computing paradigm, the Mobile Cloud Computing (MCC), also called *Mobile Cloud*, which could effectively alleviate the problems of mobile computing. More sophisticated applications, such as online gaming, augmented reality, and virtual reality, which require broad bandwidth, low response latency and large computational power have emerged. In general, such a system relies on distant cloud services to perform all the data processing tasks, which result in explicit latency. The most promising solution for the requirements of aforesaid applications was the placement of content, compute, and cloud resources on the edge of the network, closer to concentrations of users. Consequently, Edge Computing (EC), which utilizes the proximal computational and networking resources, has arisen. The concept of Edge Computing was introduced, initially in the form of Cloudlets that assist mobile devices to gain computational and storage performance benefits, which later paved the way towards the concepts of Fog computing and Mobile Edge Computing, which essentially has to be referred as Multi-Access Edge Computing (MEC). Thus, mobile cloud has

A. Dimou · C. Iliopoulos · E. Polytidou · P. Nicopolitidis (✉)
Hellenic Open University, Patras, Greece
e-mail: petros@csd.auth.gr

S. K. Dhurandher
Netaji Subhas University of Technology, New Delhi, India

G. Papadimitriou · P. Nicopolitidis
Department of Informatics, Aristotle University of Thessaloniki, Thessalonikiy, Greece

© The Author(s), under exclusive license to Springer Nature Switzerland AG 2022　　　121
P. Nicopolitidis et al. (eds.), *Advances in Computing, Informatics, Networking and Cybersecurity*, Lecture Notes in Networks and Systems 289,
https://doi.org/10.1007/978-3-030-87049-2_30

extended its computation model from centralized to distributed, from Internet cloud to edge cloud, and from mobile devices to mobile services. In this survey paper we investigate the pre-existing works in MCC to determine their role in the current trends for edge computing. We provide a holistic overview on the exploitation of EC for the realization of pervasive mobile applications and we briefly present the state-of-the art of edge computing paradigms and the key enabling technologies. We also briefly present how the crowd sourcing was based on the MCC to work efficiently and how MEC enriches the dynamics of crowd sourcing. Finally, we present the open research challenges.

Keywords Mobile computing · Cloud computing · Mobile cloud computing · Edge computing · Cloudlets · Fog computing · Multi-access edge computing · Crowdsourcing

1 Introduction

Three decades ago, two visions changed the way we live and communicate: mobile computing vision [1] and ubiquitous computing vision [2]. The essence of mobile computing vision was "information at your fingertips anywhere, anytime" [3]. The essence of ubiquitous computing, also called pervasive computing, was "the creation of environments saturated with computing and communication capability, yet gracefully integrated with human users" [4]. Tremendous developments in a multitude of technologies ranging from personalized and embedded smart devices (e.g., smartphones, sensors, wearables, IoTs, etc.) to ubiquitous connectivity, via a variety of wireless mobile communications and cognitive networking infrastructures, to advanced computing techniques and user-friendly middleware services and platforms have significantly contributed to the unprecedented advances in pervasive and mobile computing.

1.1 The Road to the Edge Computing

Figure 1 presents the evolution from first distributed systems to MEC. Mobile phones and smartphones have been one of the greatest drivers of this evolution. Smartphones provide a pervasive computing platform that is evolving almost as fast as can invent new applications and support needs for work, entertainment, and education. The current era comprises of unique powerful devices with capabilities of sensing, communication, computing and also have information on fingertips anywhere, anytime.

Mobile devices (e.g., smartphones, phablets, tablets, PDAs, and wearable devices) are now an integral part of people's everyday lives, both at a professional and a social level. According to Cisco Systems analysis [5] for the mobile network through 2022,

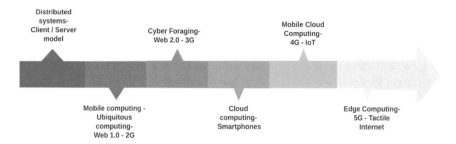

Fig. 1 Evolution to MEC

mobile data traffic will reach the following milestones. Monthly global mobile data traffic will 77 exabytes, mobile will represent 20% of total IP traffic, the number of mobile-connected devices per capita will reach 1.5, smartphones will surpass 90% of mobile traffic, nearly 59% will be offloaded from cellular networks on to Wi-Fi and 79% of the worlds mobile data traffic will be video. The trend of increase in mobile devices usage is fundamentally driven by the augmentation of mobile users, mobile applications development, e.g., iPhone apps and Google apps [6]. Also, mobile devices evolution, especially the smartphone evolution, has accelerated the proliferation of mobile internet (e.g., wireless networks) and spurred a new wave of mobile applications on smartphones, leading to an unprecedented mobile data volume generated from the mobile devices, the Mobile Big Data (MBD) [7].

Nowadays smartphones are equipped with touch screens, multiple sensors, diverse networking capacities, massive storage, high-end multi-core processors as well as the more utilized camera and microphone. Mobile applications, such as speech recognition, natural language processing, multimedia services, MBD applications (i.e. data driven activity recognition for pervasive health computing and context-aware recommendation for urban planning) and enterprise applications, are becoming increasingly ubiquitous and can provide better user experience on mobile devices. However, such applications are complex and resource-intensive. as they require a large Central Processing Unit (CPU) capable of dealing with many operations per second, a lot of Random-Access Memory (RAM) to load the code and data, extensive disk storage to store contents, and long-lasting batteries, which are not available in today's smartphones. Smartphones' small size, lightness, and mobility impose severe limitations on their processing capabilities, battery life, storage capacity, and screen size and capability, impeding execution of resource-intensive computation and bulky data storage on smartphones [8, 9].

Researchers have used the concept of Cloud Computing (CC) to address the limitations of mobile devices and fulfill users' demands, which has led to the state-of-the-art technology Mobile Cloud Computing (MCC). MCC researchers envision enhancing the computational capabilities of contemporary mobile devices to enable users to perform unrestricted computing, anywhere, anytime, from any device [10].

CC is a model for enabling ubiquitous, convenient, on-demand network access to a shared pool of configurable computing resources (e.g., networks, servers, storage, applications, and services) that can be rapidly provisioned and released with minimal management effort or service provider interaction [11]. CC provides fundamental support to address Big Data and therefore MBD challenges (e.g., store, transport, process, mine and serve the data) with shared computing resources, including computing, storage, networking and analytical software [12].

MCC involves execution of only those applications that require extensive computational and storage resources beyond native mobile devices. So, the usefulness of MCC is realized by exploiting the computational power of computing entities, including giant clouds, desktop computers in public places, and resource-rich mobile devices that inherit cloud computing technologies and principles. MCC faces many challenges because it is an amalgamation of complex technologies (mobile computing, CC and networking). Though, the dramatic increase in cybercrime and security threats within mobile devices, cloud resources and wireless transactions makes security and privacy more challenging than ever.

1.2   Trends

The rapid development of technology has undoubtedly enhanced the humans' abilities, capable of providing to them even superhuman powers. Modern trends focus on creating intelligent environments, thereby affecting the daily lives of people who interact with these either at work or in entertainment venues. As businesses adopt digital solutions, it is important for business leaders to understand the trends that differentiate how customers experience digital technology. At the same time, IT executives need to understand how people interact and experience digital technology. The combination and development of these trends is preferable to implement according to the business needs of the people as well as the human beings in general. This view leads us to adopt some trends in digital technology of the new decade. Some of the main applications and service trends that are likely to affect the MEC sector are discussed below [13, 14]:

- *Hyperautomation*: With the help of Artificial Intelligence and Machine Learning, processes are automated, and decisions are made voluntarily by the various intelligent environments while no human intervention is required. As a result, hyperautomation enhances human capabilities and automation processes are further optimized as systems are now able to discover, design, measure, monitor and combine all this information.
- *Multiexperience*: The user experience when interacting with technology is multiplied by modern multimodal interfaces, multiple sensors and virtual or augmented reality. Users need to interact with a variety of devices and applications. Therefore, a consistent and unified user experience across the web, mobile, wearables, conversational and immersive touch points must be ensured. A new category of

things and experiences is coming to the surface, such as the "environmental experience", where each user can remotely monitor the progress of a product they order.

- *Democratization*: The user's specialization with technology is becoming more and more user-friendly to provide him with all the know-how required without the need for much intellectual or financial effort.
- *Human augmentation*: Human augmentation is about helping people with physical disabilities or even enhancing their cognitive abilities by raising certain ethical and cultural issues. In summary, emphasis is placed on enhancing human senses (hearing, speech), biological needs, brain augmentation (use of implants) and genetic augments. Likewise, cognitive enhancement is attempted to enhance cognitive abilities such as better decision making, thinking ability and perception.
- *Transparency and traceability*: It is clear that the new generation of people is very familiar with technology. This gives organizations an additional burden of responsibility for the best possible, most efficient and secure collection and management of personal data, as users are now able to understand this process to a satisfactory extent. The use of artificial intelligence in conjunction with machine learning to make decisions, without the consent of the user, poses risks and calls into question the credibility of the processes on the part of people by pushing this particular tendency to focus on six key elements of trust such as "ethics, integrity, open-mindedness, accountability, competence and consistency".
- *The empowered edge*: The development of smart spaces and IoT systems in general require network topologies capable of managing and processing data as close to the source of information as possible with the aim of maintaining traffic local and distributed achieving low latency (edge computing).
- *The distributed cloud*: The idea of distributed cloud focuses on creating cloud services in remote locations relative to the physical data centers of each provider but under their own supervision and management as this implies for service architecture, delivery, functionality and any upgrades (cloud computing). The purpose of this trend is to resolve any technical issues and regulatory challenges such as latency as well as an attempt to combine the advantages of public and private cloud services.
- *Autonomous things*: The tendency of autonomous objects refers to the creation of objects such as drones, robots, and many others capable of acting autonomously or almost autonomously in different environments without human intervention but with the help of artificial intelligence. However, even today, technology combined with artificial intelligence and machine learning algorithms are unable to create objects capable of responding to all conditions as effectively as a human.
- *Practical blockchain*: Blockchain allows secure communication between unknown users in a common digital world as well as the exchange of information without a centralized authority. Users can trace the origin of their assets while using this technology they can also track common information with the relevant access permission. Completion of this model is still incomplete in the business sector due to the challenges of scalability and interoperability, but it aims to fully integrate with technologies such as artificial intelligence and IoT.

- *AI security*: In addition to the benefits of the above trends, they raise security issues. Nevertheless, this trend emphasizes the influence of artificial intelligence on security issues, focusing on the protection of AI systems, the more effective use of AI to enhance security defense as well as the prevention of possible malicious use of AI by attackers.
- *Interfaceless machines*: Applications running on mobile and portable devices have led manufacturers to adopt interface models quite different from traditional ones. The evolution of these devices in terms of hardware and rich device APIs requires a very different approach from what can be achieved with on-machine interfaces.
- *Inclusive design* is the way to serve the needs of the wider community considering the specific needs of all potential communities. Inclusive design should be based on the principle that we should keep in mind all potential users, products, and services we design. As for our data, it must respond accurately to potential user segments and be complete.

1.3   Motivation and Contribution

The motivation of this primary research was the convergence of mobile and cloud computing technologies. This led to the investigation of MCC paradigm. According to [15], a list of highly cited papers was found and the investigation for recent papers in the domain according to the data was started in the aforementioned research, regarding the most productive authors, organizations and journals. From the early stages of the investigation the emergence of edge computing and related computing paradigms were identified. Thus, search for surveys in these computing paradigms were carried out. Unfortunately, it was found that in many studies there was a confusion about terminology and definitions for MCC, EC, Cloudlets, Fog Computing and MEC. On the other hand, because edge computing is in the early stages of research the last two years observed new synergies, mergers and renames (or example MEC change its scope and renamed from Mobile Edge Computing to Multi-Access Edge Computing, the Industrial Internet Consortium and OpenFog Consortium unite and so on) resulting in greater confusion about the implementation of each EC paradigm and for the content and contribution of each individual survey. So, we came up with our research questions:

- What was the main motivation for the creation of MCC?
- Edge computing is one of the solutions already proposed by MCC or it's a different computing paradigm?
- Which mobile applications will benefit from EC?
- What are the differences and similarities between Cloudlets, Fog Computing and MEC?
- How these computing paradigms relate to 5G?

There are many excellent surveys in the field of MCC, EC, Cloudlets, Fog Computing and MEC. So, the focus of this paper is not to propose another survey about a single computing paradigm. Our scope is to provide a comprehensive review of the current state-of-the-art and the latest development on edge computing, which would help early-stage researchers to have an overview of the existing technologies, applications and implementations and investigate research issues and future challenges in this domain.

The organization and structure of the paper has a sequential form. Section 2, presenting some basic elements, definitions, and concepts of mobile and pervasive computing as well as cloud computing and virtualization. Through these concepts we lead to the analysis of MCC and EC. This is followed by Sect. 3, a brief description of Cloudlets, Fog Computing and Multi-Access Edge Computing (MEC) is given, focusing our attention on their advantages / disadvantages. Follows, in Sect. 4, which lists the most important areas where MEC is implemented as regards mobile crowd-sourcing (MCS), as well as how cutting-edge laptop technology has influenced and continues to influence these technology industries, analyzing the architecture that is applied as well as the characteristics that appear. Section 5 refers to open research issues that concern us and finally we come to a text of conclusions in Sect. 6, which emerges from the analysis of all that we have discussed.

2 Overview of Basic Concepts

In this section, we present several important definitions and basic concepts for EC. First, we describe the concepts of mobile and pervasive computing, then we present cloud computing and virtualization basics, we summarize the terms and concepts of MCC and finally of EC.

2.1 Mobile and Pervasive Computing

Mobile computing is a significant contributor to the pervasiveness of computing resources. In concert with the proliferation of stationary and embedded computer technology, smartphones and other handheld or wearable computing technologies have created a state of ubiquitous and pervasive computing where we are surrounded by more computational devices than people. It is almost universally acceptable to consider mobile computing as the root of pervasive computing and even today's IoT technologies [16]. Thus, it is important to understand the difference between mobile computing, pervasive computing, and IoT technologies. The essence of the pervasive computing vision is a world saturated with sensing and computing, yet so gracefully integrated with humans that all this technology remains below their consciousness. The essence of the IoT vision is a world full of intelligent devices that sense the physical world and integrate their observations to meet some higher-level goal - such as the energy efficiency of a building. In terms of technical content relating to sensing, computing, and communication, these are virtually identical visions. A substantial difference lies in the role of the human. Mark Weiser placed the human at the center of his vision, making sensing, computing, and communications to be "disappeared". In contrast, the IoT vision is silent about the role of humans. Pervasive computing subsumes IoT-it covers all that IoT covers and more [17]. IoT applications can be seen as extensions of using mobile computing solutions, in which devices are heterogeneous, and they may belong to different administrative domains, where their computation and networking models are more distributed, and the scale of the IoT system can be much larger than mobile computing application scenarios [18, 19].

2.2 Cloud Computing, VMs and Containers

Before we describe MEC, we dive into the basic concepts of CC and key virtualization technologies of MEC, such as VMs and containers.

2.2.1 Cloud Computing

It would not be an exaggeration to say that one of the most important keys to technological development in the field of ICT is the CC as many organizations have adopted it today, thus contributing to its growing adoption. CC is now a guideline for many IT resources that provide on-demand access to the consumer, while significantly boosting the economy by contributing to the rapid growth of other businesses with all that entails [20, 21].

According to the National Institute of Standards and Technology (NIST), CC is defined as follows. "*Cloud Computing is a model for enabling ubiquitous, convenient, on-demand network access to a shared pool of configurable computing resources (e.g., networks, servers, storage, applications, and services) that can be rapidly provisioned and released with minimal management effort or service provider interaction. This cloud model is composed of five essential characteristics, three service models, and four deployment models*" [11].

The CC Essential characteristics are the following [22]:

- On-demand self-service.
- Broad network access. Users can access data centers resources online using devices.
- Resource pooling. Users share the cloud service provider's resources in a multi-tenancy manner, where each user can run and stop these resources as needed.
- Rapid elasticity. The cloud resources capabilities are often unlimited and can be used anytime.
- Measured service. The service provider offers resources on Pay as you go pricing model.

The CC Service models *are* the following:

- Infrastructure as a Service (IaaS) offers on demand components for building IT infrastructure, such a storage, bandwidth and virtual servers (i.e. Amazon EC2).
- Platform as a Service (PaaS) offers development and runtime environments for applications that are hosted in the cloud (i.e. Microsoft Azure).
- Software as a Service (SaaS) offers multi-tenant applications and service on demand (i.e. Google Docs).

The deployment models are classified into four types: private, community, public and hybrid clouds [11].

2.2.2 VMs and Containers

In order to clarify the cloud meaning we can just think that some physical computers (machines) undertake to carry out different user's needs forming with this way a logical entity. So, from a user's perspective these personal operations are performed from a unique machine, which in fact is a virtual machine (VM) using a hypervisor. In this way the underlying VMs are isolated giving the user the illusion that he is using a particular machine. Virtual machines offer a great deal of flexibility in the use of resources, enabling the user to start, download and install various programs without actually interfering with the underlying hardware [23, 24].

Abstraction is a widespread term using in great extend in ICT systems. Contrastingly, this technique can be used equally efficient from containers. Containers act inside the OS level, executing an application or an operation using without problem the underlying libraries or system resources. At this point is the differential part of this method as it can manage partially physical machine's resources achieving a multitude number of instances in a smaller size as opposed to VMs. This enables a single OS to use many containers positively affecting the CPU performance, memory, disk or networking capabilities. Undoubtedly, the container is a lighter version of VMs but it is believed that it is less secure than the latter technique [23].

In short, application deployments are more efficient using hypervisors only when they need different OS at the same cloud, whereas containers perform better in an application which share the same OS, resulting in deployments with smaller size, as can be seen in Fig. 2 [25].

2.3 Mobile Cloud Computing

2.3.1 Architectures for Mobile Cloud Computing

To fulfill the diverse computational and quality of service (QoS) requirements of numerous different mobile applications and end users, several Cloud-based Mobile Augmentation (CMA) solutions have been undertaken that suggest architectures for

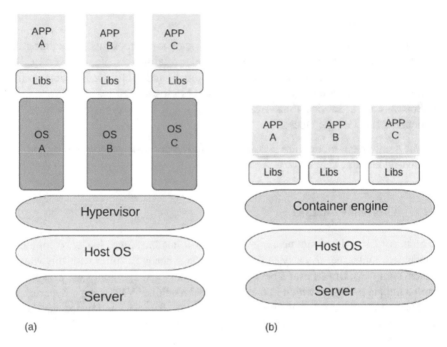

Fig. 2 Comparison of **a** hypervisor and **b** container-based deployments

MCC. The major differences in these MCC architectures derive from various cloud-based resources with different features, namely multiplicity, elasticity, mobility, and proximity to the mobile users [9]. Multiplicity refers to the abundance and volume of cloud-based resources and mobility is unrestricted movement of the computing device while its wireless communication is maintained, uninterrupted. Four types of cloud-based resources are identified in [9] and were generally accepted in recent studies by other researchers and with a different taxonomy in [26, 27, 86] (which essentially follow a similar approach). We adopt the taxonomy as described first by the researchers in [9] and we give a brief description for each of them as follows.

- *Distant Immobile Cloud (DIC)*: A general abstract architecture for a typical MCC system consists of Mobile Devices (MD) users that consume computational resources of public or private clouds using the Internet. Computational tasks in this model are executed inside the DIC resources (the term Remote Cloud is also used) and the results are sent back to the mobile client. This architecture as shown in Fig. 3a depicts the conventional two-tiered MCC architecture with the MD at one end and DIC at the other. Exploiting DIC to augment mobile devices leads to long WAN latency due to many intermediate hops and high data communication overheads in the intermittent wireless networks. Generally, heterogeneities between DIC resources and MDs complicate code and data portability and interoperability among them and excess overheads by employing handling techniques (i.e. virtualization) [10].

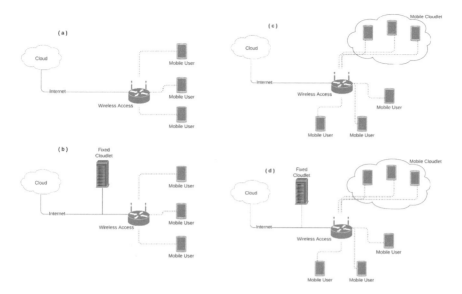

Fig. 3 **a** Distant immobile cloud; **b** Proximate immobile cloud; **c** Proximate mobil cloud; **d** Hybrid

- *Proximate Immobile Cloud (PIC)*: To mitigate the effects of long WAN latency, researchers [28] have endeavored to access computing resources with the least number of intermediate hops. Instead of travelling through numerous hops to performing intensive computations in DIC, tasks are executed inside the one hop distance computing entities in the vicinity, shown in Fig. 3b. These are medium location-granular (compared to the coarse grain resources, PICs are more numerous and are located nearer to mobile service consumers) and feature moderate computational power that provides less scalability and elasticity. In [28] the Cloudlet term is mentioned for the first time, as "trusted, resource-rich form factor that is well connected to the Internet and available for use by nearby mobile devices". One of the authors of [28] is one of the founders of pervasive computing and perhaps the most famous researcher in this field, Mahadev Satyanarayanan. Satyanarayanan further develops this concept by regarding the cloudlet as an intermediary step between the mobile device and the cloud, in a three-tier hierarchy in which the cloudlet is deemed to be a "Data Centre in a box" set-up to "bring the cloud closer" to the device, therefore reducing latency. Conceptually, the idea of cloudlet is the building blog that sustains edge computing.
- *Proximate-Mobile-Cloud (PMC)*: This architecture, as shown in Fig. 3c, has been proposed to employ a cloud of nearby resource-rich mobile devices that are willing to share resources with proximate resource-constraint mobile devices. Rapidly increasing popularity and ever-increasing numbers of contemporary mobile devices are enabling the vision of building PMC to be realized. Two different computing models are feasible: peer-to-peer and client-server. In peer-to-peer, service consumers and providers can communicate with each other directly

to negotiate and initiate the augmentation. For this architecture recently, the term Mobile Ad hoc Cloud Computing (MAC) is also used [27].

- *Hybrid (H)*: Each of the three architectures features advantages and disadvantages that hinder optimal exploitation for efficient mobile computation augmentation. Hybrid multi-tiered infrastructures, shown in Fig. 3d, are comprised of various proximate and distant computing nodes, either mobile or immobile. The main idea behind building hybrid resources is to employ heterogeneous computing resources to create a balance between user requirements (mainly latency and computation power) and available options [9].

2.3.2 Computation Offloading

Generally, the software-oriented Cloud-based Mobile Augmentation (CMA) techniques in literature are referred to as offloading techniques. Offloading works as a fundamental feature that enables MMC to relieve task load and extend data storage through cloud-based resources [29]. To meet the goals of saving energy consumption of mobile devices, improving application performance, or achieving both, offloading decisions can be made based on multiple perspectives: *when, what, where* and *how* to offload [9, 30]. Specifically, *when* is to decide whether to offload or not, according to the knowledge on the amount of computation and communication data, the wireless network conditions and dynamic changes of context, since sometimes offloading is not worthwhile at all; *what* is to decide how much and what should be offloaded, it defines the name of the candidate tasks to be offloaded through application partitioning; *where* describes the type of surrogate and choosing the appropriate offloading target, e.g., cloud-based resources in which the application has to be offloaded; *how* introduces offloading plans that enable the device to schedule offloading operations. Hence, many research efforts have been devoted to determining via algorithms the right time, the right component, the right place and the right way to offload. In [30] the authors present a survey on these efforts and explore methods of multi-objective decision making for time- and energy aware task offloading.

Offloading decisions in MCC may involve multiple factors, such as the resource heterogeneity of mobile devices and cloud-based resources, the complexity of mobile applications, the interruption of heterogeneous wireless networks and the characteristic of transferring a large amount of data, which may significantly impede the improvement of service quality. In [30] the authors categorized these factors into three main modules:

- *Profiling*: On the mobile side, upon receipt of an offloading request, resource information about the device and network characteristics are gathered by the profiling module. There are three different kinds of profilers, namely, program profiler, network profiler and energy profiler. Among them, program profiler (static or dynamic) collects characteristics of applications, such as the execution time, the memory usage and the size of data. Network profiler collects information about

the network bandwidth and the wireless connection status (connected or unconnected). Energy profiler is used to collect the energy characteristics of mobile
devices through software and hardware monitors.

- *Metrics*: Offloading decisions are usually made based on a selected cost criterion.
  On the one side, energy, monetary cost and storage are cost criteria which are the
  less the better, and on the other side, performance, robustness and security are
  benefit criteria which need to be maximized. Among such criteria, energy and
  performance are the two most important aspects mobile users concern about.
- *Application partitioning*: On the basis of the collected information, the offloading
  decision making module takes the decision according to the metrics module (i.e.
  minimizing or maximizing some criteria), and then the partitioning module is
  invoked to cut the classes that make up an application into local and remote
  partitions, where the former is executed locally on the mobile device and the
  latter will be offloaded to a dedicated cloud server. The application partitioning
  can be done either statically or dynamically.

2.4   Edge Computing

In this paper we take the approach to *Edge Computing* as defined by the Linux Foundation Open Edge Computing Glossary [31] which defines Edge Computing as "*the
delivery of computing capabilities to the logical extremes of a network in order to
improve the performance, operating cost and reliability of applications and services.
By shortening the distance between devices and the cloud resources that serve them,
and also reducing network hops, edge computing mitigates the latency and bandwidth constraints of today's Internet, ushering in new classes of applications. In
practical terms, this means distributing new resources and software stacks along the
path between today's centralized data centers and the increasingly large number of
devices in the field, concentrated, in particular, but not exclusively, in close proximity to the last mile network, on both the infrastructure and device sides*". Also, the
glossary defines Edge Cloud as "*Cloud-like capabilities located at the infrastructure
edge, including from the user perspective access to elastically-allocated compute,
data storage and network resources. Often operated as a seamless extension of a
centralized public or private cloud, constructed from micro data centers deployed at
the infrastructure edge.*"

The need to utilize services on the edge of the network is as old as the need created
in the IT world in the past to utilize the cloud by requiring very low latency services
compared to human response as well as saving both computing and energy resources.
This situation on the one hand necessarily leads to the execution of the computing
load in the part of the cloud; on the other hand the speed of light reminds us that the
cloud or otherwise the edge of the network to which the service is to be offloaded
may be too far away [32].

2.4.1 Edge Computing Characteristics

Concerning the features that the EC includes, which are of particular reference as they determine how the EC integrates with others structures, we should focus on some of them, like transmission latency, data storage, privacy and security as well as energy consumption. Thus, these are the key features that characterize EC as a structure that improves IoT performance. We refer to them based on the potential they have over other features [33].

- Regarding latency and delay: edge computing has been developed to solve the bottleneck problem of network resources. By offloading the data computation and storage to end users, the response time and traffic flow will be significantly reduced. The hierarchical distributed edge nodes are able to satisfy the demands of time-sensitive applications. Obviously, computation offloading from the central cloud to the network edge can help reduce transmission delay via a proper offloading strategy. There are some efforts focused on how to make optimal computing offloading decisions. For example, Deng et al. in [34] proposed an adaptive sequential offloading game scheme for a multi-cell MEC scenario, and then designed a multi-user computation offloading algorithm. In their designed scheme, the mobile users make offloading decisions by considering the current interference as well as available computation resources. In this way, reduced latency and energy consumption can be realized by mobile users. Due to the locations of edge computing nodes being close to end users, the peak in traffic flows will be alleviated. Edge computing platform allows edge nodes to respond to service demands, reducing bandwidth consumption and network latency [33, 35, 36].
- *Regarding privacy and security*: as is the case with many new technologies, solving one problem can create others. From a security standpoint, data at the edge can be troublesome, especially when it's being handled by different devices that might not be as secure as a centralized or cloud-based system. The edge nodes are running in different organizations, it is difficult to ensure the integrity, information protection, anonymity assessment, non-repudiation, and freshness of the original data [37]. As the number of IoT devices grow, it's imperative that IT understand the potential security issues around these devices, and to make sure those systems can be secured. Furthermore, differing device requirements for processing power, electricity and network connectivity can have an impact on the reliability of an edge device. This makes redundancy and failover management crucial for devices that process data at the edge to ensure that the data is delivered and processed correctly when a single node goes down. Edge computing-based storage assistance, sensitive data can be replicated and the different pieces of data stored in different geological locations. This remarkably mitigates the risk of data loss [35, 36, 38].
- Regarding storage: Compared to large data centers of cloud, it is fact that edge nodes have less storage space resulting in limited large-scale and long-lived storage. Thus, uploading data means employment and coordination of various

edge nodes for storing as a result grater complexity of data management. The advantages of EC for storing and disposing of data consist of storage balancing and data reliability. Concerning the storage balancing as Edge computing encompasses data computing and storage that is being performed at the network "edge", edge computing-based storage can reduce the storage time by selecting the nearest edge storage nodes, or some storage processing rating and weighting schemes. Based on the requirements that typically have structures such as Industrial IoT, which use sensors or cameras, sending data to cloud-based storage is not the most efficient solution. On the contrary, according to EC functions, if the data is sent to different storage nodes, network traffic over long distances will be reduced [36, 39]. At the same time data reliability, which consists of availability and data replication, is a very important element of Edge computing. Unlike cloud-based systems, when no storage services are available, backup storage servers are facing the problem, in computer edge storage systems, this role performed by other available edge nodes, that will act as redundant storage. This is a very important feature, especially in environments with a large number of devices and with a constant demand for data [35].

- *Regarding energy consumption*: this is very important when it comes to devices that have limited energy capabilities but must at the same time operate continuously and accurately. Such devices are widely used on IoT platforms, which underscores the importance of EP for the operation of such a structure [35]. The Edge computer can transfer computing and basic communication, from nodes with a limited battery or power supply to nodes with significant power resources [35, 36]. In this way, the battery life of the nodes with a limited battery will be extended. It is worth mentioning that energy consumption in use a cloud service, according to [40] usually depends on six factors:

(1) the power consumption of the end-user device accessing the service.
(2) the power consumption of the data center, including the power consumption of the data center
(3) the volume of traffic being transferred between the user and the cloud,
(4) the complexity of the project implementation;
(5) factors such as the number of users sharing a computing resource and.
(6) the energy consumption of the transport network.

At the same time, according to [41] 14% of Internet energy consumption is due to data transfer. Therefore, processing and transferring data near the end user, in addition to reducing transmission time, also contributes to a reduction in energy consumption and, consequently, to increased device or sensor autonomy [36].

- Regarding location awareness: The current cloud computing standard enhances the limited resources capabilities, but this cannot meet the location awareness requirements. This is also one of the main differences of CC from EC where Edge technology (and to a greater extent fog computing) is coming to cover. Based on servers installed at the edge of the network, EC makes the best use of location awareness by enabling users and applications to know their mobility as well as their position in relation to real-world objects. Furthermore, applications utilizing this information are able to adapt their functionality while greatly optimizing user navigation for various purposes. It is understood that the EC is taking full advantage of the location awareness compared with other related technologies such as CC [36].

The combination of all the above features could be said to give EC technology characteristics such as *Intelligence, Smartness and Autonomy*. Applications are now able to understand the environment, people's needs and the way they interact with computer systems (machines) so they can be adapted accordingly while the reduction of energy consumption and the optimized use of resources offers high levels of autonomy [35, 36].

2.4.2   The Hierarchical Structure of Edge Computing

The common denominator in edge paradigms is the deployment of cloud computing-like capabilities at the edge of the network. Most edge paradigms follow the structure shown in Fig. 4 [87].

Edge computing paradigms are often divided into three aspects, the front-end, near-end, and far-end. Edge clouds, which are owned and deployed by infrastructure providers, implement a multi-tenant virtualization infrastructure. Any customer— from third-party service providers to end users and the infrastructure providers themselves—can make use of these edge clouds' services. Notice that the edge computing servers are closer to the end user than cloud servers resulting in decreasing latency to a great extent, providing sufficient QoS although they lack computation power. Additionally, while edge clouds can act autonomously and cooperate with each other, they are not disconnected from the traditional cloud. In this way they manage interconnection with the whole network infrastructure creating an architecture with multiple tiers. Besides, we have to consider the potential existence of an underlying infrastructure, or core infrastructure (e.g., mobile core networks, centralized cloud services), that provide various support mechanisms, such as management platforms and user registration services. A hierarchical model proposed in [42], which integrates the Mobile Edge Computing (MEC) servers and cloudlet infrastructures. In this model, the mobile users can obtain their requested services as MEC provides the ability to meet their computing and storage needs. Finally, one trust domain (i.e. edge infrastructure that is owned by an infrastructure provider) can cooperate with other trust domains, creating an open ecosystem where multitude of customers can be served [33, 43].

Although intended for different parts of the overall network, Edge Computing (EC) and Cloud Computing (CC) are interrelated. Evaluating the features and hierarchical structure applied to Edge Computing, a comparison between CC and EC is depicted in Table 1.

Edge computing has further advantages over cloud computing for IoT, even though it has more limited computational capacity and storage. Edge computing offers a

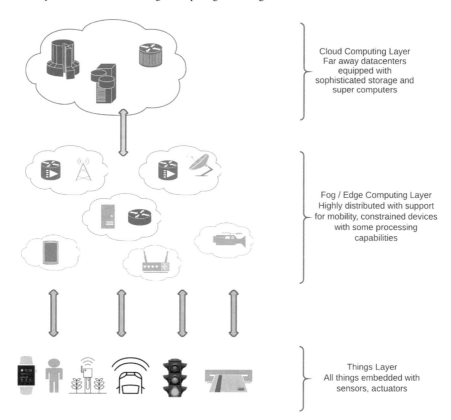

Fig. 4 The hierarchical structure of edge paradigms

Table 1 Differences between CC and edge computing

Characteristics/features	Cloud computing	Edge computing
Response time/latency	Slow/high	Fast/low
Distribution	Centralized	Distributed
Storage	Unlimited	Limited
Network access type	Mostly WAN	LAN (WLAN)
Components	Virtual resources	Edge nodes
Server location	Anywhere within network	At the edge
User device	PC, mobile devices (limited)	Mobile-smart-wearable devices
Mobility support	Low	High
Computational	Unlimited	Limited
Number of servers	High	Low
Task/app needs	Higher computation power	Lower latency

tolerable computational capacity, enough storage space, and fast response time to satisfy IoT application requirements. Nonetheless, the computational capacity of individual edge nodes is limited, and thus the scalability of computational capacity for edge computing is a challenging problem [44].

In cloud computing-based systems, redundant storage servers are deployed to handle the problem of unavailability (error to storage hardware, remove or change the authority of the disks, maintenance of the system etc.) Nonetheless, in edge computing storage systems, the other available edge nodes will act as redundant storage [45].

3 The State-of-the Art of Edge Computing Paradigms

3.1 Cloudlets

A Cloudlet is a form of cloud that refers mainly to private or public cloud applications with an emphasis on enhancing mobility (mainly for mobile devices) and in fact it is often paralleled with the terms Edge Cloud, Edge Data Center or Edge Node. In more detail, imagine a computing architecture consisting of 3 tiers, so cloudlet is going to represent the middle tier (tier 2) with tier 1 being the cloud and tier 3 User Equipment's devices (e.g., smartwatch, smartphones) [46]. In a Content Delivery Network, cloudlet plays the role of deploying applications at these nodes with self-serviceable capability [47]. Logically, the above architecture significantly reduces communication latency with the ability to increase the execution speed of various demanding applications. In addition, benefits are observed in localized cloud services with negative elements of lower bandwidth and issues of high economic impact of utilization [48, 49, 50].

3.2 Fog Computing

Fog computing almost "resides in" at the edge network and except that it offers the appropriate network at the edge nodes to Cloud, it also provides storage and computing services as every edge computing paradigm does [46]. We would say that Fog Computing has the ability of distributing computing, data storage and application services between end user and Cloud optimally emphasizing on performance and redundancy meaning that utilized if needed the whole "resource line" from Cloud to infrastructure edge, rightly representing the term of decentralization [47]. Fog Computing is inextricably linked with Internet of Everything (IoE) since its creation was inspired from this outgoing technology. Using Fog Computing we manage extreme low latency, context and location awareness and streaming services with high bandwidth. In contrast with cloud data centers, fog nodes are shared across multiple parts of the network hardening the appropriate maintenance. Also, as a continuation of the previous one, many nodes lead to complexity, security and authentication problems [45, 51].

3.3 Multi-access Edge Computing

As its name suggests, Multi-Access Edge Computing has the potential to offer powerful computing capabilities as well as a rich cloud information services environment to any type of user on the edge of the network such as from a simple user to a developer. With the term rich environment, we refer mainly to the offered abilities such as ultra-low latency with wide bandwidth, real time processing and various choices of connection to the remote world focusing mainly to radio access networking such as cellular [50]. All the above led the ETSI (European Telecommunications Standards Institute) to rename what we knew as Mobile Edge Computing to Multi-Access Edge Computing [48]. Then wherever we mention MEC we mean Multi-Access Edge Computing. One of the key characteristics of the MEC specification is that services requiring high context and real time awareness of their local surroundings benefit significantly from this architecture. As we move into the twenty-first century, mobile users demand connectivity everywhere so cellular connectivity seems to be the most optimal way with capabilities enhancing tremendously with the help of MEC [23, 52].

3.4 A High-Level Comparison

There are various differences among edge paradigms, such as the focus on mobile network operators as infrastructure providers in MEC, the existence of user-owned edge data centers in Cloudlets a Fog, and the use of different underlying protocols and interfaces, among others. Nonetheless, there remain numerous similarities. Still, little of the research in these fields takes into consideration these similarities. Most architectures, protocols, services, and mechanisms are designed with only one edge paradigm in mind, and they do not consider the state of the art of other edge paradigms. At this initial stage, researchers should consider that research findings in relation to one edge paradigm might also be applied or adapted to other edge paradigms. Some of the main properties that have to be considered before using a specific architecture are listed in Table 2 [46, 47, 48].

Table 2 A comparison of edge computing paradigms

Properties	Cloudlets	Fog	MEC
Latency reduction	✓	✓	✓
Multi-tenancy	✓	✓	✓
Co-location	✓	✓	✓
Geographical Distributed	✓	✓	✓
Mobility support	✓	✓	✓
Extended from cloud	✓	✓	May or may not
Wireless access usage	✓	✓	✓
Located between data centers and device	✓ (or may run on a device)	✓	✓
User experience improvement	Less than the others	✓	✓
Use cases	Resource hungry and interactive	IoT, IoE, IoM	Video analytics location services, IoT, AR, VR, V2X, Smart Cities, Healthcare, Disaster Management, Agriculture
Resources closer to RAN	No	Maybe	✓
Context awareness	No	✓	✓

4 Applications Categories and MEC Relation to Crowdsourcing

Given recent advances in pervasive computing, there are currently a myriad of diverse mobile applications for many different domains, as shown in Fig. 5, which are expected to enhance and improve the quality of everyday-life for the end-user. The variety of these applications dictate that there is no *one-fits-for-all* solution, as each of these applications have different characteristics, and they can be broadly categorized into a number of different fields, since they also have different latency and data rate requirements.

Most of these applications are based on integrated sensors of contemporary smartphones that collect user's data. As a result, mobile crowdsourcing (MCS) has emerged as an efficient method of collecting and processing this data [53]. Smartphone integrated sensors offer advanced services, which when combined with Web-enabled real-world devices located near the mobile user (e.g., body area networks, RFIDs, energy monitors, environmental sensors, etc.), have the potential of enhancing the overall user knowledge, perception and experience, encouraging more informed choices and better decisions. In [54] authors give the definition as the domain of mobile computing and IoT as "the research area that involves case studies, prototypes,

Fig. 5 Mobile apps domain

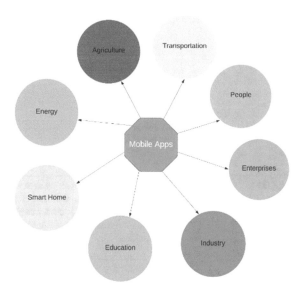

demonstrations, applications and business cases of the IoT/WoT (*Web of Things*), through mobile phones, where the user of the mobile phone interacts with cyber-physical things that are enabled to the Internet/Web, through his/her phone device, exploiting at the same time the sensing capabilities of the phone".

4.1 Categories

Emphasizing on how the EC is integrated into the technological environment, we should refer to the major areas where Mobile Cloud/Edge computing is applied. These areas relate to both daily use such as sports, urban transport, and interaction with everyday devices, as well as commercial / industrial levels such as agricultural, health and product transport. Each of them contains a multitude of subcategories and ways of use, with each requiring the collection of different types of data and using different types of technology for this purpose as depicted in Table 3. At the same time some commercial applications may refer to simple daily use and thus lose their commercial character or some non-commercial may through their role in the daily lives of people transform into commercial. Our perspective is entirely based on how technology is integrated and not on the commercial or non-commercial way of using it. We focus primarily on the way mobile edge computing technology has influenced and continues to affect below technological sectors.

- *Participatory Sensing*: When we refer to the concept of participatory sensing, this involves mobile systems encouraging users to record and share information, towards the co-creation of advanced knowledge, that affects both the social and the

Table 3 MEC application areas

Category	Description	Used technologies	Type of data
Sensing participatory	Involves mobile systems, encouraging users to record and share information	GPG, microphone, camera	Location-based search, map visualization, information sharing
Actuation and control	Mobile applications that control certain physical devices	Smart electricity meters, smart appliances, light switches, smart factory sensors	Rule-based inference, adaptive reasoning, optimization
Health	Receives user health information through sensors for better analysis, comparisons and feedback	Body area sensors, GPG, microphone, accelerometer, communication, and conversation sensing	Personalization and profiling, anomaly detection, emotions detection, stream data processing, machine learning
Sports	Device built-in sensors record various measurements about sport activities	Body area sensors, motion sensors, GPS, pressure sensors, pedometers	Personalization and profiling, historical comparisons, performance visualization, information sharing, statistics
Agriculture	Improve productivity, management of livestock and increase consumer satisfaction and transparency	Wearable collars, GPS, barcodes, RFID tags, soil sensors	Historical comparisons, information retrieval, stream data processing, optimization
Gaming	Enhance the gaming experience through VR, AR features	Accelerometer, gyrometer, cameras, pedometers, barcodes, RFID tags	Activity recognition, machine learning, information sharing, image and video processing
Transportation	Sensing features of mobile phones to enhance transport industry features	GPS, compass, carrier connectivity, RFID tags	Location-based search, Big Data analysis, stream data processing, anomaly detection, information sharing, machine learning, image and video processing

public level [54]. With the popularity of smart phones with various sensors such as camera, sound, accelerometer, GPS, gyroscope, compass, proximity and ambient light, a new model of common sense is called mobile crowdsensing (MCS) [55]. MCS represents a class of IoT applications based on the collection of data from many mobile monitoring devices such as smartphones [56]. In addition, Industry

4.0 benefits of using mobile crowdsensing and mobile crowdsourcing in many areas such as Quality control, Supply chain management and Product monitoring [57].

- *Actuation and control*: Cloud Computing (CC) provides ubiquitous on-demand access to end users and this feature is used for actuation and control ion smart environments [58]. Through mobile applications, it is possible to manage devices and activate events related to home automation. As home automation evolves, it becomes more and more necessary in our daily lives. Real-time response is required even if it is not vital in some cases. Especially the perception of the user's position is an important point as it activates the various situations and is required to be accurate. In addition, it must have a good response time because the change of user location can be continuous. Accuracy in small radius is also important because in small areas such as the home or office, position change is frequent and continuous due to movement within this area. Edge Computing (EC) speeds up and improves CC performance by bringing processing closer to the data source. This reduces response time and dramatically increases accuracy [38].

- *Health*: The edge computing has been employed in healthcare to meet the real-time response requirements of applications. With the emergence of wearable computing and gadgets, cognitive assistance-based applications are becoming a reality [35]. Therefore, new cloudlet-based architectures are proposed for data collection and processing by Body Area Networks (BANs), to minimize packet-to-cloud energy and packet delay [59, 60]. Simultaneously, high definition medical data is remotely located and monitored by many healthcare industries. 5G is widely used in medical applications. Low latency and high-resolution imaging are very important as it ensures uninterrupted and lossless data delivery, regardless of the high frequency asynchronous channel bandwidth and relatively high bit rates [61]. Therefore, in the future, 6G-supported innovations such as AR/VR, holographic communications and AI [62] will be a milestone in the evolution of a smart healthcare system and MCC-based healthcare group applications like Electronic Health Record (EHR), Picture Archiving and Communication System (PACS), Telemonitoring and Biosignal Processing and Multi-Agent Medical Consultation [63].

- *Sports*: A various combination of physical sensors built into devices (smart-watches, smart bands etc.) and mobile applications, used for measurements and improve athlete performance. Real Time Monitoring System for physical fitness of athletes requires [64] reliable and uninterrupted operation, especially in vital measurements. Real-time monitoring through MC applications has evolved rapidly over the last decade. Commercial devices and applications such as Nike + Sports Sensor, Apple Watch, Garmin, Suunto and Samsung smartwatches make the sports experience more entertaining and fun for the user, helping him/her to improve performance through competitions and comparisons with friends [33]. At the same time, front-end systems intended for healthcare applications, such as those in Wearable 2.0 [65], can be used to measure biometric data from athletes. The mobility of the various sensors requires the advantages of Mobile edge computing, for the smooth operation of collecting large health data during exercise.

- *Agriculture*: In both agriculture and animal husbandry, productivity improvements through smart farming have achieved dramatic quality improvements and cost reductions. In addition, the three parts (hardware, web application, and mobile application) that compose the smart Agriculture implementation scenarios [66], will benefit from Mobile Cloud/Edge computing capabilities. At the same time, the MEC will enable the use of connected unmanned aerial vehicles (UAVs), which play an increasingly important role in various precision farming scenarios, such as photography, disaster management, inspection and monitoring [67]. There are some interesting applications in this area such as AGBRIDGE, RRXtend Spray, Xarvio Scouting, Cropstream, FieldCheck, AgriSync, Kugler Timing, Agsense, FieldAgent and Agrellus [68].

- *Gaming*: In recent years gaming has shifted from the local situation to the Online world. The Interconnection of Everything has given additional dynamism to the gaming community which can now enjoy its digital entertainment even using mobile phones without making concessions in quality. Multiplayer games are also gaining in popularity. But if the server is far from the player's client or the communication network is congested, then the QoE decreases dramatically. Response time is an important element of online gaming as it affects the way we play especially in competitive games and e-sports [69]. There are still significant challenges to cloud gaming such as Low Delay Requirement, High Bandwidth Consumption and Sensitive Users' QoE [70]. Edge computing and 5G connectivity offers additional benefits to meet the challenges of online gaming. Especially Edge technologies can help minimize bandwidth consumption by placing high compute processing power closer to users and devices. In addition, edge computing provides low-latency address to support high-quality and enhanced multi-player gaming experience [63].

- *Transportation*: The Internet of Vehicles (IoV) and the Intelligent Transportation Services (ITS) are two growing domains which the sensing features of the MEC are harnessed. However, for ITS to be successful for better driving experience, safety, transport productivity, travel reliability, informed travel choices, environment protection, and traffic resilience, the use, integration, and deployment of emerging technologies becomes imperative [71, 72]. An interesting cloud computing model for intelligent transportation system (ITS-Cloud) has been proposed, that based on two sub-models: conventional cloud and vehicular cloud. Therefore, IoV dynamically collects, processes and disseminates traffic information in real time, and has a huge amount of data to handle. The solution for efficient handling of this data is the cloud computing framework. Cloud computing facilitates the storage, processing and analysis of data in IoV, while the use of EC will provide the benefits needed for the safe and smooth operation of smart vehicles [73].

4.2 MEC Key Elements to Crowdsourcing Architecture

As mentioned earlier, MEC is applied in various areas of everyday life, enriching the existing possibilities and offering new properties that can be used. In addition, we observe rapid growth and evolution of portable and wearable devices that carry sensors and operate in a ubiquitous environment. Collected and processed data tend to become innumerable as even the smallest and most invisible everyday devices have multi-sensing capabilities like movement, light, audio, and visual sensing. Taking advantage of the ubiquitous computing and continuous data collection, crowdsourcing is emerging as a "phenomenon" [74]. Mahmud et al. mentions that the main catalysts of the successful implementation of crowdsourcing are the widespread availability of mobile. Therefore, through crowdsourcing, tasks demand and workload will be more manageable and can be fast performed by the crowd [75].

The basic architecture of MCS as analyzed [76] and presented [77] is a four-level architecture, which consists of the Sensing, Communication, Data and Application layers. EC capabilities of parallelization and partitioning of a problem, computation offloading, reduced computational complexity in combination with reduced latency and reduced privacy threats, give additional dynamics in the implementation of MCS. M. Marjanović et al. present the functional four-tier MCS architecture according to the key features that the MEC can offer. Table 4 summarizes this four-level architec-

Table 4 Four-level MCS architecture

Architecture layers	Protocols	Processes performed	Devices
Applications	• TCP, UDP, HTTP	• Domain specific application and data visualization	• End users
Cloud computing	• TCP, UDP, HTTP • Security and privacy • Communication protocol • Real-time processing	• Quality of service • Data analytics and storage • User/sensor reputation management	• Cloud storage services
Edge computing	• Security and privacy • Communication protocol • Real – time processing • IEEE 802.15.4/ZigBee, 6LoWPAN, MQTT, CoAP, Bluetooth	• Device discovery • Mobility management • Energy management • Sampling mode management • Sensing scope management • Data (pre)processing	• BTS • Access points
User equipment	• IEEE 802.15.4/ZigBee, 6LoWPAN, MQTT, CoAP, Bluetooth	• Data (Pre)processing • Sensing nodes • User involvement • Incentives	• Smartphone sensors • Smartwatch sensors • Wearable sensors • Smart vehicles

ture that includes User Equipment layer, EC layer, CC layer and Application layer [78]:

- **Sensing/User Equipment layer** is the umbilical cord of the MCS with the real world as here the data feed is achieved through the built-in sensors. Data collection in MCS can be based on a variety of devices and everyday objects such as mobile devices (cameras, smartphones, or smart watches), smart vehicles, laptops [79], temperature, pressure, but also wearable sensors or the latest generation sensors, such as NFC. A basic principle, if the capabilities of these devices allow it, is the initial processing of the data to be done before they are sent to the server. Through this process we have advantages in energy efficiency and reduction of bandwidth.
- **Edge Computing layer**: As the EC is closer to the user, it is responsible for worker management in certain geographical areas, data collection, data processing and filtering of sensor data. At this level is device discovery, mobility and energy management, while at the same time MEC enhances cellular network services with low latency, high bandwidth and data analysis before the cloud, and in addition Enable Real-Time Context-Aware services [80] something that which helps MCS services.
- **Cloud computing** layer is responsible for processing and storage complex data. Through data analytics and sensors reputation management offers the necessary QoS, while additionally enable interactions between multiple MCS services.
- **Applications layer:** At this layer, specifying MCS tasks is provided to those requesting analysis of the data collected and stored. Applications used to visualize the results of the collected data analysis. They are domain specific and usually include different interfaces for performing visualization and sharing knowledge between users. A key challenge at this level is the conflict between privacy and trustworthiness. As the collection and sharing of personal information related to human activity introduces the concern of privacy [81].

5   Issues and Open Research Challenges

MEC technology includes a multitude of mobile devices, cloud-based resources, networking infrastructures and various connection technologies that are managed via software systems. Also, MEC involves execution of only those applications that require extensive computational and storage resources beyond native mobile devices. So, this technology faces multidimensional challenges, and despite impressive findings, several challenges require further efforts. In this section we briefly summarize the most important challenges which emerged from our study as follows.

- Big Data: The need for mobility enhancement with ultra-low latency roaming, among others, brought to the surface what we explain earlier as Multi-Access Edge Computing (MEC). However, the ICT systems evolving rapidly in proportion to human needs giving birth to innumerable data sources and Big Data. As a result, it seems impossible for MEC servers to be able to meet this demanding need for data processing and storage with the same efficiency as the enormous cloud data centers. Numerous tries and ideas have been proposed such as the addressing of "4Cs" ("computing, caching, communication and control") at the edge of the network or the availability of relevant datasets, near to the MEC servers, but still this challenge raises concerns as it is difficult enough to predict how far these huge amounts of real-world data streams can go [46, 82]

- Optimal deployment: One of the major factors of latency is the distance of involved entities and the efficient management of given bandwidth. Considering MEC technology, the deployment of so many nodes optimally remains an unsolved puzzle where every time you have to look for another missed piece on this confusing topology [46].

- Heterogeneity—Offloading: These two factors may seem independent but in fact they can be interconnected. Offloading decisions in MEC may involve multiple factors, such as the resource heterogeneity of mobile devices and cloud-based resources, the complexity of mobile applications, the interruption of heterogeneous wireless networks (especially in MEC supporting cellular technology, Wi-Fi, Wi-Max etc.) and the characteristic of transferring a large amount of data which may significantly impede the improvement of service quality [30]. Resource-poverty in mobile computing is the major factor that necessitates the development of lightweight techniques for mobile consumers, cloud service providers, and network providers. To realize lightweight approaches in MEC, shrinking application dependency on underlying platforms (towards portability) is significantly beneficial [46].

- Seamless connectivity/Reliability: Mobility in MEC is an inseparable property of service consumers and mobile service providers that requires seamless connectivity and reliability. Establishing and maintaining continuous, consistent wireless sessions between moving service consumers and other computing entities (e.g., mobile devices and clouds) in the presence of heterogeneous networking technologies requires future research and development. Lack of seamless connectivity increases application execution time and mobile energy consumption due to frequent disconnections and interruptions, which substantially degrade user experience [9]. Consequently, seamless connectivity has to followed by reliable and accurate data processing as well as the corresponding mobile applications must be able to detect data abnormalities and errors [46].

- Availability: Availability challenge remains a constant headache for every aspect of cloud computing as it refers to the capability of each server and the medium access challenges because of frequency scarcity or congestion issues [46].

- Security and privacy: MEC security challenges lie in the fact that it is required to meet all the security challenges of its components, e.g., hardware and software. So, the modern security and privacy issues remain in the MEC. Solving a security problem usually creates a new need for security and privacy. Although it seems like a "vicious circle", but we think this is that creates the need for continuous development and improvement of the sector [83]. One of the most challenging aspects of MEC is protecting offloaded code and data inside the cloud. While securing contents inside the mobile consumes huge resources, offloading plain contents through insecure wireless medium and storing plain data inside the cloud highly violates user security and privacy. Despite of the large number of research and development in establishing trust in cloud, security and privacy is still one of the major user concerns in utilizing cloud resources [46, 84, 85].

6 Conclusion

Computing technologies have made a great step forward in the last decades. It is a fact that most of them are being key enablers for many ICT systems significantly enhancing the performance of the respective systems. In this paper we study the various architectures and the key points of some of the main computing technologies today with regard to cloud, starting from Cloud Computing (CC) which was the key pillar for what followed and that is Mobile Computing (MC), Edge Computing (EC), Fog Computing (FC) and the promising Multi-Access Edge Computing (MEC) being the evolution of Mobile Edge Computing. As it is perceived, going through the era of the IoE every user demands broadband connectivity with high QoS and ultra-low latency for optimal user experience. MEC, on the one hand is able to satisfy the above characteristics creating a very useful domain of crowd sourcing, on the other hand some important open research issues have to be moderated such as security and privacy, the constant headache of telecommunications as well as heterogeneity and big data, to name a few. By this study we hope that the reader will understand the main differences of each aforementioned Edge Computing aspects and the importance of their contribution to communication' s evolution.

References

1. Satyanarayanan, M. Fundamental challenges in mobile computing. In: Proceedings of the Fifteenth ACM Symposium on Principles of Distributed Computing. Philadelphia, PA, May (1996)
2. Weiser, M.: The computer for the 21st century. In: Scientific American, vol. 265(3), pp. 94–104. SciAm (1991)
3. Satyanarayanan, M.: Mobile computing: the next decade. In: Proceedings of the 1st ACM Workshop on Mobile Cloud Computing & #38; Services: SocialNetworks and Beyond (MCS'10), pp. 5:1–5:6. ACM, New York, NY, USA (2010)
4. Satyanarayanan, M.: Pervasive computing: vision and challenges. IEEE Pers. Commun. **8**(4), 10–17 (2001)
5. Cisco Systems. Cisco Visual Networking Index: Global Mobile Data Traffic Forecast Update, 2017–2022. White Paper, last visited May 2019. [Online]. Available: https://www.cisco.com/c/en/us/solutions/collateral/service-provider/visual-networking-index-vni/white-paper-c11-738429.html
6. Abbas, N., Zhang, Y., Taherkordi, A., Skeie, T.: Mobile edge computing: a survey. IEEE Internet Things J. **5**(1), 450–465 (2017)
7. Cheng, X., Fang, L., Yang, L., Cui, S.: Mobile big data: the fuel for data-driven wireless. IEEE Internet Things J. **4**(5), 1489–1516 (2017)

8. Abolfazli, S., Sanaei, Z., Gani, A., Xia, F., Yang, L.T.: Rich mobile application: genesis, taxonomy, and open issues. J. Netw. Comput. Appl. **40**, 345–362 (2014)
9. Abolfazli, S., Sanaei, Z., Ahmed, E., Gani, A., Buyya, R.: Cloud-based augmentation for mobile devices: motivation, taxonomies, and open challenges. IEEE Commun. Surv. Tutor. **16**(1), 337–368 (2014)
10. Sanaei, Z., Abolfazli, S., Gani, A., Buyya, R.: Heterogeneity in mobile cloud computing: taxonomy and open challenges. IEEE Commun. Surv. Tutor. **16**(1), 369–392 (2014)
11. Mell, P.M., Grance, T.: The NIST definition of cloud computing. National Institute of Standards and Technology, Gaithersburg, MD, USA, Technical Repot, pp. 800–145 (2011)
12. Skourletopoulos, G., Mavromoustakakis, C.X., Mastorakis, G., Batalla, J.B., Dobre, C., Panagiotakis, S., Pallis, E.: Big data and cloud computing: a survey of the state-of-the-art and research challenges. In: Advances in mobile cloud computing and big data in the 5G era, pp. 23–38. Springer International Publishing (2017)
13. https://www.gartner.com/smarterwithgartner/5-digital-technology-trends-for-2020/
14. https://www.gartner.com/smarterwithgartner/gartner-top-10-strategic-technology-trends-for-2020/
15. Gupta, B.M., Dhawan, S.M., Gupta, R.: Mobile cloud computing: a scientometric assessment of global publications output during 2007–16. J. Sci. Res. **6**(3), 186–194 (2017)
16. Atzori, L., Iera, A., Morabito, G.: Understanding the Internet of things: definition, potentials, and societal role of a fast evolving paradigm. Ad Hoc Netw. **56**, 122–140 (2017)
17. Ebling, M.R., Want, R.: Pervasive computing: vision and challenges. IEEE Pervasive Comput. **16**(3), 20–23 (2017)
18. Elazhary, H.: Internet of Things (IoT), mobile cloud, cloudlet, mobile IoT, IoT cloud, fog, mobile edge, and edge emerging computing paradigms: disambiguation and research directions. J. Netw. Comput. Appl. **128**, 105–140 (2019)
19. Din, I.U., Guizani, M., Hassan, S., Kim, B.S., Khan, M.K., Atiquzzaman, M., Ahmed, S.H.: The internet of things: a review of enabled technologies and future challenges. IEEE Access **7**, 7606–7640 (2018)
20. Buyya, R., Srirama, S.N., Casale, G., Calheiros, R., Simmhan, Y., Varghese, B., et al.: A manifesto for future generation cloud computing: Research directions for the next decade. ACM Comput. Surv. (CSUR), **51**(5), 1–38 (2018)
21. Buyya, R., Yeo, C.S., Venugopal, S., Broberg, J., Brandic, I.: Cloud computing and emerging IT platforms: vision, hype, and reality for delivering computing as the 5th utility. Futur. Gener. Comput. Syst. **25**(6), 599–616 (2009)
22. Kumar, R., Goyal, R.: On cloud security requirements, threats, vulnerabilities and countermeasures: a survey. Computer Sci. Rev. **33**, 1–48 (2019)
23. Taleb, T., Samdanis, K., Mada, B., Flinck, H., Dutta, S., Sabella, D.: On multi-access edge computing: a survey of the emerging 5G network edge cloud architecture and orchestration. IEEE Commun. Surv. Tutor. **19**(3), 1657–1681 (2017)
24. Varghese, B., Buyya, R.: Next generation cloud computing: New trends and research directions. Futur. Gener. Comput. Syst. **79**, 849–861 (2018)
25. Bernstein, D.: Containers and cloud: From lxc to docker to kubernetes. IEEE Cloud Comput. **1**(3), 81–84 (2014)
26. Somula, R., Ra, S.: A survey on mobile cloud computing: mobile computing + cloud computing (MCC = MC + CC). Scalable Comput. Practice Experience **19**(4), 309–337 (2018)
27. Ferrer, J., Marquès, J.M., Jorba, J.: Towards the decentralised cloud: survey on approaches and challenges for mobile, ad hoc, and edge computing. ACM Comput. Surv. **51**(6), Article 111 (2019)
28. Satyanarayanan, M., Bahl, P., Caceres, R., Davies, N.: The case for VM-based cloudlets in mobile computing. IEEE Pervasive Comput. **8**(4), 14–23
29. Flores, H., Hui, P., Tarkoma, S., Li, Y., Srirama, S., Buyya, R.: Mobile code offloading: from concept to practice and beyond. IEEE Commun. Mag. **53**(3), 80–88 (2015)
30. Wu, H.: Multi-objective decision-making for mobile cloud offloading: a survey. IEEE Access **6**, 3962–3976 (2017)

31. https://www.lfedge.org/openglossary/
32. Peterson, L., Anderson, T., Katti, S., McKeown, N., Parulkar, G., et al.: Democratizing the network edge. ACM SIGCOMM Computer Commun. Rev. **49**(2), 31–36 (2019)
33. Ahmed, E., Ahmed, A., Yaqoob, I., Shuja, J., Gani, A., Imran, M., Shoaib, M.: Bringing computation closer toward the user network: is edge computing the solution? IEEE Commun. Mag. **55**(11), 138–144 (2017)
34. Deng, M., Tian, H., Lyu, X.: Adaptive sequential offloading game for multi-cell mobile edge computing. In 2016 23rd International Conference on Telecommunications (ICT), pp. 1–5. IEEE (2016, May)
35. Bilal, K., Khalid, O., Erbad, A., Khan, S.U.: Potentials, trends, and prospects in edge technologies: fog, cloudlet, mobile edge, and micro data centers. Comput. Netw. **130**, 94–120 (2018)
36. Khan, W.Z., Ahmed, E., Hakak, S., Yaqoob, I., Ahmed, A.: Edge computing: a survey. Futur. Gener. Comput. Syst. **97**, 219–235 (2019)
37. Hossain, M.M., Fotouhi, M., Hasan, R.: Towards an analysis of security issues, challenges, and open problems in the internet of things. In: 2015 IEEE World Congress on Services, pp. 21–28. IEEE (2015)
38. Shi, W., Cao, J., Zhang, Q., Li, Y., Xu, L.: Edge computing: vision and challenges. IEEE Internet Things J. **3**(5), 637–646 (2016)
39. Yu, W., Liang, F., He, X., Hatcher, W.G., Lu, C., Lin, J., Yang, X.: A survey on the edge computing for the Internet of Things. IEEE Access **6**, 6900–6919 (2017)
40. Jalali, F.: Energy consumption of cloud computing and fog computing applications. Doctoral dissertation (2015)
41. Costenaro, D., Duer, A.: The megawatts behind your megabytes: going from data-center to desktop. In: Proceedings of the 2012 ACEEE Summer Study on Energy Efficiency in Buildings, ACEEE, Washington, pp. 13–65 (2012)
42. Jararweh, Y., Doulat, A., AlQudah, O., Ahmed, E., Al-Ayyoub, M., Benkhelifa, E.: The future of mobile cloud computing: integrating cloudlets and mobile edge computing. In: 2016 23rd International Conference on Telecommunications (ICT), pp. 1–5. IEEE (2016)
43. Toczé, K., Nadjm-Tehrani, S.: A taxonomy for management and optimization of multiple resources in edge computing. Wirel. Commun. Mob. Comput. (2018)
44. De Donno, M., Tange, K., Dragoni, N.: Foundations and evolution of modern computing paradigms: cloud, IoT, edge, and fog. IEEE Access **7**, 150936–150948 (2019)
45. Alli, A.A., Alam, M.M.: The fog cloud of things: a survey on concepts, architecture, standards, tools, and applications. Internet Things **9**, 100177 (2020)
46. Shahzadi, S., Iqbal, M., Dagiuklas, T., Qayyum, Z.U.: Multi-access edge computing: open issues, challenges and future perspectives. J. Cloud Comput. **6**(1), 1–13 (2017)
47. Hu, Y.C., Patel, M., Sabella, D., Sprecher, N., Young, V.: Mobile edge computing—a key technology towards 5G. ETSI white paper **11**(11), 1–16 (2015)
48. Bonomi, F., Milito, R., Zhu, J., Addepalli, S.: Fog computing and its role in the internet of things. In: Proceedings of the first edition of the MCC workshop on mobile cloud computing, pp. 13–16 (2012)
49. Shaukat, U., Ahmed, E., Anwar, Z., Xia, F.: Cloudlet architectures, applications, and open challenges to deployment in local area wireless networks. J. Network Computer Appl. (2015)
50. Porambage, P., Okwuibe, J., Liyanage, M., Ylianttila, M., Taleb, T.: Survey on multi-access edge computing for internet of things realization. IEEE Commun. Surv. Tutor. **20**(4), 2961–2991 (2018)
51. Puliafito, C., Mingozzi, E., Longo, F., Puliafito, A., Rana, O.: Fog computing for the internet of things: a survey. ACM Trans. Internet Technol. (TOIT) **19**(2), 1–41 (2019)
52. Pham, Q.V., Fang, F., Ha, V.N., Piran, M.J., Le, M., Le, L. B., et al.: A survey of multi-access edge computing in 5G and beyond: fundamentals, technology integration, and state-of-the-art. IEEE Access **8**, 116974–117017 (2020)
53. Feng, W., Yan, Z., Zhang, H., Zeng, K., Xiao, Y., Hou, Y.T.: A survey on security, privacy, and trust in mobile crowdsourcing. IEEE Internet Things J. **5**(4), 2971–2992 (2017)

54. Kamilaris, A., Pitsillides, A.: Mobile phone computing and the internet of things: a survey. IEEE Internet Things J. **3**(6), 885–898 (2016)
55. Ganti, R.K., Ye, F., Lei, H.: Mobile crowdsensing: current state and future challenges. IEEE Commun. Mag. **49**(11), 32–39 (2011)
56. Pouryazdan, M., Kantarci, B., Soyata, T., Song, H.: Anchor-assisted and vote-based trustworthiness assurance in smart city crowdsensing. IEEE Access **PP**(99), 1–1. https://doi.org/10.1109/ACCESS.2016.2519820
57. Pilloni, V.: How data will transform industrial processes: Crowdsensing, crowdsourcing and big data as pillars of industry 4.0. Future Internet **10**(3), 24 (2018)
58. Satpathy, S., Sahoo, B., Turuk, A.K.: Sensing and actuation as a service delivery model in cloud edge centric internet of things. Futur. Gener. Comput. Syst. **86**, 281–296 (2018)
59. Quwaider, M., Jararweh, Y.: Cloudlet-based for big data collection in body area networks. In: 8th International Conference for Internet Technology and Secured Transactions (ICITST-2013), pp. 137–141. IEEE (2014)
60. Quwaider, M., Jararweh, Y.: An efficient big data collection in body area networks. In: 2014 5th International Conference on Information and Communication Systems (ICICS), pp. 1–6. IEEE (2014)
61. Sodhro, A.H., Luo, Z., Sangaiah, A.K., Baik, S.W.: Mobile edge computing based QoS optimization in medical healthcare applications. Int. J. Inf. Manage. **45**, 308–318 (2019)
62. Zhao, Y., Yu, G., Xu, H.: 6G mobile communication network: vision, challenges and key technologies (2019). arXiv preprint arXiv:1905.04983
63. Wang, X., Jin, Z.: An overview of mobile cloud computing for pervasive healthcare. IEEE Access **7**, 66774–66791 (2019)
64. Shao, J., Yao, H., Cao, R.: A real time monitoring system for physical fitness of athletes based on internet of things and cloud computing
65. Chen, M., Ma, Y., Li, Y., Wu, D., Zhang, Y., Youn, C.H.: Wearable 2.0: Enabling human-cloud integration in next generation healthcare systems. IEEE Commun. Mag. 55(1):54–61 (2017)
66. Muangprathub, J., Boonnam, N., Kajornkasirat, S., Lekbangpong, N., Wanichsombat, A., Nillaor, P.: IoT and agriculture data analysis for smart farm. Comput. Electron. Agric. **156**, 467–474 (2019)
67. Mao, Y., You, C., Zhang, J., Huang, K., Letaief, K.B.: A survey on mobile edge computing: The communication perspective. IEEE Commun. Surv. Tutor. **19**(4), 2322–2358 (2017)
68. https://www.h2020fairshare.eu/10-best-agriculture-apps-for-2019/#
69. Bilal, K., Erbad, A.: Edge computing for interactive media and video streaming. In: 2017 Second International Conference on Fog and Mobile Edge Computing (FMEC), Valencia, Spain, 2017, pp. 68–73 (2017). https://doi.org/10.1109/FMEC.2017.7946410
70. Zhang, X., et al.: Improving cloud gaming experience through mobile edge computing. IEEE Wirel. Commun. **26**(4), 178–183 (2019). https://doi.org/10.1109/MWC.2019.1800440
71. Bitam, S., Mellouk, A.: Its-cloud: cloud computing for intelligent transportation system. In: 2012 IEEE Global Communications Conference (GLOBECOM), pp. 2054–2059. IEEE (2012)
72. Guerrero-Ibanez, J.A., Zeadally, S., Contreras-Castillo, J.: Integration challenges of intelligent transportation systems with connected vehicle, cloud computing, and internet of things technologies. IEEE Wirel. Commun. **22**(6), 122–128 (2015)
73. Sharma, S., Kaushik, B.. A survey on internet of vehicles: applications, security issues & solutions. Vehicular Commun. **20**, 100182 (2019)
74. De Vreede, T., Nguyen, C., De Vreede, G.J., Boughzala, I., Oh, O., Reiter-Palmon, R.: A theoretical model of user engagement in crowdsourcing. In: International Conference on Collaboration and Technology, pp. 94–109. Springer, Berlin, Heidelberg (2013)
75. Mahmud, F., Aris, H.: State of mobile crowdsourcing applications: a review. In: 2015 4th International Conference on Software Engineering and Computer Systems (ICSECS), Kuantan, Malaysia, pp. 27–32 (2015). https://doi.org/10.1109/ICSECS.2015.7333118
76. Guo, B., Han, Q., Chen, H., Shangguan, L., Zhou, Z., Yu, Z.: The emergence of visual crowdsensing: challenges and opportunities. IEEE Commun. Surv. Tutorials **19**(4):2526–2543

(Fourth Quarter 2017) [Phuttharak, J., Loke, S.W.: A review of mobile crowdsourcing architectures and challenges: towards crowd-empowered internet-of things. IEEE Access 1–22 (2018)

77. Capponi, A., Fiandrino, C., Kantarci, B., Foschini, L., Kliazovich, D., Bouvry, P.: A survey on mobile crowdsensing systems: challenges, solutions, and opportunities. IEEE Commun. Surv. Tutorials **21**(3), 2419–2465 (2019)

78. Marjanović, M., Antonić, A., Žarko, I.P.: Edge computing architecture for mobile crowdsensing. IEEE Access **6**, 10662–10674 (2018)

79. Kong, X., Liu, X., Jedari, B., Li, M., Wan, L., Xia, F.: Mobile crowdsourcing in smart cities: technologies, applications, and future challenges. IEEE Internet Things J. **6**(5), 8095–8113 (2019). https://doi.org/10.1109/JIOT.2019.2921879

80. Nunna, S., Kousaridas, A., Ibrahim, M., Dillinger, M., Thuemmler, C., Feussner, H., Schneider, A.: Enabling real-time context-aware collaboration through 5G and mobile edge computing. In: Proceedings of the 2015 12th International Conference on Information Technology-New Generations, Las Vegas, NV, USA, pp. 601–605, 13–15 April 2015

81. Chen, P.Y., Cheng, S.M., Ting, P.S., Lien, C.W., Chu, F.J.: When crowdsourcing meets mobile sensing: a social network perspective. IEEE Commun. Mag. **53**(10), 157–163 (2015)

82. Ndikumana, A., Tran, N.H., Ho, T.M., Han, Z., Saad, W., Niyato, D., Hong, C.S.: Joint communication, computation, caching, and control in big data multi-access edge computing. IEEE Trans. Mob. Comput. **19**(6), 1359–1374 (2019)

83. Mollah, M.B., Azad, A.K., Vasilakos, A.: Security and privacy challenges in mobile cloud computing: survey and way ahead. J. Netw. Comput. Appl. **84**, 38–54 (2017)

84. Anand, A., Muthusamy, A.: Data security and privacy-preserving in cloud computing paradigm: survey and open issues. In: Cloud Computing Applications and Techniques for E-Commerce, pp. 99–133. IGI Global (2020)

85. European Union Agency for Network and Information Security (ENISA): Privacy and data protection in mobile applications. Report, last visited May 2019. [Online]. Available: https://www.enisa.europa.eu/publications/privacy-and-data-protection-in-mobile-applications/at_download/fullReport

86. Nayyer, M.Z., Raza, I., Hussain, S.A.: A survey of cloudlet-based mobile augmentation approaches for resource optimization. ACM Comput. Surv. **51**(5), Article 107 (2018)

87. Omoniwa, B., Hussain, R., Javed, M.A., Bouk, S.H., Malik, S.A.: Fog/edge computing-based IoT (FECIoT): architecture, applications, and research issues. IEEE Internet Things J. **6**(3), 4118–4149 (2018)

Informatics

Smart Healthcare: Rough Set Theory in Predicting Heart Disease

Arpit Singh, Subhas Chandra Misra, and Sameer Kumar

Abstract The paper presents the rough set based methodologies for developing a predictive model for the occurrence of heart disease. The current study employs classical rough set theory (CRSA), which uses similarity relationships between the objects in the decision table to determine the inclusion of a certain condition in a decision class. The other variant of rough set methodology based on dominance relationship, Dominance-based Rough Set Analysis (DRSA) incorporates the monotonic relationship between the condition attributes and the desired variable to explain their respective decision classes. The data for this study is from Cleveland's heart disease database of 303 heart disease patients taken from the open source online. Classification accuracy and explanatory power of the rough set based methods are compared with that of other machine learning languages and non-rule based methods. Non-rule-based methods performed marginally better than the rule based rough set methods, which cannot shadow the benefits of rough set methods provided by the provision of allowing a large amount of inconsistency in the decision rules.

Keywords Rough sets · Machine learning · Heart disease · Classification

1 Introduction

The adoption of information technology in healthcare has revolutionized processes within the sector. Cloud based services provide an immense amount of readily available data pertaining to the healthcare industry. Not only do we have an enormous amount of information available about a plethora of diseases, their symptoms, death rates, hospital re-admissions, etc., we also are in a state where the information is continuously and instantly updated. We are indebted to the cyber revolution for

S. C. Misra (✉) · S. Kumar
Kanpur, India
e-mail: subhasm@iitk.ac.in

A. Singh
O.P. Jindal Global University, Jindal Global Business School, Sonipat, Haryana, India

© The Author(s), under exclusive license to Springer Nature Switzerland AG 2022
P. Nicopolitidis et al. (eds.), *Advances in Computing, Informatics, Networking and Cybersecurity*, Lecture Notes in Networks and Systems 289,
https://doi.org/10.1007/978-3-030-87049-2_5

producing a flood of information about an industry as critical and important as the healthcare industry [38]. There have been instances in the past where the most efficient doctors were helpless in grave situations because proper and timely data was not available. Many lives have been lost because there was a lack of proper and appropriate information on the evaluation and prediction of a disease following certain symptoms. Introduction of smart devices such as smart shirts, sensors, etc., have made it possible to monitor a patient or a disease growth at all times without requiring the patients to be near a doctor.

Since we are working with such a large amount of data, we need methods to properly use and extract the data [56]. These methods will also suggest easy ways to understand and interpret complicated data that is otherwise impossible or very difficult to comprehend. We firmly believe that by understanding the data's intricacies and underlying patterns, we would be able to decipher important and substantial information present in it and use it to facilitate better decision making by the medical practitioner, researchers, and even the patients themselves. The main purpose of this paper is to present methods and techniques that can be extremely helpful in predicting the presence or absence of any disease given the symptoms. More specifically, we are talking about decision making using the Rough-set theory.

We would like to draw the attention of the readers towards the importance and convenience that the application of techniques based on rough set methodology provides in the prediction of diseases that can become fatal if not properly diagnosed. The present study has been taken on to showcase important methods based on rough set theory that can be readily applied to any disease's information database. In this case, using the data from heart disease, the methods based on rough set theory [35, 42] can identify patterns within the data and devise rules that aid in effective decision making. The study further attempts to draw comparisons between several pre-existing machine learning tools that primarily leverage the non-rule-based algorithms in devising models for prediction [8, 19]. We chose machine learning algorithms as the comparison reference because of their popularity in the prediction domain along with the mathematical validity that these techniques provide. The subsequent sections deal with background literature, illustrate the methodology, descriptions about the techniques used, the results obtained and discussions, conclusions of the results thus obtained, and future research directions.

2   Background Literature

The healthcare industry in the past made various efforts to employ data mining tools to increase the efficiency of the predictive capability of the model. One such step in this regard was made by [32, 44]. They proposed a hybrid methodology for feature selection based on a genetic algorithm (GA), yielding the top analytic model and the best prediction of the decision variable. Interestingly, GA based techniques were exploited on each of the three exploratory methods: k-nearest neighbor (KNN), Support Vector Machines (SVM), and Artificial Neural Network (ANN). GA-SVM proved to be the best predictor with the highest accuracy [44]. CARL-sim is a graphical processing units (GPU) accelerated spiking neural network simulator, modeling spiking neurons with Spike Timing Dependent Plasticity (STDP) and homeostatic scaling that estimates heart rate from the electrocardiogram (ECG) data collected from the wearable devices. The output displayed high accuracy and low energy footprints for subjects with and without cardiac irregularities [12].

Convolutional neural networks were applied to different intracranial and electroencephalogram (EEG) datasets to yield a patient-specific seizure prediction algorithm. It used short-time Fourier transform on EEG windows to generate optimized features for each patient in order to classify preictal and interictal segments. The method was statistically found to perform better than traditional predictors; displaying high sensitivity and low false positives when compared with traditional predictors [17]. Additionally, a novel method comprising phase space reconstruction (PSR), empirical mode decomposition (EMD), and neural networks was effectively shown to produce significantly better results in classification of patients with Parkinson's disease (PD) than the other state-of-the-art methods.

Specifically, neural networks were used to distinguish between patients with PD and healthy controls on the basis of differential gait dynamics preserved in the gait features between the considered groups [61]. To increase the prediction accuracy of the presence of PD, Hierarchical Bayesian framework was employed to trace the heterogeneity among patients that was a crucial determinant of the presence of the disease. It was shown that the method based on Bayesian framework outperformed existing methods that involved complex parameter tuning and hyper-parameter estimation [62]. Recognition of daily living activities was effectively demonstrated by the use of the adaptive Bayesian inference system (BasIS): A probabilistic formulation with sequential analysis is performed to identify the gait events, which subsequently determined the state of the human body while walking.

To understand and estimate user's acceptance and willingness to buy smart technology for prevention of cardiovascular diseases, structural equation modeling was conducted with the data obtained from 212 non-hypertensive Italian individuals. Ease of use and usefulness were the key determinants that emerged from the analysis. Technology promptness, innovativeness and prevention awareness were other factors that is shown to exert influence on the decision of adopting smart technology for the prevention of heart disease [4]. Smart technology has revolutionized the health care sector in unimaginable ways. With miniaturization of sensors, various devices that are used daily such as watches, glasses and even clothes can be made smart by fixing the small sensors that records pulses, heart rate, blood pressure, cholesterol levels, etc. This information is then relayed to the medical experts that can recommend proper and timely diagnosis [10]. With the advent of smart phones, there has been an abundance in the applications that promise to offer significant help and support to people diagnosed with dementia and Alzheimer's. To investigate the applications that are directed towards addressing the problems related to mental disorders, an extensive literature review and app search was conducted specially google play store and apple app store. The study revealed the important role these applications play for the dementia community as well as to their caregivers by providing necessary, detailed and timely advices and diagnosis thus confirming the utility of smart technology in health care domain [60].

In another study on predicting retinopathy in diabetic patients, a Type-2 fuzzy regression model was suggested. This method had a two-fold advantage: It addressed the ambiguity inherent in the medical data pertaining to the diabetic patients, and the regression model dealt with the small sample size [1]. The prediction of conversion of mild cognitive impairment (MCI) to Alzheimer's disease has drawn a lot of attention in the past. Conventional classification models that require biomarkers to make predictions of the conversion suffered from some serious drawbacks of decreased accuracy and increased uncertainty. Through various discrete simulations models, hyper acute stroke systems were modeled. However, the limitations of such models to include logistics and clinical aspects of the strokes rendered these simulations of less worth.

The model studied in the current study choose domain specific framework that lead to a more pronounced understanding of hyper acute stroke systems [39]. A novel method based on sequence tree based classifier (STC) proved to be better performing than the existing methods in that it offers an optimal sequence of biomarkers that also yields that a two-sided cut offs of each biomarker that further increases prediction accuracy [54]. To alleviate the shortcoming of inaccurate and uncertain diagnosis of breast cancer, SVM based ensemble technique was employed in the prediction of breast cancer. It used the datasets elicited from the Wisconsin Breast Cancer, Wisconsin Diagnostic Breast Cancer, and the U.S. National Cancer Institute's Surveillance, Epidemiology, and End Results. Twelve SVMs were hybridized based on Weighted Area under the Receiver Operating Characteristic Curve Ensemble (WAUCE) approach. The results obtained were significantly better and offered higher accuracy in predicting breast cancer when compared to other ensemble models such as adaptive boosting and bagging classification tree [57].

To shed light on the increased usability and efficiency of Health-care domain by adopting information system was highlighted in the study where an attempt was made to showcase the utility of Healthcare information system (HIS) in improved healthcare functionality [40]. Troponin tests separated by a period of six hours are employed to detect acute myocardial infarction. To test the efficiency of an advanced version of troponin that has the potential of detecting infarction separated by a period of three hours, a discrete event simulation model was conducted on the emergency admissions following treatments with the advanced troponin [50]. The first outbreak of dengue at Madeira Island was launched by the serotype having Aedes aegypti as the vector for the virus. The impact of a future outbreak of dengue on the population was studied especially because of the potential lethal combination of two serotypes, by mathematical modeling and numerical simulations [51].

The pathway of patients suffering from cutaneous malignant melanoma was simulated using system dynamics simulation. The model produced quantified output of diagnosed patients staged by severity [25]. A general-purpose machine learning algorithm based on discriminant analysis-mixed integer program (DAMIP), was used to identify gene signatures that were instrumental in predicting the immunity provided by the vaccine and efficacy [30]. When applied to yellow fever, DAMIP demonstrated its ability to predict vaccine immunity with more than 90% accuracy. New statistical models were created with the aim of predicting the outcomes of clinical trials, testing combination chemotherapy regimens, and eventually selecting the combination chemotherapy treatments that would improve the quality of regimens tested in third phase trials.

The new models were created based on clinical datasets with an out-of-sample R^2 of 0.56 with the median overall survival and an out-of-sample area under the curve (AUC) of 0.83 [3]. In predicting the probable outcome of a disease for a patient, a new prediction technique was used, termed as isotonic prediction. This was observed to be an improvement over traditional statistical techniques used for prognosis purposes, such as Kaplan-Meiser product interval estimation and Cox's regression, where it predicts the individual survival time frame of patients [52].

Helm et al. [26] also made a vital addition to the healthcare literature. The concern about monitoring glaucoma patients during the period of their disease was discovered by integrating novel optimization techniques with dynamic state space models of trajectory of the disease progression. The algorithm proved better than the methods based on fixed-interval schedules and age-based threshold policies. Authors further added that the algorithm could be used with other ailments as well. In an interesting effort to diagnose malaria, typhoid fever, yellow fever, and dengue, authors envisaged a diagnostic tool called QAMDiagnos.

It is a model of Quantum Associative Memory (QAM) and helps the medical staff who lacks in sufficient experience and laboratory facilities. Using the tool, it is easy to detect a single infection from a poly-infection scenario [43]. In order to address the health-threatening diseases, a novel method is studied that gradually increases the generalization capabilities of a Radial Basis Probabilistic Neural Network classifier without sacrificing the sensitivity and precision. The methodology was tested on ECGs, and the results showed satisfactory performances of the classifier [2]. The identification of high risk individuals particularly related to the health conditions was effectively carried out using a prognosis driven selective sensing method. Due to the lack of sufficient data about the large number of patients, the correct and timely prognosis is hindered. An integrated approach that involved prognostic models, collaborative learning and sensing resource allocation showed tremendous potential in detecting patients with depression and Alzheimer's disease [34].

A recent work on studying the ECG patterns and deriving the cardiac dynamics is done by employing the concept of a dynamical neural learning mechanism, where the data was divided into test and training subsets. During the training phase, cardiac dynamics was extracted using a Radial Basis Function (RBF) network and was stored in constant RBF networks. Trained gait patterns were then determined by using the collection of estimators from the RBF networks. Multiple Health Information Systems (HISs) were integrated using a novel technology Department Data Depot (DDD) that demonstrated the utility of the proposed methodology in reducing the turnaround times for the radiologists at Mayo Clinic in Arizona [58]. Post-traumatic stress disorder (PTSD) is an indication of suicidal behavior that is mostly seen in the veterans. Authors developed a model that was materialized in the form of a wearable sensor that records real time heart rate data. The heart rate data was used to train some machine learning algorithms such as Naïve Bayes, Neural networks, Support vector machine and decision tree. The prediction performance was assessed using the metric area under the curve. It was suggested by the authors that it was a possibility to use the heart rate data to detect the PTSD triggers [37].

In an effort to measure the dementia degree from the locomotion data, a sensor-based technique was employed that estimates Dementia conditions from daily loco-motion data. Interestingly, the algorithm demonstrated a good predictive ability with the true positives and negatives of more than 85% [11]. In an interesting study to explore the factors responsible for the cognitive decline, certain statistical tech-niques were employed that revealed previous cognitive stroke history, night time disturbances and the presence of neuropsychiatric symptoms were the indicators of Dementia or cognitive decline. On the other hand, late onset of cognitive decline, educational level of more than 12 years and the presence of irritation marked the indication of slower cognitive decline [31].

A staggering figure of about 9.7 million US adults not receiving mental health treatment prompted the authors to study the variables causing mental disability and also assess the requirement of the sufficient medical care. Optimal hierarchical discriminant analysis revealed that several socio-economic groups have different unmet mental health treatment needs which was shown to have a prediction accu-racy of about 95% [29]. There has been immense amount of work done in the context of decision making in the screening, detecting and treatment of prostate cancer as was revealed in a detailed literature review. The methods involved number of opera-tions research tools and mathematical analysis [48]. The progression of Hepatitis B virus was calibrated using a Markov modeling process. The population mortalities was found to affect the state transitions of the Markov process [9].

Pawlak [45] had an idea that revolutionized the way uncertain or vague information is dealt with in decision theory. To be precise, it was argued by the pioneers of rough sets that we can very explicitly draw concrete conclusions from a database by investigating the patterns inscribed in it. It was suggested that every entity that is a part of the database clearly forms a granule of information that can be helpful in determining whether an element is part of a set or not, or whether it possibly is. Mathematical fundamentals of rough set theory will be laid down in the subsequent sections.

3   Methods

3.1   Classical Rough Set Approach (CRSA)

CRSA, proposed by Pawlak, is a way in which the redundant information present in a dataset is identified and treated. Ultimately, this facilitates a proper decision-making mechanism [46, 45]. Typically, there are superfluous elements in an information source that increases redundancy and complexity in the decision making process

[47]. CRSA is a relatively new approach to dealing with inconsistencies present in the dataset by grouping similar objects together on the basis of indiscernibility relation between the objects under consideration. This might appear similar to what is normally attempted in statistical techniques such as principal component analysis and factor analysis. However, CRSA has an edge over these traditional methods because it does not require statistical assumptions about the data that is being used. Thus, CRSA is much more flexible and computationally convenient. The biggest advantage, however, is that it outputs decision rules in the form of "*if–then*," which is simple to understand and is readily actionable [16].

3.2 Mathematical Preliminaries of CRSA

To work with CRSA, the dataset is presented as an information database given by $\{U_n, A_t, V_a, func\}$. U_n represents the non-empty finite set of entities. A_t indicates the attribute set which is the union of condition and decision attributes sets given as $A_t = C_o \cup Dec$, where C_o is the set of conditions features and Dec is the set of decision variables. V_a represents the values that an object assumes for each of the condition and decision attributes $a \in A$. *func* is the function represented as $func : U_n \times A_t \rightarrow V_a$ that maps the conditions A to its value. The indiscernibility relation forms the backbone of the CRSA [50]. Essentially, objects with certain attributes are grouped together. Therefore, for a given condition attribute b$\in$ A, two objects x_i and x_j are said to be indiscernible or similar if $func(x_i, b) = func(x_j, b)$. Put mathematically, the indiscernible relation is given by $R(B) = (x_i, x_j) \in U_n :$ $\forall b \in A_t, func(x_i, b) = func(x_j, b)$. The classes that are equivalent are specified by the object $[x_i]$, and with respect to R(B) is denoted as $[\chi_i]_B$ [16, 55]. This way the classes form the basic granule of knowledge, that there is always some information associated with every element in the set of objects considered for the study. Keeping these similar groups as the foundation, similar sets of objects are approximated by lower and upper approximations. This will be explained in the following section.

CRSA operates on the dataset and treats the inconsistencies present in the dataset because inconsistent data interferes with the correct inferences. The decision about whether the data is consistent or inconsistent depends on the decision class or set Y. If the objects can surely be attributed as belonging to the decision class Y, we say the information is consistent and can thus render crisp decision rules. However, if there is any ambiguity about the belonging of an element to decision class Y, then inconsistencies creep into the data. This is handled by CRSA by approximating the sets of such objects. The group of entities that can be said to be included in a particular set with utmost certainty are said to be in lower approximation, whereas entities that are possibly included in a set are said to be a part of upper approximation, which, in terms of set notation, can be written as $R_{Blow}(Y) = x \in U : [x]_B \subseteq Y$ and $R_{Bupp}(Y) = x \in U : [x]_B \cap Y \neq \phi$, respectively.

There are also sets of objects that can neither be grouped in absolutely belonging to set Y nor as not belonging to the set Y. These sets are referred to as the boundary regions given in set notation as follows

$$R_{Bupp}(Y) - R_{Blow}(Y) = BND_B(Y)$$

In other words, these are the elements that possibly belong to set Y but not certainly [24].

The way to assess the authenticity of the output given by CRSA is to look at the measure known as *quality of approximation*. Let us consider a set of decision attributes given by $M = Y_1, Y_2, Y_3, \ldots, Y_n$. Then, the quality of approximation is the ratio of the number of objects in the lower approximation by the decision class Y_i to the total number of objects in the universe of discourse [45, 55].

3.3   Dominance-Based Rough Set Approach (DRSA)

In 1982, rough set theory emerged in response to the difficulty of interpreting vague and inconsistent data [19]. Though immensely useful in an uncertain environment, rough set theory's usefulness is limited when dealing with inconsistencies in preference ordered data. Intense effort and thought brought about a new technique which evolved from the rough set theory that eliminated inconsistencies present in the preference ordered data [22]. The replacement of the similarity relationship in rough set theory with the preference-ordered dominance relationship gives rise to DRSA. The output released with this algorithm is extremely lucid and easy to comprehend [5]. The detailed information about the DRSA will be presented in the subsequent paragraphs and has been leveraged from [5, 23 20 21 22].

For the algorithmic viewpoint, the data collected is arranged in the form of a rectangular array where the rows represent the entities being observed, and the columns represent the condition attributes, the conditions on which the entities will be evaluated [14]. When these condition attributes are in a preference-ordered domain, they are referred to as criteria. The entries in each cell represent the actual evaluation the entity receives on every attribute. Mathematically, the information is in the form of a table that is given by $InfSys = \{U_n, Q_t, V_a, func\}$, where finite set of entities belongs to U_n, finite set of attributes or criteria belong to $Q_t = q_1, q_2, \ldots, q_m, V_{q_t}$ is the domain of the criteria q, $V_a = \cup_{q \in Q_t} V_{q_t}$ and $func : U_n \times V_a \to Q_t$ is a total function such that $func(x, q_t) \in V_{q_t}$ for each $q_t \in Q_t$; $x \in U_n$, called the information function.

3.3.1   Rough Approximation by Means of Dominance Relation

Considering an outranking relationship $\succeq_{q_t}$, we have the following

$$x \succeq_{q_t} y \Rightarrow func(x, q_t) \succeq func(y, q_t) \tag{1}$$

The above relation holds for gain type situations that imply a "the more the better" kind of a scenario. The cost type where "the less the better" is obtained by negating the values of V_q.

3.3.2 Dominance Relation

When entity x is dominating entity y, it means that x is better than y on every $P \subseteq C$ represented as $x D_P y$, $x \succcurlyeq_{q_t} y$, $\forall q_t \in P$. Given $P \subseteq Cond$ and $x \epsilon Univ$, let

$$PD^+(x) = (y \epsilon Univ : y D_P x)$$
$$PD^-(x) = (y \epsilon Univ : x D_P y) \tag{2}$$

represent the set of elements dominating and being dominated for each $x \in Univ$, respectively.

Next, we consider the representation and nomenclature of the decision classes. The way of depicting decision classes is by the notation Cl_t, which, expressed mathematically, is $Cl_t = \{x \in Univ : func(x, d) = t\}$. The x should be assigned to one decision class only, and the Cl_t classes are all preference-ordered. Thus, the resultant decisions are of the form "at least" and "at most," shown as

$$Cl_t^{\succcurlyeq} = \bigcup_{s \succcurlyeq t} Cl_s$$
$$Cl_t^{\preccurlyeq} = \bigcup_{s \preccurlyeq t} Cl_s \tag{3}$$

respectively for each $t \in T$.

The backbone of investigating inconsistencies in the dataset is assisted by approximations of the "at least" and "at most" decision classes, given as $P_{low}(Cl_t^{\succcurlyeq})$, $P_{upp}(Cl_t^{\succcurlyeq})$, $P_{low}(Cl_t^{\preccurlyeq})$ and $P_{upp}(Cl_t^{\preccurlyeq})$, respectively. The various approximations are given as follows:

$$P_{low}(Cl_t^{\succcurlyeq}) = (x \in Univ : PD^+(x) \subseteq Cl_t^{\succcurlyeq})$$
$$P_{upp}(Cl_t^{\succcurlyeq}) = \bigcup_{x \epsilon Cl_t^{\succcurlyeq}} PD^+(x)$$
$$P_{low}(Cl_t^{\preccurlyeq}) = (x \in Univ : PD^-(x) \subseteq Cl_t^{\preccurlyeq}) \tag{4}$$

The boundaries (doubtful regions) produced by P are given by

$$PBound(Cl_t^{\succcurlyeq}) = P_{low}(Cl_t^{\succcurlyeq}) - P_{upp}(Cl_t^{\succcurlyeq})$$
$$PBound(Cl_t^{\preccurlyeq}) = P_{low}(Cl_t^{\preccurlyeq}) - P_{upp}(Cl_t^{\preccurlyeq}) \tag{5}$$

Since these approximations are uncertain quantities, we define the accuracy of the approximations as follows:

$$\alpha_p(Cl_t^{\succeq}) = \frac{|P_{low}(Cl_t^{\succeq})|}{|P_{upp}(Cl_t^{\succeq})|}$$

$$\alpha_p(Cl_t^{\preceq}) = \frac{|P_{low}(Cl_t^{\preceq})|}{|P_{upp}(Cl_t^{\preceq})|} \tag{6}$$

Also, the quality of approximation is given by

$$\gamma_p(Cl) = \frac{|Univ - \bigcup_{t\in[2,...,n]} PBoun(Cl_t^{\succeq})|}{|Univ|} = \frac{|Univ - \bigcup_{t\in[1,...,n-1]} PBoun(Cl_t^{\preceq})|}{|Univ|} \tag{7}$$

This represents the ratio that describes the number of P—correctly classified objects to the total number of objects considered and shows the efficiency of the classification result using the technique. The subset $P \subseteq C$ such that $\gamma_p(Cl) = \gamma_c(Cl)$ is called a reduct of C with respect to Cl. Again, a data table may have more than one reduct. Since we may be having more than one reduct of a particular data set, we take the common elements of the reducts and refer to the set of the common elements as the core set [20].

3.3.3 The Variable Consistency (VC) DRSA

When the strictness of the dominance relation is relaxed in DRSA, a limited number of inconsistent objects are included in the lower approximation of the sets of objects according to the object consistency level threshold $c \in [0, 1]$. That is equivalent to saying that an object x belongs to the lower approximation of the upward union if at least $l \times 100\%$ of all the objects dominating x also belong to the lower approximation.

Mathematically, the lower approximations of VC-DRSA is represented as

$$Plow_c(Cl_t^{\succeq}) = set\ of\ all\ elements\ \in Cl_t^{\succeq} : \frac{|PD_+(x) \cap P(Cl_t^{\succeq})|}{|PD_+(x)|} \geq cPlow_c(Cl_t^{\preceq})$$

$$= set\ of\ all\ elements\ \in Cl_t^{\preceq} : \frac{|PD_+(x) \cap P(Cl_t^{\preceq})|}{|PD_-(x)|} \geq c \tag{8}$$

3.3.4 Decision Rules

It is necessary to understand the origination of decision rules from all the afore-mentioned techniques as it will assist in informed decision making. We resort to

MODLEM and VC-MODLEM techniques for the induction of decision rules from the CRSA and DRSA methods. With these decision rules, we would be able to classify any object with reasonable accuracy, resulting in an appreciably correct inference.

As already mentioned, the decision rules are of the form "*if A then D.*" A represents the set of condition attributes wherein if more than one attribute is considered, then logical union is considered, and D represents the decision classes.

For traditional CRSA the decision rules take the form of:

If $func(x, ai) = (V_{a_i})$ then $x \in Y_t$, where V_{a_i} represents the value that the object takes for the attribute, Y_t represents the decision class, and *func* represents the function.

The decision rules from DRSA can be complicated, since they incorporate the attributes as well as the criteria and the upward and downward unions of the decision classes. There are different decision rules for the lower and upper approximations of the upward and downward unions of the decision classes, as shown below:

$$\text{If } (\cap_i (func(x, b_i) \preccurlyeq V_{a_i}) \cup (\cap_j (func(x, a_j) = V_{a_j}) \text{ then } x \in Cl_t^{\preccurlyeq}$$

where b_i belongs to criteria and a_j belongs to attributes. On similar grounds, we obtain the decision rules for the downward union of the decision classes where $Cl_t^{\succcurlyeq}$ is considered in place of $Cl_t^{\preccurlyeq}$.

The decision rules for the boundary regions are given as:

$$If \cap_i (func(x, a_i) \succcurlyeq V_{a_i}) \cap (\cap_i (func(x, b_i) \preccurlyeq V_{b_i}) \cap (\cap_i (func(x, r_k) = V_{r_k})$$
$$then$$
$$x \in Y_t \cup Y_{t+1} \cup \ldots, \cup Y$$

The evaluation of the decision rules is established using MODLEM and VC-DOMLEM methods for CRSA and VC-DRSA, respectively. The advantage of MODLEM and VC-DOMLEM is the way they bring out decision rules. The two methods ensure the elimination of the redundant attributes and thus assist in data dimensionality reduction. Fundamentally, it uses an iterative procedure where, at each iteration, a certain number of attributes are chosen to create the decision rules, following that in the next iteration those attributes are excluded and the next set of attributes is chosen. And the process is repeated until all the attributes are utilized. However, once an attribute is chosen for a decision rule, it can still be taken for other rules [18].

4   Problem Dataset

The dataset used to conduct the study has been taken from UCI repository which discusses the heart disease of humans [15]. A detailed description of the attributes considered is shown in Appendix.

4.1 Data Preprocessing and Discretization

The dataset contained a set of 14 attributes with 303 instances. There were certain missing values for some instances for the attribute "*thal.*" We suggested a value of 0 for those missing values so that they are not ascribed to unnecessarily wrong information. After imputing responses to the missing values, we subjected the data information to the process of discretization.

Through discretization, the continuous variables are converted to discrete or ordinal variables. Discretization is necessary particularly when using classification techniques. Moreover, when using DRSA based rule induction, it is customary to have the variables in an ordinal category [20]. The continuous variables in our dataset are age, restbps, chol, thalach, and oldpeak.

Since the dataset contained observations on people with the mean age of about 54 years and a small standard deviation, we kept a binary variable for the people who are above 54 and below 54 years of age. The other continuous variables are medical related phenomena for which the information was gathered from the medical related sites and then corresponding ratings were assigned. This was done to be informed about the correct ranges of the values of the variables that would assist in correct decision making.

The subsequent sections focus on displaying the methods employed to derive decision rules from the CRSA and VC DRSA techniques, followed by a brief description of the comparative machine learning techniques and their performance metrics.

4.2 Rule Induction from CRSA and VC DRSA

We used MODLEM algorithm for deriving decision rules from the CRSA and used VC-DOMLEM for the DRSA methods. MODLEM works closely in relation with the rough set theory. MODLEM exploits the sequential covering algorithm, which searches for the reduced set of attributes without the need to discretize the data, which is an important step in the rough set analysis. In principle, MODLEM treats the nominal and numeric attributes alike [27, 36]. VC-DOMLEM is carried out in jMAF (Dominance-based Rough Set Approach Data Analysis Framework). It operates on ordinal data such as preference-ordered data, exploiting the dominance relationship between objects in place of the indiscernibility relationship. Because of this, VC-DOMLEM is an exquisite tool for carrying out the analysis of DRSA.

5   Brief Overview of Machine Learning Tools

5.1   Statistical Classifier C4.5

The Id3 algorithm is exploited in this method of classification in machine learning. The foundation of the technique lies on information entropy. We choose a p-dimensional vector representing the attributes as our training data. The node that is split based on maximum information gain is the criteria for splitting [7, 13].

C4.5 is implemented using the J48 classifier in WEKA 3.8.3 following the configuration: {weka.classifiers.trees.J48 -C 0.25 -M 2}, where C 0.25 is the confidence factor of 0.25 which is used for pruning purposes and M 2 represents the minimum number of instances per leaf which is 2 in this case.

5.2   Random Forest

Random forests work on the premise that many decision trees are constructed for a given classification problem. The output is the mode of the decision classes when we are dealing with a classification problem or a mean prediction of the decision classes when the model under consideration is a regression analysis [53]. Random forest is a tool for addressing the problem of overfitting data when working with machine learning tools.

Random forest is accomplished using the following configuration in WEKA: {weka.classifiers.trees.RandomForest -P 100 -I 100 -num-slots 1 -K 0 -M 1.0 -V 0.001 -S 1}, where P 100 represents the size of each bag as a percentage of training set size, I 100 is the number of iterations performed, num-slots 1 are the number of execution slots or threads to use for constructing the ensemble, K sets the number of features to consider where the value less than 1 implies Log M + 1 inputs where M is the number of inputs, S is the random number seed that is to be used and V is the minimum variance for split that is it sets the minimum numeric class proportion of train variance for split (default is 0.001).

5.3   Naïve Bayes

Naïve Bayes is a collection of algorithms that assigns labels to the problem instances. The labels assigned to the entities are drawn from a finite feature set and it is assumed that the features selected from a highlighted set are independent of each other. It is important to note that Naïve Bayes is not a single algorithm but a collection of algorithms [28, 33, 41].

5.4 Support Vector Machines (SVM)

Support vector machines (SVMs) accomplish the regression and classification through machine learning theory. The biggest advantage of SVMs is that it identifies the relevant, important variables using the supervised learning methods where the classifier is made to learn so the accuracy is maximized and overfitting is avoided [49]. We implemented John Platt's sequential minimal optimization algorithm for training a support vector classifier using the following configuration: {weka.classifiers.functions.LibSVM -S 0 -K 2 -D 3 -G 0.001 -R 0.0 -N 0.5 -M 40.0 -C 250.0 -E 0.001 -P 0.1 -model}, where P 0.1 represents the epsilon for round-off error, M represents the fitting of logistic models to SVM outputs, S is the random number seed for cross validation, K is the Kernel to use, D is the number of decimal after the outputs in the model, R sets the ridge in the log-likelihood, G is the gamma for RBF Kernel, N represents whether to normalize, standardize or neither of the two, M fits calibration models to SVM outputs, E is the exponent value, C is the cache size for the poly-kernel and model is the SVM chosen for the classification.

5.5 Logistic Regression

Logistic regression is a member of the nonlinear family of regressions where the regressand is dichotomous or binary in nature and regressors can be of any type (i.e. continuous or discrete). There are other forms of logistic regressions available, depending upon whether the dependent variable is nominal, ordinal, or categorical. In WEKA it is implemented using the configuration: {weka.classifiers.functions.Logistic -R 1.0E-8 -M -1 -num-decimal-places 4}, where R 1.0E-8 is the ridge value in the log-likelihood, M represents the maximum number of iterations to be performed and num-decimal-places 4 represents the number of decimal places for the output numbers.

5.6 Multilayer Perceptron

To be able to solve nonlinearly separable problems, a number of neurons are connected in layers to build a multilayer perceptron. Each perceptron is used to identify small linearly separable sections of the inputs. Outputs of perceptrons are combined into another perceptron to produce the final output. The hard-limiting (step) function used for producing the output prevents information on the initial set of inputs from flowing on to inner neurons. To solve this problem, the step function is replaced with a continuous function—usually the *sigmoid function* [17]. In a multilayer perceptron, the neurons are arranged in an input layer, an output layer, and one or more hidden layers. The configuration that is employed in WEKA to carry out the multilayer perceptron analysis is given by the following: {weka.classifiers.functions.MultilayerPerceptron -L 0.3 -M 0.2 -N 500 -V 0 -S 0 -E 20 -H a}, where L 0.3 is the learning rate for weight updates, M 0.2 is the momentum applied to the weight updates, N 500 represents the number of epochs to train through, E 20 represents validation threshold that is used to terminated validation testing, S 0 is used to set the seed to initialize the random number generator, H a represents the number of hidden layers and V is the minimum variance for split that is it sets the minimum numeric class proportion of train variance for split.

6 Performance Evaluation Metrics

To evaluate the performance of various models we describe some metrics below. These metrics have proved to suggest a sound model for addressing the problem at hand [59]. The metrics are used for all models despite the nature of the technique used. However, due to the binary classification nature of the problem, these metrics are much easier to understand and interpret and, therefore, their use in other classifications should not be ruled out [19].

Confusion Matrix

To completely appreciate the confusion matrix, it is essential to understand certain terminologies as mentioned below:

TrPo: true positives: instances shown positive that indeed are positive
FaPo: false positives: instances shown positive that indeed are positive
TrNe: true negatives: instances shown negative that indeed are negative
FaNe: false negatives: instances shown negative that indeed are positive.
So, confusion matrix is defined as in Table 1 where Predicted is given by Pred, Actual is given by Act, Positive by POS, and Negative by NEG.

Sensitivity

Sensitivity is also known as the True Positive Rate (TPR) or the recall value. It demonstrates how successfully the classifier does not miss a positive case.
Mathematically, it is represented as

$$Sensitivity = \frac{TrPo}{TrPo + FaNe}$$

Table 1 Confusion matrix

Pred	Act	
	POS	NEG
POS	TrPo	FaPo
NEG	FaNe	TrPo

Specificity

The fraction of false positives yielded by the classifier is given by specificity, also known as the true negative rate, represented as shown below

$$Specificity = \frac{TrNe}{FaPo + TrNe}$$

Precision

When the classifier results in less false positives, it is considered to be more precise. Precision is given mathematically by

$$Precision = \frac{TrPo}{TrPo + FaPo}$$

Accuracy

The fraction of correctly classified objects is called the accuracy of the classifier given by

$$Accuracy = \frac{TrPo + TrNe}{TrPo + FaPo + TrNe + FaNe}$$

Receiver Operating Characteristics (ROC) Curve

The classifiers' potential to discriminate the entities contained in different decision classes is measured in the form of Receiver Operating Characteristic (ROC) Curve or as an approximation of the Area Under the Curve (AUC). It is a plot of sensitivity versus 1—specificity in the range 0–1, where the straight line from (0, 0) to (1, 1) represents no discriminatory power and a line from (0, 0) to (0, 1) to (1, 1) represents perfect discrimination [7].

The cross-validation procedure is utilized to verify the accuracy of the classifiers. In particular, the tenfold method is used for estimating the generalized error of the prediction model.

7   Results

In this section we will present the results obtained using the aforementioned algorithms.

Table 2 lists the correct and incorrect classified instances for all methods described above. Amongst the decision and rule-based methods, MODLEM and DRSA produce similar results, whereas in the non-rule based category, SVM outperforms other methods.

Table 2 List of correctly and incorrectly classified instances

	Correctly classified instances (% of total)	Incorrectly classified instances (% of total)
Rule based methods		
Rough sets based MODLEM	245 (80.85)	58(19.14)
DRSA	243 (80.20)	60 (19.20)
VC DOMLEM (consistency factor0.6)	201 (66.34)	36 (11.88)
Non rule based classification methods		
C4.5	73 (70.87)	30 (29.13)
Logistic Regression	251 (82.84)	52 (17.16)
Naïve Bayes	87 (84.47)	16 (15.53)
Random forest	80 (77.67)	23 (22.33)
SVM	88 (85.44)	15 (14.56)
Multilayer Perceptron	251 (82.8)	52 (17.2)

The various performance metrics for the methods are listed in Table 3.

By assessing the quality of the outputs, we observe that the approximation quality comes out to be 0.9835 for CRSA, 0.8555 for DRSA, and 0.9210 VC-DRSA. The object consistency parameter was considered at the value of 0.6. The approximation quality of an algorithm gives an indication of the appropriateness of the chosen attributes for the classification process. A higher approximation quality indicates

Table 3 Performance metrics

	Precision	Sensitivity (recall)	AUC	K-coefficient
Rule based methods				
Rough sets based MODLEM	0.813	0.809	0.7720	0.5326
DRSA	0.855	0.868	0.7538	0.4481
VC DRSA (consistency factor 0.6)	0.878	0.901	0.7492	0.5501
Non rule based classification methods				
C4.5	0.700	0.709	0.6219	0.3552
Logistic regression	0.825	0.828	0.7749	0.5621
Naïve Bayes	0.845	0.845	0.8292	0.5963
Random forest	0.774	0.777	0.7191	0.4870
SVM	0.858	0.854	0.8109	0.6047
Multilayer perceptron	0.828	0.828	0.8750	0.5718

that the chosen attributes for a given classifier performs well in the classification process.

We obtained 57 rules from the MODLEM CRSA approach by employing the entropy measure, 39 rules from monotonic VC-DOMLEM DRSA approach, and 45 rules from VC DRSA. The core obtained using the MODLEM algorithm comprises 9 elements instead of the original 13 elements as is shown in the outputs. On the other hand, with DRSA we obtained all the attributes as the core. We observe similar performance metrics from all rule based methods with VC-DRSA performed best while SVM performs better than the other methods in the non-rule based category.

For SVM, we chose the parameters gamma and cost coefficient as 0.001 and 250. Since the software jMAF used for DRSA and VC-DRSA algorithms does not output Cohen's kappa coefficient (K-coeff.), we calculated it using excel and the corresponding K-coeff. is shown in Table 3. The highest kappa value is obtained for the SVM method, and suggests a decent inter-rater agreement for the items. We also observe that the value of the K-coeff. for Logistic regression, Naïve Bayes, and VC-DRSA are close to SVM. However, the worst value of the K-coeff. is shown by C4.5, accompanied by only marginally better values of DRSA and Random forest. The area under the ROC curve is highest for the multilayer perceptron method with the value of 0.8750, followed very closely by Naïve Bayes and SVM with the values 0.8292 and 0.8109, respectively.

8   Discussion

If we assess the classification ability of the classifiers in terms of the area under the curve (AUC) values obtained from ROC, then we observe all classifiers perform satisfactorily [6]. The ROC curves for various classifiers are shown in Fig. 1.

The best performance of the non-rule-based methods in terms of AUC values seems to be a multilayer perceptron with the value of 0.8750, and the worst performance is C4.5 with the value 0.6219. The rule-based methods, namely rough sets and DRSA, seem to be performing reasonably well in classification with the values of AUC being 0.7720 and 0.7538. We also see that VC-DRSA outperforms DRSA in terms of performance metrics. This is explained by a better quality of approximation of 0.922, as compared to 0.8555 for DRSA. This happens primarily because the conditions of including the objects in the lower approximation is relaxed with the consistency factor of 0.6. The quality of approximation of CRSA is better than VC-DRSA with the value of 0.9835, yet, it underperforms compared to VC-DRSA, as is reflected from the higher sensitivity and AUC values of VC-DRSA. In fact, if we consider all the metrics together to gauge the performance of the classifier, we see that the Naïve Bayes algorithm performs best in the non-rule based category and VC-DRSA is the best performer in the rule based category. However, we notice the difference between the best performers is not strikingly large, as is seen in Fig. 2. This indicates that we obtain comparable results with both methods on average. The rule based methods should have an edge in the choice for classification models

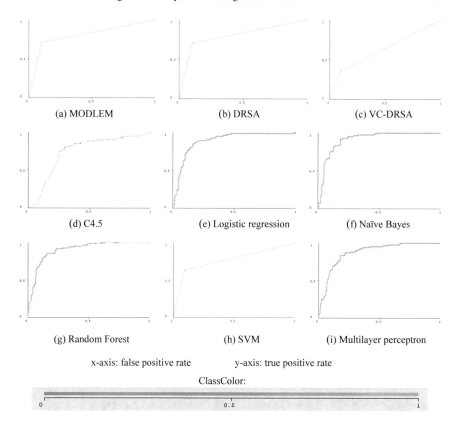

Fig. 1 Receiver operating characteristics (ROC) curves

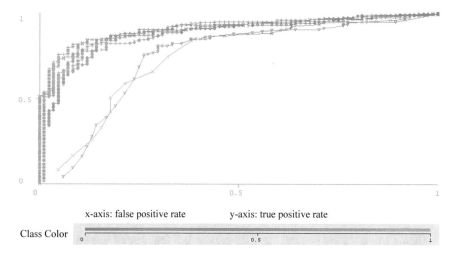

Fig. 2 Consolidated ROC curves of the classifiers

because it does not consider certain strict assumptions necessary for analysis with non-rule based methods. Figure 2 illustrates that almost all classifiers demonstrate good classification performance.

The ease in understandability and accessibility of the decision rules obtained from rough sets also makes it a superior choice over the non-rule-based methods.

The main purpose of this paper is to showcase the methods based on Rough set theory, which have been validated quantitatively as illustrated in the afore mentioned sections [53]. It can be time consuming to follow the pre-specified protocol in disease diagnosis. It can also, in some cases, become chaotic due to the multitude of tests, non-availability of suitable testing equipment, scarcity of staff, and most often, the time constraints. It is in these situations that we propose using data mining methods namely C4.5, logistic regression, Naïve Bayes, Random Forest, SVM and Multi-layer perceptrons. At least, the presence of some specific symptoms can point medical professionals towards the presence of specific diseases that are mapped in the models that we used. We also think that the algorithms presented can be recreated and coded in the form of downloadable apps that can prove handy to the healthcare professionals and the patients, who would be given a preliminary diagnosis based on their symptoms. In this case, they will be notified by the potential presence of the disease even before seeing the doctor. This would possibly probe patients to seek medical attention before their condition worsens, preventing them from serious consequences.

8.1 Applicability of Decision Rules in Detecting Heart Disease

The results obtained from the rough sets method and its preference-based variant dominance-based rough sets can be used as a plug-and-play software for detecting and predicting the presence of heart disease. To exemplify the outputs, a few rules are mentioned below that were obtained using the rough sets methodologies.

if sex $= 0$ & ca $= 3$ then num $> = 2$
if exang $= 1$ & ca $= 3$ then num $> = 2$
if age $= 1$ & slope $> = 2$ & ca $= 3$ then num $> = 2$
if sex $= 0$ & ca $= 3$ then num $> = 2$
if oldpeak $> = 4$ & ca $= 2$ then num $> = 2$
if sex $= 0$ & ca $= 3$ then num $> = 2$
if trestbps $> = 4$ & slope $> = 2$ & ca $= 2$ & thal $> = 6$ then num $> = 2$
if age $= 1$ & cp $< = 3$ & thal $< = 3$ then num $< = 1$
if exang $= 0$ & slope $< = 1$ & ca $= 0$ then num $< = 1$.

The first rule should be read as: if the patient is a female and there are 3 vessels colored by fluoroscopy, the patient is likely to have heart disease. Similarly, the last rule suggests: if there is no exercise-induced angina, an upslope of the peak exercise segment with no colored vessels, then the patient does not have heart disease.

We observe that the rules are an easy way to know which factors are potentially causing heart disease, and it serves as a tool to determine the future occurrence of heart disease given these conditions. Further, the rules exploit the information present in the database and thus the rules are intuitive to understand. Since the database is appreciably large, we see the information contained in the database as a representative picture of the situation that we are trying to model in our problem. Also, rough set methods allow us to figure out which factors are essential to deriving rules that are present in all the rules obtained. These essential factors are termed as the core of the information table. Since rule-based methods clearly and explicitly demonstrate how each variable is contributing in the decision making, the methods are much more accessible and transparent than the non-rule-based methods.

9 Conclusions

The current study attempts to draw comparisons between the rule based and non-rule-based classification methods in determining the presence or absence of heart disease. Clearly, the non-rule based classifiers have demonstrated a greater ability in accomplishing the aforementioned objective than the rule based methods. However, we believe that the marginal performance difference between the two sets of methods can be ruled out by the benefits the rule based methods offer over non-rule based methods. The application of the above-mentioned algorithms was found to be consistent with the findings of [18], where the CRSA and DRSA were employed for the classification purposes in the case of hospice referral. The current study creates opportunities for exploration in machine learning to obtain more intuitive results by considering the rule based and non-rule methods in conjunction rather than separately. The current data table used for the analysis consisted of just a few variables. If we can collect more information on some more relevant data points for the patients then we firmly believe that the results will look more precise and relevant. Collecting this information is one part of our future research.

We would like to enlist certain limitations the case poses. For all classification purposes, we assumed a binary output of either the existence or non-existence of heart disease. This assumption is too restrictive—the intensity of heart disease varies in different levels of fatality. Next, all the non-rule-based classification methods assume a linear relationship between the conditions and the decision variables, which is too realistic of an assumption to be true. Finally, there could have been more data mining methods used to conduct the comparative study that might have been more useful in establishing the findings.

The classifiers in this study displayed reasonable performance in the classification of the patients with heart disease. This is demonstrated by performance metrics such as precision, sensitivity, and AUC. The central theme of the current work revolves around the application of rule-based methods in the classification process. In this study, we applied CRSA, variable consistency rough set approach, and DRSA in classifying patients with heart disease. Variable consistency dominance-based approach performs better because it considers a high variance in the patients' characteristics that are considered for the study. Hence, improved performance by increasing the variance in the characteristics of patients is a subject for further research. Another scope for future research is building an integrated framework where the non-rule-based classifiers may be combined or integrated with the rule based classifiers. This would increase the accuracy of the classification process by analyzing the granularity of the information contained in the data in much more depth.

Though the current study focuses on predicting the presence of heart disease that is considered critical, it clearly has the potential of dealing with diseases that are not readily detectable and not easily diagnosed. Therefore, further research in those categories of diseases is highly recommended.

Acknowledgements The data used in this study are third party and are publicly available from the UCI repository of machine learning databases: http://archive.ics.uci.edu/ml/datasets/Heart+Dis ease.

Competing Interests The authors declare that they have no competing interests.

Appendix: Description of the Variables in the Information Table

Variable name	Description	Values
Age	Age in years	1: > 54 0: otherwise
Sex	Sex	1: male 0: female
cp	Chest pain type	1: typical angina 2: atypical angina 3: non-anginal pain 4: asymptomatic
trestbps	Resting blood pressure (in mm Hg on admission to the hospital)	0: 90–120 (normal) 4: 70–90 (low) & 130 - above (high)
chol	Serum cholesterol in mg/dl	0: 125–200 (normal) 4: otherwise
fbs	Fasting blood sugar > 120 mg/dl	0: False 1: True
restecg	Resting electrocardiographic results	0: normal 1: having ST-T wave abnormality 2: showing hypertrophy

(continued)

(continued)

Variable name	Description	Values
thalach	Maximum heart rate achieved	3: in the range of 140–179 AND 40–54 4: > = 180 AND < 49
exang	Exercise induced angina	0: No 1: Yes
oldpeak	ST depression induced by exercise relative to rest	> 0.5 = 1, > 1 = 3, > 2 = 4
slope	The slope of the peak exercise ST segment	1: up sloping 2: Flat 3: down sloping
ca	Number of major vessels (0–3) colored by fluoroscopy	–
thal	Defect	3: normal 6: fixed defect 7: reversible defect
num	Diagnosis of heart disease (angiographic disease status)	1: absence 2: presence

References

1. Bajestani, N.S., Kamyad, A.V., Esfahani, E.N., Zare, A.: Prediction of retinopathy in diabetic patients using type-2 fuzzy regression model. Eur. J. Oper. Res. **264**(3), 859–869 (2018)
2. Beritelli, F., Capizz, G., Lo Sciuto, G., Napoli, C., Woźniak, M.: A novel training method to preserve generalization of RBPNN classifiers applied to ECG signals diagnosis. Neural Netw. **108**, 331–338 (2018). https://doi.org/10.1016/j.neunet.2018.08.023
3. Bertsimas, D., O'Hair, A., Relyea, S., Silberholz, J.: An analytics approach to designing combination chemotherapy regimens for cancer. Manag. Sci. **62**(5), 1511–1531 (2016)
4. Bettiga, D., Lamberti, L., Lettieri, E.: Individuals' adoption of smart technologies for preventive health care: a structural equation modeling approach. Health Care Manag. Sci., 1–12 (2019)
5. Blaszczynski, J., Greco, S., Slowinski, R.: Multi-criteria classification—a new scheme for application of dominance-based decision rules. Eur. J. Oper. Res. **181**, 1030–1044 (2007)
6. Bradley, A.P.: The use of the area under the ROC curve in the evaluation of machine learning algorithms. Pattern Recogn. **30**(7), 1145–1159 (1997)
7. Caruana, R.,Niculescu-Mizil, A.: An empirical comparison of supervised learning algorithms. In: Proceedings of the 23rd International Conference on Machine Learning, pp. 161–168. ACM (2006)
8. Castro-Lopez, O., Lopez-Barron, D.E., Vega-Lopez, I.F.: Next-generation heartbeat classification with a column-store DBMS and UDFs. J. Intell. Inf. Syst. (2019). https://doi.org/10.1007/s10844-019-00557-w
9. Chen, J.H., Chen, S.Y., Luh, H.P., Chien, R.N.: Modeling chronic hepatitis B virus infections with survival probability metrics. Oper. Res. Health Care **12**, 29–42 (2017)
10. Chen, T.C.T., Chaovalitwongse, W.A., O'Grady, M.J., Honda, K.: Smart technologies for improving the quality of mobile health care. Health Care Manag. Sci., 1–2 (2019)
11. Cheng, C., Yang, H.: Multi-scale graph modeling and analysis of locomotion dynamics towards sensor-based dementia assessment. IISE Trans. Healthc. Syst. Eng., 1–18 (2018) (in press)

12. Das, A., Pradhapan, P., Groenendaal, W., Adiraju, P., Thilak Rajan, R., Catthoor, F., Schaafsma, S., Krichmar, J.L., Dutt, N., Hoof, C.V.: Unsupervised heart-rate estimation in wearables with liquid states and a probabilistic readout. Neural Netw. **99**, 134–147 (2018). https://doi.org/10.1016/j.neunet.2017.12.015

13. Disha, T., Jeevan, N., Varsha, N.J., Kavi, M.: Terrorism analytics: learning to predict the perpetrator. In: International Conference on Advances in Computing, Communications and Informatics (ICACCI) (2017)

14. Dominance Based Rough Set Approach: Data Analysis Framework. Accessed Oct 2018. http://www.cs.put.poznan.ple/jblaszczynski/Site/jRS_files/jMAFmanual.pdf

15. Dua, D., Karra Taniskidou, E.: UCI Machine Learning Repository. University of California. School of Information and Computer Science, Irvine, CA (2017). http://archive.ics.uci.edu/ml

16. Dubois, D., Prade, H.: Foreword. In: Pawlak, Z. (ed.) Rough Sets: Theoretical Aspects of Reasoning About Data. Kluwer, Dordrecht, The Netherlands (1991)

17. Duy Truong., N., Duy Nguyen, A., Kuhlmann, L., Reza Bonyadi, M., Yang, J., Ippolito, S., Kavehei, O.: Convolutional neural networks for seizure prediction using intracranial and scalp electroencephalogram. Neural Netw. **105**, 104–111 (2018). https://doi.org/10.1016/j.neunet.2018.04.018

18. Gil-Herrera, E., et al.: Rough set theory based prognostic classification models for hospice referral. BMC Med. Inf. Decis. Making. **15**, 98 (2015). https://doi.org/10.1186/s12911-015-0216-9

19. Goldberg, D.E., Holland, J.H.: Genetic algorithms and machine learning. Mach. Learn. **3**(2), 95–99 (1988)

20. Greco, S., Matarazzo, B., Slowinski, R.: Rough sets methodology for sorting problems in presence of multiple attributes and criteria. Eur. J. Oper. Res. **138**, 247–259 (2002)

21. Greco, S., Matarazzo, B., Slowinski, R.: Rough set analysis of preference-ordered data. In: Alpigini, J.J., Peters, J.F., Skowron, A., Zhong, N. (eds.) Rough Sets and Current Trends in Computing, pp. 44–59. Springer, Berlin (2002)

22. Greco, S., Matarazzo, B., Slowinski, R.: A new rough set approach to evaluation of bankruptcy risk. In: Zopounidis, C. (ed.) Rough Fuzzy and Fuzzy Rough Sets, pp. 1121–1136. Kluwer, Dordrecht (1998)

23. Greco, S., Matarazzo, B., Slowinski, R., Stefanowski, J.: An algorithm for induction of decision rules consistent with dominance principle. In: Ziarko, W., Yao, Y. (eds.) Rough Sets and Current Trends in Computing. LNAI. 2005, pp. 304–313. Springer, Berlin (2001)

24. Grzymala-Busse, J.W.: Knowledge acquisition under uncertainty—a rough set approach. J. Intel. Rob. Syst. **1**(1), 3–16 (1988). Grzymala-Busse, J.W.: Managing Uncertainty in Expert Systems. Kluwer, Dordrecht, The Netherlands (1991)

25. Hallberg, S., Claeson, M., Holmström, P., Paoli, J., Larkö, A.M.W., Gonzalez, H.: Developing a simulation model for the patient pathway of cutaneous malignant melanoma. Oper. Res. Health Care **6**, 23–30 (2015)

26. Helm, J.E., Lavieri, M.S., Van Oyen Mark, P., Stein, J.D., Musch, D.C.: Dynamic forecasting and control algorithms of glaucoma progression for clinician decision support. Oper. Res. **63**(5), 979–999 (2015)

27. Induction of rules. Accessed Nov 2018. http://www.cs.put.poznan.pl/jstefanowski/sed/DM-6rulesnew.pdf

28. Jiang, L.: Learning instance weighted Naive Bayes from labeled and unlabeled data. J. Intell. Inf. Syst. **38**(1), 257–268 (2012)

29. Kumar, S., Luo, C.: US adults with unmet mental health treatment needs–profiling and underlying causes using machine learning techniques. IISE Trans. Healthc. Syst. Eng., 1–13 (2019) (in press)

30. Lee, E.K., Nakaya, H.I., Fan, Y., Querec, T.D., Greg, B., Pietz, F.H., Benecke, B.A., Bali, P.: Machine learning for predicting vaccine immunogenicity. Interfaces **46**(5), 368–390 (2016)

31. Li, X., Bilen-Green, C., Farahmand, K., Langley, L.: A semiparametric method for estimating the progression of cognitive decline in dementia. IISE Trans. Healthc. Syst. Eng. **8**(4), 303–314 (2018)

32. Li, B., Chow, T.W., Huang, D.: A novel feature selection method and its application. J. Intell. Inf. Syst. **41**(2), 235–268 (2013)
33. Liaw, A., Wiener, M.: Classification and regression by RandomForest. R News. **2**(3), 18–22 (2002)
34. Lin, Y., Liu, S., Huang, S.: Selective sensing of a heterogeneous population of units with dynamic health conditions. IISE Trans. **50**(12), 1076–1088 (2018)
35. Lingras, P.: Unsupervised rough set classification using GAs. J. Intell. Inf. Syst. **16**(3), 215–228 (2001)
36. MODLEM: MODLEM rule algorithm. Accessed Nov 2018. http://weka.sourceforge.net/pac kageMetaData/MODLEM/index.html
37. McDonald, A.D., Sasangohar, F., Jatav, A., Rao, A.H.: Continuous monitoring and detection of post-traumatic stress disorder (PTSD) triggers among veterans: a supervised machine learning approach. IISE Trans. Healthc. Syst. Eng. 1–15 (2019) (in press)
38. Moghaddasi, H., Tabatabaei Tabrizi, A.: Applications of cloud computing in health systems. Glob. J. Health Sci. **9**. (2016). https://doi.org/10.5539/gjhs.v9n6p33
39. Monks, T., Van der Zee, D.J., Lahr, M.M., Allen, M., Pearn, K., James, M.A., Luijckx, G.J.: A framework to accelerate simulation studies of hyperacute stroke systems. Oper. Res. Health Care **15**, 57–67 (2017)
40. Morana, S., Dehling, T., Reuter-Oppermann, M., Sunyaev, A.: User assistance for health care information systems. In: SIG-Health Pre-ICIS Workshop, Seoul, South Korea, Dec 2017
41. Murphy, K.P.: Naive bayes classifiers. University of British Columbia, p. 18 (2006)
42. Ning Zhong, N., Dong, J., Ohsuga, S.: Using rough sets with heuristics for feature selection. J. Intell. Inf. Syst. **16**(3), 199–214 (2001)
43. Njafa, J.-P., Tchapet, E., Nana, S.G.: Quantum associative memory with linear and non-linear algorithms for the diagnosis of some tropical diseases. Neural Netw. **97**, 1–10 (2018). https://doi.org/10.1016/j.neunet.2017.09.002
44. Oztekin, A., Al-Ebbini, L., Sevkli, Z., Delen, D.: A decision analytic approach to predicting quality of life for lung transplant recipients: a hybrid genetic algorithms-based methodology. Eur. J. Oper. Res. **266**(2), 639–651 (2018)
45. Pawlak, Z.: Rough Sets: Theoretical Aspects of Reasoning About Data. Springer, Norwell, MA (1982)
46. Pawlak, Z., Skowron, A.: Rudiments of rough sets. Inf. Sci. **177**(1), 3–27 (2007)
47. Pawlak, Z.: Vagueness a rough set view. In: Mycielski, J., Rozenberg, G., Salomaa, A. (eds.) Structures in Logic and Computer Science, vol. 1261, pp. 106–117. Springer, Berlin, Heidelberg (1997). https://doi.org/10.1007/3-540-63246-8_7
48. Price, S., Golden, B., Wasil, E., Denton, B.T.: Operations research models and methods in the screening, detection, and treatment of prostate cancer: a categorized, annotated review. Oper. Res. Health Care **8**, 9–21 (2016)
49. Quinlan, J.R.: C4. 5: Programs for Machine Learning. Elsevier (2014)
50. Rachuba, S., Salmon, A., Zhelev, Z., Pitt, M.: Redesigning the diagnostic pathway for chest pain patients in emergency departments. Health Care Manag. Sci. **21**(2), 177–191 (2018)
51. Rocha, F.P., Rodrigues, H.S., Monteiro, M.T.T., Torres, D.F.: Coexistence of two dengue virus serotypes and forecasting for Madeira Island. Oper. Res. Health Care **7**, 122–131 (2015)
52. Ryu, Y.U., Chandrasekaran, R., Jacob, V.: Prognosis using an isotonic prediction technique. Manage. Sci. **50**(6), 777–785 (2004)
53. Schuldt, C., Laptev, I., Caputo, B.: Recognizing human actions: a local SVM approach. In: Pattern Recognition. 2004. ICPR 2004. Proceedings of the 17th International Conference, vol. 3, pp. 32–36. IEEE (2004)
54. Si, B., Yakushev, I., Li, J.: A sequential tree-based classifier for personalized biomarker testing of Alzheimer's disease risk. IISE Trans. Healthc. Syst. Eng. **7**(4), 248–260 (2017)
55. Skowron, A., Rauszer, C.: The discernibility matrices and functions in information systems. In: Slowinski, R. (ed.) Intelligent Decision Support. Handbook of Advances and Applications of the Rough Set Theory, pp. 331–362. Kluwer, Dordrecht, The Netherlands (1992)

56. Srinivas, K., Kavitha Rani, B., Govrdhan, A.: Applications of data mining techniques in healthcare and prediction of heart attacks. Int. J. Comput. Sci. Eng. (IJCSE) **2**(2), 250–255 (2010)
57. Wang, H., Zheng, B., Yoon, S.W., Ko, H.S.: A support vector machine-based ensemble algorithm for breast cancer diagnosis. Eur. J. Oper. Res. **267**(2), 687–699 (2018)
58. Wang, K., Zwart, C., Wellnitz, C., Wu, T., Li, J.: Integration of multiple health information systems for quality improvement of radiologic care. IISE Trans. Healthc. Syst. Eng. **7**(3), 169–180 (2017)
59. Williams, N., Zander, S., Armitage, G.: A preliminary performance comparison of five machine learning algorithms for practical IP traffic flow classification. ACM SIGCOMM Comput. Commun. Rev. **36**(5), 5–16 (2006)
60. Yousaf, K., Mehmood, Z., Awan, I. A., Saba, T., Alharbey, R., Qadah, T., Alrige, M.A.: A comprehensive study of mobile-health based assistive technology for the healthcare of dementia and Alzheimer's disease (AD). Health Care Manag. Sci. 1–23 (2019)
61. Zeng, W., Yuan, C., Wang, Q., Liu, F., Wang, Y.: Classification of gait patterns between patients with Parkinson's disease and healthy controls using phase space reconstruction (PSR), empirical mode decomposition (EMD) and neural networks. Neural Netw. (2019). https://doi.org/10.1016/j.neunet.2018.12.012
62. Zou, N., Huang, X.: Empirical Bayes transfer learning for uncertainty characterization in predicting Parkinson's disease severity. IISE Trans. Healthc. Syst. Eng. **8**(3), 209–219 (2018)

Healthcare Patient Flow Modeling and Analysis with Timed Petri Nets

Jiacun Wang

Abstract Patients moving though a healthcare facility to receive medical service form patient flow. Optimizing patient flow is crucial to improving the efficiency of healthcare practice, providing a positive experience for patients and healthcare teams. Patient flow is event-driven, and thus Petri nets are a good fit for its modeling and analysis. In this chapter, the fundamental concepts of Petri nets, including transition firing rules, modeling power, structural and behavioral properties are briefly introduced. Petri net modeling practice in healthcare systems and patient flows, in particular, is discussed. Details of patient flow performance analysis based on stochastic timed Petri nets, such as the average patient waiting time, resource constraints modeling, task durations modeling, and patient flow hierarchical modeling that handles complexity are discussed.

Keywords Patient flow · Petri nets · Modeling · Performance analysis · Resource constraints

Patients moving though a healthcare facility to receive medical service form patient flow. Optimizing patient flow is crucial to improving the efficiency of healthcare practice, providing a positive experience for patients and healthcare teams. Patient flow is event-driven, and thus Petri nets are a good fit for its modeling and analysis. In this chapter, the fundamental concepts of Petri nets, including transition firing rules, modeling power, structural and behavioral properties are briefly introduced. Petri net modeling practice in healthcare systems and patient flows, in particular, is discussed. Details of patient flow performance analysis based on stochastic timed Petri nets, such as the average patient waiting time, resource constraints modeling, task durations modeling, and patient flow hierarchical modeling that handles complexity are discussed.

J. Wang (✉)
Department of Computer Science and Software Engineering, Monmouth University, West Long Branch, NJ, USA
e-mail: jwang@monmouth.edu

© The Author(s), under exclusive license to Springer Nature Switzerland AG 2022
P. Nicopolitidis et al. (eds.), *Advances in Computing, Informatics, Networking and Cybersecurity*, Lecture Notes in Networks and Systems 289,
https://doi.org/10.1007/978-3-030-87049-2_6

1 Introduction

Patient flow is the movement of patients through a healthcare facility [12]. It involves the medical care, physical resources, and internal systems needed to get patients from the point of admission to the point of discharge while maintaining quality and patient and provider satisfaction. Different from typical predefined industrial workflow, patient flow is full of uncertainty and sometime chaos. Significant change in patient visit rate over a given time period often causes large variation of patient length of stay. Improving patient flow and reducing patient waiting time is a critical component in providing better health care service and higher patient satisfaction.

It has been well recognized that applying formal methods in system design and development can increase the correctness and robustness of the system [9, 28]. Petri nets are a powerful formal tool and have foundation applications in various discrete event systems. Patient flow is event-driven, and thus Petri nets are a good fit for the modeling and analysis of patient flow [10, 13, 18, 24].

Patient flow is a type of workflow. Workflow technology has received considerable attention in the healthcare filed in recent years for the automation of healthcare processes [8]. For example, Song et al. classified computer-aided healthcare workflows, including their approaches, goals, and major characteristics in [22]. Browne et al. proposed a two tier, goal-driven workflow model for the healthcare domain in [7]. Dwivedi et al. discussed how workflow management systems enable the achievement of value driven healthcare delivery in [8]. In [21], Poulymenopoulou and Vassilacopoulos introduced a web-based workflow system for healthcare. Patient flow modeling and analysis can also play a great role in hospital staffing [11, 16].

A number of formal modeling techniques that work for workflow were proposed in the past decades. In our previous research, we introduced a new Workflows Intuitive and Formal Approach (WIFA) for the modeling and analysis of workflows [25]. In addition to the abilities of supporting automatic workflow validation and enactment, WIFA possesses the distinguishing feature of allowing users who are not proficient in formal methods to construct and dynamically modify the workflows that address their needs. In [26], a timed workflow model based on WIFA was developed to support the emergency response workflow modeling and real-time property analysis. Petri nets, because of its generality in application, are also among the most popular ones [1]. As a variation of Petri nets, Resource Oriented Workflow Nets (ROWNs) were introduced in [23] that supports workflow resource requirements analysis.

This chapter attempts to apply timed Petri nets in modeling and evaluating patient flow. The fundamental concepts of Petri nets, including transition firing rules, modeling power, structural and behavioral properties are briefly introduced. Petri net modeling practice in patient flow is discussed. Details of patient flow performance analysis based on stochastic timed Petri nets, such as the average patient waiting time, resource constraints modeling, task durations modeling, and patient flow hierarchical modeling that handles complexity are discussed.

The rest of the chapter is organized as follows: Sect. 2 introduces Petri nets and timed Petri nets. Section 3 describes patient flows and their modeling with Petri nets.

Section 4 presents STPN based patient flow simulation and evaluation. Section 5 discusses hierarchical modeling of patient flow. Section 6 summarizes the work.

2 Petri Nets

This section will introduce fundamental concepts of Petri nets and timed Petri nets.

2.1 Definition

A Petri net is a particular kind of bipartite directed graph, composed of four types of objects: places, transitions, directed arcs, and tokens [20]. Places are passive elements, representing status, conditions, resources, etc. Transitions, on the other hand, are active elements, representing actions, events, computation, etc. Directed arcs connect places to transitions or transitions to places. In its simplest form, a Petri net can be represented by a transition together with an input place and an output place. This elementary net may be used to represent various aspects of the modeled systems. For example, a transition and its input place and output place can be used to represent a data processing event, its input data and output data, respectively, in a data processing system. They can also represent a machining process, a raw part, and a process part in a manufacturing system. In order to use Petri nets to study a system's dynamic behavior, in terms of states and state changes, each place may potentially hold either none or a positive number of tokens. Tokens are a primitive concept for Petri nets in addition to places and transitions. The presence or absence of a token in a place can indicate whether a condition associated with this place is true or false, for instance.

A Petri net is formally defined as a 5-tuple $N = (P, T, I, O, M_0)$, where

- $P = \{p_1, p_2, ..., p_m\}$ is a finite set of places;
- $T = \{t_1, t_2, ..., t_n\}$ is a finite set of transitions, where $P \cap T = \varnothing$, and $P \cup T \neq \varnothing$;
- $I: P \times T \rightarrow N$ is an input function that defines directed arcs from places to transitions, where N is the set of all nonnegative integers;
- $O: T \times P \rightarrow N$ is an output function that defines directed arcs from places to transitions;
- $M: P \rightarrow N$ is the initial marking.

A *marking* in a Petri net is an assignment of tokens to the places of a Petri net. Tokens reside in the places of a Petri net. The number and location of tokens may change during the execution of a Petri net. The tokens are used to define the execution of a Petri net. A place containing one or more tokens is said to be *marked*.

Most theoretical work on Petri nets is based on the formal definition of Petri nets. However, a graphical representation of a Petri net is much more useful for illustrating the structure and dynamics of the modeled system. A Petri net graph is a Petri net depicted as a bipartite directed multigraph. Corresponding to the definition of Petri nets, a Petri net graph has two types of nodes: a circle represents a place, and a bar or box represents a transition. Directed arcs (arrows) connect places and transitions, with some arcs directed from places to transitions and other arcs directed from transitions to places. An arc directed from a place p to a transition t indicates that p is an input place of t, denoted by $I(p, t) = 1$. An arc directed from a transition t to a place p indicates that p is an output place of t, denoted by $O(p, t) = 1$. If $I(p, t) = k$ for some $k > 1$, then there exist k directed (parallel) arcs connecting p to t. Usually, in the graphical representation, parallel arcs connecting a place (transition) to a transition (place) are represented by a single directed arc labeled with its multiplicity, or weight k. A circle containing a dot represents a place containing a token.

2.2   Transition Firing Rules

The execution of a Petri net is controlled by the number and distribution of tokens in the Petri net [27]. By changing distribution of tokens in places, which may reflect the occurrence of events or execution of operations, for instance, one can study the dynamic behavior of the modeled system. A Petri net is executed by firing transitions. Denote by $M(p)$ the number of tokens in a place p in the marking M. We now introduce the enabling rule and firing rule of a transition, which govern the flows of tokens:

- Enabling Rule: A transition t is said to be enabled if each input place p of t contains at least the number of tokens equal to the weight of the directed arc connecting p to t, i.e., $M(p) \geq I(t, p)$ for all p in P. If $I(t, p) = 0$, then t and p are not connected, so we don't care about the marking of p when considering the firing of t.
- Firing Rule: Only enabled transitions can fire. The firing of an enabled transition t removes from each input place p the number of tokens equal to $I(t, p)$, and deposits in each output place p the number of tokens equal to $O(t, p)$.

Mathematically, firing t in M yields a new marking

$$M'(p) = M(p) - I(t, p) + O(t, p), \quad \forall p \in P$$

Notice that since only enabled transitions can fire, the number of tokens in each place always remains non-negative when a transition is fired. Firing a transition can never try to remove a token that is not there.

A transition without any input place is called a *source* transition, and one without any output place is called a *sink* transition. Note that a source transition is unconditionally enabled, and that the firing of a sink transition consumes tokens, but doesn't produce tokens.

A pair of a place p and transition t is called a self-loop, if p is both an input place and output place of t. A Petri net is said to be *pure* if it contains no self-loop.

Consider the simple Petri net shown in Fig. 1. The initial marking is

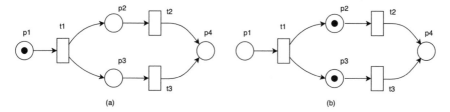

Fig. 1 A simple Petri net and transition firing

$$M_0 = (1, 0, 0, 0)$$

In the initial marking only t_1 is enabled. Firing of t_1 results in a new marking, say M_1. It follows from the firing rule that

$$M_1 = (0, 1, 1, 0)$$

The new token distribution of this Petri net is shown in Fig. 1b. Then in the marking M_1, both transitions t_2 and t_3 are enabled. If t_2 fires, the new marking, say M_2, is:

$$M_2 = (0, 0, 1, 1)$$

If t_3 fires instead, the new marking, say M_3, is:

$$M_3 = (0, 1, 0, 1)$$

2.3 Modeling Power

One of the major reasons that Petri nets are so popular in applications is its great power in modeling system operational dynamics. Here are some most common patterns of activities in discrete event systems and their Petri net modeling constructs.

1. Sequential execution. In Fig. 2, the transition t_2 can fire only after the firing of t_1. This imposes the precedence constraint "t_2 fires after t_1." Such precedence constraints are typical among tasks in discrete event systems. Also, this Petri net construct models the causal relationship among activities.

Fig. 2 Two sequential transitions

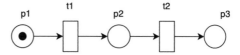

Fig. 3 Transitions t_2 and t_3 are in conflict

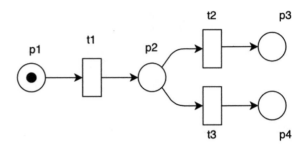

2. *Conflict*. Transitions t_2 and t_3 are in conflict in Fig. 3. Both are enabled; however, the firing of any transition leads to the disabling of the other one. Such a situation will arise, for example, when two computation tasks compete for the CPU or any other shared resource. The resulting conflict may be resolved in a purely non-deterministic way or in a probabilistic way, by assigning appropriate probabilities to the conflicting transitions.

3. *Concurrency*. In Fig. 4, t_2 and t_3 are concurrent, because after firing t_1 both p_2 and p_3 will get a token, and thus t_2 and t_3 are bother enabled and can fire without disabling each other. Concurrency is an important attribute of system interactions.

4. *Synchronization*. It is quite normal in a dynamic system that an event requires multiple resources. The resulting synchronization of resources can be captured by a transition with multiple input places. In Fig. 4, t_4 is enabled only when each of its two input places receives a token. In general cases, the arrival of a token into each input place could be the result of a complex sequence of operations elsewhere in the rest of the Petri net model. Essentially, a transition of synchronization models a joining operation that merges multiple stream events into a single sequence.

5. *Resource contention*. Resource contention is a common phenomena in event-driven systems. Petri nets are well suited to model resource contention. Figure 5 shows such a case where three patients are waiting for being serviced by one CT scanner. Because there is only one scanner, so at any moment only one patient

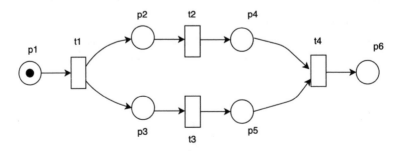

Fig. 4 Transitions t_2 and t_3 are concurrent. Transition t_4 synchronizes two sequences

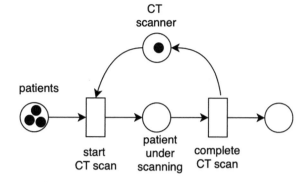

Fig. 5 Resource contention: patients are waiting for CT scan service

can be serviced. When the transition "start CT scan" fires, it cannot fire again until the transition "complete CT scan" fires, which resturns a token to the place "CT scanner."

2.4   System Analysis with Petri Net Models

Petri nets are a mathematical tool that can be used to analyze the properties of a system that it models. Two types of properties can be distinguished: behavioral and structural. The behavioral properties depend on the initial state or marking of a Petri net. The structural properties, on the other hand, do not depend on the initial marking of; they only depend on its topology or structure instead.

The most important behavioral property of Petri nets is reachability. By enumerating all reachable markings, or states, of a Petri net model, one can tell whether a system can reach a specific state, or exhibit a particular functional behavior. For example, when we design patient flow with a Petri net, we want to make sure a patient, in any case, can find a path to move (or be moved) out of the healthcare facility. With the model, we can also predict if any undesired property, such as dead lock, can be researched. If it is, then we need to analyze why—is it because there is a flaw in the system operation logic, or because the resource configuration has a problem? Fig. 6 shows such an example, in which R_1 and R_2 are containers of two different types of resources. With this model, if the transition t_1 fires first and then t_4 fires, then p_1 and p_3 will each have a token. This is a deadlock situation because no transitions are enabled at this marking. Other behavioral properties included boundedness, safety and liveness.

T-invariants and S-invariants are two structural properties of Petri nets. A T-invariant is a set of transitions that, if they fire in certain order, would take the Petri net to the starting state. Often times we expect a system to be cyclic, that is, it can repeat certain operations. Such a property can be checked by searching the system's Petri net model for T-invariants. Take Fig. 6 as an example, firing transition t_1, t_2 and t_3 in a row will bring the Petri net back to the initial marking. Therefore, t_1, t_2 and t_3 constitute a T-invariant. By the same token, t_4, t_5 and t_6 also form a T-invariant.

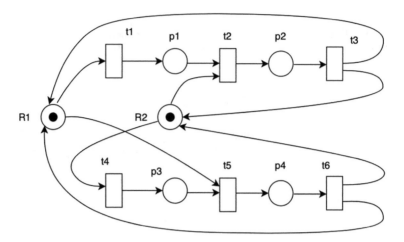

Fig. 6 A Petri net that can run into a deadlock

An S-invariant is a set of places such that the weighted total of tokens in these places are a constant, regardless of how the marking changes. This property can be used to check if there is any resource leak or accumulation in a system, for example. The two place "CT scanner" and "patient under scanning" in Fig. 5 constitute an S-invariant: the total number of tokens in these two places is always one. In fact, these two places represent the single scanner is either idle or busy. There is no other possibility.

2.5 Timed Petri Nets

Regular Petri nets do not have the notion of time built into the model. As such, they are not able to model and analyze a system's time related properties. Timed Petri nets are Petri nets introduced with time parameters [2, 19, 27]. There are three different ways to introducing time to a Petri net, namely associating time to tokens, associating time to places, or associating time to transitions. As far as time, it can be fixed, an interval, or random.

The simplest timed Petri nets are deterministic timed Petri nets (DTPN), in which each transition is associated with a fixed time parameter. A DTPN is a 6-tuples (P, T, I, O, M_0, ST), where (P, T, I, O, M_0) is a regular Petri net, ST is a mapping $T \to R^+$, R^+ being the set of non-negative real numbers. For a transition t, $ST(t)$ is called the *static firing time* of the transition; it represents the delay from the transition being enabled to firing.

Consider the simple DTPN shown in Fig. 7. As a convention, we put a transition's static firing time in a pair of parentheses, next to the label of the transition. In this model, $ST(t_1) = 2$, $ST(t_2) = 7$, $ST(t_3) = 3$, $ST(t_4) = 3$. Moreover, without loss

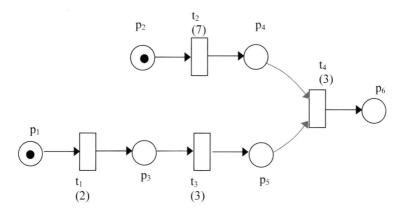

Fig. 7 A deterministic timed Petri net

of generality, we assume a timed Petri net starts running at time 0 from the initial marking. The initial marking of this DTPN is

$$M_0 = (1, 1, 0, 0, 0, 0).$$

Enabled transitions in the initial marking are t_1 and t_2. At time 0, we know t_1 will fire at time 2 and t_2 will fire at time 7, if they are not disabled before their static firing times are up. Because $ST(t_1) < ST(t_2)$, t_1 will fire before t_2. At time 2, t_1 fires and the new marking is

$$M_1 = (0, 1, 1, 0, 0, 0),$$

in which t_2 continues to be enabled and meanwhile t_3 also becomes enabled. Because t_3 just becomes enabled, it is due to fire 3 time unites later, at time 5. For t_2, because it became enabled at time 0, thus at time 2, it has already been enabled for 2 time units; It is due to fire in 5 time units at time 7. Therefore, the situation of the two enabled transitions is as follows:

- t_2 fires in 3 time units, which is equal to its static firing time.
- t_3 fires in 5 time units, which is its *remaining* static firing time in M_1.

At time 5, t_2 fires and the Petri net reaches the marking

$$M_2 = (0, 1, 0, 0, 1, 0).$$

At time 7, t_3 fires and the Petri net reaches the marking

$$M_3 = (0, 0, 0, 1, 1, 0)$$

At time 9, by firing t_4 the Petri net reaches the final marking

$$M_4 = (0, 0, 0, 0, 0, 1)$$

From the above analysis, we can conclude that it will take 9 time units for the Petri net to run to completion. It is worth to mention that the reachable set of a DTPN is a subset of its underlying Petri nets. For example, from the initial marking of the model in Fig. 7, its underlying Petri net can reach $(1, 0, 0, 1, 0, 1)$, but the DTPN cannot, because the transition t_1 has to fire at time 2.

Stochastic timed Petri nets (STPNs) are Petri nets extended with random event times [27]. Event times can be associated with arcs, places or transitions. Most STPN models introduce times to transitions, called *transition firing times*. The time can be deterministic or stochastic. Stochastic transition firing time model is widely adopted because it covers general cases. STPN-based system performance evaluation starts with the modeling of the system with Oetri nets. The STPN model has an underline process that governs the system behavior. This stochastic process is then analyzed using known analytical techniques or through computer simulation.

Formally, an STPN is a tuple (P, T, I, O, M_0, Φ), where

- (P, T, I, O, M_0) is a regular Petri net;
- $\Phi = (\Phi_1, \Phi_2, ..., \Phi_n)$, where Φ_i is the random distribution of the firing time of the transition t_i. We denote by τ_i the firing time of the transition t_i. Then

$$\Phi_i(x) = \Pr\{\tau_i < x\}$$

For a regular Petri net, the set of all possible transition firing sequences starting from the initial marking M_0 is denoted by $L(M_0)$. The timed execution instance of an STPN can be specified by a sequence in $L(M_0)$ augmented with a non-decreasing sequence of real nonnegative values, representing the instants of transition firings, such that consecutive transitions t_i and t_{i+1} correspond to ordered firing epochs τ_i and τ_{i+1}, $\tau_i \leq \tau_{i+1}$. That is, if $\sigma \in L(M_0)$ and

$$\sigma = t_1 t_2 \dots t_i t_{i+1} \dots$$

Then the augmented sequence is

$$\sigma' = (t_1, \tau_1)(t_2, \tau_2) \dots (t_i, \tau_i)(t_{i+1}, \tau_{i+1}) \dots$$

Assume the marking of the STPN after firing transition t_i is M_i. The time intervals $[\tau_i, \tau_{i+1})$ between consecutive epochs represent the periods in which the STPN sojourns in M_i. A *history* of an STPN up to the kth epoch τ_i, or the sequence of σ' up to τ_i, is denoted by $Z(k)$. We assume that for all k, Z, and M, the following distribution functions can be uniquely determined [2]:

$$F(i, x|M, Z) = \Pr\{t_i \text{ fires}, d \leq x|M, Z\}$$

where the random variable d represents the time elapsed from the STPN's entry to M up to the next transition firing epoch, i.e., the time difference $\tau_{i+1} - \tau_i$.

In modeling customer arrival processes, a common assumption is that they follow the Poisson distribution. Also, customer service time is normally treated as exponentially distributed. The biggest advantage of the two assumptions is the memoryless property associated with Poisson process and exponential distribution. If in a marking all enabled transitions are associated with exponentially distributed transition firing times, the above distribution function will only depend on the current marking. That is,

$$F(i, x|M) = \Pr\{t_i \text{ fires}, \ d \le x|M\}$$

When multiple transitions are enabled in a marking, a policy is needed to decide which transition to fire next. This is particularly important when we investigate the performance of an STPN model through simulation. We consider two policies [2, 17]:

Preselection. With this policy, transitions that are enabled in a marking are assigned fixed probabilities to fire.

Race. Let $E(M)$ be the set of all transitions enabled by M. In the race policy, for each transition it in the set $E(M)$, a random sample τ is extracted from the joint distribution:

$$\Phi(x_1, x_2, \ldots, x_n|M, Z) = \Pr\{_1 \le x_{1,2} \le x_2, \ldots,_n \le x_n|M, Z\}$$

The minimum of these samples determines both the transition to actually fire and the sojourn time in M.

In a STPN, concurrency is not only exhibited when different transitions are enabled and can run in parallel, but also by the possibility that a single transition with multiple instances. For example, we can use a single transition to model three machines of the same type. If the transition is triple-enabled, i.e., the transition can fire three times continuously, then the transition actually works like three transitions. We call this transition a *multi-server*. On the other hand, if a transition only represents a single machine or processor or any other task performer, then the transition is called a *single-server* [3, 5].

3 Patient Flow and Modeling

Management of patient flow in a hospital is of major importance, for its impact both on the quality of care and on the associated costs. Uncertainty is the dominant character of patient flow. Patient visit rate varies significantly a day. The care demand of each individual patient is also different. When care demand exceeds capacity at any service point, patient queue forms. When a queue forms, service bottleneck forms. The prevention of the building up of bottlenecks is important.

Patient flow problems can be caused by multiple reasons. Unexpected surge of patient visits is something hard to control. Poor flow design, unreasonable service scheduling and insufficient staffing are all root causes. Patient flow is the movement of patients, information and sometimes equipment between various care service departments, staff groups or organizations as part of a patient's care pathway. Poorly managed patient flow in hospitals can lead to adverse health outcomes, including decreased patient satisfaction, increased re-admissions and mortality rates. By improving patient flow, a hospital can optimize staffing levels to meet patient demand, decrease wait times and boost patient and clinician satisfaction [12].

This section will provide insight into patient flow in hospitals and potential causes for delays. It will also discuss the formal modeling and analysis of patient flows with Petri nets.

3.1   Patient Flow

Figure 8 depicts typical patient flow in hospitals. It shows that patients can experience several different care paths throughout their hospital visit. When a patient arrives at a hospital, he goes through registration first, and a nurse takes his vitals by soliciting

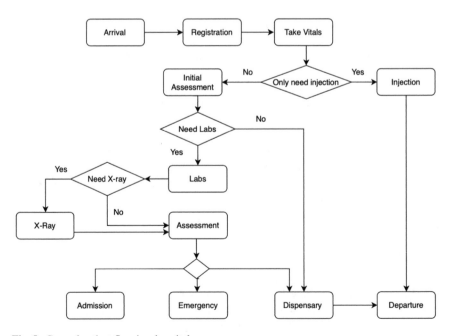

Fig. 8   General patient flow in a hospital

chief complaint and obtaining his temperature, blood pressure and oxygen level. The patient may just need an injection for chronical diseases and then leave the hospital. If it is not the case, an initial assessment is performed and then a lab and/or X-ray may be ordered, or the patient is asked to get some medication and leave for home. In case lab and/or X-ray are ordered, when their results available, a second assessment is performed and, depending on the assessment results, he may be admitted into the hospital, taken to emergence department, or get some medication and then go home. In a real hospital, a patient may be asked to take lab test or radiology test first. That is, there is no particular order between the two tests. In this model we assume lab test always goes first, just to simplify the model a little bit.

3.2 Initial Petri Net Model of the Patient Flow

Figure 9 shows the Petri net model of the patient flow depicted in Fig. 8. It is pretty much a direct translation of the patient flow to the language of Petri nets. The self-loop composed of the place *Patient* and transition *Arrival* models the continuous arrival of patients. The place *End of flow* in the right-bottom corner of the model represents that patients are done with the service. There are two transitions named *Dispensary*, which represents the service of acquiring medication at different service stages of the care flow. The model is easy to understand. However, some important information is missing from the model.

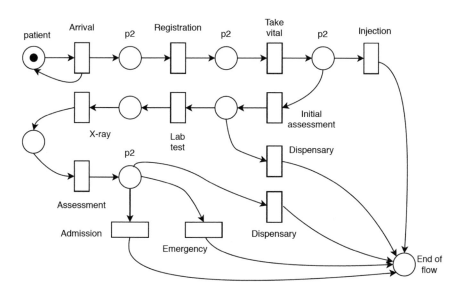

Fig. 9 Petri net model of the patient flow of Fig. 8

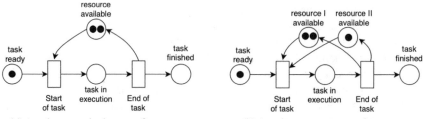

(a) A task uses a single type of resource.   (b) A task uses two types of resources.

Fig. 10 Resource modeling of tasks

3.3 Modeling Resource Requirements

In a hospital or other health care facility, each caring activity needs resources. For example, registration needs a registrar, triage is performed by a doctor or registered nurse, consultations are performed by doctors, a CT machine must be available before a patient is taken to do CT scan, etc. Here, registrars, nurses, doctors, CT machines are all resources. Resources are constraints. Patients have to compete for resources to move on. When related resources are not available, they have to wait.

Countable resources are modeled with tokens in places in a Petri net. Figure 5 shows the model of the CT scan job or task. In general, to model the use of a resource by a task, we use the Petri net construct shown in Fig. 10a. Each task is modeled with two transitions, representing task starts and task ends. A place with tokens models the resource required to execute the task. Another place models the state that the task is in execution. If a task requires multiple types of resources to execute, then the task model can be constructed in the way shown in Fig. 10b.

3.4 Modeling Time

Task execution time is associated with the transition that represents the end of the task. The transition for the beginning of a task does not take time. The importance of this modeling practice is that when a task is ready to execute, it will grab all resources needed and make them unavailable to other tasks immediately. Because the transition for end of task is associated with time, when it is enabled, it will have to wait until the associated time has elapsed before it can fire, which models the time taken to complete the task.

In general, task durations are random numbers. That is, each instance of a task may take different amount of time. For example, it is quite possible the registration of patient A takes 5 min, while that of patient B takes 3 min. Task durations change in certain range and follow some distributions. It is a common practice to associate a probability distribution with a task to characterize the random property of the task duration. Uniform distribution, normal distribution and experiential distribution are

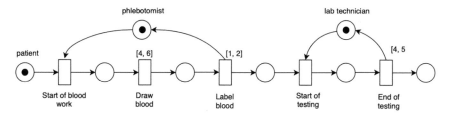

Fig. 11 Timing in blood test events

the ones wildly used because they match real random properties of task durations very well.

Because each task (registration, triage, lab work, patient assessment, etc.) takes time to complete in patient flow, queues are always formed before some tasks if the task execution rate is slower than patient arrival rate. As an example, Fig. 11 shows time spent in blood work. It assumes that drawing blood takes 4–6 min, labeling blood takes 1–2 min, and blood testing takes 4–5 min. There is only one phlebotomist to take blood samples from patients and one lab technician to analyze blood samples. The place *Patient* will receive tokens continuously (imagine there is an input transition to this place). Tokens may accumulate in the place.

3.5 Identify Resource Usage and Task Durations

To build a resource constrained and timed Petri net model for patient flow, we need to identify the resource usage and duration for each task. One must have a good understanding and rich experience of healthcare process and practice in order to get this job done. Also, observations and historic records are necessary in order to determine task durations.

Let consider the patient flow of Fig. 8. Main tasks are

- Registration
- Take vitals
- Initial assessment
- Injection
- Lab test
- X-ray test
- Assessment
- Dispensary
- Paperwork for admission into hospital
- Paperwork for checking into emergency room.

We need to analyze resource usage and duration property for each of them. In a healthcare unit, there are various types of resources. Some are trivial, such as printer paper, tissue paper and water bottles. Because there is rarely a shortage of these kind

Table 1 Tasks, resource usages and durations of a patient flow

Task name	Resource required	Resource released	Duration property	Transition name
Registration	Registration staff	Registration staff	Uniform, 3–6 min	End reg
Take vitals	Nurse	Nurse	Uniform, 6–10 min	End vital
Initial assessment	Doctor	Doctor	Exponential mean: 10 min	End IA
Injection	Nurse	Nurse	Uniform, 10–12 min	End injection
X-ray	X-ray technician, X-ray device	X-ray technician, X-ray device	Uniform, 5–7 min	End X-ray
Lab	Lab technician	Lab technician	Uniform, 20–30 min	End lab
Assessment	Doctor	Doctor	Exponential, mean: 10 min	End asmt
Dispensary	Pharmacist	Pharmacist	Uniform, 10–15 min	End dispensary
Admission	Nurse	Nurse	Uniform, 8–12 min	End adm
Emergency	Nurse	Nurse	Uniform, 8–12 min	End emerg

of resources, we can ignore them in patient flow modeling. Other resources may be critical, such as doctors, nurses, examination beds, lab test machines, and so on. The quantity of these resources can have a significant impact on patient flow performance. Therefore, we have to consider them in the model. Table 1 listed tasks, their resource requirement and durations for the patient flow in Fig. 8. Notice that this table is only meant to illustrate our approach. The accuracy is not a big concern here.

The introduction of resources and event durations to the patient flow results in a new STPN model as shown in Fig. 12.

Transitions. Transitions' firing durations are listed Table 1. All other transitions, except Arrive, are immediate transitions and have 0 firing times. There are two pairs of (*Start Dispensary*, *End Dispensary*) transitions and they represent that patients visit dispensary at different stages of the patient flow. There are also three groups of special transitions in the model that are graphically depicted with thin boxes. They represent *decisions* and are where pre-selection firing policy applies. For example, $t11$ and $t12$ are a pair. When they are enabled, firing one will disable the other. After taking the vital of patient, the nurse may decide that the patient only needs a regular injection and then leaves the hospital (firing of $t11$), or the patient needs to go through more assessment and test (firing of $t12$). The firing probability distribution of these three groups of pre-selection transitions will be discussed in the next section.

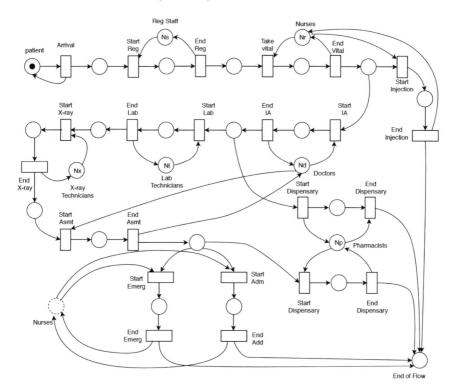

Fig. 12 Patient flow model with resource constraints

Places. Places with names represent resources. For example, the place *Reg Staff* models registration staff, and the number of tokens in this place, *Ns*, represent the number of registration staff. Only most critical resources are considered in the model. The dashed box marks the construct to be refined later. The dotted circle place *Nurse* and the solid circle place *Nurse* are actually one place. Plotting it in two is to avoid too many directed arcs linked to/from one place that will inevitably make the model look messy. This model clearly shows resource constraints and competition. For example, many tasks require nurses to execute.

4   Patient Flow Evaluation

4.1   STPN Simulation

Based on the STPN model in Fig. 12, we can evaluate the patient flow's performance. A long list of Petri net software tools have been developed over the past three decades. The majority of these tools are for specific classes of STPNs, such as GreatSPN [6],

Fig. 13 Architecture of the STPN simulation system

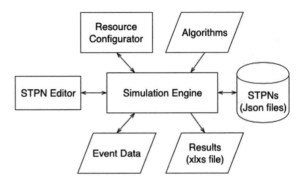

TimeNET [31], PIPE [4], SPNP [15]. In [14], a tool called TiPeNeSS (Timed Petri Net Simulator Software) is presented, which supports the simulation of timed Petri nets containing transitions with generally distributed firing delays.

We designed and developed a resource oriented STPN simulation system [32]. As shown in Fig. 13, the simulation system is composed of seven function units: (1) A simulation engine, which is an STPN based state transition system and supports non-Markovian transition firing times, race and preselection transition firing policies, and both single-server and multi-server transitions; (2) An STPN editor, which is a graphic tool for users to construct STPN models; (3) An STPN bank, which stores all STPN models built with the tool in the JSON file format. (4) A resource configurator, which is designed to provision, manually or automatically, the resources of a service system modeled by a STPN; (5) An algorithm bank, which provides different algorithms to the simulation engine. (6) An event data set, which stores event names and all timing parameters; and (7) A file to store simulation results.

The tool is written in Java and is a standalone application.

4.2 Performance Evaluation

There are many different ways to define a system's performance matrix. There are also many factors that affect a system's performance. When a system is modelled with an STPN, the STPN logic structure, initial marking, each transition's firing time, transition firing policies, etc., are all parameters of the system's performance. Theoretically, our simulation system can analyze the impact of any system parameters on the system's performance. In this chapter, however, we focus on how a patients average waiting time in a hospital changes with the following factors:

- Patient arrival rate
- Key resource provisioning
- Mixture of patient acuity.

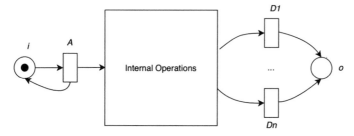

Fig. 14 Service system input and output

We can view the patient flow as a queuing system that customers arrive, go through all steps of services, and then leave the system. An external view of the STPN model is illustrated in Fig. 14, where the transition A is a *source* transition, representing customer arrival and it is normally assumed to be a Poisson process. The token in the place i represents a customer. A and i constitute a self-loop, modeling consecutive arrival of customers. The sink place o is for customers who have served by the system. The sink transitions $D_1, D_2, ..., D_n$ model different ways that customers depart the service system.

During simulation, we record when each token in the place i is generated, denoted by $a_1, a_2, ..., a_N$, and when each token in the place o is created, denoted by $d_1, d_2, ..., d_N$, where N is the simulation times. The average sojourn time, s, of customers in the system can be calculated by

$$s = \frac{\sum_{k=1}^{N}(d_i - a_i)}{N}$$

If the average customer waiting time is a concern, then we can get the average service time of the system by simulating the system operation with a single customer. The average waiting is the difference between the average sojourn time and average service time.

4.3 Simulation Set Up and Results

Recall in Sect. 2.5 we mentioned two transition firing policies in a STPN, namely pre-selection policy and race policy. Transitions in Fig. 12 that follow the race firing policy are listed in the right-most column of Table 1. The firing rate of transition *arrival* reflect the patient volume. All other transitions are immediate transitions whose firing delays are 0. Among these immediate transitions, there are three groups of transitions following the pre-selection firing policy. They are listed in Table 2. All transitions in the model are multi-server ones.

To perform the simulation, we set related parameters as follows:

Table 2 Pre-selection transitions

Group	Transition	Firing probability
I	t11	α
	t12	$1 - \alpha$
II	t21	β
	t22	$1 - \beta$
III	t31	η
	t32	$1 - \eta$
IV	t41	γ_1
	t42	γ_2
	t43	$1 - \gamma_1 - \gamma_2$

- Resources: $Ns = 2$; $Nl = Nx = Np = 1$; $Nr = 2, 6$; $Nd = 2, 4, 6, 8, 10$.
- Pre-selection probabilities: $\alpha = 0.3$, $\beta = 0.7$, $\eta = 0.5$, $\gamma_1 = 0.2$, $\gamma_2 = 0.4$.
- Average patient arrival interval (minutes): 5, 10, 15, 20, 20.

The simulation results are shown in Fig. 15. Figure 15a is the case when the numbers is set to 2 and the number of doctors is changing, and Fig. 15b is the case when the numbers is set to 6 and the number of doctors is changing. Two trends can be observed easily from the figure: when the number of doctors or nurses is increased, the average patient waiting time is decreased; when the average patient arrival interval is increased, the average patient waiting time is decreased. One can observe another important fact, that is, when the number of doctors is increased from 8 to 10, the path flow's performance changed very little, which means that the number of doctors is no longer the bottleneck of the performance when more than 8 doctors are provisioned. Of course, more experiments can be performed with different resource provisioning, such registration staff, lab technicians and pharmacists, and pre-selection probabilities which indicates the dynamics of patients.

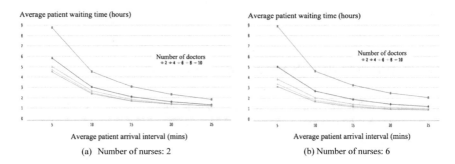

(a) Number of nurses: 2   (b) Number of nurses: 6

Fig. 15 Performance evaluation of the patient flow model of Fig. 12

5   Refinement of Patient Flow Model

The patient flow depicted in Fig. 8 or Fig. 12 is quite high-level. It skipped some details of the care service. For example, that task of lab test can actually be an aggregation of three tasks:

- CT scan
- Blood test
- Urine test.

These three tests can be conducted in an arbitrary order. However, if you build a Petri net model with too many details, the model may look very complex and hard to read and verify. A practical approach to modeling a complex system is hierarchical modeling, that is, one first models a system at a high level, then starts to refine the model by introducing more details. You don't need to refine the entire model at a time; instead, refine a small component or even just a task a time. This way, you can manage the complexity at different level of abstractions. The refining process stops when a desire level of details is reached.

As an example, we refine the task of lab test by replacing the small construct composed of transitions *Start Lab* and *End Lab* and the two places between them in the STPN model of Fig. 12 with the model of Fig. 16. The model shows that CT scan, urine test and blood can be performed in any order. However, they cannot run concurrently because for any individual patient they have to be performed one by one. That is the role the place *patient* plays. Transitions *End CT*, *End Urine*, *Draw Blood*, *Label Blood*, and *End Test* have tangible durations; all others are immediate transitions with 0 firing times.

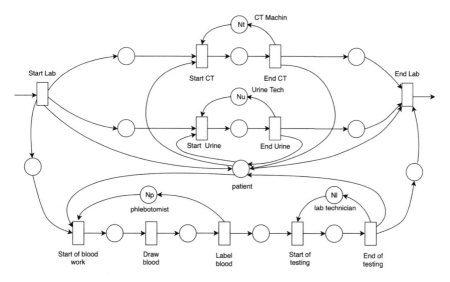

Fig. 16 Refinement of the task of lab test

We can refine other tasks as well. After all the desired refinement, we can plug the refined blocks into the original model and perform the performance simulation.

6 Conclusion

Petri nets are a powerful tool in modeling and analyzing discrete event systems. When time parameters are introduced into Petri nets, they can also be used for system performance analysis. Patient flow is composed of various steps of patient movement and care services. Each individual service is an event and thus patient flow is event-driven. This chapter applies timed Petri nets to the modeling and performance analysis of patient flow. Firstly, a Petri net tool is designed and developed, which supports Petri net edit, simulation, and file operations. Secondly, the patient flow and patient flow modeling are discussed. A Petri net model directly converted from an informal description of patient flow, such as flow chart, often misses some details of patient flow, particularly resource constraints. This chapter discussed how to model resource constraints in patient flow with Petri nets. To capture task duration, we showed how to introduce time to the Petri net model. Hierarchical modeling is also discussed to handle the complexity of patient flow when details of the flow is considered in the model.

Future work includes introducing more details to the patient flow STPN model, which will make the simulation results more accurate. Also, more healthcare resources, such as various types of technicians and testing devices, should also be taken into account in the patient flow modeling. Another thing to do is consider applying the patient flow modeling and analysis to hospital staffing, which is a topic drawing lots of attention from healthcare practitioners [29, 30].

References

1. Ajmone Marsan, M., Balbo, G.., Bobbio, A., Chiola, G., Conte, G., Cumani, A.: The effect of execution policies on the semantics and analysis of stochastic Petri nets. IEEE Trans. Softw. Eng. **SE-15**(7), 832–846 (1989)
2. Boyer, M., Diaz, M.: Multiple enabledness of transitions in Petri nets with time. In: Proceedings of 9th International Workshop on Petri Nets and Performance Models. Aachen, Germany, Sept 2001
3. Bonet, P., Lladó, C.M., Puijaner, R., Knottenbelt, W.J.: Pipe v2. 5: a Petri net tool for performance modelling. In: Proceedings of the 23rd Latin American Conference on Informatics (2007)
4. Boucheneb, H., Lime, D., Roux, O.H.: On multi-enabledness in time Petri nets. In: PETRI NETS 2013: Application and Theory of Petri Nets and Concurrency, pp. 130–149 (2013)
5. Baarir, S., Beccuti, M., Cerotti, D., De Pierro, M., Donatelli, S., Franceschinis, G.: The GreatSPN tool: recent enhancements. ACM SIGMETRICS Perform. Eval. Rev. **36**(4), 4–9 (2009)

6. Browne, E., Schrefl, M., Warren, J.: A two tier, goal-driven workflow model for the healthcare domain. In: Proceedings of the 5th International Conference on Enterprise Information Systems. Angers, France (2003)
7. Dwivedi, A., Bali, R.K., Wickramasinghe, N., Nagui, R.N.G.: How workflow management systems enable the achievement of value driven healthcare delivery. Int. J. Electron. Healthc. **3**, 382–393 (2007)
8. Deng, Y., Wang, J., Tsai, J.: Formal analysis of software security architectures. In: Proceedings 5th International Symposium on Autonomous Decentralized Systems, pp. 426–434. Dallas (2001)
9. Dotoli, M., Fanti, M.P., Iacobellis, G., Martino, L., Moretti, A.M., Ukovich, W.: Modeling and management of a hospital department via Petri nets. In: IEEE Workshop on Health Care Management (2010)
10. Dixon, C.A., Punguyire, D., Mahabee-Gittens, M., Ho, M., Lindsell, C.J.: Patient flow analysis in resource-limited settings: a practical tutorial and case study. Glob. Health: Sci. Pract. **3**(1), 126–134 (2015). https://doi.org/10.9745/GHSP-D-14-00121
11. Envision Healthcare: Emergency Department Operations Management and Patient Flow. An Envision Playbook—Best Practices, Tools and Timelines. https://www.envisionphysicianservices.com/campaigns/breakthrough-series/presentation-materials/playbooks/ed-operations-management-and-patient-flow-playbook.pdf (2017)
12. Fanti, M., Mangni, A., Dotoli, M., Ukovich, W.: A three-level strategy for the design and performance evaluation of hospital departments. IEEE Trans. Syst. Man Cybernet. Syst. **43**(4), 742–756 (2013)
13. Horváth, Á., Molnár, A.: TiPeNeSS: a timed Petri net simulator software with generally distributed firing delays. In: SIMUTOOLS 2015. Athens, Greece (2015)
14. Hirel, C., Tuffin, B., Trivedi, K.S.: SPNP: stochastic Petri nets. Version 6.0. In: Computer Performance Evaluation. Modelling Techniques and Tools, pp. 354–357. Springer (2000)
15. Jensen, K.: Staffing your emergency department efficiently, effectively and safely: core concepts. https://www.envisionphysicianservices.com/campaigns/breakthrough-series/presentation-materials/presentations/09-staffing-your-ed-core-concepts.pdf (2017)
16. Marsan, M.A., Chiola, G.: On Petri nets with deterministic and exponentially distributed firing times. In: Advances in Petri Nets 1987, pp. 132–145. Springer, Berlin, Heidelberg (1987)
17. Mahulea, C., Mahulea, L., García-Soriano, J., Colom, J.: Modular Petri net modeling of healthcare systems. Flex. Serv. Manuf. J. Spec. Issue Anal. Des. Manag. Health Care Syst. (2017)
18. Molloy, M.K.: On the integration of delay and throughput measures in distributed processing models. Ph.D. thesis. UCLA, Los Angeles, CA (1981)
19. Murata, T.: Petri nets: properties, analysis and applications. Proc. IEEE **77**(4), 541–580 (1989)
20. Poulymenopoulou, M., Malamateniou, F., Vassilacopoulos, G.: Emergency healthcare process automation using workflow technology and web services. Inform. Health Soc. Care **28**(3), 195–207 (2003)
21. Song, X., Hwong, B., Matos, G., Rudorfer, A., Nelson, C., Han, M., Girenkov, A.: Understanding requirements for computer-aided healthcare workflows. In: Proceedings of International Conference on Software Engineering: Experiences and Challenges, pp. 930–934 (2006)
22. Wang, J., Li, D.: Resource oriented workflow nets and workflow resource requirement analysis. Int. J. Software Eng. Knowl. Eng. **23**(5), 667–693 (2013)
23. Wang, J., Tian, J., Sun, R.:. Emergency healthcare resource requirement analysis: a stochastic timed Petri net approach. In: 2018 IEEE 15th International Conference on Networking, Sensing and Control, IEEE, Mar 2018
24. Wang, J., Tepfenhart, W., Rosca, D.: Emergency response workflow resource requirements modeling and analysis. IEEE Trans. Syst. Man Cybern. Part C **39**(3), 270–283 (2009)
25. Wang, J.: Emergency healthcare workflow modeling and timeliness analysis. IEEE Trans. Syst. Man Cybern. Part A **42**(6), 1323–1331 (2012)
26. Wang, J.: Timed Petri Nets: Theory and Application. Kluwer Academic Publishers (1998)

27. Wang, J.: Formal Methods in Computer Science. CRC Press (2019)
28. Wolf, L., Perhats, C., Delao, A., Clark, P., Moon, M.: On the threshold of safety: a qualitative exploration of nurses' perceptions of factors involved in safe staffing levels in emergency departments. J. Emerg. Nurs. **43**, 150–157 (2017)
29. Yankovic, N., Greenn, L.: Identifying good nursing levels: a queuing approach. Oper. Approach **59**, 942–955 (2011)
30. Zimmermann, A., Knoke, M.: TimeNET 4.0: a software tool for the performability evaluation with stochastic and colored Petri nets; user manual. TU, Professoren der Fak. IV (2007)
31. Zhou. J., Wang, J., Wang, J.: A simulation engine for stochastic timed petri nets and application to emergency healthcare systems. IEEE/CAA J. Autom. Sin. **6**(4), 969–980 (2019)
32. van der Aalst, W.M.P.: Verification of workflow nets. In: Proceedings of application and theory of Petri nets, vol. 1248. Lecture Notes in Computer Science, pp. 407–426 (1997)

Enabling and Enforcing Social Distancing Measures at Smart Parking Infrastructures Using Blockchain Technology in COVID-19

Amtul Waheed, Jana Shafi, and P. Venkata Krishna

Abstract Social distancing is the only preventive nonpharmaceutical measure to limit the transmission of Coronavirus disease (covid-19). By reducing the closeness and frequency of human physical gatherings and maintaining adequate distance among the individuals can minimize the spread of disease. IoT and AI revolutionary domain technologies can support in implementing and measuring minimum social distancing set-up by using the smart application. Though Smart City and Smart parking System are innovative, smart, and safe practices for end-users, the outbreaks of COVID-19 have revealed limitations of the existing deployments. Hence timely enforcements, framework, and architecture have to be developed accordingly. The focus of this paper is to provide a practical model where emerging blockchain technology is enforcing in finding many solutions and dealing with the problems in maintaining social distancing at smart parking in such pandemic outbreaks. The research aims to provide an architecture and framework in developing such applications using for smart parking concerning important issues and challenges of privacy and cybersecurity in implementing social distancing in practice.

Keywords Smart parking · COVID-19 · Social distancing · IoT · Blockchain technologies

1 Introduction

Ever since the coronavirus has emerged and widely blow-out, the whole world is traumatized and shaken by WHO declaring it as a global pandemic. Coronavirus starts with common flu and cold but can lead to Severe Acute Respiratory Syndrome.

A. Waheed · J. Shafi
Prince Sattam Bin Abdul Aziz University, Al-Kharj, Saudi Arabia

P. Venkata Krishna (✉)
Sri Padmavati Mahila Visvavidyalayam, Tirupati, Andhra Pradesh, India
e-mail: pvk@spmvv.ac.in

This virus is unrestrained and can be easily spread from an infected person to others whoever comes in physical contact without protection [1].

Health organizations, front-end worriers, and medical fraternities are struggling high to stop the spread of the virus. At the same time, researchers and scientists are struggling for the invention of vaccines, yet no vaccines are developed. Hence the primary solution to prevent this disease is to maintain social distancing with people in public places at least for a period of the next few months. Social distancing states the actions that reduce the disease spread by reducing the closeness and frequency of people's physical contact, by taking actions such as closing public places such as schools, and workplaces avoiding mass gatherings and keeping a sufficient distance amongst people. The physical distancing of 6 feet in public places is an effective nonpharmaceutical approach to limit the pandemic outrage. If executed properly, social distancing measures can play a key role in reducing the infection rate. Smart governances are extensively using satellite images, CCTV cameras, Drones, IoT devices-based surveillance, and smart applications to track and maintain social distancing at public places within cities such as highways, tourist destinations, industrial areas, and shopping locations, and parking lots [2].

With the increase in corona cases, smart parking system has also progressed contactless parking solutions at parking spaces. Keeping a view of social distancing, smart Parking will provide digital payment methods through which customers can enter their car number, link their wallet, and pay anytime during their stay at parking lots. Mobile applications are used for the reservation and digital payment for hassle-free and smooth entry and exit from smart parking lots. In doing so, the smart parking lots can ensure a contactless and ticketless parking system to further intensify the focus on containing the spread of the virus. While keeping a view of all the aspects of safety and hygiene and also enforce social distancing practices at smart parking lots, managing customer expectations, and following government regulations.

In this research, we present comprehensive details on enabling and emerging blockchain technology used for social distancing at the smart parking system. In this article, we have proposed blockchain which is a distributed decentralized, and proactive solution for smart parking with social distancing. In Sect. 2 related work of covid-19, social distancing, blockchain technology, smart parking are discussed. Section 3 proposed a blockchain-based framework for a smart parking system with social distancing. In Sect. 4 represents features of using blockchain technologies for social distancing in Smart Parking. Section 5 analyzes the security and financial measures achieved through the proposed blockchain framework for smart parking systems for distancing and finally Sect. 6 is the conclusion.

2 Related Work

2.1 COVID-19 Pandemic

COVID-19 is caused by severe acute respiratory syndrome coronavirus [3]. Initially, COVID-19 infected case was reported in Wuhan City, China in December 2019, and has rapidly spread out all over the world. With huge destruction and still there is no sign that the numbers of infected people and dead cases will decrease and the condition will be under control. Due to the worst situation and high level of risk assessment of corona throughout the world, the WHO (World Health Organisation) has declared it as a global pandemic. Even though mass research and attempts are carried out by researchers and scientists to find the solution to fight against the covid-19 pandemic, but still no proper vaccination is introduced and the only solution by governments afford to stop the pandemic is to follow a nonpharmaceutical approach that is by social distancing, partial lockdown in crowded places (such as markets, schools, universities, and workplaces, etc.), wearing the mask at public places, washing or sanitizing hands frequently.

2.2 Social Distancing

Maintaining a physical distance of a minimum of 6-ft or one meter from the people in public places, To avoid direct contact with other people that can help in the reduction of direct transmission of corona infection. Many advancements and innovations are taking place in different technologies such as IoT, BigData, blockchain, and AI are used to maintain social distancing. AI Software is used to scan and create a profile of every citizen, it can also be used to detect the proximity of people to enforce distancing rules in a particular area of hot infectious spot [2].

2.3 Smart Parking

A smart parking system plays a vital role in imposing social distancing at any parking lot [4]. To find out vacant or occupied parking spaces on street or in garage parking lots, AI and sensors technology are used [5]. Such information can be used to calculate the number of vehicles parked in parking lots, rules and restrictions can also be applied to vehicles of such particular parking lots. Different rules at different parking locations are challenging for users, to overcome these difficulties such rules can be updated every day by informing law enforcement agencies. Smart parking lots can automatically implement these rules while notifying the Vehicles through transmitted messages [6].

2.4 Blockchain Technology

With the innovation of Blockchain technology redefining financial prospects and economics are emerging in the public system. The primary advantage of blockchain technology is prominent as a decentralized, distributed, and trusted model. The three main components of blockchain technology are data blocks, distributed ledger, and consensus algorithms. Blockchain is distributed decentralized hierarchical chain of blocks where secure public ledges store records of the transaction [7]. Cryptography techniques and hash values are used to secure each block of the blockchain to ensure the integrity of the transaction. For creating a cryptographically secure chain of blocks every block adds the hash value of the previous block [8]. After the block is mined and validated by every node, it cannot be manipulated or altered. By this data, integrity is sustained and provides assurance for tamper resistance secure data storage in distributed model without a centralized third party. Since each transaction is signed by the user so authentication is also ensured. The distributed ledger is a database that records and stores unique transactions created by users with a cryptographic signature that resists alterations. The consensus algorithm process establishes a contract between entities concerning the validation of each data block. This can be achieved by nodes connecting in the mining procedure and competing with one another to verify the block.

We summarize some of the related work done in the field of blockchain technology to aid in this covid-19 pandemic, according to Ting et al. [9] many cutting edge technologies such as IoT, blockchain, big data, and AI are used in creating simulation models that predict and mitigating the spread of the covid-19 disease. And also explains how these technologies can aid in establishing a screening tool that can facilitate in terms of diagnosis and monitoring of the disease's spread. The author also defines the use of blockchain to track and drop the medications to the patient's doorsteps to avoid physical contact and to maintain social distancing to stop the spread of Covid-19. Another author Mashamba-Thompson [10] projected blockchain and AI to achieve COVID-19 self-testing. The use of blockchain to automatically detect infected cases and evaluate the infection risk of the COVID-19 in society was approached by Torky et al. [11]. Nguyen et al. [12] proposed an approach using AI along with blockchain to process a large volume of medical data that has a complex pattern. As well as the introduced blockchain-based approach to help with donation tracking and the healthcare supply chain. Another author Bansal et al. [13] proposed the use of immutable blockchain technology to avoid the spread of false reports and information by creating immunity certificates. Kumar et al. [14] research use blockchain as a means of sharing data while maintaining privacy by using an approach to improve the deep recognition of a deep learning model to rec recognize COVID-19 patients based on tomography (CT) slices. Finally, Rosiere [15] proposed the use of blockchain technology to attain collective scientific research and medical cooperation to fight against the spread of COVID-19 infectious disease.

In almost all research blockchain technology is used to help in stopping the spread of fake information, or to track and trace covid-19 patients for testing, diagnosing, and medication. None of the research proposed the implementation of social distancing in smart parking systems using blockchain technology. Unlike other existing work, our main aim is to propose a framework based on blockchain technology for social distancing at smart parking lots in the covid-19 pandemic. Additional contributions presented are features of the blockchain platform. We analyzed challenges related to security and financial cost for our framework concerning reliability and feasibility.

3 Blockchain-Based Framework with Social Distancing in Smart Parking

In this section, we present a detailed description of the proposed framework of Blockchain-based social distancing in smart parking. In this framework, ethereum platform is used for smart contracts of parking lot users and parking lot logs for identifying occupied and vacant parking spaces in parking lots. We aim to provide parking space in parking lots with social distancing using blockchain technology. This can helps in avoiding direct contact with other people and also help in mitigating the spread of the COVID-19 virus.

Figure 1 shows the framework of the proposed blockchain-based solution for maintaining social distancing at smart parking lots. It represents the user participation and registration in smart parking applications with a request for smart parking space by using smart contracts, blockchain clients, integrated parking services, and

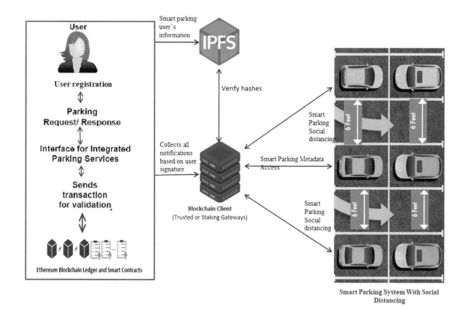

Fig. 1 Blockchain-based Model with social distancing in smart parking

distributed storage. If the user decides to park the vehicle in the parking lot the information related to the user and vehicle is updated using IPFS hash functions. IPFS storage is a distributed and decentralized way of storing documents related to smart parking systems and user identification. Therefore securely storing data is mandatory as it is public to everyone. Consequently, the information stored on IPFS should be encrypted and only authorized entities should be able to read the plaintext content, our system allows multiple people to access the content on the servers while maintaining confidentiality [16, 17].

In our framework, a user smart contract can generate an event to request a smart parking space. With this parking, management communicates in an event and parking lots to ensure vacant parking lots. Private user information is disclosed using an IPFS hash and this information is further encrypted to ensure secure data storage. Blockchain clients are considered an important part of this framework as it works as an interface between events and the blockchain network. Blockchain clients do not enhance any security features to the blockchain network. The main objective of this client is to carry a source of information in the blockchain network in a tamper-proof manner.

4 Features of Using Blockchain Technologies for Social Distancing in Smart Parking

The blockchain platform is a decentralized and distributed technology with advanced features such as transparency, cryptographic encryption, and impenetrable information infrastructure which help to maintain smart parking user's information secure with tamper-proof assures of data and authenticate the user in the blockchain network. For is very difficult to modify the transaction in a blockchain network or requires lots of resources to modify because once a transaction is validated and verified then it gets chained with a previous transaction with unique hash values. Furthermore, data stored on the blockchain is made available to all members of the network ensuring transparency among participants. Table 1. Define the features of using blockchain technology in smart parking with social distancing measures.

5 Analysis of Proposed Model

Analysis of our proposed model concerning financial and security concerns within a real-world setting.

Table 1 Features of using blockchain technology in smart parking with social distancing measure

S. No.	Features	Blockchain platform
1	Authority	Decentralized even in the private network
2	Data handling	Merely read and write preferences are available
3	Data Integrity	Data is immutable and auditable
4	Data Privacy	Data is stored using cryptography technology
5	Transparency	Data is stored in a distributed network
6	Quality assurance	Data can be tracked and traced right from its origin using cryptography technology
7	Fault tolerance	The distributed ledger is highly faulted tolerant
8	Cost	Uncertainty in the operating and maintenance costs
9	Performance	Handle's minimal transactions and scalability is a challenge

5.1   Financial Analysis

A certain transaction fee is incurred on ethereum blockchain transactions for information storage on the blockchain network. The execution cost and transaction cost are always part of the logs in the Remix environment. The miners line up transactions that hold a higher amount of Loops, arrays, mappings, variable storage, and manipulation as well as data types play a major role in transaction costs. The extremely important aspect is the efficiency and feasibility of the solution. Therefore, our proposed model influences the blockchain's immutable properties and depends on the events and information then on-chain storage.

5.2   Security Analysis

Blockchain platforms are highly robust secure and resilient based on cryptographic essentials. But if smart contracts are not written carefully they can be exploited. Smart contracts should be reviewed for infinite loops, software bugs, errors, and other common slips which can make the smart contract highly vulnerable to hackers. Blockchain integrates key security mechanism that makes many applications ideal to use blockchain platform. It enables secure and trusted solutions that are tamper-proof and strong. Additionally, blockchain provides many security characteristics such as availability, integrity, authorization, confidentiality, and nonrepudiation.

6 Conclusion

In this paper, we presented a framework for a blockchain-based smart parking system with social distancing as a primary objective to combat the COVID-19 pandemic. We proposed and evaluated a blockchain-based parking system with social distancing for validating the user information without modifying data. Our proposed blockchain-based solution promotes integrity, confidentiality, transparency, traceability and streamlines the communication between the user in the network by requesting and responding to parking space considering social distancing as an important aspect, our main aim is to propose a framework based on blockchain technology for social distancing at smart parking lots in a covid-19 pandemic. Additional contributions presented are features of the blockchain platform. We analyzed challenges related to security and financial cost for our framework regarding reliability and feasibility.

Acknowledgements We are thankful to the Deanship of Scientific Research of Prince Sattam bin Abdul Aziz University, KSA for their kind support.

References

1. World Health Organization: Coronavirus (COVID-19), World Health Organization, 22 Apr 2020. Accessed on: 22 Apr 2020 [Online]. Available: https://covid19.who.int/
2. Waheed, A., Shafi, J.: Successful role of smart technology to combat COVID-19, In: 2020 Fourth International Conference on I-SMAC (IoT in Social, Mobile, Analytics, and Cloud) (I-SMAC), pp. 772–777. Palladam, India (2020). https://doi.org/10.1109/I-SMAC49090.2020. 9243444
3. Sohrabi, C., Alsafi, Z., O'Neill, N., Khan, M., Kerwan, A., Al-Jabir, A., Ios Fidis, C., Agha, R.: World Health Organization declares global emergency: a review of the 2019 novel coronavirus (COVID-19). Int. J. Surg. **76**, 71–76 (2020)
4. Khanna, A., Anand, R.: IoT based smart parking system. In: 2016 International Conference on Internet of Things and Applications (IOTA), pp. 266–270. IEEE (2016)
5. Waheed, A., Venkata Krishna, P.: Detecting predominance of on-street parking payment schemes by means of linear regression. Int. J. Rec. Technol. Eng. (IJRTE) **8**(6) (2020). ISSN: 2277-3878
6. Waheed, A., Shafi, J., Venkata Krishna, P.: Analyzing significant reduction in traffic by using restricted smart parking. In: Venkata Krishna, P., Obaidat, M. (eds.) Emerging Research in Data Engineering Systems and Computer Communications. Advances in Intelligent Systems and Computing, vol. 1054. Springer, Singapore (2020). https://doi.org/10.1007/978-981-15-0135-7_4
7. Waheed, A., Venkata Krishna, P.: Comparing biometric and blockchain security mechanisms in smart parking system, In: 2020 International Conference on Inventive Computation Technologies (ICICT), pp. 634–638. Coimbatore, India (2020). https://doi.org/10.1109/ICICT48043. 2020.9112483.
8. Zheng, Z., Xie, S., Dai, H., Chen, X., Wang, H.: An overview of blockchain technology: architecture, consensus, and future trends. In: IEEE 6th International Congress on Big Data, Honolulu (2017)
9. Ting, D.S.W., Carin, L., Dzau, V., Wong, T.Y.: Digital technology and covid-19. Nat. Med. **26**(4), 459–461 (2020)

10. Mashamba-Thompson, T.P., Crayton, E.D., Blockchain and artificial intelligence technology for novel coronavirus disease-19 self-testing. Diagnostics (2020)
11. Torky, M., Hassanien, A.E.: Covid-19 blockchain framework innovative approach. arXiv preprint arXiv:2004.06081 (2020)
12. Nguyen, D., Ding, M., Pathirana, P.N., Seneviratne, A.: Blockchain and AI-based solutions to combat coronavirus (covid-19)-like epidemics: a survey (2020). https://doi.org/10.36227/tec hrxiv.12121962.v1
13. Bansal, A., Garg, C., Padappayil, R.P.: Optimizing the implementation of covid-19 "immunity certificates" using blockchain. J. Med. Syst. **44**(9), 1–2 (2020)
14. Kumar, R., Khan, A.A., Zhang, S., Wang, W., Abuidris, Y., Amin, W., Kumar, J.: Blockchain-federated-learning and deep learning models for covid-19 detection using CT imaging. arXiv preprint arXiv:2007.06537 (2020)
15. Resiere, D., Resiere, D., Kallel, H.: Implementation of medical and scientific cooperation in the caribbean using blockchain technology in coronavirus (covid-19) pandemics. J. Med. Syst. **44**, 1–2 (2020)
16. Ateniese, G., Fu, K., Green, M., Hohenberger, S.: Improved proxy re-encryption schemes with applications to secure distributed storage. ACM Trans. Inf. Syst. Secur. (TISSEC) **9**(1), 1–30 (2006)
17. Green, M., Ateniese, G.: Identity-based proxy re-encryption. In: International Conference on Applied Cryptography and Network Security, pp. 288–306. Springer (2007)

Using DEVS for Full Life Cycle Model-Based System Engineering in Complex Network Design

Abdurrahman Alshareef, Maria Julia Blas, Matias Bonaventura, Thomas Paris, Aznam Yacoub, and Bernard P. Zeigler

Abstract The Discrete Event System Specification (DEVS) is a modeling formalism that supports a general methodology for describing discrete event systems with the capability to represent continuous, discrete, and hybrid systems due to its system theoretic basis. In this chapter, we discuss the use of DEVS as the basic modeling and simulation framework for Model-Based System Engineering methodology that supports the critical stages in a top down design of complex networks. Focusing on the design of communication networks for emergency response, we show how such networks pose challenges to current technologies that current simulators cannot address. This sets the stage for considering how the DEVS formalism supports the required phases of top down design and the transitions from one phase to the next. After describing the proposed DEVS-based system engineering methodology in depth, we conclude with a discussion of the current state of its application, also mentioning open research needed to bring it into general practice.

A. Alshareef
King Saud University, P.O. Box 4545, Riyadh 11451, Saudi Arabia
e-mail: ashareef@ksu.edu.sa

M. J. Blas
Instituto de Desarrollo y Diseño INGAR (UTN-CONICET), Avellaneda 3657, Santa Fe, Argentina
e-mail: mariajuliablas@santafe-conicet.gov.ar

M. Bonaventura
Instituto de Ciencias de la Computación (Universidad de Buenos Aires-CONICET), Pabellón I, Ciudad Universitaria, Buenos Aires, Argentina
e-mail: mbonaventura@dc.uba.ar

T. Paris
Université de Lorraine, CNRS, LORIA, 54000 Nancy, France
e-mail: thomas.paris@loria.fr

A. Yacoub
School of Computer Science, University of Windsor, Windsor, ON N9B 3P4, Canada
e-mail: aznam.yacoub@uwindsor.ca

B. P. Zeigler (✉)
RTSync Corp, Chandler, AZ, USA
e-mail: zeigler@rtsync.com

© The Author(s), under exclusive license to Springer Nature Switzerland AG 2022   215
P. Nicopolitidis et al. (eds.), *Advances in Computing, Informatics, Networking and Cybersecurity*, Lecture Notes in Networks and Systems 289,
https://doi.org/10.1007/978-3-030-87049-2_8

Keywords DEVS · Complex network behavior · Routing mechanisms ·
Co-simulation · Network simulation infrastructure · Simulation-based testing ·
Modeling and simulation · Discrete event simulation · Hybrid simulation · Activity
diagrams · Model-based system engineering · High-level system specification ·
MBSE · Emergency disaster response

1 Introduction

The design of complex networks with current technologies must meet several challenges that current simulators cannot address. This calls for new modeling and simulation (M&S) approach that can support a full life-cycle Model-Based System Engineering methodology. In this chapter, we discuss such an M&S approach based on the Discrete Event System Specification (DEVS) formalism and supporting computational environments.

Our focus here is particularly on the design of complex networks in the case of emergency response, and the problems posed by the validation of these networks, and more specifically, why we need to support the coexistence of simulation and network execution.

The approach includes specification of required behaviors using UML/SysML metamodels that map to DEVS simulation models, with particular attention paid to the high level specification of routing mechanisms in such models. We go on to discuss infrastructure design to implement the behaviors of such models. The emergency response then requires us to consider interoperability co-simulation support for heterogeneous node devices. Finally, we discuss simulation-based testing and evaluation of such designs using powerful hybrid fluid-flow/packet-level mechanisms.

These together provide a comprehensive, unified system engineering methodology with modeling expressiveness and high simulation performance. This chapter outlines each facet and shows how they work together in a real world example of packet routing in data networks.

Our discussion takes the following form:

- Section 2: Motivation of the problem of complex network design applied to emergency response,
- Section 3: Review of background on DEVS in complex network design,
- Section 4: Specification of required behaviors using UML/SysML metamodels that map to DEVS simulation models,
- Section 5: High level specification of routing mechanisms in such models,
- Section 6: Co-simulation of complex networks using DEVS as the formal basis for co-simulation support,
- Section 7: DEVS-Based architectures to design, model, and validate complex network simulation infrastructures,
- Section 8: Simulation-based testing of such designs using powerful DEVS-based hybrid-flow/packet-level mechanisms, and

- Section 9: A summary of the proposed DEVS-based system engineering methodology and of the current state of its application with open research needed to bring it into general practice.

2 Complex Network Design in Emergency Decision Making

2.1 Emergency Decision-Making and Requirements

Emergency response is a critical problem, especially with the increase in catastrophic natural disasters such as earthquakes, typhoons, diseases, fire-spreading, etc. Nowadays, old traditional centralized approaches of emergency management have been replaced by new decentralized decision-making processes which put the emphasis on collaboration [41]. Although the analysis of the psychological mechanisms underlying decision-making processes has become more complex, decision-making psychologists intensively studied different approaches to understand how to minimize errors and support the various domain experts, especially in situations in which lives are at stake [95]. Particularly, modern theoretical frameworks identify at different levels of analysis specific recurrent factors that greatly impact decision-making in emergencies: complexity arising from the collaboration between different organizations and teams, uncertainty related to the lack of reliability of the available information, complexity related to the situation itself and how the experts understand it in a short time, previous experience and training. Moreover, the literature on Emergency Decision Making (EDM) and Natural Disaster EDM highlights the importance of mathematical models, simulation, and knowledge management in methodologies and decision support systems. Data management and communication are therefore crucial, and the base of the entire decision-making process, including in situational awareness [49, 96]. Indeed, in order to make the best decision, information must be presented according to the role of the different actors concisely and with accuracy.

At a lower level, quickly sharing and spreading good information is also critical in the context of emergency response. This aspect imposes specific challenges to information and communication technologies (ICT) [64]: information and actors in the processes of emergency management are heterogeneous by nature. From an operational point of view, this greatly impacts the design of software and networks which support communication in order to make them flexible, safe, and reliable. As an example, requirements may impose that the information collected by sensors or drones deployed in a disaster area with limited communication capabilities are transmitted to a decision center several kilometers away. This information has to be processed and presented to decision makers who should be able to communicate their instructions to actors in the incident scene. This scenario already implies two levels of network analysis: the communication network itself, but also the network created by the actors and agents involved in the emergency management processes.

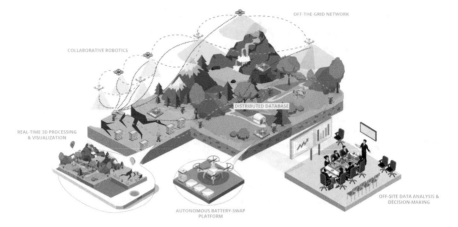

Fig. 1 An example of heterogeneous emergency ecosystem

Concerning the communication network itself, researchers identify four main types of network solutions to support emergency response [52, 64]:

- Wireless ad hoc networks (WANET);
- Cellular networks;
- Special emergency networks;
- Router-based networks.

Generally, WANETs are popular because they can be operated without any prior infrastructure or any central system. The deployment of such a kind of network begins with two nodes and finishes when the last node is connected. Moreover, they allow the deployment of the agents on the same device, which is a node component of the network, meaning it is possible to make plans early in the design of solutions. However, using WANETs has an impact on the requirements of the software which will use these networks [64]:

- Applications should inform the users about the availability and status of the network;
- Applications should obtain information in alternative ways in case of unavailability, which supposes the existence of a second network;
- Applications should have current information according to the role of the users, to keep their attention on their primary goals;
- Applications should work across different platforms, modalities, and screen sizes; moreover, they should have adaptive behaviors and take into account the heterogeneity of the devices and nodes that constitutes the network (Fig. 1);
- The network itself should be flexible regarding the number of nodes, the capacity of interconnection, replication, speed and latency.

This analysis concludes that simulation and analysis tools should take into account both applications and network requirements and should not focus only on one of these two aspects as most of the existing papers on this topic do [64]. Furthermore, in the context of emergency response, nodes of these ad hoc networks are dynamic and move in space. Mobile Ad Hoc Networks (MANETs) [43, 61, 70] thus bring challenges related to classical WANET (architecture, topology, routing, protocols, layers), but also new issues related to mobility: computing a route efficiently depends on the ability of planning dynamics of the different actors in time and space.

Because of the critical nature of these missions, analysis of all these aspects before any technological deployment is crucial. However, we readily admit that accurate modeling of this entire ecosystem, including the aspects of physical phenomena, is a challenging, perhaps intractable, task.

2.2   Problems with Current MANET Modeling and Simulation

In this section, we detail the kinds of problems that arise in current MANET modeling and simulation that our DEVS-based design approach is intended to address. We will summarize the problems into two categories.

- First, the improper abstraction of the problem to solve, especially in emergency response, leads to strong assumptions; these assumptions are the main reason for the lack of accuracy of the simulators.
- Second, heterogeneity of the components and aspects, implementations, and paradigms lead to a lack of integration of existing tools. This makes it difficult to cross-validate results between different simulators in a context in which experimental validation is almost impossible or unrealistic.

In the following, we delve more deeply into these problems to give readers a sense of the choices that simulation modelers have to make. Indeed, the literature has proposed numerous MANET infrastructure and protocols [7, 70, 77] aimed at secure and reliable networks. Because analysis is difficult, network designers have tried to improve their algorithms by adding redundancy or improving route computation. As a result, these architectures pose challenges for the simulation community [6, 23], especially because:

- the topology of MANETs changes more frequently than do usual networks. Indeed, the network structure depends on the mobility of each node, and these moves are hardly predictable. However, integrating dynamic models into network models is a hard task. Consequently, mobility models are studied in separate works [42, 76], leading to oversimplified MANET models [75];
- modeling physical phenomena involved in wireless communication, like radio propagation, is also a challenging task. Primarily, accurate models are generally expressed using continuous representations at a level of abstraction different from the packet-level, used for modeling routing protocols [37]. Hence designers tend to focus only on the routing aspects in MANET simulators and their performance and provide only oversimplified models for physical interactions [6].

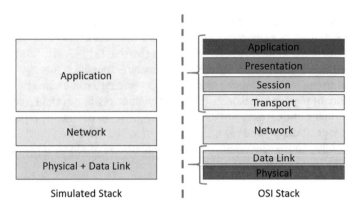

Fig. 2 Simulation stack (left) versus real OSI stack (right)

Existing surveys about MANET simulation [37, 50] also addressed two other issues. The first one concerns the extreme divergence and inconsistencies between existing simulators[20] regarding routing protocols. Suppose we already highlighted the heterogeneity of the different aspects of communication through MANET as the main reason. In that case, we can also underline the use of unrealistic implementation of applications in simulation. For example, Andel et al. [6] showed that the protocol stack is modified in the simulator because MANET tests occur in the network layer. As a result, designers and modelers combine all the layers above the network layer in a single one. In contrast, the data link layer and physical layer are randomized (Fig. 2).

Here we question the relevance of this approach. We agree that a model cannot focus on every aspect of a system, and analysis should focus on interesting properties to be validated with applicable experimental frames [93]. However, in network communication, the physical layer and data link layer are not just abstract conceptual models. We mean that these two layers depend on their actual implementation in operating systems and drivers. Abstract models of the computerized aspect of these two layers can be easily derived from their actual implementation. Similarly, the upper layers of the OSI model depend on their software implementation. As a result, we argue that strong assumptions are made while accuracy could have been easily increased. Readers may reply that actual implementation depends on the real nodes (and we cannot simulate all the existing products) or that generally, source code is not available. However, it does not justify the statement made by Andel et al. [6]: most of the studies in MANET simulation literature use overly simplified application instances like constant-bitrate traffic generators as experimental frames. These applications' complexity is far from the complexity of real applications and do not allow efficient validation. If we do not question the principle of making an abstraction of the application layer, we can question the impact on this simplification's accuracy. Especially in a fully connected environment with heterogeneous components, misuses of protocols can harmfully reduce the performance and stress network robustness. By pushing the reasoning further, interactions between stacks are even more complicated in virtualized environments, in which several stacks can be combined. For instance, in virtualized operating systems, there are at least two communication stacks in the environment, which can conflict, resulting in a drastic change in communication speed. Therefore, simulating such an environment using one simplified stack can not guarantee the modeled network efficiency.

Other studies [78] showed the effect of inaccurate models on the simulation: a small variation of parameters in the physical model underlying the network layer drastically changed the results of the comparative analysis between two evaluated protocols [40, 68]. In other words, routing protocols are extremely sensitive to physical phenomena. In the context of emergency response, this might lead to increase uncertainty, which is not acceptable. If the result is expected and understandable, it leads to an important question: the level of details is essential in MANET simulation [37] and abstraction or refinement can lead to erroneous outcomes. This opens the question of the interpretation of the simulation results and the verification and validation of models.

This result comes from the lack of definition of the excellent level of abstraction when a model is developed, what Hogie et al. [37] name the *granularity*. Verification and Validation of Simulation Models [73] is also related to the paradigms and the nature of the different models: discrete-event models, discrete-time models, continuous models, without clearly using a specified formalism. Discrete-event paradigms are especially well-used for modeling the computational aspects, but accurate physical modeling needs a continuous approach. Using the wrong paradigm without motivating it by a serious analysis leads to increasing the models heterogeneity. While existing simulators have all their purposes [56, 57], researchers generally focus on one existing simulator and try to implement models in the paradigm of this simulator [37], whereas integrating many simulators is a hard challenge. The outcome of such a practice is to blur the choice of the right level of abstraction. By extension, it also brings problems where the heterogeneity also exists at the level of implementation because all of these simulators have their programming languages.

To summarize, improper abstractions are the main reason for the lack of accuracy displayed by currently available MANET simulators. Moreover, the heterogeneity of designs leads to a lack of integration of existing tools rendering the necessary cross-validation difficult in the emergency application context.

2.3   Towards DEVS-Based Methodology for Complex Network Modeling and Simulation

Now that we have identified the problems, the question becomes: what can we do to solve them? Before we show how DEVS-based methodology helps to address them, we briefly review some attempts that have already been proposed in the literature. For instance, DIANEmu [46] tries to improve the application model by providing an environment for simulating applications communicating through a network. This architecture can be transposed directly on an actual device. However, DIANEmu does not simulate the four first layers of the network stacks. JANE [26] combines the advantages of a simulator, an emulator, and testbeds by providing a simulator able to work in hybrid mode thanks to a simulation environment and an execution platform. An application can easily swap between the simulated network and the real devices while the communication interface is the same from the software point of view. However, the simulation models are themselves defined at a high level of abstraction. Consequently, JANE is not well-suited for a complex heterogeneous environment. Testbeds [63] and emulators [25, 39] can help researchers to overcome these two problems by doing real experiments and considering these two models as controlled parameters. However, testbeds remain not scalable and cost expensive. Various literature areas addressed the problem of making heterogeneous simulators coexisting and working together to improve the accuracy of simulations and deal with repeatability and consistency. Essentially techniques like co-simulation [30, 85] have been proved as good approaches that allow modelers to take into account specifics of different subsystems.

Despite these attempts, there were few attempts to fully integrate heterogeneous simulators and execution in the same environment to take advantage of hybrid modeling and simulation. However, a close look at the simulation paradigm used by the well-used MANET simulators [23] shows that most of them implement a discrete approach to reduce the intrinsic complexity of the MANET analytic models. More precisely, some of them like OMNeT++ [84] and ns-3 [71] use a discrete-event-based architecture without explicitly following the DEVS formalism [92].

The use of discrete event simulation is particularly interesting because, as we show in the next section, the Theory of Modeling and Simulation (TMS) [93], and its associated DEVS formalism offer important recommendations to resolve the problem stated previously. Indeed, this chapter will show that the DEVS methodology helps in more accurate modeling and simulation of MANET and how we can provide a methodology to integrate existing MANET simulators into a DEVS-compliant environment.

2.4 Summary

Emergency decision-making is a source of issues and challenges for engineers and researchers. Especially, decentralized approaches must rely on reliable complex networks that have to be deployed quickly in a hostile environment. Information acquisition and communication are critical, especially in situations in which lives are at stake. In order to overcome these problems, some technologies and tools have been proposed and studied in different scientific fields: research on ad hoc networks provided protocols and simulators focused on technical network issues like topology, routing, security; research on ergonomics in emergency response focuses on the end-user interface of application and software, or provides some learning platforms in order to train actors; research on cognitive science and psychology focuses on cognitive process behind decisions, exemplified by the impact of multimodalities.

More specifically, studies about MANET lead to creating many simulators and tools that focus on specific aspects of networks (like routing protocols), with no interoperability between these tools and no integration. As a result, simulation is applied to unrealistic scenarios that do not cover an emergency mission complexity, heterogeneous by nature. Therefore, we argue that complex network design in emergency decision-making goes far beyond design applications or networks. We have to provide a framework in which designers should specify, verify, and validate requirements related to the network itself and the final end-user applications. They should also choose the right level of abstraction and integrate different simulation tools in a hybrid approach. This perspective proposes in this chapter to explore a DEVS-based network design methodology to achieve these objectives.

3 DEVS in Complex Network Design

DEVS is a modeling formalism that was developed in the late 70s [92] as a general methodology for describing discrete event systems. In this regard, the DEVS formalism fits the general structure of deterministic, causal systems in classical systems theory [82].

The core of DEVS [88] consists of:

- a set of *basic concepts* as DEVS, DEVS Simulation and System Entity Structure;
- the *M & S framework* defined as an ontology of the M&S domain;
- the *systems-theory basis* used to formulate the framework entities in terms of system specifications, and the framework relations in terms of morphisms among system specifications.

Centered in the *M&S framework*, such an ontology provides a set of entities and their relationships as illustrated in Fig. 3. The basic entities of the framework are *source system*, *model*, *simulator*, and *experimental frame*. The *source system* is the real/virtual environment to be modeled. The *model* describes the set of instructions for generating data comparable to the data observable in the *system*. The *simulator* refers to the computational system that executes the set of instructions specified in the *model*. The *experimental frame* represents the conditions under which the *source system* is observed or experimented with. On the other hand, the basic interrelationships between entities are the *modeling* and *simulation* relationships. The *modeling relation* determines when a model can be said to be a valid representation of the *source system* within an *experimental frame*. The *simulation relation* specifies what constitutes a correct simulation of a *model* by a *simulator*.

The design of complex networks must meet several challenges that current simulators cannot address. Such challenges call for new M&S approaches that can support a full life-cycle Model-Based System Engineering methodology. Regarding the M&S framework entities and relationships, the basic issues encountered in performing M&S activities in *complex network design* based on DEVS can be better understood.

Fig. 3 M&S framework

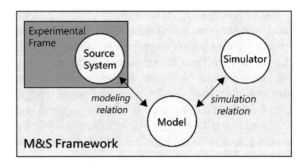

Hence, using such a framework as a foundation, coherent solutions can be developed for the M&S of complex network designs.

In the next sections, new DEVS-centered solutions are developed that together provide a comprehensive, unified system engineering methodology with modeling expressiveness and high simulation performance for studying complex networks.

4 Specification of Required Behaviors Using UML/SysML Metamodels that Map to Hierarchical DEVS Simulation Models

In order to address specific design and analysis needs, especially for complex systems, we use a DEVS specification for fundamental elements in UML/SysML behavioral diagrams, i.e., actions and control nodes. Such account considers the practices of the Model-Based System Engineering (MBSE) while confining the developed artifacts into a simulation modeling formalism (i.e., MBSE/M&S in the sequel). The specification conforms to the semantics of their corresponding element in the abstraction. Additional specifications are necessary for the simulation to be attainable such as precise temporal and I/O descriptions. As a result, various network dynamics can be simulated and examined with various configurations while conforming to the modular and hierarchical DEVS formalism. The resulting simulations can serve as a basis for studying and understanding different aspects of the design and dynamics of behavior in complex networks. Initially, a restricted simulation corresponds to a basic execution of the behavioral model is generated and then can be extended and refined further in different settings to accommodate for a variety of System Entity Structure (SES) attributions, including the ones at the hardware level and real-time environments [94].

The following scenarios provide a high-level yet rigorously grounded with DEVS specification and code generation facilities to attain the initial simulation models. Using activity diagrams can describe different flow scenarios, highlighting two sets of elements, actions, and control nodes. Different synchronization and routing schemes can be specified to describe various possible flow scenarios within a network. More-

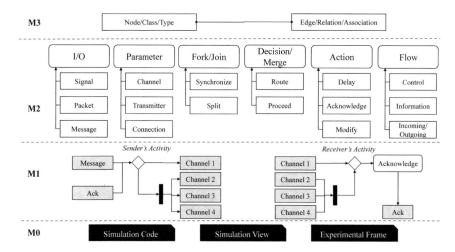

Fig. 4 The elements in the top level of the M2 layer are major activity elements and below them further specializations to describe some network specification and dynamics. From these specializations, concrete instances of activity diagrams are drawn at M1 to be then transformed into Parallel DEVS models in the simulation environment

over, it is possible to create multiple layers of activities relying on the previous specification of hierarchical activities [5] inspired by the modular and hierarchical Parallel DEVS formalism [93] and subsequently transformed to it.

Figure 4 illustrate essential artifacts across layers for using activities to describe some network dynamics in general with various input/output and flow schemes. The architecture highlights major elements to satisfy specific needs throughout the flow in the activity, such as routing and synchronization. The general specification of the activity allows for generic I/O types, while further specializations take place at lower levels closer to the concrete simulation model. These layers can vary in numbers in the model-driven architecture. Higher layers capture the network specification in a generalized manner to be suitable to serve different applications, among which computer networks and neural networks. Moreover, this generalized manner corresponds and later transformed to a generalized Parallel DEVS specification while accounting for as many details and the capacity to be refined for hybrid models or accommodating for a hardware specification.

The routing and synchronization offered with the proposed specialization can facilitate a high-level specification of different processing regimes of packets or signals in various networks. The fork/join elements are specialized into synchronization nodes. The *SYNC* atomic model corresponds to the dynamics where multiple outgoing flows are ought to be dispatched simultaneously. The same atomic model can also correspond to various splitting schemes of inputs or outputs in network flows. On the other end, another *SYNC* model corresponding to another fork/join element can be used to model the synchronizing of the incoming flows at a later point with another decision node to send back possible acknowledgments or receipts. Different routing scenarios can use decision or merging nodes that are mapped to *SELECT* atomic models in the DEVS formalism. As of now, automatic code generation facilities have been developed for DEVS-Suite [1] and MS4 Me [62] simulation environments with the support for generating codes for Markov DEVS models in the latter environment [4]. The formal specification of the atomic models and the coupled models, along with other DEVS specification, can be found in [3].

Figure 5 shows a counterpart design of the network topology for activities within networks or network nodes. Each activity corresponds to a router in Fig. 11 in Sect. 5 to describe the routing process. For example, Activity 1 describes the routing process in router 1. An example of such activity is shown at M1 layer in Fig. 4. Additional to the routing process, different synchronization regimes can also be specified and included in the process through *fork* and *join* node, which is then transformed into *SYNC* atomic model. The benefit of allowing such design at the behavioral level allows for more complex accounts of interaction within/between heterogeneous network/networks encountered in the discrete event system without settling to the numerical analysis and calibrations only.

4.1 Enabling Implementation of Various Routing Infrastructures

Activity diagrams mapped into DEVS enable the implementation of various routing infrastructures with different behavioral specifications. The control *SELECT* (see Listing 8.1) element that captures the semantics of the activity decision node contains a general routing behavior and, by extension, more specific and desired routing behaviors. Instead of focusing on the atomic level, the activity diagram mapped into DEVS establishes a higher view of the behavior specified at both atomic and coupled levels. Different dynamics can take place by using different structures of the activity nodes. The different structures result in different implementations generated using the code generation facility conforming to the proposed DEVS specification.

Listing 8.1 The formal specification of *SELECT* element which captures the semantics of the decision nodes in activities used for routing processes

$SELECT = \langle X^b, Y^b, S, \delta_{ext}, \delta_{int}, \delta_{con}, \lambda, ta \rangle$, where

$X^b = \{(p, v) : p \in IPorts, \ v \in X_p,$
$\quad IPorts = \{in_1, \ldots, in_n\}, \ X_p = V(\text{an arbitrary set})\}$
$\quad$ is the set of input ports and values
$Y^b = \{(p, v) : p \in OPorts, \ v \in Y_p,$
$\quad OPorts = \{out_1, \ldots, out_m\}, \ Y_p = V(\text{an arbitrary set})\}$
$\quad$ is the set of output ports and values
$S = phase \times \sigma \times task \times C$, where
$\quad phase = \{passive, sending\}, \ \sigma = \mathbb{R}^+_{0,\infty}, \ task \subseteq X^b$

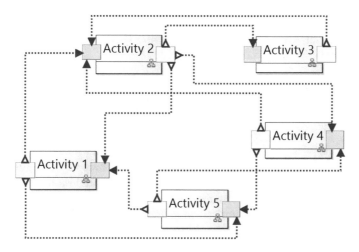

Fig. 5 Some activities interacting in correspondence to a certain topology

$C = \{(p, c) : p \in OPorts \ and \ c \in \{true, false\}\}$
  is the set of output ports and conditions
$\delta_{ext}((phase, \sigma, C, task), e, X^b) =$
  $(sending, 0, C, (in, v_1), \cdots, (in, v_n))$
$\delta_{int}(phase, \sigma, C, task) = (passive, \infty, (p, false), task)$
$\delta_{con}(s, ta(s), x) = \delta_{ext}(\delta_{int}(s), 0, x)$
$\lambda(sending, \sigma, C, task) =$
  $(p_i, task) \ if \ (p_i, true)$
$ta(phase, \sigma) = \sigma.$

The decision node current implementation includes a table identifying which output port should be activated to carry out the flow acting as a routing table in the Routed DEVS adaptation to be discussed in the next section. Allowing the specification to be encoded allows for lifting some restrictions imposed on the behavioral specification at the atomic level. The resulting DEVS components describe the interacting elements' collective behavior in the activity as a behavioral diagram. The activity gives an overall view of the routing process behavior and a more refined mechanism enabling the specification of more complex behavior in complex networks with as much modeling transparency.

5 High-Level Specification of Routing Mechanisms in DEVS

This section focuses on the study of DEVS as M&S formalism for routing processes as networks with dependent nodes and routings. In this regard, DEVS modeling levels are revised to show how the formalism provides a solid basis to build routing simulation models. Such models are defined with a DEVS adaptation named Routed DEVS.

5.1 Routing Processes as Network Models

A routing process is a system of interacting components in which the operation of a component and the routing of its outputs depend on what is happening throughout the process. Hence, interactions between components depend on local information but also external data derived from the process structure.

The routing process definition can contain multiple types of components. Each type of component operates independently. When routing depends equally on the component operative description and process structure, the component decides the output destinations. Hence, for example, components can take decisions about: (i) alternate the routing of its outputs to avoid congestion, (ii) block the routing of its outputs from entering into a sector of the nodes, and (iii) accelerate/decelerate the processing of its inputs (to produce faster/slower outputs) when knowing that downstream nodes are free/busy.

A reasonable conceptualization of *routing processes* is through the use of constrained *network models*. Figure 6 depicts a domain ontology that unifies both domains using the concepts of network model domain (*network, network element, edge, node, operation, starts at, ends at, composed of,* and *holds*) as stereotypes of the classes and relationships associated with the routing process domain (*Routing Process, Routing Process Element, Interaction, Component, Functionality, defined as,* and *exhibits*). Then, concepts of the routing process domain are enriched with the definition of the network model domain. This ontology is represented with UML [31] following the conceptual modeling guidelines proposed by Guizzardi [35]. The modeling primitives of the UML profile [32] used to represent this ontology are outside the scope of this chapter.

From a broad point of view, a *network* is a composition of *nodes* and *edges* between these *nodes*. Both *nodes* and *edges* are *network elements* defined in terms of their properties. Each *edge starts at* a *node*. Moreover, it also *ends at* a *node* (possibly the same that the one at the beginning). If the model is composed of different types of nodes, each *node* holds an *operation*. Several *nodes* can share the same *operation* (multiplicity 1...*).

When *networks* are used as supporting models for *Routing Processes*, the model is *defined as* a set of *Routing Process Elements*. The *nodes* denote *Components*, and *edges* express directed *Interactions* between *Components*. Two *Components* are linked in an *Interaction*. One *Component* acts as the *source from* which *Interaction* takes place. The other *Component* acts as the *destination to* where the *Interaction* is directed. Each *Component* exhibits *Functionality* as *behavior*. The *Component* acts as the *container* of such *Functionality*. Several *Components* can *exhibit* the same *Functionality* (multiplicity 1...*).

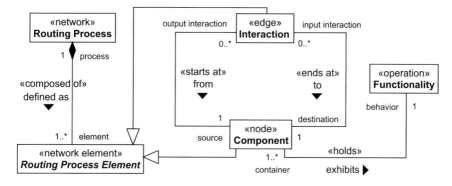

Fig. 6 Routing processes as networks (domain abstraction)

Even when all concepts of the routing process domain are included in the network model domain, constraints are required to ensure model correctness. For example, *i) Components* (i.e., the *nodes*) of the *Routing Process* cannot be isolated, and *ii) Interactions* cannot use the same pair of *Components* as *source* and *destination*. These kinds of constraints restrict the original model giving a constrained network model for representing routing processes.

5.2   DEVS Modeling Levels

DEVS is a system theoretic-based formalism that provides representation for systems whose sequences of events can describe input/output behavior. It provides a general methodology for the hierarchical construction of reusable models in a modular way [93].

Regarding system theory, DEVS models can be interpreted at two distinct modeling levels: *behavior* and *structure* [82]. An *atomic DEVS* describes the autonomous behavior of a discrete-event system as a sequence of deterministic transitions between sequential states. It also specifies how the system reacts to external input (events) and generates output (events). Instead, a *coupled DEVS* describes a system as a structure of coupled components (that can be defined as *atomic DEVS* models or *coupled DEVS* in their own right). Hence, DEVS embodies a set of concepts related to systems theory and modeling to describe discrete event models in terms of their *behavior* and *structure*.

When routing processes are studied as discrete-event systems, the DEVS formalism provides a solid basis for their M&S. From this point of view, routing mechanisms can be structured over DEVS models with aims to define routing processes. Such routing mechanisms can be interpreted as *routing functionalities embedded in the system behavior*.

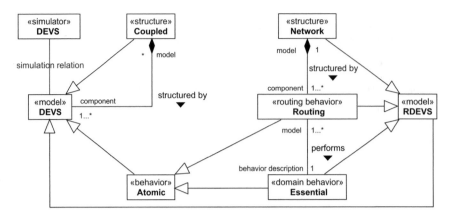

Fig. 7 Main relationships between DEVS and RDEVS

The Routed DEVS formalism (RDEVS) employs the "embedding routing functionality" strategy over DEVS models to provide routing capability from the simulation model conception. Such a DEVS extension was presented in [9] as an adaptation of the classic DEVS that adds routing features to the models by introducing a new modeling level named *routing behavior*. Based on the use of routing policies, RDEVS models provide a formal definition for the M&S of general routing processes as discrete-event systems. Such a definition allows the use of the RDEVS formalism as a "layer" above the DEVS formalism that provides routing functionality without requiring the user to "dip down" to DEVS itself for any functions [93].

The UML class diagram depicted in Fig. 7 illustrates how DEVS and RDEVS models are related. Each concept is described in a class. Two types of classes are included in the diagram: *high-level M&S classes* and *modeling-level classes*. The first set of classes is used to describe concepts related to the M&S framework. These classes are stereotyped as ≪*model*≫ and ≪*simulator* ≫. From the basic classes stereotyped as ≪*model*≫, DEVS and RDEVS models are derived. For DEVS models, *DEVS* is specialized in *Atomic* and *Coupled*. Instead, for RDEVS models, the *RDEVS* class is specialized in *Essential*, *Routing*, and *Network*. Hence, each concrete type of discrete-event model is derived from a ≪*model*≫ class. For the classes that represent concrete models, the stereotype is used to detail the modeling-level. The *modeling-level classes* are stereotyped as ≪*behavior*≫, ≪*domain behavior*≫, ≪*routing behavior*≫ and ≪*structure*≫ to contrast models.

5.2.1   Understanding the Relationships Between DEVS and RDEVS

DEVS models are designed to provide *behavior* and *structure* definitions through *Atomic* and *Coupled* models, respectively. The RDEVS formalism uses three types of models: *Essential*, *Routing*, and *Network*. These models take advantage of the DEVS modeling levels by refining the *behavior* level in two new kinds of modeling levels: *domain behavior* and *routing behavior*. Hence, each RDEVS model belongs to a specific modeling level as follows: The *Essential* model is at the *domain behavior* level, the *Routing* model is at the *routing behavior* level, and the *Network* model is at the *structure* level.

In RDEVS, the *behavior* level of DEVS is now divided into two modeling levels, *routing behavior* and *domain behavior*. Formally, the definition of the *Essential* model is equivalent to the *Atomic* model specification. Then, the *behavior* level of DEVS is conceptually equivalent to the *domain behavior* level of RDEVS. The new modeling level introduced by RDEVS formalism is the *routing behavior* level. Such a modeling level can be seen as a layer above the *behavior* modeling level of DEVS that embeds the routing functionality in *domain behavior* models. To conduct the routing functionality, *Routing* models use routing policies to authenticate senders and receivers before executing their *domain behavior*. Hence, the RDEVS formalism provides a suitable separation of concerns between *domain* and *routing behaviors* using *Atomic* DEVS models an underlying M&S layer.

At the top level, the *structure* modeling levels of DEVS and RDEVS are defined in the same way. Both *Coupled* and *Network* support the construction of models by coupling together *components*. Hence, connections among the models denote how such models (playing the role *component*) influence each other. In a model playing the role of *model*, the output events of one *component* can become, via a coupling definition, input events of another *component*. However, instead of using a modeler design-centered approach to define couplings, the *Network* model structures the connections statically. That is the *Network* definition prefix couplings as all-to-all connections between the *Routing* models that compose it. This leaves the routing task to the routing policies. Then, the events that flow inside a *Network* are defined as identified events. Such events contain data from their *Routing* source and the set of admissible *Routing* destinations. The main theoretical properties of the RDEVS formalism are treated in [9].

The *simulation relation* depicted in Fig. 7 is the one proposed in the M&S Framework 3. A *DEVS model* is executed by a *DEVS simulator*. Now, *RDEVS models* are defined as an inheritance of *DEVS models*. Inheritance enables the construction of chains of derivations forming class inheritance hierarchies. Hence, *RDEVS models* are executable using a *DEVS simulator*. This feature allows building powerful discrete-event models combining DEVS and RDEVS formalisms.

5.3   RDEVS as a Formalization of the Network Model

Formalization makes it easier to work out the implications of an *abstraction* and implement them in reality [93]. In this context, Fig. 6 provides an *abstraction* of the routing process domain through a constrained network model. With aims to show how the routing capabilities introduced by RDEVS formalism provide a reasonable formalization for the M&S of routing processes, Fig. 8 links the conceptual models detailed above. Such a diagram exhibits how the main components of *RDEVS models*

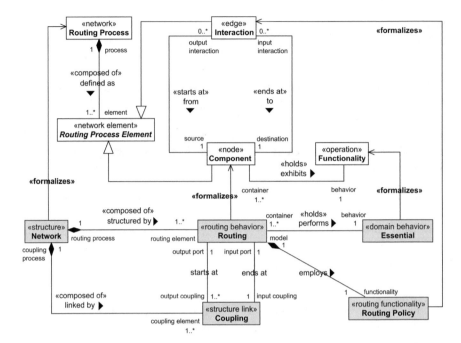

Fig. 8 Conceptual mapping of RDEVS simulation models to network models (discrete-event formalization)

(concepts in gray) can be used to formalize simulation models for routing processes. Such processes are defined as *constrained network models* (concepts in white). The relationship stereotyped as ≪*formalizes*≫ is used to link both sets of concepts. Due to both sets are defined at different levels (abstraction vs. formalization), the association stereotypes are used to clarify the dependencies.

Each type of RDEVS model act as a formalization of some routing process entity. The *Essential* model *formalizes* a *Functionality* defined as an *operation exhibited* by a *node*. Then, the *Functionality* is interpreted as the *domain behavior* of the *Component*. The *Routing* model *formalizes* the *routing behavior* of a *Component*. Regarding the *Routing* model definition, the *routing behavior performs* a *domain behavior* (i.e., an *Essential* model). Moreover, a *Routing* model employs a *Routing Policy* that acts as *routing functionality*. Such a *functionality formalizes* the *Interactions* among *Components*. These *Components* are defined as the *nodes* that compose the constrained *network*. Hence, the *Network* model *formalizes* the *structure* of a *Routing Process* through *Routing* models and *Couplings*. While the *Routing* models act as *nodes*, the *Couplings* act as *links* among *nodes*.

In a DEVS-based solution, coupling establishes output-to-input pathways. Hence, the system modeler is free to specify how events flow through such couplings. Instead, in a RDEVS-based solution, the *Couplings* are defined as part of the *Network* model definition. Such a definition establishes output-to-input pathways as all-to-all connections among *Routing* models. Now, the interaction among *Routing* models depends only on the *Routing Policies*. Such policies are defined considering the *Interactions* among *Components*. The *Routing Policy* is used to define destinations of outgoing events but, also, to verify the incoming events before passing on the functional content defined in the *Essential* model. Hence, the *Network* model is conceived as an extension of the original *Network* that defines the *Routing Process*. Such an extension defines all-to-all *edges* over the *nodes* to make *routing functionality* independent of the *domain behavior*.

Since DEVS is most naturally implemented in a computational form in an object-oriented framework [93], the model detailed in Fig. 8 gives a suitable formalization that can be used to design RDEVS implementations over an object-oriented framework. Regarding such an object-oriented framework, an example is introduced in the following subsection.

5.4 Computer Networks as a Domain-Specific Example

A *Computer Network* is a collection of autonomous computers interconnected by a single technology [79]. Such a network is composed of computer *nodes* and their *interconnections*. Two computers are interconnected if they can exchange information. Intermediate nodes are typically *hardware devices* (e.g., *routers*, *switches*, or *firewalls*) identified by their *hostnames* and *network addresses*. *Interconnections* among *nodes* are defined from a broad spectrum of telecommunication network technologies based, for example, on *physically wired*, *optical*, and *wireless connections*.

Regarding the information exchange that takes place inside a *Computer Network*, *Routing* is the process of selecting suitable paths to carry network traffic through the nodes. The *routing process* usually directs forwarding using *routing tables*. Usually, multiple routes can be taken to reach a destination. Most routing algorithms use only one path at a time. However, there are multipath routing techniques that enable the use of multiple alternative paths to reach a destination. With aims to get a representation of *Routing Processes* in the *Computer Network* domain, Fig. 9 presents a domain model that uses the *routing process abstraction* (Fig. 6) as the foundational metamodel.

Now, following the discrete-event formalization of network models into RDEVS simulation models (Fig. 8), Fig. 10 presents a RDEVS-based solution for the abstraction model depicted in Fig. 9. Operations are defined with aims to provide a basis for further implementations. Given that such a formalization is designed as a conceptual routing framework, multipath routing is allowed. The routing technique to be used as routing behavior depends on network constraints. Hence, the routing technique is attached to the *Computer Network* scenario. When the formalization be applied to solve a particular scenario, the routing technique will be defined as part of the routing function specification.

The formalization framework defines an *Essential* model for each type of *Networking Hardware*. Such formalizations are defined in terms of the model state (i.e., the attribute named *state*) and the discrete-event functions that describe the domain

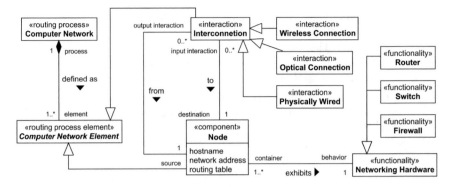

Fig. 9 Abstraction model of the Computer Network domain

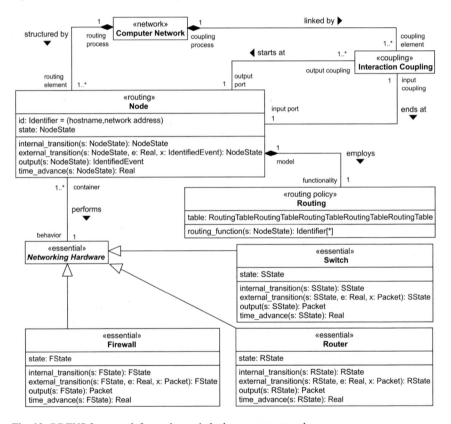

Fig. 10 RDEVS framework for packet-switched computer networks

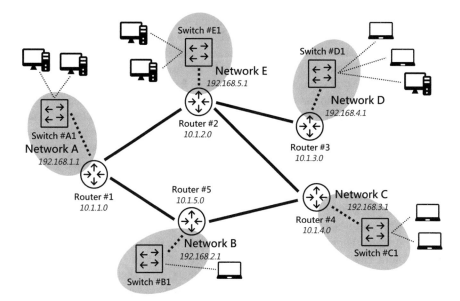

Fig. 11 Computer Network (example)

behavior (i.e., the set of available operations). Each model employs its own state definition. For example, the *Router* model employs the *RState* (*Router State*) type, while the *Firewall* model uses the *FState* (*Firewall State*) type. Then, the formalization allows defining a specific state for each *Networking Hardware*. In all models, the output function produces a *packet*.

Each *Node* that composes the *Computer Network* is defined as a *Routing* model with *id* = *(hostname, network address)*. The model state is detailed in the attribute *state* (that is based on the *NodeState* type). The *Routing* is performed by the *routing_function* operation using the routing *table* as a *predefined parameter.*[1] The domain behavior (*Essential* model) attached to the nodes is accessed through the *behavior* property. Each *Node* performs the *routing behavior* over the *functionality* defined in the *Essential* embedded in it. For example, when the *Essential* produces an output event (a *packet*), the *Node* produces an *Identified Event*[2] (that surrounds the *packet* with its *routing data*). The same approach is used for transition functions.

With aims to provide an example for further implementation, take the *Computer Network* depicted in Fig. 11 as a situation where multipath routing is not allowed. The scenario is divided into five networks named *Network A* to *Network E*. The intermediate nodes are defined as routers and switches. The routing directs packet forwarding through intermediate nodes considering the transmission capacity.[3]

---

[1] Each entry is defined as the tuple *(destination, intermediate, capacity)*.

[2] An *Identified Event* is defined as the tuple *(packet, origin, destination)*.

[3] Maximum capacity is used to choose between several interconnections.

The hostnames and network addresses are defined as part of the scenario description. Moreover, the routing table of each node can be statically defined by analyzing the structure proposed in the scenario. Then, all the parameters required to configure the RDEVS models can be automatically obtained from the scenario description.

It is easy to see that each component presented in Fig. 11 can be mapped into the *Routing Process* description of the *Computer Network* domain proposed in Fig. 9. Hence, a RDEVS-based formalization can be obtained following the formalization framework detailed in Fig. 10. According to such a framework, the scenario employs 2 essential models (*Router* and *Switch*), 5 routing models that represent *Nodes* that perform the *Router* functionality, 5 routing models that represent *Nodes* that perform the *Switch* functionality, and 1 network model that structures the overall *Computer Network* scenario.

Moreover, to define how the *Nodes* perform their behavior, the following set of OCL [33] constraints need to be defined. Such constraints are used to specify (in a declarative language) the operations related to the routing behavior. These specifications can be (later) implemented in any M&S tool (based on a general-purpose programming language) that supports the DEVS abstract simulator. The domain behaviors of the *Networking Hardware* are outside the scope of this chapter.

context Node::internal_transition(s:NodeState): NodeState
—*the result of the internal transition function of a Node is the result of the internal transition function of the Networking Hardware embedded in it.*
post: result = self.behavior.internal_transition(s)

context Node::external_transition(s:NodeState,e:**Real**,x:IdentifiedEvent)
    : NodeState
—*the result of the external transition function of a Node depends on the acceptance of the input event.*
post:
    if x.destination–>includes(self.id) **then** result = self.behavior.
        external_transition(s,s.elapsedtime,x.packet)
    else result = s.update(e) **endif**

context Node::output(s:NodeState): IdentifiedEvent
—*the result of the output function of a Node is a new Identified Event that contains the packet produced by the Networking Hardware embedded in it along with the routing data.*
post: result.oclIsNew() **and** result.packet = self.behavior.output(s) **and**
    result.origin = self.id **and** result.destination = self.
    functionality.routing_function(s)

context Node::time_advance(s:NodeState): **Real**
—*the time advance function of a Node returns the sigma value.*
body: s.sigma

context Routing::routing_function(s:NodeState):**Set**(Identifier)
—*the routing function of a Node returns the packet destination following the transmission capacity primitive.*
body: let paths:**Set**(Tuple{destination:Identifier,intermediate: Identifier,capacity:**Real**})=self.table–>select(r|r.destination=s. destination) in paths–>select(r|r.capacity=paths.capacity–>max()). intermediate–>asSet()

Hence, the formalization framework presented in this section allows defining domain-specific formalizations for routing processes using network models as structure. The final simulation models are based on the RDEVS formalism. Once the domain-specific formalization is defined, new scenarios can be modeled. Even when all scenarios are based on the same domain, each situation can be defined to use its routing policy. Such a definition is detailed in the operations. The domain-specific formalization is then complemented with the specification of the required operations that perform the routing behavior for each case. These specifications can be used as inputs during the implementation process.

6 Co-simulation of Complex Networks

After the presentation of complex network challenges, this section will highlight how co-simulation can be a support for their M&S. It presents the basics of co-simulation in comparison with simulation and positions its challenges and how the DEVS formalism can be helpful. As an example, the design of a co-simulation with a network dedicated M&S software is exemplified using a coupling example between OMNeT++ [84] and MECSYCO [16].

6.1 Co-simulation Principles

We present in the following the general co-simulation principles and challenges.

6.1.1 Simulation to Co-simulation

In standard M&S approaches, the simulation of a system is done using a single model, simulated using one simulator implemented in one software (or library, language…). This activity can be split in four main steps [28]:

- **Conceptualization**: formulation of a set of hypotheses and assumptions on the target system to build the model.
- **Formalization**: description of the system, often in a mathematical rigorous notation. This process is twofold: (1) the structure and behavior of the model and (2) the algorithm to advance it through time (the simulator) have to be defined [74].

- **Operational specification**: it consists in preparing the transformation of the model and its simulator into something that can be programmed on a computer.
- **Computing**: writing the corresponding code to execute and collecting the results.

These steps are time-consuming and most of the time, modelers use dedicated M&S software to earn time, e.g. OMNeT++ or ns-3 in network M&S.

Complications come when one wants to consider phenomenon from other domains and modeled with other formalisms and tools (emergency networks which couple mobility and IP networks [89], electrical grid behavior coupled with economic market [11], smart-grid modeling [86]…). A way to deal with this complexity is to split the overall system into interacting sub parts and to design a set of interacting models (multi-modeling). Co-simulation consists then in simulating each model separately and ensuring the synchronization of the simulators (often implemented in different software). We pass from simple simulation with one model, one simulator in one software to a co-simulation of a set of interacting models simulated by their simulators and possibly implemented in several software.

Co-simulation is gaining interest because, for many reasons (different developing teams, different suited formalisms or domains), it is profitable to implement each heterogeneous model in the most appropriate M&S software, and co-simulation is the way to put them in interaction [30].

6.1.2 Co-simulation Challenges

Putting in interaction models from various software implies to solve several interoperability issues. Three main levels can be identified [80]:

- *technical*: can the different software be connected? This level is linked to both the computing and the operational specification steps of the M&S activity.
- *formal*: can the simulators be supervised to enable data exchange and synchronization?
- *conceptual*: are the models consistent with each other and is it meaningful to couple them?

The technical challenge is related to the way M&S software are implemented. If there is no way to interact with them, e.g. no API, then their usage in co-simulation can be very hard and request high re-implementation costs.

The formal level is linked to the formalism chosen during the software implementation (if there is one). It will notably influence the representation of models and the implementation of simulators. If there is no bridge between abstract simulators, interaction between their concrete implementation cannot be done in a rigorous way.

On the conceptual level, it is often impossible to reuse already implemented models from another software to use them in a co-simulation. Such models were implemented for a specific target system and were not meant to be coupled with others [67]. The interest of co-simulation is rather to let a modeler benefits from the dedicated features of a specific M&S software to write a model designed to be coupled and co-simulated with others. Each software comes with a set of assumptions on the models that can be written within them and how they can be simulated. These assumptions can also be linked to the technical level. This is why the consistency between the models we want to couple must be carefully checked.

In addition to the interoperability challenges, other side questions must be dealt with such as the performances (accuracy of results, computation time) linked to the synchronization algorithm, the modularity (changing one model by another), the extensibility (adding models from other simulators/software) etc. Providing these M&S features requires relying on standard means to perform co-simulations. The next part highlights how DEVS can be a step to that as a pivot formalism for the interaction between several simulators.

6.2 DEVS as Co-simulation Support

DEVS offers a clear distinction between models and simulators and its abstract simulation protocol can be used as a co-simulation algorithm. In addition, its universality property means that every discrete event system can be represented in DEVS, i.e. other discrete event formalisms can be mapped in DEVS as we can see on the formalism transformation graph in [83].

These advantages have been used for several DEVS-based network co-simulation works such as the ones presented below.

6.2.1 DEVS/NS2

DEVS/NS2 [44] is a framework that couples DEVS models with NS2 (network simulator 2). The purpose was to benefit from the model library of NS2 in a DEVS framework without having to re-implement it in DEVS. At the technical level, NS2 is an open source C++ library in which new components and simulators can be implemented. The core idea at the formal level was the design of the *NS2 event queue agent*, this atomic model wrapped around the NS2 event queue enables to handle it like another DEVS atomic model.

6.2.2 DEVS/BUS

DEVS BUS [45] is a co-simulation middleware architecture used to couple several discrete event software. The purpose was to propose a generic simulation infrastructure to couple heterogeneous simulators. DEVS BUS focuses on the standardization of the data exchange and synchronization strategy to couple simulators. The key point is at the formal level, DEVS converters are defined for every simulators to make them DEVS-compliant. Then the DEVS BUS controller, a centralized entity based on the DEVS simulation protocol, ensures their synchronization and data exchange. At the technical level, the DEVS BUS controller uses a communication infrastructure whose implementation is left to the user's choice.

6.2.3 MECSYCO

MECSYCO (Multi-agent Environment for Complex SYstem CO-simulation) is a DEVS-based co-simulation middleware [16]. It was notably used to couple network simulators, OMNeT++ [84] and ns-3 [36], with equations-based simulator, Modelica [27] exported as Functional Mockup Units [10], for cyber-physical system simulation [85, 86]. MECSYCO is on license AGPLv3, its source code can be found on gitlab.inria.fr/Simbiot/mecsyco.

This approach is similar to DEVS BUS, it is a DEVS-based middleware relying on a DEVS wrapping strategy. The difference lies in the choice of implementation. The particularity of MECSYCO is its multi-agent architecture that offers a decentralized co-simulation algorithm and means to resolve data representation heterogeneities (different time scales or data types).

6.2.4 Synthesis

DEVS/NS2 highlights the integration of one network simulator into a DEVS-based one but is limited to this integration. DEVS BUS and MECSYCO offer solutions to implement co-simulations of heterogeneous simulators and software. They are meant to be generic and mainly focus on the formal level. Regarding the co-simulation challenges, all these works highlight the integration ability of DEVS and its usage as pivot at the formal level.

6.3 Network Co-simulation Example

To complete the discussion about complex network co-simulation, the section below gives insights on how to connect an existing network simulator, namely OMNeT++, in a DEVS-based platform, namely MECSYCO. This example is directly based on the work done in [85, 86].

6.3.1 OMNeT++ Integration Principle

The Fig. 12 shows a simplified scheme of OMNeT++ simulation architecture, see 12, and the elements added or modified for the co-simulation, see 12.

To give a short overview, OMNeT++ simulation involved three main entities: scheduler ('.cc'), network model ('.ned') and configuration file ('.ini'). The scheduler is responsible for the event-based simulation and provides methods to run it (get

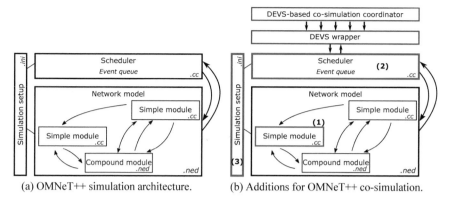

(a) OMNeT++ simulation architecture.   (b) Additions for OMNeT++ co-simulation.

Fig. 12 From OMNeT++ simulation to OMNeT++ co-simulation

next event time, get next event from the event queue...). The network model is composed of modules that are either simple modules whose behaviors are described in C files, or compound modules composed of connected submodules. Simple and compound modules are the OMNeT++ equivalent of DEVS atomic and coupled models. Modules interact through messages described with '.msg' files. Finally, the configuration file sets up the simulation by selecting the network, its parameters, and notably the scheduler to use.

The implementation steps are detailed below:

1. **Open OMNeT++ simulator**: The approach used in [85] overrides OMNeT++ sequential scheduler to make MECSYCO able to handle the simulation process. The new scheduler bypasses the normal OMNeT++ simulation. By choosing it, MECSYCO can control the OMNeT++ event-queue and simulation and synchronize it with other simulators. The new scheduler connects itself to a MECSYCO wrapper that ensures the link with the co-simulation middleware.

2. **Open OMNeT++ network model**: The next step is to define input and output events of the network model. This part is the most use case specific one. Whatever one wants to do, it will require the implementation of simple modules whose behaviors will take into account external events and which will select the events to be extracted and sent to other models. A network model has then to be built with these modules. Note that these new simple modules can be integrated in network based on existing OMNeT++ model libraries. For example, the Inet TCP library enables to implement new TCP applications and to link them with the standard host module. This possibility is exploited in the co-simulation below.

3. **Configure OMNeT++ simulation**: Finally, an OMNeT++ configuration file has to be set up with the new scheduler and a network containing the new modules.

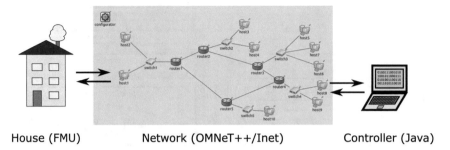

House (FMU)   Network (OMNeT++/Inet)   Controller (Java)

Fig. 13 Co-simulation with an Inet TCP network model

6.3.2   Co-simulation Using Inet TCP Library

The following co-simulation involves a network model connecting the air conditioner of a smart-house to a remote controller. The house is built on Modelica and exported as FMU, the external weather is from an SQL database and the external controller is designed here in Java.

This example is from [66], it is used for teaching in a course about co-simulation principles, challenges and limits. In [66], the network is not taken into account. Here, a network model written in OMNeT++ using the Inet TCP library is added using the MECSYCO/OMNeT++ wrapper described previously. The Fig. 13 presents the interaction between the house, the controller and the network.

The network is based on the Computer Network defined Fig. 11. All the communication (routing, encapsulation etc.) is handled by TCP library of Inet. Data travels through the network in TCP chunks. From the network point of view, the house is a TCP application on the first host (on the left) and the controller is a TCP application on the eighth host (on the right).

6.3.3   Perspectives and Limits

The previous example shows how a network model can be integrated in a co-simulation. The interest is that we can benefit from a wide network library by implementing a wrapper instead of re-implementing all the network models. Using DEVS as a formal basis to develop a generic middleware (MECSYCO here), the network wrapper joins a set of existing wrappers for various M&S tools (FMU, MatLab etc.) that can interact with each other, enhancing the modeling possibilities. However, co-simulation has several side effects and limits.

In terms of update, the wrapper is greatly dependent on the evolution of model libraries. The wrapper coded for the previous example may be outdated with new versions of Inet.

In terms of integration, OMNeT++ framework enables to write network libraries in a flexible way leading to several heterogeneities. The developed wrapper is specific to the TCP library of Inet. Using another library of models will require coding a new specific wrapper (following the same principles). The current wrapper has also its limits, e.g. the time resolution is assumed to be in nanoseconds for the simulation. Better integration is possible but would request more time.

In terms of representation, the network wrapper offers a highly detailed representation of data that travel through communication networks. This can be interesting for cyber-physical system modeling when one wants to study in detail the impact of the interaction of several smart-systems on the network (and vice-versa). However, taking into account other phenomena (dynamic structure changes, breakdowns...) will depend on the capabilities of the integrated library and simulator. In all cases, it will require an update of the wrapper to define these kinds of events at the co-simulation level.

6.4 Synthesis on Co-simulation

Along this section, we highlight both how co-simulation can be used for complex network M&S and the interests of using DEVS as formal basis. Co-simulation has two main advantages: (1) using several simulators allows mixing several perspectives and (2) it is compliant with a wrapping-based software reuse strategy which make possible to benefit from existing M&S software capabilities and libraries avoiding high re-implementation costs.

Nevertheless, co-simulation requests a rigorous handling of the software heterogeneities to ensure their interoperability. This can lead to a high complexity. Furthermore, reusing a software allows to take advantage of its functionalities, but this also puts us within its boundaries. This means that we have to be careful about what we want to model and be familiar with the software we want to use so that we do not get stuck by theirs limits and assumptions.

To conclude, co-simulation is gaining interest and is becoming easier to use with the development of technical standards such as FMI [10], or the use of a rigorous formal basis like DEVS. This can lead to new studies using co-simulation but which focus on the inter-influence between the models handled by the simulators. Regarding complex networks involved in emergency situations (Sect. 2), co-simulation can be used to design frameworks where we can cross-test and validate the routing infrastructures (Sect. 5) and the end-user applications modeled in distinct dedicated M&S software.

7 DEVS-Based Infrastructure to Design Complex Networks

Understanding the exact mechanisms behind decision-making or training efficiently, actors require accurate simulators. However, integrate all these different aspects increases the complexity of the simulation environment, because of the heterogeneity of the models.

Previous sections addressed problems related to network design in emergency response. We defined what kind of properties and behaviors are important for situational awareness, and we highlighted the importance of reliability in the process of decision-making. We saw that communication and information sharing are critical, and we presented different ICT solutions studied in the literature.

Most of the challenges addressed by network design in emergency management are related to specifications of the requirements, routing, multiplicity of tools, and oversimplified models. Existing architectures and MANET simulators suffer from a lack of accuracy induced by simplified scenarios, a lack of integration between existing technologies, which often leads to inconsistent behaviors, and difficulties to focus on the right aspects of complex applications like emergency response. As an example, traditional MANET simulators focus on the verification and validation of routing protocols and algorithms. However, they do not take into account the specificities of the environment in which the real technology is deployed. For instance, real routing protocols and underlying layers are generally implemented and handled by the operating systems and drivers, not accessible to developers; communications depend on physical phenomena in case of wireless transmission; literature provides many examples of emergency situations like aircraft emergency in which the pilots minimize some important signals because they were overtrained in unrealistic conditions and simulation scenarios.

Then, we talked about specifications of behaviors using UML metamodels, which map to hierarchical DEVS models, and we gave an example of high-level specifications of routing mechanisms in RDEVS. Next, we highlighted some co-simulation architectures based on DEVS.

In this section, we present a software and hardware infrastructure to support the design and development of complex networks [89, 90] and we show how M&S approaches based on DEVS formalisms enhance computing environments. Especially, we introduce the concept of an integrated heterogeneous simulation environment for the verification and validation of the MANET-based ecosystem. This integrated architecture helps designers to implement a co-simulation environment enabling hybrid modeling. It also supports validation processes by providing a way to execute real applications directly in simulation environments.

7.1 DEVS-Based Architecture for MANET Simulation

Using DEVS formalism as a basis of MANET simulation to bring scalability, adaptability, reusability to mobile network applications is not new. Some work has been proposed, for example, to use existing MANET simulators like NS2 for modeling of low-level network protocols and components, while DEVS simulators are used as a controller for high-level behaviors and as a handler of the interactions between actors and components [44, 81]. From this approach arises the idea that it is possible to create a DEVS-based heterogeneous simulation framework [45, 91] in which MANET simulators and DEVS-compliant simulators may coexist.

Heterogeneous simulation [22, 24] concerns the use of a collection of simulators developed in different simulation languages, environments, and paradigms, and which work in an interoperable way to achieve a simulation. The main addressed problem is that such interoperations need data exchange and time synchronization between the simulators but also the definition of an interaction semantics. While data exchange can be easily resolved through a standard messaging protocol between the simulators, time synchronization is a hard task due to the different possible natures of models: untimed, continuous, discrete-time, or discrete-event. Furthermore, computerized models may differ because of the language used to implement these conceptual models and because of the compiler or the interpreter, which translates the source code into executable. Indeed, time representation strongly depends on the programming language used for implementing the simulators, and possible operations on time also depend on the hardware architecture of the simulation host. Moreover, the same statement can be made if we consider parallel and distributed simulations in which hardware architecture can also bring errors because of approximations and abstraction. From this point of view, heterogeneous simulation is a kind of co-simulation as defined in Sect. 6. However, heterogeneous simulation also encompasses two aspects: hardware-in-the-loop, which integrates real hardware into the simulation as a source of data, and software-in-the-loop, which recreates a piece of software, abstracted or not, in a simulated environment. In other words, a heterogeneous simulation makes simulators and real products coexist and interact in the same ecosystem.

Achieving this level of integration in the case of emergency management requires a framework that combines MANET simulators focusing on protocols evaluation, environmental simulators, and models of real applications. As we stated in Sect. 2, using real applications or models derived from real applications, allow designers to validate the routing protocols under development. At this step, it is also important to remind that a computerized DEVS simulator is a program as any other software and can be described as a classical state machine derived from a conceptual DEVS model. A DEVS-compliant simulator is defined as a model whose conceptual model can be defined using a DEVS algebraic formalism and whose the computerized simulation follows the DEVS abstract simulator algorithm [89]. From this perspective, it is possible to turn any application into a DEVS-compliant simulator thanks to the hierarchy of simulation formalisms [29].

7.2   OSI Model and Communication Stack

Interconnection and communication have been conceptualized and modeled mainly through two models: the OSI model [38] and the Internet Protocol Suite [69] as described in Fig. 14. A quick comparison between the two models shows that:

- layers 5, 6 and 7 correspond to the upper layer of the TCP/IP model;

Layer			TCP/IP Layer	Protocol data unit (PDU)	Role	Implementation Level
Host layers	7	Application	Application Protocols	Data	High-level API & Software.	End-user software & application
	6	Presentation			Translation of data using an application protocol for character encoding, data compression and encryption/decryption	Network Library
	5	Session	Security		Managing sessions between two nodes	
	4	Transport	Transport	Segment, Datagram	Reliable transmission between points on a network (segmentation, acknowledgement and multiplexing)	OS Kernel, Socket Library
Media layers	3	Network	Network	Packet	Network topology management (addressing, routing and traffic control)	OS Kernel & Drivers
	2	Data link	Link	Frame	Reliable transmission between two nodes connected by a physical layer	
	1	Physical	Network access	Bit, Symbol, Waves	Transmission and reception of raw bit streams over a physical medium	Electrical / Hardware

Fig. 14 The OSI model and TCP/IP communication model, and their classical implementation. Colors represent the possible abstracted layer in simulation model

Fig. 15 Layer-to-layer communication in OSI model

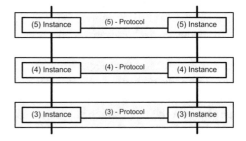

- layers 4, 5, 6, and 7 operate at the host level;
- layer 3 is present in both models;
- layers 1 and 2 operate at the media level.

Therefore, this structure justifies the abstract model of communication described in Sect. 2. Indeed, the host layers generally correspond to the host that serves as a software and a set of accessible algorithms to the designers, while the media layers correspond to what designers consider as hardware or hidden implementation. From this perspective, it seems logical to abstract layers 4, 5, 6, and 7 in one application layer on the one hand, and abstract layers 1 and 2 into one unique physical layer on the other hand. While routing protocols take place in layer 3, simulation models keep it as an independent layer (Fig. 2).

Nevertheless, if we analyze the real implementation of these layers, we notice that actually layers 3, 4, 5, 6, and 7 are implemented in software (kernel or application), while only the physical layer is handled by hardware. Responsibility of Layer 2 is shared between kernel and hardware. From this observation, we can deduce that layer 1 is the only *non-computerized* non-accessible model. Moreover, OSI model and TCP/IP model describes layer-to-layer communication (Fig. 15). This communication mode is represented by the fact that each protocol data unit (PDU) at an upper level is encapsulated in the PDU at the lower level. Therefore when a message is sent (top to down), each PDU contains information from the upper layer. When a node receives a message (bottom to up), internal PDU are decapsulated and sent to the upper layer.

Given these elements, we propose a new classification:

- Virtual layers correspond to computerized layers, meaning layers 2–7;
- Physical layers correspond to physical phenomena, which occurs in layer 1.

Using this representation, we argue that any virtual layers can be replaced by any other instance of layers at the same level independently of their implementation (abstracted or real) or any combination of these implementations. On the other hand, the physical layer can be either real hardware or an abstract model, but not both. This new model is the base of the Virtual Communication Stack model [89, 90].

7.3 *Virtual Communication Stack*

The virtual communication stack (VCS) is a model which is inspired by and extends ns-3 TAP mechanism [71], and Virtual Private Network (VPN) which adds other layers above the traditional media layers through TUN/TAP adapters: VPN link/network layers are implemented at the kernel level or handled by a third party (e.g., OpenVPN). Packets are carried using virtual kernel network interfaces and injected into the operating system stack, emulating communication. From the transport and application layers perspective, communication is entirely transparent (Fig. 16).

The VCS acts exactly as a VPN middleware (Fig. 17) by creating a new stack below the application layer and above the OS transport layer. VCS replicates all the virtual layers described in the previous section. VCS also carries a message between virtual space and operating system space. Concretely, this middleware handles connections and communication between nodes, devices, and with any external, real or virtual, networks.

Using this principle, the VCS model allows the designer to swap at any moment between a real implementation handled by the operating systems or real hardware and an abstracted space in which simulation models can live. This is possible because:

- from the perspective of the application layer, VCS is a part of the transport layer. Therefore, there is no need to define a specific implementation for simulation;
- from the perspective of the operating system space and simulation space, VCS is a part of the application layer.

Moreover, the exact implementation of the VCS (e.g., using one or several services, distributed or embedded, nature of devices, OS restrictions) has no impact on the global design of the communication. Therefore, a packet can follow any path between the OS space and the virtual space (i.e., a final message can be a combination of PDU from both OS space and virtual space). Communication between software through the network is done thanks to virtual sockets that connect a client and a server as usual. Data transmission is totally transparent for the end-user application.

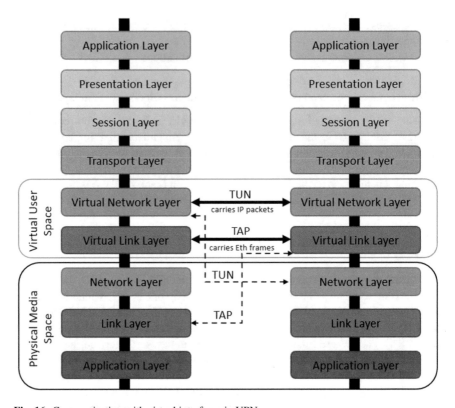

Fig. 16 Communication with virtual interfaces in VPN

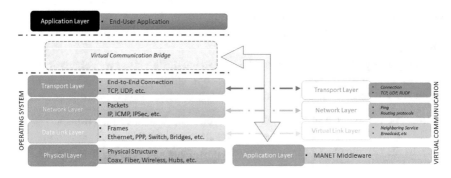

Fig. 17 Virtual communication stack

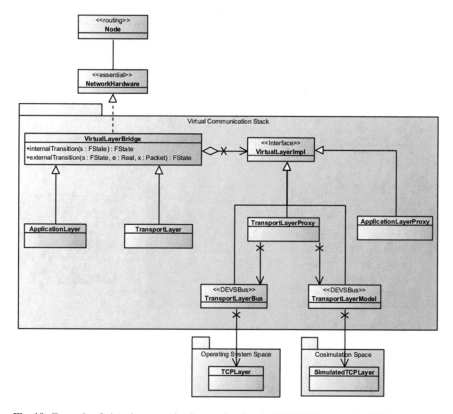

Fig. 18 Example of virtual communication stack using the RDEVS framework. All layers are not represented

7.4 Co-simulation Using the VCS and DEVS

The property of the VCS enables the possibility to combine a real-world space with a DEVS-based simulated space. This is possible thanks to the RDEVS formalism described in Sect. 4 and the co-simulation principles described in Sect. 6.

First, any automata of application may be defined as a DEVS model that behaves exactly as this automata. While the idea is to embed the VCS in any application - we remind that from the application point of view, the VCS is a part of the transport layer -, we can define a conceptual model of the VCS as a coupling of automata with synchronization performing two actions: send and receive data, and update its internal routing table. Consequently, a layer of the VCS may be considered a Node in the RDEVS framework (Fig. 18). In this design, the virtual communication is a set of *VirtualLayerBridge* which acts as *NetworkHardware*. *VirtualLayerBridge* can be refined into specific layers (*ApplicationLayer*, *TransportLayer*, etc.) implementing any corresponding protocols. From this perspective, we denote that communication through the VCS is seen as communication through a network. In other words, the

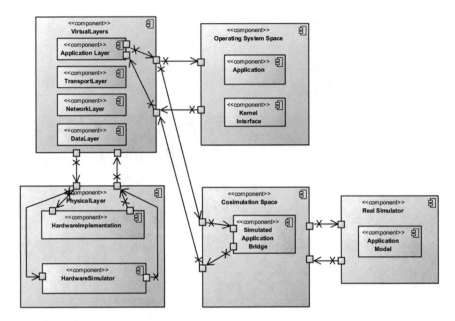

Fig. 19 Coupled DEVS model representing the final heterogeneous simulation architecture. All the layers are not represented

VCS itself is seen as a complex network that supports the MANET protocols. This model may be then abstracted or refined: at a more abstract level, *VirtualLayerBridge* may be associated with an *Essential* component as described in Sect. 4, or to a more complex DEVS model. It is also interesting to note that this conceptual model is completely independent of the implementation of the VCS. While a layer is a node of a network, this model describes a monolithic implementation of the VCS as well as a distributed version. In the second one, we can also refine the model to make the protocol communication between each layer appears using RDEVS. This example shows then that RDEVS allows the designer to model any algorithm which requires a routing protocol.

Second, co-simulation is achieved using a DEVS-bus strategy. We stated that any computerized algorithms might be mapped to a DEVS-compliant simulator by implementing the DEVS protocol. Therefore, coupling a DEVS-bus with any algorithm through a proxy should allow the transformation of any computerized algorithm (seen as a finite automaton) to a DEVS-compliant model. The critical point concerns the synchronization between simulators. This synchronization may be achieved using a DEVS-Bus between the VCS and the real application on the hand, and between the VCS and the other co-simulation environment on the other hand. Finally, we get the DEVS structure described in Fig. 19: Virtual Layers and Operating System Space are seen as a DEVS coupled model named Virtual Domain, while the Physical Layers compose the Physical Domain as expected in the initial requirements of our commu-

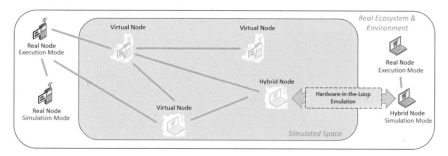

Fig. 20 Heterogeneous simulation infrastructure and ecosystem

nication model. This property allows the realization of the two important aspects of heterogeneous simulation: Software-in-the-loop and Hardware-in-the-loop.

7.5 Integrated Software-in-the-Loop and Hardware-in-the-Loop Paradigms

Software-in-the-loop (SIL) basically uses a software model to simulate a component of a system. It may be basically a simple simulator or a complete emulator. In contrast, Hardware-in-the-loop (HIL) consists of using real piece of hardware in a closed simulation environment. In both cases, the models work in "execution time," while the co-simulation environment works in simulated time. Consequently, the crucial problem is the ability to synchronize all these components. We already stated that VCS architecture has this ability. Indeed, DEVS mechanisms simulate time using the time advance function. Using this function, the VCS itself makes retention of packets (if needed) if the date of the next packet transmission is late in the future. To illustrate this approach, if a physical simulation model computes that a given packet will take 3 s to be transmitted to another node if this computation takes ten milliseconds, the VCS buffers this packet and waits before transmitting it to the upper layer, if this layer is in the operating system space, and conversely. This property allows the connection of real hardware and real software to a simulated environment as shown in Fig. 20

We can distinguish three types of nodes:

- Real nodes are devices which are executing the real communication pipeline through a VCS;
- Virtual nodes are fully simulated devices; a real implementation of VCS is running to handle communication, but a part or the entire devices is modeled;
- Hybrid nodes are devices whose model is composed of real implementation and simulated model. For instance, mobility can be fully simulated while the physical layer can be partly simulated. Simulation can help to extrapolate values in this case.

Again, from the point of view of an end-user application, the communication is entirely transparent through a socket mechanism. This means a normal application can be used to test all the protocols supporting a network and then compared with the result obtained in real conditions in order to perform triple validation: validation of the simulation model, validation of the MANET-based infrastructure, and validation of the application.

7.6 Summary

This section presented a software architecture based on RDEVS and virtualization that supports the implementation of co-simulation environments for the complex network. The routing infrastructure proposed by the RDEVS framework allows specifications of routing mechanisms which are crucial in the context of simulation of complex networks for two reasons: the co-simulation environment is itself a complex network on the one hand, with the ability of messages routing from real space to virtual space and vice-versa; on the other hand, this resulting network simulates a real communication network. Naturally, RDEVS proposes an elegant way to solve a complex problem encountered in all the previous attempts of MANET modeling and simulations: it makes the topology of the simulation structure match with the topology of the simulated network.

Virtualization also allows integration in the same environment, execution, and simulation. Instead of abstracting implemented components of a system under study or reimplementing software specifically for a specific simulation tool, the virtual communication stack allows the designers to reuse real parts of their system and choose what they want to abstract. Hence the experimental frame is closer to the real world. This increases the reliability and accuracy of the simulation results, which is critical in the case of emergency response. Putting together, co-simulation and virtualization allow the implementation of different simulators or emulators to verify and validate properties related to different kinds of networks in the case of emergency response (communication network, decision-making processes, swarm intelligence interactions, etc.)

Finally, the VCS architecture proposed in this section solves two problems presented in Sect. 2: the choice of the good level of abstraction by subclassing the *essentials* components and the possibility to integrate SIL/HIL testbeds. DEVS-based co-simulation associated with DEVS-based VCS achieves to define a comprehensive framework in which different simulators can be integrated. An example of this approach has been proposed to integrate OMNeT++, and ns-3 [89]. Heterogeneity is completed by hybrid modeling and simulation of complex data networks.

8   Hybrid Modeling and Simulation of Complex Data Networks

In this section, we will discuss how hybrid packet-level [14] and fluid-flow [13] simulations can interact within the same unifying DEVS formalism, reducing execution times while retaining individual packet traces.

The *packet-level* network simulation approach represents data flow packet-by-packet as it was seen in previous sections. It is the most widely used approach as it provides fine-grained results yielding results comparable to the real systems. Unfortunately, simulation time grows linearly with the number of nodes and link speeds, making them unsuitable for large high-speed networks due to the high execution times. In the future, the gap between real-world networks and the performance capabilities of network simulations is expected to grow, as network technologies have witnessed an exponential growth in terms of the bandwidth [2] and topology size.

On the other hand, the *fluid-flow* network simulation approach uses Ordinary Differential Equations (ODEs) to represent network traffic, trading execution speedups for coarser-grained accuracy [13, 55, 65]. ODEs are faster to solve, execution time is independent of link speed, but they only represent averaged network behavior. Tools for solving ODEs require knowledge of continuous dynamic systems which are radically different from the packet-level simulation tools. Also, ODEs are classically solved using discrete-time numerical methods, like Euler or Runge-Kutta. ODEs can also be solved within DEVS using Quantized State Systems (QSS) numerical methods [21, 48, 93], which result in discrete-time simulations and models can be developed with the same tools as packet-level simulations.

A third *hybrid* approach tries to retain the benefits of the *packet-level* and the *fluid-flow* approaches by treating the system as an interaction between the discrete and fluid parts [18, 34, 54]. Hybrid models aim at retaining the performance advantages of fluid-flow models while providing detailed simulation traces of selected packet-level models. One possible solution is to execute a co-simulation between classic discrete-time ODE solvers and discrete-events packet simulators. Unfortunately, ODE solvers and packet simulators require substantially different backgrounds and tools, so network experts need a strong mathematical background to define continuous dynamic equations and numerical solving methods as well as to master network protocols and control logic. Moreover, the disparate time management of discrete-event simulators and discrete-time time solvers must be synchronized. This leads to the need for clever ad hoc synchronization algorithms, constraints in choosing step sizes, smoothing of packet-level traffic, and handling discontinuities in numerical solvers. At the heart of the problem is the issue of trying to force one paradigm to fit into another [34, 54].

In this section, we will discuss the hybrid approach where both packet-level and fluid-flow simulations are represented under the DEVS discrete-event formalism yielding hybrid simulations that do not require time synchronizations and interact smoothly. Contrary to other approaches [34, 54], there is no need to define boundaries or regions of the network that are represented uniquely by either packet-level or fluid-flow model. Moreover, it simplifies the adoption from network experts as fluid and discrete models are developed using the same PowerDEVS [8] tool. DEVS hierarchical compositions are used to abstract away details of modeling dynamic equations. At the same time, model and simulation are kept in separated domains removing the need for knowledge on numerical solvers. In this approach, ODEs are solved by QSS methods where asynchronous events are the natural way for the simulation to advance so that stochastic packet arrivals can be treated within normal integration steps. Moreover, QSS integration results in dense polynomial approximations with errors guaranteed to be bounded at all time intervals; thus, discontinuities associated with state events, for example, when routers queues are filled, can be correctly detected and handled naturally, influencing discrete packet control logic.

Finally, we will discuss the differences, advantages, and disadvantages of the co-simulation and hybrid approach.

8.1 Combining Packet and Fluid DEVS Network Models

The proposed hybrid approach is an extension to the DEVS hybrid model proposed by Castro [17, 18]. Packet-level model libraries were developed and used at a large scale to model CERN data acquisition networks [14, 15, 51]. Fluid-flow ODEs are based on well-known dynamic TCP models originally proposed by Towsley [60, 65] which were implemented in QSS and extended to allow additional features (e.g., UDP flows, tail-drop queues, etc.) [13].

The hybrid model use packet-level and fluid-flow models without modifications, and flows interact within the output ports of new *hybrid router* models as shown in Fig. 21. Additionally, packet-level sources are connected to hybrid routers by new *hybrid links* that incorporate a piece-wise continuous signal representation of packets. We now discuss these two new models in detail.

The *hybrid link* model shown in Fig. 21 represents packets traversing a link as a sustained stream of bits. Discrete packet arrivals are transformed into piece-wise continuous signals without loss of accuracy by representing the link's ON/OFF state (ON: sending bits, OFF: not sending). Like in packet-level links, when packets traverse hybrid links, they are scheduled to arrive at the other end according to the configured propagation delay D and link capacity C. Additionally, a continuous outgoing rate $d^p(t)$ is incorporated as follows: upon receiving a packet of size P at time t, the outgoing rate is set to the link capacity $d^p(t') = C$ for the period of time the link would take to send the packet $t + D <= t' <= t + D + P/C$; if there are no scheduled packets, the outgoing rate is reset to zero. This way, the outgoing rate $d^p(t)$ models the link state either sending bit streams at the link capacity (when $d^p(t) = C$) or not sending (when $d^p(t) = 0$). This approach relies heavily on DEVS ability to represent both discrete event packet arrivals and QSS continuous signals within the same formal framework, as well as the inherent discrete event nature of QSS and its efficiency in treating discontinuities in piece-wise signals. Smoothing continuous signals can greatly help to enhance performance on hybrid models, although it is not

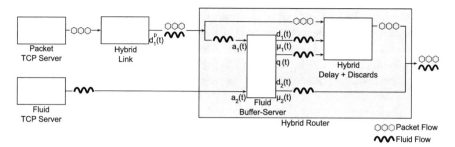

Fig. 21 New link and router DEVS models to support hybrid network simulation. The hybrid router is depicted with 2 inports, 1 outport, and the routing table is omitted for simplicity

strictly necessary. A simple time window smoothing is already available, but more complex algorithms could be implemented.

As shown in Fig. 21 The *hybrid router* coupled model has a number of input and output ports where both packet-level and fluid-flow can be connected and encapsulates the logic for their interaction. It contains a routing table, a *fluid-flow buffer-server* queue for each output port, and applies fluid-flow queue metrics to outgoing packets (*Delay+Discards* model). At the routing table, incoming flows are directed to the corresponding outgoing port where packet-level and fluid-flows can converge at the fluid-flow queue.

The *buffer-server* queue model is reused from the fluid-flow library without modifications. This model encapsulates the ODE equations described in [12] handling multiple flows, detection of buffer over/under flow, and Retarded Functional Differential Equations with implicit delays [19]. There exist versions for tail-drop behavior and Random Early Discard (RED) policies. The model takes a maximum buffer size Q_{max} and an output capacity C as parameters. The **inputs** $a_1(t)$, ..., $a_n(t)$ are continuous QSS signals representing incoming bit streams at time t of flows $1, \ldots, n$. The outputs are also continuous QSS signals representing at time t: $q(t)$, the **buffer size** ; $\mu_1(t)$, ..., $\mu_n(t)$, the **discard rates** of flows $1, \ldots, n$; and $d_1(t)$, ...,$d_n(t)$, the **departure (output) rates** of flows $1, \ldots, n$.

At the hybrid router, packet-level and fluid-flows converge that the fluid-flow queue determined by the routing table. Fluid-flows are treated just like in fluid-flow routers, i.e., they are routed to an outgoing port and fed as input a_i in the corresponding fluid-flow queue that determines its new departure rate $d_i(t)$ and discards rate $\mu_i(t)$. Similarly, the ON/OFF continuous piece-wise rates d_i^p generated by the hybrid link for each packet-level flow are also fed as inputs a_j to the fluid-flow queue determined by the routing table. Thus, the hybrid router output ports model the queuing effects of the interacting packet-level and fluid-flow. At the same time, these continuous queuing metrics are used to schedule outgoing discrete packets. QSS methods produce dense polynomial output trajectories that describe dynamics at any point in time with errors guaranteed to be bounded by QSS quantums ΔQ_{rel} and ΔQ_{min}. In particular, QSS output trajectories from the fluid-flow queue are evaluated on every discrete packet arrival. For a packet arriving at time t, it is discarded with probability $\frac{\mu_i(t)}{a_i(t)}$ taken from an uniform distribution. If not discarded, the packet is scheduled to leave the outgoing port at time $t + q(t)/C$, i.e. after the queuing delay $q(t)/C$ determined by the fluid-flow. This way, discrete packets are impacted by the queuing effects of the interacting packet-level and fluid-flow seamlessly and avoiding overflow and underflow issues present in other hybrid approaches [54].

8.2   Performance and Accuracy of Hybrid Topologies in PowerDEVS

Packet-level, fluid-flow and hybrid models were developed in PowerDEVS. PowerDEVS [8] is a general purpose simulator specifically conceived for the simulation of hybrid systems implementing DEVS and all QSS integration methods [47, 47, 58, 59]. It provides a block-oriented graphical user interface (GUI) and python bindings for model coupling, a wide library of pre-developed models, and atomic models are implemented in C++.

Hybrid topologies are constructed similarly as packet-level and fluid-flow topologies, either by connecting high-level blocks in the PowerDEVS GUI, pragmatically using Py2PDEVS python bindings [72], or could be automatically generated by tools like TopoGen [51]. In this case, routers and links are represented by their hybrid models. Figure 22 shows the graphical definition of a pair of sender and receiver packet-level and fluid-flow hosts, interconnected by two hybrid routers and interacting at the output port of the first router.

Network experts can describe topologies with drag&drop in the GUI or programmatically in Python without knowledge on the underlying mathematical equations

$$\frac{dW_i(t)}{dt} = \frac{1}{\tau_i(t)} - \frac{W_i(t)}{2}\frac{\mu_i}{aN}(t - \tau_i(t))$$

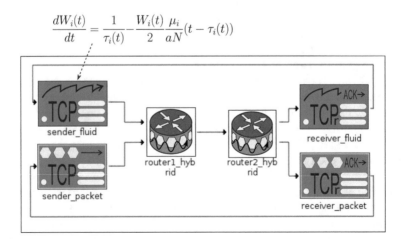

Fig. 22 Hybrid topology in the PowerDEVS GUI with a fluid-flow and a packet-level host sharing a hybrid bottleneck queue

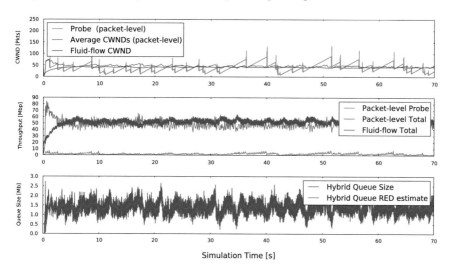

Fig. 23 Hybrid simulation metrics with 20 packet-level and 20 fluid-flow TCP sessions using a smoothing window of 10ms

describing network dynamics nor of the complex numerical methods to solve them. Additionally, the same tools can be used to describe packet-level, fluid-flow, and hybrid models.

We now turn our attention to the execution performance and accuracy of the hybrid model. Using the topology from Fig. 22, we evaluate the trade-off between precision (packet-level) and performance (fluid-flow) by increasing packet-level traffic in each execution. For execution N, we configure $N_p = N$ packet-level TCP sessions and $N_f = 40 - N$ fluid flow TCP sessions. All links are configured with a 10ms propagation delay and bandwidth fixed at 200 Mps except for the routers' bottleneck links which are fixed at 100Mbps. For comparison, an additional equivalent packet-level only topology is used, as well as an additional hybrid simulation with a hybrid link smoother configured with $delta_{min} = 1$. 70s are simulated and the total throughput of both traffic flows is measured as from 10s.

Figure 23 shows the smooth hybrid simulation metrics when there are 20 packet-level and 20 fluid-flow TCP sessions. On the first plot, the 20 packet-level windows are averaged (in red) and a single probe packet-level (in green) is also included to verify the behavior of detailed foreground traffic. The fluid-flow congestion window (in blue) approximates very closely the packet-level average. The plot on the second row shows the throughput of both types of traffic that interact to quickly share the 100 Mbps link. The fluid-flow rate shows fast fluctuations regardless of the smoothing and mean-value foundations. This is a consequence of the queue size, shown in the third row, which is affected by the hybridized traffic which changes its rate according to the smoothing.

Figure 24 compares the execution times of increasing packet-level TCP sessions N_p for the hybrid simulation with smoothing (in green) and without smoothing (in

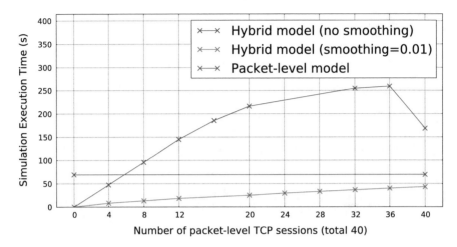

Fig. 24 Execution times of the packet-level and hybrid model with and without smoothing for increasing percentage of packet-level traffic, 70 s of virtual simulation time, and aggregated throughput of ∼ 7 Gb

blue). The full packet-packet model execution time with 40 TCP sessions is also included for comparison (in red). As expected, the hybrid simulation performance is driven by the number of packets traversing the network. In this scenario, the hybrid simulation without smoothing quickly becomes slower than the full packet-level simulation making it useless (slower and less detailed). When percentage of packet-level traffic is high compared to the fluid-flow traffic it is worth to smooth the ON/OFF state changes of the hybridized flow. When smoothing is active, the execution time greatly reduces and remains lower than the packet-level only simulation at all executions.

These results show that hybrid simulation is able to retain detailed network flows while drastically reducing execution times. A set of individual flows can be accurately simulated with packet-level models, while background traffic is approximated with fluid-flow representations. There is a trade-off between performance and the number of packet-level traces simulated which can be easily tuned using this hybrid approach. Compared to the packet-level simulation, the experiment shows that the hybrid simulation presents a speed up of 5.9 when there is 10% of packet-level traffic and 90% of fluid-flow traffic.

8.3 Synthesis on Hybrid Simulation

So far we have discussed how a hybrid modeling approach can drastically improve simulation performance and produce detailed packet trace results. This is achieved thanks to a unified approach where both the fluid-flow (dynamic ODEs) and the packet-level (discrete packets) models are represented under DEVS, resulting in hybrid simulations that do not require time synchronizations and interact smoothly. On the other hand, this approach relies on mathematical models that accurately represent network dynamics. In this regard, although there exists a well known literature, such models are not trivial to calibrate and do not adequately represent all practical scenarios and protocols. Moreover, further study and characterization should be performed on tandems of queues which could degrade performance if changes in one queue causes a "ripple effect" on other queues downstream [53].

Regarding hybrid and co-simulation approaches, they are not antagonistic, they have to deal with similar challenges but are still significantly different. The main difference is that co-simulation enables reusing and enabling inter-operation of different simulators possibly implemented in distinct software. Conversely, the hybrid packet-fluid approach focuses on the efficient and accurate simulation within a single formalism, model and tool. Hybrid packet-fluid simulations have been implemented as a co-simulation of a packet network simulator and an ODE solver [34], with pros and cons as discussed in this section. On the other hand, at the technical level both need to consistently mix different perspectives in terms of model representation and deal with data exchange and synchronization issues. As discussed along this chapter, these challenges can be tackled using DEVS as a unifying formalism.

Regarding emergency response scenarios described in Sect. 2, hybrid modeling can be used to reduce simulation execution time as well as integrating emergency models defined as ODEs. Routing protocols are an integral part of packet-level and fluid-flow models which can be readily defined using the infrastructure defined in Sect. 5.

9 Summary and Conclusions

The proposed DEVS-based system engineering methodology is motivated by complex network design applied to emergency response. Design of such networks with current technologies must meet several challenges that current simulators cannot address. The proposed methodology offers a new M&S approach that can support a full life-cycle Model-Based System Engineering. In this chapter, we discussed such an M&S approach based on the DEVS formalism and supporting computational environments. The proposed methodology advocates:

- High level specification of the behaviors required in emergency response management using UML/SysML metamodels that map to DEVS simulation models. We discussed concepts and tools to support such specification and illustrated their application with high level specification of routing mechanisms with an example of packet routing in communication networks.
- The development of DEVS-Based architectures to design, model, and validate complex network simulation infrastructures. We provided a framework that combines simulators focusing on protocols evaluation, environmental representations, and models of real applications.

- For verification, validation, and testing such models we discussed two alternatives, both enabled by the DEVS formalism:

  1. Co-simulation of complex networks using DEVS as the formal basis for co-simulation support,
  2. Simulation-based testing of DEVS-based designs using powerful hybrid fluid-flow/packet-level mechanisms.

We also showed that the hybrid and co-simulation approaches offer complementary methods that together can overcome the limitations of each one.

As is evident from the motivation discussion, in the current state of the art, the proposed framework is quite ready for application to real problem situations but has not yet been widely adopted. We conclude with the following discussion of open research needed to bring it into general practice.

We have advocated using the DEVS formalism as the basic modeling and simulation framework for Model-Based System Engineering methodology to support the critical stages in the top-down design of complex networks. However, it must be recognized that while DEVS provides a suitable solution in all these stages, challenges remain in the development of M&S software tools that can improve the modeling activities necessary in such complex network designs. Furthermore, while DEVS-based software tools can be used as general-purpose computational environments for building the required simulation models, specific software tools might need to be developed when domain-specific abstraction models are brought in to complement such general tools.

Despite the complexity of complex network designs, several types of networks can be modeled following the same approach, provided that, as we have discussed, such networks share their primary structures that are based on the same underlying conceptual models. Although we have primarily addressed the high-level specification of the behaviors required in emergency response management, our approaches can be used to support new types of networking problems. The development of semi-automatic procedures to construct implementations from the defined abstractions is a challenge that can be addressed when MBSE approaches are combined with Software Engineering methods. In this case, modeling languages can be used to support modeling activity and reduce implementation times. Moreover, interoperability models among different stages can be used to support traceability during the M&S tasks.

Integration of models with inherently different behaviors results in substantial limitations encountered with commonly used monolithic approaches. Discrete event systems and formalisms, mainly the DEVS formalism due to its expressive capacity and behavioral abstractness, are candidates to solve, or at least mitigate, such a challenging problem. In contrast to the DEVS well-defined system theory-based formalism, there are semi-formal higher-level modeling languages (e.g., UML/SysML) that can facilitate coming up with useful abstractions and guide the production of concrete artifacts. However, in using such approaches, modelers should be provided clear guidance in realizing their models in computational environments when necessary, which is usually the case in simulation modeling. In this chapter, we have seen that the behavioral specification aligned with DEVS modeling provides an initial basis for fulfilling knowledge gaps and semantics within entangling modular components with heterogeneous behaviors instead of simplified ones, especially in complex networks. Research is needed to establish how to couple together the overlaying higher-level activity diagrams, metamodels, and the collective behavior with simulation experimental design to allow decision-makers and modelers to realize behavioral specifications with far more control and transparency than found in ad hoc approaches. The theory of modeling and simulation with its concepts of levels of system specification and associated morphisms [93] can inform the DEVS formalism's support for phases of top down design and the transitions from one phase to the next.

Co-simulation has demonstrated its ability to mix perspectives from several fields by coupling simulators that may be implemented in different software. The developments of co-simulation middleware based on standards and formal foundations like DEVS tend to reduce the extra complexity due to the interoperability challenge and offer opportunities for further use-case specific research projects. Next developments must therefore continue the standardization work [87] on middleware and put it into practice in various examples such as complex networks.

Concerning the VCS architecture, its main strength is also its major drawback: the addition of new layers in the communication process increases the complexity of the simulation and may lead to significant overhead. More than that, although the VCS is transparent from the application layer perspective, its implementation has to be verified and validated independently because the VCS depends on the final architecture of the network. With this approach, an error or a defect in the VCS itself cannot be detected under the current framework and could lead to undesired changes in the other layers. Hence, research is needed to develop guidelines for safe implementation of the VCS. Other interesting research concerns the use of dynamic structure [93] to adequately model the mobility and dynamic network reconfiguration likely to occur in a response management setting.

The hybrid approach proposed in this chapter simplifies the adoption from network experts as fluid, and discrete models are developed using the same formalism and tool (with graphical and/or programmatic model definitions). Still, as a generic environment for simulation, it can lack particular features present in domain specific tools. Moreover, the literature on fluid network modeling and simulation is significantly less abundant compared to packet network simulation; thus, there are limitations on the scenarios that can be represented by fluid (and consequently by hybrid) models. Future developments could include models for WiFi networks and the study of fluid models for a broader set of network protocols.

References

1. ACIMS: DEVS-Suite Simulator version 5.0.0 (2019). https://sourceforge.net/projects/devs-suitesim/. Available at https://sourceforge.net/projects/devs-suitesim/. Accessed 1 July 2020

2. Alliance, E.: Ethernet roadmap. www.ethernetalliance.org/roadmap (2019). Accessed: 10 June 2020
3. Alshareef, A.: Activity specification for time-based discrete event simulation models. Ph.D. dissertation, Arizona State University, Tempe (2019)
4. Alshareef, A., Kim, D., Seo, C., Zeigler, B.P.: Activity diagrams between DEVS-based modeling & simulation and fUML-based model execution. In: Proceedings of the 2020 Summer Simulation Conference. Society for Computer Simulation International (2020)
5. Alshareef, A., Sarjoughian, H.: Metamodeling activities for hierarchical component-based models. In: 2019 Spring Simulation Conference (SpringSim), pp. 1–12. IEEE (2019)
6. Andel, T.R., Yasinsac, A.: On the credibility of manet simulations. Computer **39**(7), 48–54 (2006). https://doi.org/10.1109/MC.2006.242
7. Anjum, S.S., Noor, R.M., Anisi, M.H.: Review on manet based communication for search and rescue operations. Wirel. Pers. Commun. **94**(1), 31–52 (2017)
8. Bergero, F., Kofman, E.: Powerdevs: a tool for hybrid system modeling and real-time simulation. Simulation **87**(1–2), 113–132 (2011)
9. Blas, M.J., Gonnet, S., Leone, H.: Routing structure over discrete event system specification: a DEVS adaptation to develop smart routing in simulation models. In: 2017 Winter Simulation Conference (WSC), pp. 774–785. IEEE (2017)
10. Blochwitz, T., Otter, M., Arnold, M., Bausch, C., Clauß, C., Elmqvist, H., Junghanns, A., Mauss, J., Monteiro, M., Neidhold, T., et al.: The Functional Mockup Interface for tool independent exchange of simulation models. In: Proceedings of the 8th International Modelica Conference, pp. 105–114. Linköping University Press (2011). http://elib.dlr.de/74668/
11. Bollinger, L.A., van Blijswijk, M.J., Dijkema, G.P., Nikolic, I.: An energy systems modelling tool for the social simulation community. J. Artif. Soc. Social Simul. **19**(1) (2016). https://doi.org/10.18564/jasss.2971. http://jasss.soc.surrey.ac.uk/19/1/1.html
12. Bonaventura, M.: Hybrid modeling and simulation of complex data networks. Ph.D. thesis, University of Buenos Aires, Argentina (2019). https://ri.conicet.gov.ar/handle/11336/83438
13. Bonaventura, M., Castro, R.: Fluid-flow and packet-level models of data networks unified under a modular/hierarchical framework: Speedups and simplicity, combined. In: Proceedings of 2018 Winter Simulation Conference (WSC) (2018)
14. Bonaventura, M., Foguelman, D., Castro, R.: Discrete event modeling and simulation-driven engineering for the ATLAS data acquisition network. Comp. cSci. Eng. **18**(3), 70–83 (2016)
15. Bonaventura, M., Jonckheere, M., Castro, R.: Simulation study of dynamic load balancing for processor sharing servers with finite capacity under generalized halfin-whitt-jagerman regimes. In: Proceedings of 2018 Winter Simulation Conference (WSC) (2018)
16. Camus, B., Paris, T., Vaubourg, J., Presse, Y., Bourjot, C., Ciarletta, L., Chevrier, V.: Co-simulation of cyber-physical systems using a DEVS wrapping strategy in the MECSYCO middleware. SIMULATION (2018)
17. Castro, R.: Integrative tools for modeling, simulation and control of data networks. Ph.D. thesis, National University of Rosario, Argentina (2010). Spanish, extended summary in English
18. Castro, R., Kofman, E.: An integrative approach for hybrid modeling, simulation and control of data networks based on the devs formalism. In: Modeling and Simulation of Computer Networks and Systems: Methodologies and Applications, chap. 18. Morgan Kaufmann (2015)
19. Castro, R., Kofman, E., Cellier, F.E.: Quantization-based integration methods for delay-differential equations. Simul. Model. Practice Theory **19**(1), 314–336 (2011)
20. Cavin, D., Sasson, Y., Schiper, A.: On the accuracy of MANET simulators. In: Proceedings of the Second ACM International Workshop on Principles of Mobile Computing, POMC '02, pp. 38–43. ACM, New York, NY, USA (2002). https://doi.org/10.1145/584490.584499. http://doi.acm.org/10.1145/584490.584499
21. Cellier, F.E., Kofman, E.: Continuous System Simulation. Springer Science & Business Media (2006)
22. Chang, W.T., Ha, S., Lee, E.A.: Heterogeneous simulation-mixing discrete-event models with dataflow. Journal of VLSI signal processing systems for signal, image and video technology **15**(1–2), 127–144 (1997)

23. Dorathy, I., Chandrasekaran, M.: Simulation tools for mobile ad hoc networks: a survey. J. Appl. Res. Technol. **16**, 437–445 (2019)
24. Fennibay, D., Yurdakul, A., Sen, A.: A heterogeneous simulation and modeling framework for automation systems. IEEE Trans. Computer-Aided Design Integr. Circ. Syst. **31**(11), 1642–1655 (2012). https://doi.org/10.1109/TCAD.2012.2199116
25. Flynn, J., Tewari, H., O'Mahony, D.: Jemu: A real time emulation system for mobile ad hoc networks. In: Proceedings of the first joint IEI/IEE Symposium on Telecommunications Systems Research, pp. 262–267 (2001)
26. Frey, H., Görgen, D., Lehnert, J.K., Sturm, P.: A java-based uniform workbench for simulating and executing distributed mobile applications. In: Guelfi, N., Astesiano, E., Reggio, G. (eds.) Scientific Engineering of Distributed Java Applications, pp. 116–127. Springer, Berlin (2004)
27. Fritzson, P., Engelson, V.: Modelica—A unified object-oriented language for system modeling and simulation. In: ECOOP'98-Object-Oriented Programming, pp. 67–90. Springer (1998). http://link.springer.com/chapter/10.1007/BFb0054087
28. Galán, J.M., Izquierdo, L.R., Izquierdo, S.S., Santos, J.I., Del Olmo, R., López-Paredes, A., Edmonds, B.: Errors and artefacts in agent-based modelling. J. Artifi. Soc. Social Simul.**12**(1) (2009). http://jasss.soc.surrey.ac.uk/12/1/1.Html
29. Giambiasi, N.: From sequential machines to DEVS formalism. In: Proceedings of the 2009 Summer Computer Simulation Conference, SCSC '09, pp. 216–222. Society for Modeling; Simulation International, Vista, CA (2009)
30. Gomes, C., Thule, C., Broman, D., Larsen, P.G., Vangheluwe, H.: Co-simulation: A survey. ACM Comput. Surv. **51**(3), 49:1–49:33 (2018). https://doi.org/10.1145/3179993. http://doi.acm.org/10.1145/3179993
31. Group, O.M.: Ontology definition metamodel request for proposal (2003)
32. Group, O.M.: UML 2.0 Infrastructure Specification (2003)
33. Group, O.M.: Object constraint language (2014)
34. Gu, Y., Liu, Y., Towsley, D.: On integrating fluid models with packet simulation. In: INFO-COM 2004. Twenty-third Annual Joint Conference of the IEEE Computer and Communications Societies, vol. 4, pp. 2856–2866. IEEE (2004)
35. Guizzardi, G.: Ontological Foundations for Structural Conceptual Models. No. 15 in Telematica Institute Fundamental Research Series. Telematica Instituut, Enschede, The Netherlands (2005). http://www.researchgate.net/publication/215697579_Ontological_Foundations_for_Structural_Conceptual_Models
36. Henderson, T.R., Roy, S., Floyd, S., Riley, G.F.: ns-3 Project Goals. In: Proceedings of WNS2'06, p. 13. ACM (2006)
37. Hogie, L., Bouvry, P., Guinand, F.: An overview of MANETs simulation. Electron. Notes Theor. Comput. Sci. **150**(1), 81–101 (2006). https://doi.org/10.1016/j.entcs.2005.12.025
38. Information technology—Open Systems Interconnection—Basic Reference Model: The Basic Model (1994)
39. Ivanic, N., Rivera, B., Adamson, B.: Mobile ad hoc network emulation environment. In: MIL-COM 2009 - 2009 IEEE Military Communications Conference, pp. 1–6 (2009). https://doi.org/10.1109/MILCOM.2009.5379781
40. Johnson, D.B., Maltz, D.A., Broch, J.: Ad hoc networking. chap. DSR: The Dynamic Source Routing Protocol for Multihop Wireless Ad Hoc Networks, pp. 139–172. Addison-Wesley Longman Publishing Co., Inc., Boston, MA, USA (2001). http://dl.acm.org/citation.cfm?id=374547.374552
41. Kapucu, N., Garayev, V.: Collaborative decision-making in emergency and disaster management. Int. J. Public Admin. **34**(6), 366–375 (2011). https://doi.org/10.1080/01900692.2011.561477
42. Khairnar, V.D., Pradhan, S.N.: Mobility models for vehicular ad-hoc network simulation. In: 2011 IEEE Symposium on Computers Informatics, pp. 460–465 (2011). https://doi.org/10.1109/ISCI.2011.5958959
43. Kiess, W., Mauve, M.: A survey on real-world implementations of mobile ad-hoc networks. Ad Hoc Networks **5**(3), 324–339 (2007)

44. Kim, T., Hwang, M.H., Kim, D., Zeigler, B.P.: DEVS/NS-2 environment; integrated tool for efficient networks modeling and simulation. In: Proceedings of the 2007 Spring Simulation Multiconference, vol. 2, p. 8. SCS/ACM, Norfolk, Virginia, USA (2007)
45. Kim, Y.J., Kim, J.H., Kim, T.G.: Heterogeneous simulation framework using DEVS bus. SIMULATION **79**(1), 3–18 (2003). https://doi.org/10.1177/0037549703253543
46. Klein, M.: Dianemu: A java based generic simulation environment for distributed protocols (2003)
47. Kofman, E.: A third order discrete event method for continuous system simulation. Latin Am. Appl. Res. **36**(2), 101–108 (2006)
48. Kofman, E., Junco, S.: Quantized-state systems: a Devs approach for continuous system simulation. Trans. Soc. Model. Simul. Int. **18**(3), 123–132 (2001)
49. Komazec, N., Bozanic, D., Pamucar, D.: Aspects of decision-making in emergency situations. In: ICT Forum Nis, pp. 55–59 (2014)
50. Kurkowski, S., Camp, T., Colagrosso, M.: MANET simulation studies: The incredibles. SIGMOBILE Mob. Comput. Commun. Rev. **9**(4), 50–61 (2005)
51. Laurito, A., Bonaventura, M., Eukeni Pozo Astigarraga, M., Castro, R.: Topogen: A network topology generation architecture with application to automating simulations of software defined networks. In: 2017 Winter Simulation Conference (WSC), pp. 1049–1060 (2017). https://doi.org/10.1109/WSC.2017.8247854
52. Lien, Y., Jang, H., Tsai, T.: A MANET based emergency communication and information system for catastrophic natural disasters. In: 2009 29th IEEE International Conference on Distributed Computing Systems Workshops, pp. 412–417 (2009). https://doi.org/10.1109/ICDCSW.2009.72
53. Liu, B., Figueiredo, D.R., Guo, Y., Kurose, J., Towsley, D.: A study of networks simulation efficiency: Fluid simulation vs. packet-level simulation. In: Proceedings of the 20th Annual Joint Conference of the IEEE Computer and Communications Societies, vol. 3, pp. 1244–1253. IEEE (2001)
54. Liu, J., Liu, Y., Du, Z., Li, T.: Gpu-assisted hybrid network traffic model. In: Proceedings of the 2nd ACM SIGSIM Conference on Principles of Advanced Discrete Simulation, pp. 63–74. ACM (2014)
55. Liu, Y., Lo Presti, F., Misra, V., Towsley, D., Gu, Y.: Fluid models and solutions for large-scale ip networks. In: ACM SIGMETRICS Performance Evaluation Review, vol. 31, pp. 91–101. ACM (2003)
56. Mallapur, S.V., Patil, S.R.: Survey on simulation tools for mobile ad-hoc networks. Int. J. Comput. Networks Wirel. Commun. (IJCNWC) **2**(2) (2012)
57. Manpreet, Malhotra, J.: A survey on MANET simulation tools. In: 2014 Innovative Applications of Computational Intelligence on Power, Energy and Controls with their impact on Humanity (CIPECH), pp. 495–498 (2014). https://doi.org/10.1109/CIPECH.2014.7019120
58. Migoni, G., Bortolotto, M., Kofman, E., Cellier, F.E.: Linearly implicit quantization-based integration methods for stiff ordinary differential equations. Simul. Model. Practice Theory **35**, 118–136 (2013)
59. Migoni, G., Kofman, E., Cellier, F.: Quantization-based new integration methods for stiff ordinary differential equations. Simulation **88**(4), 387–407 (2012)
60. Misra, V., Gong, W.B., Towsley, D.: Fluid-based analysis of a network of aqm routers supporting tcp flows with an application to red. In: ACM SIGCOMM Computer Communication Review, vol. 30, pp. 151–160. ACM (2000)
61. Mohammed, A., Al-Ghrairi, A.: Differences between ad hoc networks and mobile ad hoc networks: A survey. Xinan Jiaotong Daxue Xuebao/J. Southwest Jiaotong Univ. **54**, 12 (2019). https://doi.org/10.35741/issn.0258-2724.54.4.20
62. MS4 Systems: MS4 Me Simulator version 3.0 (2018). http://ms4systems.com/pages/ms4me.php. Available at http://ms4systems.com/pages/ms4me.php (Accessed July 1, 2020)
63. Muchtar, F., Abdullah, A.H., Latiff, M.S.A., Hassan, S., Wahab, M.H.A., Abdul-Salaam, G.: A technical review of MANET testbed using mobile robot technology. J. Phys. Conf. Ser. **1049**, 012001. IOP Publishing (2018)

64. Nilsson, E.G., Stølen, K.: Ad hoc networks and mobile devices in emergency response—a perfect match? In: Zheng, J., Simplot-Ryl, D., Leung, V.C.M. (eds.) Ad Hoc Networks, pp. 17–33. Springer, Berlin Heidelberg, Berlin, Heidelberg (2010)

65. Padhye, J., Firoiu, V., Towsley, D., Kurose, J.: Modeling TCP throughput: a simple model and its empirical validation. ACM SIGCOMM Computer Commun. Rev. **28**(4), 303–314 (1998)

66. Paris, T., Wiart, J.B., Netter, D., Chevrier, V.: Teaching co-simulation basics through practice. In: Proceedings of the 51th Computer Simulation Conference. Society for Computer Simulation International, Berlin, Germany (2019). https://hal.archives-ouvertes.fr/hal-02268350/file/TeachingCosimulationBasicsThroughPractice_HAL.pdf

67. Pennock, M.J., Rouse, W.B.: Why connecting theories together may not work: How to address complex paradigm-spanning questions. In: 2014 IEEE International Conference on Systems, Man, and Cybernetics (SMC), pp. 373–378. IEEE (2014). http://ieeexplore.ieee.org/xpls/abs_all.jsp?arnumber=6973936

68. Perkins, C.E., Royer, E.M.: Ad-hoc on-demand distance vector routing. In: Proceedings WMCSA'99. Second IEEE Workshop on Mobile Computing Systems and Applications, pp. 90–100 (1999). https://doi.org/10.1109/MCSA.1999.749281

69. R., B.: Requirements for Internet Hosts—Communication Layers (1989). https://tools.ietf.org/html/rfc1122

70. Reina, D.G., Askalani, M., Toral, S.L., Barrero, F., Asimakopoulou, E., Bessis, N.: A survey on multihop ad hoc networks for disaster response scenarios. Int. J. Distrib. Sensor Networks **11**(10), 647037 (2015). https://doi.org/10.1155/2015/647037

71. Riley, G.F., Henderson, T.R.: The ns-3 Network Simulator, pp. 15–34. Springer, Berlin (2010). https://doi.org/10.1007/978-3-642-12331-3_2. https://doi.org/10.1007/978-3-642-12331-3_2

72. Santi, L., Bonaventura, M.: py2pdevs: a python to powerdevs interface. https://gitlab.cern.ch/tdaq-simulation/powerdevs/ (2018)

73. Sargent, R.G.: Some approaches and paradigms for verifying and validating simulation models. In: Simulation Conference, 2001. Proceedings of the Winter, vol. 1, pp. 106–114 (2001)

74. Sarjoughian, H.S.: Model composability. In: Proceedings of the 38th Conference on Winter Simulation, pp. 149–158. Winter Simulation Conference (2006). http://dl.acm.org/citation.cfm?id=1218144

75. Schindelhauer, C., Lukovszki, T., Rührup, S., Volbert, K.: Worst case mobility in ad hoc networks. In: Proceedings of the Fifteenth Annual ACM Symposium on Parallel Algorithms and Architectures, SPAA '03, pp. 230–239. ACM, New York, NY, USA (2003). https://doi.org/10.1145/777412.777448

76. Sichitiu, M.L.: Mobility Models for Ad Hoc Networks, pp. 237–254. Springer, London (2009). https://doi.org/10.1007/978-1-84800-328-6_10.

77. Sikora, A., Niewiadomska-Szynkiewicz, E., Krzysztoń, M.: Simulation of mobile wireless ad hoc networks for emergency situation awareness. In: 2015 Federated Conference on Computer Science and Information Systems (FedCSIS), pp. 1087–1095 (2015). https://doi.org/10.15439/2015F52

78. Takai, M., Martin, J., Bagrodia, R.: Effects of wireless physical layer modeling in mobile ad hoc networks. In: Proceedings of the 2Nd ACM International Symposium on Mobile Ad Hoc Networking &Amp; Computing, MobiHoc '01, pp. 87–94. ACM, New York, NY, USA (2001). https://doi.org/10.1145/501426.501429.

79. Tannenbaum, A., Wetherall, D.: Computer Networks, 5th edn. (2010)

80. Tolk, A., Diallo, S., Padilla, J., Turnitsa, C.: How is M&S Interoperability different from other Interoperability Domains? GUEST EDITORIAL p. 5 (2012). http://www.msco.mil/documents/MSJournal2012-2013Winter.pdf#page=7

81. Tüncel, S., Ekiz, H., Zengin, A.: Design and implementation of a new MANET simulator model for AODV simulation (2016). https://doi.org/10.3906/elk-1311-120

82. Vangheluwe, H.: Multi-formalism modelling and simulation. Ph.D. thesis, Ghent University (2000)

83. Vangheluwe, H., De Lara, J., Mosterman, P.J.: An introduction to multi-paradigm modelling and simulation. In: Proceedings of the AIS'2002 conference (AI, Simulation and Planning in High Autonomy Systems), Lisboa, Portugal, pp. 9–20 (2002)

84. Varga, A., Hornig, R.: An overview of the OMNeT++ simulation environment. In: Proceedings of the 1st International Conference on Simulation Tools and Techniques for Communications, Networks and Systems & Workshops, Simutools '08, pp. 60:1–60:10. ICST (Institute for Computer Sciences, Social-Informatics and Telecommunications Engineering), ICST, Brussels, Belgium, Belgium (2008). http://dl.acm.org/citation.cfm?id=1416222.1416290

85. Vaubourg, J., Chevrier, V., Ciarletta, L., Camus, B.: Co-simulation of IP network models in the Cyber-Physical systems context, using a DEVS-based platform. In: Proceedings of the 19th Communications & Networking Symposium, p. 2. Society for Computer Simulation International (2016). http://dl.acm.org/citation.cfm?id=2962688

86. Vaubourg, J., Presse, Y., Camus, B., Bourjot, C., Ciarletta, L., Chevrier, V., Tavella, J.P., Morais, H.: Multi-agent Multi-Model Simulation of Smart Grids in the MS4SG Project. In: Y. Demazeau, K.S. Decker, J. Bajo Pérez, F. de la Prieta (eds.) Advances in Practical Applications of Agents, Multi-Agent Systems, and Sustainability: The PAAMS Collection, vol. 9086, pp. 240–251. Springer International Publishing, Cham (2015). http://link.springer.com/10.1007/978-3-319-18944-4_20

87. Wainer, G., Al-Zoubi, K., Dalle, O., Hill, D.R.C., Mittal, S., Martin, J.L.R., Sarjoughian, H., Touraille, L., Traoré, M.K., Zeigler, B.P.: Standardizing DEVS Simulation Middleware. In: Discrete-Event Modeling and Simulation: Theory and Applications, p. 459 (2010)

88. Wainer, G.A., Mosterman, P.J.: Discrete-Event Modeling and Simulation: Theory and Applications. CRC Press, Boca Raton (2016)

89. Yacoub, A.: Integrated Simulator of Mobile Ad-hoc Network-based Infrastructure : A Case Study. In: Spring Simulation Conference (SpringSim 2020). Society for Modeling and Simulation International (SCS), Fairfax, VA, USA (2020). https://doi.org/10.22360/SpringSim.2020.CNS.006. https://dl.acm.org/doi/abs/10.5555/3408207.3408230

90. Yacoub, A.: Virtual Communication Stack: Towards Building Integrated Simulator of Mobile Ad Hoc Network-based Infrastructure for Disaster Response Scenarios. arXiv:2004.14093 [cs] (2020). http://arxiv.org/abs/2004.14093. ArXiv: 2004.14093

91. Yong, J. K., Tag, G. K.: A heterogeneous simulation framework based on the DEVS bus and the high level architecture. In: 1998 Winter Simulation Conference. Proceedings (Cat. No.98CH36274), vol. 1, pp. 421–428, vol.1 (1998). https://doi.org/10.1109/WSC.1998.745017

92. Zeigler, B.P.: Theory of modeling and simulation. Wiley, New York, NY (1976)

93. Zeigler, B.P., Muzy, A., Kofman, E.: Theory of Modeling and Simulation: Discrete Event & Iterative System Computational Foundations, 3rd edn. Academic (2018)

94. Zeigler, B.P., Sarjoughian, H.S.: Guide to Modeling and Simulation of Systems of Systems, 2nd edn. Springer, Berlin (2017)

95. Zhou, L., Wu, X., Xu, Z., Fujita, H.: Emergency decision making for natural disasters: an overview. Int. J. Disaster Risk Reduc. **27**, 567–576 (2018)

96. Zsambok, C.E., Klein, G.: Naturalistic Decision Making. Psychology Press (2014)

Touchless Palmprint and Fingerprint Recognition

Ruggero Donida Labati, Angelo Genovese, Vincenzo Piuri, and Fabio Scotti

Abstract Biometric systems based on hand traits captured using touchless acquisition procedures are increasingly being used for the automatic recognition of individuals due to their favorable trade-off between accuracy and acceptability by users. Among hand traits, palmprint and fingerprints are the most studied modalities because they offer higher recognition accuracy than other hand-based traits such as finger texture, knuckle prints, or hand geometry. For capturing palmprints and fingerprints, touchless and less-constrained acquisition procedures have the advantage of mitigating the problems caused by latent prints, dirty sensors, and skin distortions. However, touchless acquisition systems for palmprints and fingerprints face several challenges caused by the need to capture the hand while it is moving and under varying illumination conditions. Moreover, images captured using touchless acquisition procedures tend to exhibit complex backgrounds, nonuniform reflections, and perspective distortions. Recently, methods such as adaptive filtering, three-dimensional reconstruction, local texture descriptors, and deep learning have been proposed to compensate for the nonidealities of touchless acquisition procedures, thereby increasing the recognition accuracy while maintaining high usability. This chapter presents an overview of the various methods reported in the literature for touchless palmprint and fingerprint recognition, describing the corresponding acquisition methodologies and processing methods.

Keywords Biometrics · Touchless · Palmprint · Fingerprint

R. Donida Labati · A. Genovese · V. Piuri (✉) · F. Scotti
Department of Computer Science, Università Degli Studi di Milano,
Via Celoria 18, 20133 Milan, MI, Italy
e-mail: vincenzo.piuri@unimi.it

R. Donida Labati
e-mail: ruggero.donida@unimi.it

A. Genovese
e-mail: angelo.genovese@unimi.it

F. Scotti
e-mail: fabio.scotti@unimi.it

© The Author(s), under exclusive license to Springer Nature Switzerland AG 2022   267
P. Nicopolitidis et al. (eds.), *Advances in Computing, Informatics, Networking
and Cybersecurity*, Lecture Notes in Networks and Systems 289,
https://doi.org/10.1007/978-3-030-87049-2_9

1 Introduction

Biometric systems based on hand characteristics are widely used in both private and governmental applications. The main reasons for their popularity are their high accuracy, simplicity of use, low cost for hardware devices, compatibility with governmental and forensic applications, and availability of ad-hoc techniques for protecting the privacy of biometric data [16, 36]. The main biometric traits that can be extracted from a hand are the palmprint [56], fingerprints [80], hand geometry [2], finger texture [1], knuckle prints [62], hand veins [124], and finger veins [102]. Among these traits, palmprints and fingerprints are the ones characterized by the most mature technologies and are the most widely used in real-world applications [51].

Traditionally, palmprint and fingerprint recognition systems require the user to touch a sensor platen to acquire a biometric sample. However, the touch-based acquisition process presents several disadvantages:

- the acquired samples can exhibit nonlinear and unpredictable distortions due to the skin deformations induced by touching a surface;
- touching a sensor previously used by unknown persons can present hygiene issues;
- to obtain samples of sufficient quality, users should be trained in the proper way to apply pressure to the acquisition surface; and
- the quality of the acquired samples can deteriorate over time due to the accumulation of grease and dirt released by the hands of users on the sensor platen.

To overcome the mentioned problems and to enhance the usability and acceptability of palmprint- and fingerprint-based biometric systems, the research community has proposed various technological solutions for realizing touchless biometric systems based on cameras placed at a distance from the biometric trait to be captured. The design of such touchless acquisition systems faces several challenges due to the need to capture the hand while it is moving and under varying illumination conditions. Furthermore, touchless samples exhibit relevant differences with respect to those collected through touch-based acquisitions. In particular, images captured touchlessly (Fig. 1) tend to exhibit complex backgrounds, nonuniform reflections, and perspective distortions. Therefore, touchless palmprint and fingerprint recognition systems need to adopt different techniques compared with touch-based technologies for all modules of the biometric recognition chain [51].

Touchless palmprint and fingerprint recognition systems can be designed for heterogeneous application contexts, present important differences in their acquisition setups, use two-dimensional (2-D) or three-dimensional (3-D) data, use dedicated preprocessing algorithms, and be based on different feature extraction and matching methods. Their accuracy and robustness have recently increased considerably by virtue of the introduction of deep learning (DL) techniques into every step of the computational chain.

This chapter presents a review of the state of the art in touchless palmprint and fingerprint recognition systems from a technological point of view. Specifically, it describes recent acquisition methods, preprocessing techniques, and feature extraction and matching methods developed for touchless biometric systems based on palmprints and fingerprints. To the best of our knowledge, this is the first literature review providing a systematic analysis of touchless technologies for both palmprint and fingerprint recognition, elucidating their differences and commonalities. While most recent surveys in the literature have focused only on either 2-D or 3-D approaches [24, 30, 94], this work considers both 2-D and 3-D technologies, offering a detailed and comprehensive review.

The chapter is organized as follows. Sec. 2 analyzes touchless palmprint recognition technologies. Sec. 3 reviews touchless fingerprint recognition systems. Sec. 4 concludes the work.

2 Palmprint Recognition

The palm is defined as the region of the palmar side of the hand that extends from the wrist to the base of the fingers (Fig. 1). In this region, the skin covering the hand is of the same type as the skin that covers the fingertips and therefore possesses several distinctive characteristics that enable high-accuracy biometric recognition [38]. Compared to fingerprints, palmprints have the advantages that they can be captured even using low-cost acquisition devices with a low resolution (< 100 dpi), they enable high-accuracy recognition even in the case of damaged hands (e.g., the hands of manual workers or elderly people) since biometric recognition algorithms can exploit features at different levels of detail, and their acquisition is generally

Fig. 1 Positions of the palmprint and fingerprints in a touchless acquisition of a hand.

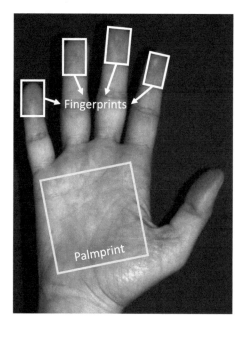

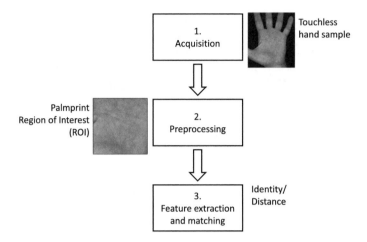

Fig. 2 Outline of the biometric recognition process based on touchless palmprint acquisition.

well accepted by users, who regard palmprints as less-invasive biometric traits than fingerprints or irises [27, 36, 65, 66, 89].

Because of the high usability and favorable user acceptability commonly associated with palmprint recognition, several recent biometric recognition approaches based on the palm consider the use of touchless and less-constrained acquisition procedures, without the need for a fixed position of the hand or a requirement to touch any surface [3, 38, 55, 64, 85, 134].

In this section, we present the most recent approaches for touchless palmprint recognition, detailing the acquisition procedure, preprocessing algorithms, and methods for biometric feature extraction and matching. Specifically, the recognition procedure of a touchless palmprint recognition system usually consists of the following phases: (i) acquisition, (ii) preprocessing, and (iii) feature extraction and matching. Figure 2 shows the outline of the recognition process.

2.1 Acquisition

The purpose of the acquisition phase is to capture a 2-D image or 3-D model of the hand in which the details of the palmprint are sufficiently visible to perform a biometric recognition (Fig. 3). However, unlike for fingerprints, there is no standard set of features for palmprints, and various acquisition systems have been proposed that capture different types of details at different resolutions. Therefore, there is no standard set of requirements for palmprint acquisition devices. For this reason, most public palmprint databases that are currently available have been captured with different devices and feature high variability in terms of image resolution, dynamic range, and quality [27].

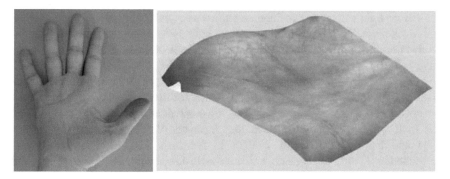

Fig. 3 Examples of touchless palmprint acquisition: **a** a 2-D image captured with an off-the-shelf webcam and a controlled background and **b** a 3-D model computed using a multiple-view acquisition setup. In both cases, although the details of the fingerprints are not visible, the features of the palmprint (e.g., the principal lines) are visible.

Another drawback of palmprint databases is that the majority of the methods proposed in the literature describe the collection of datasets containing a limited number of samples that have not been made public, and therefore, these datasets do not enable the research community to assess the validity of acquisition methodologies by comparing the accuracy of different recognition algorithms [38]. In this section, we do not consider such approaches; instead, to provide the research community with an overview of the most significant acquisition methods in the literature, we consider only the approaches for which a public database has been made available on the internet.

Based on the approaches described in the literature, it is possible to classify the current methods for touchless palmprint acquisition based on the dimensionality of the processed samples. In particular, we can distinguish approaches based on 2-D images from approaches based on 3-D models [38].

2.1.1 Two-Dimensional Approaches

Touchless palmprint recognition methods based on 2-D images use acquisition setups that do not require the contact of the palm of the hand with any surface. However, some acquisition systems may use partially constrained setups, in which the back of the hand must be placed against a fixed support. In contrast, other acquisition systems use less-constrained setups that do not require the hand to touch any surface [38]. The acquisition setups used for 2-D touchless palmprint recognition usually include a charge-coupled device (CCD) camera, an enclosure to guide the position of the hand, and an illumination source. Depending on the wavelengths of the illumination used, these acquisition methods can be divided into two categories: (i) methods based on visible light and (ii) multispectral methods.

Visible-light acquisition methods include the approach proposed in [47], which describes a touchless acquisition procedure with a controlled background and controlled illumination, in which the user places the back of the hand against a fixed surface inside an enclosure. A method with a similar acquisition procedure is described in [10], which proposes an ad hoc device for capturing the palm of a user while the back of the hand rests on a fixed surface. An enclosure is also used in the approach described in [113]. However, in this case, the hand does not touch any surface, and the enclosure is used only to restrict the placement of the hand inside the field of view and depth of focus of the camera. In contrast to [10, 47, 113], the method described in [8] does not consider either an enclosure or controlled illumination, and proposes a database in which the samples are captured with uncontrolled rotations of the hand. However, the back of the hand is placed against a fixed surface. The method described in [59, 93] considers a similar acquisition procedure used to collect a database from people of different ethnicities, occupations, and ages. While the majority of touchless palmprint databases consider a controlled background, the methods described in [87, 115, 138] are based on a procedure for capturing palmprint samples with uncontrolled backgrounds under visible light using a smartphone. The corresponding samples exhibit high variation in terms of pose, rotation, and distance from the camera. A similar database is described in [82, 86], with the difference that the images are collected from the internet rather than directly captured by the authors.

Multispectral acquisition methods use illuminators at different wavelengths, with the purpose of enhancing different details of the skin. One example is the method described in [9], which relies on a uniform illumination setup composed of six different illuminators ranging from violet to near-infrared wavelengths. In contrast to [9], the approach proposed in [116] involves a simpler acquisition setup, consisting of one visible-light illuminator and one infrared illuminator. The infrared illuminator is obtained by replacing the infrared filter in an off-the-shelf webcam with a visible-light filter.

2.1.2 Three-Dimensional Approaches

Touchless palmprint recognition methods based on 3-D models, similar to those based on 2-D images, use acquisition setups that do not require contact of the palm of the hand with any surface. Their main advantage over methods based on 2-D images is the possibility of reconstructing a 3-D model of the hand that describes the position and orientation of the hand in a 3-D metric space. By using such a 3-D model, it is possible to measure and compensate for variations in the distance and rotation of the hand, which do not need to be fixed, allowing less-constrained acquisition compared with the acquisition setups for 2-D images [21]. Another major advantage is the possibility of using information derived from the 3-D model as additional biometric features, thus increasing the recognition accuracy [30]. However, the acquisition setups for 3-D models are more complex than those for 2-D images since they require devices that are able to determine the position of the hand in 3-D space. More specifically, methods based on 3-D models can be divided into two categories: (i) methods based on laser scanners and (ii) approaches using multiple-view acquisitions [38].

Methods based on laser scanners include the approach proposed in [55, 110]. In this approach, the reflection of an illumination beam on the surface of the hand is first detected, and triangulation is then applied to determine the position of the hand and reconstruct the 3-D model. The illumination and background conditions are both controlled. The approach described in [54, 111] uses the same acquisition setup described in [55] and additionally introduces a method of increasing the recognition accuracy by compensating for the pose and orientation of the hand at the moment of acquisition.

Approaches using multiple-view acquisition setups include the method proposed in [38],[1] which describes a two-view acquisition setup composed only of red-green-blue (RGB) cameras and visible-light illuminators. Notably, while laser scanners enable high-accuracy 3-D reconstruction, the devices are expensive and possibly difficult to obtain. In contrast, the method described in [38] is able to capture two synchronized images of the hand and then reconstruct a 3-D model of the palm using only off-the-shelf components, with a precision sufficient to enable high-accuracy biometric recognition.

2.2 Preprocessing

The purpose of the preprocessing phase is to extract the region of interest (ROI) of the palmprint from a touchless hand sample (Fig. 4). In most methods in the literature, this phase is divided into four steps: (i) segmentation of the hand, (ii) extraction of valley points, (iii) computation of the ROI, and (iv) enhancement.

The purpose of the hand segmentation step is to remove the background from the captured sample. This step differs depending on whether the background is controlled or uncontrolled. In the case of a controlled background, the majority of palmprint recognition approaches use algorithms based on gray-level thresholding or edge detection, such as the method described in [37]. In the case of an uncontrolled background, most methods in the literature apply segmentation procedures designed to isolate skin-color pixels in an RGB image [43, 63, 84]

The purpose of valley point extraction is to establish a reference system for the subsequent extraction of the ROI, and this step is usually performed by analyzing the local minima of the contour of the segmented hand [14, 84]. For example, in Fig. 4, the ring-little finger valley and the index-middle finger valley are used as reference points to extract the palmprint ROI. However, methods that analyze local minima are robust only for hands with all fingers separated. To compensate for this problem and enable successful valley point extraction in the case of poor separation between fingers, the method proposed in [48] extends the algorithms described in [14, 84] by adding a step based on edge detection.

---

[1] https://homes.di.unimi.it/genovese/3dpalm/.

Fig. 4 Example of a hand
image, with the extracted
valley points, the
corresponding reference
system, and the resulting
region of interest (ROI) of
the palmprint

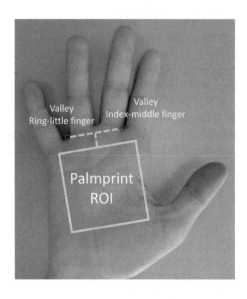

The aim of ROI computation is to extract a square region capturing the details of the palmprint, and this step is performed by using the extracted valley points to robustly estimate a reference system on the hand (as shown in Fig. 4). Different methods in the literature have considered variants of such reference systems based on the database and the procedure used to extract the features. For example, the procedure described in [3] extracts a rectangular ROI that spans most of the actual palm area by considering the ring-little finger valley and the index-middle finger valley. Recently, methods reported in the literature have been increasingly relying on the use of DL and convolutional neural networks (CNNs) because of their ability to automatically learn data representations by processing the spatial relationships between pixels in an image [40]. In particular, CNNs have been used for the automatic preprocessing of touchless hand images, as in the method described in [82], which extracts the ROI by using a CNN trained on the positions of landmarks.

Finally, palmprint images may be enhanced to increase the visibility of the details used for recognition; however, the enhancement step is seldom performed in the methods reported in the literature, especially with the growing popularity of methods based on machine learning and, in particular, DL. In fact, DL methods based on CNNs can automatically learn a filter structure that can be optimally adapted to each image, as in the approach proposed in [37]. However, the study described in [133] does present an enhancement method for palmprint samples and demonstrates that a particular range of image sharpness levels is correlated with a higher recognition accuracy.

2.3 Feature Extraction and Matching

The purpose of the feature extraction and matching phase is to process the ROI to extract a discriminant representation of the individual and then match the extracted representation to determine the identity associated with the palmprint sample. Traditionally, methods for feature extraction and matching can be divided into line-based, texture-based, subspace-based, coding-based, and local-texture-descriptor-based methods [56]. However, recent approaches are increasingly relying on DL, while line-based, texture-based, and subspace-based approaches are less commonly studied [27, 37]. Therefore, to offer the research community useful insight into the most studied research directions, this chapter will focus on coding-based, local-texture-descriptor-based, and DL-based methods.

2.3.1 Coding-Based Approaches

In methods based on coding, the feature extraction step is performed by first using a set of filters to process the image, quantizing the magnitude or phase of the response for each pixel, and finally encoding the results to compute a biometric template. Then, this template is matched using procedures based on the Hamming distance [38]. Based on the type and number of filters used to process the image, coding-based methods can be divided into three categories: (i) methods based on a single orientation, (ii) methods based on multiple orientations, and (iii) methods based on 3-D shapes. Table 1 presents an overview of such methods.

Methods based on a single orientation encode only one orientation for each pixel in an image. For example, the PalmCode approach [131] uses a single Gabor filter to process the image and then encodes the response for each pixel. Improving on PalmCode, the competitive code method [139] uses several Gabor filters with different orientations to process the image and then, for each pixel, encodes only the index of the filter corresponding to the minimum magnitude response. Such encoding creates a map of the main orientations of the palmprint lines in the image. Similar coding-based methods have subsequently been proposed in the literature to further increase the recognition accuracy by improving the set of filters as well as the coding scheme, such as the double-orientation code [29] and robust line orientation code [53] methods.

Methods based on multiple orientations, in addition to considering the principal orientation of each palmprint line, also consider secondary orientations at each pixel to compute the biometric template. For example, the binary orientation co-occurrence vector approach [41] encodes the responses of all Gabor filters for each pixel. Similarly, for each pixel, the neighboring direction indicator (NDI) method [33] encodes both the principal orientation and the relations with the orientations of neighboring regions. More recently, the robust competitive code approach [127] has been proposed by combining the competitive code algorithm with the NDI approach. Specifically, the robust competitive code method consists of encoding, for each pixel,

Table 1 Summary of coding-based approaches for palmprint recognition

Year	Method	Class	Approach
2003	PalmCode [131]	(i) Single orientation	Uses a single Gabor filter to process the image, then encodes the magnitude response for each pixel
2008	Robust line orientation code [53]	(i) Single orientation	Uses a modified finite Radon transform (MFRAT) to filter the image, encodes the most relevant response for each pixel, and matches encodings based on pixel-to-area comparison
2010	Competitive code [139]	(i) Single orientation	Uses multiple Gabor filters with different orientations to filter the image, computes the encoding for each pixel as a number indicating the filter for which the minimum response is obtained, and performs matching using the angular distance
2016	Double-orientation code [29]	(i) Single orientation	Uses multiple Gabor filters with different orientations to filter the image, encodes numbers indicating the two most representative filters for each pixel, and then performs matching using the nonlinear angular distance
2009	Binary orientation co-occurrence vector [41]	(ii) Multiple orientations	Uses multiple Gabor filters with different orientations to filter the image, then performs encoding for each pixel by considering the responses of all filters
2016	Neighboring direction indicator [33]	(ii) Multiple orientations	Uses multiple Gabor filters with different orientations to filter the image, then computes the encoding for each pixel by considering both the orientation of the most relevant filter and the relations with the orientations in adjacent regions
2018	Robust competitive code [127]	(ii) Multiple orientations	Uses multiple Gabor filters with different orientations to filter the image, encodes the representation by considering the orientation of the filter with the most relevant response as well as the weighted responses for adjacent regions, and then matches the representation using the angular distance
2011	SurfaceCode [55]	(iii) 3-D shape	Applies surface interpolation to local areas, computes a shape index for each pixel, and encodes the results using 4 bits for each pixel

the most relevant orientation and a weighted combination of the orientations in adjacent regions.

Methods based on 3-D shapes encode the 3-D shape of a palmprint. For example, the SurfaceCode approach [55] applies surface interpolation to the point cloud obtained using a laser scanner; then, for each pixel, it computes a shape index describing the local 3-D model and encodes the result using 4 bits.

2.3.2 Local-Texture-Descriptor-Based Approaches

Recent methods for touchless palmprint recognition have widely considered local texture descriptors since they have been proven to be robust to local variations in rotation, translation, scale, and illumination [12, 129], which are more likely to be present in touchless samples than in samples captured using a touch-based procedure [37]. Therefore, methods based on local texture descriptors are better suited than coding-based methods for achieving high-accuracy recognition based on palmprint samples captured using touchless acquisition procedures [27]. Specifically, recent approaches for touchless palmprint recognition based on local texture descriptors involve computing, for each local region of the ROI image, a blockwise histogram describing the orientations of the lines on the palm [67, 77, 134]. Then, a biometric template is computed by concatenating all of these blockwise histograms, thereby obtaining a global feature descriptor for the whole image. Finally, various distance measures (e.g., the Euclidean or chi-squared distance) are used to compare different templates generated in this way [5]. Approaches based on local texture descriptors can be divided into three categories: (i) methods using general-purpose descriptors applied to palmprint images, (ii) methods using texture descriptors encoding the main orientation for each pixel, and (iii) methods using texture descriptors encoding multiple orientations for each pixel. Table 2 presents an overview of such methods.

Methods using general-purpose descriptors consider local texture descriptors that have been previously proposed in the literature and apply them for touchless palmprint recognition, such as the local binary patterns (LBP) descriptor [119], the scale-invariant feature transform (SIFT) [125], the local directional patterns (LDP) descriptor [28], the histograms of oriented gradients (HOG) descriptor [52], and the local tetra patterns (LTrP) descriptor [67].

Methods using texture descriptors encoding the main orientation for each pixel, in contrast to methods using general-purpose descriptors, rely on local texture descriptors designed especially for palmprint recognition, such as the histograms of oriented lines (HOL) descriptor [52], which is a variant of the HOG descriptor based on Gabor filters, or the modified finite Radon transform (MFRAT), to better enhance the palmprint lines. Similarly, the collaborative representation competitive code (CR-CompCode) method [134] is a modification of the competitive code approach [139] in which a technique of computing a template based on blockwise histograms is introduced and then a sparse representation classifier [135] is used to compare templates.

Table 2 Summary of local-texture-descriptor-based approaches for palmprint recognition

Year	Method	Class	Approach
2014	SIFT [125]	(i) General purpose	Combines the scale-invariant feature transform (SIFT) for feature extraction with random sample consensus (RANSAC) for filtering outliers
2006	LBP [119]	(i) General purpose	Uses the local binary patterns (LBP) descriptor to compute a template, then performs matching using an AdaBoost classifier
2014	HOG [52]	(i) General purpose	Uses the histograms of oriented gradients (HOG) descriptor to compute a template
2016	LDP [28]	(i) General purpose	Applies the local directional patterns (LDP) descriptor to compute a template, then matches the template using the chi-square distance
2017	LTrP [67]	(i) General purpose	Uses the local tetra patterns (LTrP) descriptor to compute a template
2014	HOL [52]	(ii) Texture descriptors encoding the main orientation	Uses a variant of the HOG descriptor to preprocess the input image by applying either Gabor filters or the MFRAT
2017	CR-CompCode [134]	(ii) Texture descriptors encoding the main orientation	Uses a combination of competitive code with blockwise histograms to compute a template, then performs matching using a sparse representation classifier
2016	LLDP [77]	(iii) Texture descriptors encoding multiple orientations	Computes the most relevant orientation for each pixel by using either Gabor filters or the MFRAT, creates a template by encoding the corresponding responses, and then performs matching using either the Manhattan distance or the chi-square distance
2016	LMDP [28]	(iii) Texture descriptors encoding multiple orientations	Uses an encoding scheme that considers multiple relevant orientations for each pixel as well as their confidence and the relations with neighboring regions
2017	LMTrP [67]	(iii) Texture descriptors encoding multiple orientations	Computes the most relevant orientation for each pixel by using either Gabor filters or the MFRAT, then extracts the derivatives at each pixel in both the horizontal and vertical directions while also considering adjacent pixels to account for the thickness of the lines
2019	DDBC [31]	(iii) Texture descriptors encoding multiple orientations	Uses a filter-based approach to compute local convolutions for different directions, then learns a feature mapping to extract a feature vector
2020	LDDBP [32]	(iii) Texture descriptors encoding multiple orientations	Applies a method based on a combination of LBP and an analysis of the most discriminative directions for each pixel

Methods using texture descriptors encoding multiple orientations for each pixel, in contrast to methods that encode a single orientation for each pixel, consider a feature descriptor that encodes multiple orientations. For example, the local line directional patterns (LLDP) descriptor [77] is an extension of the LDP descriptor [50] that computes the line responses at each pixel using several Gabor filters with different orientations or the MFRAT. Then, both the minimum and maximum responses are encoded for each pixel, the corresponding blockwise histograms are calculated, and a distance measure is used to compare the resulting templates. Improving on the LLDP descriptor, the local multiple directional patterns (LMDP) descriptor [28] considers multiple dominant directions for each pixel, the confidence associated with each direction, and the relations with directions in adjacent regions. Similarly, the discriminant direction binary code (DDBC) [31] considers different directions by using a filter-based approach to compute the convolution differences between neighboring directions and then learns a feature mapping to project the convolution results into a feature vector. To further improve the accuracy by gaining insight into which directions are the most representative, the local discriminant direction binary pattern (LDDBP) approach [32] is based on an analysis of the discriminative power of each different direction in combination with the LBP descriptor. In contrast to the majority of the methods of class *iii)*, which achieve increased accuracy by encoding the most representative orientations, the local microstructure tetra patterns (LMTrP) descriptor [67] improves the recognition accuracy by considering the line thickness at each pixel in addition to describing the different local orientations.

2.3.3 Deep-Learning-Based Approaches

Currently, the majority of approaches for pattern recognition, including biometric systems, consider techniques based on DL and CNNs [105]. Approaches using CNNs are capable of extracting knowledge from data affected by noise, such as perspective distortions and local changes in rotation, translation, and scale, which are typical of biometric samples captured using touchless or less-constrained procedures [15, 19]. Moreover, CNNs can adapt to samples captured in heterogeneous environments [18]. Because of the advantages of DL for biometric recognition, several approaches in the literature consider CNNs for touchless and less-constrained palmprint recognition [37, 82, 83, 97, 106]. These approaches usually involve applying a CNN to ROI images to extract discriminative features and then computing a distance measure to compare the resulting templates. DL-based approaches for touchless palmprint recognition can be divided into three categories: (i) methods using pretrained CNNs, (ii) methods using CNNs fine-tuned on palmprint images, and (iii) methods using CNNs trained on palmprint images. Table 3 presents an overview of such methods.

Methods using pretrained CNNs extract features from palmprint images using CNNs previously trained on a general-purpose dataset, such as the method introduced in [109], which compares the results obtained using AlexNet [57], VGG-16, and VGG-19 [103]. Then, this method uses a support vector machine (SVM) to perform

Table 3 Summary of DL-based approaches for palmprint recognition

Year	Method	Class	Approach
2018	AlexNet, VGG-16, VGG-19 [109]	(i) Pretrained CNNs	Uses pretrained CNNs to extract features, then classifies them using a support vector machine (SVM)
2018	AlexNet [97]	(i) Pretrained CNNs	Uses pretrained CNNs to extract features, then classifies the feature vectors using an SVM
2016	AlexNet, discriminative index learning [106]	(ii) CNNs fine-tuned on palmprint images	Uses a CNN based on the AlexNet architecture, trained using a loss function that considers the separation between genuine and impostor distributions
2020	C-LMCL [137]	(ii) CNNs fine-tuned on palmprint images	Uses a CNN based on the ResNet architecture, with a loss function designed for uniformly clustering feature vectors of different classes
2020	GoogLeNet, adversarial metric learning [138]	(ii) CNNs fine-tuned on palmprint images	Uses a CNN based on the GoogLeNet architecture, trained using a technique based on adversarial metric learning
2020	EE-PRNet [82]	(ii) CNNs fine-tuned on palmprint images	Uses a CNN trained to segment and classify palmprint images using an end-to-end learning algorithm
2017	PCANet [83]	(iii) CNNs trained on palmprint images	Uses a CNN in which the filters are learned using an unsupervised procedure based on principal component analysis (PCA)
2019	PalmNet [37]	(iii) CNNs trained on palmprint images	Uses a CNN in which the filters are learned and adapted to the database using an unsupervised procedure based on Gabor analysis and PCA
2019	FusionNet [39]	(iii) CNNs trained on palmprint images	Uses PCANet for the fusion of palmprint and inner finger texture features

classification. A similar procedure is described in [97] for recognizing the palmprints of newborns as captured using a touchless acquisition procedure.

Methods using CNNs fine-tuned on palmprint images also rely on pretrained CNNs, but only after these CNNs have been fine-tuned on palmprint images. Such methods adapt the pretrained neural models to palmprint samples and can achieve a greater recognition accuracy than methods using only pretrained CNNs. For example, the work proposed in [106] starts from a CNN architecture based on the AlexNet model and then trains the CNN using a loss function based on the separation between genuine and impostor scores. Similarly, the centralized large margin cosine loss (C-LMCL) method proposed in [137] uses a CNN based on the ResNet architecture [44] and introduces a loss function designed to uniformly cluster classes in the feature space by separating the feature vectors of different individuals while ensuring that the feature vectors of the same individual remain close to each other. The work described in [138] extends this concept by introducing a CNN based on the GoogLeNet architecture [107], trained using a technique based on adversarial metric learning, with the purpose of further improving the division of templates in the feature space in accordance with their classes. Rather than using segmented ROIs, the work proposed in [82] describes EE-PRNet, a CNN based on the VGG-16 architecture fine-tuned to directly process touchless hand images, extract the palmprint ROI, and then perform individual classification.

Methods using CNNs trained on palmprint images rely on training CNNs from scratch on palmprint images. In particular, the approach described in [83] considers PCANet [7], a CNN trained using an unsupervised procedure based on principal component analysis (PCA). This method uses PCANet to extract a feature vector from the palmprint ROI and then classifies the resulting template using an SVM. Similarly, PalmNet,[2] proposed in [37], is a CNN in which the filters are learned using an unsupervised procedure based on Gabor analysis and PCA. Gabor analysis is performed to preliminarily select the Gabor filters that are best adapted to the palmprint images based on the palm size, rotation, and scale. The filters are then further adapted to the images using a PCA-based procedure. PCANet is also used in FusionNet[3] [39], which fuses the feature vectors obtained by applying PCANet to both palmprint and inner finger texture image regions.

3 Fingerprint Recognition

Fingerprints are reproductions of the surface pattern of the fingertip epidermis. This pattern is a characteristic sequence of interleaved ridges and valleys, usually considered unique for each individual. Biometric systems based on touchless fingerprint samples attempt to perform recognition by extracting and processing the discriminative information present in traditional fingerprint images from touchless finger images

---

[2] http://iebil.di.unimi.it/palmnet/index.htm.
[3] http://iebil.di.unimi.it/fusionnet/index.htm.

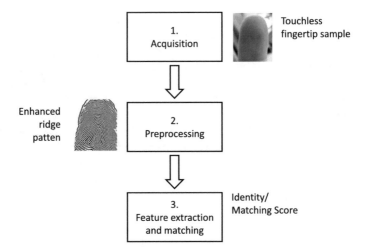

Fig. 5 Outline of the biometric recognition process based on touchless fingerprint acquisition.

[94] or 3-D models acquired using touchless technologies [22, 26]. Compared with palmprints, fingerprints offer the following advantages: users are frequently more familiar with this biometric recognition approach since fingerprint recognition systems are the most mature and widely used biometric technologies [80], and touchless fingerprint recognition technologies can produce templates compatible with existing governmental and investigative databases, such as those adopted in the Automated Fingerprint Identification System [81].

With the aim of designing systems that are more usable and acceptable than traditional touch-based technologies for fingerprint recognition, several recent studies have focused on touchless and less-constrained acquisition procedures, which can be based on either the integrated cameras in mobile devices [24] or dedicated acquisition systems [58].

In this section, we present the most recent approaches for touchless fingerprint recognition, describing state-of-the-art methods designed for every step of the biometric recognition process. The recognition procedure of a touchless fingerprint recognition system usually consists of the following phases: (i) acquisition, (ii) preprocessing, and (iii) feature extraction and matching. Figure 5 shows the outline of the recognition process.

3.1 Acquisition

The purpose of the acquisition phase is acquire an image, a 3-D model of the last finger phalanx, or a 3-D model of the ridge pattern with sufficient distinguishability of the distinctive characteristics to allow biometric recognition to be performed (Fig. 6).

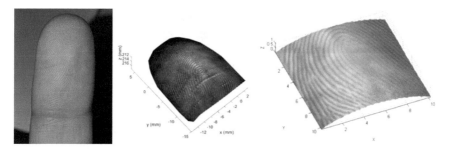

Fig. 6 Examples of touchless fingerprint acquisition: **a** a single image acquired using a smartphone with uncontrolled background and illumination conditions, **b** a 3-D model of the finger shape, and **c** a 3-D model of the ridge pattern.

The approaches proposed in the literature for acquiring touchless fingerprint samples exhibit important differences in terms of finger placement guides, the number of cameras, illumination techniques, and the use of either uniform or uncontrolled backgrounds. Therefore, there is no standard approach, and the types of acquisition constraints imposed depend on the application scenario for which the biometric system has been designed.

In the following, we divide state-of-the-art touchless fingerprint acquisition techniques into approaches based on 2-D images and approaches based on 3-D models [26].

3.1.1 Two-Dimensional Approaches

The state-of-the-art approaches designed for the acquisition of 2-D samples can be divided into three main groups: (i) methods for acquiring a single image using a frontal illumination source, (ii) approaches designed to compensate for the nonidealities of single touchless images using hardware solutions, and (iii) multimodal acquisition approaches.

Methods for acquiring a single image using a frontal illumination source are the most common ones. They can be based on heterogeneous kinds of cameras, such as the integrated cameras in smartphones [24], webcams [92], cameras designed for industrial applications [23], and consumer cameras [39]. Various constraints may also be imposed during the acquisition process depending on the scenario for which the biometric system is designed. Depending on the acquisition constraints, the following classes of acquisition setups can be distinguished: (a) setups for collecting single fingerprints with controlled finger positioning as well as controlled background and illumination conditions [17], (b) setups for collecting single fingerprints with uncontrolled finger positioning but controlled background and illumination conditions [112], (c) setups for collecting single fingerprints with uncontrolled finger positioning as well as uncontrolled background and illumination conditions [96], (d) setups for collecting multiple fingerprints with controlled finger positioning but uncontrolled background and illumination conditions [46], and (e) setups for collecting multiple fingerprints with uncontrolled finger positioning as well as uncontrolled background and illumination conditions [13].

Approaches designed to compensate for the nonidealities of single touchless images using hardware solutions are based on more complex and expensive acquisition setups than methods for acquiring a single image using a frontal illumination source. The approaches pertaining to this class can be divided into methods of compensating for perspective rotations, increasing the depth of field, and mitigating the detrimental effects of damaged finger skin. To compensate for perspective rotations, some studies have investigated capture devices capable of acquiring the nail-to-nail finger surface by means of rotating line scan cameras [90]. Other approaches rely on capturing multiple images from different viewpoints to compute a fingerprint image representing the complete ridge pattern by means of image stitching techniques [11]. To increase the depth of field of traditional cameras and thus enhance the ability of biometric systems to process fingerprint images acquired from nonfrontal positions, some studies have investigated the use of digital variable-focus liquid lenses [114]. To mitigate the detrimental effects of damaged finger skin, some systems are able to capture ridge patterns in the internal layers of the finger by using a red-light illumination source placed against the back of the fingernail [100].

Multimodal acquisition approaches are designed to acquire heterogeneous biometric traits using a single hardware device. For example, there are methods for simultaneously acquiring fingerprints and finger vein patterns [90, 118] as well as handheld embedded devices that can capture multiple touchless fingerprints and face images [123].

3.1.2   Three-Dimensional Approaches

The advantage of 3-D acquisition systems compared with 2-D acquisition systems is that they can capture additional information, thereby overcoming some of the problems related to perspective distortions, providing additional data for processing, and enhancing the compatibility between touchless and touch-based technologies. However, 3-D acquisition systems also require more complex and expensive hardware than 2-D acquisition technologies.

Systems for 3-D fingerprint acquisition can be divided into two classes: (i) systems that acquire models describing the 3D shape of the finger and (ii) systems that acquire models describing the 3-D characteristics of the ridge pattern.

Systems that acquire models describing the 3D shape of the finger are usually based on multiple-view acquisition setups and use multiple images acquired from different viewpoints to compute the 3-D coordinates of corresponding points in the images by applying the triangulation principle. Most of these methods require the use of guides for finger placement to control the orientation of the finger in 3-D space [74, 91]. However, there are also methods that are able to acquire 3-D samples on the move, without any guide for finger placement [23]. The 3-D reconstruction process requires a search for corresponding pairs of points in the images, which can be performed using correlation-based strategies [23] or methods for computing denser representations by using more complex feature sets [73, 75].

Systems that acquire models describing the 3-D characteristics of the ridge pattern have the advantage of collecting additional information that can be used for feature extraction and matching. However, they usually acquire multiple frames over time and thus require the use of finger placement guides to keep the finger still for the required acquisition time. Systems that acquire models describing the 3-D characteristics of the ridge pattern can be based on various techniques: photometric stereo 3-D reconstruction, ultrasonic sensors, structured light imaging, or laser sensing. Systems based on photometric stereo 3-D reconstruction capture multiple images under variable lighting conditions from a fixed viewpoint [70, 126, 136]. Such a reconstruction system assumes that the finger is illuminated only by the light sources of the sensor itself. Ultrasonic sensors may be used to acquire either 3-D or 2-D models of the internal skin layers [49]. Technologies for 3-D fingerprint acquisition based on ultrasonic sensors are currently in the prototype stage [78, 98], and there are not yet any studies on complete biometric recognition systems based on this technology. Systems based on structured light imaging project successive light patterns of different frequencies onto the finger. A fixed camera is used to acquire a set of images, which are then processed to estimate the shape of each single pattern and compute the distance of every point in the field to determine its deformation with respect to the original pattern [72, 120, 121]. Laser sensing permits accurate 3-D reconstruction with limited processing resources. As an example, the 3-D fingerprint reconstruction systems presented in [34, 35] use a laser line scanner to estimate the depth of the ridges.

3.2 Preprocessing

The touchless fingerprint recognition systems reported in the literature exhibit important differences in the preprocessing phase, which can be due to the acquisition techniques used or to the application scenario for which a system has been designed. Nevertheless, it is possible to distinguish six main computational tasks: (i) segmentation, (ii) texture enhancement, (iii) 3-D model enhancement, (iv) resolution normalization, (v) compensation of perspective deformations, and (vi) mapping of 3-D models to 2-D images.

The purpose of segmenting a touchless fingerprint sample is to remove the background and select an ROI corresponding to the last phalanx of a single finger. The segmentation methods presented in the literature differ depending on whether the acquisition system uses a fixed or uncontrolled background. In the first case, it is often possible to use general-purpose segmentation approaches, such as thresholding techniques [112] or background subtraction [6]. In the second case, segmenting the finger region is a more challenging task and requires more complex techniques, such as skin detection algorithms [4, 99] or methods based on CNNs [68, 79]. In the case of acquisitions of multiple fingers, the segmentation task requires separating every finger, frequently using image processing algorithms based on edge detectors [13].

The purpose of texture enhancement is to reduce noise and improve the distinguishability of the distinctive characteristics of the ridge pattern. This task can be performed by both 2-D and 3-D touchless fingerprint recognition systems that acquire samples using CCD cameras. The enhancement techniques for touchless fingerprint images can be divided into two classes: algorithms that enhance the visibility of the ridges using reduced computational resources and methods that compute an enhanced representation of the ridge pattern similar to those obtained from touch-based images. Techniques of the first class usually increase the contrast between ridges and valleys using algorithms such as Wiener filtering [96] and adaptive histogram equalization [112]. Techniques of the second class are more computationally expensive and usually consist of a noise reduction algorithm and a method for enhancing the ridge pattern [76, 99, 104]. Some methods for computing ridge pattern images similar to touch-based samples can also be performed in a single computational step, for example, using a bank of wavelets [4].

The purpose of 3-D model enhancement is to reduce the presence of noise and outliers. Systems designed to reconstruct the 3-D finger shape frequently refine the computed point clouds by applying techniques for approximating the finger shape as a 3-D surface. Some techniques approximate the finger shape as a previously defined shape [74]. Other methods perform noise reduction by approximating the finger shape using thin plate splines [23]. Recent studies have obtained remarkable results using binary quadratic functions [132]. Systems designed to reconstruct the 3-D ridge pattern adopt fewer assumptions on the finger shape, and the preprocessing phase frequently consists of the application of frequency filters [120].

Resolution normalization may be necessary to enable accurate biometric recognition since most state-of-the-art matchers require samples of fixed resolution. In touchless fingerprint recognition systems based on 3-D models and in systems based on 2-D samples that impose a fixed placement of the finger, the sample resolution is known a priori. In contrast, 2-D touchless systems that do not impose any constraint on finger placement frequently need to estimate and normalize the image resolution. This task can be performed by imposing a constant size for each finger [92], assuming that the ridge frequency is constant for each finger [104], or identifying the thick valley between the intermediate phalanges and proximal phalanges to estimate the finger size [96].

The aim of compensating for perspective deformations is to mitigate the detrimental effects of 3-D rotations of the finger samples. Methods in the literature estimate the finger pose from a single touchless fingerprint image [21, 104, 130] and subsequently apply compensation techniques based on synthetic 3-D models approximating the finger shape [21, 108].

Methods for mapping 3-D models to 2-D images attempt to enhance the compatibility between 3-D touchless fingerprint recognition systems and touch-based technologies. The mapping process may consist of unwrapping the 3-D models to obtain 2-D images similar to rolled fingerprints [42, 42, 95, 101, 120, 121] or may use geometrical models to compensate for both the perspective deformations of touchless samples and the nonlinear deformations introduced by touching an acquisition surface [69]. Unwrapping approaches can be further classified into parametric and nonparametric methods. Parametric methods use geometrical shapes that are known a priori to approximate the finger shape. They can convert the coordinates of a fingerprint into cylindrical coordinates [42], apply local conversions in polar coordinates by using sets of rings [120], or perform conversion into cylindrical coordinates followed by a refinement algorithm [121]. Nonparametric methods use more complex techniques that aim to preserve the distances between distinctive points of the ridge pattern. They can be based on various heuristics [42, 101] and include algorithms intended to enhance the compatibility with touch-based fingerprint databases by simulating the pressure of the finger on the acquisition sensor [95].

3.3 Feature Extraction and Matching

The goal of the feature extraction and matching phase is to extract distinctive characteristics from the fingerprint samples and compute the result of the recognition process. Most touch-based fingerprint recognition systems are based on minutiae features and are designed for identity verification [80]. Similarly, most studies on touchless fingerprint recognition rely on minutiae points. There also exist methods based on different features as well as machine learning approaches and DL strategies.

3.3.1 Minutiae-Based Approaches

The minutiae are distinctive points of the ridge pattern corresponding to bifurcations and terminations of the ridges [80]. Feature extraction and matching approaches based on minutiae features can be divided into three classes: (i) methods designed for touch-based samples, (ii) methods designed for 2-D touchless samples, and (iii) methods designed for 3-D touchless samples. Table 4 presents an overview of such methods.

Methods designed for touch-based samples are the most commonly used approaches in touchless fingerprint recognition systems based on both 2-D and 3-D samples. Specifically, many touchless systems adopt commercial feature extractors and matchers designed for touch-based samples [88], achieving impressive accuracy [104]. Open-source libraries designed for touch-based samples, such as the National Institute of Standards and Technology Biometric Image Software (NIST NBIS) [122], can also achieve satisfactory performance [20] when applied to enhanced representations of ridge patterns obtained from touchless images.

Methods designed for 2-D touchless samples attempt to overcome the nonidealities specific to touchless fingerprint images. Some minutiae extractors for touchless samples have been designed based on DL strategies [108]. Furthermore, minutiae matchers designed for touchless fingerprint images have been reported based on genetic algorithms [128] and artificial neural networks [25].

Methods designed for 3-D touchless samples attempt to achieve improved recognition accuracy compared to 2-D minutiae-based approaches. One approach consists of compensating for the perspective deformations of 2-D images by computing the best match score obtained by applying a set of 3-D rotations to a 3-D model of the probe sample [23]. To exploit the additional information provided by 3-D fingerprint models, some biometric systems use matching algorithms that compare the minutiae coordinates in 3-D space [60]. In this case, a pair of minutiae is considered to correspond if the Euclidean distance between their 3-D coordinates in the spatially aligned samples is less than a certain threshold and if the differences between their angles in 3-D space are less than certain fixed values.

3.3.2   Approaches Based on Non-minutiae Features

In the literature, there are some biometric recognition approaches designed for touchless fingerprint recognition systems that use features different from minutiae points to achieve accurate results in heterogeneous application scenarios. Specifically, these biometric recognition approaches can be divided into the following classes: (i) methods based on algorithmic matchers, (ii) methods based on computational intelligence techniques, and (iii) methods based on DL strategies. Table 5 presents an overview of such methods.

Methods based on algorithmic matchers exhibit relevant differences depending on the application scenario for which they have been designed. Some methods are designed to perform biometric recognition based on low-resolution touchless fingerprint images using level zero features, such as local texture patterns, which are matched by using the Hamming distance operator [61]. Another method consists of computing Speeded-Up Robust Features (SURF) and evaluating the number of corresponding pairs of points [112].

Methods based on computational intelligence learn distinctive characteristics of samples with the aim of overcoming the nonidealities that can detrimentally affect touchless samples, such as perspective distortions, reflections, and low visibility of the ridge patterns. Some of these methods are based on feature extractors applied for heterogeneous machine learning applications, such as sets of local Gabor filters [45], or a combination of the LBP and local gradient code (LGC) descriptors [117]. Other methods are based on more descriptive feature extractors such as scattering networks [79, 99], which are filter banks of wavelets able to compute distinctive representations that are stable with respect to local affine transformations.

Methods that rely on DL strategies use feature extraction functions learned and optimized for touchless fingerprint samples. Some methods use feature extractors learned by deep neural networks to compare touchless samples, such as the recognition method presented in [13], which uses a competitive coding algorithm in conjunction with a residual network to extract templates that are compared using the Hamming function. There are also studies on matching partial fingerprints with 3-D fingerprint acquisitions by using multiple Siamese CNNs [68]. Multiple Siamese CNNs are also used to perform cross-domain matching between touchless and touch-

Table 4 Summary of minutiae-based approaches for fingerprint recognition

Year	Method	Class	Approach
Since 1998	Neurotechnology VeriFinger [88]	(i) Methods designed for touch-based samples	Is the most widely used software development kit for feature extraction and matching of touchless fingerprint samples (commercial software)
2007	NIST NBIS [122]	(i) Methods designed for touch-based samples	Is a software development kit that can achieve satisfactory performance when applied to enhanced representations of ridge patterns obtained from touchless images (open-source software)
2020	Deep minutiae [108]	(ii) Methods designed for 2-D touchless samples	Uses a 3-D-based method to compensate for perspective distortions, applies a deep neural network to extract minutiae without any preprocessing, and performs matching using a method designed for touch-based samples
2020	Genetic matcher [108]	(ii) Methods designed for 2-D touchless samples	Uses genetic algorithms to match minutiae-based templates
2011	Neural matcher [25]	(ii) Methods designed for 2-D touchless samples	Uses artificial neural networks to compare pairs of minutiae
2016	Neural matcher [23]	(iii) Methods designed for 3-D touchless samples	Compensates for perspective deformations of 2-D images by computing the best matching score obtained by applying a set of 3-D rotations to a 3-D model of the probe sample
2015	Neural matcher [60]	(iii) Methods designed for 3-D touchless samples	Is a minutiae-based matcher that compares minutiae coordinates in 3-D space

based samples [71]. Deep neural networks can also be used to analyze ultrathin details of the fingertip (pores, incipient ridges, and local ridge characteristics). For example, the method described in [18] estimates and refines pore coordinates by using multiple CNNs.

Table 5 Summary of approaches based on non-minutiae features for fingerprint recognition

Year	Method	Class	Approach
2011	Level zero features [61]	(i) Methods based on algorithmic matchers	Uses level zero features (such as local texture patterns) and performs matching by using the Hamming distance operator
2015	SURF [112]	(i) Methods based on algorithmic matchers	Uses Speeded-Up Robust Features (SURF) to compute a template and performs matching by evaluating corresponding pairs of points
2007	Local Gabor filters [45]	(ii) Methods based on computational intelligence techniques	Uses a set of Gabor filters to compute biometric templates and performs matching by using an SVM classifier
2016	LBP and LGC [117]	(ii) Methods based on computational intelligence techniques	Uses the LBP and local gradient code (LGC) descriptors to compute a template and performs matching by using a nearest neighbor classifier
2015	Scattering networks A [99]	(ii) Methods based on computational intelligence techniques	Uses scattering networks to compute a template and performs matching by using a random forest classifier
2017	Scattering networks B [79]	(ii) Methods based on computational intelligence techniques	Uses scattering networks to compute a template and performs matching by using trained machine learning classifiers
2018	Deep feature vector [13]	(iii) Methods based on DL strategies	Uses a competitive coding algorithm in conjunction with a residual network and performs matching by using the Hamming distance operator
2018	Partial fingerprints [68]	(iii) Methods based on DL strategies	Performs the matching of partial fingerprints with 3-D fingerprint acquisitions by using multiple Siamese CNNs
2019	Cross-domain [71]	(iii) Methods based on DL strategies	Performs cross-domain matching between touchless and touch-based samples by using multiple Siamese CNNs
2018	Pore extraction [18]	(iii) Methods based on DL strategies	Uses multiple CNNs to estimate and refine pore coordinates

4 Conclusions

Palmprints and fingerprints are the most commonly used hand characteristics for biometric recognition. Recent studies have introduced accurate touchless technologies that offer enhanced usability, acceptability, hygiene, and robustness to grease and dirt compared with traditional touch-based technologies.

This chapter has presented a literature review on touchless palmprint and fingerprint recognition systems, focusing on the technological perspective. In particular, it has analyzed every phase of the biometric recognition chain, considering acquisition systems, preprocessing techniques, and feature extraction and matching methods. This review has focused on both two- and three-dimensional technologies, highlighting recent advances enabled by computational intelligence approaches and deep neural networks.

From the presented analysis of the state of the art, it is evident that deep neural networks enable marked improvement in the robustness and accuracy of touchless palmprint and fingerprint recognition systems. However, there are still open problems to be solved in order to develop highly usable and accurate systems that are fully compatible with touch-based biometric databases.

Acknowledgements This work was supported in part by the EC within the H2020 Program under projects MOSAICrOWN and MARSAL, by the Italian Ministry of Research within the PRIN program under project HOPE, by the Universitá degli Studi di Milano under project AI4FAO, and by JPMorgan Chase & Co. We thank the NVIDIA Corporation for the GPU donated.

References

1. Al-Nima, R., Abdullah, M., Al-Kaltakchi, M., Dlay, S., Woo, W., Chambers, J.: Finger texture biometric verification exploiting multi-scale sobel angles local binary pattern features and score-based fusion. Digital Signal Process. **70**, 178–189 (2017)
2. Barra, S., De Marsico, M., Nappi, M., Narducci, F., Riccio, D.: A hand-based biometric system in visible light for mobile environments. Inform. Sci. **479**, 472–485 (2019)
3. Bingöl, Ö., Ekinci, M.: Stereo-based palmprint recognition in various 3D postures. Expert Syst. Appl. **78**, 74–88 (2017)
4. Birajadar, P., Gupta, S., Shirvalkar, P., Patidar, V., Sharma, U., Naik, A., Gadre, V.: Touchless fingerphoto feature extraction, analysis and matching using monogenic wavelets. In: Proceedings of the 2016 International Conference on Signal and Information Processing (IConSIP), pp. 1–6 (2016)
5. Brahnam, S., Jain, L.C., Nanni, L., Lumini, A.: Local Binary Patterns: New Variants and Applications. Springer (2013)
6. Carney, L.A., Kane, J., Mather, J.F., Othman, A., Simpson, A.G., Tavanai, A., Tyson, R.A., Xue, Y.: A multi-finger touchless fingerprinting system: mobile fingerphoto and legacy database interoperability. In: Proceedings of the 2017 4th International Conference on Biomedical and Bioinformatics Engineering (ICBBE), pp. 139–147. ACM, New York, NY, USA (2017)
7. Chan, T., Jia, K., Gao, S., Lu, J., Zeng, Z., Ma, Y.: PCANet: a simple deep learning baseline for image classification? IEEE Trans. Image Process. **24**(12), 5017–5032 (2015)

8. Charfi, N., Trichili, H., Alimi, A.M., Solaiman, B.: Local invariant representation for multi-instance touchless palmprint identification. In: Proceedings of 2016 IEEE International Conference on Systems, Man, and Cybernetics (SMC), pp. 3522–3527 (2016)

9. Chinese Academy of Sciences, Institute of Automation: CASIA multi-spectral palmprint database (2007). http://www.cbsr.ia.ac.cn/english/MS_PalmprintDatabases.asp

10. Chinese Academy of Sciences, Institute of Automation: CASIA Palmprint Image Database (2009). http://english.ia.cas.cn/db/201611/t20161101_169936.html

11. Choi, H., Choi, K., Kim, J.: Mosaicing touchless and mirror-reflected fingerprint images. IEEE Trans. Inform. Forensi. Secur. **5**(1), 52–61 (2010)

12. Choi, J.Y., Ro, Y.M., Plataniotis, K.N.: Color local texture features for color face recognition. IEEE Trans. Image Process. **21**(3), 1366–1380 (2012)

13. Chopra, S., Malhotra, A., Vatsa, M., Singh, R.: Unconstrained fingerphoto database (2018)

14. Connie, T., Teoh, A.B.J., Ong, M.G.K., Ling, D.N.C.: An automated palmprint recognition system. Image Vis Comput. **23**(5), 501–515 (2005)

15. Das, R., Piciucco, E., Maiorana, E., Campisi, P.: Convolutional neural network for finger-vein-based biometric identification. IEEE Trans. Inf. Forensic. Secur. **14**(2), 360–373 (2019)

16. De Capitani di Vimercati, S., Foresti, S., Livraga, G., Samarati, P.: Data privacy: definitions and techniques. Int. J. Uncertain. Fuzziness Knowl. Based Syst. **20**(6), 793–817 (2012)

17. Derawi, M.O., Yang, B., Busch, C.: Fingerprint recognition with embedded cameras on mobile phones. In: Prasad, R., Farkas, K., Schmidt, A.U., Lioy, A., Russello, G., Luccio, F.L. (eds.) Security and Privacy in Mobile Information and Communication Systems, pp. 136–147. Springer, Berlin (2012)

18. Donida Labati, R., Genovese, A., Muñoz, E., Piuri, V., Scotti, F.: A novel pore extraction method for heterogeneous fingerprint images using Convolutional Neural Networks. Pattern Recognit. Lett. (2017)

19. Donida Labati, R., Genovese, A., Muñoz, E., Piuri, V., Scotti, F., Sforza, G.: Computational intelligence for biometric applications: a survey. Int. J. Comput. **15**(1), 40–49 (2016)

20. Donida Labati, R., Genovese, A., Piuri, V., Scotti, F.: Fast 3-D fingertip reconstruction using a single two-view structured light acquisition. In: Proceedings of the IEEE Workshop on Biometric Measurements and Systems for Security and Medical Applications (BIOMS), pp. 1–8 (2011)

21. Donida Labati, R., Genovese, A., Piuri, V., Scotti, F.: Contactless fingerprint recognition: a neural approach for perspective and rotation effects reduction. In: Proceedings of the IEEE Workshop on Computational Intelligence in Biometrics and Identity Management (CIBIM), pp. 22–30 (2013)

22. Donida Labati, R., Genovese, A., Piuri, V., Scotti, F.: Touchless fingerprint biometrics: a survey on 2D and 3D technologies. J. Internet Technol. **15**(3), 325–332 (2014)

23. Donida Labati, R., Genovese, A., Piuri, V., Scotti, F.: Toward unconstrained fingerprint recognition: A fully touchless 3-D system based on two views on the move. IEEE Trans. Syst. Man Cybern. Syst. **46**(2), 202–219 (2016)

24. Donida Labati, R., Genovese, A., Piuri, V., Scotti, F.: A scheme for fingerphoto recognition in smartphones. In: Rattani, A., Derakhshani, R., Ross, A. (eds.) Selfie Biometrics: Advances and Challenges, pp. 49–66. Springer International Publishing, Cham (2019)

25. Donida Labati, R., Piuri, V., Scotti, F.: A neural-based minutiae pair identification method for touch-less fingerprint images. In: Proceedings of the IEEE Workshop on Computational Intelligence in Biometrics and Identity Management (CIBIM), pp. 96–102 (2011)

26. Donida Labati, R., Piuri, V., Scotti, F.: Touchless Fingerprint Biometrics. Series in Security, Privacy and Trust. CRC Press, Boca Raton (2015)

27. Fei, L., Lu, G., Jia, W., Teng, S., Zhang, D.: Feature extraction methods for palmprint recognition: A survey and evaluation. IEEE Trans. Syst., Man Cybern. Syst. 1–18 (2018)

28. Fei, L., Wen, J., Zhang, Z., Yan, K., Zhong, Z.: Local multiple directional pattern of palmprint image. In: Proceedings of 2016 23rd International Conference on Pattern Recognition (ICPR), pp. 3013–3018 (2016)

29. Fei, L., Xu, Y., Tang, W., Zhang, D.: Double-orientation code and nonlinear matching scheme for palmprint recognition. Pattern Recognit. **49**, 89–101 (2016)
30. Fei, L., Zhang, B., Jia, W., Wen, J., Zhang, D.: Feature extraction for 3-D palmprint recognition: a survey. IEEE Trans. Instrum. Measure. **69**(3), 645–656 (2020)
31. Fei, L., Zhang, B., Xu, Y., Guo, Z., Wen, J., Jia, W.: Learning discriminant direction binary palmprint descriptor. IEEE Trans. Image Process. **28**(8), 3808–3820 (2019)
32. Fei, L., Zhang, B., Xu, Y., Huang, D., Jia, W., Wen, J.: Local discriminant direction binary pattern for palmprint representation and recognition. IEEE Trans. Circ. Syst. Video Technol. **30**(2), 468–481 (2020)
33. Fei, L., Zhang, B., Xu, Y., Yan, L.: Palmprint recognition using neighboring direction indicator. IEEE Trans. Human-Mach. Syst. **46**(6), 787–798 (2016)
34. Galbally, J., Beslay, L., Böstrom, G.: 3D-flare: a touchless full-3D fingerprint recognition system based on laser sensing. IEEE Access **8**, 145513–145534 (2020)
35. Galbally, J., Bostrom, G., Beslay, L.: Full 3D touchless fingerprint recognition: Sensor, database and baseline performance. In: Proceedings of the IEEE International Joint Conference on Biometrics (IJCB), pp. 225–233 (2017)
36. Genovese, A., Muñoz, E., Piuri, V., Scotti, F.: Advanced biometric technologies: emerging scenarios and research trends. In: Samarati, P., Ray, I., Ray, I. (eds.) From Database to Cyber Security: Essays Dedicated to Sushil Jajodia on the Occasion of His 70th Birthday. Lecture Notes in Computer Science, vol. 11170, pp. 324–352. Springer International Publishing, Cham (2018)
37. Genovese, A., Piuri, V., Plataniotis, K.N., Scotti, F.: PalmNet: Gabor-PCA convolutional networks for touchless palmprint recognition. IEEE Trans. Inform. Forens. Secur. **14**(12), 3160–3174 (2019)
38. Genovese, A., Piuri, V., Scotti, F.: Touchless palmprint recognition systems. In: Advances in Information Security, vol. 60. Springer, Berlin (2014)
39. Genovese, A., Piuri, V., Scotti, F., Vishwakarma, S.: Touchless palmprint and finger texture recognition: a deep learning fusion approach. In: Proceedings of the 2019 IEEE International Conference on Computational Intelligence and Virtual Environments for Measurement Systems and Applications (CIVEMSA), pp. 1–6 (2019)
40. Gu, J., Wang, Z., Kuen, J., Ma, L., Shahroudy, A., Shuai, B., Liu, T., Wang, X., Wang, G., Cai, J., Chen, T.: Recent advances in convolutional neural networks. Pattern Recogn. **77**, 354–377 (2018)
41. Guo, Z., Zhang, D., Zhang, L., Zuo, W.: Palmprint verification using binary orientation co-occurrence vector. Pattern Recognit. Lett. **30**(13), 1219–1227 (2009)
42. Han, F., Hu, J., Alkhathami, M., Xi, K.: Compatibility of photographed images with touch-based fingerprint verification software. In: Proceedings of the 6th IEEE Conference on Industrial Electronics and Applications, pp. 1034–1039 (2011)
43. Han, Y., Sun, Z., Wang, F., Tan, T.: Palmprint recognition under unconstrained scenes. In: Proceedings 8th Asian Conference on Computer Vision (AACV), pp. 1–11 (2007)
44. He, K., Zhang, X., Ren, S., Sun, J.: Deep residual learning for image recognition. In: Proceedings of the 2016 IEEE Conference on Computer Vision and Pattern Recognition (CVPR), pp. 770–778 (2016)
45. Hiew, B.Y., Teoh, A.B.J., Pang, Y.H.: Touch-less fingerprint recognition system. In: Proceedings of the 2007 IEEE Workshop on Automatic Identification Advanced Technologies, pp. 24–29 (2007)
46. IIIT Delhi: IIITD SmartPhone Fingerphoto Database v1 (ISPFDv1). http://iab-rubric.org/resources/spfd.html
47. Indian Institute of Technology Delhi: IIT Delhi Touchless Palmprint Database (Version 1.0) (2008). http://www4.comp.polyu.edu.hk/~csajaykr/IITD/Database_Palm.htm
48. Ito, K., Sato, T., Aoyama, S., Sakai, S., Yusa, S., Aoki, T.: Palm region extraction for contactless palmprint recognition. In: Proceedings of 2015 International Conference on Biometrics (ICB), pp. 334–340 (2015)
49. Iula, A.: Ultrasound systems for biometric recognition. Sensors **19**(10) (2019)

50. Jabid, T., Kabir, M.H., Chae, O.: Robust facial expression recognition based on Local Directional Pattern. ETRI J. **32**(5), 784–794 (2010)
51. Jain, A.K., Flynn, P., Ross, A.A.: Handbook of Biometrics, 1st edn. Springer (2010)
52. Jia, W., Hu, R., Lei, Y., Zhao, Y., Gui, J.: Histogram of Oriented Lines for palmprint recognition. IEEE Trans. Syst. Man Cybern. Syst. **44**(3), 385–395 (2014)
53. Jia, W., Huang, D.S., Zhang, D.: Palmprint verification based on robust line orientation code. Pattern Recognit. **41**(5), 1504–1513 (2008)
54. Kanhangad, V., Kumar, A., Zhang, D.: Contactless and pose invariant biometric identification using hand surface. IEEE Trans. Image Process. **20**(5), 1415–1424 (2011)
55. Kanhangad, V., Kumar, A., Zhang, D.: A unified framework for contactless hand verification. IEEE Trans. Inf. Forens. Secur. **6**(3), 1014–1027 (2011)
56. Kong, A., Zhang, D., Kamel, M.: A survey of palmprint recognition. Pattern Recogn. **42**(7), 1408–1418 (2009)
57. Krizhevsky, A., Sutskever, I., Hinton, G.E.: ImageNet classification with Deep Convolutional Neural Networks. In: Proceedings of 25th International Conference on Neural Information Processing Systems (NIPS), pp. 1097–1105 (2012)
58. Kumar, A.: Introduction to Trends in Fingerprint Identification. Springer International Publishing, Cham (2018)
59. Kumar, A.: Toward more accurate matching of contactless palmprint images under less constrained environments. IEEE Trans. Inform. Forens. Secur. **14**(1), 34–47 (2019)
60. Kumar, A., Kwong, C.: Towards contactless, low-cost and accurate 3D fingerprint identification. IEEE Trans. Pattern Anal. Mach. Intell. **37**(3), 681–696 (2015)
61. Kumar, A., Zhou, Y.: Contactless fingerprint identification using level zero features. In: Proceedings of the Conference on Computer Vision and Pattern Recognition Workshops (CVPRW), pp. 114–119 (2011)
62. L. Sathiya, V.P.: A survey on finger knuckle print based biometric authentication. Int. J. Computer Sci. Eng. **6**, 236–240 (2018)
63. Leng, L., Gao, F., Chen, Q., Kim, C.: Palmprint recognition system on mobile devices with double-line-single-point assistance. Personal Ubiquitous Comput. **22**(1), 93–104 (2018)
64. Leng, L., Li, M., Kim, C., Bi, X.: Dual-source discrimination power analysis for multi-instance contactless palmprint recognition. Multimed. Tools Appl. **76**(1), 333–354 (2017)
65. Leng, L., Li, M., Leng, L., Teoh, A.B.J.: Conjugate 2DPalmHash code for secure palm-print-vein verification. In: Proceedings of 2013 6th International Congress on Image and Signal Processing (CISP), pp. 1705–1710 (2013)
66. Leng, L., Zhang, J., Khan, M.K., Chen, X., Alghathbar, K.: Dynamic weighted discrimination power analysis: a novel approach for face and palmprint recognition in DCT domain. Int. J. Phys. Sci. **5**(17), 2543–2554 (2010)
67. Li, G., Kim, J.: Palmprint recognition with Local Micro-structure Tetra Pattern. Pattern Recognit. **61**, 29–46 (2017)
68. Lin, C., Kumar, A.: Contactless and partial 3D fingerprint recognition using multi-view deep representation. Pattern Recogn. **83**, 314–327 (2018)
69. Lin, C., Kumar, A.: Matching contactless and contact-based conventional fingerprint images for biometrics identification. IEEE Trans. Image Process. **27**(4), 2008–2021 (2018)
70. Lin, C., Kumar, A.: Tetrahedron based fast 3D fingerprint identification using colored leds illumination. IEEE Trans. Pattern Anal. Mach. Intell. **40**(12), 3022–3033 (2018)
71. Lin, C., Kumar, A.: A CNN-based framework for comparison of contactless to contact-based fingerprints. IEEE Trans. Inform. Forens. Secur. **14**(3), 662–676 (2019)
72. Liu, F., Liang, J., Shen, L., Yang, M., Zhang, D., Lai, Z.: Case study of 3D fingerprints applications. PLOS ONE **12**(4), 1–15 (2017)
73. Liu, F., Shen, L., Zhang, D.: Feature-based 3D reconstruction model for close-range objects and its application to human finger. In: Zha, H., Chen, X., Wang, L., Miao, Q. (eds.) Computer Vis., pp. 379–393. Springer, Berlin Heidelberg, Berlin, Heidelberg (2015)
74. Liu, F., Zhang, D.: 3D fingerprint reconstruction system using feature correspondences and prior estimated finger model. Pattern Recogn. **47**(1), 178–193 (2014)

75. Liu, F., Zhao, Q., Zhang, D.: 3D fingerprint generation. In: Advanced Fingerprint Recognition: From 3D Shape to Ridge Detail, pp. 15–32. Springer, Singapore (2020)

76. Liu, X., Pedersen, M., Charrier, C., Cheikh, F.A., Bours, P.: An improved 3-step contactless fingerprint image enhancement approach for minutiae detection. In: Proceedings of the 2016 6th European Workshop on Visual Information Processing (EUVIP), pp. 1–6 (2016)

77. Luo, Y.T., Zhao, L.Y., Zhang, B., Jia, W., Xue, F., Lu, J.T., Zhu, Y.H., Xu, B.Q.: Local line directional pattern for palmprint recognition. Pattern Recognit. **50**, 26–44 (2016)

78. Maev, R., Bakulin, E., Maeva, E., Severin, F.: High resolution ultrasonic method for 3D fingerprint representation in biometrics. In: Akiyama, I. (ed.) Acoust. Imaging, pp. 279–285. Springer, Netherlands, Dordrecht (2009)

79. Malhotra, A., Sankaran, A., Mittal, A., Vatsa, M., Singh, R.: Fingerphoto authentication using smartphone camera captured under varying environmental conditions. In: De Marsico, M., Nappi, M., Proença, H. (eds.) Human Recognition in Unconstrained Environments, pp. 119–144. Academic, London (2017)

80. Maltoni, D., Maio, D., Jain, A.K., Prabhakar, S.: Handbook of Fingerprint Recognition, 2nd edn. Springer Publishing Company, Berlin (2009)

81. Mather, F.: 4F allows the use of smartphone finger photos as a contactless fingerprint identification system to match with legacy databases (2016). http://www.biometricupdate. com/201601/4f-allows-the-use-of-smartphone-finger-photos-as-a-contactless-fingerprint-identification-system-to-match-with-legacy-databases

82. Matkowski, W.M., Chai, T., Kong, A.W.K.: Palmprint recognition in uncontrolled and unco-operative environment. IEEE Trans. Inform. Forens. Secur. **15**, 1601–1615 (2020)

83. Meraoumia, A., Kadri, F., Bendjenna, H., Chitroub, S., Bouridane, A.: Improving biometric identification performance using PCANet deep learning and multispectral palmprint. In: Jiang, R., Al-maadeed, S., Bouridane, A., Crookes, D., Beghdadi, A. (eds.) Biometric Security and Privacy: Opportunities & Challenges in the Big Data Era, pp. 51–69. Springer, Cham (2017)

84. Michael, G.K.O., Connie, T., Teoh, A.B.J.: Touch-less palm print biometrics: novel design and implementation. Image Vis. Comput. **26**(12), 1551–1560 (2008)

85. Michael, G.K.O., Connie, T., Teoh, A.B.J.: An innovative contactless palm print and knuckle print recognition system. Pattern Recognit. Lett. **31**(12), 1708–1719 (2010)

86. Nanyang Technological University: NTU Palmprints from the Internet (NTU-PI-v1) (2019). https://github.com/matkowski-voy/Palmprint-Recognition-in-the-Wild

87. National University of Ireland: NUIG_Palm2 database of palmprints (2020). https://github. com/AdrianUng/NUIG-Palm2-palmprint-database

88. Neurotechnology: VeriFinger SDK. http://www.neurotechnology.com/verifinger.html

89. Palma, D., Montessoro, P.L., Giordano, G., Blanchini, F.: Biometric palmprint verification: a dynamical system approach. IEEE Trans. Syst. Man Cybern. Syst. **49**(12), 2676–2687 (2019)

90. Palma, J., Liessner, C., Mil'Shtein, S.: Contactless optical scanning of fingerprints with 180° view. Scanning **28**(6), 301–304 (2006)

91. Parziale, G., Diaz-Santana, E., Hauke, R.: The surround imagerTM: A multi-camera touchless device to acquire 3D rolled-equivalent fingerprints. In: Zhang, D., Jain, A.K. (eds.) Advances in Biometrics, pp. 244–250. Springer, Berlin Heidelberg, Berlin, Heidelberg (2005)

92. Piuri, V., Scotti, F.: Fingerprint biometrics via low-cost sensors and webcams. In: Proceedings of the 2008 IEEE International Conference on Biometrics: Theory, Applications and Systems (BTAS), pp. 1–6. Washington, D.C., USA (2008)

93. PolyU-IITD: Contactless Palmprint Images Database (Version 3.0) (2011). https://www4. comp.polyu.edu.hk/~csajaykr/palmprint3.htm

94. Priesnitz, J., Rathgeb, C., Buchmann, N., Busch, C., Margraf, M.: An overview of touchless 2D fingerprint recognition. EURASIP J. Image Video Process. **2021** (2021)

95. Qijun Zhao, Jain, A., Abramovich, G.: 3D to 2D fingerprints: unrolling and distortion correction. In: Proceedings of the International Joint Conference on Biometrics (IJCB), pp. 1–8 (2011)

96. Raghavendra, R., Busch, C., Yang, B.: Scaling-robust fingerprint verification with smartphone camera in real-life scenarios. In: Proc. of the 2013 IEEE 6th International Conference on Biometrics: Theory, Applications and Systems (BTAS), pp. 1–8 (2013)

97. Ramachandra, R., Raja, K.B., Venkatesh, S., Hegde, S., Dandappanavar, S.D., Busch, C.: Verifying the newborns without infection risks using contactless palmprints. In: Proceedings of 2018 International Conference on Biometrics (ICB), pp. 209–216 (2018)

98. Saijo, Y., Kobayashi, K., Okada, N., Hozumi, N., Yoshihiro Hagiwara, Tanaka, A., Iwamoto, T.: High frequency ultrasound imaging of surface and subsurface structures of fingerprints. In: Proceedings of the 2008 30th Annual Int. Conf. of the IEEE Engineering in Medicine and Biology Society, pp. 2173–2176 (2008)

99. Sankaran, A., Malhotra, A., Mittal, A., Vatsa, M., Singh, R.: On smartphone camera based fingerphoto authentication. In: Proceedings of the 2015 IEEE 7th International Conference on Biometrics Theory, Applications and Systems (BTAS), pp. 1–7 (2015)

100. Sano, E., Maeda, T., Nakamura, T., Shikai, M., Sakata, K., Matsushita, M., Sasakawa, K.: Fingerprint authentication device based on optical characteristics inside a finger. In: Proceedings of the Conference on Computer Vision and Pattern Recognition Workshop (CVPRW), p. 27 (2006)

101. Shafaei, S., Inanc, T., Hassebrook, L.G.: A new approach to unwrap a 3-D fingerprint to a 2-D rolled equivalent fingerprint. In: Proceedings of the 3rd IEEE International Conference on Biometrics: Theory, Applications, and Systems, pp. 1–5 (2009)

102. Shaheed, K., Liu, H., Yang, G., Qureshi, I., Gou, J., Yin, Y.: A systematic review of finger vein recognition techniques. Information 9(9) (2018)

103. Simonyan, K., Zisserman, A.: Very deep convolutional networks for large-scale image recognition. In: Proceedings of International Conference on Learning Representations (ICLR) (2015)

104. Stein, C., Nickel, C., Busch, C.: Fingerphoto recognition with smartphone cameras. In: Proceedings of the 2012 International Conference of Biometrics Special Interest Group (BIOSIG), pp. 1–12 (2012)

105. Sundararajan, K., Woodard, D.L.: Deep Learning for biometrics: a survey. ACM Comput. Surv. 51(3), 65:1–65:34 (2018)

106. Svoboda, J., Masci, J., Bronstein, M.M.: Palmprint recognition via discriminative index learning. In: Proceedings of 2016 23rd International Conference on Pattern Recognition (ICPR), pp. 4232–4237 (2016)

107. Szegedy, C., Wei Liu, Yangqing Jia, Sermanet, P., Reed, S., Anguelov, D., Erhan, D., Vanhoucke, V., Rabinovich, A.: Going deeper with convolutions. In: Proceedings of the 2015 IEEE Conference on Computer Vision and Pattern Recognition (CVPR), pp. 1–9 (2015)

108. Tan, H., Kumar, A.: Towards more accurate contactless fingerprint minutiae extraction and pose-invariant matching. IEEE Trans. Inform. Forens. Secur. 15, 3924–3937 (2020)

109. Tarawneh, A.S., Chetverikov, D., Hassanat, A.B.: Pilot comparative study of different Deep features for palmprint identification in low-quality images. CoRR abs/1804.04602 (2018)

110. The Hong Kong Polytechnic University: Contact-free 3D/2D Hand Images Database (Ver 1.0) (2011). http://www4.comp.polyu.edu.hk/csajaykr/myhome/database_request/3dhand/Hand3D.htm

111. The Hong Kong Polytechnic University: Contact-free 3D/2D Hand Images Database (Version 2.0) (2011). http://www4.comp.polyu.edu.hk/~csajaykr/Database/3Dhand/Hand3DPose.htm

112. Tiwari, K., Gupta, P.: A touch-less fingerphoto recognition system for mobile hand-held devices. In: Proceedings of the 2015 International Conference on Biometrics (ICB), pp. 151–156 (2015)

113. Tongji University: Tongji Contactless Palmprint Dataset (2017). https://cslinzhang.github.io/ContactlessPalm/

114. Tsai, C.W., Wang, P.J., Yeh, J.A.: Compact touchless fingerprint reader based on digital variable-focus liquid lens. In: Gregory, G.G., Davis, A.J. (eds.) Novel Optical Systems Design and Optimization XVII, vol. 9193, pp. 173–178. International Society for Optics and Photonics, SPIE (2014)

115. Ungureanu, A., Thavalengal, S., Cognard, T.E., Costache, C., Corcoran, P.: Unconstrained palmprint as a smartphone biometric. IEEE Trans. Consum. Electron. 63(3), 334–342 (2017)

116. University of Las Palmas de Gran Canaria: Grupo de Procesado Digital de la Señal (GPDS) GPDS100Contactlesshands2Band database (2011). http://www.gpds.ulpgc.es/
117. Wang, K., Jiang, J., Cao, Y., Xing, X., Zhang, R.: Preprocessing algorithm research of touchless fingerprint feature extraction and matching. In: Tan, T., Li, X., Chen, X., Zhou, J., Yang, J., Cheng, H. (eds.) Pattern Recogn., pp. 436–450. Springer Singapore, Singapore (2016)
118. Wang, L., El-Maksoud, R.H.A., Sasian, J.M., Kuhn, W.P., Gee, K., Valencia, V.S.: A novel contactless aliveness-testing (CAT) fingerprint sensor. In: Koshel, R.J., Gregory, G.G. (eds.) Novel Optical Systems Design and Optimization XII, vol. 7429, pp. 333–343. International Society for Optics and Photonics, SPIE (2009)
119. Wang, X., Gong, H., Zhang, H., Li, B., Zhuang, Z.: Palmprint identification using boosting Local Binary Pattern. In: Proceedings 18th International Conference on Pattern Recognition (ICPR), vol. 3, pp. 503–506 (2006)
120. Wang, Y., Hassebrook, L.G., Lau, D.L.: Data acquisition and processing of 3-D fingerprints. IEEE Trans. Inform. Forens. Secur. 5(4), 750–760 (2010)
121. Wang, Y., Lau, D.L., Hassebrook, L.G.: Fit-sphere unwrapping and performance analysis of 3D fingerprints. Appl. Opt. 49(4), 592–600 (2010)
122. Watson, C.I., Garris, M.D., Tabassi, E., Wilson, C.L., Mccabe, R.M., Janet, S., Ko, K.: User's guide to NIST biometric image software (NBIS) (2007)
123. Weissenfeld, A., Strobl, B., Daubner, F.: Contactless finger and face capturing on a secure handheld embedded device. In: Proceedings of the Design, Automation Test in Europe Conf. Exhibition (DATE), pp. 1321–1326 (2018)
124. Wu, W., Elliott, S.J., Lin, S., Sun, S., Tang, Y.: Review of palm vein recognition. IET Biometrics 9(1), 1–10 (2020)
125. Wu, X., Zhao, Q., Bu, W.: A SIFT-based contactless palmprint verification approach using iterative RANSAC and local palmprint descriptors. Pattern Recognit. 47(10), 3314–3326 (2014)
126. Xie, W., Song, Z., Chung, R.C.: Real-time three-dimensional fingerprint acquisition via a new photometric stereo means. Opt. Eng. 52(10), 1–11 (2013)
127. Xu, Y., Fei, L., Wen, J., Zhang, D.: Discriminative and robust competitive code for palmprint recognition. IEEE Trans. Syst. Man Cybern. Syst. 48(2), 232–241 (2018)
128. Yin, X., Zhu, Y., Hu, J.: Contactless fingerprint recognition based on global minutia topology and loose genetic algorithm. IEEE Trans. Inform. Forens. Secur. 15, 28–41 (2020)
129. Zaghetto, C., Mendelson, M., Zaghetto, A., d. B. Vidal, F.: Liveness detection on touchless fingerprint devices using texture descriptors and artificial neural networks. In: Proceedings of 2017 IEEE International Joint Conference on Biometrics (IJCB), pp. 406–412 (2017)
130. Zaghetto, C., Zaghetto, A., d. B. Vidal, F., Aguiar, L.H.M.: Touchless multiview fingerprint quality assessment: rotational bad-positioning detection using artificial neural networks. In: Proceedings of the International Conference on Biometrics (ICB), pp. 394–399 (2015)
131. Zhang, D., Kong, W.K., You, J., Wong, M.: Online palmprint identification. IEEE Trans. Pattern Anal. Mach. Intell. 25(9), 1041–1050 (2003)
132. Zhang, D., Lu, G., Zhang, L.: 3D fingerprint reconstruction and recognition. In: Advanced Biometrics, pp. 177–212. Springer International Publishing, Cham (2018)
133. Zhang, K., Huang, D., Zhang, D.: An optimized palmprint recognition approach based on image sharpness. Pattern Recogn. Lett. 85, 65–71 (2017)
134. Zhang, L., Li, L., Yang, A., Shen, Y., Yang, M.: Towards contactless palmprint recognition: a novel device, a new benchmark, and a collaborative representation based identification approach. Pattern Recognit. 69, 199–212 (2017)
135. Zhang, L., Yang, M., Feng, X.: Sparse representation or collaborative representation: which helps face recognition? In: Proceedings 2011 International Conference on Computer Vision (ICCV), pp. 471–478 (2011)
136. Zheng, Q., Kumar, A., Pan, G.: Contactless 3D fingerprint identification without 3D reconstruction. In: Proceedings of the 2018 International Workshop on Biometrics and Forensics (IWBF), pp. 1–6 (2018)

137. Zhong, D., Zhu, J.: Centralized large margin cosine loss for open-set deep palmprint recognition. IEEE Trans. Circ. Syst. Video Technol. **30**(6), 1559–1568 (2020)
138. Zhu, J., Zhong, D., Luo, K.: Boosting unconstrained palmprint recognition with adversarial metric learning. IEEE Trans.Biometrics Behavior Identity Sci. **2**(4), 388–398 (2020)
139. Zuo, W., Lin, Z., Guo, Z., Zhang, D.: The multiscale competitive code via sparse representation for palmprint verification. In: Proc. 2010 IEEE Computer Society Conference on Computer Vision and Pattern Recognition (CVPR), pp. 2265–2272 (2010)

A Survey of IoT Software Platforms

Konstantinos Astropekakis, Emmanouil Drakakis, Konstantinos Grammatikakis, and Christos Goumopoulos

Abstract In the last few years an outbreak of Internet of Things (IoT) solutions have been placed in service around the globe giving rise to a multi-billion market size. IoT software platforms are evolving as the main pillar for the rapid development of scalable and efficient IoT applications. Such platforms are offering end-to-end services including connectivity between smart objects and cloud infrastructures, scalable storage as well as programming frameworks and data analytics to create valuable insights for new business use cases. This paper proposes a broad evaluation scheme for IoT software platforms based on an analysis of the characteristics found in contemporary platforms that are surveyed. The evaluation criteria are organized into four categories: core (Connectivity, Device Management, Integration APIs, Security, Endorsed Hardware, Hosting), data management and processing (Storage, Multi-Modal Support, Analytics, Visualization), application empowerment (Event-driven Actions, Push Notifications, Programming Frameworks, Third Party Integration), and accessibility (Pricing, Support). A comparative analysis of state of-the-art IoT software platforms is conducted based on the defined criteria followed by a critical presentation of fundamental features, commonalities and differences between them. Furthermore, challenges in this domain that are open to address in the future are discussed.

Keywords IoT · Platform · Applications · Survey · Comparative analysis

K. Astropekakis · E. Drakakis · K. Grammatikakis · C. Goumopoulos (✉)
School of Science and Technology, Hellenic Open University, Patras, Greece
e-mail: goumop@aegean.gr

C. Goumopoulos
Information and Communication Systems Engineering Department, University of the Aegean, Mytilene, Greece

P. Nicopolitidis et al. (eds.), *Advances in Computing, Informatics, Networking and Cybersecurity*, Lecture Notes in Networks and Systems 289,
https://doi.org/10.1007/978-3-030-87049-2_10

1 Introduction

The Internet of Things (IoT) represents a vision where the Internet extends to the physical world including everyday objects that can be remotely controlled and act as access points to smart services [1]. In this context, the IoT paradigm promotes the augmentation of the physical world with computing, sensing and wireless communication capabilities, which enables the collection of massive data and their connection to intelligent cloud services [2].

According to IEEE [3], IoT is defined as a self-configuring, adaptive, complex network that interconnects things to the Internet through standard communication protocols. The interconnected things have physical or virtual representation, sensing/actuation capability, programmability and are uniquely identifiable. The things offer services available anywhere, anytime, and for anything, taking security into consideration.

A number of developments in various fields have allowed the exponential growth of the IoT ecosystem:

- *Sensors*: The price of sensors has decreased over the last few years and their value is expected to fall further. This fact allows the mass use of sensors in a cost-effective way [4].
- *Ubiquitous wireless connection*: Sensors require access to wireless networks to forward available data for processing. Connectivity is now available everywhere, with different types of networks (WiFi, BLE, 3G/4G/5G and LoRa) [5].
- *Cloud Computing*: The cloud computing flourishing meets the processing and storage requirements of IoT applications. Big data in particular, generated by billions of sensors can find a place to be stored and processed [6].
- *IPv6 Networking*: The latest version of the IP protocol allows an almost unlimited number of unique IP addresses that can be mapped to identify any potential IoT device [7].

Since the participating IoT nodes may generate data in high rates, the new trend in the field of IoT applications is the transfer of computing to end devices to increase response speed and reduce costs [8]. These devices are located at the edge of the network, so it is usually preferable to process the data before sending them to the cloud. The term edge computing has been introduced to refer to the enabling technologies allowing computation to be performed at the edge of the network and the term "edge" is defined as "any computing and network resources along the path between data sources and cloud data centers" [9].

Over the last few years, IoT has emerged as a key driver for digital transformation in several application domains among which manufacturing, business, health, smart cities and smart farming by emphasizing the internetworking of physical and virtual resources in order to gather and exchange data. Some researchers advocate that applications like predictive maintenance in industrial IoT have the potential to boost the IoT technology in this niche market [10]. On the other hand, more voices argue that there is no clear killer application and that there is no point in chasing for such applications in IoT [11, 12]. Instead the IoT industry should put more emphasis on comprehending customer needs and assessing how the IoT technology can solve identified problems efficiently, rapidly, and with a lower cost. In that way, any company can build the right IoT application that addresses the requirements of their particular customer, vertically in a specific domain.

Almost all IoT applications require some form of communication, external storage, processing, and display of the data they generate. Due to the complexity of IoT application development numerous software platforms have been emerged to facilitate the engineering process. Such platforms facilitate the development and operation of IoT applications by ensuring the seamless interaction of different components and by confronting the inherent heterogeneity of the environments and the dynamic changes that occur in the lifetime of applications. For convenience in the context of this paper the use of the term *IoT platform* will be equivalent to *IoT software platform* unless otherwise noted.

To state the main contributions, this paper: (a) surveys contemporary software platforms for IoT application development; (b) proposes an evaluation scheme for such platforms based on an analysis of the features found on modern IoT platforms; (c) compares IoT platforms based on the proposed evaluation criteria; and (d) identifies a number of challenges and open issues that future studies need to address in this IoT research area.

The remaining of the paper is organized as follows. Section 2 describes the research methodology followed regarding the selection and evaluation process of the IoT platforms. Section 3 clarifies the role of an IoT platform, gives a generic IoT platform architecture scheme and defines the evaluation criteria. Section 4 gives a brief description of each platform selected, followed by a comparative presentation and a discussion of fundamental features, commonalities and differences between the platforms. In Sect. 5 challenges concerning this field are discussed, whereas conclusions are outlined in Sect. 6.

2   Research Methodology

2.1   IoT Platforms Selection

In this survey, a meticulous approach was followed to select IoT platforms among the current state-of-the-art solutions in this technology field by distinguishing which ones have the greatest potential and popularity. To achieve this, a composite weighted formula was defined by taking into account the relative rating of each platform as found in four sources, namely, Alexa Site Ranking, Google Search Results, Google Scholar Results (after 2016), and IEEE Digital Library (after 2016) in order to arrive at an overall ranking of IoT platform popularity.

For each platform the ranking of its website was observed in Alexa Site Ranking (Alexa.com), where a smaller ranking number can show higher popularity in users.

Then all platforms were ranked from highest to lowest in traffic. In rankings based on Google Search Results, Google Scholar Results and IEEE Digital Library, the number of search results for each platform was observed. For each information source a decreasing classification for the number of results was made. The first platform for each of the ranking sources, received 1 point, the second platform in ranking received 2 points and so on.

Furthermore, IoT-Analytics evaluation was adopted [13]. This evaluation divides the platforms into three categories "Leaders", "Challengers" and "Followers". Depending on the categorization 10, 20, 30 points were awarded depending on the categorization of each platform and unspecified platforms received 40 points.

The total ranking was calculated by summing each individual ranking source result, weighted depending on each source accuracy. The IoT platforms with the lowest sum of points were selected.

2.2   IoT Platforms Evaluation

This paper proposes a broad evaluation scheme for IoT software platforms based on an analysis of the characteristics found in contemporary platforms that are surveyed. The evaluation criteria are organized into four categories: *core* (Connectivity, Device Management, Integration APIs, Security, Endorsed Hardware, Hosting), *data management and processing* (Storage, Multi-Modal Support, Analytics, Visualization), *application empowerment* (Event-driven Actions, Push Notifications, Programming Frameworks, Third Party Integration), and *accessibility* (Pricing, Support).

For the evaluation of each platform information was sought in the website of each platform. Related research articles on the subject were explored using Google Scholar search engine. The following keywords were used: "IoT platforms", "Edge devices", "IoT", "Device Management", "Integration API", "IoT Security", "Databases", "IoT Analytics", "IoT Visualization", "Event-driven Actions". Articles from relevant scientific journals in the IoT field were examined, including:

- ACM Transactions on Internet of Things
- Computer Networks
- IEEE Communications
- IEEE Internet of Things Journal
- IEEE Pervasive Computing
- IEEE Wireless Communications
- International Journal of Internet of Things and Cyber-Assurance
- Internet of Things Journal
- Pervasive and Mobile Computing Sensors.

A total of 42 relevant papers were reviewed and carefully analyzed to extract the required information. In addition to studying the scientific literature from journals, information was obtained from the official sites and sources of each platform due to the rapid changes that occur in this specific field.

3 Background and Evaluation Scheme

3.1 The Role of IoT Platforms

The number of devices that generate data is increasing exponentially. As a result, conventional systems have a difficulty to record and analyze efficiently the data generated, especially if real time performance is required. These difficulties have piqued the interest of research centers and industry and the IoT platforms have emerged as a solution to such challenges. IoT platforms play an important role in many aspects of the new reality by connecting devices and things, data networks, and IoT gateways to cloud applications and services [14].

IoT platforms provide ultimately the link between the physical and digital worlds. A very large number of devices and things are connected and interact with the IoT platforms. Despite the fact that these things may be far apart, they can be managed and controlled and their data can be processed seamlessly by the IoT platform [14]. One of the most important service of an IoT platform is the interconnection of things in a secure and reliable manner, ensuring the authentication and authorization of users and things, the integrity of the content and the security of the transferred data.

Other features of IoT platforms include the device management, the perception of the environment, the transfer of data from edge devices to cloud or local storage, event and action management, and the data analysis and visualization in real time. Thus, the end user is able to take advantage of the generated information through applications. IoT platform providers are numerous and offer different types of architecture depending on the capabilities that the platforms offer. A general structure of IoT platform architecture is shown in Fig. 1 as a guide to understanding the main components involved in IoT platforms.

3.2 Evaluation Criteria

In order to compare the IoT platforms, specific features were selected which were grouped into four categories. The criteria of the first group related to the **core** of the platform operation are listed thereafter:

- **Connectivity** includes features related to how the platform communicates with IoT devices and protocols for data transfer. Commonly used protocols for machine-to-machine (M2M) communication at the application level are HTTP, MQTT, XMPP, AMQP and CoAP. HTTP is the oldest and one of the most common protocols but is not optimized for IoT devices and it does not provide Quality of

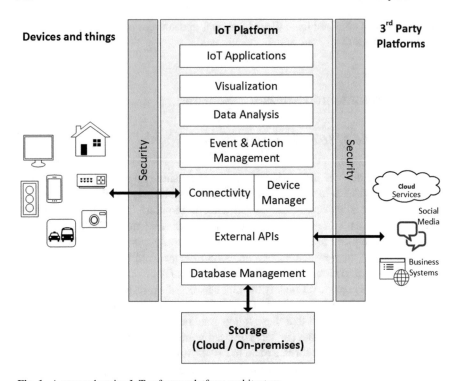

Fig. 1 A comprehensive IoT software platform architecture

Service (QoS). On the other hand, MQTT and CoAP [15, 16] are particularly suitable for IoT gadgets and use comparatively fewer resources, both from the device and the network side as they have a smaller overhead due to the smaller header size at 2 and 4 Bytes respectively [17]. In addition, they do provide QoS support [18].

- **Device Management** allows the platform to remotely control, authenticate and configure device and sensors as well as upgrade their software and firmware [19]. Desirable device management features include scaling and automation capabilities, compatibility, openness, context awareness as well as multi-role service [20].
- **Integration APIs** are responsible for interconnecting all the components that make up an IoT ecosystem. They are points of interaction between an IoT device and other components of the IoT environment. Particularly important features are openness and ease of use. An example of such an API is the REST API which defines a set of functions that allows developers to execute requests and receive responses via HTTP [21]. It can be implemented in almost any programming language and due to its convenience, it is supported by most available platforms, as shown by other research studies on IoT platforms [22, 23].

- **Security** is one of the most important features in an IoT Platform. Key elements of security are data privacy, confidentiality and integrity. IoT devices are prone to attacks that can allow the attacker to affect their data and integrity. Authentication, authorization capabilities and account management are essential elements to ensure security on IoT devices. At the same time, the great heterogeneity of IoT platforms and the diversity of authentication mechanisms is a particular challenge [24]. In this work the existing authentication, authorization, access management and communications security mechanisms of the IoT platforms under consideration are examined. Security techniques are implemented with protocols such as SSL/TLS and DTLS, as well as authentication tokens.
- **Endorsed Hardware** is defined as the compatible hardware to which the various sensors are connected and function with the specific IoT platform. They can either be supported without the need for any intervention or conversion, or require special firmware and settings provided by the IoT platform.
- **Hosting** is related to how the IoT platform implementation is deployed. Most commercial IoT platforms use their own cloud infrastructure. They are more flexible in computing needs and data management [25]. They allow easier access to shared computing resources, such as networks, servers, storage, applications and services, which can be quickly provided and require little effort in managing and interacting with the provider's services [26]. In addition, they provide better scalability depending on the use. Other methods for hosting are the use of cloud infrastructure different from third party chosen by the user or the development of a local system.

The second group of criteria is related to **data management and processing**. The relevant features that were examined are the following:

- **Storage** can be realized by different technologies. It can be relational databases like SQL or non-relational like NoSQL. SQL databases have capabilities for vertical scalability by increasing their system resources, while NoSQL have greater capabilities for horizontal scalability by partitioning among multiple servers. There are also differences in structure and compatibility. All these differences can be important factors in choosing one platform over another [27].
- **Multi-Modal Support** is related to the ability of the IoT platform to manage and process various types of files that are not easily manageable, i.e., large files such as blob, video and audio files.
- **Analytics** allows the analysis and data process with the aid of algorithms that perform complex calculations and machine learning. The analysis gives more value to the collected data [28]. Real time data analysis and process is considered as an important feature. As it can be observed from relevant research [23], the analysis is done by third party tools, such as MATLAB, Google Analytics, etc.
- **Visualization** is the graphical display and presentation of data from sensors, as well as other information, such as statistics, metadata and others. Many platforms have built-in techniques to support this feature, while others rely on third-party systems, such as MATLAB or Grafana.

For the third group we analyze the **application empowerment**. More specifically, four criteria are included:

- **Event-driven Actions** is the mechanism by which the user can define rules upon the data collected by the platform, to trigger specific actions. These actions may involve one or more smart objects by changing their behavior or by sending alerts and messages to users or other interconnected devices. The definition of rules for the implementation of event-driven actions is usually done with the help of a separate subsystem called the rule engine [29].
- **Push Notifications** refers to the provision of mechanisms through which notifications are sent to users about various events that occurred and are considered critical to the smooth operation of an IoT system. Notifications are triggered by the rules engine and its event-driven actions. Common forms of notifications include, SMS, email or social media messages.
- **Programming Frameworks** is the provision of appropriate development tools that allow the developers to create applications for specific platform, such as Android or iOS. IoT platforms usually provide a set of libraries that implement device interconnection.
- **Third Party Integration** is related to the interconnection of external platforms linked to business systems, like ERP and CRM, web services, such as Amazon web services, as well as social media platforms, such as LinkedIn, Twitter and Facebook.

Finally, the fourth group of criteria is about **accessibility**, which examines the pricing and support features:

- **Pricing** is related to the supported pricing model. This feature indicates whether the platform is open source, or proprietary available for free with limited functionality or with tiered paid plan options. A platform can be provided free of charge as open source software and at the same time provides support packages.
- **Support** is related to the different ways this can be done. The existence of official forums and wiki for answering questions, use case scenarios for use as examples, official user guides or provided by third parties, as well as the provision of paid assistance packages, are included in this feature.

4 Survey of IoT Platforms

4.1 Research Sample

After evaluating the information sources specified in the research methodology for the impact of each platform and its general position in the global IoT platform list the following fifteen (15) platforms were selected in the research sample and are introduced briefly in alphabetical order.

Alibaba IoT Cloud [30] is part of Alibaba cloud services founded in 2009 and provides services to enterprises, developers, governments and organizations. The platform provides 4G/5G connectivity and seamless networking on heterogeneous networks. Regarding its functions, provides device connection and management, message communication and security. It also makes use of a rule engine for SQL parsing and data forwarding. The documentation of the platform is open source and anyone can contribute.

Amazon AWS IoT [31] is an IoT platform, which through its services, offers solutions to comply with industrial, consumer and commercial requirements. Through easy-to-use services, it provides data management and analytics, collects data through sensors and responds to events. Services are also provided, for all layers of security including mechanisms such as encryption and access control. The platform has been built on a secure cloud environment thus significantly improving its scalability. AI integration, as well as the ability to integrate other AWS services enhances the building of smart and complete IoT applications.

Carriots [32] is a platform provided in the form of Platform as a Service (PaaS) or on-premises. It specializes in creating and hosting IoT services. Provides tools for fast prototyping to reduce development time and it is scalable. With the use of SmartSight, SmartCore and SmartEdge developed by Altair, the platform is able to provide the necessary tools for IoT applications development and deployment.

Cisco Kinetic [33] is a platform implemented in a modular architecture that combines different modules to provide secure device connectivity, data transfer and processing. With the "Edge and Fog Processing Module" the processing is done from the distributed nodes for efficient use of network resources while the "Data Control Module" is responsible for the most efficient transfer to cloud-based applications.

Google Cloud IoT Core [34] uses other Google Cloud services to collect, process, analyze, and display IoT data in real time. Its services include, among others, advanced analytics, visualizations and machine learning. It allows the use of existing devices with minimal firmware changes and provides secure connectivity and two-way communications.

IBM Watson IoT [35] is a fully managed, cloud-hosted service with features like registration, secure connection and control of devices, as well as visualization, AI-driven analysis and storage of collected data. Regarding the security, through a blockchain service, it enables shared and secured information across private networks [36].

Microsoft Azure IoT [37] offers a wide range of services and capabilities for integration into customer's IoT solutions. Worth mentioning are the multiple programming languages and frameworks support, the hybrid implementation in cloud and edge and its security features. Some of the services are provided free of charge and there are demos and learning modules to help familiarize users with the platform.

Oracle IoT Cloud [38] is based on a cloud service and is offered with the Platform as a Service (PaaS) model. Allows customers to connect their devices to the cloud, collect and analyze device data in real time and integrate them into their applications. In the area of security, it provides a secure environment for the participating devices and communications applying end-to-end security. It is worth mentioning that the platform has built-in machine learning and artificial intelligence capabilities.

PTC Thingworx [39], is an IoT platform created to be used primarily in the field of Industrial IoT. Exploring the basic features it provides easy connection of devices / objects on the platform, easy and fast development of applications through the platform, while there is a built-in machine learning feature to automate complex data analysis tasks. In terms of data storage, local as well as cloud service storage is supported.

SiteWhere [40] is a microservice-based infrastructure targeting the industrial IoT domain, with capabilities for independently scale. It is deployed in Kubernetes, providing deployment capabilities for any cloud structure. It is important to mention its flexible modular structure and that it is open source.

Thinger.io [41] is an open source IoT platform where the user can install it either on his own infrastructure or on the provided hosted cloud infrastructure thus gaining in scaling, speed and security. Its simplicity in writing code, without limiting its capabilities, makes it an ideal choice for developers of any level. At the same time, it is hardware agnostic, enabling device integration from any manufacturer.

ThingsBoard [42] is an open-source IoT platform. It enables device connectivity via industry standard IoT protocols, so that monitoring and controlling of the devices is possible. It allows the collection, processing, analysis and visualization of data while at the same time reacting accordingly to the rules that are set by the user through the rule engine. The platform supports both cloud and on-premises deployments.

ThingSpeak [43] is provided by MathWorks known for the MATLAB tool. A special feature is the ability to execute MATLAB code in ThingSpeak to perform real time analysis and processing of the data. It also accelerates the development of proof-of-concept IoT systems and can support small and medium-sized IoT solutions.

Ubidots [44] is a platform provided for use with the Platform as a Service model. It supports the connection of a large number of devices to the Ubidots cloud through libraries and device kits. It also enables real-time data processing and provides data protection and encryption. From 2018 the Ubidots Stem version is provided which is suitable for students. Lastly, the Gateway Management, manages and deploys on Cisco gateways through a secure VPN connection.

WSO₂ [45] is an open source IoT platform that enables users to manage and monitor their devices. It is designed with the security of both devices and data in mind. Regarding the capabilities in data processing and analysis, it is able to collect sensor data, visualize them, recognize patterns and act on them. The ability to process and analyze data can also be deployed on edge devices in order to optimize network traffic, but also to preserve data privacy.

4.2 Comparative Analysis

The evaluation criteria as defined and presented in Sect. 3.2 have been the pillars for conducting the comparative analysis of the IoT platforms. The results of the evaluation process are presented in the following paragraphs based on the data collected and reported in the associated tables.

By examining the data related to the core operation of the platform (Table 1) several insights can be given. All IoT platforms support the MQTT communication protocol, which is to be expected. MQTT is the standard messaging protocol for the IoT connectivity and is suitable for communicating IoT devices with small code footprint and minimal network bandwidth [15]. Also, all platforms support the HTTP(S) protocol either for connecting only to Gateways (e.g. Cisco Kinetic) or for connecting devices. HTTP(S) although not customized for IoT has the advantage of being supported by most devices as it is the earliest communication protocol. Other protocols that appear to be used by the sample platforms are CoAP (33%) and AMQP (26%), which have features that make them suitable for use in the IoT. It is followed by the multi-purposed XMPP with 20%. Figure 2 summarizes the occurrence rate of the various communication protocols in the surveyed IoT platforms. Device Management is supported by 13 of the platforms, which corresponds to the 87% of the sample. In terms of Integration APIs, all platforms support REST API, and some of them point out the support for MQTT API, which is an alternative to REST API [46]. The use of the appropriate Integration API for each use is important, as they have their own unique properties and differences in throughput [47, 48]. There is no homogeneity in the technologies and techniques used for security, but every platform follows some approach. Most platforms support authentication and use SSL/TLS at a ratio of 67% and 87% respectively. Popular hardware platforms such as Arduino and Raspberry Pi are supported by the majority of IoT platforms either natively or with minor modifications. Some IoT platforms make use of sensors connected to gateways and smart objects designed for a particular use (e.g., smart city things). Regarding hosting, there are solutions that are based exclusively on cloud infrastructure. For almost half of the sample platforms, and especially for open source there is usually the capability of on-premises deployment.

The following observations can be made by examining the data management and processing data reported in Table 2. Pertaining to storage, it is observed that 14 out of 15 platforms support non-relational (NoSQL) databases and 11 out of 15 relational (SQL). Additional 3 platforms support hybrid storage implementations. This was to be anticipated, since due to the type of data and the needs of the IoT platforms, the use of NoSQL is preferable (Fig. 4, left side). Multimodal data support is available on 11 of the platforms. In most cases there is support for BLOB files. All platforms support some form of real-time analytics and 40% of them can also make use of 3rd

Table 1 Core criteria

Platform	Connectivity	Device management	Integration APIs	security	Endorsed hardware	Hosting
Alibaba IoT Cloud	MQTT, CoAP, HTTP(S)	Yes	REST APIs	Authentication tokens, Hardware crypto, TLS	AliOS Things devices, Arduino, Raspberry Pi	Cloud
Amazon AWS IoT	MQTT, Websockets, HTTP(S)	Yes	REST APIs	Authentication (X.509), TLS	Arduino, Raspberry Pi; SigFox	Cloud
Carriots	MQTT, HTTP(S)	Yes	REST APIs	Authentication tokens, TLS	Arduino, Raspberry Pi; Beagle Bone, Cubie Board, FEZ Cerbuino, TST Gate, TST Mote	Cloud, On- Premises
Cisco Kinetic	MQTT, AMQP, HTTP(S)*	N/A	Kinetic API (REST)	Authentiction, Secure websockets, TLS	Smart city things, Industrial environments, Raspberry Pi, Panduit Sensors, Gateways	Cloud
Google Cloud IoT	MQTT, HTTP(S)	Yes		Authentication (X.509), Authorization (OAuth 2.0), TLS	Android things, various devices including Raspberry Pi	Cloud
IBM Watson IoT	MQTT, HTTP(S)	Yes	REST APIs	Authentication Tokens, HTTPS, TLS	Arduino, Raspberry Pi	Cloud

(continued)

Table 1 (continued)

Platform	Connectivity	Device management	Integration APIs	security	Endorsed hardware	Hosting
Microsoft Azure IoT	MQTT, AMQP, HTTP(S)	Yes	REST APIs	Hardware, multi-layer software, renewable security, SSL/TLS, X.509 certificates	Windows 10 family, iOS, Linux, Android, Mbed, Azure Sphere, Arduino, Espressif, Qualcomm IoT Modem, ST Microelectronics, TI LaunchPad	Cloud
Oracle IoT Cloud	MQTT, CoAP, AMQP, XMPP, Websockets, HTTP(S)	Yes	REST APIs	Authentication Tokens, Access Tokens	Arduino, Raspberry Pi	Cloud
PTC ThingWorx	MQTT, CoAP, XMPP, Websockets, HTTP(S)	Yes	REST APIs	HTTPS, SSL, TLS	Arduino, Raspberry Pi, SigFox, iOS, Androidora	Cloud, On- Premises
SiteWhere	MQTT, AMQP, Stomp, HTTP(S)	Yes	REST APIs	Authentication tokens, Authorization, User Permission Management	Arduino, Appliances (printers, air conditioners, PCs etc.)	Cloud, On- Premises
Thinger.io	MQTT, HTTP(S)	Yes	REST APIs, Websocket API	SSL, TLS	Hardware Agnostic (Arduino, Raspberry Pi, SigFox etc.)	Cloud, On- Premises
ThingsBoard	MQTT, CoAP, HTTP(S)	Yes	REST APIs	Authentication (X.509), HTTPS, SSL	Arduino, Raspberry Pi, SigFox	Cloud, On- Premises

(continued)

Table 1 (continued)

Platform	Connectivity	Device management	Integration APIs	security	Endorsed hardware	Hosting
ThingSpeak	MQTT, HTTP	N/A	ThingSpeak API (REST & MQTT)	HTTPS, SSL, TLS	Arduino, Particle Photon Core, Raspberry Pi, Electric Imp, ESP-8266	Cloud, On- Premises
Ubidots	MQTT, CoAP, HTTP(S)	Yes (limited)	REST APIs	Authentication tokens, SSL, User Permission Management	Arduino, Raspberry Pi, Spark Core, Microchip, WCM, Adafruit, FONA	Cloud
WSO$_2$	MQTT, XMPP, HTTP	Yes	REST APIs	Authorization (OAuth 2.0), Basic Auth, Mutual SSL, SCEP, JWT, Scopes	Windows, Android, iOS, Arduino, Raspberry Pi,	Cloud, On- Premises

Fig. 2 Communication protocols adopted by the surveyed IoT platforms

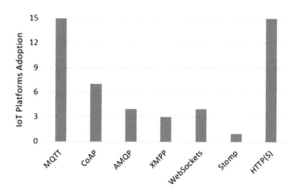

party data analysis platforms. Finally, they all support some form of visualization. SiteWhere in particular provides visualization exclusively from 3rd party tools, such as Grafana.

Table 3, concerning application empowerment, shows that all platforms support event driven actions with the use of a rule-engine. Regarding push notifications, most of the platforms can send e-mail and SMS. Additional ways, such as HTTP responses are supported depending on the platform. There is a variety in programming frameworks, with 12 out of the 15 platforms supporting Java through SDKs. Other popular languages supported are C, Python and JavaScript. Figure 3 provides additional data regarding programming languages and frameworks. Third Party Integration is supported in some form by all platforms.

Finally, the data in Table 4 manifest that 4 of the platforms are completely open source, 1 platform had most of its modules open source and 3 were partially open source providing libraries and tools for building IoT applications and connecting devices. That means that more than half of the platforms (53%) were, to some extent, offered as open source (Fig. 4, right side). Out of 15 platforms the 14 provide a free version. On closed source platforms, the free version comes with restrictions either in terms of usage time or in terms of functionality. All platforms have basic user help tools such as providing a wiki, examples and usage scenarios. SiteWhere is the only platform that does not allow to purchase support and doesn't offer official forum for asynchronous communication between users.

4.3 Discussion

The IoT development and the growing needs of customers, in terms of new prospecting applications, has led to the creation of a large number of IoT platforms. Today, more than 600 IoT platforms are offered by various companies and organizations according to IoT-Analytics [49]. These platforms often differ considerably in their structure and deployment approach. More specifically, based on the study of

Table 2 Data management and processing criteria

Platform	Storage	Multi-modal support	Analytics	Visualization
Alibaba IoT Cloud	SQL (*ApsaraDB*) NoSQL (HiTSDB) OSS (Object Storage Service)	BLOB, Big data, Multimedia	Quick BI, MaxCompute	DataV (Dashboard)
Amazon AWS IoT	SQL (*PostgreSQL*), NoSQL (*DynamoDB, MongoDB*) Amazon S3	Amazon Kinesis Video Streams	Real-Time Analytics (Amazon IoT Analytics)	AWS IoT Dashboard
Carriots	NoSQL (*Big data DB*)	Raw data (BiG Data DB)	Altair SmartSight 3rd party: Microsoft Azure, IBM, Ducksboard, Nibodha	Altair SmartWorks (deprecated), 3rd party: Geckoboard, Freeboard
Cisco Kinetic	SQL (*ParStream, JDBC*)	BLOB	Real-Time Analytics	Cisco Kinetic dashboard
Google Cloud IoT	SQL (*Cloud Spanner, Cloud SQL*), NoSQL (*Cloud Bigtable, Cloud Firestore, Firebase Realtime Database, Cloud Memorystore*)	Large Files, ML services, BigQuery	Google Big data analytics, Cloud Dataflow, BigQuery, Cloud Bigtable, ML, Google Data Studio, BI tools	Stackdriver Monitoring
IBM Watson IoT	SQL (*Db2, PostgreSQL*), NoSQL (*MongoDB*),		IBM Watson Analytics	Dashboard
Microsoft Azure IoT	SQL (*MySQL, MariaDB, PostgreSQL*), NoSQL (*Cosmos DB*)	BLOB	Real-Time Analytics (Azure Stream analytics)	Web Portal Dashboard
Oracle IoT Cloud	SQL(*Oracle*), NoSQL(*Oracle NoSQL*), Oracle DBaaS	BLOB	Oracle IoT Cloud Enterprise Platform	Dashboard
PTC ThingWorx	SQL(*H2, MS SQL Server, Azure SQL, PostgreSQL*), NoSQL(*InfluxDB*) Hybrid (*Datastax Enterprise*)	BLOB	Thingworx Analytics	Dashboard
SiteWhere	NoSQL (*InfluxDB, MongoDB, Apache HBase*)	BJSON, big data event storage	Amazon SOS, Wattics	3rd party Dashboard
Thinger.io	NoSQL (*InfluxDB, MongoDB, DynamoDB*)	N/A	Real-Time Analytics	

(continued)

Table 2 (continued)

Platform	Storage	Multi-modal support	Analytics	Visualization
ThingsBoard	SQL(*PostgreSQL*), NoSQL (*deprecated*), Hybrid (*PostgreSQL + Cassandra, PostgreSQL + TimescaleDB*)	BLOB	Trendz Analytics, Kafka Streams	Dashboard
ThingSpeak	SQL (*MySQL*) NoSQL (*Timeseries*)	N/A	MATLAB	Dashboard
Ubidots	NoSQL (*Timestamps*)	N/A	Distimo, Google Analytics	Dashboard
WSO$_2$	SQL(*MySQL, SQL Server, Oracle, PostgreSQL*) NoSQL(*Cassandra, MongoDB*)	BLOB, Streaming support	WSO$_2$ Analytics Platform	Dashboard

the IoT platforms explored in this survey, three approaches can be identified, each of them having specific advantages and concerns.

The first approach refers to *IoT platforms as dedicated servers* build either from scratch, or by using open source solutions. For example, the very popular Grafana platform can be used to analyze and visualize the collected data and InfluxDB can store the time-series data directly received by the device, without the need for an additional server. Additionally the Google Firebase cloud service support environment can be used successfully. This category can also include open source IoT platforms which are mainly structured in terms of their basic features and use many third-party open source software solutions. Examples of such platforms and tools are SiteWhere and, to a lesser extent, Node-RED.

The main advantage of this approach is that the system is highly customized for the purpose that it is created and each system component can be ideally selected in the optimal way. In addition, depending on its size, it can have zero initial cost even for commercial use as this is usually built with open source software. The only investment required in this case concerns actually the hardware cost.

On the other hand, developing an IoT application combining different technologies requires time, programming skills knowledge and familiarity with the programming requirements of each solution while the code for the IoT devices should be constructed from scratch. At an organizational level, there is certain difficulty in creating, installing, and managing servers for use with IoT devices from scratch.

The second approach is *IoT platforms with highly integrated services* provided typically by large IT and software companies. These solutions are usually an evolution of pre-existing cloud systems and services that provided items to costumers and then adapted to support the IoT sector. Examples of such IoT platforms are Microsoft Azure IoT, Amazon AWS IoT Core, Google Cloud IoT, and IBM Watson.

Table 3 Application empowerment criteria

Platform	Event-driven actions	Push notifications	Programming frameworks	Third party integration
Alibaba IoT Cloud	Rule Engine	Email, SMS	Java, Pythin, C,.NET, PHP, Gateway	MaxCompute Lightning, Quick BI
Amazon AWS IoT	AWS IoT Events (Trigger Action)	SNS (Simple Notification Service)	Java, Javascript, Pythin, C#,.NET, PHP, Ruby, Go	Yes (AWS IoT Greengrass Core)
Carriots	Rule Engine	Google Cloud Messaging, SMS	Java Carriots SDK	PUSH or PULL data to/from CRMs, ERPs or HTTP API
Cisco Kinetic	Cisco Kinetic Rule Engine	Email—SMS alerts	Java, JavaScript, Python, C, C++, Ruby, Dart, Scala	3rd party products and services, OAuth 2.0 framework
Google Cloud IoT	Google Cloud Functions	FIREBASE, Email, Cloud Mobile App, PagerButy, SMS, Slack, Webhooks, Pub/Sub, HTTP responces	Java, Python, C, Embedded-C, Node.js, Go	Collaborative Support, Workload centric Support, Third-party Support
IBM Watson IoT	Rule Engine	Email	Java, JavaScript, Python, C, PHP	Yes
Microsoft Azure IoT	SQL-like rules	API for all major platforms (iOS, Android, Windows, Kindle, Baidu), SMS & E-mail notifications	Java, Python, C,.NET (C#), Node.js	Yes
Oracle IoT Cloud	IoT Production Monitoring Cloud Service	Email, HTTP	Java, JavaScript, Python, C, C POSIX, Swift	3rd party Apps, Oracle Cloud Services (ERP, CRM)
PTC ThingWorx	Built-In, Custom	SMS, Voice messages	Java, Python,.NET, Android, iOS	Yes
SiteWhere	OpenHAB	SMS, SiteWhere Java Agent notifications	Java, JavaScript, Objective-C, Android	Cross-Platform Deployment using Kubernetes, Apache Kafka

(continued)

Table 3 (continued)

Platform	Event-driven actions	Push notifications	Programming frameworks	Third party integration
Thinger.io	Rule Engine	Email, SMS	Arduino IDE, Visual Studio Code	AWS, Digital Ocean, Google Cloud, Microsoft Azure
ThingsBoard	Rule Engine	Telegram, Email	Java, Node.js	Yes
ThingSpeak	React App, ThingTweet App	Social Networks feedback, control devices, HTTP POST	Python, Node.js, Ruby, Phusion Passengeer Enterprise	Yes
Ubidots	Conditional Events	Email, SMS, Telegram, Slack, Voice Call	Ubidots Libraries for Supported devices (ex. Android SDK)	Yes
WSO2	Rule Engine	FCM (Firebase Cloud Messaging) Server, APNS (Apple Push Notification Service) Server	Java, JavaScript, Python, C, C++, PHP	Yes

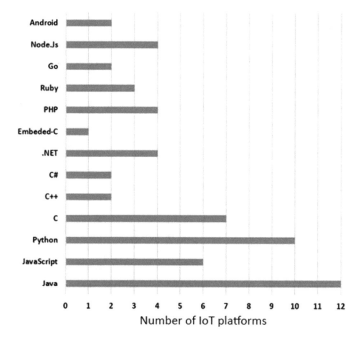

Fig. 3 Programming languages and frameworks supported by the surveyed IoT platforms

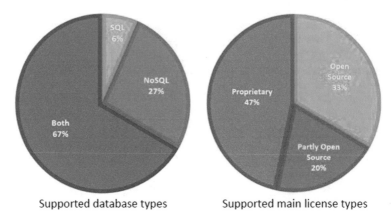

Supported database types   Supported main license types

Fig. 4 Database types and source code availability supported by the surveyed IoT platforms

A key advantage of this approach is the ability to integrate other in-house services, provided under the same technology umbrella, for the implementation of the final IoT application. Such services may be related to cloud services, data analytics, machine learning etc. In addition, ready-made solutions of large companies usually have increased security features and high availability. Finally, there are often comprehensive tutorials and user guides available that make this kind of IoT platforms use very effective.

On the downside, for this category it should be mentioned that although free packages are given for limited use the cost can increase rapidly depending on the amount of information transmitted and by the number of devices that are interconnected. As far as the developers are concerned, they require a significant amount of time to familiarize themselves with the IoT platform, and often need to write some low-level code to communicate with the devices. Finally, very low-cost devices are often not directly supported, due to the security requirements posed by such IoT platforms, and the lack of processing power of the devices for encryption.

The latter approach concerns *IoT platforms for rapid prototyping*. These platforms are provided usually by smaller start-up companies and organizations and have a clear focus on IoT application development and are not an evolution of other services. Examples of such platforms are ThingWorx, Thinger and ThingsBoard.

On the positive side, the solutions that demonstrate a clear orientation to the IoT, are providing usually a flexible data model (e.g. ThingWorx object-oriented model) which enforces flexibility and scalability. Ready-made libraries and code examples for the most popular IoT devices are also provided and enable increased customization capabilities. There are also numerous user guides available, oriented for certain purposes. Such features enable the ability for the rapid prototyping of IoT applications even by users who are not particularly familiar with the technology.

On the downside, perhaps the most important element is the risk of discontinuation or not having a long-term support. Also, when building professional applications, license cost should be taken into account.

Table 4 Accessibility criteria

Platform	Pricing		Support		
	Source licensing	Subscription plans	Forum	User guides/wiki	Paid plan
Alibaba IoT Cloud	N/A	Free, Paid	Yes	Yes	Yes
Amazon AWS IoT	Partially—open source libraries	Free, Paid			
Carriots	N/A	Free, Paid	Yes	Yes	Yes
Cisco Kinetic	N/A	Paid	Yes	Yes	Yes
Google Cloud IoT	Partially—open source libraries	Free, Paid			
IBM Watson IoT	N/A	Free, Paid	Yes	Yes	Yes
Microsoft Azure IoT	Partially—Azure IoT Edge—Open Source Microsoft Software License	Free, Paid	Yes	Yes	Yes
Oracle IoT Cloud	N/A	Free, Paid	Yes	Yes	Yes
PTC ThingWorx	N/A	Free, Paid	Yes	Yes	Yes
SiteWhere	Open Source CPAL 1.0	Free	N/A	Yes	N/A
Thinger.io	Open Source MIT License for most modules	Free, Paid	Yes	Yes	Yes
ThingsBoard	Open Source Apache 2.0	Free, Paid (professional edition)	Yes	Yes	Yes
ThingSpeak	Open Source GNU General Public License v3.0	Free, Paid	Yes	Yes	Yes
Ubidots	N/A	Free, Paid	Yes	Yes	Yes
WSO_2	Open Source Apache 2.0	Free, Paid	Yes	Yes	Yes

A summary of the advantages and concerns for each category of the IoT platforms is given in Table 5.

5 Challenges

IoT is proposed as a single, seamless and pervasive technology. Nonetheless, systems based on this technology involve many heterogeneous technologies, in order to enable

Table 5 Synopsis of IoT platforms categorization with main characteristics

	IoT platforms as dedicated servers	IoT platforms with highly integrated services	IoT platforms for rapid prototyping
Advantages	Highly customized and optimized, Zero initial cost	Easy integration with other services, Increased security features, High availability, Guides and Tutorials	Clear IoT orientation Flexibility Scalability Easy Prototyping
Concerns	Time investment, Requires advanced programing skills	Higher initial cost, Steep learning curve, Low-end devices inclusion Due to security cost	Long-term support risk Usually high cost for licensing

the autonomous exchange of data between devices and objects, while at the same time having a significant impact on all aspects of human life (safety, health, energy efficiency, environment, etc.). Therefore, these systems are usually complex, and this complexity brings issues and challenges that need to be ad-dressed. The large number of connected devices, the increased bandwidth and Quality of Service (QoS) requirements, the heterogeneous environments, the need for larger data warehouses, the large amount of processing power required, as well as security and privacy issues, are just some of the challenges that need to be addressed and are briefly discussed below.

Standardization

IoT is a technology that involves a wide range of disciplines and each has its own standards and regulations [50]. This makes the standardization process particularly complex. Standards play an important role in IoT design and development and their coordination is an important resource for the technology advancement, infrastructure improvement and application and service development [51]. The standardization of IoT architecture, communication and networking, are considered critical for the IoT development [52].

Energy Management

IoT has its own special energy requirements, such as powering small embedded devices. These devices usually have limited resources, while at the same time there is a need for power and energy efficiency. Thus, it is important to find a way to improve device's capabilities and reduce their cost of construction. Existing communication protocols are not optimized for IoT, due to the fact that they are not intended to be used in low power systems [53]. Energy management should be addressed by improving the energy consumption of the devices both in active and in idle state, in order to provide high processing power with low energy consumption at the same time.

Security and Privacy

Compared to traditional networks, maintaining security and privacy is even more important, as all IoT nodes often manage end-user personal data.

An IoT system is easy to be compromised due to the fact that it consists of many devices (typically of limited resources) and components (objects, networks, services, etc.) and consequently it possesses many levels of management [24]. Consequently, maintaining IoT security and privacy is a demanding and complex process. Failure to maintain a high level sense of security often endangers end users and their privacy. Issues such as inadequate authentication and authorization, lack of encryption and vulnerabilities in the software and hardware, are indicative cases of an IoT system inadequacy regarding security and privacy [54]. Moreover, IoT communication is often not conventional (between user and system) but it takes place also between devices or systems (M2M). For this reason, the existing human-oriented security architecture may not be appropriate when implemented in an IoT system.

Hence, solutions oriented to communication between machines, such as encryption algorithms with low latency, low cost and low power requirements are needed [22].

Scalability

In an IoT system, scaling is the ability to add new devices and/or services without compromising service quality. The challenge in terms of scaling is to support a large number of devices that may have different hardware and software but also limited resources (processing power, memory, power, etc.) [55]. At the same time, an IoT system should be able to provide seamless connection to all new devices and objects, but also be able to support possible network topology changes.

It is consequently, an issue concerning the architecture of the system and rises the need to develop mechanisms that will help in the scaling of the system but also in the interoperability of its devices. A possible solution to this problem is considered to be the use cloud service based platforms that will provide scaling capabilities and the ability to store and manage large amount of collected data [56].

Management and Self-Configuration

Device and application management is a very important factor for the successful development and operation of an IoT system.

However, the complexity of such a system, the heterogeneity of its components, but also the very large number of devices, makes its monitoring, control and regulation a very demanding process. In addition to managing its components, an IoT system must have self-configurable capabilities, i.e. be able to dynamically adapt its operation to changes occurring in the environment [56]. The data management that an IoT system has to perform is also a challenge, as a huge amount of data, coming from objects and devices (e.g. sensors), is stored, processed and analyzed by the system [57].

Current data center architectures cannot meet the increased needs of an IoT system, so companies will either switch to data sharing across multiple data centers for increased data processing efficiency and lower system response times, or in storing data in cloud services.

Data Storage and Processing

The large number of connected devices in an IoT system causes increased large-volume data traffic. This fact creates the need for a single platform that will allow data analysis, big data support, but also data mining methods using AI, machine learning and other advanced decision algorithms [58].

The above-mentioned techniques have a particularly high processing cost, which leads to the extend use of cloud technology, as it is much more capable than a conventional storage system and responds better to the ever-increasing volume of data produced. This redefines the need to store and process data and requires the use of a cloud platform.

Nevertheless, even the use of a cloud service does not completely solve all storage and processing issues, as data transfer from devices/objects to the cloud infra-structure can suffer from network performance and processing power issues, due to the fact that any action (network transfer, storage, encryption, analysis, etc.) has a corresponding cost [59].

Availability and Reliability of Services

The availability and reliability of services are two of the most important challenges that an IoT system may face and are considered to be of paramount importance. A system is considered as available, when it provides its services on demand, at any time of the day, to any authorized user or service/object. The IoT system must be constantly available regardless of changes either in the network topology (it often happens to users who have high mobility, e.g. the use of an IoT system in a vehicle in motion) or in the technologies used [60]. In order for this to be possible, there must be mechanisms responsible for interoperability, handover, and recovery of the system in case of any malfunction.

Network Performance and QoS

Network performance is related to features such as latency, throughput, bandwidth and data rate. These features can be affected by the number of IoT devices, the communication protocols used, the bandwidth of the transmission medium as well as the type of network traffic. In an IoT system, wireless networks are usually used more intensively. A major problem is that in wireless networks, bandwidth will not be sufficient as the number of devices operating in a limited space increases [61]. This leads to the need to develop models for predicting the expansion of wireless networks. Regarding QoS, its management is a challenge due to the fact that IoT devices operate under bandwidth, processing, and power limitations [62].

The use of cloud computing, despite the convenience it offers, seems to have a negative effect on network performance (latency, congestion, reliability), but also on location awareness when transferring data from Edge network devices (sensors, smartphones, IoT devices, etc.) in the cloud. The contribution of Edge Computing technologies is quite promising in addressing some of the challenges posed by cloud computing, but in turn could create new challenges, as they use devices with limited resources [56].

Architecture

An IoT system includes a large number of technologies, as well as an increasing number of devices and sensors. At the same time, it must support the heterogeneous nature of objects and networks so that they can be connected seamlessly, in a wireless and autonomous way [51]. One of the biggest challenges in terms of the architecture of an IoT system, is to ensure interoperability, scalability and the smooth operation of services. This need leads to the use of open architecture using standards and separation of application logic from the hardware infrastructure [56].

Interoperability

Interoperability, is the ability to operate multiple devices and systems together without being affected by the use of different software or hardware. It is directly related to standardization and system architecture, though it is negatively affected by the integration of different heterogeneous technologies from different vendors, who often use the lack of interoperability for their advantage [25]. The challenge of supporting multi-level interoperability (technical, syntactic, semantic and organizational) in IoT frameworks has been also explored [63]. In addition to features such as object connectivity, device management, data processing, analysis and visualization, IoT platforms should also improve interoperability in a heterogeneous environment.

New Opportunities

IoT provides many new opportunities to end users and industry, with a wide range of applications and services. However, as each new technology faces a number of challenges arising from the involvement of many manufacturers, the integration of many heterogeneous technologies, potential vulnerabilities in hardware and software, the need to manage and report personal data and several more. When the issues, or a big part of them, are resolved, it will be another big step towards the evolution of technology to the next level, but also its universal acceptance by society.

6 Conclusions

IoT platforms represent the software that allows the IoT to be functional. These platforms contribute to the management, processing, analysis and display of data. The platforms also aim to ensure secure communication between objects and are considered an integral part of any IoT architecture. IoT platforms can be differentiated by various criteria such as how they analyze data, the databases they use to store data, communication protocols, service pricing, supported hardware, the ability to activate events based on changes in data and much more.

In this paper, a broad evaluation scheme for IoT software platforms is proposed. It is based on an analysis of the characteristics found in 15 leader IoT software platforms. The characteristics analysis is followed by a comparative analysis of these platforms based on the defined criteria and a critical presentation of fundamental features. In addition to the analysis, challenges and open issues of IoT platforms were discussed.

References

1. Borgia, E.: The internet of things vision: Key features, applications and open issues. Comput. Commun. **54**, 1–31 (2014)
2. Atzori, L., Iera, A., Morabito, G.: The internet of things: a survey. Comput. Netw. **54**(15), 2787–2805 (2010)
3. Minerva, R., Biru, A., Rotondi, D.: Towards a definition of the internet of things (IoT). IEEE Internet Initiative **1**(1), 1–86 (2015)
4. Mahdi, M.A., Hasson, S.T.: A contribution to the role of the wireless sensors in the IoT era. J. Telecommun. Electron. Comput. Eng. (JTEC) **9**(2–11), 1–6 (2017)
5. Yanagida, R., Bhatti, S.N.: Seamless internet connectivity for ubiquitous communication. In: Adjunct Proceedings of the 2019 ACM International Joint Conference on Pervasive and Ubiquitous Computing and Proceedings of the 2019 ACM International Symposium on Wearable Computers, pp. 1022–1033 (2019)
6. Yang, C., Huang, Q., Li, Z., Liu, K., Hu, F.: Big Data and cloud computing: innovation opportunities and challenges. Int. J. Dig. Earth **10**(1), 13–53 (2017)
7. Babik, M., Prelz, F., Froy, T., Grigoras, C., Chudoba, J., Finnern, T., et al.: IOP: IPv6 Security. J. Phys.: Conf. Ser. **898**, 102008 (2017)
8. Premsankar, G., Di Francesco, M., Taleb, T.: Edge computing for the internet of things: a case study. IEEE Internet Things J. **5**(2), 1275–1284 (2018)
9. Shi, W., Cao, J., Zhang, Q., Li, Y., Xu, L.: Edge computing: vision and challenges. IEEE Internet Things J. **3**(5), 637–646 (2016)
10. Karagiorgou, S., Vafeiadis, G., Ntalaperas, D., Lykousas, N., Vergeti, D., Alexandrou, D.: Unveiling trends and predictions in digital factories. In: 2019 15th International Conference on Distributed Computing in Sensor Systems (DCOSS), pp. 326–332. IEEE (2019)
11. Elizalde, D.: Why There's No Killer App for IoT (2017). Available online: https://medium. com/iotforall/why-theres-no-killer-app-for-iot-5646fda5a0fe
12. Lamarre, E., May, B.: Ten Trends Shaping the Internet of Things Business Landscape. McKinsey Digital (2019). Available online: https://www.mckinsey.com/business-functions/ digital-mckinsey/our-insights/ten-trends-shaping-the-internet-of-things-business-landscape
13. Lueth, K.L.: The 25 Best IoT Platforms 2019—Based on Customer Reviews (2019). Available online: https://iot-analytics.com/the-25-best-iot-platforms-2019/
14. Hejazi, H., Rajab, H., Cinkler, T., Lengyel, L.: Survey of platforms for massive IoT. In: 2018 IEEE International Conference on Future IoT Technologies (Future IoT), pp. 1–8. IEEE (2018)
15. Banks, A., Briggs, E., Borgendale, K., Gupta, R.: MQTT Version 5.0. OASIS Standard (2019). Available online: https://docs.oasis-open.org/mqtt/mqtt/v5.0/mqtt-v5.0.html
16. Shelby, Z., Hartke, K., Bormann, C., Frank, B.: RFC 7252: Constrained Application Protocol (CoAP). Asma A. Elmangoush, pp. 156 (2014). Available online: https://coap.technology/
17. Naik, N.: Choice of effective messaging protocols for IoT systems: MQTT, CoAP, AMQP and HTTP. In: 2017 IEEE international systems engineering symposium (ISSE), pp. 1–7. IEEE (2017)
18. Karagiannis, V., Chatzimisios, P., Vazquez-Gallego, F., Alonso-Zarate, J.: A survey on application layer protocols for the internet of things. Trans. IoT Cloud Comp. **3**(1), 11–17 (2015)
19. Meddeb, M., Alaya, M.B., Monteil, T., Dhraief, A., Drira, K.: M2M platform with autonomic device management service. Procedia Comput. Sci. **32**, 1063–1070 (2014)

20. Microsoft Azure IoT Hub: Overview of Device Management with Azure IoT Hub (2017). Available online: https://docs.microsoft.com/en-us/azure/iot-hub/iot-hub-device-management-overview
21. Zdravković, M., Trajanović, M., Sarraipa, J., Jardim-Gonçalves, R., Lezoche, M., Aubry, A., Panetto, H.: Survey of internet-of-things platforms. In: 6th International Conference on Information Society and Techology, ICIST 2016, vol. 1, pp. 216–220 (2017)
22. Solapure, S.S., Kenchannavar, H.: Internet of things: a survey related to various recent architectures and platforms available. In: 2016 International Conference on Advances in Computing, Communications and Informatics (ICACCI), pp. 2296–2301. IEEE (2016)
23. Singh, K.J., Kapoor, D.S.: Create your own internet of things: a survey of IoT platforms. IEEE Consumer Electron. Mag. 6(2), 57–68 (2017)
24. Khan, M.A., Salah, K.: IoT security: review, blockchain solutions, and open challenges. Futur. Gener. Comput. Syst. 82, 395–411 (2018)
25. Díaz, M., Martín, C., Rubio, B.: State-of-the-art, challenges, and open issues in the integration of Internet of things and cloud computing. J. Netw. Comput. Appl. 67, 99–117 (2016)
26. Truong, H.L., Dustdar, S.: Principles for engineering IoT cloud systems. IEEE Cloud Comput. 2(2), 68–76 (2015)
27. NoSQL vs Relational Databases: Available online: https://www.mongodb.com/scale/nosql-vs-relational-databases
28. Mohammadi, M., Al-Fuqaha, A., Sorour, S., Guizani, M.: Deep learning for IoT big data and streaming analytics: a survey. IEEE Commun. Surv. Tutor. 20(4), 2923–2960 (2018)
29. Ammar, M., Russello, G., Crispo, B.: Internet of things: a survey on the security of IoT frameworks. J. Inf. Secur. Appl. 38, 8–27 (2018)
30. Introduction to IoT Platform: Alibaba Cloud Document Center. Available online: https://www.alibabacloud.com/help/product/30520.htm
31. AWS IoT: Overview. IOT. Available online: https://aws.amazon.com/iot/
32. Altair SmartWorks™: Carriots. Available online: https://www.carriots.com/what-is-carriots
33. Cisco IoT Solutions: Cisco Kinetic IoT Platform. Available online: https://www.cisco.com/c/en/us/solutions/internet-of-things/iot-kinetic.html
34. Google Cloud Platform: Google Cloud IoT Core. Available online: https://cloud.google.com/iot-core
35. Watson IoT Platform—Overview. Available online: https://www.ibm.com/cloud/watson-iot-platform
36. IBM, Securely connect, manage and analyze IoT data with Watson IoT Platform. Available online: https://www.ibm.com/business-operations/iot-platform
37. Microsoft Azure: Internet of Things Platform. Available online: https://azure.microsoft.com/en-us/overview/iot/
38. Oracle: IoT Intelligent Applications. Available online: https://www.oracle.com/internet-of-things/
39. Industrial Internet of Things (IIoT): ThingWorx IoT Platform. Available online: https://www.ptc.com/en/products/iiot/thingworx-platform
40. SiteWhere. Available online: https://www.sitewhere.com/
41. Luis Bustamante, A., Patricio, M.A., Molina, J.M.: Thinger. io: An open source platform for deploying data fusion applications in IoT environments. Sensors 19(5), 1044 (2019)
42. Thingsboard. Open-source IoT Platform. Available online: https://thingsboard.io/
43. ThingSpeak for IoT Projects. Available online: https://thingspeak.com/
44. Ubidots. Ubidots. Available online: https://ubidots.com/
45. WSO2 IoT Server—Flexible Open Source IoT Platform. Available online: https://wso2.com/iot/
46. Flespi MQTT API: (n.d.) Available online: https://flespi.com/mqtt-api
47. MathWorks. Choose Between REST API and MQTT API.: (n.d.). Available online: https://www.mathworks.com/help/thingspeak/choose-between-rest-and-mqtt.html
48. Ismail, A.A., Hamza, H.S., Kotb, A.M.: Performance evaluation of open source iot platforms. In: 2018 IEEE Global Conference on Internet of Things (GCIoT), pp. 1–5. IEEE (2018)

49. IoT Analytics (2020) List of 620 IoT Platform Companies. Available online: https://iot-analyt ics.com/product/list-of-620-iot-platform-companies/
50. Farahani, B., Firouzi, F., Chang, V., Badaroglu, M., Constant, N., Mankodiya, K.: Towards fog-driven IoT eHealth: promises and challenges of IoT in medicine and healthcare. Futur. Gener. Comput. Syst. **78**, 659–676 (2018)
51. Chen, S., Xu, H., Liu, D., Hu, B., Wang, H.: A vision of IoT: applications, challenges, and opportunities with china perspective. IEEE Internet Things J. **1**(4), 349–359 (2014)
52. Palattella, M.R., Accettura, N., Vilajosana, X., Watteyne, T., Grieco, L.A., Boggia, G., Dohler, M.: Standardized protocol stack for the internet of (important) things. IEEE Commun. Surv. Tutor. **15**(3), 1389–1406 (2012)
53. Gubbi, J., Buyya, R., Marusic, S., Palaniswami, M.: Internet of Things (IoT): a vision, architectural elements, and future directions. Futur. Gener. Comput. Syst. **29**(7), 1645–1660 (2013)
54. Meneghello, F., Calore, M., Zucchetto, D., Polese, M., Zanella, A.: IoT: internet of threats? A survey of practical security vulnerabilities in real IoT devices. IEEE Internet Things J. **6**(5), 8182–8201 (2019)
55. Pereira, C., Aguiar, A.: Towards efficient mobile M2M communications: survey and open challenges. Sensors **14**(10), 19582–19608 (2014)
56. Čolaković, A., Hadžialić, M.: Internet of Things (IoT): a review of enabling technologies, challenges, and open research issues. Comput. Netw. (2018)
57. Lee, I., Lee, K.: The internet of things (IoT): applications, investments, and challenges for enterprises. Bus. Horiz. **58**(4), 431–440 (2015)
58. L'heureux, A., Grolinger, K., Elyamany, H.F., Capretz, M.A.: Machine learning with big data: challenges and approaches. IEEE Access **5**, 7776–7797 (2017)
59. Gou, Z., Yamaguchi, S., Gupta, B.B.: Analysis of various security issues and challenges in cloud computing environment: a survey. In: Identity Theft: Breakthroughs in Research and Practice, pp. 221–247. IGI Global (2017)
60. Choi, S.I., Koh, S.J.: Use of proxy mobile IPv6 for mobility management in CoAP-based internet-of-things networks. IEEE Commun. Lett. **20**(11), 2284–2287 (2016)
61. Wu, Q., Ding, G., Du, Z., Sun, Y., Jo, M., Vasilakos, A.V.: A cloud-based architecture for the Internet of spectrum devices over future wireless networks. IEEE access **4**, 2854–2862 (2016)
62. Samie, F., Tsoutsouras, V., Xydis, S., Bauer, L., Soudris, D., Henkel, J.: Distributed QoS management for internet of things under resource constraints. In: Proceedings of the Eleventh IEEE/ACM/IFIP International Conference on Hardware/Software Codesign and System Synthesis, pp. 1–10 (2017)
63. Pliatsios, A., Goumopoulos, C., Kotis, K.: A Review on IoT frameworks supporting multi-level interoperability—the semantic social network of things framework. Int. J. Adv. Internet Technol. **13**(1), 46–64 (2020)

Networking

FLER: Fuzzy Logic-Based Energy Efficient Routing for Wireless Sensor Networks

Thompson Stephan, S. Punitha, Achyut Shankar, and Naveen Chilamkurti

Abstract The Routing of data packets in a wireless sensor network (WSN) is affected by a variety of physical factors including poor node-link quality, insufficient buffer capacity, and insufficient energy levels. These factors, lead to failures in data packet delivery, as a consequence of which packets need to be retransmitted. Thus, a higher frequency of retransmissions makes the system unstable by consuming more energy and causing more delays in packet delivery. To avoid the aforementioned routing challenges, this paper implements the concept of fuzzy logic, taking into consideration five major factors namely, residual energy, degree of closeness to shortest path, degree of closeness to sink node, buffer capacity, and trust degree. In this proposed system, the foremost aim is to choose the optimal forwarder node which is predicted with low energy consumption, high stability, low retransmission rate, and low average transmission delay. The experimental results show that the proposed algorithm shows better performance than some existing similar algorithms in terms of minimizing the average delay in packet delivery, network energy consumption, and frequency of retransmissions.

Keywords WSN · Fuzzy logic · Routing · Energy efficiency

T. Stephan · A. Shankar · N. Chilamkurti (✉)
Department of Computer Science and Engineering, M. S. Ramaiah University of Applied Sciences, Bangalore 560054, Karnataka, India
e-mail: n.chilamkurti@latrobe.edu.au

S. Punitha
Department of Computer Science and Engineering, Karunya Institute of Technology and Sciences, Coimbatore 641114, India

N. Chilamkurti
Department of Computer Science and Engineering, Amity University Uttar Pradesh, Sector-125, Noida (U.P.) 201313, India

© The Author(s), under exclusive license to Springer Nature Switzerland AG 2022
P. Nicopolitidis et al. (eds.), *Advances in Computing, Informatics, Networking and Cybersecurity*, Lecture Notes in Networks and Systems 289,
https://doi.org/10.1007/978-3-030-87049-2_11

1 Introduction

Wireless Sensor Network (WSN) technology is an essential component of IoT [1] and the recent years has seen the tremendous growth of its application in different technologies [2]. Wireless sensor and actor networks (WSANs) [3–5] may be defined as a cluster of sensors and actors connected to each other through wireless media. In WSAN, a sensor performs the task of detection of changes in the designated environment, whereas an actor makes it possible to act upon the changes accordingly. Out of the wide range of application areas of WSAN, some of the major real life applications include collection of data in the field of robotics [6], monitoring the areas of high end securities, agricultural fields, wildlife [7], industrial environments [8] etc. Each sensor or actor involved in the network is called a node, and the connected network of these nodes builds up the wireless sensor and actuator network. These networks are vastly used in many real life applications such as collection of data in the field of robotics [6], wildlife, meteorology etc. These networks are subjected to harsh conditions including physical terrain, climate issues, and inaccessibility to manual care and repair. Such situations pose technological challenges, such as availability of sufficient energy resources like batteries, reliable communication technology, reduced transmission delays, and processing capabilities.

A principle problem which is encountered is the appropriate delivery [9] of the message/information from sensor to the destination. On one hand sensors are energy deficient and are considered to be lacking resources and on the other hand actors are considered to be of enormous resource. A routing system [10, 11] needs to be built which is capable of countering the requirements of energy efficiency [12] (at sensor's end) and the collection (at actor's end). Each recipient actor is considered as a node at its site. The very first challenge is managing the energy consumption as the message transmission incurs energy loss in the sensor. Furthermore, packet loss is observed due to the random selection of the recipient nodes which is due to the factors of change in the wireless networks. Another factor leading to loss of packets could be insufficient memory of buffers [13]. Both of these challenges lead to retransmission of the data packet which is necessary to assure the dependability over the network. But in return it increases the consumption of time as well as energy, thus the system becomes inefficient. The most common issues in any sensor based networks are energy efficiency [14, 15] and reliable data collection [16]. These issues can be dealt with by designing a suitable routing protocol that is able to satisfy and improve the above mentioned problems. Also, the WSANs are not easily attainable and must conserve energy as well as store important and essential data for retrieval and transmission. To account for this, the routing algorithm shall also incorporate the factors of buffer capacity and residual energy. Moreover, it is important to consider the trustworthiness of a node before making the routing decision as a less reliable node may not deliver the data packets properly and may trigger retransmissions.

An efficient and energy aware routing algorithms aims to select an optimized node in each step. Thus, for each hop to the next node, the algorithm should minimize the cost, energy consumption and transmission delay. This can be achieved by selecting a node, closest to the current node, which also ultimately leads to the destination node. This can be achieved by taking these factors into consideration: Degree of closeness to the sink node and degree of closeness to the shortest path. Since, these WSANs are deployed in unpredictable and continually changing environment the reliability of the links between the nodes is affected. Therefore the routing protocol should have a robust mechanism that takes into consideration the node's reliability. Thus, the trustworthiness [17] of a node should be considered since selecting a low reliable node may not forward the packets and in turn triggers retransmission [18]. Many of the existing research work in WSNs have focused on clustering issues [10], mobility [19, 20], security [21] and privacy issues [11], and channel assignment issues [1, 2, 22]. However, routing issues [23] still remain the primary focus.

The major contributions of this research work are as follows. This work proposes a Fuzzy Logic-Based Energy Efficient Routing (FLER) which is a smart routing mechanism based on the Fuzzy-Logic. This mechanism uses an energy-efficient approach, thereby enhancing the sustainability of the network. Discovering the best energy efficient next-hop neighbor node is the major objective of this research work which is based on the five parameters, namely Residual Energy, Degree of closeness to shortest path, Degree of closeness to sink node, Buffer capacity, and Trust degree, along with the rule sets described in the fuzzy-Logic-System. Finally, the involvement of Dijkstra's algorithm [21, 24] allows tracing the best routes where the cost metric to each connected node in the graph is determined by the above five link selection metrics.

The organization of the paper is as follows. Section 2 presents the related work. The proposed work is described in the Sect. 3. Section 4 presents an evaluation of the proposed system's performance. Finally, the paper is concluded in Sect. 5.

2   Related Work

This section revisits the existing studies on WSANs and fuzzy logic system. Different techniques for calculating the routing metrics in WSNs are discussed. One of the studies of Bhardwaj et al. [3] has fixed upper bounds on lifetime in a WSN network by taking into consideration Energy Model. Rout et al. [4], this study showed the calculated energy consumptions for tree based rechargeable sensor networks. In another study, Lee and Lee [5], derived the upper bound of a cluster based sensor network based on network lifetime. Neamatollahi et al. [6] derived a 'fuzzy based hyper round policy mechanism' to solve the problem of re-clustering overhead. A fuzzy based logic mechanism of sensor nodes in an industrial WSN, was devised in order to improve the energy efficiency based on sensor sleeping time by Collatta et al. [7]. Lee and Cheng [8], proposed a fuzzy logic based clustering approach, where a cluster head is selected based on the residual energy of the node. Pantazis et al. [9] showed a survey on energy efficient routing protocols in a WSAN. Rout et al. [4] proposed a routing protocol for WSNs based on probabilities and network coding. Zhang et al. [12] proposed a protocol for routing where a node is selected as the next node on the basis of link weight and forward energy density. Lai et al. [13] devised routing mechanism aware of link-delay that balances the amount of energy consumed among the sensor nodes. Sun et al. [14] proposed a routing algorithm that incorporates the application of ant colony optimization by taking into account the transmission distance between two nodes, the direction of transmission, and the remaining energy. Our proposed algorithm differs from the aforementioned algorithms by taking a fuzzy based approach with five input parameters—Residual Energy, Degree of closeness to shortest path, Degree of closeness to sink node, Buffer capacity, and Trust degree.

3 FLER Protocol

In this section we introduce the network model of the system, the energy consumption model, the FLER algorithm for routing through the network and fuzzy logic system is explained which is used to formulate the fuzzy rules that are utilized for decision making.

3.1 Network Model

As shown in Fig. 1, a heterogeneous sensor actor network is represented as a graph G (V ∪ L, K). Where, V is a set of n number of nodes ranging from $\{V_1, V_2, V_3, V_4, \ldots, V_n\}$, K represents the set of links and L denotes a local actor node. This model describes different factors which are applicable for assessing the real time network characteristics. These are: residual energy (R_e), degree of closeness to shortest path (D_{sp}), degree of closeness to sink node (D_{sn}), buffer capacity (B_c), and trust degree (T_d). In this network model, the link quality between two nodes is calculated based on the values of R_e, D_{sp}, D_{sn}, B_c, and T_d. The determined values

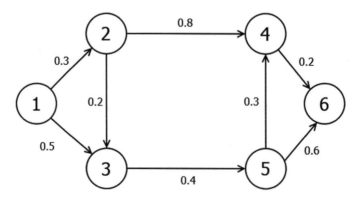

Fig. 1 Network model

are then used for determining the chance of a node to become the next hop node by applying fuzzy rules. In Fig. 1, {1, 2, 3, 4, 5, 6} represent the vertices (i.e., nodes) and {0.3, 0.8, 0.2, 0.6, 0.3, 0.4, 0.2, 0.5} represents the link quality between two nodes based on which the next best hop for data forwarding is selected.

3.2 Energy Consumption Model

FLER uses the same energy consumption model as used in [15]. This section discusses the energy consumption model considered in the network. The energy used for sensing data, receiving information, and transmitting packets, are denoted as E_{sd}, E_{ri}, E_{tp}, respectively. Furthermore, deriving from the path loss model of energy consumption, the energy required for these operations can be mathematically summarized as follows:

$$E_{sd} = \gamma_s$$
$$E_{ri} = \gamma_r$$
$$E_{tp} = \gamma_t + \gamma_a d^g \tag{1}$$

After each iteration, the signal suffers a loss in strength by a factor of $1/d^g$ where g denotes the path loss exponent and d denotes the distance. Here, γ_s represents the energy consumed for sensing data in the network in joules/bit unit. γ_r represents the energy consumed for receiving information from the network in joules/bit unit. γ_t represents the energy consumed for transmitting packets from the network in joules/bit unit. γ_a represents the energy dissipated in transmit op-amp in joules/bit unit.

3.3 Proposed Routing Mechanism

Algorithm 1 describes the optimal routing procedure for an efficient transmission. Algorithm 1 determines the next hop node for transmission in such a manner that the path formed from source node to destination node is shortest, energy aware, buffer conservative, and trusted path. The decision for next hop selection is taken by calculating values for each of the possible subsequent nodes. This value is estimated by using fuzzy logic with input value for each quality parameter. This strategy is built over Dijkstra's algorithm to find the most suitable path for data transmission.

For each possible node, the algorithm uses fuzzy logic to estimate a probabilistic value which represents the suitability of a node to be selected as the next hop node. The probability is determined by the fuzzy logic using the five primary values namely R_e, D_{sp}, D_{sn}, B_c, and T_d. In this manner, the algorithm finally finds an optimized path

Table 1 Summary of notations

R_e	Residual energy
D_{sp}	Degree of closeness to shortest path
D_{sn}	Degree of closeness to sink node
B_c	Buffer capacity
T_d	Trust degree
n	Number of nodes
L	Local actor node
d[u]	Cost of path to node u
z[v]	Predecessor of node v
P_q	Fuzzy output value of node q
$\mu(X)$	Membership value of input X

from the source node to the sink node. The summary of notations used in Algorithm 1 is shown in Table 1.

Algorithm 1: FLER	
Input:	G (V, K), S, n, R_e, D_{sp}, D_{sn}, B_c, T_d
Output:	Probability of choosing node q as the next-hop (Pq), energy-aware routing path
1.	for i = 1 to n do
2.	for each vertex j ∈ Adj[i] do
3.	Empty the list f < value, membership Wind >
4	Find the membership values, $\mu(R_e)$, $\mu(D_{sp})$, $\mu(D_{sn})$, $\mu(B_c)$ and $\mu(T_d)$ and linguistic levels using Triangular membership function
5.	DR = {A rule set with all possible combinations of determined linguistic levels}
6.	for each rule in DR do
7.	if $\mu(R_e)$, $\mu(D_{sp})$, $\mu(D_{sn})$, $\mu(B_c)$ and $\mu(T_d)$ fit the membership levels of this rule then
8.	Add an entry to the list t with
9.	value = maximum ($\mu(R_e)$, $\mu(D_{sp})$, $\mu(D_{sn})$, $\mu(B_c)$ and $\mu(T_d)$)
10.	*membership plevel* = output membership level of this rule
11.	end if
12.	end for
13.	Pq = Defuzzify(f)
14.	end for
15.	end for
16.	for each vertex u ∈ V [G] do
17.	d[u] = 0
18.	zlu] = NULL

(continued)

(continued)

19.	end for
20.	d[S] = 1
21.	S ← Ø
22.	Q ← V[G]
23.	while Q ≠ Ø do
24.	u ← MAX (Q)
25.	S ← S ∪{u}
26.	for each vertex v ∈ Adj[u] do
27.	if d[v] < d[u] • p[u][v] then
28.	d[v] ← d[u] * p[u][v]
29.	z[v] ← u
30.	end if
31.	end for
32.	end while

3.4 Fuzzy Logic System

The primary objective of our proposed algorithm is to find the best path for data packet transmission which is achieved through the application of fuzzy logic. The fuzzy logic system computes the probability of a node to become the next best hop. The fuzzy logic system takes the quality characteristics as input and provides a crisp output value that determines the probability (P_d) of a node as next hop. As shown in Fig. 2, the fuzzy logic system is composed of the following key components:

(1) Fuzzifier: It considers the input values and fixes them accordingly into various linguistic sets with the help of membership functions. These linguistic values go to the fuzzy rules set.

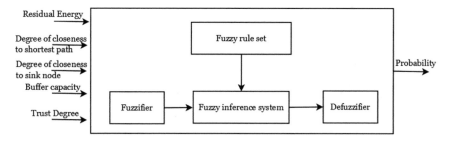

Fig. 2 Fuzzy logic system

Table 2 Fuzzy parameters and their values

Parameter name	Linguistic value	Numeric value
Residual energy (R_e)	Low	0.0–0.5
	Medium	0.2–0.8
	High	0.6–1.0
Degree of closeness to shortest path (D_{sp})	Close	0.0–0.35
	Medium	0.2–0.9
	Far	0.85–1.0
Degree of closeness to sink node (D_{sn})	Close	0.0–0.3
	Medium	0.1–0.9
	Far	0.7–1.0
Buffer capacity (B_c)	Low	0.0–0.5
	Medium	0.2–0.8
	High	0.5–1.0
Trust degree (T_d)	Low	0.0–0.5
	Medium	0.2–0.8
	High	0.5–1.0

(2) Fuzzy rules set: The combination of the linguistic values are fed into this portion of the system and based on these combinations several if–then rules are formulated. These if–then rules compose the fuzzy rules set.

(3) Fuzzy inference system: This portion of the system is responsible for carefully selecting and applying the suitable rules based on the values fed into the system.

(4) Defuzzifier: It is responsible for transforming the linguistic level values into crisp output values.

The quality characteristics govern the selection of input parameters for the fuzzy logic system. Fuzzy parameters and their values are shown in Table 2. These fuzzy parameters are as follows:

$$R_e, D_{sp}, D_{sn}, B_c, \text{ and } T_d$$

(1) Residual Energy (R_e): The first parameter is residual energy which increases the network lifetime by providing efficient rotation of cluster heads. This is necessary for the formation of clusters, which in turn helps in the fuzzy algorithm as an important parameter. The threshold values of this energy are divided into 3 linguistic categories—Low (0.0–0.5), Medium (0.2–0.8) and High (0.6–1.0).

(2) Degree of closeness to shortest path (D_{sp}): The degree of closeness to shortest path provides with the vital information of the shortest distance path. This data helps to find the optimal shortest path to reach to a particular node. The values range from—Close (0.0–0.35), Medium (0.2–0.9) and Far (0.85–1.0).

(3) Degree of closeness to sink node (D_{sn}): The fifth parameter is degree of closeness of node to sink. This is helpful to find the shortest distance from a particular node to the destination node. It plays a very vital role among all the parameters. The linguistic values range from: Close (0.0–0.3), Medium (0.1–0.9), Far (0.7–1.0).

(4) Buffer capacity (B_c): The next parameter is used for the buffer available in the nodes. This parameter is called the buffer capacity and is useful for storage of relevant information regarding the transmission sequences. The buffer capacity linguistic values range from Low (0.0–0.5), Medium (0.2–0.8) and High (0.5–1.0).

(5) Trust Degree (T_d): Trust degree measures the trustworthiness of a node in the network. The reliability of a particular node is computed according to the trust model [17]. Linguistic values ranges from: low (0.0–0.5), Medium (0.2–0.8) and Poor (0.5–1.0).

The combination of the linguistic values is fed into this portion of the system and based on these combinations several if–then rules are formulated. These if–then rules compose the fuzzy rules set. Table 3 shows the Fuzzy rule set.

Using this fuzzy rule set, the fuzzy rules for the fuzzy logic system are formulated as follows:

Rule 1: If R_e is Low, D_{sp} is Medium, D_{sn} is Medium, B_c is Medium, and T_d is Average, then P_d is Low.

Rule 2: If R_e is Low, D_{sp} is Medium, D_{sn} is Medium, B_c is Medium, and T_d is Good, then P_d is Rather Low.

Rule 3: If R_e is Low, D_{sp} is Medium, D_{sn} is Medium, B_c is High, and T_d is Average, then P_d is Low.

Rule 4: If R_e is Low, D_{sp} is Medium, D_{sn} is Medium, B_c is High, and T_d is Good, then P_d is Rather Low.

Rule 5: If R_e is Medium, D_{sp} is Medium, D_{sn} is Medium, B_c is Medium, and T_d is Average, then P_d is Medium.

Rule 6: If R_e is Medium, D_{sp} is Medium, D_{sn} is Medium, B_c is Medium, and T_d is Good, then P_d is High Medium.

Rule 7: If R_e is Medium, D_{sp} is Medium, D_{sn} is Medium, B_c is High, and T_d is Average, then P_d is Medium.

Rule 8: If R_e is Medium, D_{sp} is Medium, D_{sn} is Medium, B_c is High, and T_d is Good, then P_d is High Medium.

These rules are used by the fuzzy inference system. The fuzzy inference system carefully examines the input and finds the fuzzy rules corresponding to the respective input set. The so-found rule is then applied to calculate the estimate value for the output next hop node. This linguistic value is then transformed to its crisp output

Table 3 Fuzzy rule set

Rule	Input					Output
	R_e	D_{sp}	D_{sn}	B_c	T_d	P_d
1.	High	Close	Close	High	High	Very high
2.	High	Close	Close	High	Medium	High
3.	High	Close	Close	High	Low	Rather high
4.	High	Close	Close	Medium	High	Very high
5.	High	Close	Close	Medium	Medium	High
6.	High	Close	Close	Medium	Low	Rather high
7.	High	Close	Close	Low	High	Very high
8.	High	Close	Close	Low	Medium	High
9.	High	Close	Close	Low	Low	Rather high
10.	High	Close	Medium	High	High	Very high
11.	High	Close	Medium	High	Medium	High
12.	High	Close	Medium	High	Low	Rather high
13.	High	Close	Medium	Medium	High	Very high
14.	High	Close	Medium	Medium	Medium	High
15.	High	Close	Medium	Medium	Low	Rather high
16.	High	Close	Medium	Low	High	Very high
17.	High	Close	Medium	Low	Medium	High
18.	Low	Far	Far	Low	High	Rather low
19.	Low	Far	Far	Low	Medium	Low
20.	Low	Far	Far	Low	Low	Very low

form by the defuzzifier. The final output is used to determine the path chosen to the next hop node.

4   Performance Evaluation

This section evaluates the performance of the proposed algorithm in terms of quality characteristics of transmission. These characteristics are then compared with the existing works of Fuzzy-based energy aware routing mechanism (FEARM) algorithm, predicted remaining deliveries (PRD) algorithm, and a retransmission-based algorithm on the basis of stability of network, energy consumed for operations, number of retransmissions and packet delay.

Table 4 Simulation parameters

Number of sensor nodes	100–1000
Range of transmission	30 m
Initial energy of a node I (kJ)	20–25
Sensing energy	50×10^{-9} J/bit
Receiving energy	0.787×10^{-6} J/bit 0.937×10^{-6} J/bit
Transmitting energy	10×10^{-12} J/bit d $= 85$ m
Path loss exponent	2
Data generated per event at a node	960 bits
Average event rate per unit time	100

4.1 Simulation Environment

The simulation results are presented by calculating and graphically presenting four quality factors aforementioned in a simulated environment. This simulated environment is created with the help of Network Simulator 3. Table 4 summarizes the simulation characteristics which are used to test all four algorithms. The simulation.

4.2 Simulation Results and Discussion

In this section all four algorithms are compared on the basis of transmission quality characteristics. These characteristics are listed below:

(1) Average delay in packet delivery
(2) Number of retransmissions
(3) Stability of the network [15]
(4) Network energy consumption.

These transmission characteristics are compared between the four algorithms— 'Proposed work', 'FEARM', 'PRD' and 'Retransmission based approach'. Figure 3 shows the comparison between all algorithms in terms of average delay in packet delivery. Figure 4 shows the comparison between all algorithms in terms of energy consumption of the whole network. Figure 5 depicts the comparison between all algorithms in terms of stability of the network. Figure 6 shows the comparison between all algorithms in terms of number of retransmissions.

These results clearly show an improvement over the previously proposed algorithm for routing of data packets through wireless sensor networks.

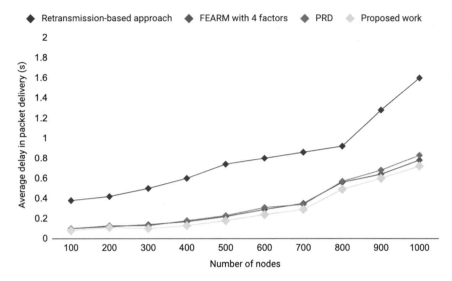

Fig. 3 Average delay in packet delivery versus number of nodes

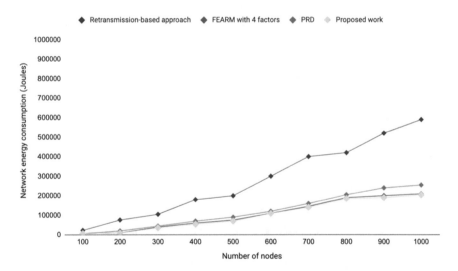

Fig. 4 Network energy consumption (Joules) versus number of nodes

5   Conclusion

This paper presents a routing algorithm that allows data packets to be transmitted in an optimized manner yielding the best path a data packet shall travel from its source to destination. The FLER algorithm makes this process highly efficient and sustainable. The main idea behind the algorithm is to apply fuzzy rules taking the

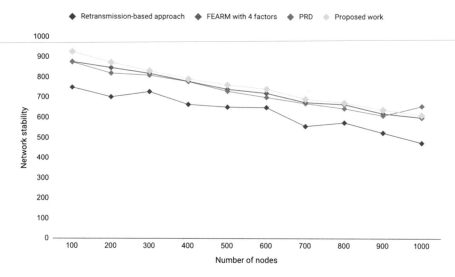

Fig. 5 Network stability versus number of nodes

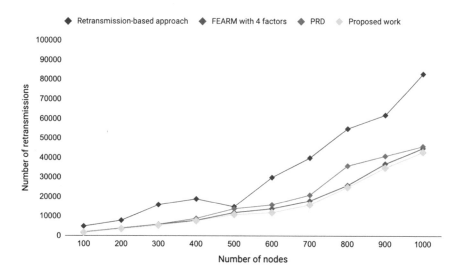

Fig. 6 Number of retransmissions versus number of nodes

quality characteristics of nodes as crisp input such as residual energy, degree of closeness to shortest path, degree of closeness to sink node, buffer capacity, and trust degree. This is used to give as output the node that shall be selected as the next hop node. The findings reveal a slight improvement over the existing transmission protocols. The proposed system manages to reduce the number of retransmissions required to safely and completely transfer data, reduce the amount of energy required for transmission, sensing and receiving operations, reduce the packet delay between two nodes in any transmission, and improves the overall stability of the network.

References

1. Stephan, T., Al-Turjman, F., Suresh Joseph, K., Balusamy, B., Srivastava, S.: Artificial intelligence inspired energy and spectrum aware cluster based routing protocol for cognitive radio sensor networks. J. Parallel Distrib. Comput. (2020). https://doi.org/10.1016/j.jpdc.2020.04.007
2. Stephan, T., Al-Turjman, F., Suresh Joseph, K., Balusamy, B.: Energy and spectrum aware unequal clustering with deep learning based primary user classification in cognitive radio sensor networks. Int. J. Mach. Learn. Cybern. (2020). https://doi.org/10.1007/s13042-020-01154-y
3. Bhardwaj, M., Garnett, T., Chandrakasan, A.P.: Upper bounds on the lifetime of wireless sensor networks. In: Proceedings of IEEE International Conference on Communications (ICC), pp. 785–790. Helsinki, Finland (2001)
4. Rout, R.R., Ghosh, S.K., Chakrabarti, S.: Co-operative routing for wireless sensor networks using network coding. IET Wireless Sens. Syst. 2(2), 75–85 (2012)
5. Lee, S., Lee, H.S.: Analysis of network lifetime in cluster-based sensor networks. IEEE Commun. Lett. 14(10), 900–902 (2010)
6. Neamatollahi, P., Naghibzadeh, M., Abrishami, S.: Fuzzy-based clustering-task scheduling for lifetime enhancement in wireless sensor networks. IEEE Sens. J. 17(20), 6837–6844 (2017)
7. Collotta, M., Pau, G., Maniscalco, V.: A Fuzzy logic approach by using particle swarm optimization for effective energy management in IWSNs. IEEE Trans. Industr. Electron. 64(12), 9496–9506 (2017)
8. Lee, J.-S., Cheng, W.-L.: Fuzzy-logic-based clustering approach for wireless sensor networks using energy predication. IEEE Sens. J. 12(9), 2891–2897 (2012)
9. Pantazis, N.A., Nikolidakis, S.A., Vergados, D.D.: Energy-efficient routing protocols in wireless sensor networks: a survey. IEEE Commun. Surv. Tutor. 15(2), 551–591 (2013)
10. Chithaluru, P., Al-Turjman, F., Kumar, M., Stephan, T.: I-AREOR: an energy-balanced clustering protocol for implementing green IoT in smart cities. Sustain. Cities Soc. 102254 (2020). https://doi.org/10.1016/j.scs.2020.102254
11. Shankar, A., Pandiaraja, P., Sumathi, K., Stephan, T., Sharma, P.: Privacy preserving E-voting cloud system based on ID based encryption. In: Peer-to-Peer Networking and Applications (2020). https://doi.org/10.1007/s12083-020-00977-4
12. Zhang, D., Li, G., Zheng, K., Ming, X., Pan, Z.-H.: An energy-balanced routing method based on forward-aware factor for wireless sensor networks. IEEE Trans. Industr. Inf. 10(1), 766–773 (2014)
13. Lai, X., Ji, X., Zhou, X., Chen, L.: Energy efficient link-delay aware routing in wireless sensor networks. IEEE Sens. J. 18(2), 837–848 (2017)
14. Sun, Y., Dong, W., Chen, Y.: An improved routing algorithm based on ant colony optimization in wireless sensor networks. IEEE Commun. Lett. 21(6), 1317–1320 (2017)
15. Mothku, S.K., Rout, R.R.: Adaptive fuzzy-based energy and delay-aware routing protocol for a heterogeneous sensor network. J. Comput. Netw. Commun. (2019)
16. Rout, R.R., Ghosh, S.K.: Adaptive data aggregation and energy efficiency using network coding in a clustered wireless sensor network: an analytical approach. Comput. Commun. 40, 65–75 (2014)
17. Salhi, I., Doudane, Y.G., Lohier, S., Roussel, G.: Network coding for event-centric wireless sensor networks. In: Proceedings of IEEE International Conference on Communications (ICC 2010), Cape Town, South Africa (2010)

18. Cormen, T.H., Leiserson, C.E., Rivest, R.L., Stein, C.: Introduction to Algorithms, 3rd edn. The MIT Press, Cambridge, MA, USA (2009)
19. Stephan, T., Suresh Joseph, K.: PSO assisted OLSR routing for cognitive radio vehicular sensor networks. In: Proceedings of the International Conference on Informatics and Analytics—ICIA-16 (2016). https://doi.org/10.1145/2980258.2980457
20. Stephan, T., Karuppanan, K.: Cognitive inspired optimal routing of OLSR in VANET. IEEE Xplore, 283–289 (2013). https://doi.org/10.1109/ICRTIT.2013.6844217
21. Bhardwaj, A., Al-Turjman, F., Kumar, M., Stephan, T., Mostarda, L.: Capturing-the-invisible (CTI): behavior-based attacks recognition in IoT-oriented industrial control systems. IEEE Access 1–1 (2020). https://doi.org/10.1109/ACCESS.2020.2998983
22. Stephan, T., Suresh Joseph, K.: Particle swarm optimization-based energy efficient channel assignment technique for clustered cognitive radio sensor networks. Comput. J. 61(6), 926–936 (2017). https://doi.org/10.1093/comjnl/bxx119
23. Stephan, T., Suresh Joseph, K.: Cognitive radio assisted OLSR routing for vehicular sensor networks. Procedia Comput. Sci. 89(2016), 271–282 (2016). https://doi.org/10.1016/j.procs.2016.06.058
24. Chanak, P., Banerjee, I.: Fuzzy rule-based faulty node classification and management scheme for large scale wireless sensor networks. Exp. Syst. Appl. 45, 307–321 (2016)

Application of Device-to-Device Communication in Video Streaming for 5G Wireless Networks

Ala'a Al-Habashna and Gabriel A. Wainer

Abstract We present one of the key-enabling technologies in 5G wireless networks, i.e., Device-to-Device (D2D) communication. Furthermore, we discus some of our work in utilizing D2D communication in one of the most bandwidth-demanding applications in 5G networks, i.e., video streaming. D2D communication, introduced by the LTE-Advanced (LTE-A) standard, allows direct communication between devices in cellular networks. We review some of the work in the literature on D2D communication and its applications in video streaming. We also discus an architecture that provides Dynamic Adaptive Streaming over HTTP (DASH)—based Peer-to-Peer (P2P) video streaming in cellular networks. The architecture employs Base-Station (BS)—assisted D2D video transmission in cellular networks for direct exchange of video contents among users. We evaluate the performance of the proposed architecture in terms of many video streaming Quality-of-Experience (QoE) metrics.

Keywords Device-to-device communication · Video streaming · 5G networks · DABAST

1 Introduction

With the advancement of smart devices and the evolution of cellular networks, video streaming service has become immensely popular. Video content is the main contributor to data traffic over cellular networks nowadays. According to [1], video traffic accounted for 60% of total mobile data traffic in 2016. Furthermore, over three-fourths (78%) of the world's mobile data traffic is expected to be video in 2021. Another reason for this increasing popularity of video streaming is the emergence of new platforms for video streaming such as YouTube and Netflix. As such, providing

A. Al-Habashna · G. A. Wainer (✉)
Department of Systems and Computer Engineering, Carleton University, 1125 Colonel By Dr, Ottawa, ON K12 5B6, Canada
e-mail: gwainer@sce.carleton.ca

© The Author(s), under exclusive license to Springer Nature Switzerland AG 2022   345
P. Nicopolitidis et al. (eds.), *Advances in Computing, Informatics, Networking and Cybersecurity*, Lecture Notes in Networks and Systems 289,
https://doi.org/10.1007/978-3-030-87049-2_12

high Quality of Experience (QoE) video streaming services has become a main concern for cellular networks operators.

Due to the limited capacity of cellular networks, it is difficult to provide the users with the data rates needed to achieve high QoE video streaming, especially in the case of high user density. The continuous increase in resource-demanding video applications is outpacing the improvements on the cellular network capacity. As per [2], video traffic is the main reason for congestion in mobile networks. The facts above made it necessary to develop new approaches that help serving the increasing video traffic over cellular networks and improve the QoE of video streaming.

In [3–5], we proposed two algorithms for BS-controlled progressive caching of video contents and Device-to-Device (D2D) video transmission in cellular networks. D2D communication, introduced by the LTE-Advanced (LTE-A) standard, allows direct communication between devices in cellular networks [6]. The algorithms are called Cached and Segmented Video Download (CSVD) and DIStributed Cached and Segmented video download (DISCS). The algorithms are employed by the BS to control progressive video content caching in selected User Equipment (UEs) in the cell, referred to as Storage Members (SMs). Furthermore, the algorithms are employed by the BS to control D2D communication between UEs in the cell for Peer-to-Peer (P2P) video content distribution to requesting UEs.

Here, we present an architecture to improve the QoE of video streaming in cellular networks. The proposed architecture employs the aforementioned cached and segmented video download algorithms. The architecture also employs Dynamic Adaptive Streaming over HTTP (DASH). DASH is an adaptive video streaming technique which allows changing the video bit rate during video streaming to adapt to the available throughput [7]. The proposed architecture is called DASH-based BS-Assisted P2P/D2D video STreaming in cellular networks (DABAST).

In [3–5], we investigated the performance of CSVD and DISCS in terms of the aggregate and average data rates. Results have shown that DISCS significantly improves the data rates over CSVD. Here, present some performance evaluation results for DABAST with CSVD (DABAST-CSVD) in various scenarios. Results have shown that DABAST-CSVD achieves significant gains and improves all the measured QoE metrics. We also show performance evaluation results for DABAST with DISCS (DABAST-DISCS) in terms of video streaming QoE metrics. We compare the results of both DABAST-CSVD and DABAST-DISCS to see if the improvement in the data rates achieved by DISCS would translate into significant improvements in terms of video streaming QoE. We provide analysis of the results and present the findings on DABAST with both algorithms.

We use the Discrete EVent System Specification (DEVS) formalism [8] to build a model for DABAST-CSVD and DABAST-DISCS. DEVS provides a formal framework for modeling generic dynamic systems. It has formal specifications for defining the structure and behavior of a discrete event model. We implement our DEVS model with the CD++ toolkit [8] and use this implementation for performance evaluation.

The rest of this chapter is organized as follows, in Sect. 2 we review the background and related work. In Sect. 3, we present CSVD, DISCS, and DABAST. We discus modeling of DABAST with DEVS in Sect. 4 and present our DEVS model. We present the simulation scenarios and results in Sect. 5. Finally, we conclude the chapter in Sect. 6.

2 Background

D2D communication is one of the main technologies in the Fifth Generation (5G) cellular networks [9] due to the improvements it provides. With D2D communication, two UEs within proximity of each other can exchange data over direct links, without the need to relay the traffic over the BS. This can improve the data rate between the two UEs due to transmission over one-hop and shorter distance. Moreover, the capacity of the cellular network can be increased by coordination of multiple short distance transmissions to achieve spatial frequency reuse. D2D communication can also extend the coverage area of the cell and improve the received signal for users at the cell edge. As such, much work has been conducted to develop applications for D2D communications in cellular networks and improve its performance [10–12]. D2D communication allows collaboration of users in cellular networks to share contents they have. However, approaches are needed to motivate participation of users in D2D communication. Incentivizing users to participate in D2D communication is a topic that has received much interest in the last couple of years. The interested reader in this area is referred to [13, 14].

Video streaming was considered ever since the early stages of the Internet, and nowadays, it is the most popular application on the web. According to [15], during the peak hours, YouTube traffic only accounts for 27% of the mobile downlink (DL) video traffic in North America. With video streaming, a user can start playing the video before the entire video file is downloaded. Most videos on the web nowadays are accessed via streaming. Contents such as movies, video news clips, and YouTube videos are watched by millions of people every day.

HTTP video streaming is the most popular form of video streaming nowadays, and it has been adopted by major video streaming solutions such as YouTube, Netflix, and Hulu. This is due to the convenience of using HTTP [16], which eliminates the need to install and use a dedicated streaming application and helps to get the streaming traffic past firewalls. HTTP video streaming works by breaking the overall video stream into a sequence of small HTTP-based file downloads, referred to as video segments. Users progressively download these small segments, while the video is being played. Playout usually starts after receiving a certain "sufficient" number of video segments. The received segments are buffered in a video/application buffer. The application that plays the video is usually referred to as the client. The streaming client requests video segments from the server. Since each segment has a fixed duration, the size of the segment depends on its duration and the video bit rate. The client receives the pieces from the video buffer. The duration of video content available for playout is called the playout buffer length, measured in seconds of video. Furthermore, every second, one second of video is removed from the buffer and played to the user.

Bad network conditions (insufficient bandwidth, delay, etc.) may cause the playout buffer to get empty, as the video bit rate is higher than the video streaming rate, which causes video playout interruptions. These interruptions are referred to as video stalling or rebufferings. When stalling occurs, playout stops until sufficient data is buffered again.

Although HTTP video streaming provided a convenient way for video streaming over the Internet, it was still challenging to stream video to wireless and mobile devices due to the high bandwidth variability of the wireless links. DASH provided a promising technique to improve video streaming over networks with varying bandwidth [17], as it allows changing the quality of video streaming to adapt to network conditions.

DASH provides two features that helped improving video streaming. First, it breaks down the video into small, easy to download segments (for example 5 s chunks). Second, each segment is encoded at multiple bit rates, providing multiple quality levels for each segment, which allows adaptive streaming. Clients will choose between various bit rates to adapt to the network conditions. As such, DASH helps improving the bandwidth utilization and reducing the interruptions of the video playback, which results in a higher streaming quality. Due to these advantages of DASH over classical HTTP video streaming, DASH has been employed by big video streaming platforms, such as YouTube and Netflix, and it is being adopted by an increasing number of video applications.

There are various adaptation strategies that can be used to determine how the client selects the streaming quality to adapt to the varying network conditions. These strategies usually try to balance between two factors. They try to maximize the video quality by selecting the highest video rate the network can support, and at the same time minimize rebufferings. We refer to the component in the client that runs the adaptation strategy as the DASH controller.

With the increasing demand for video applications, providing high quality video service as perceived by the end user has become an important concern for cellular network operators. As such, quality measure has shifted from Quality of Service (QoS) to QoE. The ITU defines QoE as the overall acceptability of the service as perceived by the end user. Video streaming QoE is especially important because users pay their operator, and they expect to get video service with good QoE in return. If the user is not satisfied, they may look for other options and switch to another provider. As such, video streaming QoE must be considered in network design and management in order to maintain user satisfaction. There are many factors that are used to measure video streaming QoE, here we present the most important ones [18].

- Video stalling (rebuffering): the stopping of video playback as the playout buffer gets empty. Increasing video stalling decreases the QoE. Many studies [18] have shown that video stalling has the biggest impact on QoE, and thus, should be avoided as much as possible.
- Video continuity index: a measure of the extent by which rebuffering pauses are avoided [19]. The continuity index is measured as follows,

$$\eta_c = 1 - \frac{\Delta T_{rb}}{\Delta T},\tag{1}$$

where ΔT_{rb} is the total time the client remains paused due to rebuffering events and ΔT is the duration of the experiment (playing time and rebuffering time).

- Initial (startup) delay: the delay from the request to stream the video until the playback starts. Initial delay is always present as certain number of video segments should be received before decoding and playback starts.
- Video bit rate: it is a measurement of the amount of data in one second of the video. Video bit rate is determined by many quality factors of the video such as video frame rate, resolution, and quantization parameters. As the video bit rate increases, the video quality increases, which increases the QoE.

Direct communication between nodes in wireless networks has been studied in the context of wireless ad hoc networks [20]. However, it has not been considered in cellular networks until the emergence of D2D communications [10–12, 21]. The introduction of D2D communications in the LTE-A standard has opened the door for direct P2P communications between UEs in cellular networks [6]. With D2D communication, two UEs that are within proximity of each other communicate directly without going through the BS or the core network.

D2D communication provides a promising solution to increase the capacity in cellular networks by allowing frequency reuse, and by increasing the data rates due to potentially improved transmission over shorter distance and fewer hops. As such, it has been investigated in the last few years [10–12, 21].

There has been some work on P2P video streaming in cellular networks. A protocol for P2P video streaming on mobile phones, called RapidStream, was proposed in [22]. It is similar to many of the P2P streaming protocols on wired networks that involve the dissemination of buffer maps and video chunks between peers. While such protocols work well in wired networks, they involve too much signaling and transmission (dissemination of buffer maps) to be appropriate for UEs that has limited power, processing, and transmission resources (especially on a large scale). In [23], multi-source video streaming was proposed where mobile users can connect through Wi-Fi direct to other users to get some of the video content. Such system requires the device to perform device discovery to find neighbors, and service discovery to find services offered by neighboring devices. These requirements along with the signaling needed to exchange content consume significant amount of resources.

In [24], a system, called MicroCast, was designed and evaluated using a testbed. MicroCast is used by a group of smart phone users who trust each other, are interested in watching the same video at the same time, and who are within proximity of each other. Users employ their cellular connection to download segments of the video and use their Wi-Fi connections to share among each other the downloaded content, to improve the streaming experience. While this could result in some improvement for a group of users, the scope of the system is limited. Furthermore, users usually do not use their cellular connection for downloading video segments when Wi-Fi is available.

In [25] the authors proposed a D2D communication system where multiple helpers collaborate to send a video segment to the requesting UE. The video, which is assumed to be in scalable video coding standard, is encoded by applying multiple description coding by each helper, and each helper sends a different description to the requesting UE. The authors analytically studied the problem of optimizing the number of transmitted descriptions to the requesting UE to maximize the video quality and efficiently consume the helpers' energy. However, the work only considers the energy consumed by the helpers to send the segments without considering the processing power and energy needed to encode the video segments. Encoding the video segments is a big favor to ask for, considering the limited energy and processing power of UEs.

None of the research studies above on P2P video streaming in cellular networks considers how the video segments are actually cached. When evaluating the performance, they consider that requested segments are already available in helper UEs. With DABAST, we provide a framework that takes care of video content caching and distribution [26]. Moreover, the employed CSVD and DISCS algorithms provide approaches for inter-cluster as well as inter-cell interference mitigation. Furthermore, the work above considers small-scale networks, i.e., up to 10 UEs including the helpers. We show that using clustering and BS assistance, the potential of collaborative D2D communication between UEs is significant.

Here, we present DABAST [27, 28] and discus its operation and implementation. We also discuss the performance evaluation of DABAST-CSVD under various scenarios, to provide quantitative evaluation of the impact of many parameters on the performance of DABAST. Furthermore, we compare the results of both DABAST-CSVD and DABAST-DISCS to see if the improvement in the data rates achieved by DISCS would translate into significant improvements in terms of video streaming QoE.

We used the DEVS formalism [29] to build a model for the DABAST architecture, and used that model to test and evaluate the performance of DABAST using various simulations. DEVS provides a formal framework for modeling generic dynamic systems. It has formal specifications for defining the structure and behavior of a discrete event model. A DEVS model is composed of structural (Coupled) and behavioral (Atomic) components, in which the coupled component maintains the hierarchical structure of the system, while each atomic component represents a behavior of a part of the system. The CD++ toolkit [8] was used to implement our model of DABAST. CD++ is an open-source simulation software written in C++ that implements the DEVS abstract simulation technique. The simulation engine tool of CD++ is built as a class hierarchy. C++ is used to develop the atomic components of the model. These components can be incorporated into the class hierarchy. Passive classes can be also used to model components of the system. Coupled models can be created using a language built in the simulation engine.

3   DABAST: The Architecture

In this section, we discuss DABAST and its operation. Thereafter, we present the employed cached and segmented video download algorithms.

3.1   DABAST

By employing the cached and segmented video download algorithms [28], DABAST relaxes the bottleneck of the Radio Access Network (RAN), and hence, improves video streaming QoE for end users.

Figure 1 illustrates the main structure for DABAST. At the bottom, we have the LTE-A network that provides the infrastructure for communication between the BS and UEs over cellular links, and the communication between UEs over D2D links where the UEs exchange data directly without going through the BS.

A BS assisted P2P/D2D communication protocol (e.g., CSVD) is implemented on top of that which uses both the cellular and D2D communication. DASH-based video streaming takes place on top of these layers, as the transmission of video segments is implemented as per the communication protocol at the layer below.

Figure 2 depicts the implementation of DABAST in cellular networks. A CSVD server/proxy is used in the RAN at the BS. This provides the processing and networking capabilities needed to implement CSVD or DISCS. The CSVD server can also be used to provide caching capabilities to store popular files at the BS. Streaming clients at the UEs send their requests asking for video segments. These requests are

Fig. 1 DABAST: the architecture

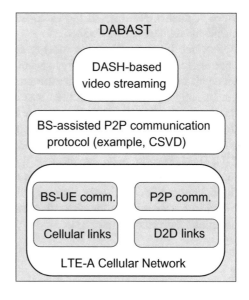

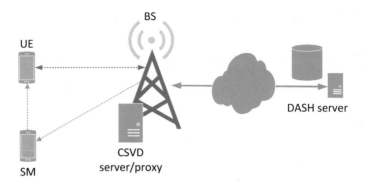

Fig. 2 Illustration of the implementation of DABAST

processed by the CSVD server. Based on the employed algorithm (CSVD or DISCS), the server decides whether to send the segment from the distributed cache over the D2D channel or get it from the content server and send it over the cellular channel. If the video segment is to be delivered from the distributed cache, an assistance request will be sent to an SM. Otherwise, the request will be forwarded to the DASH server. Under high traffic load, the BS sends a video segment from the distributed cache (when found) even if the segment found in the distributed cache does not match the video bit rate requested by the UE. This is to maximize the exploitation of the distributed cache and D2D channel.

Another feature of DABAST is that it can operate in proactive mode. In this mode, DABAST can send up to a certain number of video segments to the user when found in the distributed cache before the segments are requested. This reduces the signaling and latency between the BS and the UE and speeds up transmission of video segments. However, If the video segment is not available in the distributed cache, the request will be awaited.

A video bit rate adaptation strategy is used by DASH to determine how the client selects the streaming quality to adapt to the varying network conditions. Many video bit rate adaptation strategies are proposed for DASH [30, 31]. These strategies usually try to balance between two factors. They try to maximize the video quality by selecting the highest video rate the network can support, and at the same time minimize rebufferings. We refer to the component in the client that runs the adaptation strategy as the DASH controller. We adopt the buffer-based approach in [31]. This is because in our architecture, the UE could receive a video segment from the BS or from any SM in the cluster. As such, it would be difficult to estimate the throughput at which the next segment will be received. The adaptation algorithm used is a piecewise function, $f(B)$, that uses the length of the playout buffer, B, to determine the video bit rate [31].

3.2   The CSVD and DISCS Algorithms

CSVD and DISCS relax the RAN bottleneck by providing BS-assisted progressive video content caching and P2P video segment distribution in cellular networks. When we use the term P2P here, we refer to direct transmission of video segments between UEs in the cell over D2D links.

With both algorithms, the cell is divided into clusters. To do that, the coverage area of the cell is divided into non-overlapping subareas by the BS. Each one of these subareas will be a cluster. The BS assigns UEs to clusters based on their locations, and it selects the UEs in the central area of each cluster as SMs of that cluster. SMs are UEs that are used as helpers in the cluster. To prevent inter-cluster as well as inter-cell interference, only the UEs in the central area of each cluster are selected as SMs.

After clustering, when a UE requests a video file from the BS, the BS processes the request and responds as follows:

- Send With Assistance (SWA): if the video (or parts of it) is available in any of the SMs of the cluster, the BS will ask the SMs to send the video segments to the requesting UE over D2D links.
- Send To an SM (STSM): if the requested video is not available in the distributed cache of the cluster, and the requesting UE is an SM, the BS will send the video to that UE over a cellular link, and it will ask the UE to cache it. This case allows the SMs to cache video files. These files will be available for UEs in the cluster when requested later.
- Distribute to SMs (DTSMs): this case is only used in DISCS. In this case, if a requested video is popular and it is not available in the distributed cache of the cluster, the BS will distribute the segments of the video among the SMs. The BS asks the SMs to cache the pieces (as the video is popular) and forward the received pieces to the requesting UE.
- Send To a UE (STUE): otherwise, the BS will send the video directly to the requesting UE over a cellular link.

The difference between CSVD and DISCS is the DTSMs case. The DTSMs case is only used in DISCS. The other 3 cases are used by both CSVD and DISCS. For further detail on the operation of CSVD and DISCS, the reader is referred to [5].

4   Modeling DABAST with DEVS

As discussed earlier, we used DEVS to build a model of DABAST in an LTE-A network. Figure 3 shows the coupled DEVS model of the top-level architecture. As can be seen, we have defined a Cell coupled model that contains many UE coupled models, a BS coupled model, and a Transmission Medium coupled model. The Cell coupled model also contains Cell Manager and Log Manager atomic models.

Fig. 3 Coupled DEVS
model of DABAST

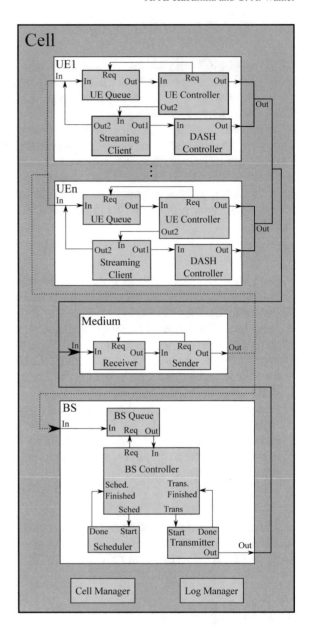

A UE coupled model contains four atomic models: UE Queue, UE Controller,
Streaming Client, and DASH controller. Messages received are buffered at the
UE Queue. The UE Controller is where the UE part of the CSVD algorithm is
implemented.

The DASH-based streaming client is implemented in the Streaming Client and
DASH controller atomic models. The streaming client manages the video buffer.

It adds video segments received to the video buffer and removes video segments that were played from the buffer. As the video buffer usually has a certain length that could be shorter than the video length, it is implemented as a sliding window. Video segments that were already played will be removed from the video buffer and the buffer slides to cover the next segments in the stream. The DASH controller implements the adaptation algorithm. It monitors the video playout buffer and updates the video bit rate accordingly. When the video bit rate is to be updated, a request is sent to the BS with the new video bit rate.

The BS coupled model includes four atomic models: BS Queue, BS Controller, Scheduler, and Transmitter. Received messages are buffered at the BS Queue. The BS Controller is where the BS part of the CSVD algorithm is implemented. The Scheduler schedules the messages to be transmitted in the next Transmission Time Interval (TTI), which is 1 ms. Every TTI, the BS Controller also asks the Transmitter to send messages that were scheduled for transmission during this TTI.

The Medium model simulates the transmission medium and the Cell Manager atomic model initializes and sets the parameters of the cellular DLs and uplinks (ULs) between the BS and the UEs, as well as the D2D links between the UEs. For further details on the communication models used for simulation of the LTE-A cellular links and D2D links, the reader is referred to [5]. In addition to the atomic models above, many other passive classes where developed to model other components of the system such as classes to model the cellular and D2D links, download sessions the BS has with UEs, etc.

5 Simulation Scenarios and Results

We executed simulations to evaluate the performance of DABAST in terms of the QoE metrics that were discussed in Sect. 2. We ran various simulations to study the impact of many parameters on DABAST such as files' popularity, the number of UEs in the cell, etc. In this section, we consider DABAST with CSVD (DABAST-CSVD). In the next section, we consider both DABAST-CSVD and DABAST-DISCS and compare their performance.

The simulation setup is shown in Table 1. The simulations consider a single LTE-A cell. The urban macro propagation model [32] was used for cellular links with a DL operating carrier frequency of 900 MHz, and a transmission bandwidth of 10 MHz. The D2D channel model at 24 GHz is used for D2D transmission [33].

In each iteration of the simulation, the UEs are uniformly distributed throughout the cell. Clustering takes place in the beginning in case of DABAST where the cell is divided into 9 clusters. The UEs then start requesting video streams. During each iteration of the simulation, each UE will request two video streams. A UE requests a video stream, and after finishing the playout, it will request a second video. The arrival of requests is generated according a Poisson arrival process. The popularity of videos is generated according to a Zipf distribution to simulate the variable popularity of the videos, as it has been established this is a good model for this purpose [34].

Table 1 Simulation setup

Parameter	Value
Cellular channel BW (MHz)	10
Cell range (m)	500
Number of clusters	9
BS antenna gain (dB)	12
BS transmission power (dBm)	43
UE antenna gain (dB)	0
UE transmission power (dBm)	21
Noise spectral density (dBm)	− 174
Antenna height (m)	15
Transmission model	UTRA-FDD
DL carrier frequency (MHz)	900
Number of requests by a UE	2
Area configuration	Urban
D2D channel BW (MHz)	60
D2D carrier frequency (GHz)	24
D2D transmitter TX power (dBm)	23
D2D large-scale fading std deviation (dB)	4.3
D2D receiver noise figure (dB)	9
D2D TX/RX height from ground (m)	1.5
Segment length (s)	10
Number of buffered segments to start playout	4
Video bit rate levels (kbps)	384, 768, 2000, 4000
Videos length (s)	441

Using this distribution, some videos are requested more often than others. The length of the videos is 441 s, which is the mean length of a YouTube video [35]. Four video bit rate levels where used as shown in Table 1. These are adapted from the H.264/AVC video coding standard [36].

Regarding the DASH controller, the buffer-based approach in [31] was employed. This is because in our architecture, the UE could receive a video segment from the BS or from any SM in the cluster. As such, it would be difficult to estimate the bit rate at which the next segment will be received. The adaptation algorithm used is a piecewise function, $f(L)$, that uses the length of the playout buffer, L, to determine the video bit rate.

We assume that the video buffer is long enough to accommodate all received segments. The playout buffer to video rate mapping is shown in Table 2. We measured the number of rebufferings, video continuity index, initial delay, and video bit rate levels of the received video segments. Table 3 shows the mean values for these

Table 2 Playout buffer length to video rate mapping

Playout buffer length (s)	Video bit rate (kbps)
$0 \leq L \leq 90$	384
$90 < L \leq 150$	768
$150 < L \leq 200$	2000
$200 < L$	4000

Table 3 Simulation results

	Conventional DASH		DABAST	
	Mean	MoE	Mean	MoE
Rebufferings	3.4448	0.0179	1.7272	0.0164
Cont. index	0.7447	0.0009	0.8699	0.0001
Initial delay (s)	56.881	0.2200	28.654	0.4821
Video bit rate (kbps)	397.27	0.3267	430.16	1.3694

measurements. The results in Table 3 are for 500 UEs in the cell, Zipf exponent of 1.5, and 500 videos. The average value for each simulation run was calculated. The values below show the mean of all the average values from 50 simulation runs. In addition to the mean, we show the Margin of Error (MoE) for 95% confidence interval.

Table 3 shows that DABAST achieves improvements over conventional DASH in terms of all the measured metrics above. Regarding the average number of rebufferings, DABAST achieved 50% decrease in the average number of rebufferings, which is a significant improvement. The continuity index is also improved with DABAST due to decreasing the average number of rebufferings as well as the rebuffering time. It is worth mentioning that the average initial delay for conventional DASH is high because in this scenario, there are 500 UEs in the cell requesting video streams and sharing fixed cellular frequency resources (10 MHz). This is also because video playout starts after receiving 4 video segments.

Table 3 shows that DABAST also achieves a 50% decrease in the average initial delay, which is also a significant improvement. In addition to the improvements above, DABAST also achieved an improvement in terms of the average video bit rate. Due to the increase in the transmission rates achieved by DABAST, video segments are delivered to UEs much faster than in the case of conventional DASH. This is because in the case of DABAST, the CSVD algorithm is employed, where video segments are sent to many UEs from both the BS (over cellular links) and SMs (over D2D links) as opposed to only from the BS. This reduces the transmission delay of the first 4 segments needed to start playout, and consequently, reduces initial delay. This also reduces the possibility of video buffer stalling, and hence, reduces the number of rebufferings. There is only a small improvement achieved by DABAST in terms of average video bit rate. This is due to two reasons. First, with both conventional DASH and DABAST, the DASH controller resorts to choosing a lower video bit rate level to increase the video playout buffer length and reduce the number of rebufferings. This means that video bit rate is the last metric that is improved when transmission rate improves. As such, the improvement in terms of the number of rebufferings is usually achieved on the expense of video bit rate. Second, as mentioned in the previous section, under high traffic load, the BS will send a video segment from the distributed cache (when found) even if the segment found in the distributed cache does not match the video bit rate requested by the UE. This is to increase the utilization of the available D2D channel and to speed up the transmission of video segments, as sending from the distributed cache is faster. This means that although DABAST might increase the transmission rate and playout buffer length for some clients (which increases the requested video rate), such users might still receive segments with low video bit rate (from the distributed cache).

Figure 4 shows the histogram of the number of rebufferings for both conventional DASH and DABAST. The figure shows that over 96% of the streaming requests have 3 or 4 rebufferings in the case of conventional DASH. With DABAST, on the other hand, half of the streaming requests have 0 rebufferings, and slightly less than half of the streams have 3 or 4 rebufferings. After videos accumulate in the clusters' caches, many video segments will be delivered from the distributed cache in the cell. These segments will be transmitted faster to the requesting UEs. Moreover, as many of the segments will be sent over D2D links, there will be more cellular resources available for segments that are transmitted over cellular resources, which also reduces the transmission delay for such segments and reduces the possibility of playout interruption, decreasing the number of rebufferings by 50%.

Figure 5 shows the histogram of the continuity index for both conventional DASH and DABAST. The continuity index results match these in Fig. 4 for the number of

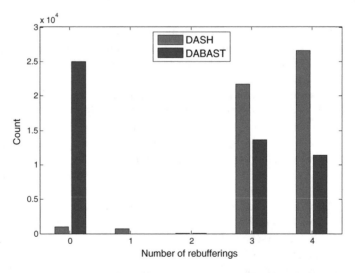

Fig. 4 Histogram of the number of rebufferings for conventional DASH and DABAST

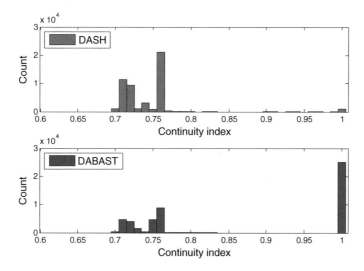

Fig. 5 Histogram of the continuity-index for conventional DASH and DABAST

rebufferings. Half of the requests with DABAST have a continuity index of 1, which corresponds to zero rebufferings. Less than half of the video streams have a continuity index less than 0.76 with DABAST (corresponds to 3 and 4 rebufferings). However, in the case of conventional DASH, over 96% of the video streams have a continuity index less than 0.76.

Figure 6 shows the ECDF of the initial delay for both conventional DASH and DABAST. We can see that the ECDF of DABAST is always higher than that of

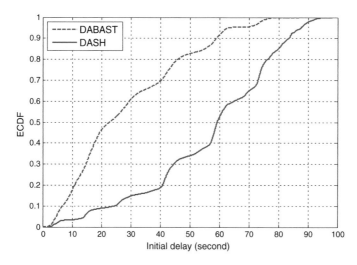

Fig. 6 ECDF of the initial delay for conventional DASH and DABAST

Table 4 Count of the received segments with each video bit rate

Video bit rate (kbps)	Count	
	Conventional DASH	DABAST
384	2,207,508	2,077,111
768	31,494	149,365
2000	10,998	19,273
4000	0	4251

conventional DASH. For example, the probability of having a stream with initial delay of 20 s or less is 0.46 with DABAST, and only 0.09 with conventional DASH. Figure 6 also shows that 50% of the streams have initial delay of 22.76 s or less with DABAST while 50% of the streams have 59.21 s or less with conventional DASH. As previously mentioned, with conventional DASH, all the UEs in the cell share the fixed frequency resources (10 MHz cellular channel), while with DABAST, the D2D channel is exploited for P2P communication in addition to cellular resources. The transmission of video segments from the distributed cache in the cell speeds up the delivery of video segments and significantly reduces initial delay.

Table 4 shows the count for the received video segments with each video bit rate level, for both conventional DASH and DABAST. The results show that with DABAST, fewer video segments with 384 kbps were received and more video segments with higher levels (768, 2000, and 4000 kbps) were received. This explains why the average video bit rate (Table 3) for DABAST is higher than that for conventional DASH. With DABAST, video segments are delivered faster to the requesting UEs as explained above. As such, clients will have more video segments in the playout buffer, i.e., higher playout buffer length. Consequently, the requested video bit rate will be higher in the case of DABAST.

The results presented in this section show that DABAST provides improvements over conventional DASH in terms of all the measured metrics, which significantly improves the QoE of video streaming in cellular networks. In the following, we investigate the impact of different parameters on the performance of DABAST.

Figure 7 shows the average number of rebufferings for both conventional DASH and DABAST versus the number of UEs in the cell. Figure 7 shows that at 300 UEs, the average number of rebufferings is 0 for both conventional DASH and DABAST. This means that the cellular resources are enough with conventional DASH to avoid rebufferings for all the video streams. As the number of UEs increases, the available cellular resources will be shared by more UEs, which reduces the average data rate and increases the transmission delay of video segments. This increases the possibility of video buffer depletion and increases the average number of rebufferings.

Even with DABAST, the average number of rebufferings increases with increasing the number of UEs. This is because we study the case of progressive caching, where in the beginning there are no videos cached, and videos are cached as requested. Figure 7 shows that the improvement achieved by DABAST increases by increasing the number of UEs in the cell. Increasing the number of UEs increases the number of requests and also increases the number of SMs in each cluster. This increases the

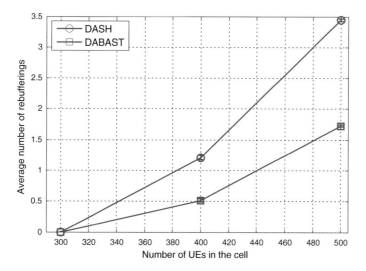

Fig. 7 Average number of rebufferings versus number of UEs in the cell Zipf exponent = 1.5 and 500 videos

number of cached videos in a cluster and the percentage of requests that would be satisfied from the cluster's cache, increasing the improvement achieved by DABAST over conventional DASH.

Figure 8 shows the average initial delay for both conventional DASH and DABAST versus the number of UEs in the cell. Figure 8 shows that although the average number of rebufferings is zero at 300 UEs for both approaches (as in Fig. 7)

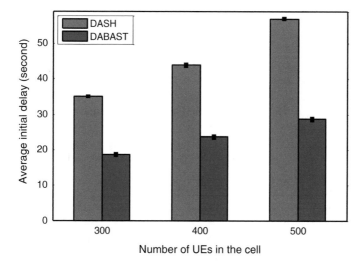

Fig. 8 Average initial delay versus number of UEs in the cell. Zipf exponent = 1.5 and 500 videos

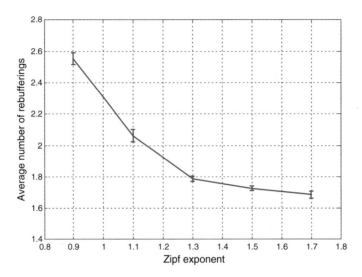

Fig. 9 Average number of rebufferings versus the Zipf exponent. 500 UEs, 500 videos

DABAST still achieves improvement in terms of the average initial delay at 300 UEs. Furthermore, it can be noticed that the gain achieved by DABAST over conventional DASH increases with the number of UEs. This is for the same reason explained above for the average number of rebufferings.

As mentioned above, the Zipf distribution was used to model the popularity of videos. The Zipf distribution has one parameter, namely the Zipf exponent. This exponent controls the relative popularity of the videos. When the value of the Zipf exponent increases, the relative popularity of some videos in the list increases. This increases the possibility of requesting these videos.

Figure 9 shows the average number of rebufferings versus the Zipf exponent. As can be seen, the average number of rebufferings decreases as the Zipf exponent increases. As the popularity of some videos increases, higher percentage of the requests will be found in the distributed cache and delivered over the D2D channel (rather than the cellular channel). This speeds up the transmission of many segments and decreases the possibility of playout buffer depletion, which decreases the average number of rebufferings.

Figure 9 shows that the improvement achieved with increasing the Zipf exponent eventually slows down. This is because in our scenario, each UE requests only two videos. Increasing the number of requests made by each UE further increases the exploitation of cached contents and increases the improvement achieved by DABAST. This is because cached videos will be further used by the later requests.

Figure 10 shows the average number of rebufferings for DABAST versus the number of videos available to request from. As can be seen, there is no considerable effect for the number of videos on the average number of rebufferings. As per the Zipf distribution, having more videos will not cause a noticeable impact on the probability

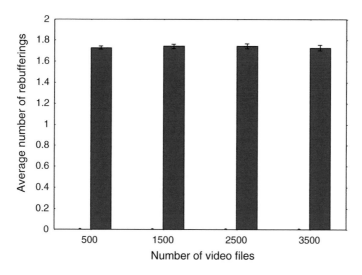

Fig. 10 Average number of rebufferings versus the number of videos 500 UEs and Zipf exponent = 1.5

of requesting the popular files. This means that for a certain Zipf exponent, although the number of videos increases, cached contents will still be exploited as long as there are popular videos.

It is worth mentioning that DABAST should be even more beneficial when considering background traffic. To get a feel of the improvement achieved by DABAST in the case background traffic is present, we ran simulations with 300 UEs in the cell and with the same setup in Table 1. Unlike the previous simulations above, we consider that background traffic is present in the cell, and that the BS dedicates 5 MHz of the channel for background traffic and 5 MHz for video streaming traffic. Let us recall that in the case background traffic is absent and the whole channel is available for video streaming traffic (300 UEs), users experienced 0 rebuffering with both DASH and DABAST. On the other hand, in the case background traffic is present, the results have shown that with conventional DASH, the average number of rebufferings is 4.47, while with DABAST, the average number of rebufferings is 2.35. This shows that DABAST in this case has achieved 47.4% reduction in the average number of rebufferings. This shows how DABAST is even more beneficial and can achieve further gains in the case background traffic is present.

In [5], we have studied the performance of CSVD and DISCS in terms of data rates. Results have shown that DISCS achieves significant improvement over CSVD in terms of the average data rate. Furthermore, in the previous section, we have studied the performance improvement achieved by DABAST that employs CSVD (DABAST-CSVD) in terms of video streaming QoE. Results have also shown that significant improvements can be achieved using DABAST-CSVD in terms of all the measured QoE metrics. In this section, we study how much improvement DABAST can further achieve if DISCS is employed instead of CSVD, i.e., we want to study

Table 5 Simulation results

	DABAST-CSVD		DABAST-DISCS	
	Mean	MoE	Mean	MoE
Rebufferings	1.7272	0.0164	1.6294	0.0169
Cont. index	0.8699	0.0001	0.8763	0.0011
Initial delay (s)	28.654	0.4821	21.192	0.4163
Video bit rate (kbps)	430.16	1.3694	448.57	0.9607

if the significant improvement achieved by DISCS over CSVD in terms of data rate translates into significant improvement in terms of QoE.

Remember that with DISCS, in addition to the 3 cases considered in CSVD, that are listed in 3.2, an additional case is employed (DTSMs). In this case, if a requested video is popular (requested n times) and it is not available in the distributed cache of the cluster, the BS will distribute the pieces among the SMs in the cluster. The BS asks the SMs to cache the received pieces (as the file is popular) and forward them to the requesting UE. This case helps speeding up accumulation of popular video files in the distributed cache of the cluster. It also allows for more parallelism and load balancing among SMs when sending video files from the distributed cache of the cluster. This should increase the utilization of the D2D channel and speeds up the transmission, and consequently increase the average data rate.

We executed simulations to evaluate the performance of DABAST that employs DISCS (DABAST-DISCS) in terms of the same QoE metrics. The simulation setup in the previous section was also used here. Table 5 shows results for both DABAST-CSVD and DABAST-DISCS. The results in Table 5 are for 500 UEs in the cell, Zipf exponent of 1.5, and 500 videos. The average value for each simulation run was calculated. The values below show the mean of all the average values from 50 simulation runs. In addition to the mean, we show the MoE for 95% confidence interval.

Before discussing the results in Table 5 for the measured QoE metrics, it is worth mentioning that the average data rates achieved with DABAST-CSVD and DABAST-DISCS are 3.89 and 8.32 Mbps, respectively. DISCS significantly improves the achieved average data rate because it speeds up video caching which increases the percentage of requests that are satisfied from the distributed cache which speeds up the transmission. This is also because in the case of DISCS, many files will be sent in parallel from multiple SMs (as opposed to one SM), thanks to the DTSMs case which distributes segments of a cached video file among multiple SMs. This causes further parallelism in sending video files and better load balancing among SMs, which speeds up the transmission of video files and increases the average data rate. From Table 5, we can see that as expected, DABAST-DISCS provides improvement over DABAST-CSVD in terms of the measured QoE metrics. Table 5 shows that DABAST-DISCS reduced the average number of rebufferings from 1.73 to 1.63, which increased the continuity index. With DABAST-DISCS, the initial delay is

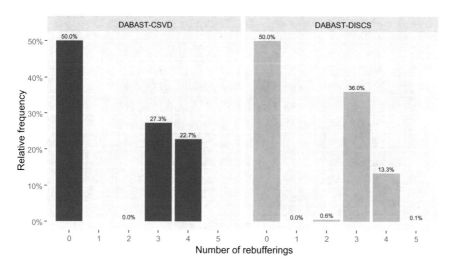

Fig. 11 Relative frequency histogram of the number of rebufferings for DABAST-CSVD and DABAST-DISCS

also reduced from 28.7 to 21.2 s, which is a significant improvement. Furthermore, the average video bit rate increased by 18 kbps with DABAST-DISCS.

Although the results above show that DABAST-DISCS improves the QoE metrics, one can see that the only significant improvement achieved by DABAST-DISCS is in terms of the initial delay. Only slight improvement is achieved in terms of the average number of rebuffering and continuity index. One would expect higher gains by DABAST-DISCS over DABAST-CSVD given the that DISCS achieves more than double the average data rate obtained with CSVD. In the following, we present further analysis of the above results to understand this behavior.

Figure 11 depicts the relative frequency histogram for the number of rebufferings of DABAST-CSVD and DABAST-DISCS. From the figure, one can see that the main difference is that with DABAST-DISCS, higher number of video streams have 3 rebufferings and fewer number of video streams have 4 rebufferings. This explains why the average number of rebufferings with DABAST-DISCS is less than that with DABAST-CSVD. The figure shows that with both DABAST-CSVD and DABAST-DISCS, 50% of the video streams have 0 rebufferings. With DABAST-CSVD, 27.3% of the video streams have 3 rebufferings, and 22.7% of the streams have 4 rebufferings. With DABAST-DISCS, 36.0% of the video streams have 3 rebufferings, and 13.3% of the streams have 4 rebufferings. This improvement in the number of rebufferings is expected, as DABAST speeds up video caching, and improves the rate at which video segments are delivered to requesting UEs. Figure 2 also shows that with DABAST-DISCS, a small percentage of the video streams (0.1%) have 5 rebufferings. This increase in the number of rebufferings experienced by a slight percentage of the streams is a result of the DTSMs case, where video segments are sent to the requesting UEs in two steps, i.e., the segment is sent to

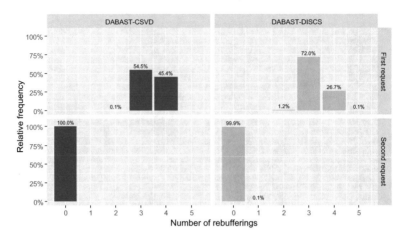

Fig. 12 Relative frequency histogram of the number of rebufferings in each request for *DABAST-CSVD and DABAST-DISCS*

the SM first, and then sent to the UE by the SM. Despite of this increase to a small percentage of the UEs, DABAST-DISCS still achieves lower initial delay and number of rebufferings, on average. This means that the DTSMs case is beneficial to the cell, as expected.

Although the results show that DABAST-DISCS decreases the number of rebufferings, one can argue that DABAST-DISCS is expected to achieve higher gains in terms of the number of rebufferings, as it significantly improves the average data rate. To further investigate the results and explain this behavior, we separate the results for each request. As previously mentioned, each UE in the simulations makes 2 video streaming requests. After playout of the first video, a UE would stay idle for a random period of time, and then generates another request for a video stream.

Figure 12 shows the relative frequency histogram for the number of rebufferings of each request for both DABAST-CSVD and DABAST-DISCS. As with the previous figure, the histograms on the right side are for DABAST-CSVD, and the ones on the left are for DABAST-DISCS. In this figure, however, the histograms on the top are for the first requests, while the ones at the bottom are for the second requests. Figure 12 shows that for DABAST-CSVD, all the rebufferings take place during the first set of video streams. All the second video streams have 0 rebufferings. This is because by the time most of the video streams start, there are no video segments cached in the distributed caches. Hence, most of the video segments will be delivered over the cellular channel. As such, the limited cellular channel will be shared by the large number of users, which means the average data rate at which these segments are delivered is low and explains the high number of rebufferings. The figure shows that with DABAST-CSVD, all the second set of streams have 0 rebufferings. By the time the second set of streams starts, there will be many video segments cached in the clusters. Hence, many of the segments will be delivered over D2D links, which eliminates rebufferings for those streams. This also relaxes the bottleneck of the

RAN because, at this time, only a portion of the video segments will be sent over the cellular channel. Because the cellular channel is now shared by a much lower number of UEs, the data rate will increase, and rebufferings will be avoided for these video streams as well.

With DABAST-DISCS, almost all rebufferings occur in the first set of video streams, as 99.9% of the second set of video streams have 0 rebufferings, and only 0.1% of the second set of video streams have 1 rebufferings. This is for the same reason all the rebufferings with DABAST-CSVD occur during the first set of video streams. Initially, there are no video segments available in the distributed cache, and hence, all video streams will experience multiple rebufferings. By the time the second set of streams starts, there will be many video segments cached in the clusters. Hence, many of the segments will be delivered over D2D links, which eliminates rebufferings for these streams, and relaxes the RAN bottleneck for segments delivered over cellular resources. The one difference here is that there is a small portion of the second set of video streams that still get its segments with the DTSMs case, which causes 1 rebuffering to 0.1% for the second set of video streams.

The results in Fig. 12 also explain why DABAST-DISCS does not achieve significant improvement in terms of the average number of rebufferings over DABAST-CSVD. As discussed above, DISCS achieves significant improvement in terms of the average data rate over CSVD. This is because in the case of DISCS, many files will be sent in parallel from multiple SMs (as opposed to one SM). This causes further parallelism in sending video segments and better load balancing between SMs, which speeds up the transmission of video files and increases the average data rate. However, we have seen from Fig. 11 that video streams with segments delivered over D2D links already have 0 rebufferings, even in the case of DABAST-CSVD. As such, the increase in the average data rate achieved by DABAST-DISCS will not translate into reduction in the average number of rebufferings, as all rebufferings take place in the first set of streams when video segments are not cached. This means that the improvement in the average number of rebufferings gained by DABAST-DISCS over DABAST-CSVD is due to the fact that DISCS speeds up video caching and achieves better hit ratio, which increases the percentage of requests that are satisfied from the cluster's cache and speeds up the relaxation of the RAN bottleneck.

Figure 13 shows the histogram of the continuity-index for both DABAST-CSVD and DABAST-DISCS. As expected, 50% of the streams have a continuity index of 1. With DABAST-DISCS there is more concentration of the values around 0.76 and less concentration of values between 0.71 and 0.75. This agrees with the results for the number of rebufferings. However, with DABAST-DISCS, there are few continuity index values less than 0.7. These values correspond to streams with 5 rebufferings.

Regarding the average video bit rate, we can also see that DABAST-DISCS did not achieve a significant improvement over DABAST-CSVD (only 4.3% improvement). This can be explained as follows. As most of the cached video segments are downloaded during high traffic load. These segments are usually downloaded with low video bit rate. By the time the second set of video streams starts, there will be many segments available in the clusters' caches. However, most of these segments have low video bit rate. As DABAST sends available segments from the clusters'

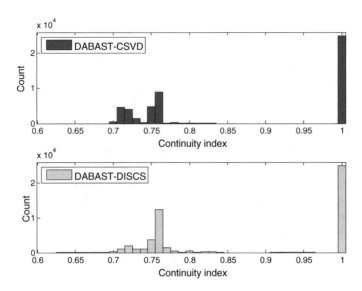

Fig. 13 Histogram of the continuity-index (DABAST-CSVD and DABAST-DISCS)

caches over D2D links (in the case an SM is available), most of the segments transmitted over the D2D channel will be sent from the distributed cache with low video bit rate. Although these segments are transmitted with higher data rates in the case of DABAST-DISCS, this does not increase the video bit rate for these segments. DABAST operates in this fashion to save the valuable cellular resources and exploit them for sending video segments that are not available in the clusters' caches to avoid rebufferings as much as possible.

A cached video segment is sent over the cellular channel only when there are no SMs available to send the segment. These video segments that are sent over cellular resources (despite being cached) will usually be sent with high video bit rate as such segments are usually requested with high video bit rate. This is because streams with cached video segments usually have long playout buffer length, as most of their segments are sent over the D2D channel with higher data rates. Furthermore, such video streams have even longer playout buffer length in the case of DABAST-DISCS, as video segments are sent with higher data rate than these with DABAST-CSVD. Because of that, video segments that are sent over cellular resources (in spite of being cached) will be sent with higher video bit rate in the case of DABAST-DISCS when compared to DABAST-CSVD. This explains the small improvement achieved by DABAST-DISCS over DABAST-CSVD in terms of the average video bit rate.

This behavior, explained above, can be seen in Table 6, which shows the number of video segments received with each video bit rate, for DABAST-CSVD and DABAST-DISCS. Table 6 shows that in the case of DABAST-DISCS, fewer segments are received with video bit rate of 768 kbps, and more segments (about the double) are received with 2Mbps and 4Mbps video bit rates, when compared to DABAST-CSVD. While this is beneficial for the streams that receive video segments with high video

Table 6 Count of the received segments with each video bit rate

Video bit rate (kbps)	Count	
	DABAST-CSVD	DABAST-DISCS
384	2,077,111	2,089,649
768	149,365	105,874
2000	19,273	46,179
4000	4251	8298

bit rates, it increases the RAN bottleneck and decreases the average data rate for other UEs receiving video segments exclusively over cellular resources.

6 Conclusion

In this chapter, we give a brief introduction on Device-to-Device (D2D) communication; a technology that allows direct communication between devices in cellular networks. Moreover, we review the work in literature on utilizing D2D communication in one of the most bandwidth-demanding applications in 5G networks, i.e., video streaming. We also discus an architecture we proposed that provides Dynamic Adaptive Streaming over HTTP (DASH)—based Peer-to-Peer (P2P) video streaming in cellular networks. The architecture employs Base-Station (BS) assisted D2D video transmission in cellular networks for direct exchange of video contents among users. We discuss in detail some performance evaluation results for the proposed architecture, which show that the proposed architecture can achieve significant performance gains in terms of video streaming Quality of Experience (QoE).

Acknowledgements The authors would like to thank Professor Stenio Fernandes from the Federal University of Pernambuco (UFPE), Brazil, for his valuable input during this work.

References

1. Cisco: Cisco Visual Networking Index: Global Mobile Data Traffic Forecast Update (2017). https://www.cisco.com/c/en/us/solutions/collateral/service-provider/visual-networking-index-vni/mobile-white-paper-c11-520862.html. Accessed 28 Feb 2018
2. Sandvine: 2016 Global Internet Phenomena: Latin America and North America. Technical report (2016)
3. Al-Habashna, A., Wainer, G., Boudreau, G., Casselman, R.: Cached and segmented video download for wireless video transmission. In: Proceedings of the 49th Annual Simulation Symposium, pp 18–25. Pasadena, USA (2016)
4. Al-Habashna, A., Wainer, G., Boudreau, G., Casselman, R.: Distributed cached and segmented video download for video transmission in cellular networks. In 2016 International Symposium on Performance Evaluation of Computer and Telecommunication Systems, pp. 473–480. IEEE, Montreal, Canada (2016)

5. Al-Habashna, A., Wainer, G.: Improving video transmission in cellular networks with cached and segmented video download algorithms. Mob. Netw. Appl **23**, 543–559 (2017). https://doi.org/10.1007/s11036-017-0906-x
6. Parkvall, S., Astely, D.: The evolution of LTE towards IMT-advanced. J. Commun. **4**, 146–154 (2009)
7. IOS: Information Technology—Dynamic Adaptive Streaming over HTTP (DASH)-Part 1: Media Presentation Description and Segment Formats. ISO/IEC 23009-1:2012 (2012)
8. Wainer, G.A.: Discrete-Event Modeling and Simulation: A Practitioner's Approach. CRC Press, Boca Raton (2009)
9. Agiwal, M., Roy, A., Saxena, N.: Next generation 5G wireless networks: a comprehensive survey. IEEE Commun. Surv. Tutorials **18**, 1617–1655 (2016). https://doi.org/10.1109/COMST.2016.2532458
10. Asadi, A., Wang, Q., Mancuso, V.: A survey on device-to-device communication in cellular networks. IEEE Commun. Surv. Tutorials **16**, 1801–1819 (2014). https://doi.org/10.1109/COMST.2014.2319555
11. Kaufman, B., Aazhang, B.: Cellular networks with an overlaid device to device network. In The 42nd Asilomar Conference on Signals, Systems and Computers, pp. 1537–1541. IEEE, Pacific Grove, USA (2008)
12. Doppler, K., Rinne, M.P., Janis, P., et al.: Device-to-device communications; functional prospects for LTE-advanced networks. In: IEEE International Conference on Communications Workshops, pp. 1–6. IEEE, Dresden, Germany (2009)
13. Duan, L., Gao, L., Huang, J.: Cooperative spectrum sharing: a contract-based approach. IEEE Trans. Mob. Comput. **13**, 174–187 (2014). https://doi.org/10.1109/TMC.2012.231
14. Zhang, Y., Song, L., Saad, W., et al.: Contract-based incentive mechanisms for device-to-device communications in cellular networks. IEEE J. Sel. Areas Commun. **33**, 2144–2155 (2015). https://doi.org/10.1109/JSAC.2015.2435356
15. 3GPP: Policy and Charging Control Signaling Flows and Quality of Service (QoS) Parameter Mapping. Technical report TS 29.213 (2016)
16. Li, B., Wang, Z., Liu, J., Zhu, W.: Two decades of internet video streaming. ACM Trans. Multimed. Comput. Commun. Appl. **9**, 1–20 (2013). https://doi.org/10.1145/2505805
17. Stockhammer, T.: Dynamic adaptive streaming over HTTP: Standards and design principles. In: Proceedings of the 2nd Annual ACM Conference on Multimedia Systems, pp. 133–144. ACM Press, New York, USA (2011)
18. Seufert, M., Egger, S., Slanina, M., et al.: A survey on quality of experience of HTTP adaptive streaming. IEEE Commun. Surv. Tutorials **17**, 469–492 (2015). https://doi.org/10.1109/COMST.2014.2360940
19. De Cicco, L., Mascolo, S.: An adaptive video streaming control system: modeling, validation, and performance evaluation. IEEE/ACM Trans. Netw. **22**, 526–539 (2014). https://doi.org/10.1109/TNET.2013.2253797
20. Zhao, J., Zhang, P., Cao, G., Das, C.R.: Cooperative caching in wireless P2P networks: design, implementation, and evaluation. IEEE Trans. Parallel Distrib. Syst. **21**, 229–241 (2010). https://doi.org/10.1109/TPDS.2009.50
21. Doppler, K., Rinne, M., Wijting, C., et al.: Device-to-device communication as an underlay to LTE-advanced networks. IEEE Commun. Mag. **47**, 42–49 (2009). https://doi.org/10.1109/MCOM.2009.5350367
22. Eittenberger, P.M., Herbst, M., Krieger, U.R.: RapidStream: P2P streaming on android. In: 2012 19th International Packet Video Workshop, pp. 125–130. IEEE, Munich, Germany (2012)
23. Siris, V.A., Dimopoulos, D.: Multi-source mobile video streaming with proactive caching and D2D communication. In: IEEE 16th International Symposium on A World of Wireless, Mobile and Multimedia Networks, pp. 1–6. IEEE, Boston, USA (2015)
24. Keller, L., Le, A., Cici, B., et al.: MicroCast: cooperative video streaming on smartphones. In: The 10th International Conference on Mobile Systems, Applications, and Services, pp. 57–70. New York, USA (2012)

25. Duong, T.Q., Vo, N.-S., Nguyen, T.-H., et al.: Energy-aware rate and description allocation optimized video streaming for mobile D2D communications. In: 2015 IEEE International Conference on Communications, pp. 6791–6796. IEEE, London, UK (2015)

26. Al-Habashna, A., Wainer, G.: QoE awareness in progressive caching and DASH-based D2D video streaming in cellular networks. Wirel. Netw. **26**, 2051–2073 (2020). https://doi.org/10.1007/s11276-019-02055-x

27. Al-Habashna, A., Fernandes, S., Wainer, G.: DASH-based peer-to-peer video streaming in cellular networks. In: 2016 International Symposium on Performance Evaluation of Computer and Telecommunication Systems, pp. 481–488. IEEE, Montreal, Canada (2016)

28. Al-Habashna, A., Wainer, G., Fernandes, S.: Improving video streaming over cellular networks with DASH-based device-to-device streaming. In: 2017 International Symposium on Performance Evaluation of Computer and Telecommunication Systems, pp. 468–475. IEEE, Seattle, USA (2017)

29. Zeigler, B.P., Praehofer, H., Kim, T.G.: Theory of Modeling and Simulation: Integrating Discrete Event and Continuous Complex Dynamic Systems. Academic Press, San Diego (2000)

30. De Cicco, L., Caldaralo, V., Palmisano, V., Mascolo, S.: ELASTIC: a client-side controller for dynamic adaptive streaming over HTTP (DASH). In: 20th International Packet Video Workshop, pp. 1–8. IEEE, San Jose, USA (2013)

31. Huang, T.-Y., Johari, R., McKeown, N., et al.: A buffer-based approach to rate adaptation: evidence from a large video streaming service. In: Proceedings of the 2014 ACM SIGCOMM, pp. 187–198. ACM, New York, USA (2014)

32. 3GPP: Evolved Universal Terrestrial Radio Access; RF System Scenarios. Technical report TR36.942 (2015)

33. Al-Hourani, A., Chandrasekharan, S., Kandeepan, S.: Path loss study for millimeter wave device-to-device communications in urban environment. In: 2014 IEEE International Conference on Communications Workshops, pp. 102–107. IEEE, Sydney, Australia (2014)

34. Cha, M., Kwak, H., Rodriguez, P., et al.: I tube, you tube, everybody tubes. In: Proceedings of the 7th ACM SIGCOMM Conference on Internet Measurement, pp. 1–14. ACM Press, New York, USA (2007)

35. Ahsan, S., Singh, V., Ott, J.: Characterizing Internet Video for Large-Scale Active Measurements. arXiv Prepr arXiv:14085777v1 (2014)

36. ITU-T: Infrastructure of Audiovisual Servicescoding of Moving Video. ITU-T Recommendation H.264 (2012)

5G Green Network

X. Ge, J. Yang, and J. Ye

Abstract Compared with conventional regular hexagonal cellular models, random cellular network models resemble real cellular networks much more closely. However, most studies of random cellular networks are based on the Poisson point process (PPP) and do not take into account the fact that adjacent base stations (BSs) should be separated with a minimum distance to avoid strong interference among each other BSs. Moreover, the user distribution in ultra-dense networks (UDNs) plays a crucial role in affecting the performance of UDNs due to the essential coupling between the traffic and the service provided by the networks. Existing studies are mostly based on the assumption that users are uniformly distributed in space. The non-uniform user distribution has not been widely considered despite that it is much closer to the real scenario. This chapter proposes a multi-user multi-antenna random cellular network model with the aforementioned minimum distance constraint for adjacent BSs, based on the hardcore point process (HCPP). A spectrum efficiency model and an energy efficiency model are presented based on the random cellular network model, and the maximum achievable energy efficiency of the considered multi-user multi-antenna HCPP random cellular networks is investigated. Moreover, a radiation and absorbing model (R&A model) is first adopted to analyze the impact of the nonuniformly distributed users on the performance of 5G UDNs. Based on the R&A model and queueing network theory, the stationary user density in each hot area is investigated. Simulation results demonstrate that the energy efficiency of conventional PPP cellular networks is underestimated when the minimum distance between adjacent BSs is ignored. Furthermore, the simulation results indicate that non-uniform user distribution has a significant impact on the performance of UDNs, compared with the uniformly distributed assumption.

Keywords Energy efficiency · Random cellular networks · Ultra-dense networks · Mobility

X. Ge (✉) · J. Yang · J. Ye
Huazhong University of Science and Technology, Wuhan, China
e-mail: xhge@mail.hust.edu.cn

© The Author(s), under exclusive license to Springer Nature Switzerland AG 2022
P. Nicopolitidis et al. (eds.), *Advances in Computing, Informatics, Networking and Cybersecurity*, Lecture Notes in Networks and Systems 289,
https://doi.org/10.1007/978-3-030-87049-2_13

1 Introduction

With the rapid growth of wireless traffic over the last decade, multiple-input multi-output (MIMO) antenna technology has been widely adopted to satisfy the high traffic requirement in the fourth generation (4G) and future fifth generation (5G) cellular networks [1, 2]. On the other hand, energy consumption in cellular networks has been increasing dramatically because of the increasing number of antennas and the increasing wireless traffic [3, 4]. By 2011, there were more than 4 million base stations (BSs) operating in cellular networks, each consuming an average of 25 MWh per year [5]. Therefore, it is important to investigate and improve the energy efficiency of multi-user multi-antenna cellular networks.

Moreover, 5G mobile communication systems are envisaged to provide a 1000 times enhancement of the network capacity while achieving a much higher energy efficiency than the fourth generation (4G) mobile communication systems. The ambitious aims of 5G mobile communication systems bring both opportunities and challenges to researchers all over the world [6, 7]. The UDNs are regarded as one of the key technologies for 5G mobile communication systems [8]. The main difference between UDNs and heterogeneous networks (HetNets) lies in the dramatic increase of small cell base station (SBS) density. The distances between users and SBSs are greatly reduced with the increase of the SBS density. Hence more wireless links are available for users in wireless networks to enhance the quality of service (QoS) [9, 10]. On the other hand, UDNs also suffer from the increasing of energy consumption with the massive deployment of SBSs. Therefore, one of the core problems for deploying UDNs is optimizing SBS density to meet the traffic demand in an energy-efficient way in hot spot areas.

2 Energy Efficiency of Random Cellular Networks

2.1 Related Work

Numerous energy efficiency models for MIMO wireless communication systems have been proposed in the literature [11–19]. A closed-form approximation for the energy efficiency-spectrum efficiency trade-off has been derived for the MIMO Rayleigh fading channel in [11]. The simulation results in [11]indicated that the energy efficiency can be effectively improved through receive diversity in the very low spectrum efficiency regime and that MIMO systems are more energy efficient than single-input single-output (SISO) systems in the high spectrum efficiency regime. Furthermore, the energy efficiency gain of MIMO over SISO systems was analyzed for various power consumption models at transmitters in [12]. The MIMO transmission energy efficiency was analyzed for wireless sensor networks considering both the diversity gain and the multiplexing gain in [13]. A precoding matrix was optimized to maximize the energy efficiency of wireless communication systems for single-user MIMO channels in [14]. The energy efficiency optimization was investigated by adaptively adjusting the bandwidth, transmission power, and precoding mode in downlink MIMO systems in [15]. Optimizations on the transmit covariance precoding matrix and active transmit antenna selection were proposed to improve the energy efficiency for MIMO broadcast channels in [16]. Based on the distributed singular value decomposition (SVD) of multi-user channels, a power allocation scheme was presented to achieve the optimal energy efficiency in multi-user MIMO systems in [17]. The trade-off between energy efficiency and spectrum efficiency was quantified for very-large multi-user MIMO systems with small-scale fading channels in [18]. However, most energy efficiency studies of MIMO systems have focused on the link level and single cell scenarios. The case of multi-cell MIMO was also studied in the literature (see, e.g. [18], where the conventional regular hexagonal cell model was adopted). Nonetheless, the investigation of the energy efficiency of realistic multi-user multi-antenna networks by following the well-known stochastic geometry approach still remains unexplored.

It is well known that the locations of the transmitters and receivers are very important for the performance of wireless communication systems. In the literature, there are published works that consider a location model in random cellular networks. In [20, 21], a QoE-driven approach based on a novel mobile cloud computing architecture is proposed to extensively improve the energy efficiency of wireless communication systems. The most popular random process used for wireless network models is the Poisson point process (PPP) [22]. Pioneering results on random wireless networks were reported in [23, 24]. The detailed mathematical definition and the properties of the PPP in random wireless networks were discussed in [25]. In [26], a density of success transmissions was derived for the downlink cellular network where the locations of the BSs were governed by a homogeneous PPP. Based on the stochastic geometry and the PPP theory, a simple single integral model for the average rate of random cellular networks was derived in [27], which is useful for performance analysis. In [28], a comprehensive mathematical framework was proposed for the analysis of the average rate of multi-tier cellular networks whose BSs are assumed to follow a PPP distribution. A signal-to-interference-plus-noise ratio (SINR) model was derived for multi-tier cellular networks where the locations of the BSs followed a PPP [29]. For Poisson distributed multi-antenna BSs, an approximation for the area-averaged spectral efficiency of a representative link was derived in [30]. Assuming that each tier of BSs was modeled by an independent homogeneous PPP, a tractable downlink model for multi-antenna heterogeneous cellular networks was proposed in [31]. Adopting the PPP for the distribution of the BSs, success probability and energy efficiency models for homogeneous single-tier macrocell and heterogeneous multi-tier small cell networks were derived under different sleeping policies in [32].

In addition to the PPP, other random point processes have also been used for modeling and performance analysis of wireless networks in [33–36]. For a wireless network with a finite and fixed number of nodes, a closed-form analytical expression for the moment generation function of the interference was derived based on the binomial point process (BPP) in [33]. For the case where node locations of clustered wireless ad hoc networks are assumed to form a Poisson cluster process (PCP), in

[34] the distribution properties of interference were derived for analyzing the outage probability. Furthermore, considering a minimal distance constraint between nodes in random carrier-sense multiple access (CSMA) wireless networks, a modified hard core point process (HCPP) was used for modeling the spatial distribution of the simultaneously active users in [35]. Different spatial stochastic models including the PPP, the HCPP, the Strauss process (SP) and the perturbed triangular lattice were compared for modeling the spatial distribution of BSs in cellular networks and it was proven that the HCPP is more realistic than the PPP for modeling the spatial structure of BSs [36]. However, a detailed investigation of the performance of multi-user multi-antenna cellular networks under the HCPP is not available in the literature. Therefore, motivated by the above gaps in the literature, in this paper, we derive the average energy efficiency of multi-user multi-antenna cellular networks with HCPP distributed BSs.

2.2 System Model

Compared with the regular hexagonal cell structure, random cellular networks are more coincident with real deployments of cellular networks. The PPP theory has been widely used for modeling of random cellular networks [23]. However, there is no constraint for the distance between two points in PPPs. Consequently, there exist scenarios where two adjacent BSs are infinitesimally close to each other for PPP random cellular networks. In this case, the interference from adjacent BSs will approach infinity when the interfering BSs are infinitesimally close to the desired BS in a PPP random cellular network. In realistic BSs deployments in cellular networks, two arbitrary BSs cannot be infinitesimally closed to each other. In general, telecommunication providers always ask that the location of two adjacent BSs must keep a protect distance or a minimum distance to avoid obvious interference. Hence, there exists a conflict for the minimum distance constraint between two arbitrary BSs in realistic BSs deployment and PPP random cellular networks. This result will conduce to the deviation for interference and energy efficiency analysis for random cellular networks, which are illustrated in Figs. 2 and 11. To solve this drawback of PPP random cellular networks, we propose to model random cellular networks based on the HCPP theory. In the following, we introduce the channel model, the traffic model and the HCPP, which has been shown to be more realistic in modeling the deployments of cellular networks than the PPP.

In the literature, the PPP is widely used for modeling random cellular networks, mainly because it leads to a mathematically tractable analysis [26–32, 37]. However, some aspects associated with the PPP analysis may render it inadequate for modeling certain realistic cellular deployments. For example, in downlink interference models of PPP random cellular networks, the locations of the interfering BSs can approach that of the desired BS arbitrarily close. As a result, the mean of the aggregated interference approaches infinity [37]. This result not only increases the model complexity but also deviates from reality. To avoid this extreme result, we base our random cellular network model on the HCPP theory [38–40]. We note that this theory has been used for the modeling of carrier sense multiple access (CSMA) networks with a specified minimum distance between adjacent wireless nodes [41]. HCPP generates patterns produced by points that have a minimum distance δ from each other. The Matern hard-core process of Type II, which represents a special case of HCPP, is essentially a stationary PPP Π_{PPP}, i.e., the Poisson point process of intensity λ_P, to which a dependent thinning is applied [40]. The thinned process Π_{HCPP}, i.e., the Matérn hard-core process is expressed as [40]

$$\Pi_{HCPP} = \left\{ x \in \Pi_{PPP} : \Phi(x) < \Phi(x^*) \text{ for all } x^* \text{ in } \Pi_{PPP} \cap d(x, \delta) \right\} \quad (1)$$

The points of Π_{PPP} are marked with random numbers uniformly distributed in $[0, 1]$ independently. The dependent thinning retains the point x of Π_{PPP} with mark $\Phi(x)$ if the disk $d(x, \delta)$ contains no points of Π_{PPP} with marks smaller than $\Phi(x)$, where $d(x, \delta)$ is a disk area with central point x and the radius δ.

We assume that both BSs and user equipments (UEs) are randomly located in the infinite plane $\mathbb{R}^2$. Moreover, the UEs' motions are isotropic and relatively slow, such that during an observation period, e.g., a time slot, the relative positions of BSs and UEs are stationary. The distribution of the BSs is assumed to be governed by a thinned process Π_{HCPP} applied to a stationary PPP Π_{PPP} of intensity λ_P. The locations of the BSs are denoted by $\Pi_{BS} = \{x_{BS_i} : i = 1, 2, 3, \ldots\}$, where x_{BS_i} are the two-dimensional Cartesian coordinates that denote the location of BS BS_i. The distribution of UEs is assumed to be governed by a PPP with intensity λ_M.

We assume that the BS and UE are equipped with N_T and N_R antennas, respectively. We also assume that each UE connects to the closest BS, which corresponds to the smallest path loss during wireless transmission. In this paper, our studies are focused on the downlink of random cellular networks. The large scale fading coefficients of the UEs in a cell are assumed to be identical to each other. The channel matrix between UEs and BS is modelled as

$$\mathbf{H} = \sqrt{\frac{\beta}{\mathfrak{R}^\alpha}} w \mathbf{h} \quad (2)$$

where $\mathbf{H}$ is the channel matrix, β is a constant depending on the antenna gain, $\mathfrak{R}$ is the distance between the transmitter and receiver, and α is the path loss coefficient. Furthermore, w models the log-normal shadowing effect in wireless channels and is given by $w = 10^{s/10}$, where s is a Gaussian distributed random variable with zero mean and variance σ_s^2, $s \sim N(0, \sigma_s^2)$. Additionally, $\mathbf{h}$ is the small scale fading channel matrix, whose elements are modelled as independent and identically distributed (i.i.d.) Gaussian random variables with zero mean and unit variance.

In early studies [42], the Poisson distribution was adopted for traffic modeling of cellular networks. Based on empirical measurement results [43], the traffic load of cellular networks has been demonstrated to have the self-similar characteristic which means the variance of similar network traffic approaches to infinity. To model the cellular network traffic with self-similar characteristic, several mathematical distributions with the infinite variance have been proposed to fit the self-similar network traffic [44–46]. Considering the analytical expression and intuitionistic engineering implication of function parameters, e.g., the traffic rate ρ, the Pareto distribution has been widely used for similar cellular network traffic modeling [47]. Without loss of generality, the Pareto distribution has been adopted for the cellular network traffic in this study. Moreover, the traffic rates of all UEs are assumed to be *i.i.d.* The probability density function (PDF) of traffic rate in cellular networks is given by

$$f_\rho(\chi) = \frac{\theta \rho_{\min}^\theta}{\chi^{\theta+1}}, \chi \ge \rho_{\min} > 0 \tag{3}$$

where $\theta \in (1, 2]$ is the heaviness index which reflects the heaviness of the distribution tail, and $\rho_{\min}$ denotes the minimum traffic rate which is configured to guarantee the user requirement in data transmission rate. We note that the heaviness index, θ, affects the distribution tail of the traffic rate, so that when θ approaches 1 the tail becomes the dominant part of the distribution. The average traffic rate at UEs is obtained as

$$\mathbb{E}(\rho) = \frac{\theta \rho_{\min}}{\theta - 1} \tag{4}$$

where $\mathbb{E}(\cdot)$ denotes the expectation operation.

2.3   Hard-Core Point Process Cellular Networks

The interference from BSs affects the transmission rate between the BS and UE. Furthermore, the interference from BSs has to be considered to evaluate wireless propagation environments. To simplify the analysis, an average interference model has been proposed for HCPP cellular networks in this section. Moreover, the impacts of the distance between the UE and desired BS, the minimum BSs distance and path loss coefficient on the average interference of HCPP cellular networks are analyzed by numerical simulations.

In Fig. 1a, based on the HCPP model, a simulation-based illustration of the distribution of the BSs for the considered system is shown. The blue nodes represent BSs and are distributed according to a PPP with intensity $\lambda_P = 1/(\pi * 800^2)$. The minimum distance is set to $\delta = 500$ m. The nodes whose marked values do not satisfy the condition in (1) are marked by red circles. As mentioned in the HCPP model in section II, these red circle points are discarded from the analysis. As a consequence, the BSs that are included in the HCPP model are the nodes marked with blue, whose distance from adjacent nodes is larger than or equal to δ.

In a hard core point distribution, a point with mark t, $t \in [0, 1]$, is retained only when there are no other points with a smaller mark at a distance less than the hard core distance δ. In a Poisson point distribution, given that the mean of point number

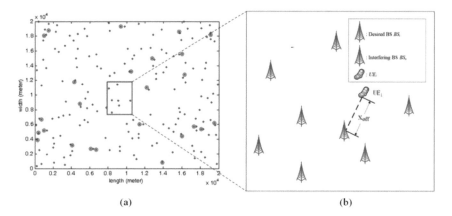

Fig. 1 HCPP BSs distribution

is $\lambda_P \pi \delta^2$ in a circle with radius δ, the probability of points is expressed as

$$\Pr\{\text{poisson point number in } \pi\delta^2 \text{ area is } \kappa\} = \frac{(\lambda_P \pi \delta^2)^\kappa e^{-\lambda_P \pi \delta^2}}{\kappa!}. \tag{5}$$

The probability that the point with mark t is retained in a Poisson point distribution is derived as

$$\Pr\{\text{A point with mark } t \text{ is retained}\}$$

$$= \sum_{\kappa=0}^{\infty} \frac{(\lambda_P \pi \delta^2)^\kappa e^{-\lambda_P \pi \delta^2}}{\kappa!} (1-t)^\kappa$$

$$= \sum_{\kappa=0}^{\infty} \frac{[\lambda_P \pi \delta^2 (1-t)]^\kappa e^{-\lambda_P \pi \delta^2}}{\kappa!} = e^{-\lambda_P \pi \delta^2 t}. \tag{6}$$

As a result, the retaining probability for a typical point can be calculated by integrating $e^{-\lambda_P \pi \delta^2 t}$ in the interval $t \in [0, 1]$. At the location x, the probability that there is a point in the infinitesimal small region dx is $\zeta^{(1)} dx$. The first moment of HCPP is expressed by

$$\zeta^{(1)} = \lambda_P \int_0^1 e^{-\lambda_P \pi \delta^2 t} dt = \frac{1 - e^{-\lambda_P \pi \delta^2}}{\pi \delta^2}. \tag{7}$$

When two points with mark t_1 and t_2 are located at two differential regions dx_1 and dx_2 in a PPP distribution, the probability that two points are retained depends only on the distance r between two points. Moreover, if two circles with the same radius δ are separated by r, the area of the union of two circles $V_\delta(r)$ is derived as

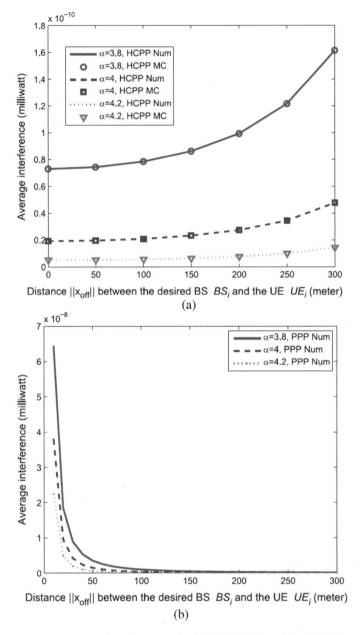

Fig. 2 Impact of the distance $\|x_{off}\|$ between the desired BS, BS_i, and the UE of interest, UE_i, on the average interference of HCPP cellular networks. **a** Average interference in HCPP model. **b** Average interference in PPP model

[38]

$$V_\delta(r) = \begin{cases} 2\pi\delta^2 - 2\delta^2 \arccos(\frac{r}{2\delta}) + r\sqrt{\delta^2 - \frac{r^2}{4}}, & 2\delta > r > 0 \\ 2\pi\delta^2, & r \geq 2\delta \end{cases} \tag{8}$$

Using (6) and (7), the probability that two points with stamp marks t_1 and t_2 are retained can be derived by considering the case of $r > \delta$ and $r \leq \delta$ independently, i.e.,

$$\begin{aligned} \varphi(r) &= \int_0^1 e^{-\lambda_P t_1 \pi \delta^2} \int_0^{t_1} e^{-\lambda_P t_2 [V_\delta(r) - \pi\delta^2]} dt_2 dt_1 \\ &+ \int_0^1 e^{-\lambda_P t_2 \pi \delta^2} \int_0^{t_2} e^{-\lambda_P t_1 [V_\delta(r) - \pi\delta^2]} dt_1 dt_2 \\ &= \begin{cases} \frac{2V_\delta(r)(1-e^{-\lambda_P \pi\delta^2}) - 2\pi\delta^2(1-e^{-\lambda_P V_\delta(r)})}{\lambda_P^2 \pi\delta^2 V_\delta(r)[V_\delta(r) - \pi\delta^2]}, & r > \delta \\ 0, & r \leq \delta. \end{cases} \end{aligned} \tag{9}$$

Furthermore, the second moment of HCPP is expressed by

$$\zeta^{(2)}(r) = \lambda_p^2 \varphi(r). \tag{10}$$

As a consequence, the probability that the distance between two points which are located in the infinitesimally small regions, dx_1 and dx_2, respectively, equals r, is expressed as $\zeta^{(2)}(r)dx_1 dx_2$.

Let us denote the desired BS and UE by BS_i and UE_i, respectively. Based on the system model in Fig. 1b, the location vector of desired BS BS_i is denoted by x_{BS_i}, the distance vector between UE_i and BS_i is denoted by x_{off}. Moreover, x_{BS_u} is the two-dimensional Cartesian coordinate of interfering BS, denoted by BS_u. Considering the impact of the distance between the desired BS and the received UE [48], the aggregated interference at UE_i is expressed as

$$I_i(x_{off}) = \sum_{u \neq i, BS_u \in \Pi_{BS}} g(||x_{BS_u} - x_{BS_i} - x_{off}||, \psi_{iu}), \tag{11}$$

where g is defined as the interference function between BS_u and UE_i, $|| \cdot ||$ is the modular operation, i.e., the Euclid distance operation. Furthermore, ψ_{iu} is the fading factor over wireless channels, given as $\psi_{iu} = \{w_{iu}, \mathbf{h}_{iu}, P_u\}$, which includes the shadowing effect w_{iu}, the small scale fading matrix $\mathbf{h}_{iu}$ and the transmission power P_u at BS_u. Based on the channel model introduced in Section II, the aggregated interference at UE_i can be extended as

$$I_i(x_{\text{off}}) = \sum_{u \neq i, \text{BS}_u \in \Pi_{\text{BS}}} \frac{\beta w_{iu} |z_{iu}|^2 P_u}{\|x_{\text{BS}_u} - x_{\text{BS}_i} - x_{\text{off}}\|^\alpha}, \tag{12}$$

where z_{iu} is the small scale fading between the received UE UE_i and the interfering BS BS_u with single antenna and is governed by a complex Gaussian distribution with mean value that equals 1.

Considering every active UE is associated with a BS and all UEs are traversed in the plane $\mathbb{R}^2$, the total interference in the plane $\mathbb{R}^2$ is given by

$$\sum_{BS_i \in \Pi_{\text{BS}}} I_i(x_{\text{off}}) = \sum_{\text{BS}_i \in \Pi_{\text{BS}}} \sum_{u \neq i, \text{BS}_u \in \Pi_{\text{BS}}} g(\|x_{\text{BS}_u} - x_{\text{BS}_i} - x_{\text{off}}\|, \psi_{iu}). \tag{13}$$

Based on the second moment of the HCPP in (10) and the corresponding properties [49], the expectation of the total interference in the plane $\mathbb{R}^2$ is derived as

$$\mathbb{E}[\sum_{\text{BS}_i \in \Pi_{BS}} I_i(x_{\text{off}})]$$

$$= \int_{\mathbb{R}^2} \int_{\mathbb{R}^2} \{E_{\psi_{iu}}[g(\|x_1 - x_2 - x_{\text{off}}\|, \psi_{iu})] \times \zeta^{(2)}(\|x_1 - x_2 - x_{\text{off}}\|)\} dx_1 dx_2. \tag{14}$$

Based on the first moment of HCPP in (7), the average BS number in the plane $\mathbb{R}^2$ is expressed as $\int_{\mathbb{R}^2} \zeta^{(1)} dx$. Without loss of generality, the aggregated interference $\sum_{\text{BS}_i \in \Pi_{\text{BS}}} I_i(x_{\text{off}})$ can be calculated by the distance among the BSs $\|x_{\text{BS}_u} - x_{\text{BS}_i}\|$ and the distance between the desired BS and the received UE $\|x_{\text{off}}\|$. Let $x = x_{\text{BS}_u} - x_{\text{BS}_i}$ be the distance vector among the BSs in HCPP cellular networks, the average interference of HCPP cellular networks is derived as

$$I_{i\_avg} = \frac{\int_{\mathbb{R}^2} \int_{\mathbb{R}^2} \{\mathbb{E}_{\psi_{iu}}[g(\|x_1 - x_2 - x_{\text{off}}\|), \psi_{iu}] \times \zeta^{(2)}(\|x_1 - x_2 - x_{\text{off}}\|)\} dx_1 dx_2}{\int_{\mathbb{R}^2} \zeta^{(1)} dx}$$

$$= \frac{\int_{\mathbb{R}^2} dx_1 \int_{\mathbb{R}^2} \{\mathbb{E}_{\psi_{iu}}[g(\| - x_2 - x_{\text{off}}\|), \psi_{iu}] \times \zeta^{(2)}(\| - x_2 - x_{\text{off}}\|)\} dx_2}{\zeta^{(1)} \int_{\mathbb{R}^2} dx}$$

$$= \frac{1}{\zeta^{(1)}} \int_{\mathbb{R}^2} \mathbb{E}_{\psi_{iu}}[g(\|x + x_{\text{off}}\|), \psi_{iu}] \zeta^{(2)}(\|x + x_{\text{off}}\|) dx. \tag{15}$$

Substituting (11) and (12) into (15), the average interference of HCPP cellular networks is derived as

$$I_{i\_avg}(x_{\text{off}}) = \frac{\beta \mathbb{E}(w_{iu}) \mathbb{E}(|z_{iu}|^2) \mathbb{E}(P_u)}{\zeta^{(1)}} \times \int_{\mathbb{R}^2} \frac{1}{\|x + x_{\text{off}}\|^\alpha} \zeta^{(2)}(\|x\|) dx, \tag{16}$$

where the distance x_{off} between the transmitter and the receiver is considered to evaluate the average interference of HCPP cellular networks.

Based on (16), the impacts of the distance between the user and the desired BS, the path loss coefficient and the minimum distance on the average interference of HCPP cellular networks are numerically analyzed in detail. In the following analysis, the default parameters used for interference model are configured as follows: $\sigma_s = 6$, which usually ranges from 4 to 9 in practice [50]; $\alpha = 3.8$ and $\beta = -31.54$ dB, which correspond to an urban area with a rich scattering environment [47]; the average transmission power of interfering BS is set as $\mathbb{E}(P_u) = 2$ Watt (W) when the interference link bandwidth is configured as 10 kHz [51, 52]. Moreover, analysis results are confirmed by Monte-Carlo (MC) simulations in HCPP cellular networks.

Figure 2 illustrates the impact of the distance $\|x_{\mathrm{off}}\|$ between the desired BS BS$_i$ and the UE UE$_i$ on the average interference of HCPP and PPP cellular networks. When the path loss coefficient α is fixed, the average interference of HCPP cellular networks increases with increasing the distance $\|x_{\mathrm{off}}\|$ and the average interference of PPP cellular networks decreases with increasing the distance $\|x_{\mathrm{off}}\|$. When the distance $\|x_{\mathrm{off}}\|$ between the desired BS BS$_i$ and the UE UE$_i$ is fixed, both HCPP and PPP cellular networks indicate that the average interference increases with decreasing the path loss coefficient α.

Figure 3 shows the impact of the minimum distance on the average interference of HCPP cellular networks, in which "Num" labels numerical results and "MC" represents MC simulation results. When the distance $\|x_{\mathrm{off}}\|$ between the desired BS BS$_i$ and the UE UE$_i$ is fixed, the average interference decreases with increasing the minimum distance δ in HCPP cellular networks. When the minimum distance is

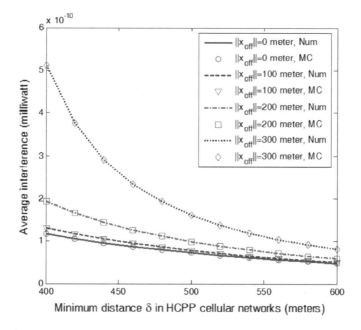

Fig. 3 Impact of the minimum distance δ on the average interference of HCPP cellular networks

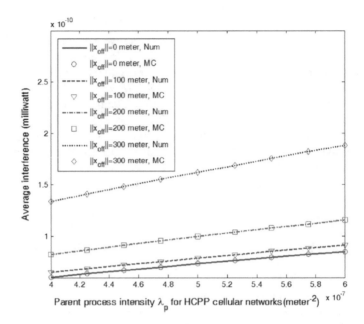

Fig. 4 Impact of the parent process intensity λ_P on the average interference of HCPP cellular networks

fixed, the average interference increases with increasing the distance $\|x_{\text{off}}\|$ in HCPP cellular networks. Based on simulation results in Fig. 3, the average interference is underestimated when the distance $\|x_{\text{off}}\|$ is ignored in HCPP cellular networks.

Figure 4 analyzes the impact of the parent process intensity λ_P on the average interference of multi-antenna HCPP cellular network. When the distance $\|x_{\text{off}}\|$ between the desired BS BS_i and the UE UE_i is fixed, the average interference increases with increasing the parent process intensity λ_P in multi-antenna HCPP cellular networks. When the parent process intensity λ_P is fixed, the average interference increases with increasing the distance $\|x_{\text{off}}\|$ between the desired BS BS_i and the UE UE_i in multi-antenna HCPP cellular networks. Simulation results in Fig. 4 indicate that the average interference is underestimated when the distance $\|x_{\text{off}}\|$ is ignored in HCPP cellular networks.

2.4 Spectrum and Energy Efficiency

In this paper, the zero-forcing precoding method is adopted for multi-user multi-antenna downlink systems. Based on the zero-forcing precoding method, the BS integrated with N_T antennas can simultaneously transmit signals to S active UEs in a cell, as illustrated in Fig. 5. All S UEs in a cell are grouped as a user equipment group (UEG). When each UE is equipped with a single antenna, the antenna number

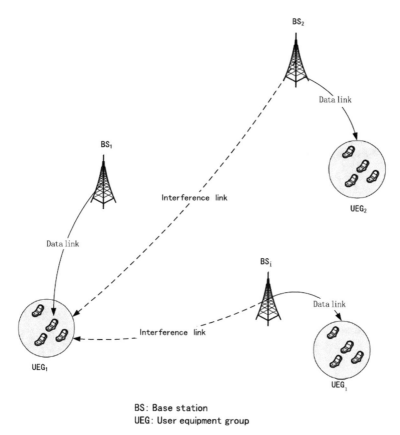

BS: Base station
UEG: User equipment group

Fig. 5 Downlink model of multi-user multi-antenna HCPP cellular networks

of UEG is S. We assume that the antenna number of BS is larger than or equal to the antenna number of UEG, i.e., $N_T \geq S$. The radius of cellular coverage in current cellular networks is about 200–400 m in the urban regions. When the protect distance, i.e., the minimum distance is configured in HCPP cellular networks, all active USs in a cell are located at a confined circular ring where has an approximated same distance to the desired BS. Considering that the large scale fading depends on the distances between all active UEs in a cell and the desired BS, the large scale fading in a cell can be regarded to be almost identical in HCPP cellular networks. This scenario where UEs in a cell are susceptible to the same large scale fading, was also adopted in [53] and identified as the homogeneous scenario. The location vector of UEG UEG_i associated with the BS BS_i is denoted as $x_{\text{BS}_i} + x_{\text{off}}$.

In Fig. 5, the signal vector $\mathbf{y}_i$ received by UEG_i can be expressed as

$$\mathbf{y}_i = \mathbf{H}_{ii}\mathbf{x}_i + \sum_{u \neq i, BS_u \in \Pi_{BS}} \mathbf{H}_{iu}\mathbf{x}_u + \mathbf{n}, \tag{17}$$

with $\mathbf{x}_i = \mathbf{F}_i\mathbf{s}_i$ and $\mathbf{x}_u = \mathbf{F}_u\mathbf{s}_u$, where $\mathbf{H}_{ii}$ is the channel matrix between BS_i and UEG_i, $\mathbf{x}_i$ is the signal vector from BS_i, $\mathbf{H}_{iu}$ is the channel matrix between BS_u and UEG_i, $\mathbf{x}_u$ is the signal vector from BS_u, and $\mathbf{n}$ is the noise vector. The noise power at each antenna is denoted by σ_n^2, the covariance matrix of noise vectors at UEG_i is denoted by $\mathbb{E}(\mathbf{n}\mathbf{n}^+) = \sigma_n^2 \mathbf{I}_{S\times S}$, where $\mathbf{I}_{S\times S}$ is the $S \times S$ identity matrix. Considering that zero-forcing precoding is adopted for all BSs, $\mathbf{F}_i$ and $\mathbf{F}_u$ are the $N_T \times S$ precoding matrixes used at BS_i and BS_u, respectively. $\mathbf{s}_i$ and $\mathbf{s}_u$ are the $S \times 1$ signal vectors at BS_i and BS_u, respectively.

For zero-forcing precoding, precoding matrix $\mathbf{F}_i$ can be expressed as

$$\mathbf{F}_i = \mathbf{H}_{ii}^+(\mathbf{H}_{ii}\mathbf{H}_{ii}^+)^{-1}, \tag{18}$$

where $(\cdot)^+$ and $(\cdot)^{-1}$ denote the conjugate transpose operation and inverse operation, respectively. Furthermore, the signal vector $\mathbf{y}_i$ received by UEG_i can be rewritten as

$$\mathbf{y}_i = \mathbf{s}_i + \sum_{u\neq i, BS_u \in \Pi_{BS}} \mathbf{H}_{iu}\mathbf{F}_u\mathbf{s}_u + \mathbf{n}. \tag{19}$$

Based on the zero-forcing precoding method [54], the transmission power of BS_i is expressed as

$$
\begin{aligned}
P_i &= \mathbb{E}(\mathbf{x}_i^+\mathbf{x}_i) \\
&= \mathbb{E}(\mathbf{s}_i^+\mathbf{F}_i^+\mathbf{F}_i\mathbf{s}_i) \\
&= \mathbb{E}\{\mathbf{s}_i^+[(\mathbf{H}_{ii}\mathbf{H}_{ii}^+)^{-1}]^+\mathbf{s}_i\} \\
&= \sum_{k=1}^{S} \mathbb{E}(\mathbf{s}_{i(k)}^+\mathbf{s}_{i(k)})(\mathbf{H}_{ii}\mathbf{H}_{ii}^+)_{(kk)}^{-1}, \\
&= \sum_{k=1}^{S} Q_{i(k)}(\mathbf{H}_{ii}\mathbf{H}_{ii}^+)_{(kk)}^{-1} \\
&= \sum_{k=1}^{S} P_{i(k)}
\end{aligned}
\tag{20}
$$

where $P_{i(k)}$ is the transmission power transmitted by BS_i over the kth sub-channel, $Q_{i(k)}$ is the UE received power transmitted from BS_i over the kth sub-channel, $(\mathbf{H}_{ii}\mathbf{H}_{ii}^+)_{(kk)}^{-1}$ is the element located at the kth row and the kth column in the matrix $(\mathbf{H}_{ii}\mathbf{H}_{ii}^+)^{-1}$. Considering $P_{i(k)} = \mathbb{E}(\mathbf{s}_{i(k)}^+\mathbf{s}_{i(k)})(\mathbf{H}_{ii}\mathbf{H}_{ii}^+)_{(kk)}^{-1}$, the UE received power transmitted from BS_i over the kth sub-channel is expressed by

$$
\begin{aligned}
Q_{i(k)} &= \mathbb{E}(\mathbf{s}_{i(k)}^+\mathbf{s}_{i(k)}) \\
&= \frac{P_{i(k)}}{(\mathbf{H}_{ii}\mathbf{H}_{ii}^+)_{(kk)}^{-1}}.
\end{aligned}
\tag{21}
$$

Assuming that the zero-forcing precoding is also adopted by the interfering BSs, the interference power over the kth sub-channel in UEG_i is obtained as

$$
\begin{aligned}
I_{i(k)}^{\text{ZF}} &= \mathbb{E}[(\sum_{u\neq i, \text{BS}_u \in \Pi_{\text{BS}}} \mathbf{H}_{iu}^{(k)} \mathbf{F}_u \mathbf{s}_u)^{+}(\sum_{u\neq i, \text{BS}_u \in \Pi_{\text{BS}}} \mathbf{H}_{iu}^{(k)} \mathbf{F}_u \mathbf{s}_u)] \\
&= \sum_{u\neq i, \text{BS}_u \in \Pi_{\text{BS}}} \sum_{k=1}^{S} \frac{P_{u(k)}}{(\mathbf{H}_{uu}\mathbf{H}_{uu}^{+})_{(kk)}^{-1}} (\mathbf{F}_u^{+}\mathbf{H}_{iu}^{(k)+}\mathbf{H}_{iu}^{(k)}\mathbf{F}_u)_{(kk)}, \\
&= \sum_{u\neq i, \text{BS}_u \in \Pi_{\text{BS}}} \sum_{k=1}^{S} \left[\frac{P_{u(k)}}{(\mathbf{H}_{uu}\mathbf{H}_{uu}^{+})_{(kk)}^{-1}} \right. \\
&\quad \left. \times \frac{\beta w_{iu}}{||x_{\text{BS}_u} - x_{\text{BS}_i} - x_{\text{off}}||^{\alpha}} \times (\mathbf{F}_u^{+}\mathbf{h}_{iu}^{(k)+}\mathbf{h}_{iu}^{(k)}\mathbf{F}_u)_{(kk)} \right]
\end{aligned}
\tag{22}
$$

where $\mathbf{H}_{iu}^{(k)}$ is the kth row of channel matrix $\mathbf{H}_{iu}$, which corresponds to the kth sub-channel in UEG_i; $P_{u(k)}$ is the transmission power transmitted by BS_u over the kth sub-channel. $\mathbf{h}_{iu}$ is the $S \times N_T$ small scale fading matrix between BS_u and UEG_i. Each element of $\mathbf{h}_{iu}$ is assumed to follow an *i.i.d.* complex Gaussian distribution with zero mean and unit variance.

Based on the derivation in (16), the average interference over the kth sub-channel in UEG_i is derived as

$$
\begin{aligned}
\mathbb{E}(I_{i(k)}^{\text{ZF}}) &= \mathbb{E}\left\{ \sum_{u\neq i, \text{BS}_u \in \Pi_{\text{BS}}} \sum_{k=1}^{S} \left[\frac{P_{u(k)}}{(\mathbf{H}_{uu}\mathbf{H}_{uu}^{+})_{(kk)}^{-1}} \times \frac{\beta w_{iu}}{||x_{\text{BS}_u} - x_{\text{BS}_i} - x_{\text{off}}||^{\alpha}} \times (\mathbf{F}_u^{+}\mathbf{h}_{iu}^{(k)+}\mathbf{h}_{iu}^{(k)}\mathbf{F}_u)_{kk} \right] \right\} \\
&= \mathbb{E}\left\{ \sum_{u\neq i, \text{BS}_u \in \Pi_{\text{BS}}} \sum_{k=1}^{S} \left\{ \frac{P_{u(k)}}{(\mathbf{H}_{uu}\mathbf{H}_{uu}^{+})_{(kk)}^{-1}} \times \frac{\beta w_{iu}}{||x_{\text{BS}_u} - x_{\text{BS}_i} - x_{\text{off}}||^{\alpha}} \times [\mathbf{F}_u^{+}\mathbb{E}(\mathbf{h}_{iu}^{(k)+}\mathbf{h}_{iu}^{(k)})\mathbf{F}_u]_{kk} \right\} \right\}.
\end{aligned}
\tag{23}
$$

Based on the result in [55], we have the following property

$$
\mathbb{E}(\mathbf{h}_{iu}^{(k)+}\mathbf{h}_{iu}^{(k)}) = \mathbf{I}_{S\times S}.
\tag{24}
$$

Therefore, combining (23) and (24), the average interference over the kth sub-channel in UEG_i is expressed by

$$\mathbb{E}(I_{i(k)}^{ZF}) = \mathbb{E}\left\{\sum_{u\neq i, BS_u \in \Pi_{BS}} \sum_{k=1}^{S} \left[\frac{P_{u(k)}}{(\mathbf{H}_{uu}\mathbf{H}_{uu}^{+})_{(kk)}^{-1}}\right.\right.$$

$$\left.\left.\times \frac{\beta w_{iu}}{||x_{BS_u} - x_{BS_i} - x_{off}||^{\alpha}} \times (\mathbf{F}_u^{+}\mathbf{F}_u)_{kk}\right]\right\}$$

$$= \mathbb{E}\left(\sum_{u\neq i, BS_u \in \Pi_{BS}} \sum_{k=1}^{S} P_{u(k)} \frac{\beta w_{iu}}{||x_{BS_u} - x_{BS_i} - x_{off}||^{\alpha}}\right),$$

$$= \mathbb{E}\left(\sum_{u\neq i, BS_u \in \Pi_{BS}} P_u \frac{\beta w_{iu}}{||x_{BS_u} - x_{BS_i} - x_{off}||^{\alpha}}\right) \qquad (25)$$

where $P_u = \sum_{k=1}^{S} P_{u(k)}$ is the total transmission power transmitted by BS_u. Capitalizing on (7), (10), (16) and (25), the average interference over the kth sub-channel in UEG_i is derived as

$$I_{i(k)\_avg}^{ZF}(x_{off}) = \frac{\beta \mathbb{E}(w_{iu})\mathbb{E}(P_u)}{\zeta^{(1)}} \int_{\mathbb{R}^2} \frac{1}{||x + x_{off}||^{\alpha}} \zeta^{(2)}(||x||)dx \qquad (26)$$

In this paper, every UE equipped with a single antenna is assumed to be allocated with the bandwidth B_W for data transmission. The total UEG bandwidth used for the data transmission is thus $S \cdot B_W$. Considering the spatial multiplexing scheme of multi-antenna systems, the data stream transmitted by a sub-channel is extended over the total UEG bandwidth. It is assumed that the noise is negligible in this paper [56]. To simplify the derivation, the average interference in (26) is used to calculate the capacity over the kth sub-channel in UEG_i. As a consequence, the capacity of the kth sub-channel in the UEG UEG_i is expressed as

$$C_{i(k)} = S \cdot B_W \log_2\left\{1 + \frac{P_{i(k)}\frac{\beta w_{ii}}{||x_{off}||^{\alpha}}[(\mathbf{h}_{ii}\mathbf{h}_{ii}^{+})_{(kk)}^{-1}]^{-1}}{I_{i(k)\_avg}^{ZF}(x_{off})}\right\} \qquad (27)$$

The term of $[(\mathbf{h}_{ii}\mathbf{h}_{ii}^{+})_{(kk)}^{-1}]^{-1}$ is the random variable which governed by a Chi-square distribution. Moreover, the PDF of $[(\mathbf{h}_{ii}\mathbf{h}_{ii}^{+})_{(kk)}^{-1}]^{-1}$ is expressed as [54]

$$\gamma_{[(\mathbf{h}_{jj}\mathbf{h}_{jj}^{+})_{(kk)}^{-1}]^{-1}}(\ell) = \frac{\ell^{N_T-S}e^{-\ell}}{(N_T - S)!}, \quad \ell \geq 0. \qquad (28)$$

Furthermore, the spectrum efficiency of UEG_i in a typical cell is derived as

$$SE_{UEG_i} = \frac{\sum_{k=1}^{S} C_{i(k)}}{SB_W} = \sum_{k=1}^{S}\left\{\log_2\left\{1 + \frac{\xi}{S}[(\mathbf{h}_{ii}\mathbf{h}_{ii}^{+})_{(kk)}^{-1}]^{-1}\right\}\right\} \qquad (29)$$

where ξ is defined as the large scale SINR environment factor over wireless channels which is given as $\xi = \left(\beta w_{ii} / \|x_{off}\|^\alpha\right) P_{i(k)} S / I_{i(k)\_avg}^{ZF}(x_{off})$.

Using Jensen's inequality and the mean of Chi-square distribution [57, 58], an upper bound for the average spectrum efficiency of UEG_i in a typical cell is derived as

$$
\begin{aligned}
\mathbb{E}(SE_{\mathrm{UEG}_i}) &= S\mathbb{E}\log_2\{1 + \frac{\xi}{S}[(\mathbf{h}_{ii}\mathbf{h}_{ii}^+)_{(kk)}^{-1}]^{-1}\} \\
&\leq S\log_2\{1 + \frac{\xi}{S}\mathbb{E}[(\mathbf{h}_{ii}\mathbf{h}_{ii}^+)_{(kk)}^{-1}]^{-1}\}. \\
&= S\log_2[1 + \frac{\xi}{S}(N_T - S + 1)]
\end{aligned}
\tag{30}
$$

The homogeneous PPP and its associated hard core Matern point process are both stationary and isotropic [59]. Based on the Palm theory [40], this feature implies that the analytical results for a typical multi-user multi-antenna HCPP cell can be extended to the whole multi-user multi-antenna HCPP cellular network.

The energy efficiency of wireless communication systems is defined as the ratio of the throughput over the consumed transmission power [18]. In this paper, the wireless link transmission power is assumed to be adaptively adjusted to satisfy the wireless traffic requirement [60], i.e., $C_{i(k)} = \rho$. For a given traffic rate ρ and the result in (27), the wireless link transmission power in the multi-user multi-antenna HCPP cellular network is derived by

$$
P_{i(k)} = \frac{\|x_{\mathrm{off}}\|^\alpha I_{i(k)\_avg}^{ZF}(2^{\frac{\rho}{SB_W}} - 1)}{\beta w_{ii}[(\mathbf{h}_{ii}\mathbf{h}_{ii}^+)_{(kk)}^{-1}]^{-1}}
\tag{31}
$$

For a wireless link of cellular networks, the multi-antenna system will consume additional transmission circuit block energy $P_{\mathrm{RF\_chain}}$ per antenna [51]. The total BS power is thus composed of the transmission power and the stationary power [52]. Therefore, the average of total BS power in multi-user multi-antenna HCPP cellular networks is derived as

$$
\mathbb{E}(P_{\mathrm{BS}}) = N_{\mathrm{link}}\left[\frac{\mathbb{E}(P_{i(k)})}{\eta} + N_T P_{\mathrm{RF\_chain}}\right] + P_{\mathrm{sta}}
\tag{32}
$$

where N_{link} is the average number of active links in multi-user multi-antenna HCPP cellular networks. Considering the BS equipped with multi-antenna and the UE equipped with single antenna, the average number of active links is configured as the average active UEs in a unit BS coverage region in HCPP cellular networks, which is calculated by the density of UEs over the density of BSs, i.e., $N_{\mathrm{link}} = \lambda_M/\zeta^{(1)}$. $\mathbb{E}(P_{i(k)})$ is the average link transmission power. η is the average efficiency of RF circuit for a BS; P_{sta} is the stationary power for a BS.

Without loss of generality, the total BS traffic is summed by the traffic rate over all wireless links. The average of total BS traffic is derived as

$$\mathbb{E}(T_{BS}) = N_{link}\mathbb{E}(\rho) = N_{link}\frac{\theta\rho_{min}}{\theta - 1} \tag{33}$$

Therefore, the average energy efficiency of multi-user multi-antenna HCPP cellular networks is derived as

$$EE = \frac{\mathbb{E}(T_{BS})}{\mathbb{E}(P_{BS})} = \frac{\frac{\theta\rho_{min}}{\theta-1}}{\left[\frac{\mathbb{E}(P_{i(k)})}{\eta} + N_T P_{RF\_chain}\right] + \frac{\zeta^{(1)}}{\lambda_M}P_{sta}}. \tag{34}$$

2.5 Simulation Results

Based on the proposed spectrum efficiency and energy efficiency models in Sections IV and V, numerical results are analyzed in detail. Moreover, analysis results are validated through MC simulations in multi-user multi-antenna HCPP cellular networks. The default parameters used for the simulations are as follows [47, 51, 52]: $\lambda_P = 1/(\pi * 800^2)$, $\delta = 500$ m, $\beta = -31.54$ dB, $\alpha = 3.8$, $\sigma_s = 6$, $\mathbb{E}(P_{u(k)}) = 2$ W, $\theta = 1.8$, $\eta = 0.38$, $P_{RF\_chain} = 50$ milliwatt (mW), $P_{sta} = 45.5$ W, $N_{link} = 30$. The maximum BS transmission power is assumed to be 40 W over 200 kHz carrier bandwidth [47]. Furthermore, the maximum transmission power over a wireless link with 10 kHz carrier bandwidth is configured as 2 W in the multi-user multi-antenna HCPP cellular networks [51, 52]. We assume that the wireless link is interrupted if the corresponding transmission power is larger than 2 W.

Figure 6 illustrates the impact of the large scale SINR environment factor ξ on the spectrum efficiency of multi-user multi-antenna HCPP cellular networks for different numbers of transmission antennas at the BS. In this case, the number of antennas at the UEG is $S = 1$. We observe that when the number of transmit antennas is fixed, the spectrum efficiency of multi-user multi-antenna HCPP cellular networks increases with increasing ξ. When ξ is fixed, the spectrum efficiency of multi-user multi-antenna HCPP cellular networks increases with increasing the number of transmit antennas at the BS. The MC simulation curves agree well with the numerical curves in Fig. 6. Since numerical curves are plotted by the upper bound of the spectrum efficiency, MC simulation values are always less than or equal to numerical values in Fig. 6.

Figure 7 analyzes the impact of ξ on the spectrum efficiency of multi-user multi-antenna HCPP cellular networks considering different receive antenna numbers in the UEG, for the case where the number of transmit antennas at the BS equals 8. We observe that when number of receive antennas at the UEG is fixed, the spectrum efficiency of multi-user multi-antenna HCPP cellular networks increases with the increasing of ξ. When ξ is large, the spectrum efficiency of multi-user multi-antenna HCPP cellular networks increases with increasing the number of receive antennas at the UEG. When the large scale SINR environment factor ξ is low, the

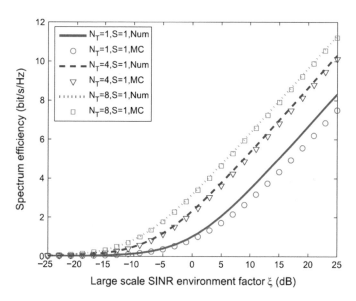

Fig. 6 Impact of the large scale SINR environment factor on the spectrum efficiency of multi-user multi-antenna HCPP cellular networks considering different number of transmit antennas at the BS

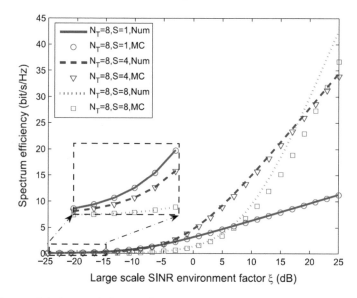

Fig. 7 Impact of the large scale SINR environment factor on the spectrum efficiency of multi-user multi-antenna HCPP cellular networks considering different number of receive antennas at the UEG

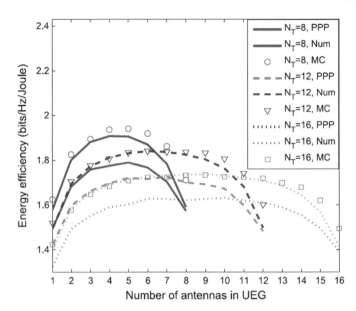

Fig. 8 Energy efficiency of multi-user multi-antenna HCPP and PPP cellular networks with respect to the number of antennas at the UEG and the BS

spectrum efficiency of multi-user multi-antenna HCPP cellular networks increases with decreasing the number of receive antennas at the UEG. The MC simulation curves exhibit a good match with the numerical curves in Fig. 7.

Figure 8 illustrates the energy efficiency of multi-user multi-antenna HCPP and PPP cellular networks with respect to the number of antennas at the UEG and the BS. When the number of antennas at the BS is fixed, both numerical and MC simulation results illustrate that the energy efficiency of multi-user multi-antenna HCPP and PPP cellular networks first increases with increasing the number of antennas at the UEG. When the number of antennas at the UEG is larger than the threshold which corresponds the maximal value of the energy efficiency in this curve, both numerical and MC simulation results show that the energy efficiency of cellular networks decrease with increasing the number of antennas at the UEG. There exist different maximal energy efficiency values of multi-user multi-antenna HCPP cellular networks when BSs are integrated with different antenna numbers. In numerical results, the maximal energy efficiency values are 1.9, 1.84 and 1.72 bits/Hz/Joule, which corresponds to the number of antennas at the BS as 8, 12 and 16, respectively. Moreover, both numerical and MC simulation results indicate that the available maximal energy efficiency values of multi-user multi-antenna HCPP cellular networks decreases with increasing the number of antennas at the BS. Meanwhile, the energy efficiency of PPP cellular networks is less than the energy efficiency of HCPP cellular networks. This result is also validated in Figs. 9, 10 and 11.

Without loss of generality, the number of antennas at the BSs and the UEGs in the multi-user multi-antenna HCPP and PPP cellular networks are configured as equal

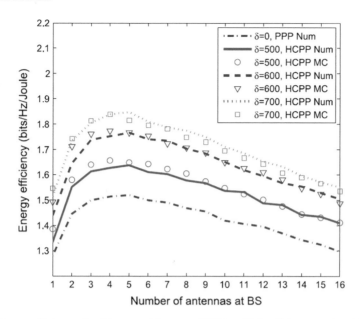

Fig. 9 Energy efficiency of multi-user multi-antenna HCPP and PPP cellular networks with respect to the number of antennas at the BS and the minimum distance δ in adjacent BSs

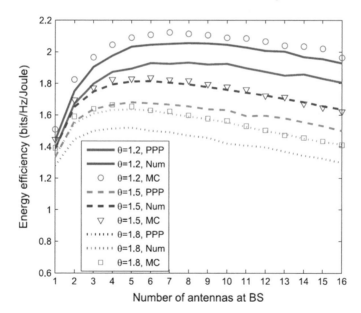

Fig. 10 Energy efficiency of multi-user multi-antenna HCPP and PPP cellular networks with respect to the number of antennas at the BS and the traffic heaviness index θ

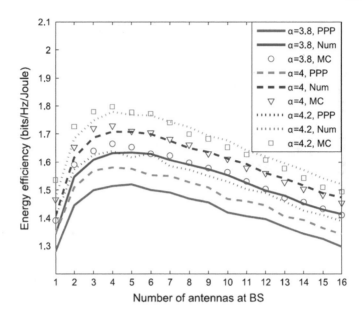

Fig. 11 Energy efficiency of multi-user multi-antenna HCPP and PPP cellular networks with respect to the number of antennas at the BS and the path loss coefficient α

in Figs. 9 and 11. Figure 9 shows the energy efficiency of multi-user multi-antenna HCPP and PPP cellular networks with respect to the number of antennas at BS and the minimum distance δ. When the minimum distance δ is fixed, both numerical and MC simulation results illustrate that the energy efficiency of multi-user multi-antenna HCPP and PPP cellular networks first increases with increasing the number of antennas at the BS. When the number of antennas at the BS is larger than the threshold, both numerical and MC simulation results show that the energy efficiency of multi-user multi-antenna HCPP and PPP cellular networks decreases with increasing the number of antennas at the BS. This result is different with the energy efficiency respect to the number of antennas at the BS in massive MIMO systems [18]. One of reasons is that the small scale fading effect is ignored for wireless channels in massive MIMO systems [61]. On the contrary, the small scale fading effect is considered for the capacity and the interference modeling in this paper. We observe that there exist different maximal energy efficiency values of multi-user multi-antenna HCPP cellular networks under different minimum distances. The maximal energy efficiency values are 1.85, 1.73 and 1.63 bits/Hz/Joule, corresponding to the minimum distances of 300, 400 and 500 m, respectively. When the number of antennas at the BS is fixed, both numerical and MC simulation results consistently demonstrate that the available maximal energy efficiency of multi-user multi-antenna HCPP cellular networks increases with increasing the minimum distance δ. Based on our previous results in [47, 62], there exist an optimal value of cell size, e.g. 1200 m, corresponding to the maximal energy efficiency of cellular networks when the stationary power P_{sta}, i.e., the embodied power, is considered for the BS power consumption. When the minimum distance δ is less than the optimal value of cell size, the maximal energy efficiency of HCPP cellular networks increases with increasing the minimum distance.

Figure 10 analyzes the energy efficiency of multi-user multi-antenna HCPP and PPP cellular networks with respect to the number of antennas at the BS and the traffic heaviness index θ, where the minimum traffic rate over the unit bandwidth is fixed as $\rho_{min}/B_W = 2$. When the number of antennas at the BS is fixed, both numerical and MC simulation results illustrate that the energy efficiency of multi-user multi-antenna HCPP and PPP cellular networks increases with decreasing the traffic heaviness index θ. There exist different maximal energy efficiency values of multi-user multi-antenna HCPP cellular networks under different traffic heaviness indices. The maximal energy efficiency values are 2.06, 1.81 and 1.64 bits/Hz/Joule, which corresponds to the traffic heaviness index as 1.2, 1.5 and 1.8, respectively. When the number of antennas at the BS is fixed, both numerical and MC simulation results demonstrate that available maximal energy efficiency of multi-user multi-antenna HCPP and PPP cellular networks decreases with increasing the traffic heaviness index θ. The increasing of the traffic heaviness index implies that the burst of traffic is increased. The increased burst of traffic will decrease the utilization efficiency of the wireless channel capacity. As a result, the available maximal energy efficiency is decreased with increasing the traffic burst in HCPP cellular networks.

Finally, the impact of path loss coefficient on the energy efficiency of multi-user multi-antenna HCPP and PPP cellular networks is evaluated in Fig. 11. When the number of antennas at the BS is fixed, both numerical and MC simulation results show that the energy efficiency of multi-user multi-antenna HCPP and PPP cellular networks increases with increasing of the path loss coefficient α. Moreover, the maximum energy efficiency with three different path loss coefficients are 1.78, 1.71 and 1.63 bits/Hz/Joule, which correspond to the path loss coefficient as 4.2, 4.0 and 3.8, respectively. The interference fading becomes severer when the path loss coefficient is increased, which leads to the higher spectrum efficiency in wireless channels. As a consequence, the available maximal energy efficiency is increased with increasing the spectrum efficiency in HCPP cellular networks.

2.6   Conclusions

We proposed an energy efficiency assessment for multi-user multi-antenna HCPP cellular networks considering the minimum distance constraint in adjacent BSs. This assessment was obtained by considering an average interference model for multi-antenna HCPP cellular networks with the shadowing and small scale fading over wireless channels. Based on the zero-forcing precoding method, a spectrum efficiency assessment was also obtained for multi-user multi-antenna HCPP cellular networks. Based on the proposed energy efficiency model of multi-user multi-antenna HCPP cellular networks, numerical results have shown that there exists the maximal energy efficiency in multi-user multi-antenna HCPP cellular networks. Our analysis indicates that the maximal energy efficiency of multi-user multi-antenna HCPP cellular networks decreases with increasing the number of transmit antennas at the BSs. Moreover, the maximal energy efficiency of multi-user multi-antenna HCPP cellular networks was shown to depend on the wireless traffic distribution, the wireless channel and the minimum distance in adjacent BSs. Furthermore, the comparison between HCPP and PPP cellular networks implies that the energy efficiency of the conventional PPP cellular networks is underestimated when the minimum distance in adjacent BSs is ignored. Interesting topics for future work include the investigation of the energy efficiency of random cellular networks under massive MIMO scenarios and the UE association based on channel conditions.

3 5G Ultra-Dense Networks with Nonuniform Distributed Users

3.1 Related Work

Some recently studies were conducted to evaluate the performance of UDNs. A basic question was investigated in [1]: how to choose the SBS density for a 5G UDN when the backhaul capacity and energy efficiency are jointly considered. A comprehensive introduction of researches about the energy efficiency of UDNs was given by G. Wu et al. and the tradeoff between energy efficiency and spectrum efficiency was highlighted in [63]. A resource allocation scheme was proposed to analyze trade-offs between the spectrum efficiency and energy efficiency of 5G UDNs [63]. The influence of SBS density on the outage probability was discussed and different multiple access technologies were compared to obtain the optimal solutions for allocating subcarriers with interference constraints [64]. The Lyapunov method and mean field game were utilized to optimize the energy efficiency with the QoS constraints [65]. The impact of the spectrum bandwidth and SBS density on the capacity and the spectrum efficiency was analyzed in [66]. Comparisons between different spectrum bandwidths and SBS densities were made by D. López-Pérez et al. to propose a technique for increasing the average user capacity to 1 gigabit per second (Gbps) [66]. Cognitive radio technologies were adopted to enhance the throughput of 5G UDNs [67], and energy efficient optimal power allocation strategies subject to constraints on the average interference power are proposed in [68]. On the other hand, the stochastic geometry theory is widely used to evaluate the performance of UDNs. The stability and the delay of the wireless networks with spatial and temporal fluctuation in the traffic was investigated by using stochastic geometry in [69, 70] respectively. Closed-form formulas of network capacity and energy efficiency were derived in [71] by modeling the wireless network using the stochastic geometry tools. The difference between line of sight (LoS) channel and non-line of sight (NLoS) channel was considered to evaluate the impact of channel fading on the spatial spectrum efficiency [72]. It was shown in [72] that the network spatial spectrum efficiency will not monotonously increase with the increase of the SBS density when LoS and NLoS channels are considered. A new type of UDNs including femtocell base stations (FBSs), macro cell base stations (MBSs) with distributed antennas was proposed in [73], where the energy efficiency, spectrum efficiency and spatial spectrum efficiency of UDNs were improved. By utilizing an innovative millimeter-wave (mmWave) decoupling method, mmWave can be used by network users for uplink transmissions to traditional microwave base stations (BSs) [74]. Based on the mmWave decoupling method, a closed-form formula for spectrum efficiency of UDNs was derived and a resource management method was proposed to optimize downlink transmission rate of UDNs with a minimum uplink rate constraint. The backhaul traffic of UDNs based on the mmWave and massive MIMO technologies was analyzed and the impact of different pre-coding methods was also investigated in [75]. When both the long-term evolution (LTE) and the wireless fidelity (WiFi) were deployed in UDNs, a Markov chain model was utilized to analyze the performance of the proposed LTE-WIFI UDNs [76]. Based on the energy harvesting technology, the trade-off between the energy efficiency and the QoS was analyzed for UDNs [77]. The orbital angular momentum spatial modulation (OAM-SM) technique was investigated for millimeter wave based UDNs in [78]. Furthermore, a fractional programming method was adopted to investigate the energy efficiency maximization problem of wireless networks under a minimum system throughput constraint [79].

Though the aforementioned studies have investigated many aspects of UDNs, most of them only focused on studying the influence of SBS density on the capacity and energy efficiency. Few effort has been spent on the effect of the SBS density on both the network performance and the user experiences simultaneously. Recently, the virtual cell technology has been widely used to improve the user experience of UDNs [80–85]. By adopting the virtual cell technology, a new type of UDNs termed user-centric UDNs were introduced in [80]. The main idea of user-centric UDNs is to cluster SBSs dynamically and intelligently to provide better services for users. Based on virtual cells in UDNs, a new type of beam forming technology named the balanced beam forming algorithm was developed to optimize the network capacity [81]. A type of virtual cell networks based on distributed antennas were presented in [82], and the influence of cell size on the users' maximum downlink achievable rate was investigated. Based on stochastic geometry theory, the trade-off between the energy efficiency and spectrum efficiency of UDNs was investigated to optimize the energy efficiency with the minimum network capacity constraints [83]. By the joint optimization of virtual cell clustering method and the beam forming algorithm, the sum capacity was improved for UDNs [84]. By optimizing the interference nulling range, the user-centric interference nulling method was proposed in [85], where the outage probability is reduced by 35–40%.

From the above discussion, one can observe that the focus of the researches on UDNs turns from network performance to user experience. Moreover, the design and optimization of UDNs develops from network-centric to user-centric. Due to uncertainties of users' activities, few work have studied the impact of users' activities on the network performance. By measuring the entropy of each individual's moving trajectory, Song et al. found a 93% potential predictability in user mobility [86]. Based on the human mobility trajectory measured from real data, the individual mobility model considering the human mobility tendency habit was proposed to describe the human mobility in the real world [87]. The gravity law model was established to analyze the number of commuters between different areas [88]. The theoretical result of [88] was improved in [89]. By proposing the Radiation and Absorbing (R&A) model and comparing it with the empirical data, the R&A model was proved to be highly reliable in large range of spatial scale.

Although there exist a large amount of meaningful and important studies for UDNs, one basic question is still not well answered: how many SBSs are required to meet the greatly increased traffic demand while guaranteeing a high energy efficiency? To answer this question, a theoretical model including user density, BS deployment, interference analysis and energy consumption need to be established for 5G UDNs.

3.2   System Model

Definitions of some default parameters are shown in the following Table 1.

In this paper, the R&A model is used to analyze the distribution of mobile users in a two-dimension plane $\mathbb{R}^2$. Let $\mathbb{A}_T$ be a region with area S_A on the plane $\mathbb{R}^2$. The places with large number of users are denoted as hot spots in $\mathbb{A}_T$. The set of hot spots is denoted by Υ, the cardinality of Υ is N_p and elements of Υ are denoted by $\{HS_i | 1 \leq i \leq N_p\}$. The system QN is defined as a set formed by all users, hot spots and SBSs in $\mathbb{A}_T$. The coverage region of the hot spot HS_i is denoted by CR_i which is assumed to be a circle with a radius l_i. Thus, the area of CR_i is $S_i = \pi l_i^2$. According to the definition of the R&A model, users can commute with all other hot spots in $\mathbb{A}_T$. The probability that one user moves from HS_i to HS_j is denoted as r_{ij} and expressed as follow [89].

Table 1 Definitions of parameters

Default parameters	Definitions
HS_i	Hot spot i
CR_i	Coverage region of HS_i
S_i	Coverage area of CR_i
l_i	Radius of CR_i
P_i	Number of users in CR_i
$N_S(i)$	Number of SBSs in CR_i
m_i	Attraction exponent of HS_i
T_{ij}	Number of moving users from HS_i to HS_j
U_{HS_i}	Typical user of HS_i
S_{HS_i}	Typical SBS of HS_i

$$r_{ij} = \frac{m_i m_j}{(m_i + s_{ij})(m_i + m_j + s_{ij})}, \tag{35}$$

where m_i and m_j are attraction exponents (AEs) of HS_i and HS_j, respectively. Based on the definition in the supplementary material of [89], AEs are configured as fixed values determined by several parameters like the number of available jobs, average salary and consumption level of the corresponding hot spot. s_{ij} denotes the total value of AEs (excluding AEs of HS_i and HS_j) in the circle centered at HS_i with radius l_{ij}, where l_{ij} is the distance between the centers of HS_i and HS_j.

The total number of users commuting from HS_i to HS_j during Δt is denoted by T_{ij}. Then, T_{ij} is a binomial random variable by assuming moving characteristics of different users inside the QN system are independent from each other. The expectation of T_{ij} is

$$\mathbb{E}(T_{ij}) = \zeta P_i \frac{m_i m_j}{(m_i + s_{ij})(m_i + m_j + s_{ij})}, \tag{36}$$

where $\mathbb{E}(\cdot)$ is the expectation operation. The ratio of moving users in the total population is configured as a fixed value denoted by ζ. An example of this system is shown in Fig. 12.

Based on the definition in [89], the original R&A model is only used for the theoretical analysis in the spatial domain. Thus, to evaluate the stationary number of users in the coverage region of each hot spot in the QN system, the queueing network model is adopted in this paper to extend the R&A model to the temporal domain. We assume that all users in the coverage region of a hot spot are stayed in a queue so that the number of users in the coverage region is equal to the length of the corresponding queue. Furthermore, users in the queue are assumed to be served by a server in the corresponding hot spot, and a user will leave the hot spot after the user is served. Thus, the arrival rate and serving rate of a queue are evaluated by the expected number of users that moving in and out of the hot spot respectively during a given time slot Δt.

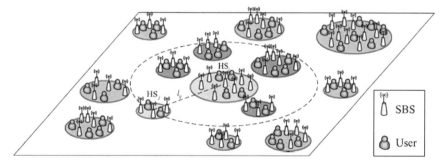

Fig. 12 System model. Coverage areas of different hot spots are shown as circles with different radius, s_{ij} is the total number of AEs in circles marked as the purple regions

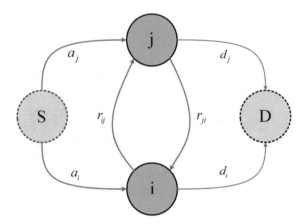

Fig. 13 An example of JNS that contains two queues i and j. Users outside of JNS are denoted by S, and users exiting from JNS are denoted by D. The probability that a user transfers from S to i and j is a_i and a_j. The probability that a user leaves the JNS from i (j) is d_i (d_j). The probabilities that a user moves between i and j are denoted by r_{ij} and r_{ji}, respectively

The Jackson network (JN) is a basic type of queueing networks. Based on the definition of JN [90], users outside of a queueing network system may move into a hot spot contained in the queueing network system. Figure 13 is a Jackson network system (JNS) formed by two queues.

In this paper, the number of users in CR_i at time t is denoted as P_i. Based on the definition of JN in [90], P_i is independent from the length of other queues in the QN system. The serving rate of the queue in HS_i is denoted as μ_i. In the QN system, the stationary value of P_i exists if the condition $\rho_i = \frac{\lambda_i}{\mu_i} < 1$ is satisfied. The QN system keeps a stationary state when lengths of all queues in the QN system remain stationary.

In this paper, all SBSs and users of HS_i are assumed to be randomly and uniformly distributed in CR_i. Furthermore, an typical user in CR_i is donated by U_{HS_i} and an typical SBS in CR_i is denoted as S_{HS_i}. The distance between U_{HS_i} and S_{HS_i} is denoted by l_{mc}. Based on the result in [91], we have

$$f_{l_{mc}}(x) = \frac{4x}{\pi l_i^2}\left(\arccos\frac{x}{2l_i} - \frac{x}{2l_i}\sqrt{1 - \frac{x^2}{4l_i^2}}\right), \tag{37}$$

where $f_{l_{mc}}(x)$ is the probability density function (PDF) of the random variable l_{mc}. In this paper, the function $f_X(\cdot)$ is used to denote the PDF of a random variable X.

The orthogonal frequency division multiplexing (OFDM) technology is assumed to be adopted by all SBSs in the QN system. In this paper, SBSs that transmit useful signals to users are called the associated SBSs while SBSs that generate interference are called interfering SBSs. Users are assumed to receive downlink interference only from SBSs within the same hot spot with them. The small scale fading of all channels are assumed to be governed by independent and identically distributed (*i.i.d.*) Rayleigh distributions. By ignoring the shadowing, the coverage probability of S_{HS_i} to U_{HS_i} is expressed by

$$P_{\text{cover}} = \Pr\left(\frac{p_t G_s R_s^{-\alpha_p}}{\sigma^2 + \sum_{i=0}^{\overline{N_{\text{int}}}} p_t G_i R_i^{-\alpha_p}} \geq \gamma_0\right), \tag{38}$$

where $\Pr(\cdot)$ is the probability corresponding to the expression in parentheses, γ_0 is the threshold of the received signal to interference plus noise ratio (SINR) at user devices, p_t is the transmit power consumption of a SBS, σ^2 is the power of the additive white Gaussian noise (AWGN), α_p is the path loss exponent, $\overline{N_{\text{int}}}$ is the average number of interfering SBSs, R_s is the distance between a user and an associated SBS, R_i is the distance between a user and an interfering SBS, G_s and G_i are the small scale fading experienced by the desired link and interfering links, respectively. Based on the assumption that small scale fading of all channels follows *i.i.d.* Rayleigh distributions, G_s and G_i are exponential distributed random variables with expectations η.

The event that a user is covered by a SBS is assumed to be independent of the event that the user is covered by another SBS. Thus, the number of SBSs that cover the same user is a binomial distributed random variable. The expectation of the binomial distributed random variable is

$$\mathbb{E}(N_{\text{cover}}) = \overline{N_{\text{cover}}} = N_S(i) P_{\text{cover}}. \tag{39}$$

All SBSs that can cover a same user are configured to form a virtual cell cluster (VC). SBSs in a VC are assumed to be able to transmit signals to the served user simultaneously without causing interference [9]. Thus, the number of interfering SBSs is

$$\overline{N_{\text{int}}} = N_{\text{SBS}} - \overline{N_{\text{cover}}} = N_S(i)(1 - P_{\text{cover}}). \tag{40}$$

By substituting (40) into (38), the coverage probability is further derived by

$$P_{\text{cover}} = \Pr\left(\frac{p_t G_s R_s^{-\alpha_p}}{\sigma^2 + \sum_{i=0}^{N_S(i)(1-P_{\text{cover}})} p_t G_i R_i^{-\alpha_p}} \geq \gamma_0\right). \tag{41}$$

A channel is called an available channel if the received SINR at the user using this channel is larger than a given threshold γ_0. A user will be blocked when all available channels of the associated SBS are occupied. Thus, a two-dimensional Markov chain is utilized to analyze channel access processes of SBSs in the QN system [47]. Two states of the two-dimensional Markov chain are denoted by (v_m, v_n), where v_m denotes the number of occupied channels in a SBS and v_n is the number of available channels in a SBS ($v_m \leq v_n$), respectively.

The probability that a user has a downlink communication request to a SBS is denoted by p_s. Communication requests of different users are assumed to be *i.i.d.*. Thus, the number of users with communication requests follows a binomial distribution with expectation $\overline{P_i} \cdot p_s$, where $\overline{P_i}$ is the stationary number of users in CR_i. Since $\overline{P_i}$ is usually a large value, the binomial distribution can be approximated by a Poisson distribution with the same expectation $\overline{P_i} \cdot p_s$. The call arriving process of S_{HS_i} is denoted by AP and the number of calls arriving at S_{HS_i} before time t is denoted by $AP(t)$. Thus, we have the following equation

$$\Pr\{AP(t + \Delta t) - AP(t) = n\} = e^{-\frac{\overline{P_i} \cdot p_s \Delta t}{N_S(i)}} \frac{\overline{P_i} \cdot p_s (\Delta t)^n}{n! N_S(i)}. \tag{42}$$

The call arriving process of S_{HS_i} is supposed to be a stochastic process with independent and stationary increments. Thus, the stochastic process AP is a Poisson process. The value of the serving duration of S_{HS_i} to U_{HS_i} is assumed to be an exponential distributed random variable with expectation μ_s. The corresponding transition diagram of the two-dimensional Markov chain is shown as follows.

The transition diagram of the two-dimensional Markov chain is explained as follows.

(1) $(v_m, v_n) \rightarrow (v_m + 1, v_n)$: A user has been successfully allocated an available idle channel of S_{HS_i}, then $v_m + 1$.
(2) $(v_m, v_n) \rightarrow (v_m, v_n + 1)$: An unavailable channel of S_{HS_i} becomes available due to the time-varying interference, thus $v_n + 1$.
(3) $(v_m, v_n) \rightarrow (v_m - 1, v_n)$: A user has been fully served and releases the occupied channel of S_{HS_i}, then $v_m - 1$.
(4) $(v_m, v_n) \rightarrow (v_m, v_n - 1)$: An available channel of S_{HS_i} becomes unavailable due to the time-varying interference, thus $v_n - 1$.

The maximum number of available channels offered by S_{HS_i} is denoted as C. The transition rate for an unavailable channel becoming available is denoted by α_s, and the transition rate for an available channel becoming unavailable is denoted by β_s. The call arriving rate $\lambda_{i,s}$ is

$$\lambda_{i,s} = \frac{\overline{P_i} \cdot p_s \cdot \overline{N_{\text{cover}}}}{N_S(i)}. \tag{43}$$

Based on the Kolmogorov criteria [47], the two-dimensional Markov chain in Fig. 14 is reversible and the stationary state distribution of the two-dimensional Markov chain exists.

3.3 Coverage Probability, Blocking Probability

The coverage probability in (41) can be further evaluated as (44).

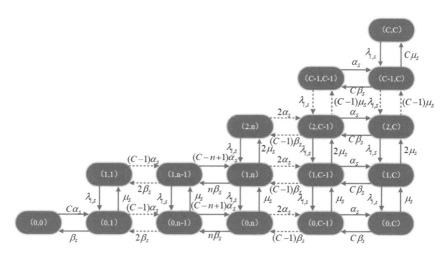

Fig. 14 Transition diagram of the two-dimensional Markov chain

$$P_{\text{cover}} = \Pr\left(\frac{p_t G_s R_s^{-\alpha_p}}{\sigma^2 + \sum_{i=0}^{N_S(i)(1-P_{\text{cover}})} p_t G_i R_i^{-\alpha_p}} \geq \gamma_0\right)$$

$$= \Pr\left(\frac{G_s}{\frac{1}{\rho_0} + \frac{\sum_{i=0}^{N_S(i)(1-P_{\text{cover}})} G_i R_i^{-\alpha_p}}{R_s^{-\alpha_p}}} \geq \gamma_0\right),$$

$$= \Pr\left(\frac{1}{\rho_0 G_s} + \frac{\sum_{i=0}^{N_S(i)(1-P_{\text{cover}})} G_i R_i^{-\alpha_p}}{G_s R_s^{-\alpha_p}} \leq \gamma_0^{-1}\right) \tag{44}$$

where ρ_0 is the received signal to noise ratio (SNR) of U_{HS_i}. The expression of ρ_0 is

$$\rho_0 = \frac{p_t R_s^{-\alpha_p}}{\sigma^2}. \tag{45}$$

Based on the result in [92], ρ_0 is configured as a determined variable in this paper. Based on (44) and (45), the coverage probability P_{cover} is finally derived as

$$P_{\text{cover}} = 2^{-B_e} \gamma_0 \exp\left(\frac{A_e}{2}\right) \sum_{b_e=0}^{B_e} \binom{B_e}{b_e} \sum_{c_e=0}^{C_e+b_e} \frac{(-1)^{c_e}}{D_e}$$

$$\times \text{Re}\left(\left(\int_0^\infty \left(\eta \exp\left(-\frac{h}{\rho_0 G_s} - \eta G_s\right)\right.\right.$$

$$\times \left(\int_0^\infty \int_0^{2l_i} \int_0^{2l_i} \exp\left(\frac{-\hbar G_i R_i^{-\alpha_p}}{G_s R_s^{-\alpha_p}}\right) \left(\frac{4R_s}{\pi l_i^2}\left(\arccos\frac{R_s}{2l_i} - \frac{R_s}{2l_i}\sqrt{1 - \frac{R_s^2}{4l_i^2}}\right)\right).\right.$$

$$\times \left(\frac{4R_i}{\pi l_i^2}\left(\arccos\frac{R_i}{2l_i} - \frac{R_i}{2l_i}\sqrt{1 - \frac{R_i^2}{4l_i^2}}\right)\right)$$

$$\left.\times (\eta \exp(-\eta G_i)) dG_i dR_s dR_i)^{\overline{N_{\text{int}}}} dG_s\right) / \hbar\right) \tag{46}$$

The corresponding derivation and explanation are shown in the following.
The Laplace transform of the random variable Z is denoted by $L_Z(\varepsilon)$, where Z is

$$Z = \frac{1}{\rho_0 G_s} + \frac{\sum_{i=0}^{N_S(i)(1-P_{\text{cover}})} G_i R_i^{-\alpha_p}}{G_s R_s^{-\alpha_p}}. \tag{47}$$

Thus, the Laplace transform $L_Z(\varepsilon)$ is derived as (41), where $\exp(\cdot)$ is the exponential function and $\mathbb{E}_{G_s, G_i, R_s, R_i}(\cdot)$ is the expectation operations on random variables G_s, G_i, R_s and R_i.

$$L_Z(\varepsilon) = \mathbb{E}_{G_s, G_i, R_s, R_i}\left(\exp\left(-\varepsilon\left(\frac{1}{\rho_0 G_s} + \frac{\sum_{i=0}^{N_S(i)(1-P_{\text{cover}})} G_i R_i^{-\alpha_p}}{G_s R_s^{-\alpha_p}}\right)\right)\right), \tag{48}$$

and the PDFs of G_s, G_i, R_s and R_i are

$$\begin{cases} f_{G_s}(x) = \eta \exp(-\eta x) \\ f_{G_i}(x) = \eta \exp(-\eta x) \\ f_{R_s}(x) = \frac{4x}{\pi l_i^2}\left(\arccos\frac{x}{2l_i} - \frac{x}{2l_i}\sqrt{1 - \frac{x^2}{4l_i^2}}\right) \\ f_{R_i}(x) = \frac{4x}{\pi l_i^2}\left(\arccos\frac{x}{2l_i} - \frac{x}{2l_i}\sqrt{1 - \frac{x^2}{4l_i^2}}\right) \end{cases} \tag{49}$$

Based on (40), (48) and (49), the Laplace transform of Z is further derived by (50).

$$L_Z(\varepsilon) = \mathbb{E}_{G_s, G_i, R_s, R_i}\left(\exp\left(-\varepsilon\left(\frac{1}{\rho_0 G_s} + \frac{\sum_{i=0}^{N_S(i)(1-P_{\text{cover}})} G_i R_i^{-\alpha_p}}{G_s R_s^{-\alpha_p}}\right)\right)\right)$$

$$= \mathbb{E}_{G_s, G_i, R_s, R_i}\left(\exp\left(\frac{-\varepsilon}{\rho_0 G_s}\right)\prod_{i=0}^{N_S(i)(1-P_{\text{cover}})}\exp\left(\frac{-\varepsilon G_i R_i^{-\alpha_p}}{G_s R_s^{-\alpha_p}}\right)\right)$$

$$= \mathbb{E}_{G_s}\left(\exp\left(\frac{-\varepsilon}{\rho_0 G_s}\right) \cdot \left(\mathbb{E}_{G_i, R_s, R_i}\left(\exp\left(\frac{-\varepsilon G_i R_i^{-\alpha_p}}{G_s R_s^{-\alpha_p}}\right)\right)\right)^{N_S(i)(1-P_{\text{cover}})}\right),$$

$$= \int_0^\infty \exp\left(\frac{-\varepsilon}{\rho_0 G_s}\right)(\eta \exp(-\eta G_s))$$

$$\cdot \left(\mathbb{E}_{G_i, R_s, R_i}\left(\exp\left(\frac{-\varepsilon G_i R_i^{-\alpha_p}}{G_s R_s^{-\alpha_p}}\right)\right)\right)^{\overline{N_{int}}} dG_s$$

$$= \int_0^\infty \eta \exp\left(\frac{-\varepsilon}{\rho_0 G_s} - \eta G_s\right)\left(\mathbb{E}_{G_i, R_s, R_i}\left(\exp\left(\frac{-\varepsilon G_i R_i^{-\alpha_p}}{G_s R_s^{-\alpha_p}}\right)\right)\right)^{\overline{N_{int}}} dG_s \quad (50)$$

with (51).

$$\mathbb{E}_{G_i, R_s, R_i}\left(\exp\left(\frac{-\varepsilon G_i R_i^{-\alpha_p}}{G_s R_s^{-\alpha_p}}\right)\right) = \int_0^{2l_0}\int_0^{2l_0}\int_0^\infty \exp\left(\frac{-\varepsilon G_i R_i^{-\alpha_p}}{G_s R_s^{-\alpha_p}}\right)$$

$$\times (f_{G_i}(G_i) f_{R_s}(R_s) f_{R_i}(R_i)) dG_i dR_s dR_i \quad (51)$$

By substituting (49) and (51) into (50), the Laplace transform in (50) becomes (52).

$$L_Z(\varepsilon) = \mathbb{E}_{G_0}\left(\exp\left(-\frac{\varepsilon}{\rho_0 G_s}\right)\left(\int_0^\infty\int_0^{2l_i}\int_0^{2l_i} \exp\left(\frac{-\varepsilon G_i R_i^{-\alpha_p}}{G_s R_s^{-\alpha_p}}\right)\right.\right.$$

$$\times \left(\frac{4R_s}{\pi l_i^2}\left(\arccos\frac{R_s}{2l_i} - \frac{R_s}{2l_i}\sqrt{1 - \frac{R_s^2}{4l_i^2}}\right)\right)$$

$$\left(\frac{4R_i}{\pi l_i^2}\left(\arccos\frac{R_i}{2l_i} - \frac{R_i}{2l_i}\sqrt{1 - \frac{R_i^2}{4l_i^2}}\right)\right)$$

$$\times \left. (\eta \exp(-\eta G_i)) dG_i dR_s dR_i\right)^{\overline{N_{int}}}\right)$$

$$= \int_0^\infty\left(\eta \exp\left(-\frac{\varepsilon}{\rho_0 G_s} - \eta G_s\right)\left(\int_0^\infty\int_0^{2l_i}\int_0^{2l_i} \exp\left(\frac{-\varepsilon G_i R_i^{-\alpha_p}}{G_s R_s^{-\alpha_p}}\right)\right.\right.$$

$$\times \left(\frac{4R_s}{\pi l_i^2}\left(\arccos\frac{R_s}{2l_i} - \frac{R_s}{2l_i}\sqrt{1 - \frac{R_s^2}{4l_i^2}}\right)\right)$$

$$\times \left(\frac{4R_i}{\pi l_i^2}\left(\arccos\frac{R_i}{2l_i} - \frac{R_i}{2l_i}\sqrt{1 - \frac{R_i^2}{4l_i^2}}\right)\right)$$

$$\left. (\eta \exp(-\eta G_i)) dG_i dR_s dR_i\right)^{\overline{N_{int}}}\right) dG_s \quad (52)$$

Similar to [93], by the Euler summation on (52), P_{cover} becomes

$$P_{\text{cover}} = 2^{-B_e} \gamma_0 \exp\left(\frac{A_e}{2}\right) \sum_{b_e=0}^{B_e} \binom{B_e}{b_e} \times \sum_{c_e=0}^{C_e+b_e} \frac{(-1)^{c_e}}{D_e} \text{Re}\left(\frac{L_Z(\hbar)}{\hbar}\right). \tag{53}$$

Based on the analysis in [94], A_e, B_e and C_e should be no less than $t \ln 10$, $1.243t - 1$ and $1.467t$ respectively to obtain a numerical accuracy of 10^{-t}. In this paper, A_e, B_e and C_e are configured to be $8 \ln 10$, 11 and 14 in order to obtain a numerical accuracy of 10^{-9} for the theoretical analysis. The parameter $\hbar$ is expressed as $\hbar = (A_e + 2\pi c_e J)/(2\gamma_0^{-1})$, where J is the imaginary unit and $\text{Re}(\cdot)$ denotes the real part of the given variable in parentheses. The parameter D_e is

$$D_e = \begin{cases} 2 \text{ if } c_e = 0 \\ 0 \quad \text{others} \end{cases} \tag{54}$$

By substituting (52) into (53), the coverage probability P_{cover} becomes (46), which finishes the derivation.

By the definition of JN in [90], a QN system should meet the following three necessary conditions for being a JN:

(1) User arriving processes in all hot spots are independent Poisson processes in the QN system
(2) Users outside of the QN system can move into the QN system.
(3) Users who are fully served in the QN system can either move to another hot spot in the QN system or leave the QN system.

Denote the total number of users moving from outside of $\mathbb{A}_T$ into CR_i during a time period Δt as e_i. By assuming e_i as a random variable with expectation a_i, we get the following lemmas.

Lemma 1 *For a large P_i, the binomial distributed random variable T_{ij} can be approximated by a Poisson distributed random variable with the expectation being* $\mathbb{E}(T_{ij})$.

The number of users in a hot spot is modeled by a queueing network model in the temporal domain. Based on the definition of JN users who have been fully served in a hot spot either choose to be added into another queue or to leave the queueing network [90]. Based on the above model, we get the following Lemma 2.

Lemma 2 *Based on Lemma 1 and the assumption that e_i is a Poisson distributed random variable, the number of users moving into CR_i during Δt is a Poisson distributed random variable. The expectation of the number of users moving into CR_i during Δt is*

$$\lambda_i = a_i + \sum_{j\in(1,N_p),j\neq i} \mathbb{E}(T_{ji}). \tag{55}$$

Lemma 3 *Assuming numbers of users moving into* CR_i *during each disjoint* Δt *in the temporal domain are i.i.d.. Thus, the user arriving process of* HS_i *is a stochastic process with independent and stationary increments.*

By combining Lemma 1, Lemma 2 and Lemma 3, the user arriving process of HS_i is proved to be a Poisson process. Then, we get the following theorem.

Theorem 1 *Assuming users' moving actions among different hot spots are governed by a R&A model. By denoting the number of users in each hot spot as the size of the corresponding queue, the* QN *system can be modeled by a Jackson network.*

The transferring matrix of the QN system is denoted by **R**. The element of the matrix **R** located at the row i and the line j is denoted as $[\mathbf{R}]_{ij}$, which is the probability that a user moves from HS_i to HS_j. Thus, $[\mathbf{R}]_{ij}$ is expressed by

$$[\mathbf{R}]_{ij} = \begin{cases} \frac{m_i m_j}{(m_i + s_{ij})(m_i + m_j + s_{ij})} & 0 < i \neq j \leq N_p \\ 0 & 0 < i = j \leq N_p \end{cases}. \tag{56}$$

The probability that a user leaves the QN system from HS_i is

$$d_i = 1 - \sum_{j=1}^{N_p} [\mathbf{R}]_{ij}. \tag{57}$$

When the QN system is in a stationary state the number of users leave the QN system from HS_i during Δt is denoted as $\overline{D_i}$ with expectation as follows.

$$\mathbb{E}(\overline{D_i}) = \zeta \overline{P_i}, \tag{58}$$

where $\overline{P_i}$ denotes the stationary number of users in CR_i. Substitute (56) and (58) into (36), the expectation of the number of users moving into CR_i during Δt is

$$\lambda_i = a_i + \sum_{j \in (1, N_p), j \neq i} \lambda_j r_{ji}. \tag{59}$$

To keep the stability of the total number of users in the QN system, a_i is configured as

$$a_i = \zeta \frac{\sum_{i=1}^{N_p} m_i d_i}{N_p} = \zeta \frac{\sum_{i=1}^{N_p} m_i \left(1 - \sum_{j=1}^{N_p} [\mathbf{R}]_{ij}\right)}{N_p}. \tag{60}$$

By denoting $\mathbf{a} = (a_1, a_2, \ldots, a_{N_p})_{1 \times N_p}$ and $\lambda = (\lambda_1, \lambda_2, \ldots, \lambda_{N_p})_{1 \times N_p}$, (60) is further derived by a matrix form as

$$\lambda = \mathbf{a}(\mathbf{I} - \mathbf{R})^{-1}, \tag{61}$$

where $\mathbf{I}$ is an identical matrix. Based on (61), the user arriving rate of each hot spot can be obtained.

The stationary serving rate of the queue in HS_i is denoted as μ_i. By substituting λ_i into $\rho_i = \frac{\lambda_i}{\mu_i} = \frac{\lambda_i}{\zeta \overline{P_i}}$, the number of users in the stable status in CR_i is

$$\overline{P_i} = \frac{\rho_i}{1 - \rho_i} = \frac{\lambda_i}{\zeta \overline{P_i} - \lambda_i}. \tag{62}$$

By solving the implicit function in (45), $\overline{P_i}$ is further derived by

$$\overline{P_i} = \sqrt{\left(\frac{\lambda_i}{2\zeta}\right)^2 + \frac{\lambda_i}{\zeta}} + \frac{\lambda_i}{2\zeta}. \tag{63}$$

The stationary distribution of the state (v_m, v_n) in the two-dimensional Markov chain in Fig. 14 is denoted by $\pi(v_m, v_n)$. Based on the result in [47], $\pi(v_m, v_n)$ is expressed by (64)

$$\pi(v_m, v_n) = \begin{cases} \frac{1}{\chi}\left(\frac{\lambda_{i,s}}{\mu_s}\right)^{v_m} \frac{1}{v_m!} \binom{C}{v_n} \left(\frac{\alpha_s}{\beta_s}\right)^{v_n} \\ \chi = \sum\limits_{v_m \leq v_n \leq C} \left(\frac{\lambda_{i,s}}{\mu_s}\right)^{v_m} \frac{1}{v_m!} \binom{C}{v_n} \left(\frac{\alpha_s}{\beta_s}\right)^{v_n} \end{cases}, \tag{64}$$

where $\binom{C}{v_n}$ is a binomial coefficient denoting the number of ways to pick v_n unordered outcomes from C possibilities. By the Gilbert-Elliott model [47], we have

$$\frac{\alpha_s}{\beta_s} = \frac{1 - \varepsilon_s}{\varepsilon_s}, \tag{65}$$

where ε_s is expressed as (35) and p_{oc} is the probability that a channel of S_{HS_i} is occupied.

$$\varepsilon_s = \sum_{\Delta_s=0}^{\overline{N_{\mathrm{int}}}} (1 - P_{\mathrm{cover}}) \binom{\overline{N_{\mathrm{int}}}}{\Delta_s} (p_{oc})^{\Delta_s} (1 - p_{oc})^{\overline{N_{\mathrm{int}}} - \Delta_s} \tag{66}$$

By assuming U_{HS_i} accesses each channel with equal probability, the probability that a channel of S_{HS_i} is occupied is

$$P_{oc} = \sum_{v_m=0}^{v_n} \sum_{v_n=0}^{C} \frac{v_m}{C} \pi(v_m, v_n). \tag{67}$$

By substituting (65), (66) and (67) into (64), the stationary probability $\pi(v_m, v_n)$ of the state (v_m, v_n) is obtained. The call blocking probability of U_{HS_i} is

$$P_B = \sum_{v_m=v_n \leq C} \pi(v_m, v_n). \tag{68}$$

By (64) and (68), the probability that U_{HS_i} is served by S_{HS_i} is derived as

$$
\begin{aligned}
P_{serve} &= 1 - P_B \\
&= 1 - \sum_{v_m=v_n \leq C} \frac{1}{\chi} \left(\frac{\lambda_{i,s}}{\mu_s} \right)^{v_m} \frac{1}{v_m!} \binom{C}{v_n} \left(\frac{\alpha_s}{\beta_s} \right)^{v_n}.
\end{aligned} \tag{69}
$$

The probability that all channels of S_{HS_i} are idle is P_{idle}, given by

$$P_{idle} = \sum_{v_n=0}^{C} \pi(0, v_n). \tag{70}$$

3.4 Throughput and Energy Efficiency

Based on the transition diagram of the two-dimensional Markov Chain in Fig. 14, the average number of occupied channels at S_{HS_i} is

$$\overline{N_{oc}} = \sum_{v_m=0}^{v_n} \sum_{v_n=0}^{C} v_m \pi(v_m, v_n) \tag{71}$$

We assume that the transmission rate is assumed to be equal to the channel capacity when the channel is occupied. Thus, the stationary throughput of S_{HS_i} is

$$
\begin{aligned}
T_s &= \overline{N_{oc}} \cdot (C_{ca}) \\
&= (C_{ca}) \sum_{v_m=0}^{v_n} \sum_{v_n=0}^{C} v_m \pi(v_m, v_n).
\end{aligned} \tag{72}
$$

By assuming the throughputs of all SBSs in the QN system are *i.i.d.*, the throughput of the small cell network restricted in CR_i is (73).

$$T_{HS_i} = N_S(i) \left(\int_{x=\gamma_0}^{\infty} \log_2(1+x) f_{\tilde{x}}(x) dx \right) \sum_{v_m=0}^{v_n} \sum_{v_n=0}^{C} v_m \pi(v_m, v_n). \qquad (73)$$

Furthermore, the throughput of the small cell network in the QN system is (74).

$$T_{QN} = \sum_{i=1}^{N_p} N_S(i) \left(\int_{x=\gamma_0}^{\infty} \log_2(1+x) f_{\tilde{x}}(x) dx \right) \sum_{m_s=0}^{n_s} \sum_{n_s=0}^{C} m_s \pi(m_s, n_s). \qquad (74)$$

The total power consumption p_{to} of a SBS s_{ori} is [95]

$$p_{to} = N_{oc} \frac{\frac{p_t}{\eta_{pa}} + p_{rf} + p_{bb}}{(1 - \sigma_{dc})(1 - \sigma_{ms})} + p_{st}, \qquad (75)$$

where p_t is the transmit power consumption of s_{ori}, η_{pa} is the efficiency coefficient of the power amplify module, p_{rf} is the power consumption of the radio frequency module, p_{bb} is the power consumption of the base band module, σ_{dc} is the loss coefficient of the digital control module, σ_{ms} is the power supply loss coefficient, p_{st} is the constant power consumption which is independent from the traffic load of s_{ori}, and N_{oc} is the number of occupied channels at s_{ori}, respectively. Based on the configuration in [95], the values of σ_{ms} and σ_{dc} are smaller than 1.

We assume that all SBSs are able to be shut down and turned on instantaneously without delay. Thus, the energy consumption of S_{HS_i} is

$$E_{to} = \left(\sum_{v_m=1}^{v_n} \sum_{v_n=1}^{C} v_m \pi(v_m, v_n) \frac{\frac{p_t}{\eta_{pa}} + p_{rf} + p_{bb}}{(1 - \sigma_{dc})(1 - \sigma_{ms})} + p_{st} \right)(1 - P_{idle}) \cdot t_{to} \qquad (76)$$

where t_{to} is the operation duration of S_{HS_i}. By assuming the energy consumptions of different SBSs to be *i.i.d.*, the energy consumed by the small cell network restricted in CR_i is

$$E_{HS_i} = \left(\sum_{v_m=1}^{v_n} \sum_{v_n=1}^{C} v_m \pi(v_m, v_n) \frac{\frac{p_t}{\eta_{pa}} + p_{rf} + p_{bb}}{(1 - \sigma_{dc})(1 - \sigma_{ms})} + p_{st} \right)$$
$$\times N_S(i) \cdot (1 - P_{idle}) \cdot t_{to}. \qquad (77)$$

Thus, the network energy consumption of the small cell network in the QN system is

$$E_{QN} = \sum_{i=1}^{N_p} \left(\sum_{v_m=1}^{v_n} \sum_{v_n=1}^{C} v_m \pi(v_m, v_n) \frac{\frac{p_t}{\eta_{pa}} + p_{rf} + p_{bb}}{(1 - \sigma_{dc})(1 - \sigma_{ms})} + p_{st} \right)$$
$$\times N_S(i) \cdot (1 - P_{idle}) \cdot t_{to}. \qquad (78)$$

By combining (72) and (76), the energy efficiency of S_{HS_i} will be (79).

$$EE_s = \frac{\left(\int_{x=\gamma_0}^{\infty} \log_2(x+1) f_{\tilde{x}}(x)dx\right) \sum_{v_m=0}^{v_n} \sum_{v_n=0}^{C} v_m \pi(v_m, v_n)}{\left(\sum_{v_m=1}^{v_n} \sum_{v_n=1}^{C} v_m \pi(v_m, v_n) \frac{\frac{p_t}{\eta_{pa}}+p_{rf}+p_{bb}}{(1-\sigma_{dc})(1-\sigma_{ms})} + p_{st}\right)(1 - P_{idle})}. \quad (79)$$

Combine (73) and (77), the network energy efficiency restricted in CR_i is (80).

$$EE_{HS_i} = \frac{N_S(i)\left(\int_{x=\gamma_0}^{\infty} \log_2(x+1) f_{\tilde{x}}(x)dx\right) \sum_{v_m=0}^{v_n} \sum_{v_n=0}^{C} v_m \pi(v_m, v_n)}{N_S(i)\left(\sum_{v_m=1}^{v_n} \sum_{v_n=1}^{C} v_m \pi(v_m, v_n) \frac{\frac{p_t}{\eta_{pa}}+p_{rf}+p_{bb}}{(1-\sigma_{dc})(1-\sigma_{ms})} + p_{st}\right)(1 - P_{idle})}. \quad (80)$$

Based on (74) and (78), the energy efficiency of the small cell network in the QN system is (81).

$$EE_{QN} = \frac{\sum_{i=1}^{N_p} N_S(i)\left(\int_{x=\gamma_0}^{\infty} \log_2(x+1) f_{\tilde{x}}(x)dx\right) \sum_{v_m=0}^{v_n} \sum_{v_n=0}^{C} v_m \pi(v_m, v_n)}{\sum_{i=1}^{N_p} N_S(i)\left(\sum_{v_m=1}^{v_n} \sum_{v_n=1}^{C} v_m \pi(v_m, v_n) \frac{\frac{p_t}{\eta_{pa}}+p_{rf}+p_{bb}}{(1-\sigma_{dc})(1-\sigma_{ms})} + p_{st}\right)(1 - P_{idle})}.$$

$$(81)$$

3.5   Simulation Results

Based on our proposed 5G ultra-dense small cell network model in this paper, the numerical and simulation results are shown in this section. Some default parameters are configured as $l_i = 1$ km, $\zeta = 0.1$, $\alpha_p = 4$, $\rho_0 = 100$, $\eta = 1$, $p_s = 0.01$, $C = 10$, $p_t = 1.6$W, $\eta_{pa} = 8$, $p_{rf} = 0.7$W, $p_{bb} = 1.6$W, $p_{st} = 6.8$W, $\sigma_{dc} = 0.08$, $\sigma_{ms} = 0.1$. Without loss of generality, the centers of hot spots are uniformly distributed in $\mathbb{A}_T$. AEs of hot spots are configured to be $i.i.d.$ Poisson distributed with expectation 3000. The ratio of the stationary user density to the SBS density in CR_i is $\theta_i = \frac{P_i}{v_n(i)}$.

Comparison of the stationary number of users in a hot spot between numerical results and Monte Carlo simulation is shown in Fig. 15. Three hot spots are selected from $N_p = 50$ hot spots to make this comparison. It is observed from the figure that gaps between Monte carlo simulation and numerical results are very small, indicating that using queueing network theory to extend the R&A model to the temporal domain for the analysis of user density is practicable.

The impact of the SINR threshold γ_0 and the ratio θ_i on the coverage probability P_{cover} is investigated in Fig. 16. Figure 16 shows that when γ_0 is fixed the coverage

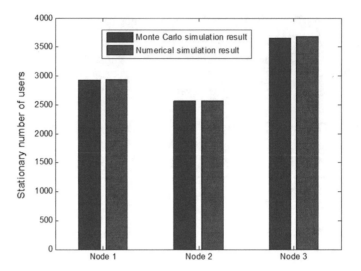

Fig. 15 Comparison between theoretical results and simulation for the stationary number of users in a hot spot

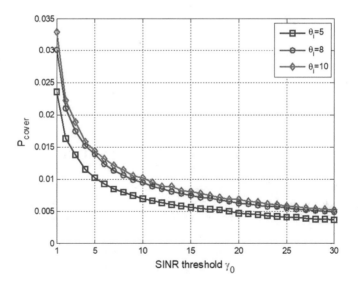

Fig. 16 Coverage probability with respect to the SINR threshold γ_0 considering different ratios of the stationary user density to the SBS density θ_i

probability P_{cover} decreases when increasing θ_i. When θ_i is fixed, the coverage probability P_{cover} decreases with the increase of γ_0. The observations is consistent with the results in [9, 47].

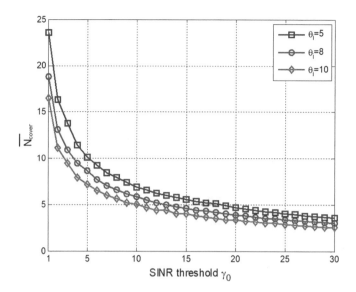

Fig. 17 Average number of SBSs in a VC with respect to the SINR threshold γ_0 considering different ratios of the stationary user density to the SBS density θ_i

The effect of the SINR threshold γ_0 and the ratio of the stationary user density to the SBS density θ_i on the average number of SBSs in a VC $\overline{N_{\text{cover}}}$ is evaluated in Fig. 17. Figure 17 reveals that when γ_0 is fixed $\overline{N_{\text{cover}}}$ decreases with the increase of θ_i. When θ_i is fixed, $\overline{N_{\text{cover}}}$ decreases with the increase of the SINR threshold γ_0. Similar results are observed in [9].

The influence of the SINR threshold γ_0 and the ratio θ_i on the number of average available SBSs in a VC $\overline{N_{\text{server}}}$ is shown in Fig. 18. As the figure shows, when γ_0 is fixed, $\overline{N_{\text{server}}}$ decreases with the increase of θ_i. When θ_i is fixed, $\overline{N_{\text{server}}}$ decreases with the increase of γ_0. By comparing Fig. 17 with Fig. 18, we find that when γ_0 and θ_i are configured to be the same, $\overline{N_{\text{server}}}$ in Fig. 17 is always smaller than $\overline{N_{\text{cover}}}$ in Fig. 16, due to the limited service capability of SBSs.

The effect of the SINR threshold γ_0 and the ratio θ_i on the blocking probability P_B is shown in Fig. 19. When γ_0 is fixed, the blocking probability P_B decreases with the increase of θ_i. When θ_i is fixed, the blocking probability P_B first increases with the increase of the SINR threshold γ_0. When the SINR threshold γ_0 is larger than a given threshold, the blocking probability P_B decreases with the increase of the SINR threshold γ_0.

The impact of the SINR threshold γ_0 and the ratio θ_i on the network energy efficiency is illustrated in Fig. 20. As the figure shows, when θ_i is fixed, the network energy efficiency EE_{HS_i} increases with the increase of γ_0. Moreover, when the SINR threshold γ_0 is fixed, the network energy efficiency EE_{HS_i} decreases with the increase of θ_i.

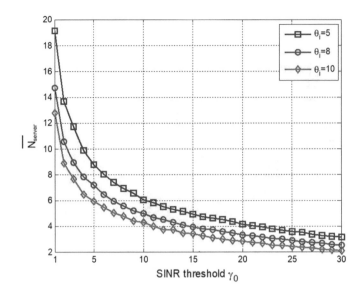

Fig. 18 Average number of available SBSs in a VC with respect to the SINR threshold γ_0 considering different ratios of the stationary user density to the SBS density θ_i

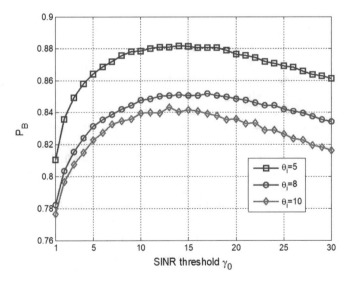

Fig. 19 Blocking probability P_B with respect to the SINR threshold γ_0 considering different ratios of the stationary user density to the SBS density θ_i

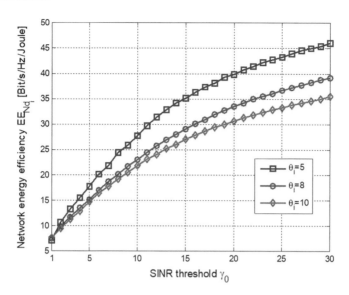

Fig. 20 Network energy efficiency EE_{HS_i} with respect to the SINR threshold γ_0 considering different ratios of the stationary user density to the SBS density θ_i

The impact of the SINR threshold γ_0 and the ratio θ_i on the energy efficiency of the small cell network contained in the QN system is shown in Fig. 21. The user density based deploying strategy (UDS) is implemented by deploying 1000 SBSs

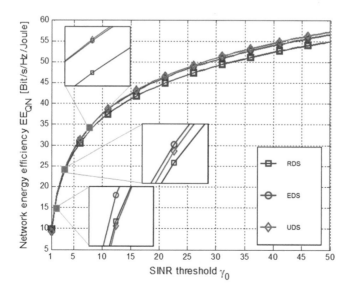

Fig. 21 Network energy efficiency EE_{QN} with respect to the SINR threshold γ_0 for different SBS deploying strategies

into 10 hot spots based on the stationary number of users in each hot spot (e. g. the ratio of user density to base station density is the same for each hot spot), and the randomly deploying strategy (RDS) is implemented by letting SBSs and users be randomly and uniformly distributed in each hot spot without considering R&A model. As the figure shows when $1 < \gamma_0 < 2.2$, energy efficiency of the network based on the RDS is larger than that based on the UDS. When $\gamma_0 > 2.2$, the UDS based network energy efficiency becomes larger than the RDS based network energy efficiency.

3.6   Conclusions

Based on the R&A model, the impact of the non-uniformly distributed users on performance of 5G ultra-dense small cell networks is firstly investigated in this paper. Moreover, the queueing network theory is firstly adopted to extend the analysis of R&A model to the temporal domain. Based on the theoretical analysis results for the stationary number of users, the coverage probability and the blocking probability are evaluated. Furthermore, by adopting virtual cell technology, the theoretical analyses of the throughput and the energy efficiency for each hot spot are proposed. By the simulations of the proposed model, we observe that when the virtual cell technology is adopted, the network energy efficiency is complicatedly coupled other than varying monotonically with the ratio of the stationary user density to the SBS density. This result provides insights into the important issues such as the optimal deployment of SBSs.

References

1. Ge, X., et al.: 5G ultra-dense cellular networks. IEEE Wirel. Commun. **23**(1), 72–79 (2016) (Art. no. 7422408)
2. Larsson, E.G., et al.: Antenna count for massive MIMO: 1.9 GHz vs. 60 GHz. IEEE Commun. Mag. **56**(9), 132–137 (2018)
3. Ge, X., et al.: Energy efficiency of small cell backhaul networks based on Gauss-Markov mobile models. IET Netw. **4**(2), 158–167 (2015)
4. Zhong, Y., et al.: Traffic matching in 5G ultra-dense networks. IEEE Commun. Mag. **56**(8), 100–105 (2018)
5. Hasan, Z., et al.: Green cellular networks: a survey, some research issues and challenges, (in English). IEEE Commun. Surv. Tutor. **13**(4), 524–540 (2011)
6. Demestichas, P., et al.: 5G on the horizon: key challenges for the radio-access network. IEEE Veh. Technol. Mag. **8**(3), 47–53 (2013) (Art. no. 6568922)
7. Shafi, M., et al.: 5G: a tutorial overview of standards, trials, challenges, deployment, and practice. IEEE J. Sel. Areas Commun. **35**(6), 1201–1221 (2017)
8. Andrews, J.G., et al.: What will 5G be? IEEE J. Select. Areas Commun. **32**(6), 1065–1082 (2014) Art. no. 6824752
9. Ge, X., et al.: User mobility evaluation for 5G small cell networks based on individual mobility model. IEEE J. Select. Areas Commun. **34**(3), 528–541 (2016) (Art. no. 7399689)

10. Zhong, Y., et al.: QoE and cost for wireless networks with mobility under spatio-temporal traffic. IEEE Access **7**, 47206–47220 (2019)
11. Héliot, F., et al.: An accurate closed-form approximation of the energy efficiency-spectral efficiency trade-off over the MIMO Rayleigh fading channel. In: IEEE International Conference on Communications (2011)
12. Héliot, F., et al.: On the energy efficiency gain of MIMO communication under various power consumption models. In: 2011 Future Network and Mobile Summit, FutureNetw 2011 (2011)
13. Liu, W., et al.: Energy efficiency of MIMO transmissions in wireless sensor networks with diversity and multiplexing gains. In: ICASSP, IEEE International Conference on Acoustics, Speech and Signal Processing—Proceedings, vol. IV, pp. 897–900 (2005)
14. Belmega, E.V., Lasaulce, S.: An information-theoretic look at MIMO energy-efficient communications. In: VALUETOOLS 2009—4th International Conference on Performance Evaluation Methodologies and Tools (2009)
15. Xu, J., et al.: Improving energy efficiency through multimode transmission in the downlink MIMO systems. Eurasip J. Wirel. Commun. Netw. **2011**(1) (Art. no. 200) (2011)
16. Xu, J., Qiu, L.: Energy efficiency optimization for MIMO broadcast channels. IEEE Trans. Wireless Commun. **12**(2), 690–701(Art. no. 6409501) (2013)
17. Miao, G., Zhang, J.: On optimal energy-efficient multi-user MIMO. In: GLOBECOM—IEEE Global Telecommunications Conference (2011)
18. Ngo, H.Q., et al.: Energy and spectral efficiency of very large multiuser MIMO systems. IEEE Trans. Commun. **61**(4), 1436–1449(Art. no. 6457363) (2013)
19. Chen, M., et al.: On the computation offloading at ad hoc cloudlet: architecture and service modes. IEEE Commun. Mag. **53**(6), 18–24 (Art. no. 7120041) (2015)
20. Chen, M., et al.: EMC: emotion-aware mobile cloud computing in 5G. IEEE Netw. **29**(2), 32–38 (Art. no. 7064900) (2015)
21. Chen, M., et al.: AIWAC: affective interaction through wearable computing and cloud technology. IEEE Wirel. Commun. **22**(1), 20–27 (Art. no. 7054715) (2015)
22. Elsawy, H., et al.: Stochastic geometry for modeling, analysis, and design of multi-tier and cognitive cellular wireless networks: a survey. IEEE Commun. Surv. Tutor. **15**(3), 996–1019 (Art. no. 6524460) (2013)
23. Andrews, J.G., et al.: A tractable approach to coverage and rate in cellular networks. IEEE Trans. Commun. **59**(11), 3122–3134 (Art. no. 6042301) (2011)
24. Ge, X., et al.: 5G wireless backhaul networks: challenges and research advances. IEEE Netw. **28**(6), 6–11 (Art. no. 6963798) (2014)
25. Chan, C.C., Hanly, S.V.: Calculating the outage probability in a CDMA network with spatial poisson traffic. IEEE Trans. Veh. Technol. **50**(1), 183–204 (2001)
26. Yu, S.M., Kim, S.L.: Downlink capacity and base station density in cellular networks. In: 2013 11th International Symposium and Workshops on Modeling and Optimization in Mobile, Ad Hoc and Wireless Networks, WiOpt 2013, pp. 119–124 (2013)
27. Guidotti, A., et al.: Simplified expression of the average rate of cellular networks using stochastic geometry. In: IEEE International Conference on Communications, pp. 2398–2403 (2012)
28. Renzo, M.D., et al.: Average rate of downlink heterogeneous cellular networks over generalized fading channels: a stochastic geometry approach. IEEE Trans. Commun. **61**(7), 3050–3071 (Art. no. 6516171) (2013)
29. Mukherjee, S.: Distribution of downlink SINR in heterogeneous cellular networks. IEEE J. Select. Areas Commun. **30**(3), 575–585 (Art. no. 6171998) (2012)
30. Govindasamy, S., et al.: Asymptotic spectral efficiency of the uplink in spatially distributed wireless networks with multi-antenna base stations. IEEE Trans. Commun. **61**(7), 100–112 (Art. no. 6528073) (2013)
31. Dhillon, H.S., et al.: Downlink MIMO HetNets: modeling, ordering results and performance analysis. IEEE Trans. Wirel. Commun. **12**(10), 5208–5222 (Art. no. 6596082) (2013)
32. Soh, Y.S., et al.: Energy efficient heterogeneous cellular networks, (in English). Ieee J. Select. Areas Commun. **31**(5), 840–850 (2013)

33. Srinivasa, S., Haenggi, M.: Modeling interference in finite uniformly random networks. In: Proceedings of International Workshop on Information Theory for Sensor Networks (WITS'07), pp. 1–12 (2007)
34. Ganti, R.K., Haenggi, M.: Interference and outage in clustered wireless ad hoc networks. IEEE Trans. Inf. Theory **55**(9), 4067–4086 (2009)
35. Elsawy, H., et al.: Characterizing random CSMA wireless networks: a stochastic geometry approach. In: IEEE International Conference on Communications, pp. 5000–5004 (2012)
36. Guo, A., Haenggi, M.: Spatial stochastic models and metrics for the structure of base stations in cellular networks. IEEE Trans. Wireless Commun. **12**(11), 5800–5812 (2013)
37. Win, M.Z., et al.: A mathematical theory of network interference and its applications. Proc. IEEE **97**(2), 205–230 (Art. no. 4802198) (2009)
38. Haenggi, M.: Mean interference in hard-core wireless networks. IEEE Commun. Lett. **15**(8), 792–794 (Art. no. 5934671) (2011)
39. Matérn, B.: Spatial Variation. Springer Science & Business Media (1986)
40. Chiu, S.N., et al.: Stochastic Geometry and Its Applications. Wiley (2013)
41. Elsawy, H., Hossain, E.: Modeling random CSMA wireless networks in general fading environments. In: IEEE International Conference on Communications, pp. 5457–5461 (2012)
42. Frost, V.S., Melamed, B.: Traffic modeling for telecommunications networks as new communications services evolve, professionals must create better models to predict system performance. IEEE Commun. Mag. **32**(3), 70–81 (1994)
43. Lilith, N., Doğançay, K.: Using reinforcement learning for call admission control in cellular environments featuring self-similar traffic. In: IEEE Region 10 Annual International Conference, Proceedings/TENCON, vol. 2007 (2005)
44. Ramakrishnan, P.: Self-similar traffic model. Technical Report CSHCN T.R.99–5 (ISR T.R. 99–12) (1997)
45. Norros, I.: On the use of fractional brownian motion in the theory of connectionless networks. IEEE J. Sel. Areas Commun. **13**(6), 953–962 (1995)
46. Karasaridis, A., Hatzinakos, D.: Network heavy traffic modeling using α-stable self-similar processes. IEEE Trans. Commun. **49**(7), 1203–1214 (2001)
47. Ge, X., et al.: Spatial spectrum and energy efficiency of random cellular networks. IEEE Trans. Commun. **63**(3), 1019–1030 (Art. no. 7015548) (2015)
48. Andrews, J.G., et al.: Overcoming interference in spatial multiplexing mimo cellular networks. IEEE Wirel. Commun. **14**(6), 95–104 (2007)
49. Cho, B., et al.: Bounding the mean interference in Matern Type II hard-core wireless networks. IEEE Wirel. Commun. Lett. **2**(5), 563–566 (Art. no. 6574907) (2013)
50. Simon, M.K., Alouini, M.S.: Digital Communication over Fading Channels: A Unified Approach to Performance Analysis. Wiley (2002)
51. Cui, S., et al.: Energy-efficiency of MIMO and cooperative MIMO techniques in sensor networks. IEEE J. Sel. Areas Commun. **22**(6), 1089–1098 (2004)
52. Arnold, O., et al.: Power consumption modeling of different base station types in heterogeneous cellular networks. In: 2010 Future Network and Mobile Summit (2010)
53. Chen, C.J., Wang, L.C.: Performance analysis of scheduling in multiuser MIMO systems with zero-forcing receivers. IEEE J. Sel. Areas Commun. **25**(7), 1435–1445 (2007)
54. Wang, L.C., Yeh, C.J.: Scheduling for multiuser MIMO broadcast systems: transmit or receive beamforming? IEEE Trans. Wirel. Commun. **9**(9), 2779–2791 (Art. no. 5529760) (2010)
55. Telatar, E.: Capacity of multi-antenna Gaussian channels. Eur. Trans. Telecommun. **10**(6), 585–595 (1999)
56. Ge, X., et al.: Capacity analysis of a multi-cell multi-antenna cooperative cellular network with co-channel interference. IEEE Trans. Wireless Commun. **10**(10), 3298–3309 (Art. no. 6064713) (2011)
57. Masouros, C., et al.: Large-scale MIMO transmitters in fixed physical spaces: the effect of transmit correlation and mutual coupling. IEEE Trans. Commun. **61**(7), 2794–2804 (Art. no. 6522419) (2013)

58. Paulraj, A., et al.: Introduction to space-time wireless communications. In: Introduction to Space-Time Wireless Communications, pp. 1–270 (2003)
59. Baccelli, F., Blaszczyszyn, B.: Stochastic Geometry and Wireless Networks-Volume I : Theory. Now Publishers Inc (2009)
60. Cioffi, J.M.: A Multicarrier Primer (1991)
61. Marzetta, T.L.: Noncooperative cellular wireless with unlimited numbers of base station antennas. IEEE Trans. Wirel. Commun. 9(11), 3590–3600 (Art. no. 5595728) (2010)
62. Humar, I., et al.: Rethinking energy efficiency models of cellular networks with embodied energy. IEEE Netw. 25(2), 40–49 (Art. no. 5730527) (2011)
63. Wu, G., et al.: Recent advances in energy-efficient networks and their application in 5G systems. IEEE Wirel. Commun. 22(2), 145–151 (Art. no. 7096297) (2015)
64. Stefanatos, S., Alexiou, A.: Access point density and bandwidth partitioning in ultra dense wireless networks. IEEE Trans. Commun. 62(9), 3376–3384 (Art. no. 6883156) (2014)
65. Samarakoon, S., et al.: Ultra dense small cell networks: turning density into energy efficiency. IEEE J. Select. Areas Commun. 34(5), 1267–1280 (Art. no. 7439746) (2016)
66. López-Pérez, D., et al.: Towards 1 Gbps/UE in cellular systems: understanding ultra-dense small cell deployments. IEEE Commun. Surv. Tutor. 17(4), 2078–2101 (Art. no. 7126919) (2015)
67. Tseng, F.H., et al.: Ultra-dense small cell planning using cognitive radio network toward 5g (in English). IEEE Wirel. Commun. 22(6), 76–83 (2015)
68. Zhou, F., et al.: Energy-efficient optimal power allocation for fading cognitive radio channels: ergodic capacity, outage capacity, and minimum-rate capacity. IEEE Trans. Wirel. Commun. 15(4), 2741–2755 (Art. no. 7358164) (2016)
69. Zhong, Y., et al.: On the stability of static poisson networks under random access. IEEE Trans. Commun. 64(7), 2985–2998 (Art. no. 7486114) (2016)
70. Zhong, Y., et al.: Heterogeneous cellular networks with spatio-temporal traffic: delay analysis and scheduling. IEEE J. Select. Areas Commun. 35(6), 1373–1386 (Art. no. 7886285) (2017)
71. Zhang, T., et al.: Energy efficiency of base station deployment in ultra dense HetNets: a stochastic geometry analysis. IEEE Wirel. Commun. Lett. 5(2), 184–187 (Art. no. 7377022) (2016)
72. Ding, M., et al.: Performance impact of LoS and NLoS transmissions in dense cellular networks. IEEE Trans. Wireless Commun. 15(3), 2365–2380 (Art. no. 7335646) (2016)
73. Yunas, S., et al.: Spectral and energy efficiency of ultra-dense networks under different deployment strategies. IEEE Commun. Mag. 53(1), 90–100 (Art. no. 7010521) (2015)
74. Park, J., et al.: Tractable resource management with uplink decoupled millimeter-wave overlay in ultra-dense cellular networks. IEEE Trans. Wirel. Commun. 15(6), 4362–4379 (Art. no. 7430349) (2016)
75. Gao, Z., et al.: MmWave massive-MIMO-based wireless backhaul for the 5G ultra-dense network. IEEE Wirel. Commun. 22(5), 13–21 (Art. no. 7306533) (2015)
76. Galinina, O., et al.: 5G multi-RAT LTE-WiFi ultra-dense small cells: performance dynamics, architecture, and trends. IEEE J. Sel. Areas Commun. 33(6), 1224–1240 (2015)
77. Ghazanfari, A., et al.: Ambient RF energy harvesting in ultra-dense small cell networks: performance and trade-offs. IEEE Wirel. Commun. 23(2), 38–45 (Art. no. 7462483) (2016)
78. Ge, X., et al.: Millimeter wave communications with OAM-SM scheme for future mobile networks. IEEE J. Select. Areas Commun. 35(9), 2163–2177 (Art. no. 7968418) (2017)
79. Wu, Q., et al.: Energy-efficient resource allocation for wireless powered communication networks. IEEE Trans. Wirel. Commun. 15(3), 2312–2327 (Art. no. 7332956) (2016)
80. Chen, S., et al.: User-centric ultra-dense networks for 5G: challenges, methodologies, and directions. IEEE Wirel. Commun. 23(2), 78–85 (Art. no. 7462488) (2016)
81. Kim, J., et al.: Virtual cell beamforming in cooperative networks. IEEE J. Select. Areas Commun. 32(6), 1126–1138 (Art. no. 6827165) (2014)
82. Wang, J., Dai, L.: Downlink rate analysis for virtual-cell based large-scale distributed antenna systems. IEEE Trans. Wirel. Commun. 15(3), 1998–2011 (Art. no. 7317799) (2016)

83. Nie, W., et al.: User-centric cross-tier base station clustering and cooperation in heterogeneous networks: rate improvement and energy saving. IEEE J. Select. Areas Commun. **34**(5), 1192–1206 (Art. no. 7448831) (2016)

84. Hong, M., et al.: Joint base station clustering and beamformer design for partial coordinated transmission in heterogeneous networks. IEEE J. Select. Areas Commun. **31**(2), 226–240 (Art. no. 6415394) (2013)

85. Feng, Z., et al.: An effective approach to 5G: wireless network virtualization. Commun. Mag. IEEE **53**(12), 53–59 (2015)

86. Song, C., et al.: Limits of predictability in human mobility. Science **327**(5968), 1018–1021 (2010)

87. Song, C., et al.: Modelling the scaling properties of human mobility. Nat. Phys. **6**(10), 818–823 (2010)

88. Zipf, G.K.: The P1P2/D hypothesis: on the intercity movement of persons. Am. Sociol. Rev. **11**(6), 677–686 (1946)

89. Simini, F., et al.: A universal model for mobility and migration patterns. Nature **484**(7392), 96–100 (2012)

90. Jackson, J.R.: Networks of waiting lines. Oper. Res. **5**(4), 518–521 (1957)

91. Santaló, L.A.: Integral geometry and geometric probability. In: Encyclopedia of mathematics and its applications, vol. 1. Cambridge University Press, Cambridge, U.K. (1976)

92. Guo, J., et al.: Outage probability in arbitrarily-shaped finite wireless networks. IEEE Trans. Commun. **62**(2), 699–712 (Art. no. 6712183) (2014)

93. Abate, J., Whitt, W.: Numerical inversion of Laplace transforms of probability distributions. ORSA J. Comput. **7**(1), 36–43 (1995)

94. O'Cinneide, C.A.: Euler summation for Fourier series and Laplace transform inversion. Commun. Stat. Part C Stochastic Mod. **13**(2), 315–337 (1997)

95. Auer, G., et al.: Energy efficiency analysis of the reference systems, areas of improvements and target breakdown. INFSO-ICT-247733 EARTH2012. Available: http://www.ict-earth.eu/

Geocommunity Based Data Forwarding in Social Delay Tolerant Networks

Jagdeep Singh, Sanjay Kumar Dhurandher, and Isaac Woungang

Abstract The Delay Tolerant Networks (*DTNs*) are types of ad hoc networks that can be used to realize a wide range of applications in a challenged environment. Despite the challenges of intermittent connectivity and unpredictable mobility characteristics, the mobile nodes need to communicate and share their valuable information without the need of an infrastructure. Socially aware networking is an emerging paradigm for high-efficiency data dissemination. Existing protocols take advantage of mobile node's social characteristics such as user interests to improve the dissemination performance. However, these protocols have not exploited enough the types of relations that are valuable between user interests and how these relations can affect the dissemination of social *DTNs*. In this context, this Chapter inverstiagtes the problem of data forwarding in social-based networks. Several realistic datasets will be explored to reveal both the geographical and social regularities of human mobility; the concepts of geocommunity and geocentrality into social network analysis will also be studied and the geocommunity characteristics as well as the envisioned application areas different from general *DTNs* will be explored. From a social network perspective, people sharing interesting properties such as common hobbies, social functions and occupations, will tend to form a community. Through a trace-based study, an interesting phenomenon in social *DTNs* was observed, which is that the community always strongly relates to its geographical location. Motivated by this fact, this Chapter will also study the existing location-based routing schemes based on the message endpoint, forwarding utility computation approach, relay node selec-

J. Singh (✉)
Department of Computer Science and Engineering,
Sant Longowal Institute of Engineering and Technology, Longowal, India

S. K. Dhurandher
Department of Information Technology, Netaji Subhas University of Technology,
New Delhi, India

I. Woungang
Department of Computer Science, Ryerson University, Toronto, ON, Canada

421

P. Nicopolitidis et al. (eds.), *Advances in Computing, Informatics, Networking and Cybersecurity*, Lecture Notes in Networks and Systems 289,
https://doi.org/10.1007/978-3-030-87049-2_14

tion, routing decision and message dissemination. The current challenges and future research direction for geocommunity-based routing in social delay tolerant networks will also be highlighted.

Keywords Social delay tolerant networks · Data dissemination · Broadcasting · Geography · Community

1 Introduction

DTNs [2] are characterized by high delays, imbalance data rates, and poor connectivity. In this kind of network, routing relies on the store-carry-and-forward mechanism. The consideration of social attributes provides a new angle of view in the design of *DTN* routing protocols. In most *DTN* applications, a multitude of mobile devices are used and carried by people, whose behaviors are better characterized by social models [3]. One of the most important applications of *DTN* is in the disaster scenarios [8, 9], where 3G and 4G mobile networks are inaccessible and there is no connection between the mobile devices. DTNs could play an important role in this kind of scenarios by establishing a network connection between mobile devices to deliver some crucial information to the end users. With the current addressing methodologies in *DTNs*, the information should be sent individually to each node or broadcasted to all the nodes in the network. However, the rescue notifications may vary between different regions. With an effective implementation of the DTN, each area could be grouped based on the location, and each user will get some notifications specific only to its own area. Also, this will reduce the network traffic due to the unnecessary message broadcasting to all unwanted users associated with broadcasting.

In Fig. 1, the user applications contact the *DTN* protocols, generating bundles that are stored in persistent storage until the bus passes by the village. Then, these bundles are transferred to the *DTN* agent in the bus and carried to a town to be delivered to an Internet gateway.

These procedures are illustrated in Fig. 2 for the case of an email service [4]. An application protocol, named Opportunistic Connection Management Protocol (OCMP), runs on top of the bundle layer. *OCMP* provides the interface between the email client and the *DTN* agent at the kiosk and between the *DTN* agent at the gateway and an email server [5].

It was found by many researchers that some *DTN* instances such as mobile social networks (MSNs) exhibit the human behaviours, where mobile users move around, communicate and share the data with each other using mobile devices such as smartphones, laptops, and tablet PCs. An extensive research in the area of DTNs have revealed that human activities have an extremely strong regularity, presenting a small-world network phenomenon in ubiquitous personal communication. The social characteristics in specific *DTN* application scenarios should be exploited accurately for relay node selection purpose. This fact has been validated by several proposals of social-based routing protocols. In social networks, a node usually has some com-

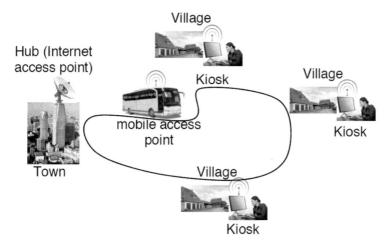

Fig. 1 A route bus and *DTN* protocols offer the gateway between the kiosks and the Internet at a neighboring town [1]

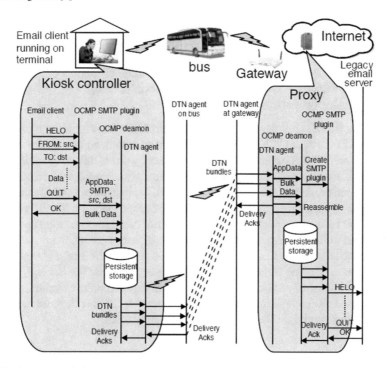

Fig. 2 The opportunistic connection management protocol runs on top of the *DTN* bundle layer to allow email transfers [1]

plicated social relationships with other nodes rather than being completely isolated because of the necessary social daily activities. This reality opens new possibilities of designing new social-based *DTN* routing protocols [6, 7], in which the knowledge of the social characteristics are utilized to make a better forwarding decision in *DTN* routing. Indeed, in *DTN* routing, some well-known social properties can be utilized for the purpose of relay node selection (e.g.: community, centrality, similarity, friendship, to name a few). In such type of networks, the social characteristics are used for data transmission in unicasting, anycasting, multicasting or geocasting scenarios. In this chapter, we have considered the geocasting scenario, in which the message is transmitted on the basis of geographical location of particular type of nodes. This group of nodes is referred to as geocommunity.

In many application scenarios of *DTNs*, location is an important factor that has been considered. Nowadays, mobile devices utilize the "location services" to specify their own location coordinates with reasonable accuracy. In the geocasting approach, the messages are delivered to a group of nodes that are located in the same geographical region. Besides, the implementation of an effective geocast addressing methodology has a significant potential in the real world use of opportunistic networks. Examples of such applications include:

1. Geographical notification for emergency situations such as fire alarms and natural disasters.
2. Location targeted advertising where a large volume of users is concentrated at specific locations to attend music festivals or sports events.
3. Geographically restricted service discovery.

In the geocasting approach [10], geocommunity is used as one of the most important factors in the routing decision. Geocommunity is defined as a group of nodes (users) that are tightly coupled to each other, either by direct link or by easily accessible users that act as relay nodes. In such community, the participating nodes share some common properties such as social functions, geolocations, and occupations. In this Chapter, we have discussed the geocasting concepts, along with different routing protocols based on geocasting. We have also simulated some existing benchmark schemes that utilized a geocast approach for message forwarding.

The remainder of the Chapter is organized as follows. In Sect. 2, some related work on geocast routing protocols for OppNets are presented. The Sects. 3, 4, 5 and 6 present the open challenges in DTNs w.r.t. security, green communication, machine learning based routing and multigeocasting respectively. In Sect. 7, some simulation results are discussed. Finally, Sect. 8 concludes the Chapter.

2 Related Work

In the literature, a significant amount of work has been done on geocast routing protocols [11–37]; the quality of each of these protocols been judged based on factors such as robustness, probability of message reaching the destination, scalability, path topol-

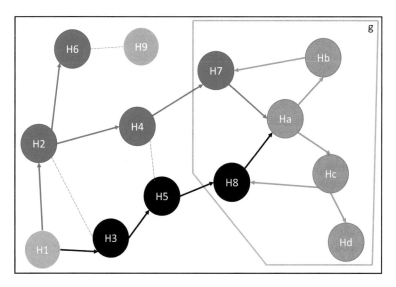

Fig. 3 Single Geocast region

ogy and message complexity. Typically, geocast routing protocols are characterized based on whether they require flooding, directed flooding, or no flooding. Some of these protocols are location-based multicast (*LBM*), *GeoNode*, *Mesh*, Voronoi-based multicast, *GeoTORA*, and *GeoGrid*.

Generally, in the geocasting approach, the message from a source node is destined to a destination in specific location, therefore some kind of destination information (such as the Endpoint Identifier (EID)) must be included in the message. For example, if the casts are pre-defined at the deployment time and known to all devices, a message may carry a cast identifier; otherwise, the cast definition (e.g. centre/radius pair or coordinates of a polygon as in our approach) must be in the message. Whenever a node receives a message, it compares its own location with the *EID* of the message. This device is a recipient of the message if it currently resides within the cast defined by the *EID* and the message has not expired yet. According to the *DTN* architecture specification [2], the EIDs can be obtained based on the IP address of the mobile device to retain their uniqueness.

The approach to design a geocasting routing scheme consists of two separate phases:

- Phase 1: Carrying the message to its destination cast.
- Phase 2: Delivering the message to all devices within the destination cast (Fig. 3).

Representative geocast routing protocols for DTNs are as follows.

In [16], the authors proposed GeoGrid, a geocast routing protocol that consists of a sender, a receiver, a forwarding zone, a destination region, intermediate nodes, a gateway, and logical grids. The basic idea of this approach is that the network is divided into logical grids, each of which has one elected gateway; and only this gateway can forward the message packets rather than all the nodes in the grid. This helps in avoiding inefficient and redundant flooding. In this approach, each grid has one gateway responsible for forwarding the packets. The sender sends the message packet to its neighbouring gateways. Gateways that are outside the forwarding zone are discarded. If the message packet goes outside the forwarding region, then it is discarded; otherwise it is retransmitted by the gateway to its neighbouring gateways until the message get to the destination region. Once the message reaches that destination region, a node transmits the packet further. This scheme is based on two phases that consists of unicast forwarding from the sender until the message packet reaches any mobile node inside the destination region; and once the packet reaches the destination region, it is flooded. Each mobile node which is a part of the message forwarding checks if it belongs to the destination region. If that is the case, then the message is flooded by the node; otherwise, the message forwarding continues.

In [17], the authors proposed GeoNode, a geocast routing protocol that consists of a GeoNode cell, a GeoRouter, a GeoHost, a sender and a destination region [9]. Here, the GeoRouters are responsible for the transmission of a message packet from a sender to the receiver. They are also aware of their service area and they operate with other routers within that service area. On the other hand, the GeoNodes act as the receivers of the message and they store the incoming packets and multicast them to their service area periodically. Besides, the GeoHost notifies the current location address of the GeoNode. It can also receive and send the geocast packets between the GeoNodes. In fact, the routing in a GeoNode involves sending the packet, commuting between routers, and receiving the packets. It starts with the GeoHost being asked for the IP address of the GeoNode. Then, the packet is forwarded to the GeoNode; which in turn forwards it to the GeoRouter. The GeoRouter then checks whether its own service area and destination polygon intersect with each other. If the service area and destination polygon do not intersect, the message packet is sent to the parent router. Meanwhile, a copy of the message packet is also sent if the service area and destination polygon slightly intersect each other. In that case, the GeoRouter scans each child's service area and then sends a duplicate of the message packet if they intersect each other. It then transmit the packet to the destination region bound (GeoNode), which in turn floods the area with the message packet.

In general, geographic routing protocols for DTNs can be categorized on the basis of the awareness of the destination. The primary objective is to achieve a coherent message replication through the process of geometric utility, without the need of tracking where the destination is at the current moment. This type of routing protocol does not take into consideration the location of the destination while routing the message at any point in time.

The main mechanism of Vector Routing [18] is to replicate the message that is to be delivered on the basis of an encounter angle among the encountered nodes that form a pair. The idea behind this protocol is that it tends to replicate the messages directly proportionally to the value of the encountered angle, meaning that if the given numeric value of this angle is small, a smaller number of messages will be replicated and vice-versa. This is attributed to the fact that pairwise encountered nodes that are moving in the same direction could cause redundant duplication, which will unnecessarily increase the network traffic. Even if the two nodes in question are moving in completely different directions, it would be redundant to create duplicates copies of the messages provided. The reason behind this is the that the encountered node is presently moving with a previous trajectory of the message carrier. Since the vector routing protocol emphasizes on the replication of a small number of messages, the results show that it has a very low routing overhead compared to other mechanisms such as the blind flooding protocol.

In [12], the authors proposed the so-called RoRo-LT, a routing protocol for DTNs, in which the location of a node at a certain time is exploited to gain further insights on its future location [10]. In this scheme, a long term (LT) observation - usually for several weeks or even months - is carried out from the spatio temporal history in order to approximate the location of the node in the future. In doing so, self-periodicity is considered as a measure of the similarity between the current routine of a node and its historical pattern. If the similarity level of the current routine and the historical pattern are high, then the current routine will serve as basis for predicting the node mobility. Besides, by using the projected future trajectories of pairwise encountered nodes, the message carried by a node, say node X, will be forwarded to an encountered node, say node Y, if and only if these nodes are estimated to be distant from each other in the near future. This is a stark contrast from the mechanism of Vector Routing, in which the number of messages that are to be replicated are estimated on the basis of the mobility of pairwise encountered nodes. Although the approach used in the design of this protocol seems intuitive - as a longer training period is taken to estimate the future location of a node from its past historical data, the simulation results showed that this protocol does not perform well. The authors reported that this may be attributed to the fact that a method to track the message destination was lacking in the design. They also reported that exploring the numerous geometric features and characteristics of a node to further select the relay nodes may have led to an improvement of the routing performance of this scheme.

In [19], the authors proposed the so-called MOVE routing protocol for DTNs, in which a message from the source node is forwarded towards the destination based on the moving direction of the node. Here, the term 'forward' implies that a duplicate copy of the message will not be generated throughout the course of the routing process. Moreover, based on the consistent moving direction among the pairwise encountered nodes, the relative distance is used to discern the node which does not have any substantial contribution to the message delivery. This implies that the said node is further away from the destination node. The major shortcoming of this scheme is that it does not take into account the nodal moving speed, noting that the node with a faster moving speed has a relatively faster proximity to the destination and as such could help reduce the delivery delay. Furthermore, this scheme does not address the problem of local maximum, where the relay node with a shorter distance to the destination may not always be present.

In [20], the authors introduced Geoopp, a routing protocol for opportunistic networks that uses the navigation system to determine the route that the message has to follow to reach its destination. The whole premise behind Geoopp is to choose the nearest point (NP) to the destination along the route that has been suggested by the navigation system. In this message forwarding phase, the utility function used to determine the relay node for delivering the message toward its destination relies on the Minimum Estimated Time of Delivery (METD), the Estimated Time of Arrival (*ETA*) (i.e. the time taken to travel from the current node location to the NP), and the distance from the NP to the final destination (so-called *DistNP*). Besides, Geoopp takes into consideration the nodal moving speed, and it was proved that it achieves a much lower delivery latency compared to the MOVE protocol [19]. The most important aspect of the *Geoopp* routing scheme is the selection of the NP, which requires the knowledge of the underlying network topology. As a result, some concerns arise due to the high level of sophistication that prevails in different road/network topologies. Some of these concerns include scalability in large scale road systems since the algorithm used to find the NP would require a greater amount of time for completion, resulting in increased computational complexity. Furthermore, the local maximum problem is also prevalent in Geoopp as since the node with a lower *METD* value may not always be present in the network. These factors directly impact the calculation of the NP, a task that requires a lot of resources. From a practical perspective, these concerns may hamper the performance of the Geoopp protocol.

In [21], the authors proposed AeroRP, a routing protocol for aeronautical networks. The premise behind this scheme is that the selection of the relay node that will forward the message toward the destination is based on a combination of distance and moving direction, while taking into account the nodal speed. This scheme only takes into account the scenario that pairwise encountered nodes are heading towards the destination node, but it fails to address the case when the message carrier is moving away from the destination node even if its paired node is moving towards the destination. As a result, the message will never be forwarded because the TTL value is negative, which is not a valid value for establishing a routing path decision. Consequently, the message is not delivered. In addition, some limitations in terms of routing performance of aeronautical networks may be attributed to the high mobility of nodes and the sparsity of the network density.

In [22], the authors proposed the so-called Delegation Geographic Routing (DGR) for DTNs, in which the shortcoming of the AeroRP routing protocol [21] have been addressed by comparing the TTL value of the node encountered historically against the TTL of the currently encountered node. As a result, any unsuccessful message delivery or failure of routing due to opposing moving directions between the pairwise encountered nodes is avoided. This mechanism is enforced through the Delegation Forwarding (*DF*) optimization policy, which always reports on the TTL of the relay node that has been selected after successful transmission of the message. Consequently, the number of message duplicated copies is shown to be reduced.

In [23], the authors proposed geoDTN, a routing protocol for DTNs, in which the historical nodal movement data is used to determine the intersection area where any two nodes have encountered each other previously. This is then used as a numerical metric to score the encountered likelihood. This is analogous to the scenario where two people may move around in the same zone because of their social habits. In this scheme, the distance model uses the minimum distance to the final destination node if the scores of the paired nodes are below a certain predefined threshold value.

Due to several predefined parameters, scalability is one of the major issues faced by *geoDTN*.

In [24], the authors proposed a routing protocol for mobile sensor networks (called MPAD), in which a node independently makes the decision to replicate the messages and forward them only to those nodes whose probability of meeting the sink node is high. This scheme involves two components: queue management and data transmission. The former is responsible for decision making on where and when the message should be transmitted or not based on the node's delivery probability. In doing so, the messages are replicated selectively to those nodes that have the higher probability of meeting the sink node. The latter is designed to help enhancing the message survival time when the message is routed toward the destination while minimizing the transmission overhead. In this part, the messages that have higher survival time are dropped, making the full use of the network bandwidth while reducing the network energy consumption. The performance of the MPAD scheme is simulated and compared against few benchmarks, showing promising results in terms of delivery ratio.

In [28], the authors proposed the so-called Distance Aware Epidemic Routing (DAER) for DTNs, which exploits the real time geographic information of nodes that are present in the network to perform the message delivery. In this scheme, it is presumed that the mobile destination's real time location, i.e. the current location of the mobile destination, can be obtained from a centralized location service system, irrespective of the amount of delay in collecting this information in a sparsely populated network. If any node that is encountered by the message carrier has a relatively shorter distance to the final destination than the message carrier, then the message is replicated with its copy. After the message transmission is successfully carried out, the original message carrier that had the message from the start, moving further away from the destination node, would discard the message in its local buffer. The main advantage of this policy is that the routing overhead will be decreased by avoiding additional message replication redundancy. The main issue with this scheme is how to procure the real time location of the final destination node given the challenges present in these networks. In addition, this scheme only depends on a distance metric and completely overlooks the prediction of the nodal mobility. As a result, the local maxima problem also prevails as a node with a closer distance to the final destination may not always be present.

In [34], the authors proposed the so-called Packet Oriented Routing (POR) for DTNs, an extension of the DAER protocol [28], in which the decision for routing a message from source to destination relies on parameters such as distance to destination and number of message replicas. The whole premise behind the POR scheme is to find a suitable longer distance route in order to replicate fewer messages, with the sole notion of bolstering transmission reliability. The simulation results have shown the superiority of the POR scheme over the DAER scheme, especially when the transmission bandwidth is limited. Some of the drawbacks of the DAER protocol still remain in the POR scheme such as the issue of dependency on the centralized location server, the prediction of nodal mobility and the local maxima problem.

In [30], the authors proposed the so-called LAROD-LoDiS protocol for DTNs, which takes into account the historical geographic information to estimate the probable location of the destination. In this scheme, the historical location data of the destination is used as a measure to judge where the destination is at the present moment. The primary involvement of the LAROID-LoDiS scheme, especially in sparse network topology, is the implementation of a location service system, which is done by keeping the records of the different nodes in a local database. The information has to be updated to keep every node well aware of the current network scenario with the aid of the broadcast gossip with routing overhearing. Based on the predicted replication zone, in order to minimize the routing redundancy, the distance metric is utilized to replicate the message towards the intended destination. This routing scheme also faces the local maximum problem with respect to the distance and mobility prediction since the speed and direction parameters are not involved in the design of the routing process. Another issue with this scheme is that the local database may not always have the most up to date record of the current node's location, which directly affects the reliability of the routing decision. Indeed, if the node's record contains an old location value, the routing decision would be completely wrong and not suitable for message transmission.

Some routing protocols referred to as hybrid class of protocols [31–33], are an amalgamation of the previous POR and LAROD-LoDiS categories of protocols in the sense that in the initial phases of their designs, when the location of the destination is not available, the routing mechanisms of the Destination Unawareness category is utilized. As the subsequent routing phases continue, the location of the destination is estimated by means of the historical geographic information using the Destination Awareness class.

In [48], the authors proposed the so-called Centrality based Geocasting in opportunistic networks, which consists of two parts: (1) transmitting the messages in the direction of their destination cast; and (2) flooding them to the nodes within the destination cast. In the first part, initially, a sequence ID S is set to the message and each time the message is duplicated and transferred to another node, S is decremented by 1. At the point when the S value reaches zero, the message can't be handed-off any longer and it is expelled when the local buffer gets occupied completely or when the message runs out. In the second part, the message is distributed to all nodes within the cast by using an intelligent centrality-based flooding strategy, in which a higher degree of centrality means that the node connects with a larger number of peers in the network.

In [50], the authors proposed a fuzzy-based geocast routing protocol for opportunistic networks, which is an extension of the GSAF protocol [10]. In this scheme, a fuzzy technique is used in the selection of the next hop for message transmission. Some fuzzy attributes such as speed, residual energy, direction, and buffer space of nodes, are used in determining the relay nodes to carry the message toward its destination.

In [49], the authors proposed an energy-efficient check and spray Geocast routing protocol for opportunistic networks scheme, which is an energy version of GSAF. In this scheme, all the messages are scheduled to a targeted area, and the information on the endpoint identifier is embedded into the messages block. Whenever a node receives a message, it matches it using its own location with the message's endpoint identifier before forwarding it to any other node. The message endpoint identification and the cast definition are used for membership check. Also, an energy-efficient controlled flooding mechanism is utilized to distribute the geomessages to all nodes within the geocast region.

The above discussed routing protocols are summarized in Table 1, where they are compared on the basis of implementation tool used (Glomosim [58], ns-2 [57], ONE [60], AQUA-GLOMO [59]), and the techniques used in phases 1 and 2. The open challenges that can be addressed are shown in Table 2. (here Yes means: This can be done, No means: This is already done).

3 Open Challenges W.r.t. Security

Security is a concern in DTNs in the sense that in such networks, the routers and the gateways should be authenticated, and the sender information should be authenticated by the forwarding nodes. In this way, the transport of the prohibited traffic would be prevented at the earliest opportunity. Security in DTNs presents some research challenges [38–40]. The requirement for out-of-band contacts; for example, checking that the public keys are not in the blacklist in the Internet is problematic in DTNs. The issue of the lack of a delay tolerant method for key management also prevails. In addition, methods for the protection of the network against traffic analysis are to be investigated, as well as a policy for introducing new nodes in the network without impacting the security level.

4 Open Challenges W.r.t. Green Communication

The scalability of the proposed DTN protocols has yet to be proved. Therefore, green communication is one of the primary research areas in DTNs [41, 42]. There are no large scale deployments of DTNs running the protocols, and all studies for a large number of nodes are currently based on simulations. Scalability of the network includes also the issues of how to effectively manage the network and how to deal with interoperability problems when several network operators get involved.

Table 1 Geocasting based routing protocols

Protocol	Phase 1 technique	Phase 2 technique	Tool
GeoGrid [16]	Unicast forwarding	Flooding	GlomoSim
GeoNode [17]	Polygon intersection technique	Flooding	GlomoSim
Vector Routing [18]	Encounter angle based forwarding	Blind Flooding	GlomoSim
RoRo-LT [12]	Forwarding based on location at corresponding time	Pair wise replication of messages	GlomoSim
MOVE [19]	Relative distance based forwarding	Flooding	GlomoSim
Geoopp [20]	Minimum Estimated Time of Delivery based forwarding	Intelligent delivery	ONE
AeroRP [21]	Aeronautical Nodal based forwarding	Flooding	GlomoSim
DGR [22]	Delegation Forwarding (DF) optimization policy	Intelligent Flooding	ONE
geoDTN [23]	Historical nodal movement data consideration	Probability	ONE
MPAD [24]	Mobility prediction based transmission	Backup metric based delivery	ns-2
DAER [28]	Real time geographic information based forwarding	Shorter distance based delivery	ONE
POR [34]	Real time geographic information based forwarding	Optimized distance based delivery	ONE
LAROD-LoDiS [30]	Gossip based forwarding	Distance based delivery	ONE
CGOPP [48]	Unicasting	Intelligent Flooding using centrality	ONE
FGSAF [50]	Fuzzy based message transmission	Intelligent Flooding	ONE
EECSG [49]	Energy efficient ticket based technique	Check and spray based delivery	ONE
Floating Content [29]	Content sharing based forwarding	Flooding by calculating center and radius of circle	ONE
EVR [34]	Expected probability of each user visiting to other nodes/cast	highest EVR rate based message forwarding in inside cast	ONE
RMTG [35]	Greedy selection	Multicast tree formation for routing	GlomoSim
ERMTG [36]	Current node module	Receiving node module	AQUA-GLOMO
RMTG-HD [37]	Hole detection	Boundary routing based routing	GlomoSim

Table 2 Open challenges that can be addressed

Open challenge crea	GeoEpidemic	GSAF	CGOPP	EECSG	FGSAF
Security	Yes	No	No	Yes	Yes
Green communication	Yes	Yes	Yes	No	Yes
Machine learning	Yes	Yes	Yes	Yes	Yes
Multigeocasting	Yes	Yes	Yes	Yes	Yes

Yes means: This can be done, No means: This is already done

5 Open Challenges W.r.t. Machine Learning Based Routing

The bundle protocol of DTNs does not describe how the routes are to be setup between the nodes. It only deal with the forwarding plane. In addition, the control plane issues remain open [43–46]. To address these issues, various works that use machine learning techniques are available in the literature. It is worth mentioning that the performance of most of the routing protocols highly depends on the level of cooperation and autonomy of nodes. By default, most of the protocols assume a full node cooperation and little attention has been devoted to the study of the effect of reduced levels of cooperation. Indeed, by applying and fine-tuning simple knowledge-based cooperation mechanisms, the routing performance can be considerably improved.

6 Open Challenges W.r.t. Multi Geocasting

The geocast-based and multi-geocast-based routing protocols for *DTNs* have been less investigated. Geocast routing protocols use the fact that the group of destinations is defined by their geographic locations whereas the multigeocast routing protocols is designed based on the principle that the geocast regions are multiple in number. Geocast is of significant importance to DTNs as frequently, the information to be disseminated is only locally useful [34, 47, 50]. In addition, the geocast approach imposes some restriction on the resource consumption. As an example, Fig. 4 illustrates a multi geocast routing scheme with three geocast regions g, g_1, and g_2.

7 Experimental Analysis

In this Section, we present an experimental analysis of selected geocasting protocols, namely *FGSAF* [50], *GSAF* [10], *CGOPP* [48], *EECSG* [49] and *GeoEpidemic* [25] using the ONE simulator [60]. The workflow of simulator is represented in Fig. 5.

The default total number of nodes in our simulations, unless otherwise specified, is 195. Also, we have experimented with different sizes of system buffers (5, 10, 15, 20, 25 and 30 MB; the default being 10 MB) and various lifespans of messages

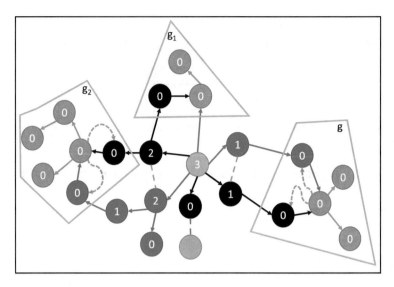

Fig. 4 Multi Geocast region

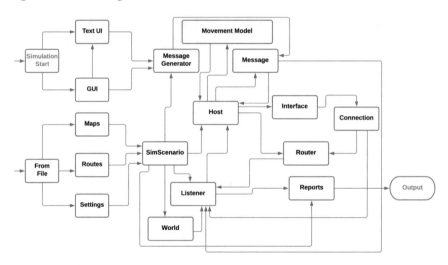

Fig. 5 ONE simulator working flow

(100, 150, 200, 250, 300, 350 min; the default being 300 min). All simulations were running for 8 h; for each simulation, the warm-up and cool-down periods were 1 h each. The messages are scheduled as follows: a sender and multiple destination casts are selected at random from the set of devices in the networks and the set of predefined casts, respectively; the message weight is fixed to 500 KB, the size of each element is 125 KB, and a new message is scheduled every 25–35 s. The buffer scheduling policy is random, which means that the messages are picked randomly from the system buffer when a node encounter another one in the network.

7.1 Real Mobility Datasets

In this subsection, we categorize the location based on the datasets and we describe the collection environment, the collection methods and the data characteristics of each dataset. The Summary of *GPS* Enabled Real Data Traces is shown in Table 3 (Table 4).

7.2 Performance Metrics

The considered performance metrics are delivery probability, average latency, and overhead ratio.

- Delivery Probability: This is the proportion of messages that have been delivered out of the total unique messages created. In the context of geocasting, a message

Table 3 Summary of GPS enabled real data traces

Dataset	Duration	Nodes	Communication distance	Environment
EPFL [51]	Over 30 days	536	–	Transportation
ZebraNet [52]	2 days	7	–	Field
Cattle [53]	9 days	5	10 m	Dairy
LocShare [54]	Over 6 months	20	Several kilometers	University
CoSphere [55]	Six weeks	12	Several kilometers	Daily routines
MIT [56]	Over 9 months	100	Several kilometers	University

Table 4 Experimental results when varying the TTL of messages

Performance metric	GeoEpidemic	GSAF	CGOPP	EECSG	FGSAF
Delivery probability	0.43	0.513	0.535	0.592	0.603
Overhead ratio	0.485	0.405	0.3	0.221	0.141
Average latency (s)	1907.5	1724.37	1804.23	1991.06	1856.40

is addressed to several nodes. Therefore, it is important to look at the fraction of nodes that received the message out of all the nodes that were present inside the destination casts of the message during its lifetime. Therefore, for every message, the delivery probability is defined as the ratio of the number of nodes that received a message by the number of nodes that were present in any one of its destination geographic casts during its lifetime. The delivery probability of the network is given by averaging the total delivered messages to all the messages created in the network.

$$delivery\ probabiltiy = \sum_{i=1}^{n} \frac{pmdr_i}{number\ of\ messages\ created} \qquad (1)$$

where, $pmdr_i$ is the total number of delivered messages. It should be noted that here we look at the number of messages created and not geomessages.

- Overhead Ratio: This is the total number of messages relayed over the total number of unique messages delivered. It can be interpreted as the number of geomessages that had to be relayed for each relay that resulted in delivery.
- Average Latency: This is the average time that is used to deliver messages to the corresponding destination. We consider the delay time of undelivered messages as the time that they have been staying in the network by the time our simulation stops.

7.3   Results

First, Fig. 6 shows the delivery probability under varying TTL (Time-To-Live) in the network). It is observed that the delivery probability of *FGSAF, GSAF, CGOPP, EECSG* and *GeoEpidemic* increases as the TTL is increased. This increase is due to the fact that the time duration allotted to each message is increased when the TTL

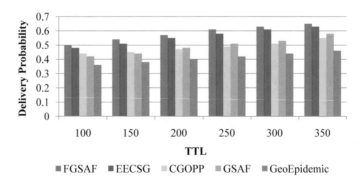

Fig. 6 Delivery probability versus TTL (in min)

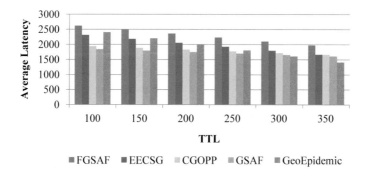

Fig. 7 Average latency (in s) versus TTL (in min)

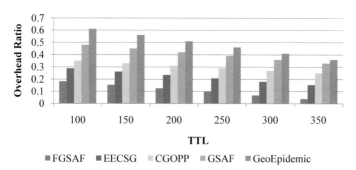

Fig. 8 Overhead ratio versus TTL (in min)

increases; and as more number of messages get stored in the buffer of nodes, the message delivery probability decreases. The performance of delivery probability is calculated when the TTL is varied for all studied protocols. It is found that *FGSAF* outperforms the other routing protocols.

Second, Fig. 7 shows the results of the latency under varying TTL for all the scenarios. It is observed that, with the increase in TTL, the average latency also increases. This occurs due to the fact that a substantial TTL value increases the stay of the message in the node's buffer. Third, Fig. 8 shows the results of overhead ratio under varying message TTL values for all the scenarios.

Fourth, Fig. 9 shows the delivery probability under varying number of nodes. It is observed that the delivery probability of *FGSAF, GSAF, CGOPP, EECSG* and *GeoEpidemic* increases as the number of hosts is increased. Other results obtained by varying the number of hosts are given in Figs. 10 and 11 respectively (Table 5).

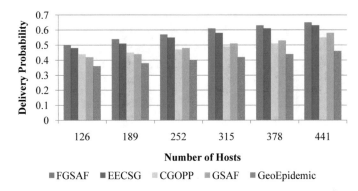

Fig. 9 Delivery probability versus number of hosts

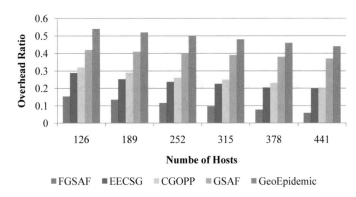

Fig. 10 Overhead ratio versus number of hosts

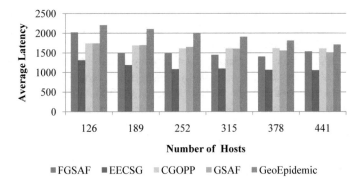

Fig. 11 Average latency (in s) versus number of hosts

Table 5 Experimental results when varying number of hosts in the network

Performance metric	GeoEpidemic	GSAF	CGOPP	EECSG	**FGSAF**
Delivery probability	0.41	0.4933	0.485	0.56	0.5833
Overhead Ratio	0.49	0.395	0.258	0.233	0.161
Average latency (s)	1957.6	1636.737	1646.326	1131.046	1565.318

8 Conclusion

In this Chapter, the recent literature regarding location-based routing protocols for delay tolerant networks has been explored, which are based on message endpoints, forwarding utility computation approach, relay node selection, routing decision approach and message dissemination. The concepts of geocommunity and geocentrality into social network analysis have been discussed and the geocommunity characteristics and their envisioned application areas different from general DTNs have been highlighted.

Through a trace-based study, an interesting phenomenon in social DTNs has been observed, namely, the fact that a community always strongly relates to its geographical location. The current challenges and future research direction for geocommunity-based routing in social delay tolerant networks have been explored. We have also presented an experimental analysis of selected geocasting protocols, namely F-GSAF, GSAF, CGOPP, EECSG, and GeoEpidemic using the ONE simulator; and the results of our simulations comparing these protocols in terms of delivery probability, overhead ratio, and average latency, under varying number of host, TTL, have been analyzed, leading to insightful conclusions.

References

1. Pereira, P.R., Casaca, A., Rodrigues, J.J., Soares, V.N., Triay, J., Cervelló-Pastor, C.: From delay-tolerant networks to vehicular delay-tolerant networks. IEEE Comm. Surv. Tutor. **14**(4), 1166–1182 (2011)
2. Lilien, L., Kamal, Z.H., Bhuse, V., Gupta, A.: Opportunistic networks: the concept and research challenges in privacy and security. In: Proceedings of International Workshop on Research Challenges in Security and Privacy for Mobile and Wireless Networks (WSPWN), Miami, FL, USA, 1–14 (2006)
3. Zhao, Y., Song, W.: Survey on social-aware data dissemination over mobile wireless networks. IEEE Access **5**, 6049–6059 (2017)
4. Tanwar, S., Tyagi, S., Kumar, N., Obaidat, M.S.: LA-MHR: learning automata based multilevel heterogeneous routing for opportunistic shared spectrum access to enhance lifetime of WSN. IEEE Syst. J. **13**(1), 313–23 (2018)

5. Alajeely, M., Doss, R., Ahmad, A.A.: Security and trust in opportunistic networks—a survey. IETE Tech. Rev. **33**(3), 256–268 (2016)
6. Li, Z., Wang, C., Yang, S., Jiang, C., Li, X.:Lass: local-activity and social-similarity based data forwarding in mobile social networks. IEEE Trans. Parallel Distrib. Syst. **26**(1), 174–184 (2014)
7. Khan, R., Kumar, P., Jayakody, D.N.K., Liyanage, M.: A survey on security and privacy of 5G technologies: potential solutions, recent advancements and future directions. IEEE Comm. Surv. Tutor. 196–248 (2019)
8. Tourani, R., Misra, S., Mick, T., Panwar, G.: Security, privacy, and access control in information-centric networking: a survey. IEEE Comm. Surv. Tutor. **20**(1), 566–600 (2017)
9. Dhurandher, S.K., Kumar, A., Obaidat, M.S.: Cryptography-based misbehavior detection and trust control mechanism for opportunistic network systems. IEEE Syst. J. **12**(4), 3191–202 (2017)
10. Rajaei, A., Chalmers, D., Wakeman, I., Parisis, G.: Efficient geocasting in opportunistic networks. Computer Commun. **127**, 105 – 121 (2018)
11. Maihofer, C.: A survey of geocast routing protocols. IEEE Comm. Surv. Tutor. **6**(2), 32–42 (2004)
12. Chen, K., Shen, H., Zhang, H.: Leveraging social networks for P2P content-based file sharing in disconnected manets. IEEE Trans. Mobile Comput. **13**(2), 235–249 (2014)
13. Blazevic, L., Le Boudec, J.Y., Giordano. S.: A location-based routing method for mobile ad hoc networks. IEEE Trans. Mobile Comput. 97–110 (2005)
14. Meghanathan, N.: Location prediction based routing protocol for mobile ad hoc networks. In: GLOBECOM 2008, pp. 1–5
15. Link, J.A.B., Schmitz, D., Wehrle, K.: Geodtn: geographic routing indisruption tolerant networks. In: GLOBECOM 2011, pp. 1–5
16. Lee, K.C., Lee, U., Gerla, M.: Geo-opportunistic routing for vehicular networks. Communi. Mag. **48**(5), 164–170 (2010)
17. Zhang, L., Yu, B., Pan, J.: GeoMob: A mobility-aware geocast scheme in metropolitans via taxicabs and buses. In: IEEE INFOCOM (2014), pp. 1279–1787
18. Zhu, K., Li, W., Fu, X.: SMART: a social-and mobile-aware routing strategy for disruption-tolerant networks. IEEE Trans. on Vehicular Technology **63**(7), 3423–3434 (2014)
19. Huang, J., Su, Y., Liu, W., Wang, F.: Adaptive modulation and coding techniques for global navigation satellite system inter-satellite communication based on the channel condition. IET Commun. **10**(16), 2091–2095 (2016)
20. Lu, S., Liu, Y.: Geoopp: geocasting for opportunistic networks. In: IEEE Wireless Communications and Networking Conference (WCNC), pp. 2582–2587 (2014)
21. Cardei, I., Liu, C., Wu, J., Yuan, Q.: Dtn routing with probabilistic trajectory prediction. Wirel. Algor. Syst. Appl. 40–51 (2008)
22. Ruhrup, S.: Theory and practice of geographic routing. In: Ad Hoc and Sensor Wireless Networks: Architectures, Algorithms and Protocols, 69 (2009)
23. Ma, Y., Jamalipour, A.: Opportunistic geocast in disruption-tolerant networks. In:IEEE GLOBECOM 2011, pp. 1–5
24. Zhu, Z., Cao, J., Liu, M., Zheng, Y.,Gong, H., Chen, G.: A mobility prediction-based adaptive data gathering protocol for delay tolerant mobile sensor Network. In: IEEE GLOBECOM 2008, pp. 1–5
25. Vahdat, A., Becker, D.: Epidemic routing for partially connected ad hoc networks. Technical Report CS-200006, Duke University (2000)
26. Nassima, H., Myoupo, J.F.: Multi-geocast algorithms for wireless sparse or dense ad-hoc sensor networks. In: Proceedings of 4th IEEE International Conference on Networking and Services (ICNS 2008), pp. 35–39
27. Cuong, T., Römer, K.: Efficient geocasting to multiple regions in large-scale wireless sensor networks. In: 37th Annual IEEE Conference on Local Computer Networks (2012), pp. 453–461
28. Linjuan, Z., Gao, D., Gao, Leung, V.C.M.: Smart geocast: dynamic abnormal traffic information dissemination to multiple regions in VANET. IEEE IWCMC 2013, pp. 1750–1755

29. Ma, Y., Jamalipour, A.: Opportunistic geocast in large scale intermittently connected mobile ad hoc networks. In: Proceedings of 17th IEEE Asia-Pacific Conference on Communications (APCC), pp. 445–449 (2011)
30. Kuiper, E., Tehrani, S.N.: Geographical routing with location service in intermittently connected MANETs. IEEE Trans. Vehicular Techn. **60**(2), 592–604 (2011)
31. Alaouia, E.A.A., Zekkorib, H., Agoujil, S.: Hybrid delay tolerant network routing protocol for heterogeneous networks. J. Network Computer Appl. **148**, 15 (2019). https://doi.org/10.1016/j.jnca.2019.102456
32. Mayer, C.P.: Hybrid Routing in Delay Tolerant Networks. KIT Scientific Publishing, 220p (2012)
33. Kang, M.W., Chung, Y.W.: An improved hybrid routing protocol combining MANET and DTN. Electronics 2020, **9**(3), 439. https://doi.org/10.3390/electronics9030439
34. Yi, Z., Wang, H.-M., Ding, Z., Ho Lee, M.: Non-orthogonal multiple access assisted multiregion geocast. IEEE Access **6** 2340–2355 (2017)
35. Dhurandher, S.K., Obaidat, M.S., Gupta, M.: An efficient technique for geocast region holes in underwater sensor networks and its performance evaluation. Simul. Model. Practice Theory **19**(9), 2102–2116 (2011)
36. Dhurandher, S.K., Obaidat, M.S., Gupta, M.: Energized geocasting model for underwater wireless sensor networks. Simul. Model. Practice Theory **37**, 125–138 (2013)
37. Dhurandher, S.K., Obaidat, M.S., Gupta, M.: Providing reliable and link stability-based geocasting model in underwater environment. Int. J. Commun. Syst. **25**(3), 356–375 (2012)
38. Jesus, E.F., Chicarino, V.R.L., de Albuquerque, C.V.N., Rocha, A.A.A.: A survey of how to use blockchain to secure internet of things and the stalker attack. Secur. Commun. Networks 9675050, 1–27 (2018). https://doi.org/10.1155/2018/9675050
39. Ramezan, Gholamreza, Leung, Cyril: A blockchain-based contractual routing protocol for the internet of things using smart contracts. Wirel. Commun. Mobile Comput. **4029591**, 1–14 (2018)
40. Jeon, J.H., Kim, K.-H., Kim, bJ.-H.: Blockchain based data security enhanced IoT server platform. In: 2018 International Conference on Information Networking (ICOIN), Kuala Lumpur, Malaysia (2018)
41. Chilipirea, C., Petre, A.C., Dobre, C.: Energy-aware social-based routing in opportunistic networks. In: 27th IEEE International Conference on Advanced Information Networking and Applications Workshops, pp. 791–796 (2013)
42. Patel, V.G., Oza, T.K., Gohil, D.M.: Vibrant energy aware spray and wait routing in delay tolerant network. J. Telematics Inform. **1**(1), 43–47 (2013)
43. Sharma, D.K., Dhurandher, S.K., Woungang, I., Srivastava, R.K., Mohananey, A., Rodrigues, J.J.: A machine learning-based protocol for efficient routing in opportunistic networks. IEEE Syst. J. **12**(3), 2207–2213 (2016)
44. Rollaand, V.G., Curado, M.: A reinforcement learning-based routing for delay tolerant networks. Eng. Appl. Artif. Intell. **26**(10), 2243–2250 (2013)
45. Sharma, D.K., Rodrigues, J.J.P.C., Vashishth, V., Khanna, A., Chhabra, A.: Rlproph: a dynamic programming based reinforcement learning approach for optimal routing in opportunistic IoT networks. Wirel. Networks 1–20 (2020)
46. Yuan, F., Wu, J., Zhou, H., Liu, L.: A double Q-learning routing in delay tolerant networks. In: IEEE ICC 2019, China, pp. 1 – 6
47. Ma, Y., Jamalipour, A.: Opportunistic geocast in disruption-tolerant networks. In: Global Telecommunications Conference (GLOBECOM 2011), IEEE, pp. 1–5
48. Singh, J., Dhurandher, S.K., Woungang, I., Takizawa, M.: Centrality based Geocasting for Opportunistic Networks. In: AINA 2018, pp. 702–712
49. Khalid, K., Woungang, I., Dhurandher, S.K., Singh, J., Rodrigues, J.J.P.C.: Energy-efficient check-and-spray geocast routing protocol for opportunistic networks. Information **11**(11), 504–514 (2020)
50. Dhurandher, S.K., Singh, J., Woungang, I., Takizawa, M., Gupta, G., Kumar, R.: Fuzzy geocasting in opportunistic networks. In: BWCCA 2019, pp. 279–292

51. Das, S.K.: Multi-periodic contact patterns in predicting future contacts over mobile networks. In: Proceedings of 18th IEEE International Symposium on a World of Wireless, Mobile and Multimedia Networks (WoWMoM 2017), pp. 1–10 (2017)
52. Yin, Z.C., Wu, Y., Winter, S., Hu, L.F., Huang, J.J.: Random encounters in probabilistic time geography. Int. J. Geogr. Inf. Sci. **32**(5), 1026–1042 (2018)
53. Xie, X., Chen, H., Wu, H.: Bargain-based stimulation mechanism for selfish mobile nodes in participatory sensing network. In: Proceedings of 6th Annual IEEE Communications Society Conference on Sensor, Mesh and Ad Hoc Communications and Networks (SECON 2009), pp. 1–9
54. Xing, X., Sun, G., Jin, Y., Tang, W., Cheng, X.: Relay selection based on social relationship prediction and information leakage reduction for mobile social networks. Math. Found. Comput. **1**(4), 369–382 (2018)
55. Fan, J., Chen, J., Du, Y., Gao, W., Wu, J., Sun, Y.: Geocommunity-based broadcasting for data dissemination in mobile social networks. IEEE Trans. Parallel Distrib. Syst. **24**(4), 734–743 (2013)
56. Lin, Z., Wu, X.: Analyzing and modeling mobility for infrastructure-less communication. J. Netw. Comput. Appl. **53**, 156–163 (2015)
57. Issariyakul, T., Hossain, E.: Introduction to Network Simulator 2 (NS2), pp. 1–18. Springer, Boston (2012)
58. Zeng, X., Bagrodia, R., Gerla, M.: GloMoSim: a library for parallel simulation of large-scale wireless networks. In: Proceedings of IEEE 12th Workshop on Parallel and Distributed Simulation (PADS'98), pp. 154–161
59. Dhurandher, S.K., Obaidat, M.S., Gupta, M.: An acoustic communication based AQUA-GLOMO simulator for underwater networks. Human-Centric Comput. Inform. Sci. **2**(1), 1–14 (2012)
60. Kernen, A., Ott, J., Krkkinen, T.: The ONE simulator for DTN protocol evaluation. In: Proceedings of the 2nd International Conference on Simulation Tools and Techniques, Rome, Italy, pp. 1–10, 2–6 Mar 2009

Resource Allocation Challenges in the Cloud and Edge Continuum

**Polyzois Soumplis, Panagiotis Kokkinos, Aristotelis Kretsis,
Petros Nicopolitidis, Georgios Papadimitriou, and Emmanouel Varvarigos**

Abstract We witness a shift of the digital infrastructures, from the current model that consists of a plethora of heterogeneous and isolated computing, storage and networking resources under centralized control, towards a cloud continuum that serves innovative applications in all sectors. This, in essence, involves the distribution of computation and storage across multiple resource types and domains. Thus, the realization of the cloud continuum through the integration of edge, fog and cloud resources under transparent orchestration is an important step in this direction. However, heterogeneous distributed infrastructures present a number of challenges regarding their management and service deployment on them. To this end, there is a movement to federations of loosely coupled autonomous or semi-autonomous systems, which incorporate various local orchestration platforms together with closed-loop control mechanisms aiming to create a self-optimizing system. We start by presenting key computing technologies of the past and the present, along with related networking technologies and continue by describing the important resource

P. Soumplis (✉) · A. Kretsis · E. Varvarigos
National and Technical University of Athens, Athens, Greece
e-mail: soumplis@mail.ntua.gr

A. Kretsis
e-mail: akretsis@mail.ntua.gr

E. Varvarigos
e-mail: vmanos@central.ntua.gr

P. Soumplis · P. Kokkinos · A. Kretsis · E. Varvarigos
Institute of Communication and Computer Systems, Athens, Greece
e-mail: kokkinop@central.ntua.gr

P. Kokkinos
Department of Digital Systems, University of Peloponnese, Tripoli, Greece

P. Nicopolitidis · G. Papadimitriou
Aristotle University of Thessaloniki, Thessaloniki, Greece
e-mail: petros@csd.auth.gr

G. Papadimitriou
e-mail: gp@csd.auth.gr

© The Author(s), under exclusive license to Springer Nature Switzerland AG 2022
P. Nicopolitidis et al. (eds.), *Advances in Computing, Informatics, Networking
and Cybersecurity*, Lecture Notes in Networks and Systems 289,
https://doi.org/10.1007/978-3-030-87049-2_15

allocation challenges that appear in this environment. We conclude, formulating and evaluating a basic resource allocation problem for assigning application's workload in an edge-fog-cloud hierarchical infrastructure.

Keywords Cloud-edge-fog computing · Optical metro and access networks · Resource allocation

1   Introduction

A wave of emerging cloud computing technologies and services that empower advanced applications from different vertical sectors, with diverse requirements has immersed in recent years. Centralized cloud computing infrastructures have been until now handling most of the processing and storage requirements of the applications, rendering cloud computing a key component of modern economy in enabling the design and the realization of novel digital services.

In addition, there is a movement from top-down-designed architectures that apply centralized resource control, towards federations of loosely coupled autonomous or semi-autonomous edge and fog systems, managed by multiple independent actors that are self-organized in a distributed manner (Fig. 1). Edge computing offers computation and storage at the very edge of the network, close to where data is produced, and has recently emerged as a way to reduce latency and limit the load that is carried to higher layers of the infrastructure hierarchy. Fog nodes operate between the edge

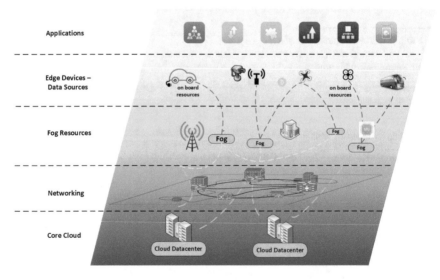

Fig. 1 Cloud-computing continuum: a hierarchy of edge, fog, cloud resources serving applications computing and storage requirements

and the cloud and are placed near highways or inside network operators' central offices, providing computing and storage services in the field.

These developments enable a cloud-computing continuum. Distributed resources form an edge-fog-cloud hierarchy (Fig. 1), where processing tasks and data can move both horizontally, across the edge/fog, and vertically, from the edge to the cloud/center, and vice versa, so as to dynamically adapt to the requirements of the end-users' applications and end-devices (e.g., cars, drones), taking into account the resources' characteristics in terms of processing latency, capacity, security, location and cost. In this context, mechanisms are required to form a data analytics pipeline infrastructure that collects continuous data, aggregates, filters, and pre-processes data at the local edge-fog level, coordinates and manages edge and fog resources and application instances and finally moves part of the processed data and executes more complex analytics in the cloud, while working in near-real-time. These resource orchestration and application management functionalities will operate centrally and in a distributed manner, creating a distributed knowledge and decision-making system.

In what follows, we initially present the basic concepts of grid computing that in some aspects led to the creation and the adoption of cloud computing. Next, we discuss about cloud and edge/fog computing and present the main characteristics of edge resources and cloud datacenters that support their operation. Resource allocation challenges in these environment are also identified. We then formulate the resource allocation problem dealing with the application workloads' allocation and deployment in an edge-fog-cloud hierarchical computing infrastructure. Finally, through simulation experiments, we illustrate the importance of the resource allocation mechanisms for the efficient service of applications and the utilization of the available computing, storage and networking infrastructures.

2 Grids

Grid networks, were designed to solve problems that require a lot of computing power, access to large volumes of data and/or use of specialized hardware tools and software, features and capabilities that are not usually locally available to a user. The Grid network technology promised faster calculations and almost unlimited storage possibilities.

Grid networks were destined to become a 'global' system that will provide access to various resources via the internet, in the same way that the World Wide Web provides access to information. In this way, the global network was transformed into an integrated virtual computer. The term "grid computing" appeared in 1999 in the work of Jan Foster and Carl Kesselman: "The Grid: Blueprint for a new computing infrastructure" [1]. The authors defined grid networks as "an infrastructure not hardware and software that provides access to high-capacity computing infrastructure". The authors envisioned that at some point in the future, users' access in terms of computing capacity will be as easy as access to electricity energy, which is generated distributed across different stations. In this way, the grid users would send their processes to the grid, without caring about exactly where they will be executed, as they receive electricity without being interested by which power station it is provided. In fact, Foster suggested one simple list of grid key features:

- It consists of resources that are not subject to central control, while at the same time covering the security issues, payments, policies, etc., that arise in this environment,
- It uses clearly defined, open, general purpose protocols and interfaces for addressing certification, authorization, resource extraction and access to them,
- It satisfies even the most complex user requirements.

Grid networks were based on the concept of "common" use of a set of distributed and heterogeneous IT resources: e.g. computer resources, data storage, network resources, operating systems, software, sensor devices, scientific instruments and any kind of resource that can be shared between multiple users. In this way, the computing power and storage capacity offered by the grid is many times higher that of a simple computer system. In addition, the model of sharing these resources, justifies and depreciates their initial purchase cost and the respective continuing maintenance and management costs, while maximizing their degree of use (e.g. resources can be used at all hours of the day and even at night). In addition, the resources of a grid network are geographically and administratively distributed and for this reason they are interconnected over a communication network. There is no central authority for managing these resources but, instead, their management is distributed, with different management policies per area or per resource.

The model for using grid networks is relatively simple. A user sets out the characteristics and requirements of the application to be executed, e.g. the data required, and sends it to the grid. The procedures applied for the execution of the application as well as the specific resources utilized are unknown to the end user. Finally, when the execution of the application is completed, the results are returned back to the user.

The grid network architecture consists of three main layers: the basic infrastructure of resources, the middleware, and the applications.

- The basic infrastructure layer contains the resources, consisting of computers (such as ordinary computers, clusters, and supercomputers), storage units, interconnecting networks, scientific instruments etc.
- The intermediate software layer provides the services and mechanisms necessary for the operation of grid networks, displaying resource entities in a unified way to the top-level applications.
- Scientific applications can be categorized in different ways, depending on their requirements, their uses, their main objective, as well as the audience to which they are addressed. We distinguish between:

  - Computationally intensive applications with high demands on computing power that need a very fast and secure data network for connecting users to resources.

- Data intensive applications that generate terabyte of data daily and use distributed storage resources. A fast data network is required for data transfers between users, storage and processing nodes.
- Community-based applications related to the cooperation between people, e.g. video conference applications between scientists from different parts of the world. Again, fast and reliable networks are required to service these applications' data.

A number of Grid Networks have been created: the TeraGrid, the Open Science Grid (OSG), the Grid'5000, the Worldwide Large Hadron Collider Computing Grid (WLCG) [2], and the Enabling Grids for E-Science (EGEE) [3]. These grid networks consisted of large capacity computing and storage resources, interconnected by a broadband data network, such as an optical network. The use of these resources was not commercial, but it was supported at a National, European or global level by respective organizations. In addition, the users of these networks were usually scientists and researchers, running specialized applications.

The above discussion highlights the range of resource optimization challenges related to grid networks, which largely determine the performance of grid networks:

- Resource allocation algorithms [4]. These algorithms assign user tasks in the computing resources of a grid network, in different locations (Fig. 2). The assignment of tasks should be performed in a way that (a) meets the requirements or characteristics (e.g., financial) of users, (b) it is fair to users and (c) makes efficient use of available resources. Scheduling is usually performed in two levels: at the

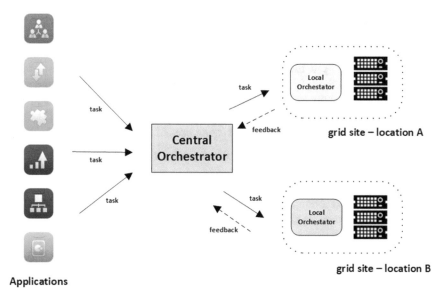

Fig. 2 Allocation of tasks to grid resources in different locations through central and local schedulers

first level a central scheduler selects the group of resources (grid site) where a process will be executed, while at the second level, a local (in the grid site) scheduler selects the exact resource to run a task. After a task's execution is completed the central scheduler is informed. This approach is followed by many intermediate systems (middleware) such as gLite [5]. In addition, fully distributed or hybrid scheduling systems have also been proposed and implemented mainly at a research level.

- Algorithms related to data management. The most common data management process is that of backup and/or caching at various nodes in the network. Through this process important data access delays can be reduced, while security and fault tolerance can be increased.
- Algorithms related to the routing of data. These algorithms route the generated data from the resources they are stored at, to the computing nodes where the respective processing will be performed. Optical networks have been identified as the technology to provide high speed data transfer between grid nodes. The combined study of issues concerning grid networks and optical networks define the research field found in the literature with the term Optical Grids or Lamda Grids. The main research problem in this field of research is the joint allocation of computing, storage and networking resources for serving a task or a set of tasks [6].

3 Clouds

Today, centralized cloud computing infrastructures and respective services are handling the processing and storage requirements of most applications, in several areas. According, to the National Institute of Standards and Technology (NIST): "Cloud computing is a model for enabling ubiquitous, convenient, on-demand network access to a shared pool of configurable computing resources (e.g., networks, servers, storage, applications, and services) that can be rapidly provisioned and released with minimal management effort or service provider interaction" [7].

According to NIST, the cloud model is composed of five essential characteristics:

1. on-demand self-service regarding the creation, configuration and use of the resources,
2. broad network access of the resources,
3. use of shared resources following a multi-tenant model,
4. dynamic and flexible allocation and release of the resources and
5. monitoring and measurement of resource usage.

Initially, three service models were identified: (i) the Infrastructure-as-a-Service (IaaS), (ii) the Platform-as-a-Service (PaaS), and (iii) the Software-as-a-Service (SaaS) models. Lately, the Container as a Service (CaaS), the Function as a Service (FaaS) and in general the "everything as a service" provisioning concept have appeared.

Amazon's Amazon Web Service [8], released in the early 2000s, was the first highly successful cloud computing service, following the Infrastructure-as-a-Service (IaaS) service model. AWS came to life when Amazon wanted to rent its excess computing capacity, in order to improve its utilization and bring more profits to the organization.

An increasing number of individual users, researchers and companies, established or startups, of any size and scope, now trust their computing and storage tasks to public and private clouds, replacing fixed Information Technology (IT) costs of ownership and operation with variable use-dependent costs. The global cloud computing market size is expected to grow from USD 371.4 billion in 2020 to USD 832.1 billion by 2025, at a Compound Annual Growth Rate (CAGR) of 17.5% during the forecast period [9]. Public Cloud providers include Amazon, Google, Microsoft, Alibaba, IBM and others.

Cloud services utilize datacenters (Sect. 5) deployed in multiple locations around the world, interconnected over private and public networking infrastructures, with the latency experienced by users varying according to the datacenter's location and networking conditions. This consolidation of computation and storage provides a variety of benefits, including economies of scale through the reduction of cost with size. Cloud computing systems also rely heavily on the virtualization technology, interconnecting dynamically a set of virtual machines, providing the ability to scale the resources according to the needs of the user and/or the application.

Cloud computing sets several algorithmic challenges in terms of using efficiently the cloud resources to serve the workload and data. The works in [10, 11] survey the respective issues and challenges. Generally, the resource allocation algorithms can be categorized into algorithms that consider: different cloud providers, or different cloud datacenters of a cloud provider, or different servers/virtual machines in a datacenter (Fig. 3).

Monitoring, analysis and optimization are important interrelated operations for cloud resource management. The sheer number of cloud resources (cloud providers, datacenters, servers) makes it difficult for a simple user or even an administrator to effectively monitor and analyze their behavior or control (optimize) the parameters that determine their proper use. A good analytic and optimization mechanism ensures that the monitored resources run uninterrupted and with acceptable performance and

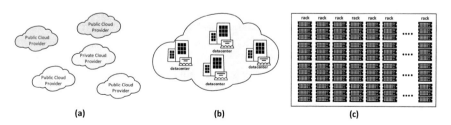

Fig. 3 **a** Set of public and private cloud providers. **b** Multiple datacenters of a cloud provider. **c** Multiple physical and virtual machines in datacenter

utilization, keeping the associated costs low. The costs encountered by a user of public clouds depends mainly on the pricing policy of the cloud provider, the number and types of resources used, and the resources' utilization. Cloud economics have emerged as an important field, studied in a number of works [12, 13].

4 Edge/Fog Computing

Internet of Things (IoT), smart cities, transportation, robots, smart-grids, industry 4.0 and other applications, running at the network edge, pose high and hard requirements on the computing, storage and networking infrastructures, while requiring low-latency and real-time processing. Day by day, these applications introduce an increasing number of new devices and data inputs that scale from thousands to millions (to billions) and reach exascale dimensions, while the sensors and cameras onboard these devices, increase in number and capabilities (e.g., resolution) further skyrocketing the volume of data produced. Generally, the handling of the emerging edge applications and devices leads to big data challenges that include: (i) high volume data, (ii) data from a variety of sources, (iii) data that are generated with high velocity and need (near-) real-time processing, (iv) the need for veracity in the interactions between data generating and data processing entities and (v) the need for identifying the business value of the data.

The centralized cloud computing and storage services, however, are distributed in a relatively small number of datacenters around the world. This means that the on the edge generated data and computations are transferred through the network to a far-away cloud datacenter, often located in a different country, and sometimes with dubious privacy guarantees. Collecting and processing all these data at a single point (even if that "point" is a distributed process in the Cloud) becomes expensive, especially when only a small fraction of the generated and aggregated data is actually useful towards decision making and high-level analytics. Also, in many situations, some of these data are temporary or trivial, and only the semantically important information they carry (e.g., a detected event or an object) has to be identified and stored. Also, the continuous transmission of data between the devices and the cloud requires a broad, constant and reliable networking infrastructure. So, in many cases, transferring and processing these edge generated data in central/fixed private and public cloud infrastructures is unnecessary and unrealistic both from an economic (e.g., cost of using the resources) and a performance (e.g., high and variable latency, available throughput) point of view.

It is clear from the above discussion that bringing computation to the edge, in the form of edge and fog computing, is a necessity and a key to the performance and success of future applications [14], as serving their data at the edge would reduce latency significantly. Edge/Fog computing paradigm is supported by the fact that many devices are becoming increasingly intelligent by adding generic or specialized computation equipment on them. Edge nodes perform processing on data collecting devices. For example, some high-end cameras are fitted with artificial intelligence

Fig. 4 Edge/fog devices range in size and capabilities: modular data centers in shipping containers, micro datacenters (mDC), computing devices combined with heaters, specialized computing devices (FPGA, GPU) and IoT computing devices (e.g., Arduino, Raspberry)

capabilities, including facial recognition technology for on device processing, drones perform on board operations like obstacle-or event detection and collision avoidance based on data collected from cameras and, of course, semi-autonomous cars perform the autonomous navigation computation operations onboard. Fog nodes, operate between the edge and the cloud, placed near highways or inside network operators' central offices (Fig. 4). These can aggregate (producing means, summaries, deviations or anomaly flags) or pre-process data, pushing to the cloud only the information required for decision making and further higher-level analysis. Fog nodes may include but are not limited to miniPCs (e.g., Raspberry Pi), hardware accelerators (e.g., FPGA and GPU) and micro-datacenters.

Of course, there are several challenges to consider. Isolated edge and fog resources are not panacea and cannot actually provide as is the required scalability, resiliency and, ultimately, the expected performance benefits. In particular, the volatile nature of the fog/edge computing and the ever-changing demands presented by the use of movable data sources (e.g., drones) may easily lead to a need for swift reorganization of the edge processing, storage and networking infrastructure. Also, swarms of independently acting edge-devices seeking access to contested resources may create dynamic bottlenecks in the system, leading to delays or drops of quality of service, at levels unacceptable for real time or critical applications. Also, the realization of any edge/fog infrastructure depends highly on its deployability. Deployment needs to be massive, in terms of the total capacity deployed and the geographical distribution of the resources, for the infrastructure to provide the desired edge/fog services. Deployability, in turn, depends on the economic benefits expected and the desired Return on Investment (ROI). The deployment of edge/fog resources is expected to be driven by network operators and content providers, but also by other players, like owners of buildings, stadiums and shopping malls that will operate edge/fog infrastructures in order to serve their own computing needs.

5 Datacenters

The increasing need for computing and storage resources "anywhere, anytime", in support of social, big data, Internet of Things (IoT), 5G and other services and applications led to the increasing development of cloud datacenters. Datacenters typically comprise of a number of interconnected servers running virtual machines.

Many of these datacenters are quite large in size, namely hyperscale datacenters, hosting hundreds of thousands of servers. Currently, hyperscale cloud operators are dominating the cloud landscape, with more and more companies moving their workloads in hyperscale-based clouds. In particular, a number of companies are deploying and operating hyperscale datacenters, like Amazon, Microsoft, IBM, Google, Salesforce, Rackspace, Facebook, Yahoo, Apple, and others. On the other hand, many large universities, institutes, and private enterprises are merging their IT services inside smaller scale datacenters (often hosting a few hundred to a few thousand servers), offering a variety of services.

Cloud datacenters are being built in several locations around the globe providing a world-wide coverage to their clients. In particular, the deployment of the datacenters depends on the computational and storage requirements in each region, while their exact location in a region has to do with the local economic conditions (like in Ireland) and capabilities of producing cheap (renewable) energy (like in Iceland). [15] identified 24 hyperscale operators that have in total 297 datacenters all over the world. North America region has the largest share, at 51%, followed by Asia Pacific region, with 29%, Western Europe region, with 17%, and Latin America region, with 3%.

Regarding the networking characteristics of the provided services, in terms of upload and download throughput, from a datacenter to the regional users and the related latency, these differ from one region to another but also between datacenters in the same region. For example, according to [15] Asia Pacific leads all regions with an average fixed download speed of 33.9 Mbps. Asia Pacific leads also all regions in average fixed network latency with 26 ms, followed by Central and Eastern Europe with 30 ms.

In 2008, peer-to-peer traffic (which is originated and terminated directly from device to device) ceased to dominate the overall Internet traffic. Network analysts report that the majority of Internet traffic is by now originated or terminated in a datacenter and it will continue to do so for the foreseeable future. This fundamental transformation in the nature of datacenter traffic is brought about by cloud applications, services, and infrastructure. Also, traffic between data centers, even though smaller, is growing faster than either traffic to end users or traffic within the data center [15]. The growth of the inter-datacenter traffic is due to the increasing prevalence of content distribution networks, the proliferation of cloud services, the migration of applications to public clouds, the size of the respective datacenters, their world-wide placement and the need to shuttle data replicated across datacenters. Some of the datacenter management operations with most relevant impact on the inter-datacenter network requirements are the following [16]:

- Storage and data replication for IT disaster recovery strategy. Although these kinds of services generate a small number of flows, they need a very high traffic throughput per flow.

- VM migration, with VM disk space asynchronously replicated at another datacenter using the network provider's WAN.
- Synchronous replication between DCs, which allows data to reside at multiple different locations and be actively accessed by VMs at all sites.

As a result, inter-data center networks have to provide high-bandwidth, low-latency network connectivity between different locations. For this reason, hyper-scale datacenters are connected using high speed optical connections both landline and subsea. The design and installation of these networking connections are closely affected and influence the position of datacenters around the globe. Thus, a data-center may be built near a subsea cable's landing point or new subsea cables are being deployed between locations with high traffic. Generally, the inter-datacenter networking performance is a critical criterion in selecting a cloud/network provider.

Network architectures and technologies inside datacenters (datacenter networks—DCNs) play a crucial role in the overall performance of the provided cloud services. High throughput, scalable, and energy/cost efficient DCN networks are required to fully harness datacenter potential. Since traffic within datacenters is much higher than incoming/outgoing traffic, the intra-datacenter network has to provide low latency and high-capacity connectivity between the numerous servers and storage nodes organized in racks. State-of-the-art intra-datacenter networks are based on elec-tronic switches connected in hierarchical topologies (e.g., fat-tree) using optical links. Also, for inter-rack communication a number of works propose overlay networking infrastructures that utilize wavelength optical switches [17].

In these intra- and inter-datacenter environments, the resource allocation deci-sions have to do with the selection of a datacenter, or the selection of a server or a virtual machine in the datacenter, to run a single or multiple tasks [18, 19]. This challenge is very often viewed in combination with the networking capabilities of the infrastructure and the networking requirements of the respective applications/tasks [17, 20]. The dynamic re-optimization of the resources allocated is also important (Fig. 5).

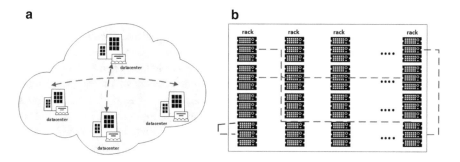

Fig. 5 a Inter-datacenter, **b** Intra-datacenter and intra-rack data transfers

6　Networking Technologies

The processing resources at the different parts of the network require an efficient communication network. To this end, optical networks offer significant advantages by increasing the transmission rate whilst satisfying latency constraints. Optical technologies were initially designed to be deployed in core networks to transport the aggregated high demands between different countries or continents. With the advent of new applications and the increase in the network traffic, optical technologies have started to find use in the metropolitan and the access part of the network. Currently, Elastic Optical Networks (EON) are the most prominent technology. It is an evolution of WDM (Wavelength Division Multiplexing) technology, providing resource flexibility and efficiency. EONs are an architecture that support variable spectrum connections as a way to increase spectral efficiency, support future transmission rates, and reduce capital costs, meeting also the strict and heterogeneous requirements of next-generation networks. An EON migrates from the fixed 50 GHz grid that traditional WDM networks utilize to a granularity of 12.5 GHz, as standardized by the International Telecommunication Union (ITU-T). Moreover, EONs can also combine spectrum units, referred to as slots, to create wider channels on an as-needed basis. EONs are built using bandwidth variable switches that are configured to create appropriately sized end-to-end optical paths of sufficient spectrum slots. Bandwidth variable switches operate in a transparent manner for transit (bypassing) traffic that is switched while remaining in the optical domain. In EONs, the traffic is served by coherent Bandwidth Variable Transponders (BVT) which can control some or all of the following parameters: (i) modulation format, (ii) baud rate, (iii) transmission power and (iv) FEC overhead.

Different architectures that leverage EONs can be employed depending on the hierarchy level. In the access part of the network, Passive Optical Networks (PONs) are worldwide adopted as the most efficient way to extend the optical fiber deep into the access network. Using PONs, multiple clients share the interface between the Central Office (CO) and the provider's metro network, as well as the optical fibers of the distribution network. PON's topology structure enables operators to deliver in a flexible manner, high bandwidth connections to multiple endpoints over long distances and with low latency. PONs also reduce the capital and operational (energy) costs of the equipment, since they are based on passive components. Different types of PONs have been proposed, handling the technological heterogeneity of the existing infrastructures and supporting their smooth upgrade. Most of the proposed technologies are based on WDM-PON. WDM-PONs consist of an optical line terminal (OLT), located at the operators' central office, and multiple optical network units (ONU) close to the end-users. The wavelengths are assigned to the ONUs exclusively, over the physical point-to-multipoint (OLT-ONUs) fiber infrastructure, based on the demanded capacity, offering also traffic isolation. NG-PON2 (also known as TWDM-PON), Next-Generation Passive Optical Network 2 is a telecommunications network standard for a passive optical network (PON), which also supports wavelength sharing for better resource utilization.

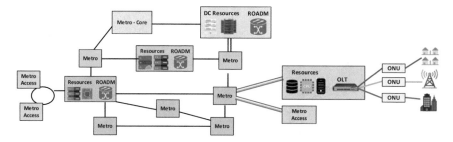

Fig. 6 A hierarchical network and computing infrastructure

Metropolitan networks that interconnect the access networks with the core are also one other focus point. In order to efficiently serve the high access demands, they are organized in hierarchical layers (Fig. 6). The exact number of these layers may depend on several parameters such as the coverage area, the number and distribution of users, the traffic characteristics, the capabilities of the networking devices and other. Thus, the nodes of the lower layer, the metro access nodes, are typically co-located with the OLT nodes inside the operators' central offices and are organized in ring and mesh topologies. In the intermediate metro layers, the supported transmission rates increase as more powerful transceivers are employed to transfer the aggregated traffic. The higher layers of the metro infrastructure consist of fewer nodes, namely the metro core nodes, and are organized in mesh topologies. To further decrease the cost of the lower parts of the metro networks filterless networks can be utilized. These networks eliminate the use of active photonic components such as Reconfigurable Optical Add Drop Multiplexers (ROADMs) based on Wavelength Selective Switches (WSS), utilizing instead passive splitters and combiners to interconnect fiber links. Their operation also requires the use of wavelength tunable optical transmitters and wavelength selective coherent receivers. In this way, the optical signal is transmitted in broadcast and select manner through the network, moving the network reconfigurability from the network nodes to the network edge that is the transceivers.

7 Close-Loop Resources Orchestration

The plethora of heterogeneous distributed computing, storage and networking infrastructures, in the cloud-computing continuum, presents a number of challenges regarding management and service/application deployment. Both the academia [21] and the industry [22, 23] are targeting solutions that enable self-optimization based on artificial intelligence and machine learning techniques that facilitate the autonomous adaptation and management of the deployed services and resources.

These properties require a hierarchical and close-loop operation of the orchestration mechanisms (Fig. 7). In the hierarchical operation, there is a central resource

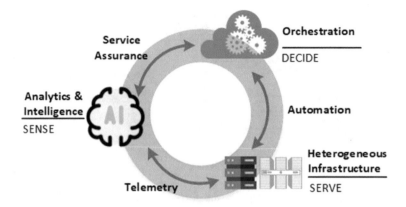

Fig. 7 Closed-loop management and application deployment for the cloud continuum

orchestrator that communicates with resource-hosted local orchestrators, like in the original grid systems (Fig. 2). The close-loop operation of the orchestration mechanisms enables the continuous adaptation of the respective decisions, based on the ability to sense (detect what is happening), discern (interpret senses), infer (understand implications), decide (choose a course of action), and act (take action). These operations run over an infinite time horizon, adjusting resources and migrating the assigned tasks based on feedback regarding the application's and the resources' state.

The intelligent and autonomous orchestration mechanisms automatically determine the most appropriate resources to be used, transparently deploying on them the workloads, coordinating the efficient movement of the required data, and providing automatic continuous adaptations. To facilitate the autonomous adaptation and management for the deployed services and provide load predictions, performance degradation detection and recommendation actions and service assurance mechanisms are also required. Autonomous, scalable and data-driven cloud and network telemetry mechanisms collect intelligently and dynamically telemetry data (regarding the resources' current state, the deployed applications status, etc.) across the distributed infrastructure. These telemetry data are utilized by the resource orchestration and service assurance mechanisms, which correlate the collected information and monitored events to efficiently allocate the available resources, triggering automatic re-optimization adjustments and performing automated data and workload movements.

Closing the control loop in the cloud-computing continuum enables real-time decisions, increases the responsiveness of the overall system and achieves high dynamicity, while yielding superior performance.

In the previous sections, we discussed about the key technologies, computing and networking, that play a major role in the shift from heterogeneous and isolated infrastructures towards a truly cloud continuum that serves innovative applications from various sectors. In the next section, we formulate the respective resource allocation problem in a hierarchical network architecture organized into layers of different processing and computing capabilities, so as to provide services that respect the demanded latency, processing and networking requirements with the aim of creating a self-optimizing system.

8 Resource Allocation Problem Formulation

The workload placement problem can be stated as follows. We let the directed graph $G = (V, E)$ denote the network topology. The set V represents the access and metro network nodes of the hierarchical topology, while the set E represents the physical connections (links) between two nodes. Each physical link $e \in E$, is characterized by its length l_e and introduces a latency τ_e due to signal propagation. Nodes are equipped with processing capacity c_v (generic or hardware accelerated), and a set of optical transceivers M_v, whose feasible transmission configurations are described by the transmission rate b over distance h. Each node with processing capacity is also associated with a processing cost ξ_v that captures the difference in the processing capabilities among the different types of processing.

Each demand $r \in R$, from a source node s to a destination node d, is characterized by its requested network capacity ζ_r, processing capacity ε_r (measured in FLOPS) and maximum latency τ_r. Putting all these parameters together, each demand is described by the tuple $r = <s, d, \zeta, \varepsilon, \tau>$. The demanded network capacity is routed through the network and processing power is allocated at the traversed nodes, with the objective is to minimize the processing cost of the utilized resources or the propagation latency of the selected network path.

Variables

$x^r_{a,\lambda,b} \in \{0, 1\}$, binary variable equal to 1 if demand request r utilizes optical path a over wavelength λ with transmission rate b.

$y^r_v \in \{0, 1\}$, binary variable equal to 1 if processing power at node v is allocated for demand r.

$z_{e,\lambda} \in \{0, 1\}$, binary variable equal to 1 if transmission over wavelength λ is performed at link $e \in E$

$0 \leq \pi^r_v \leq C_{v,t}$, an integer variable that denotes the processing power at node v allocate to demand r.

$0 \leq \gamma \leq 1$, the objective's weighting coefficient.

Objective

$$\min \gamma \sum_{r \in R} \tau_r + (1 - \gamma) \cdot \sum_{r \in R} \sum_{v \in V} \pi^r_v \cdot \xi_v$$

Subject to the following constraints

Routing of demands

$$\forall r \in R, v \in V, \sum_{a \in \delta^+(v)} \sum_{\lambda \in \Lambda} \sum_{b \in B} x^r_{a,\lambda,b} - \sum_{a \in \delta^-(v)} \sum_{\lambda \in \Lambda} \sum_{b \in B} x^r_{a,\lambda,b} = \begin{cases} -1, & if\, v = s \\ 1, & if\, v = d \\ 0, & else \end{cases}$$

Nodes where processing can be performed:

$$\forall r \in R, a \in \delta^+(v), v \in \alpha, y^r_v \le \sum_{\lambda \in \Lambda} \sum_{b \in B} x^r_{a,\lambda,b}$$

Processing capacity per node:

$$\forall r \in R, v \in V, \pi^r_v \le \left(y^r_v - 1\right) \cdot C_v$$

Satisfy the processing requirements of a demand:

$$\forall r \in R, \varepsilon_r \le \sum_{v \in V} \pi^r_v$$

Bandwidth constraints:

$$\forall r \in R, a \in \delta^+(v), \zeta_r \le \sum_{\lambda \in \Lambda} \sum_{b \in B} x^r_{a,\lambda,b} \cdot b$$

Latency Constraints

$$\forall r \in R, \tau_r \le \sum_{a \in \delta^+(v)} \sum_{\lambda \in \Lambda} \sum_{b \in B} x^r_{a,\lambda,b} \cdot \tau_a$$

Transmission reach constraint:

$$\forall r \in R, v \in V, a \in \delta^+(v) \sum_{\lambda \in \Lambda} \sum_{b \in B} x^r_{a,\lambda,b} \cdot h_b \le d_a$$

Available transceivers per node:

$$\forall v \in V, \sum_{r \in R} \sum_{a \in \delta^-(v)} \sum_{\lambda \in \Lambda} \sum_{b \in B} x^r_{a,\lambda,b} \le M_v$$

Distinct wavelength assignment constraint:

$$\forall e \in a : a \in \delta^+(v), z_{e,\lambda} \ge \sum_{r \in R} \sum_{b \in B} x^r_{a,\lambda,b},$$

$$\forall e \in E, \lambda \in \Lambda, z_{e,\lambda} \le 1$$

Table 1 The pseudocode of the heuristic algorithm for the workload placement problem	**The pseudocode of the heuristic algorithm**
	1: **Input**: Graph $G = (V, E)$
	2: Available transponders per node
	3: Available processing per node
	4: Cost of processing, transponders
	5: Number of shortest paths k
	6: Transponders' transmission configurations
	7: Set of Demands R
	8: **Output**: Path to serve each demand
	9: Allocated transponders/regenerators processing power
	10: **For** each (s, d) pair
	11: Calculate k-shortest paths
	12: **For** each transmission configuration
	13: Calculate the nodes where transponders are placed
	14: Sort requests by latency constraint in ascending order
	15: **For** each demand
	16: **For** each shortest path
	17: **If** free (processing) > demanded(processing) and free transponders(required nodes) > 0
	18: Allocate Processing in nodes of the path with respect to the objective function
	19: Allocate the required number of transponders/regenerators
	20: Update available transponders struct
	21: **end If**
	22: **end For**
	23: **If** transponders/regenerators and processing allocated for the demand
	24: Demand is served
	25: else
	26: Demand is denied
	27: **end If**
	28: **end For**

A heuristic algorithm (Table 1) can also be used to efficiently serve a large number of demands for a large size network topology significantly faster than the ILP-based optimal algorithm. The proposed heuristic serves the demands sequentially. To do so, it starts by making use of an initial phase to calculate a set of candidate paths for each demand P_r. Then, for each one of these demands and for each transmission configuration (tuple) t it calculates the regeneration points (if needed) based on the transponder's transmission configuration and the set of nodes where the demand's processing requirements can be served. Since the solution space for each demand can be vast, slowing the execution, our algorithm has an additional phase where it prunes the dominated candidate solutions. These are the configurations with more spectrum requirements and paths with available processing capacity of higher cost than other candidate solutions. For each demand, the additional cost of each solution is considered, taking into account the demands served up to that point (the current state of the network).

9　Simulation Results

In our simulations, we considered two different topologies for the computing and networking infrastructure with different characteristics in terms of the number of available resources and link lengths: a basic (Fig. 8a) and an extended (Fig. 8b) topology. For both network topologies, we assumed that the network is split into L = 3 layers, with each layer representing an edge, fog and cloud infrastructure. The basic one consists of 6 nodes: 1 cloud, 2 fog and 3 edge nodes, with link lengths varying from 100 to 500 km, while the extended one consists of 2 cloud, 3 fog and 9 edge nodes with average link lengths of 300 km that vary on the interval [100–500] km. The number, the processing capacity and the availability of the resources was assumed that increases while moving in higher layers of the infrastructure. On the other hand, the utilization cost of the processing resources decreases from the edge to the cloud.

Edge nodes are connected via lower rate links, while cloud nodes via higher speed links. The nodes of the different layers are interconnected through transponders that can transmit at 25 Gbps at 100 km reach utilizing a single wavelength, and at 50 Gbps at 200 km reach utilizing two wavelengths [24]. Each traffic request has a latency constraint, which in practice depends on the application it serves. The proposed mechanisms were evaluated based on the average cost of the utilized computing resources for the workload processing and on the experienced propagation latency of the selected network path per request.

Initially, we examined the optimality of the heuristic in the small-scale topology with a number of offline network requests that varied from 20 to 120. The demands' capacity requirements are drawn from the uniform distribution on the interval [10–50] Gbps for network, and on the interval [1–10] GFlops for processing capacity. We assumed that for the higher layer resources the cost of utilizing processing power is reduced by 20% (1.25 c.u for allocating 1 GFLOP in layer 1), in relation to lower layer resources (cloud-fog and fog-edge). Under these settings, we were able to track the optimal solution using the ILP-based algorithm.

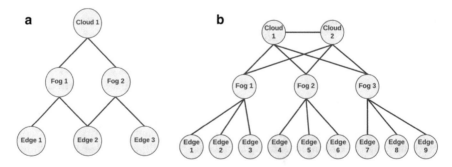

Fig. 8 **a** The basic network topology and **b** the extended network topology, split into 3 layers to represent an edge-fog-cloud infrastructure

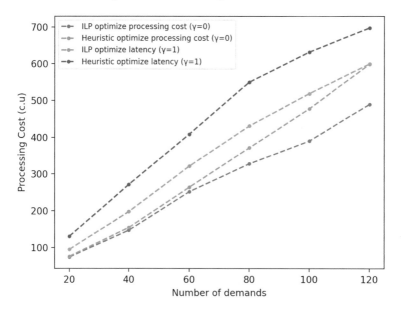

Fig. 9 Total processing cost to serve the workload demands for the ILP and heuristic for the different objective functions

As shown in Fig. 9, the lowest cost is achieved when the objective is to minimize the processing per iteration cost. In that case, the cloud resource nodes are preferred over the edge and fog nodes, due to their lower cost and higher processing capabilities. The performance of the proposed ILP and heuristic is similar for a small number of requests, while for a higher number of requests the ILP outperforms the heuristic. When the objective is the minimization of the latency for serving the requests, more edge resources are utilized, which results in increased cost when the heuristic mechanism is used.

More simulation experiments were conducted on the larger network topology and for the traffic scenarios described above, using the heuristic algorithm that provides good resource allocation solutions in reasonable time. We examined under various traffic scenarios the use of edge, fog and cloud computing nodes, for the layer selection to serve the demanded workload, considering two different objective functions: (i) minimize latency and (ii) minimize processing cost.

As shown in Fig. 10, when the objective is the minimization of the processing cost (left bar), more tasks are executed in Layer 3. The core resources are preferred over the edge ones. In this case, only the requests with most strict latency requirements offload their workload in the Layer 1, 2 nodes, while requests with relaxed latency requirements select processing nodes located in the metro core, with lower cost. Also, the number of tasks executed in Layer 3 increases as the total number of requests increase, taking advantage of the higher number and the more computational resources that are available in the cloud. On the other hand, when the objective is the minimization of the latency (right bar), Layer 1 and Layer 2 resources are preferred,

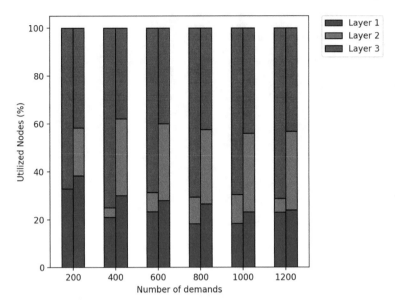

Fig. 10 The percentage of nodes selected to serve the workload demands for the (i) minimize processing cost ($\gamma = 0$ for the left bar) and the (ii) latency minimization ($\gamma = 2$ right bar) objective

while core processing resources are used only due to exhaustion of the edge nodes' capacity. More specifically, the latter ones are highly utilized due to their advantage to be allocated for workloads that cannot be served by Layer 1 resources. Hence, when the main optimization criterion is the cost, the cost and capacity advantages of cloud resources render them the most appropriate ones to be used. However, when the objective is the minimization of the latency, edge and fog resources are preferred at the expense of a higher cost.

Figure 11, illustrates the average latency and the average cost of the resource allocation decisions when faster processing equipment (e.g., hardware accelerators) is used in the edge nodes, considering the scenario with the 600 requests. We assumed that the special purpose equipment can offer better processing performance compared to the generic resources up to 50% and is available in Layer 1 nodes. Thus, we examined for the latency minimization objective, the trade-off between latency and cost, with the illustrated values being compared (as %) to the case where generic computing resources are used. As shown, the faster processing equipment in the edge resources results in a decrease in the average latency of up to 16%, while the average processing cost is increased by up to 20%.

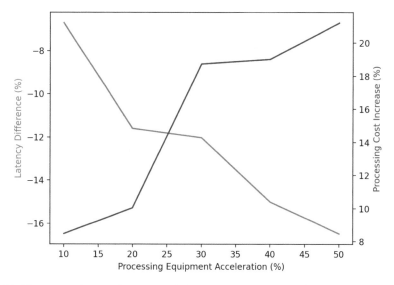

Fig. 11 The average latency decreases and the average cost increase of the resource allocation decisions when faster processing equipment is used in the edge nodes

10 Conclusions

The transparent integration of edge, fog and cloud resources is of paramount importance towards the realization of the cloud continuum. In this direction, we developed and presented an ILP-based mechanism and a heuristic for workload placement in a hierarchical computing and networking infrastructure. Our simulations indicate that the developed mechanisms serve efficiently the demands based on the objective set by the administrator, taking advantage of the processing capabilities of the resources at the different layers (edge, fog, cloud), and reducing latency at the expense of increased cost.

Acknowledgements This research has been co-financed by the European Regional Development Fund of the European Union and Greek national funds through the Operational Program Competitiveness, Entrepreneurship and Innovation, under the call RESEARCH–CREATE–INNOVATE (ARMONIA, project code: T1EDK-05061).

References

1. Foster, I., Kesselman, C.: The Grid: Blueprint for a New Computing Infrastructure. Morgan Kaufmann Publishers. ISBN 1558604758 (1999)
2. Home.cern: The grid: a system of tiers | CERN [Online] (2021). Available at: https://home.cern/science/computing/grid-system-tiers. Accessed 9 Mar 2021

3. Eu-egee-org.web.cern.ch: EGEE portal: enabling grids for E-sciencE [Online] (2021). Available at: https://eu-egee-org.web.cern.ch/index.html. Accessed 9 Mar 2021
4. Qureshi, M.B., Dehnavi, M.M., Min-Allah, N., et al.: Survey on grid resource allocation mechanisms. J. Grid Comput. **12**, 399–441 (2014)
5. Laure, E., Fisher, S.M., Frohner, A., Grandi, C., Kunszt, P., Krenek, A., Mulmo, O., Pacini, F., Prelz, F., White, J., Barroso, M., Buncic, P., Hemmer, F., Di Meglio, A., Edlund, A.: Programming the grid with gLite. Comput. Methods Sci. Technol. **12**(1), 3345 (2006)
6. Stevens, T., De Leenheer, M., Develder, C., Dhoedt, B., Christodoulopoulos, K., Kokkinos, P., Varvarigos, E.: Multi-cost job routing and scheduling in grid networks. Futur. Gener. Comput. Syst. **25**(8), 912–925 (2009)
7. The NIST Definition of Cloud Computing [Online]. Available at: https://csrc.nist.gov/public ations/detail/sp/800-145/final. Accessed 9 Mar 2021
8. Amazon Web Services, Inc.: Amazon Web Services (AWS)—Cloud Computing Services [Online] (2021). Available at: https://aws.amazon.com. Accessed 9 Mar 2021
9. https://www.globenewswire.com/news-release/2020/08/21/2081841/0/en/Cloud-Computing-Industry-to-Grow-from-371-4-Billion-in-2020-to-832-1-Billion-by-2025-at-a-CAGR-of-17-5.html
10. Zhang, J., Huang, H., Wang, W.: Resource provision algorithms in cloud computing: a survey. J. Netw. Comput. Appl. **64**, 23–42 (2016)
11. Singh, S., Chana, I.: A survey on resource scheduling in cloud computing: issues and challenges. J. Grid Comput. **14**(2), 217–264 (2016)
12. Wang, H., et al.: Distributed systems meet economics: pricing in the cloud. In: USENIX Hot Topics in Cloud Computing (HotCloud) (2010)
13. Kokkinos, P., Varvarigou, T., Kretsis, A., Soumplis, P., Varvarigos, E.: Sumo: analysis and optimization of amazon ec2 instances. J. Grid Comput. **13**(2), 255–274 (2015)
14. Yousefpour, A., et al.: All one needs to know about fog computing and related edge computing paradigms: a complete survey. J. Syst. Architect. **98**, 289–330 (2019)
15. Cisco Global Cloud Index: Forecast and Methodology, 2016–2021
16. Kokkinos, P., Kalogeras, D., Levin, A., Varvarigos, E.: Survey: live migration and disaster recovery over long-distance networks. ACM Comput. Surv. **49**(2) (2016)
17. Christodoulopoulos, K., Kontodimas, K., Siokis, A., Yiannopoulos, K., Varvarigos, E.: Efficient bandwidth allocation in the NEPHELE optical/electrical datacenter interconnect. IEEE/OSA J. Opt. Commun. Networking **9**(12) (2017)
18. Poullie, P., Bocek, T., Stiller, B.: A survey of the state-of-the-art in fair multi-resource allocations for data centers. IEEE Trans. Netw. Serv. Manage. **15**(1), 169–183 (2017)
19. Braiki, K., Youssef, H.: Resource management in cloud data centers: a survey. In: IEEE International Wireless Communications & Mobile Computing Conference (IWCMC), pp. 1007–1012 (2019)
20. Landi, G., Capitani, M., Kretsis, A., Kontodimas, K., Kokkinos, P., Gallico, D., Biancani, M., Christodoulopoulos, K., Varvarigos, E.: Inter-domain optimization and orchestration for optical datacenter networks. J. Opt. Commun. Networking **10**(7), B140–B151 (2018)
21. Ayoubi, S., Limam, N., Salahuddin, M.A., Shahriar, N., Boutaba, R., Estrada-Solano, F., Caicedo, O.M.: Machine learning for cognitive network management. IEEE Commun. Mag. **56**(1), 158–165 (2018)
22. VMware Radius: The next big thing in networking: closed loop automation [Online] (2021). Available at: https://www.vmware.com/radius/closed-loop-automation-network/. Accessed 9 Mar 2021
23. Cloudify: Closed Loop Orchestration (CLO) with Cloudify [Online] (2021). Available at: https://cloudify.co/blog/closed-loop-orchestration-clo-with-cloudify. Accessed 9 Mar 2021
24. Pavon-Marino, P., et al.: Techno-economic impact of filterless data plane and agile control plane in the 5G optical metro. J. Lightwave Technol. **38**(15), 3801–3814 (2020)

Collaborative Caching Strategy in Content-Centric Networking

Shupeng Wang and Zhaolong Ning

Abstract With the development of the mobile Internet in recent years, the amount of data in Internet has increased substantially, resulting in more and more serious delivery delay and data cost surge, which has brought great pressure to traditional host-centered networks. As a result, a new network architecture, Content-Centric Networking (CCN), has emerged. Compared with traditional networks, CCN has larger network capacity and lower delivery delay, which can better improve quality of service. Contents are cached in the routers in a CCN. Thus, when a user's interest packet is passed to routers, the router that caches the content can directly respond to the request. Efficient caching strategies in CCN can not only reduce link load and storage redundancy, but also reduce delivery delay caused by the surge of data. This chapter studies caching strategies in the CCN. In order to efficiently and reasonably place contents in routers and reduce the transmission delay, it firstly proposes a caching scheme based on on-path caching and off-path caching. Secondly, in order to further reduce the energy consumption, this thesis proposes a caching scheme to minimize energy consumption. In the caching scheme based on the on-path caching and off-path caching, we firstly analyze the factors such as the popularity of contents and the history of user's requests. Secondly, we strengthen the collaboration and the sharing of caching information between routers. Finally, an optimization problem aiming at minimizing the delivery delay is established, and an approximate algorithm is designed to obtain the approximate solution to the problem. The simulation results show that this scheme can effectively reduce delivery delay. In the caching scheme for minimizing energy consumption, we firstly set up a network content transmission model and a storage model. Secondly, through the comprehensive analysis of the content size and popularity of contents, we design the cache scheme to minimize energy consumption cost in CCN. Then, we transform the caching problem to an energy cost minimization problem, which is subject to certain constraints. Finally, an improved genetic algorithm is used to solve the problem. Simulation results show

S. Wang
Institute of Information Engineering, Chinese Academy of Sciences (CAS), Beijing, China
e-mail: wangshupeng@iie.ac.cn

Z. Ning (✉)
Dalian University of Technology, Dalian, China

© The Author(s), under exclusive license to Springer Nature Switzerland AG 2022
P. Nicopolitidis et al. (eds.), *Advances in Computing, Informatics, Networking and Cybersecurity*, Lecture Notes in Networks and Systems 289,
https://doi.org/10.1007/978-3-030-87049-2_16

that the caching scheme to minimize energy consumption can effectively reduce the energy consumption, and it is superior to other caching schemes.

Keywords Content-centric networking · Collaborative caching · Energy consumption minimization · Popularity

1 Introduction

With the rapid development of the Internet of things and the mobile Internet, data services have become an indispensable part of people's lives, and a variety of Internet applications have greatly enriched people's lives. Rich and colorful multimedia services make the network data grow exponentially, which poses a great challenge to the current network bandwidth resources, and the network is facing increasing data pressure. According to the White Paper on Global Network data Traffic Forecast [1] released by Cisco in 2019 [1], the total amount of data generated by the global network will reach 396 EB by 2022, more than triple that of 2017. Although the huge amount of data will make people's daily life more colorful, it also brings great challenges to the current network. For example, during the peak period of data traffic download and use, large-scale congestion will occur in the network link. This will lead to a series of problems, such as the user's access delay is greatly increased, the user's network experience is poor, and so on.

According to the survey of Cisco, the continuous growth of equipment in the future will have an important impact on the next generation of general communication systems. Although researchers are constantly using new technologies to improve the bandwidth of the existing network, with the increase of network bandwidth, the data demand of users is also increasing exponentially. Only relying on the traditional way of broadening the network frequency band to increase the network capacity is difficult to cope with the growing data needs of users. At the same time, users' demand for high-speed data services is also increasing. Through the analysis of the network circulation data, it is found that a small part of the network data accounts for most of the network traffic. For example, only 10% of the content in the YouTube server accounts for 80% of visits. Part of the content in the network is repeatedly requested by users, which leads to the redundancy of data circulation, which is an important reason for serious network congestion and poor user experience. With the emergence of caching technology, people begin to realize that marginalizing network resources is a very effective way to alleviate the current burden of network traffic.

With the iterative update of network technology and the deepening of the popularity of various mobile devices and Internet of things terminals, the amount of data in the network is increasing year by year, resulting in a series of problems such as network delay that make the traditional host-centric network already overwhelmed. Under this background, a new type of network architecture-Information Center Network (ICN) arises at the historic moment. The rise of ICN network makes the current network architecture change from host-centered end-to-end communication to receiver-driven content retrieval, which has been recognized by many scholars at home and abroad. As a branch of information center network [2], Content Centric Network (CCN) is regarded as the most promising network architecture in the future. Intra-network cache is considered to be the most valuable feature of CCN. As the cornerstone of CCN, cache performance directly affects the performance of the whole network. Efficient caching of content can not only reduce network traffic and relieve server pressure, but also reduce user access delay and improve the quality of service experience of users.

Network caching technology is proposed to deal with the current situation of increasingly serious network congestion and explosive growth of network traffic. According to the statistics of Cisco company, 20% of the requests in the network account for 80% of the traffic. At present, hot videos and pop music data occupy more than 80% of the IP bandwidth of the network. One of the hot areas in network research is to establish a model for network content access for quantitative and qualitative analysis, so as to implement the distribution of network resources more pertinently. At present, Zipf law is the most extensive form used to define popularity, and Zipf law is an American scholar G. K. Zipf proposed it. Zipf law can be expressed as follows: in the text library of natural language, the number of times a word appears is inversely proportional to its ranking in the frequency table. In [3], M. Joos gave a general form of Zipf distribution to define content popularity. In the aspect of cache modeling, literature [4, 5] applies Zipf distribution to cache modeling and defines content popularity as global popularity and local popularity respectively.

At present, the caching strategy in the network is mainly divided into cache decision strategy and cache replacement strategy [6, 7]. This topic studies the cache decision strategy. In a narrow sense, the definition is that when the content passes through the router, the router needs to determine whether to store the content. In a broad sense, the cache decision strategy is how to cache the content in the whole network reasonably. As a result, the delivery delay required by the user to obtain the target content is greatly reduced, and the quality of service experience of the user is improved [8]. With the emergence of the architecture with on-network cache in ICN, the cache decision strategy of CCN is also on the rise. Currently, the default caching decision strategy for CCN is the CEE (Cache Everything Everywhere) strategy [9], also known as the anywhere caching strategy. The CEE strategy reduces the distance between the user and the target content by caching all the content that passes through the routers within the network, but this also causes the same content to be stored by multiple routers, resulting in serious redundancy problems. The literature puts forward a new innovation on the basis of CEE strategy, and puts forward LCD (Leave Copy Down) caching strategy [10]. The main idea of this strategy is that after the content request packet sent by the user is responded by the router, the content is returned to the user's path, and only the next router that hits the router can store the content. other routers only forward but not store, this strategy effectively reduces the redundancy of content storage in the network. In [11, 12] proposed a caching decision strategy for Prob, which is a probabilistic caching strategy. Its main idea is that when the content passes through the router, the router will generate probability

P(x), according to its own remaining storage space and the popularity of the content. By determining the level of the P(x) value, the router decides whether to store the content or not. It is also mentioned in [13] that a probabilistic storage strategy Prob (p), is different from the Prob strategy, which stores all the content passing through the router according to probability p. A caching strategy called CL4M (Cache "Less for More") is proposed in [14]. By calculating the intermediate value centrality of the router in the network, the content is usually only stored in the location with high intermediate value centrality.

Cache replacement strategy is how to balance the relationship between new content and stored content when the available storage space of the router within the network is zero. The CCN cache replacement strategy largely draws lessons from the web cache replacement strategy, such as the common LRU and LFU cache replacement strategies [15–17]. With the deepening of CCN network research, new cache replacement strategies have been proposed. A cache replacement strategy called RUF (Recent usage Frequency), also known as recently used cache, is proposed in [18]. Through the detection of streaming content and the dynamic weighted average calculation of instantaneous traffic changes, this strategy can effectively predict the dynamic changes of users' preferences.

At present, in order to improve the utilization efficiency of CCN network storage space, researchers have proposed a large number of caching strategies, but the current storage strategy only optimizes a certain performance of the network, does not fully take into account the popularity of content, does not improve the cache cooperation between routers and other shortcomings, which will affect the overall performance of the network. Therefore, by analyzing the research status of existing network caching strategies, combining the advantages of multiple caching strategies and analyzing the influence of content popularity, this chapter puts forward the design idea of multi-factor collaborative storage. Design an efficient storage scheme to essentially improve the performance of the network.

At present, most of the cache strategies for CCN are designed to reduce network latency and other related performance, and the research on network energy consumption in the existing cache strategies has been largely ignored. However, with the enhancement of network cache capacity and the gradual increase in the number of network content, under the pressure of increasingly stringent environmental standards and rapidly rising energy consumption costs. In the design of cache strategy, the problem of network energy consumption has gradually become a problem that cannot be ignored. The energy consumption of CCN, consists of two main parts: transmission energy consumption and storage energy consumption. Transmission energy consumption includes the energy consumption of internal network, edge network and external network [19]. Storage energy consumption is mainly caused by the energy consumption generated by caching in the router, which follows the energy scale model and also depends on caching hardware technologies [20] (such as SSD, DRAM, RLDRAM, SRAM and TCAM). In this article, we will also discuss the practical problems of CCN cache scheme design from the perspective of energy consumption efficiency.

This chapter takes the Content Centric Network as the research background, aiming at some problems faced by the Content Centric Network cache, by strengthening the cooperation among the cache nodes, so as to reduce the redundant cache of the cache router to the content, reasonably and efficiently use the limited storage space, improve the cache hit rate of the content, further reduce the content transmission delay and network energy consumption, and propose an efficient Content Centric Network cache scheme.

The main contributions of this chapter are summarized as follows:

1. Design and implementation of cache scheme based on cooperation between on-path caching and off-path caching

    First of all, a network model which accords with the characteristics of CCN is established from the level of collaborative storage, and the models of content request and content placement are designed. then the factors such as content popularity, user history request data and network data transmission delay are analyzed. After a comprehensive analysis of the advantages of On-path caching and Off-path caching, the content can be reasonably placed in the limited storage space by improving the cooperation and content sharing among router nodes. finally, an optimization problem is established to minimize the communication delay caused by content acquisition in the network, and an approximate algorithm is designed to obtain the approximate solution of the problem. Simulation results show that this scheme can effectively reduce the network communication delay.

2. Design and implementation of cache scheme based on network energy consumption minimization

    Based on the composition characteristics of the energy consumption of the Content Centric Network, an energy consumption model is established, which includes the storage energy consumption and transmission energy consumption of the content. Under the premise of minimizing the energy consumption, the router nodes are urged to cooperate efficiently among nodes, so as to maximize the diversity of cached content and make users consume less energy to acquire content. This scheme analyzes and compares the influence of content popularity and content size on network energy consumption, and establishes a cache scheme for minimizing energy consumption, which is a traditional NP-hard problem. Genetic algorithm is used to solve this problem. Finally, the correctness and rationality of the scheme are proved by simulation.

2 Design and Implementation of Cache Scheme Based on Cooperation Between On-path Caching and Off-path Caching

In this chapter, we propose a caching scheme based on the cooperation of On-path caching and Off-path caching, establish an optimization problem to minimize the delivery delay caused by content acquisition in the network, and design an approximate algorithm to obtain the approximate solution of the problem. The feasibility of this scheme is verified at the end of this chapter.

2.1 Problem Description and Analysis

2.1.1 Cache Redundancy of Content Centric Network

Content Centric Network is a new architecture which is different from the traditional host-centered IP network. Starting from the characteristics of users, content-centric network pays more attention to content-centered. By perceiving users' communication behavior and obtaining relevant data information, the content within the network can be replaced and planned, so that the content can be stored more reasonably.

At present, the default caching strategy of Content Centric Network is CEE (Cache Everything Everywhere) caching strategy, that is, generalized anywhere caching. The main idea is that when the interest packet is responded by the router, all the routers passing the content packet to the user will store the content in the content packet, so as to improve the space utilization of the router. CEE is a typical non-cooperative caching mechanism. Because each router holds a copy of a packet, it faces the problem of cache redundancy. Within the network, the storage space of each router is certain, and cache redundancy will lead to limited types of content stored in the router, unable to store more content of different sizes, and various forms of network resources emerge one after another. how to deal with the diversity of users browsing content, how to reduce the network delay for users to obtain corresponding content, and how to reduce the waste of system resources caused by repeated requests. All these need to be further studied and solved.

2.1.2 Lack of Cooperation Between Router Nodes Within the Network

The Content Centric Network contains many routers, and the interest packets sent by the content requestor will be delivered to the content source along the shortest path, and then the packets will be returned to the requester along the original request path. in this process, content caching mainly occurs in the delivery path, but the delivery paths within the network are independent of each other, and each router makes its own decision cache. This causes the problem that the cooperation between routers is insufficient and the content of the router cache is single; and due to the lack of content sharing, the same content may be repeatedly stored by multiple routers, which not only results in cache redundancy, but also greatly reduces the utilization of storage space. If we increase the interconnection between routers and improve the sharing of cache content between routers, more kinds of content can be stored in the limited storage space and the utilization of cache space can be improved. when the user gets the content, it can effectively reduce the delivery delay of the content.

2.1.3 The Caching Strategy of CCN Needs to Be Optimized

As mentioned in the previous section, the default caching strategy in the Content Centric Network is CEE, With the in-depth study of the Content Centric Network, some novel and efficient caching strategies have been proposed by researchers: such as MPC caching strategy, its basic idea is to give priority to content accessed many times over a period of time in the router when making caching decisions. MPC caching strategy effectively alleviates the redundancy of content storage, but still cannot reduce the time delay for users to obtain content. Prob cache strategy, also known as probability cache, by setting a probability threshold, when a packet passes through a router, a probability value is randomly generated, and if it is greater than the set probability threshold, it will be stored in the router. Although this cache strategy can reduce the delay of network content acquisition to a certain extent, it cannot solve the problem of storage redundancy. How to design the caching strategy to make users obtain the corresponding content with less delay and reduce the redundancy of network storage is the main difficulty that we have to solve.

2.2 Construction of Collaborative Caching Model

2.2.1 On-path Caching and Off-path Caching

On-path Caching

According to the location of the cached content block, the caching mechanism can be divided into on-path caching and Off-path caching [21]. In the path cache, the interest packet is passed through the router until the response is hit, and the packet is passed back to the requester along the request path of the interest packet. During the delivery of the packet, if the content is stored in the router on the delivery path, we call it on-path caching, as shown in Fig. 1. Although content caching is an inherent feature of CCN on the delivery path, Off-path caching can also be used if there is a centralized topology manager (RH), in the network. For example, in Fig. 2, RH can decide where to cache content and forward a copy of that content to the selected router, which may be cached by the router on the delivery path or by a router outside that path, which we call Off-path caching.

In the on-path caching, when the router receives a request for content, it responds with a copy of the local cache without involving the topology manager. Although this approach reduces the computational and communication overhead of placing content on the network, it may reduce the chance of cached content hits. An important problem in On-path caching is how each node makes cache decisions to improve the cache hit rate of content. For example, popular content may need to be placed where it is next requested. In addition, the expected content popularity or time locality needs to be considered when designing intra-network caching algorithms in order to provide priority for some content. The standard of which content should be given

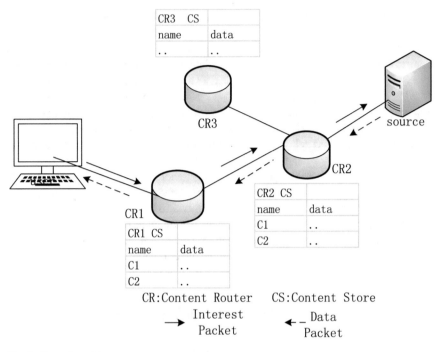

Fig. 1 On-path caching

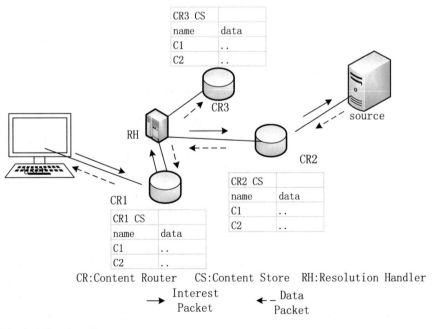

Fig. 2 Off-path caching

priority in the internal content cache of the network is also related to the business relationship between the content provider and the network operator.

Off-path Caching

Off-path caching, also known as content replication or content storage, is designed to replicate content on the network, regardless of the delivery path used, thereby improving the availability of the content. The actual number of replicas and the selection of nodes to store replicas [4, 22] are determined through monitoring and traffic analysis operations, because storage location decisions may be affected by contextual information, such as node availability, storage availability, and content popularity [23]. The balance of this large set of variables constitutes a complex system that is sensitive to short-term changes that may significantly increase the complexity of the replication mechanism. In addition, copies in the off-path caching, are usually broadcast [7, 8] in the name resolution service. The distribution of replicas and the update of the content notification service that references these replicas introduce additional traffic to the network, which needs to be taken into account when determining the location of the replicas. Therefore, off-path caching should be seen as a long-term decision. Based on this premise, content replication usually refers to copying content at the object naming granularity and performing [7] using a supplementary server located at the edge of the network. Considering that the optimal placement of content replicas in distributed storage systems has proved to be a difficult task [24, 25], optimization techniques [26, 27] are usually used to perform off-path caching. The problem of off-path caching in CCN is comparable to that of CDN content replication and web cache placement [7, 8], but only a few off-path caching strategies [28–31] have been proposed to supplement the existing optimized distributed cache [32, 33].

Compared with the existing distributed caching about off-path caching and optimization, on-path caching integrates naming and caching mechanisms into CCN networks. In more detail, the purpose of on-path caching way is to reduce traffic and latency in the network, not to increase the availability of content.

The purpose of off-path caching is to copy content in the network in order to improve the availability of content, the two complement each other and have their own emphasis.

2.2.2 Design and Implementation of Collaboration Model

In the first two sections, we introduced the advantages and disadvantages of on-path caching and off-path caching respectively. We know that the main goal of off-path caching is to reduce network traffic and network latency, and to improve the quality of service experience of users, but does not increase the availability of content.

For off-path caching, it can improve the availability of content by copying the content in the network, and can also reduce the time cost for users to obtain content to a certain extent. By analyzing the advantages and disadvantages of the two caching strategies, we consider that by combining the advantages of the two caching strategies, we can not only reduce the network traffic and network delay, but also improve the availability of content. therefore, we design a caching strategy scheme based on the cooperation of on-path caching and off-path caching.

Next, we will briefly introduce the collaborative storage model. From Fig. 3, we can see that when client1 acquires content, it first sends the interest packet to its neighboring router H, then passes the interest packet through layers, and finally to the content source, and then the packet returns along the delivery path of the interest packet. In the figure, it shows A → B → D → H → client1, in the process of packet delivery. The passing router can choose whether or not to store the content, which is on-path caching; The content is passed to router D, and through multiple copies of the content, the content can be stored in router I. this stored procedure is called off-path caching. The collaborative storage model we designed is to increase the cooperation of routers within the network, through reasonable replication and

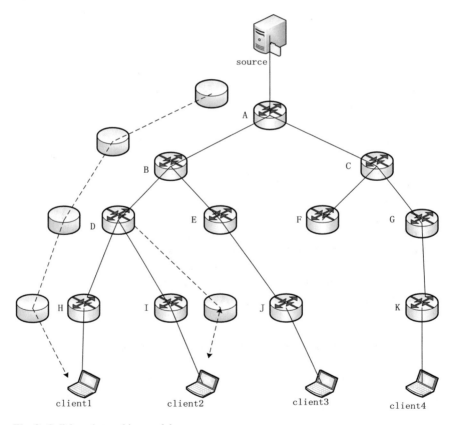

Fig. 3 Collaborative caching model

efficient placement of content in the network, on the basis that users obtain the corresponding content, obtain the target content with minimum delay and less network traffic. The specific performance in the figure is as follows: after client1 acquires the corresponding content, we make the minimum delay for client2 to obtain the same content through the reasonable placement of the content. Taking Fig. 3 as an example, content caching is carried out in the path from client1 to the content source. First, the most popular content is stored near the user side. When the client2 also wants to obtain the corresponding content, the interest packet can be hit when it is delivered to router D, then through the cooperation between the two links, the content can be stored in router D and removed from router H. at this time, through the cooperation between the two links, the content can be stored in router D and removed from router H. at this time, the content can be stored in router D and removed from router H. This not only completes the unified placement of the contents of the two links, but also reduces the storage redundancy of the router.

As you can see, introducing the cooperation between routers into the whole network and placing the content reasonably, on the one hand, it can improve the utilization of network space and reduce storage redundancy; on the other hand, it can also reduce the network delay for users to obtain content. improve the quality of service of users.

2.3 Problem Solving

In the Content Centric Network, each router has a fixed storage space by default. In view of the problems and challenges faced by the network in Sect. 2.1, this section will introduce the network topology simulation, delivery delay calculation, and knapsack problem solving.

2.3.1 Network Topology Simulation

In this section, interest packets and data packets are delivered along the shortest path from the requester to the content source. Therefore, the transmission and storage path of each content can be regarded as a linear topology, and the simulated Content Centric Network topology can be regarded as a combination of multiple linear topologies, using an undirected graph, and transforming it into a tree rooted at the content source. Routers with caching capabilities can be connected to this root node. Each router node in the network has a certain storage capacity, and the user randomly selects the router to access, as shown in Fig. 4, which is an example of the Content Centric Network topology, the figure is undirected network diagram composed of 12 router nodes.

The line topology in different colors in the figure represents the shortest transmission path, and the whole figure is composed of three paths. Interest packets are delivered and forwarded along different paths until they are hit by the router response

Fig. 4 Network topology
diagram of CCN

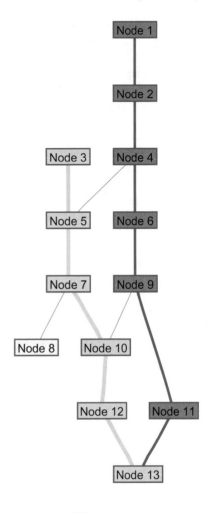

with the corresponding content, and then the data packets will be returned to the user along the original path.

2.3.2 Delivery Delay Calculation

In the Content Centric Network, for the convenience of calculation, we connect the router inside the network to fixed users, and the users sends a request to the router connected to it within a unit time. We define the product of three variables, the number of hops required to obtain content, the time required for unit hops, and the popularity of the content, as the network delay for obtaining the content, by calculating all the contents inside the network separately and then summing them up, we can get the delay in obtaining all the contents in the whole network. Next, we will take a content as an example to show the network delay required to obtain the corresponding content:

1. Initializes the number of contents within the network x, content size matrix M, number of routers y, two-dimensional matrix of interest packet routing table $Path$;enter the content popularity function in turn $C\_popularity$, network structure tree diagram matrix $Node\_fas$, content number l, and content storage table matrix W;
2. First, count the coordinates of all elements in the content storage table as 1, and assign the value of the above-mentioned coordinate point element in the network structure tree graph matrix to 0;
3. Find the forwarding path of content l in the network by traversing each router, in the subsequent transmission process of interest packets, it will be continuously transmitted along the forwarding path until it is forwarded to the node position with the element of 0 in the network structure tree diagram matrix. And assign the path of the interest package to $Path$;
4. Then analyze the two-dimensional matrix formed by the interest package transfer, and count the occurrence times of each element, and then assign the elements and the occurrence times of the elements to the matrix A in the form of a two-dimensional array;
5. We can obtain the time delay in obtaining content l based on the above data, and express it as:

$$time\_l = sum(A(:, 2)) \times C\_popularity(l) \times M\_k(l) \quad (2.1)$$

Table 1 is the pseudo code of the calculation algorithm for the time consumed from the user requesting the content to obtaining the content after any content in the network is cached to the router in the network. This section takes network delay as one of the performance indicators to evaluate the effect of cache resource allocation.

2.3.3 NP-Hard Problem Solving

Using a certain number of operations to solve a polynomial solvable problem is called an NP-hard problem. In layman's terms, how to maximize the benefits obtained through reasonable allocation of resources is the NP-hard problem. At present, Typical NP-hard problems include: selecting the smallest feature subset in the feature subset; in the induction method of the decision tree, the smallest decision tree induction method instantiates the training of the feature value subset with the largest information gain into a disjoint subset [34]; in the coverage algorithm, learn the smallest k-DNF, the smallest connection, and the most general connection that covers the most training examples. At present, ant colony algorithm is widely used to provide approximate solutions for NP-hard problems, but ant colony algorithm also has some problems, such as premature convergence and weak robustness. On this basis, the researchers proposed an ant colony pheromone matrix optimization strategy based

Table 1 Algorithm for calculating delivery delays

Steps of Algorithm: **Algorithm for calculating delivery delays**
1: Initialization: the number of contents within the network x, content size matrix M, number of routers y, router vector $Nodes$, Interest packet routing table matrix $Path$
2: Input: Content popularity function $C\_popularity$, content size vector M_k, network structure tree diagram matrix $Node\_fas$, content number l, vector of "content cached on router" W
3: $[x, y] \leftarrow size(Node\_fas)$
4: $location \leftarrow$ the coordinate of element 1 in W
5: $num \leftarrow$ calculate the dimensions of the vector $location$
6: **if** $num > 0$
7: **for all** i, $0 < i < num$ **do**
8: $Node\_fas(l, location(i)) \leftarrow 0$
9: **end for**
10: **end if**
11: Traverse each user, each user requests content l once per unit time
12: **for all** $start$, $0 < start < y$ do
13: $output(1) \leftarrow start$
14: $l \leftarrow 1$
15: $loc \leftarrow$ the index of the same element in $Nodes$ as $start$
16: **while** $Node\_fas(l, loc) \neq 0$ **do**
17: $l \leftarrow l + 1$
18: $output(l) \leftarrow Node\_fas(l, loc)$
19: $loc \leftarrow$ the index of the same element in $Nodes$ as $Node\_fas(l, loc)$
20: **end while**
21: Write the $output$ vector into the matrix $Path$ in turn, as the element in the $start$ row
22: **end for**
23: **Sum the traffic on all routes and calculate the traffic consumed by the entire network request content l as:**
24: $time\_l = sum(A(:, 2)) \times C\_popularity(l) \times M\_k(l)$
25: Get $time\_l$

on fleece bubbles to solve the traveling salesman problem and the 0/1 knapsack problem. In the mathematical model inspired by the fleece bubble, a unique feature is that the critical value can be retained during the network evolution. The optimized update strategy takes advantage of the characteristics of the ant colony algorithm, accelerates the positive feedback process of the ant colony algorithm, and makes the optimal solution converge quickly.

The backpack problem is a classic NP-hard problem, you can put n kinds of things in a backpack, the weight of item j is w_j and its value is v_j. When item j is put into the backpack, x_j is the number of items in the category that are put in the backpack, the maximum loadable weight of the backpack is $b(b > 0)$, how to place the items to maximize the total value of the items in the backpack is the traditional backpack problem, expressed as:

$$M = \max \sum_{j=1}^{n} v_j x_j, \quad v_j \geq 0 \tag{2.2}$$

$$\sum_{j=1}^{n} w_j x_j \leq b, \quad w_j \geq 0, x_j \geq 0 \tag{2.3}$$

Suppose $F_k(y)$ is the maximum value when there are only k kinds of things in the backpack and the total weight is limited to y, so:

$$F_k(y) = \max \sum_{j=1}^{k} v_j x_j, \quad (0 \leq k \leq n) \tag{2.4}$$

$$\sum_{j=1}^{k} w_j x_j \leq y, \quad (0 \leq y \leq b) \tag{2.5}$$

The knapsack problem conforms to the optimization principle, and the dynamic programming method can be used to obtain the recurrence equation and boundary conditions:

$$F_k(y) = \max\{F_{k-1}(y), F_k(y - w_k) + v_k\} \tag{2.6}$$

$$F_0(y) = 0, \quad \text{for all } y, 0 \leq y \leq b$$
$$F_k(0) = 0, \quad \text{for all } k, 0 \leq k \leq n$$

For the convenience of solving, when y is defined as a negative number, $F_k(y) \rightarrow -\infty$.

As we all know, NP-hard problems can be solved by using greedy algorithms or dynamic programming to obtain a local optimal solution to obtain a global optimal solution, for the above knapsack problem, dynamic programming can easily solve the problem, the greedy algorithm is not suitable for solving the 0/1 knapsack problem, but for the knapsack problem in special cases, the greedy algorithm can also easily find the solution of the problem.

There are many greedy strategies in the backpack problem, the realization of each strategy requires multiple steps to complete, one strategy is to put the value of the goods first, and put the most valuable goods into the backpack in turn until the maximum load-bearing capacity of the backpack is reached, but this strategy is difficult to ensure that the optimal solution is obtained. Another placement rule is to consider from the perspective of the weight of the goods, and put the smallest weight of the remaining goods into the backpack, which is also difficult to guarantee the solution obtained is the optimal solution, because these two perspectives are too unitary in thinking about the problem. The third perspective is to consider the value of the unit mass items, and put the goods with the highest unit value in the remaining items into the backpack. Among them, there is no doubt that the third solution is significantly more likely to obtain the optimal solution than the first two solutions, and it is also the most likely solution.

Through thinking about the 0/1 knapsack problem, under the premise that certain factors are restricted, the higher the unit value of the item, the more likely to obtain the optimal solution, so the improved greedy algorithm can also easily solve the 0/1 knapsack problem. To solve the problem, in this chapter, the main problem of the design of the caching scheme is how to place the content to optimize the overall performance of the network, that is, the delay required for users to obtain the content is minimized. In essence, the solution to the main problem in this chapter is actually the solution to the knapsack problem in a broad sense. In this solution, the unit income is redefined, and a storage solution based on "unit average income" is established for the content of different storage values, thus achieving reasonable placement of content. In the next section, the storage strategy design based on cache collaboration will be introduced in detail.

2.4   Caching Strategy Based on On-path Caching and Off-path Caching

In order to efficiently and reasonably place the content on the network, improve the utilization efficiency of network storage space and reduce the communication delay for users to obtain the content, taking into account the content popularity, content size and other factors, this chapter proposes the Caching strategy based on On-Path caching and Off-Path caching, OFC.

2.4.1 Mathematical Modeling

The Content Centric Network can be represented by an undirected network graph $P(V, E)$, where $V=\{v_1, ..., v_n\}$ represents a collection of router nodes within the network, and E represents a collection of link connections between router nodes. In this article, it is assumed that each user connects randomly with any router in the router set. The routers inside the network all have a certain storage capacity. When the user obtains the content, the user first sends the interest packet to the router connected to it. If the relevant content is not stored in the router, the interest packet will be passed to the next router until the response hits, then the data packet will return to the user along the original path of the interest packet.

In the cache model, we use $F = \{1, 2...k\}$ to represent the content collection. For the convenience of calculation, we uniformly define the size of each content as the unit size, and the content source server storing the content k is randomly connected to the router in the network. Each content is the smallest processing unit in the cache operation. Figure 5 is an example of the network cache of the Content Centric Network, the origin in the figure represents a router in the real network, we assume that router a is connected to the content source server, and content k is cached in router b. Then, when users under router h obtain content k, they can pass the interest packet to b through the shortest link to obtain the content k they want.

In the process just now, the user obtains the content k by connecting to the router h. The network delay can be obtained by calculating the product of the number of hops and the number of hops that are transmitted by the interest packet and the data packet and the time required for the unit hop, so we can know that the network delay required for the user to obtain content k is 4, but if the router does not cache the content k, then the interest packet will be transferred layer by layer until it is delivered to the content source, then the network delay for the user to obtain content k is 6, so 2 units of network delay can be saved by caching the content k; if we cache the content to the router f, the saved network delay will be greater, and the user's network service quality will be better, but the storage space of a router is certain,

Fig. 5 Example of caching

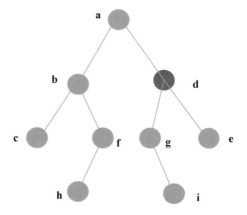

and the choice of what content to put in which router should be considered from the overall network. Therefore, we take network delay as a key factor to measure the quality of our storage strategy.

In the modeling process, we default the size of the content as the unit size, and the delay required by the unit hop between routers as the unit delay. We calculate the product of the number of times the content is requested and the single delay required to obtain the content to calculate network delay. The goal we need to optimize is how to reasonably place content under the premise of limited network storage space, so that the network delay required for users to obtain content is minimized.

The mathematical model established is as follows:

$$Min : H = \sum_{i=1}^{M} \sum_{\substack{j=1 \\ j \neq i}}^{N} \sum_{k=1}^{F} p_i^k h_{ij} x_j^k \qquad (2.7)$$

s. t.

$$X_j^k = \begin{cases} 1 & k \text{ is cached at node } j \\ 0 & others \end{cases} \qquad (2.8)$$

$$\sum_{j=1}^{N} x_j^k \geq 1 \qquad (2.9)$$

$$\sum_{k=1}^{F} s_k x_j^k \leq c_j \qquad (2.10)$$

In Formula (2.6), x_j^k is a binary constant, indicating whether content k is stored in router j, if content k is stored in router j, then the value of x_j^k is 1, otherwise it is 0. h_{ij} represents the number of hops from when user i obtains content to the router j that responds. p_i^k represents the popularity of content k. The popularity of content can also be regarded as the number of times users request in the network, because the higher the popularity, the number of times users request it, on the other hand, the lower it is. H represents the total network delay. Formula (2.8) indicates that there must be a copy of content k inside the network, and Formula (2.9) indicates that the total number of contents stored by the router cannot exceed the maximum capacity of the router itself. Therefore, the optimal content placement of the network can be transformed into the process of solving the Eq. (2.6).

2.4.2   Algorithm Design

In the previous parts, we mainly introduced the network model used in the OFC scheme and the analysis and thinking of the cache optimization problem. In this section, we mainly introduce the specific implementation details of the cache content placement algorithm. Here are the steps of the algorithm:

First, converting all routers $v_s \in V$ in the network into multiple shortest path trees rooted at the content source server v_s, and obtain the connection relationship between the routers. Through the method of storing the most popular content first, place content separately for each router in the shortest path, and uses the router close to the user to prioritize the placement decision of the most popular content. After that, the traversal algorithm is used to count the number of network hops t_s^k required by all users in the network to obtain the content after the content k is cached by the router, and record the storage router location l_a^k corresponding to the content. After caching a piece of content, place the next most popular content in turn. Similarly, count the network hops required by all users in the network to obtain the content after the content is cached by the router, record the location of the storage router corresponding to the content. and complete multiple entries. After completing the content storage of all routers on multiple shortest path trees, the minimum number of network hops for caching N contents in the tree graph rooted at $v_s \in V$ and the position where the contents are placed are obtained.

Then, because there are intersecting nodes between multiple shortest path trees, the content in the intersecting nodes is analyzed for content redundancy, the duplicated content is removed, and the stored content is sorted according to popularity, and the excess the content outside the storage capacity of the router is cleared according to the popularity ranking, and then the number of network hops required for all users to obtain the content is updated, and the storage router location corresponding to the content is recorded. After the above operations are completed, respectively place the content of each router outside the cross node of the shortest path tree twice, and still follow the principle of priority placement of the most popular content. After completing the router update in each link, the number of hops required for the user to obtain the content and the placement of all content are obtained.

Finally, when the total space occupied by the cached content is not less than the storage space of the router, complete all iterations, count and record content $F = \{1, 2, ..., k\}$ and its router location $v_b \in V$, and obtain a one-to-one mapping relationship $f_{k,b}$. After the cache content placement algorithm is completed, obtain the optimal solution of Eq. (2.6) H (Table 2).

2.5   Performance Analysis

This section will carry out simulation verification and performance analysis on the Caching strategy based on On-Path caching and Off-Path caching in this chapter. First, compare the performance of the OFC storage solution proposed in this topic

Table 2 Content placement algorithm

Steps of Algorithm: Content placement algorithm
1: Initialization: content Popularity Matrix C_ popularity, network topology structure matrix Node_ fas, content serial number j, content placement relationship matrix W, network content number F, router number N, router vector *Nodes*;
2: Count the number of rows and columns of the network topology structure relationship matrix $[x, y] \leftarrow$ size(Node_fas)
3: For all routers, v_s, $v_s \in s(F)$
4: Convert the network to a tree diagram with v_s as the root, and count the connection relationships between the routers
5: Initialize vector W, traverse and place the content according to the principle of maximum popularity storage
6: Count the number of hops and the location matrix of the content that the user needs to get
7: $N \rightarrow$ n
8: **while** $n > 0$ **do**
9: For all v_i, $v_i \in V$
10: Calculate the network hop count h_j^k after the content k is cached to j
11: $h_j\_k\_\min \leftarrow \min(h_j\_k)$
12: $location \leftarrow \arg\min(h_j\_k)$
13: $W(location) \leftarrow 1$
14: $n \leftarrow n - 1$
15: Put the value of $h_j\_k\_$ min in order to the hop count matrix H, the corresponding *location* value is stored in the router label matrix *node*, and the contents are sequentially cached according to the result of the sorting until the total cache space C is full
16: end while
17: end

with the default storage strategy of the current content center network and the original non-cached network. Then, analyze the network topology and the number of content and other factors on the network performance. We calculated according to the network delay required for users to request content after the content placement, and this performance index is used as the main measurement basis for comparison with other storage solutions.

2.5.1 Simulation Environment Design

This subject is simulated and verified by MATLAB software, in this solution, the Content Centric Network is composed of 12 routers, 1 content source server and many users and the network contains 20 different contents. It is assumed that each content has the same size and is 1 unit. In the simulation process, it will simulate the process of the user sending out the interest packet in the network through the router forwarding and transmission process, and the router will also selectively store the content during the data packet transmission process. Each user to the content source will calculate the optimal path through the shortest path algorithm, and the interest packets sent by the user will also be transmitted along the shortest path, the total amount of content stored in each router must not exceed the maximum storage capacity of the router. And popularity obeys Zipf distribution, and the exponential constant of Zipf distribution is set to 0.6. In the following subsections, other factors that affect the performance of the program are also analyzed.

2.5.2   Analysis of Simulation Result

This section mainly analyzes the impact of the use of the caching strategy based on the collaboration between the wayside cache and the cache, the absence of caching in the network, and the use of the default storage strategy in the content center network on network performance, such as network delay. The storage space of the router is 2 units of content, and the exponential constant of the Zipf distribution is set as $= 0.6$.

In this section of the simulation process, on the premise of consistent control network topology, the number of users access to different content, under the condition of different storage strategy, user access to the content of network time delay is calculated respectively, from the value of network delay, we can see that the lower the network delay, the better the quality of users' network experience, but also shows that storage strategy is more efficient. As can be seen from Fig. 6, with the increase in the number of contents acquired by users, the three curves all show an upward trend, but with different rising speeds. The OFC storage strategy proposed in this chapter is obviously superior to the other two schemes in terms of controlling network delay. When the amount of content obtained by users is small, CEE and OFC storage policy

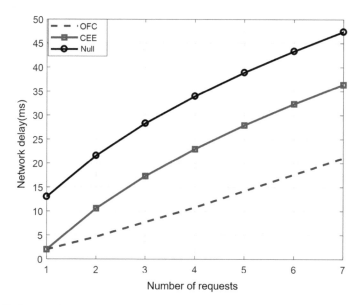

Fig. 6 OFC versus CEE versus null under different numbers of requested contents

have an intersection point. At this time, the advantages of OFC storage policy have not been fully realized. Later, with the increase in the number of content requests from users, the advantages of this scheme gradually appear. The main reasons are:

1. When the network router does not cache the content, after the user sends out the interest packet, all the interest packets need to be delivered to the content source, and then the packet returns to the user along the original request path. When CEE, the default storage strategy of content center network, is adopted, the interest package can be responded in the transmission process when the user gets the corresponding content, and there is no need to transfer the interest package to the content source. Therefore, from the perspective of network delay, we can see that CEE reduces the network delay value by about 25% compared with that without cache. Compared with the other two schemes, the OFC storage strategy proposed in this chapter significantly reduces the network delay value in terms of network delay, which is about 50% lower than the non-cache case and 25% lower than CEE storage strategy.

2. OFC storage strategy is better than CEE storage strategy in network delay because: Although CEE increased the content of the router to store, the CEE to identify the content of the stored in the router has not made the content of the stored in the router redundancy degree is serious, OFC internal storage strategy by strengthening network router, the efficient collaboration between effective solved the problem of the network router storage redundancy, increasing the routers store content type, this allows the user to access to the target content, much interest from the bag can get hit in the network router, do not need to transfer from the content source, it reduces the cost needed by users access to content the hop, fundamentally solve the problem of the network delay.

In conclusion, the simulation results are consistent with the theoretical analysis. This also indicates that our proposed OFC storage strategy is correct and efficient.

2.5.3 Influence of Relevant Parameters

Effect of Router Storage Space on Network Performance

This section compares and analyzes the impact of the storage space size of routers on the network delay caused by user acquisition of content, aiming at the use of CEE and OFC caching strategies and the network without storage conditions. The exponential constant of the Zipf distribution is set a = 0.6. As can be seen from Fig. 7, with the increase of storage space, the network delay under the condition of no storage is fixed, which is the highest among the three schemes. For a network using CEE storage strategy, with the increase of storage space, the network delay decreases gradually. With the increase of storage space, the content that can be stored increases, and the probability of interest packet being responded increases, so the network delay also decreases. With the increase of storage space, the network delay with OFC storage strategy is further reduced, and the network delay with OFC storage strategy is the

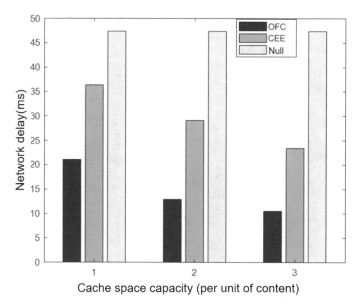

Fig. 7 OFC versus CEE versus null under different storage capacity parameters

smallest, which also verifies the high efficiency of this scheme in reducing network delay.

Therefore, we can conclude that the size of the router's storage space has an obvious impact on network performance. The larger the cache space is, the more content can be stored. However, in reality, the specific operation cost needs to be measured. In the limited storage space, OFC storage policy is more efficient. The core design principle of OFC cache policy is to improve the storage cooperation between routers.

The Effect of Topology on Network Performance

After comparing the impact of cache space size on network performance, we analyze the impact of network topology change on network performance. As is known to all, topology is the interconnection mode among routers in the network. The well-known topologies include star topology, chain topology and ring topology, etc. In this section, the chain topology and ring topology that are mainly used in the current network are simulated. Content has a popularity index of = 0.6, and each router can store 2 units of content. As can be seen from Fig. 8, although the topology has changed, the performance of OFC storage strategy is still the best, and the network delay values are all very low. The red curve represents the chain topology and the blue curve represents the ring topology. We can see that the influence of the ring topology on the network delay is less than that of the chain topology to a certain extent. The network with ring topology has slower numerical improvement in network delay

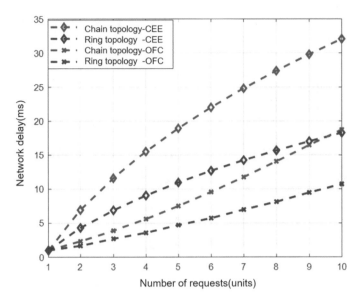

Fig. 8 OFC versus CEE under different topologies

and the growth rate is not as fast as that of chain topology. The main reasons are as follows:

First of all, in the ring topology, the cooperation probability between routers is greatly improved, so that the content can be placed efficiently between routers and the utilization rate of router storage space is improved. In the chain topology, there are few cooperation opportunities between routers, and the content storage is relatively single, which is not conducive to user acquisition.

Second, although the topological structure of network performance has the great influence, but by contrast with ring topology of CEE and adopt the chain topology OFC, we can see that the chain structure of OFC scheme in the network time delay is superior to the scheme of ring topology of CEE to some extent, this also shows the efficient storage solution is to reduce the content of the effect on the performance of the network topology structure.

After exploring the influence of network topology on network performance, we also analyze the influence of the change of router number on network performance in different topologies. The network adopts OFC storage policy.

As can be seen from Fig. 9, with the increase in the number of routers in the network, the network delay caused by content acquisition gradually decreases, because with the increase in the number of routers, the number and type of content that can be stored in the network also increases. In addition, the increase of the number of routers also improves the probability of cooperation between routers. As can be seen from the figure, with the increase in the number of requests, the network delay gradually increases, which conforms to the basic network characteristics. In

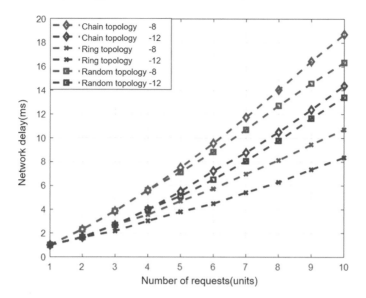

Fig. 9 OFC under different topologies and storage capacity parameters

the network with ring topology structure, the network delay caused by content acquisition is the smallest, which indicates that the network topology structure has a real influence on the network performance.

Effect of Zipf Parameters on Network Performance

This section considers the effect of CEE scheme and OFC storage scheme on network delay performance under the condition of different Zipf parameters. As can be seen from Fig. 10, in the two schemes, as the value of Zipf parameter increases, the value of network delay becomes smaller and smaller. From the overall curve, the network delay with OFC storage scheme is smaller than that with CEE storage scheme.

From the trend of the curve in the figure, we can draw a conclusion that the simulation conclusion is consistent with the theoretical analysis. Because the Zipf parameter represents the popularity of the content, different Zipf parameters have different effects on the distribution of popularity. The larger the Zipf parameter is, the more concentrated the popular content is, and the higher the probability of being requested by users is. The smaller the Zipf parameter is, the more evenly popular the content is. The OFC scheme gives priority to storing high-hot content, and through inter-network collaboration, the content can be placed in the most accessible location for users. Therefore, the higher the Zipf parameter, the more concentrated the user's requests for content, the more conducive to the storage of the network router, and the smaller the time delay for users to obtain the corresponding content.

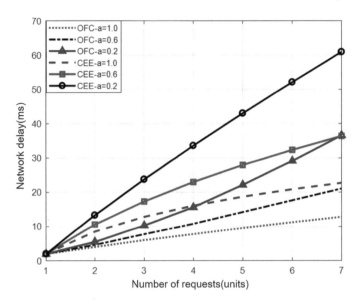

Fig. 10 OFC versus CEE under different popularity parameters

3   Design and Implementation of Cache Scheme Based on Network Energy Minimization

In the previous chapter, we proposed a caching scheme based on the collaboration between the wayside cache and the wayside cache, which greatly reduced the cache redundancy within the network and greatly reduced the delay of the network users to obtain the target content. This chapter considers further reducing the overall energy consumption of the network on the basis of maintaining the current network performance. How to minimize the overall energy consumption of the network? How can performance be further improved? This is the main problem we are going to solve.

3.1   Problem Description

With the deepening of the popularity of terminals, the number and types of content required by users are also increasing. In order to meet the huge demand for content acquisition and delivery, today's Internet is growing rapidly. At the same time, energy consumption is increasing significantly, and the Internet is becoming one of the leading players in energy consumption. According to incomplete statistics, the energy consumption of the Internet accounts for about 10% or more of the overall energy consumption in the world today, and this number is constantly climbing [35].

In the content center network, it has the advantages of lower content propagation delay and lower network transmission load due to its built-in caching function, which supports efficient and energy-saving content distribution. In the process of network operation, there is not only a certain amount of transmission energy consumption inside the network, but also extra storage energy consumption generated by each router for content storage. In the whole network, the energy consumption is mainly composed of two parts: the storage energy consumption generated by the cache content of the router and the transmission energy consumption generated by the transmission of the content within the network.

Although some excellent studies have been done on CCN caching strategies, most of the work has focused on improving network resources and related performance, and the research on energy consumption in existing caching strategies has been largely ignored. However, with the enhancement of network cache capacity and the gradual increase of network content, as well as the increasingly strict environmental standards and the rising energy cost, the caching and transmission energy consumption have gradually become a problem that cannot be ignored by people, thus leading to a new trend in solving the energy efficiency of the Internet [22]. In recent years, scholars around the world have also carried out research on network energy consumption optimization. At present, some research results have shown that optimization of caching mechanism can significantly reduce the overall energy consumption of the network. Literature [30] proposed a cache scheme EV, by considering the network energy consumption and delay optimization, caching decisions into two criteria, namely when the cache space enough cache the content whether can consume less energy and how to replace the energy consumption in the cache space more content, but because of its not considering collaboration between routers, easy to cause the cache redundancy. In the literature [36], the author thinks that network internal content popularity with its network in different areas and different time change, so an offline caching solution EE-OFD, then to the global information modeling and optimization of energy consumption of the largest gains, then puts forward the online distributed cache solution EE-OND, judging by the node the impact on the global earnings for caching decisions. In literature [31], an optimization method of energy consumption based on non-cooperative game is proposed. By creating a non-cooperative game model, each router can obtain its own optimal solution and the global energy consumption can also reach the minimum. However, this method only considers the optimization of energy consumption, but does not carry out a reasonable design of performance optimization. Literature [32] proposed the optimization algorithm of energy consumption based on network router sleeping, and designed the global centralized solution. On this basis, it proposed the distributed optimization algorithm based on directional alternating multiplier and dual decomposition. This scheme is only applicable to the environment of router sleeping, which is difficult to realize in real life. Literature [37] proposes a probabilistic caching strategy E2APC based on energy consumption optimization. It takes energy consumption as the determination condition and calculates the caching probability through the content popularity and the centrality of routers in the whole network, so as to achieve the minimum energy consumption while optimizing the network performance. However, this scheme does not consider the cooperation between routers.

The above schemes mainly have the following problems: (1) optimization of network performance only or optimization of network energy consumption only; (2) There is a lack of cooperation between routers in the network, and only the cache state of each router is considered, while the content redundancy within the network is ignored. Taking into account the advantages and disadvantages of the above schemes, this chapter proposes a Cost-oriented Caching Scheme (CCS) based on network energy minimization.

Compared with the existing studies, this scheme mainly has the following contributions:

1. The energy consumption model of CCN content transmission and storage was established to more truly show the impact of caching decisions of each router on the overall energy consumption of the network;
2. When determining whether the content is stored in the router, the size and popularity of the content should be considered in real time, and the content should be reasonably cached in the optimal storage location on the premise of minimum global energy consumption. Enhance the cooperation among router nodes, reduce the cache redundancy, and do not affect other network performance indexes on the basis of reducing the overall network energy consumption.

3.2 Caching Scheme Based on Network Energy Minimization

3.2.1 The System Model

The content center network is composed of N routers with storage capacity, and the topology structure of the content center network is modeled as undirected connection graph $G = (V, L)$, $V = \{v_1, v_2, ..., v_N\}$ represents N routers inside the network, and L represents the set of links between routers inside the network. There are K pieces of content in the network of varying sizes, their sizes are $S_k, k = 1, 2, ..., K$. The cache size of the router v_i is C_i, and $X_{i,k}$ is a binary constant indicating whether the content k is stored in the router i. The process of content acquisition and delivery in CCN is as follows: (1) when the end user wants to obtain the target content, he/she first sends an interest packet to the router connected to it, which will follow the routing path until the content source; (2) When each router receives the interest packet, it first checks whether the requested content is stored in the router. If the cache hits, the delivery of the interest packet will be terminated and the content will be returned to the end user along the original request path; Instead, the interest packet will continue to be passed until the cache hit. We define the distance between two routers as one hop, so the access distance to get the target content must be one hop or more. The access distance $H_{i,k}$ represents the number of access hops required by the router i to obtain the content k. It is an important indicator to measure the effectiveness of the cache location, and its value size mainly depends on the network topology and cache location algorithm. The network architecture is shown in Fig. 11.

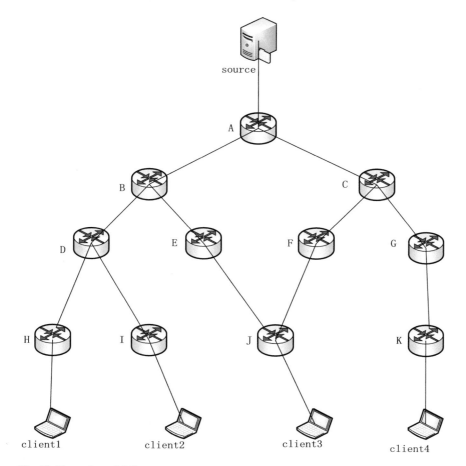

Fig. 11 Network model diagram

3.2.2 Design of Caching Mechanism

This section focuses on the design of the cache scheme, which consists of two parts: the cache policy and the replacement policy. In the design of cache strategy, energy consumption is minimized as a reference point, and two key factors, namely content size and content popularity, are considered comprehensively to reduce the competitive pressure of cache space near the user side router, effectively improve the cooperation between neighboring routers, and reduce the redundancy of content storage. In the cache replacement, it mainly replaces the stored content which has no obvious effect on the overall network energy consumption reduction, so as to minimize the energy consumption.

Design of Cache Policy

As is known to all, the process of obtaining content in the content center network is that the user sends interest packets first, and then the interest packets are transmitted between router layers until the interest packets are responded. Finally, the data packets will be transmitted step by step along the original request path and finally received by the user. When the packet is in the process of delivery, the routing routers will selectively store the content. In traditional caching policy design, content typically stored in its passes through the routers, this will cause the cache redundancy, is not conducive to the network internal content type, when the user requests a new content router due to the internal storage content types too little, interest in the user package difficult to response, which can cause for the increase of the content needed to hop, network delay is serious, poor user experience. On this basis, the researchers consider popular content store near the user's side, and reduce the redundancy phenomenon of cache of router which effectively reduces the user access to the content of network time delay, improve the user's web experience, but it also brings new problems: simply consider the prevalence of content, in the case of user experience, tend to ignore the overall energy consumption of the network. Therefore, on the basis of fully guaranteeing the overall performance of the existing network, the design of cache strategy which can effectively reduce the overall energy consumption of the network comes into being.

In routers, caching larger and more popular content saves transport costs, but larger content also takes up more cache space, increasing the cost of caching. How to reasonably store the contents with different popularity and sizes so as to cache more contents in the limited cache space and minimize the energy consumption of the whole network is a knapsack problem in the traditional sense.

So, at the beginning of design content caching decisions, first of all, according to the topology of the routers and the internal connection between the router for each router to establish its shortest path to the content source, according to the router after internal cache as much as possible the principle of popular content storage priority of popular content, in view of the content of the two attributes: popularity and size. The difference of network energy consumption is mainly reflected in the difference of content storage in the router. Based on this, the content that is most beneficial to reduce the overall energy consumption of the network is selected for storage. On this basis, the cooperation between neighboring routers is also of positive significance to the reduction of the overall network energy consumption. Routers can obtain the exact location of the content according to the content placement summary table stored in the content source, and then select the optimal content acquisition path, which further reduces the overall network energy consumption.

Design of the Replacement Strategy

Since the router has a certain amount of cache space, the router may run out of storage space, and then it will face the problem of content replacement. Therefore, in this scheme, energy consumption is mainly taken as the reference standard. When new content passes through the router, the router will face two decisions: save or not save. At this point, the storage cost needed to cache the new content is first calculated, and compared with the existing content, the storage energy consumption is greater than the storage energy consumption of the new content is screened out. Then calculate the product of content popularity and content size. If the product value of screened content is smaller than the product value of new content, the new content will be stored in the router and the selected content will be removed. Instead, the new content will not be stored in the router.

3.2.3 System Model

Content Popularity Model

Content providers have a set F containing K contents, which is sorted from 1 (most popular) to K (least popular) based on the popularity of content. According to Zipf distribution, the content popularity of k can be calculated as follows:

$p_k = \frac{k^{-\alpha}}{\sum_{j=1}^{K} j^{-\alpha}}, k \in [1, K]$, where α is the Zipf constant, its value range is $[0, 1]$. As the value of α becomes larger, the curve of the distribution of the Zipf function becomes more and more sharp, indicating that the content obtained by the user becomes more concentrated. The content popularity model reflects the probability that each content is requested by the end user.

Energy Consumption Model

In the content center network, the energy consumption mainly consists of two parts: one is the storage energy consumption generated by the content stored in each router; the other is the transmission energy consumption generated by the content passing through each link and the processing energy consumption generated by the router in the internal transmission process of the network, which is collectively referred to as the content transmission energy consumption. The sum of the two is the total energy consumption generated in the network. The symbols used in the modeling process and their meaning are shown in Table 3.

For router node i, the energy consumption of content k stored in the router, that is, the storage energy consumption is:

$$E_{ca}^{k,i} = P_{ca} S_{k,i} \tag{3.1}$$

For the whole network, the total energy consumption of the cache content of all routers within the network is E_{ca}, at this time, the cache energy consumption is:

Table 3 Symbol and its corresponding meaning

Symbol	Corresponding meaning
E_{total}	Total network energy consumption
$E_{ca}^{k,i}$	Storage energy consumption of content k stored in router i
E_{ca}	Storage energy consumption of content stored by all routers in the network
E_{tr}^{l}	Energy consumed by the content through the internal links of the network
E_{tr}^{p}	Energy consumed by the router in the network to process the passing content
E_{tr}	Transmission energy consumption of content transmitted by all routers in the network
S_k	Packet size of content k
C_i	Storage capacity of router i
P_{ca}	Energy consumed by CCN routers to store 1 bit content
P_l	Energy consumed by link transmission of 1 bit content
P_r	Energy consumed by CCN routers to process 1 bit content
p_k	Popularity of content k
X_k^i	Binary constant, the content k stored in the router i is 1, otherwise it is 0

$$E_{ca} = \sum_{i=1}^{N} \sum_{k=1}^{K} E_{ca}^{k,i} = \sum_{i=1}^{N} \sum_{k=1}^{K} P_{ca} X_k^i S_k \tag{3.2}$$

If the router node does not cache the content, then there will be no storage energy consumption, and the total network energy consumption is the sum of the content transmission energy consumption and the processing energy consumption generated by each router. In the process of content delivery, the energy consumption consumed by the content through the network link, that is, the transmission energy consumption is:

$$E_{tr}^{l} = P_l H_{i,k} S_k p_k \tag{3.3}$$

In the process of delivery, the content will pass through each router in the link, and at the same time, each router will process the content, and the energy consumption consumed is called processing energy consumption, and the processing energy consumption is:

$$E_{tr}^{p} = (H_{i,k} + 1) P_r S_k p_k \tag{3.4}$$

Therefore, the total energy consumption generated in the transmission process within the network is:

$$E_{tr} = \sum_{i=1}^{N} \sum_{k=1}^{K} [P_l H_{i,k} + (H_{i,k} + 1) P_r] S_k p_k \tag{3.5}$$

Therefore, considering the storage energy consumption and transmission energy consumption, the total energy consumption within the network is:

$$E_{total} = E_{ca} + E_{tr} = \sum_{i=1}^{N}\sum_{k=1}^{K} P_{ca} X_k^i S_k + \sum_{i=1}^{N}\sum_{k=1}^{K} [P_l H_{i,k} + (H_{i,k}+1)P_r] S_k p_k$$

$$(3.6)$$

The problem of minimizing energy consumption in a network can be expressed by the following expression:

$$\min E_{total} \tag{3.7}$$

$$s.t. \sum_{k=1}^{K} X_k^i S_k \leq C_i \tag{3.8}$$

3.2.4 Solving Model by Genetic Algorithm

Genetic Algorithm

In the previous sections, we mainly introduced the internal energy consumption of the network and the establishment of the optimization mathematical model. According to the obtained objective function, solving the optimization problem of minimizing energy consumption is an NP-hard problem, so the process of solving the mathematical model using genetic algorithm will be introduced in this section.

Genetic algorithm is a computational model created by simulating the process of biological evolution on the basis of Darwin's genetic mechanism and natural selection [35]. Genetic algorithm is a random, highly synchronized, and adaptive optimization algorithm based on "survival of the fittest". It mainly simulates the continuous exchange of biological "chromosome" groups in natural evolution, finally converge to the most qualified population, then get the suboptimal solution or the optimal solution of the problem. The reason why genetic algorithm is widely used is that its basic principle and engineering operation are simple, practical, less constrained by constraints, and has the ability to search concurrency and global solution.

The main advantages of genetic algorithm:

1. It is suitable for solving comprehensive optimization problems under complex conditions.
2. The global optimal solution of the optimization problem can be obtained.
3. The optimization result is stable and does not change with the change of initial conditions.
4. The algorithm is robust.

Table 4 Correspondence between the object and the genetic algorithm

Terminology of genetic algorithm	Research object
Population	Cache strategy solution set $c = \{c_1, c_2, ..., c_i\}$, $\forall i \in (0, \infty)$ cache strategy for all routers
Individual	$c_i = \{c_{1,i}, c_{2,i}, ..., c_{l,i}\}$, $\forall l \in [1, N]\,\forall i \in (0, \infty)$
Chromosome	Caching strategy for a single router $c_{l,i}$
Genes	Caching strategy for each content in a single router $c_{l,i}^m$
Fitness	Energy consumption of the network as a whole

The corresponding relationship between the research object in this scheme and the technical terms in genetic algorithm is mainly introduced in Table 4. In the genetic algorithm, the population is the feasibility solution set in the algorithm, and the solution set is corresponding to the content in the router in this scheme; the individual is the storage state solution of the router in the network as each feasibility solution in the algorithm; the chromosome is the coding of the feasibility solution in the algorithm, and in this scheme, each chromosome corresponds to a router content storage state solution. Gene as a feasible uncoded component in the algorithm, in this scheme, the gene corresponds to the storage of each content in the router, generally taking a value of 0 or 1. The fitness in the algorithm is regarded as the evaluation standard of the function, which corresponds to the overall energy consumption within the network in this scheme.

In the process of solving the compound problem, the genetic algorithm first calculates the fitness of the parent generation, and then selects the population with higher fitness in the parent generation to inherit to the offspring, this selective operation reflects the heredity of the algorithm. In the process of simulation, the content placement strategy that reduces the overall network energy consumption to the lowest is inherited to the offspring. The excellent parents selected by calculating the fitness, by using the mutation operation of the individual genes in the parent generation to generate offspring, which effectively ensures the diversity of the whole population as an auxiliary search operation in the algorithm. If the mutation rate of the individual in the mutation operation is too large, the algorithm will tend to random search. Through the generation of offspring after crossover operation on the genes of excellent individuals in the parent generation, the search area can be expanded on the basis of the existing solution, but with the increase of the search probability, the overall convergence speed of the algorithm will gradually decrease. If the search probability is very small, it will result in local optimization of the search results, so the reasonableness of the search probability has an important impact on the accuracy of the results. The crossover and mutation operation also reflects the evolution of the algorithm. The steps and basic flow of solving the problem by genetic algorithm are shown in Fig. 12.

First of all, all the solution sets are binary transcoded. In the simulation process of this scheme, because there are only two states of content in the router, that is, storage or non-storage, the solution set of this cache strategy is mainly composed of 0 and

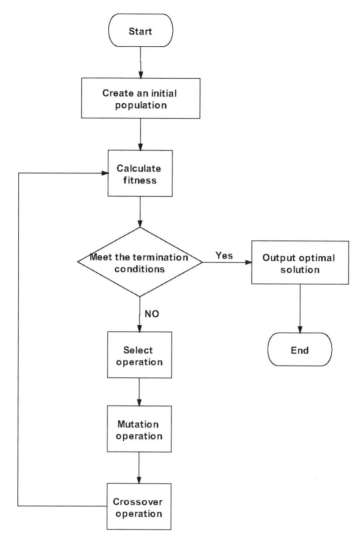

Fig. 12 Process of traditional genetic algorithm

1. In the actual simulation process, this step can be replaced by randomly generating a fixed number of feasible solutions as the original population.

Secondly, according to the randomly generated original population, the individuals in the population are evaluated and selected, and the excellent individuals who meet the cache conditions are brought into the overall energy consumption evaluation function of the network. According to the calculation of the fitness function, the individuals with high fitness are selected as the parents in the later iterative process through certain rules, and the excellent individuals can be passed on to the offspring through the operation of the parents in the later stage.

After that, the mutation operation is carried out to change the genes of some individuals in the excellent parent population with a certain probability, and then produce the offspring population. In the process of simulation, the cache state of something in a router in an excellent parent feasible solution is adjusted randomly. When the cache state of the content in the router changes, we first need to judge whether the change of the content meets the current cache conditions, and the size of the content stored in the router is smaller than the storage space of the router. If the conditions are not met, the cache policy should be adjusted in time.

Then the crossover operator is applied to the group by using the crossover operation. The crossover operator is to randomly select two individuals in the excellent parent population, then randomly select the crossing point and crossover length on a certain chromosome, cross a certain length of genes according to a certain probability, and then produce new offspring individuals. In the simulation process of this scheme, the feasible solutions of two stored states in the excellent parents are randomly selected as the crossover operator. After the completion of the crossover operation, it is also necessary to check the current storage state like the mutation operation to determine whether it meets the constraints of the objective function, and if not, it also needs to be modified and adjusted. In the process of generating new offspring through selective cross-mutation of the parent population, if the number of offspring does not reach the number of individuals of the initial offspring, the selective crossover operation will be repeated until the number of offspring meets the original requirements.

Finally, the termination condition of the algorithm is determined. When the fitness of the cache strategy meets the requirements of the objective function, the genetic iteration can be stopped. In the process of simulation, the termination condition is set as the overall energy consumption of the network tends to be minimized and stable, then the genetic iteration is stopped.

Algorithm Step

In the previous section, the theory of genetic algorithm is introduced. This section mainly introduces the process of solving the energy consumption minimization problem model by genetic algorithm. The specific steps are shown in Table 5:

1. Initialize the initial values of the initial number of individuals P, the maximum number of evolution G, the crossover rate P_{cro} and the mutation rate P_{mu} in the parameter population of genetic algorithm.
2. Randomly generate a certain number of individuals, that is, different content placement strategies that meet the constraints to form the initial population C_{ran}. The number of individuals in the population should be set reasonably, and too many individuals will increase the complexity of the algorithm and add useless calculations, waste the computing power of the computer, and affect the overall performance of the algorithm; insufficient number of individuals will lead to

Table 5 Caching algorithm based on energy minimization

Caching Algorithm based on Energy consumption minimization
1: Initialization of the number of individuals in the population P, maximum number of evolution G, crossover rate P_{cro}, mutation rate P_{mu};
2: Randomly generate initial population C_{ran};
3: $for\ k = 1 : K$
4: $for\ g = 1 : G$
5: Calculate the fitness of an individual $U(C_{ran})$;
6: Select the first L individuals with the highest fitness $C_{best} = \{C_{ran}(1), C_{ran}(2), ..., C_{ran}(L)\}$, And put their corresponding parent individuals into the preferred parent pool;
7: $if\ random[0, 1] < P_{mu}$
8: Individuals are randomly selected in the preferred parent pool, and then a chromosome on the gene is randomly mutated;
9: Check the storage status of the variation to determine whether it meets the limiting conditions, and correct it if it does not meet the conditions, until the preferred parent pool is updated after the conditions are met;
10: $end\ if$
11: $while\ length(C_{new}) < P$
12: $if\ random[0, 1] < P_{cro}$
13: Judge whether C_{new} meets the limiting conditions and correct it if it does not meet the requirements;
14: $end\ if$
15: $end\ while$
16: $C_{ran} = C_{new}$;
17: $end\ for$
18: Calculate $U_{\min}(C_{best})$;
19: $end\ for$

the algorithm cannot calculate all feasible solutions, too early to reach the local optimal solution, it is difficult to obtain the global optimal value.

3. Step 3–19 is the process of genetic algorithm to solve the function optimization. Taking the fitness function as the known condition, the optimal cache solution is gradually obtained by using the content of crossover and mutation operation.

4. Step 5 calculates the fitness of all parent individuals, and calculates the energy consumption of each individual and cache scheme.

5. The 6 step is the selection operation. According to the calculation of the fitness function, the individuals in the population are sorted according to the fitness, and the first L individuals are selected to be placed in the preferred pool of the parent generation. As a part of the offspring and as an alternative for generating offspring.

6. Step 7–10 is a mutation operation. Before mutating each individual, a random number between 0 and 1 is generated. If the random number is less than the mutation probability P_{mu}, the mutation operation will be taken for the individual, otherwise it will remain unchanged. After performing the mutation operation on the individual, we first judge the newly generated individual to check whether

it meets the requirements of the restriction, correct it if the new individual does not meet the restriction, and add it to the preferred parent pool if it does, and update the original parent pool.

7. Step 11–16 is a cross operation. If the number of offspring does not meet the requirement of population capacity, the crossover operation is repeated. Randomly select two individuals in the preferred parent pool, that is, the storage strategy to cycle, and then randomly generate an intersection and cross length, take the same intersection of the two individuals as the starting point of the crossover, and select the chromosomes of the crossover length for crossover operation. After the completion of the crossover operation, it is also necessary to judge the newly generated individuals to verify whether they meet the restrictions or not, and put them into the parent optimization pool if they are met, and vice versa.

8. The optimized parent cache strategy is substituted into the objective function of solving energy consumption, and the minimum value of network energy consumption is obtained.

3.3 Performance Analysis

This section will simulate and analyze the cache scheme of minimizing network energy consumption proposed in this chapter. Firstly, the overall network energy consumption of the networks based on Cost-oriented Caching Scheme (CCS), Most Popular Content (MPC) and CEE caching strategies are compared and analyzed, and then the effects of router storage space changes, popularity parameters and other factors on the cache strategy and network energy consumption are analyzed.

3.3.1 Simulation Parameter Setting

In this section, the MATALB simulation tool is used to analyze the performance of the caching mechanism of minimizing energy consumption proposed in this scheme. The simulation scenario includes user equipment, router with storage function, content provider and so on. The network model includes 14 router nodes as the main body of content storage, and a content source that stores 50 content blocks of different sizes. We assume that the cache capacity of each router is the same, the bandwidth of each link is equal, and the probability of users' requests for different content follows the Zipf distribution. In terms of unit energy consumption, we adopt the international energy consumption proportion model: Cache energy consumption per unit of data is $P_{ca} = 1.6 \times 10^{-9}$ J/bit, the processing energy consumption of unit data in the router is $P_r = 2 \times 10^{-8}$ J/bit, the transmission energy consumption of unit data in the link is $P_l = 1.5 \times 10^{-9}$ J/bit.

In the process of simulation, the routing mechanism of each cache strategy adopts random forwarding mechanism, and the replacement strategy of content adopts the least recently used policy. The main reference indicators of network performance in the process of simulation are as follows:

1. Energy consumption: the energy consumed to complete a certain number of content requests and data return under a certain cache strategy.
2. Energy saving rate: the difference between the energy consumed to complete a certain number of content requests and backhaul without cache and the energy consumed to complete the same task under different cache strategies, and then compared with the total energy consumption without cache.
3. Network delay: the time it takes to complete a fixed number of content requests and returns under a certain caching strategy.

3.3.2   Simulation Results and Analysis

Figure 13 shows the impact of the change of router storage space on the overall energy consumption of the network. The energy consumption of CEE caching strategy is the largest, mainly due to the serious cache redundancy of the overall content caused by a single storage of content and insufficient cooperation between neighboring routers. When users request content, because the router storage space is certain, the type of storage content is seriously limited, so the overall energy consumption remains high. With the increase of the cache space, the type and number of cache contents in the router are gradually increasing, and the energy consumption is also

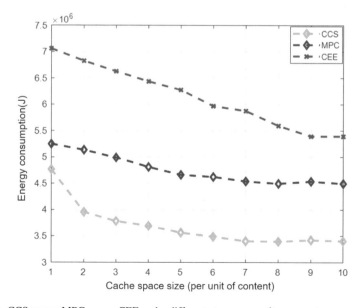

Fig. 13 CCS versus MPC versus CEE under different storage capacity parameters

gradually decreasing. When the storage space increases gradually, the storage content of each router becomes stable, and the energy consumption also reaches a fixed steady value, occasionally fluctuating slightly up and down. Compared with CEE strategy, MPC caching strategy increases the storage of popular content and improves the cooperation between neighboring routers, which effectively reduces the redundancy of router content storage. Therefore, it is obvious that the overall energy consumption of MPC policy is less than that of CEE caching strategy, and the energy consumption decreases gradually with the increase of cache space, and finally achieves stability. It can be seen from the figure that the CCS scheme mainly discussed in this chapter has a significant reduction in energy consumption compared with the default caching scheme CEE caching strategy of the Content Centric Network and the more popular MPC scheme, which also reflects the rationality and efficiency of the design of the scheme. CCS scheme not only improves the cooperation between routers and reduces the cache redundancy of content, but also plays games between the popularity and size of content.

So that the content can be placed in a position that can minimize the energy consumption of the whole network and ensure the overall performance of the network.

In Fig. 14, it is mainly analyzed that the energy saving rate of each content storage strategy varies with the storage space of the router. First of all, under the condition of different cache policies, the change of network energy consumption with the change of router storage space is calculated, and on this basis, the overall energy consumption of the network without cache in the router is calculated. Finally, the two are subtracted and compared with the energy consumption under the condition

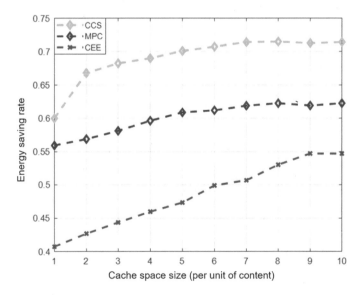

Fig. 14 CCS versus MPC versus CEE under different storage capacity parameters

of no cache, and the energy saving rate of each cache strategy under the condition of different storage space is obtained.

From Fig. 14, we can see the changes of each curve, there are two main phenomena. First, with the increase of router storage space, the energy saving rate of each cache strategy is gradually increasing, which shows that the increase of cache space has a positive significance to the overall energy consumption of the network, and can gradually reduce the energy consumption of the network. At the same time, in the curve change trend of the three cache strategies, we can also see that the overall energy saving rate of CCS cache scheme is the most prominent, the highest point is about 72%. With the increase of router storage space, the placement of content will change obviously because of the characteristics of CCS cache strategy. Therefore, the impact on the overall energy consumption of the network is also the most obvious. Second, we can observe that when the storage space of the router increases to a certain extent, the energy saving rate of the network gradually becomes stable, mainly because the buffer space of the router gradually increases at the same time, the type and number of cache content in each router are also constantly changing, so at the initial stage of the increase of storage space, the energy saving rate is gradually increasing, when the content change in the router is no longer obvious. When the storage of content becomes more and more stable, the overall energy saving rate of the network will also become stable.

Under the background of reducing the overall energy consumption of the network, we also consider the changes of network performance, so we also analyze the change of network delay with the size of router cache space. As shown in Fig. 15, we can see that when the storage space of the router is very small, the network delay is serious,

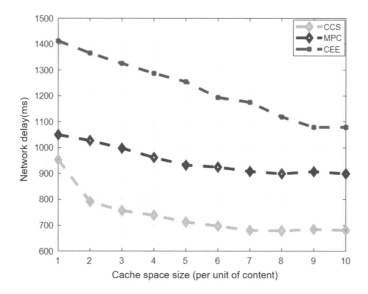

Fig. 15 CCS versus MPC versus CEE under different storage capacity parameters

and the user experience is poor, but with the increase of storage space, the network delay decreases gradually. In the change curve of CCS cache policy, we can see that the network performance has been significantly improved, and the network delay has been reduced by nearly 30%. This shows that our proposed CCS caching strategy not only effectively reduces the overall energy consumption of the network, but also improves the network performance. In the process of increasing storage space, the improvement of network performance has encountered a bottleneck period, and the network delay tends to be stable gradually, which is consistent with the change of network energy consumption. The main reason for this is that the type and number of content stored in each router will change as the cache space increases at the initial stage, but as the cache space increases to a certain extent, the content stored in the router tends to be stable in terms of type and number, and the change in the type and number of content stored in the router is no longer obvious, so the overall performance of the network is becoming more and more stable. The reduction of network delay is becoming more and more stable.

The value of the Zipf parameter directly reflects the concentration of the user's request for the content. When the Zipf parameter is small, the user's request for the content is scattered. Because the router's storage space is limited, the request for user diversity is difficult to be satisfied, and the user's request will be passed to the content source, resulting in a significant increase in transmission energy consumption and high overall energy consumption. When the Zipf parameter increases, users' requests for content are more concentrated, which are mainly popular content, so we can also see from Fig. 16 that the energy consumption under each cache mechanism decreases

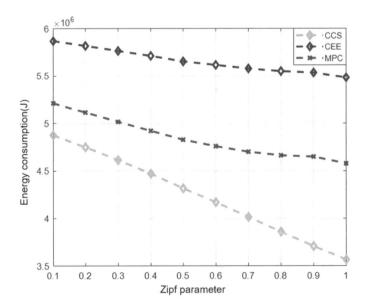

Fig. 16 CCS versus MPC versus CEE under different popularity parameters

with the increase of Zipf index. Because of the fixedness of CEE and MPC caching strategies on popular content storage, the energy consumption decreases slowly with the increase of Zipf index. On the other hand, the energy consumption curve of CCS cache strategy decreases greatly. This is because the CCS cache scheme takes into account the impact of content popularity and content size on network energy consumption. With the change of users' requests for content, the content placement under the CCS cache strategy is constantly changing, constantly changing in the direction of reducing energy consumption.

When implementing each cache strategy in the network, calculate the change of the total energy consumption of the network with the change of Zipf parameters, and calculate the total energy consumption of the network without cache, and then calculate the difference between them. We define the ratio of the difference to the total energy consumption without cache as the saving rate of energy consumption. It can be seen from Fig. 17 that the influence of Zipf parameters on energy saving rate is also obvious. From the change curve of Fig. 17, we can see that with the increase of Zipf index, the energy saving rate of each cache strategy is gradually increasing, especially the CCS cache strategy proposed in this chapter, the energy saving rate increases most rapidly, and the highest energy saving rate reaches nearly 70%, followed by MPC cache strategy. Finally, the default caching strategy of Content Centric Network is CEE caching policy.

While exploring the impact of Zipf parameters on energy saving, we also pay attention to the impact of the change of Zipf parameters on network performance, so we choose network delay as an index to measure network performance.

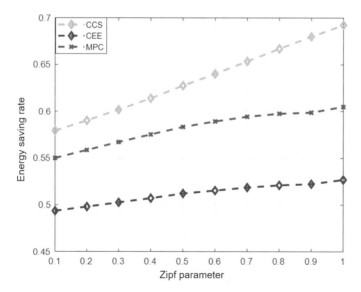

Fig. 17 CCS versus MPC versus CEE under different popularity parameters

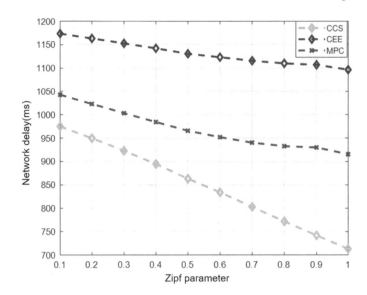

Fig. 18 CCS versus MPC versus CEE under different popularity parameters

We can see from Fig. 18 that with the increase of Zipf parameters, different policies implemented by the network lead to an overall downward trend in the delay curve of acquiring content, especially the CCS caching strategy proposed in this chapter, which results in the most obvious reduction of network latency, which is reduced by nearly 30%, followed by MPC caching strategy, and finally, the default caching strategy of the Content Centric Network, namely CEE caching strategy.

4 Conclusion

Under the background of 5G development research, aiming at the caching problem of the Content Centric Network, this chapter explores how to efficiently and reasonably place the content in the network, so as to reduce the network delay caused by users' access to content. improve the quality of service experience of users, and on this basis, further analyze the network cost caused by content delivery within the network, through the design of an efficient caching scheme. On the premise of not affecting the user experience, we can effectively reduce the network energy consumption, so we creatively propose a cache placement scheme based on collaborative caching and energy consumption minimization. The specific work is as follows:

1. We introduce the background of the generation of Content Centric Network and the current research status of the Content Centric Network cache, as well as some problems existing in the current network: how to cache the content efficiently, how to effectively reduce the network delay, how to improve the

cooperation between routers and so on. In view of the above problems, on the basis of summarizing the achievements of the current research phase, some factors that can be improved in the network are analyzed, on the basis of which the feasible scheme is studied and designed.

2. We describe the theoretical basis of the relevant technologies used in the scheme design process, including the concept of the Content Centric Network, the development of cache technology, the classical storage strategy and so on. On the basis of introducing the above basic concepts, this chapter summarizes and analyzes the related technologies, and summarizes the advantages and disadvantages of the related technologies and the reasons for using them in this chapter.

3. We propose a caching scheme based on the cooperation of on-path caching and off-path cache in content center network is designed and verified by simulation. In order to improve the utilization efficiency of router storage space, improve the rationality of content placement, and reduce the network delay caused by users when obtaining response content. By improving the collaboration between routers within the network, we give priority to the popular content close to the user side and the core of the network. On this basis, an optimization problem is established to minimize the communication delay caused by content acquisition in the network. This problem is a typical NP-hard problem, and then an approximate algorithm is designed to solve the approximate solution of the problem. According to the approximate solution of the optimization problem, what should be stored in each router is determined. In the process of simulation, this scheme is compared with the default storage scheme of the Content Centric Network, and the feasibility and efficiency of the scheme are verified from the aspects of network delay.

4. We design a caching scheme based on minimizing network energy consumption in Content Centric Network is designed and verified by simulation. In recent years, due to the increasing energy consumption cost of the network, the problem of network energy consumption has attracted great attention of researchers. The original design intention of this scheme is that how to place content efficiently can effectively reduce the energy consumption of content in the process of storage and transmission without affecting the performance of the existing network. In the design process of the scheme, we first model the process of content storage and content delivery, and then analyze the impact of content size and content popularity on network energy consumption. finally, a cache scheme for minimizing network energy consumption with low delay and low energy consumption is designed, and a submodular function optimization problem with certain constraints is established. finally, genetic algorithm is used to solve the optimization problem. The efficiency and rationality of the scheme are also verified in the subsequent simulation process.

References

1. Jose, S.: Cisco visual networking index: forecast and trends, 2017–2022. Cisco, USA, white paper (2019)
2. Kamiyama, N.: Analyzing impact of introducing CCN on profit of ISPs. IEEE Trans. Netw. Serv. Manage. **12**(2), 176–187 (2015)
3. Joos, M.: Review of G.K. Zipf. The psychobiology of language. Language **12**, 196–210 (1936)
4. Hefeeda, M., Saleh, O.: Traffic modeling and proportional partial caching for peer-to-peer systems. IEEE/ACM Trans. Networking **16**(6), 1447–1460 (2008)
5. Li, X., Wang, X., Xiao, S., Leung, V.C.M.: Delay performance analysis of cooperative cell caching in future mobile networks. In: IEEE International Conference on Communications, pp. 5652–5657 (2015)
6. Psaras, I., Chai, W.K., Pavlou, G.: In-network cache management and resource allocation for information-centric networks. IEEE Trans. Parallel Distrib. Syst. **25**(11), 2920–2931 (2013)
7. Zhang, G., Li, Y., Lin, T.: Caching in information centric networking: a survey. Comput. Netw. **57**(16), 3128–3141 (2013)
8. Rosensweig, E.J., Kurose, J.: Breadcrumbs: efficient, best-effort content location in cache networks. In: Proceedings of IEEE 28th International Conference on Computer Communication, Rio de Janeiro, Brazil, pp. 2631–2635 (2009)
9. Podlipnig, S., Böszörmenyi, L.: A survey of web cache replacement strategies. ACM Comput. Surveys **35**(4), 374–398 (2003)
10. Tyson, G., et al.: A trace-driven analysis of caching in content-centric networks. In: Proceedings of IEEE 21st International Conference on Computer Communication and Networks, Munich, Germany, pp. 1–7 (2012)
11. Kang, S.J., Lee, S.W., Ko, Y.B.: A recent popularity based dynamic cache management for content centric networking. In: International Conference on Ubiquitous and Future Networks. IEEE, pp. 219–224 (2012)
12. Rossini, G., Rossi, D.: A dive into the caching performance of content centric networking. In: Proceedings of IEEE 17th International Workshop CAMAD, Barcelona, Spain, pp. 105–109 (2012)
13. Muscariello, L., Carofiglio, G., Gallo, M.: Bandwidth and storage sharing performance in information centric networking. In: Proceedings of 1st ACM SIGCOMM Workshop ICN, Toronto, ON, Canada, pp. 26–31 (2011)
14. Katsaros, K., Xylomenos, G., Polyzos, G.C.: Multicache: an overlay architecture for Information-centric networking. Comput. Netw. **55**(4), 936–947 (2011)
15. Wang, J.M., Zhang, J., Bensaou, B.: Intra-as cooperative caching for content-centric networks. In: Proceedings of 3rd ACM SIGCOMM Workshop ICN, Hong Kong, pp. 61–66 (2013)
16. Cho, K., et al.: Wave: popularity-based and collaborative in-network caching for content-oriented networks. In: Proceedings of IEEE 31st International Conference on Computer Communication Workshops, Orlando, FL, USA, pp. 316–321 (2012)
17. Psaras, I., Chai, W.K., Pavlou, G.: Probabilistic in-network caching for information-centric networks. In: Proceedings of 2nd ACM SIGCOMM Workshop ICN, Helsinki, Finland, pp. 55–60 (2012)
18. Zhang, M., Luo, H.B., Zhang, H.K.: A survey of caching mechanisms in information centric networking. IEEE Commun. Surv. Tutorials **17**(3), 1473–1499 (2015)
19. Iio, M., Hirata, K., Yamamoto, M.: Distributed cache management considering content popularity for in-network caching. In: 23rd Asia-Pacific Conference on Communications (APCC) (2017)
20. Chai, W.K., He, D., Psaras, I., Pavlou, G.: Cache less for more in information-centric networks. In: Proceedings of 11th International IFIP TC 6 Networking Conference, Prague, Czech Republic, pp. 27–40 (2012)
21. Chao, F., Yu, F.R., Tao, H., et al.: A distributed energy-efficient algorithm in green content-centric networks. In: Proceedings of the IEEE International Conference on Communications(ICC), London, UK, pp. 5546–5551 (2015)

22. Borst, S., Gupta, V., Walid, A.: Distributed caching algorithms for content distribution networks. In: IEEE INFOCOM, pp. 1478–1486 (2010)
23. Jmal, R., Fourati, L.: Content-centric networking management based-on software defined networks: survey. IEEE Trans. Netw. Serv. Manage. **14**(4), 1–13 (2017)
24. Zhang, M., Luo, H., Zhang, H.: A survey of caching mechanisms in information-centric networking. IEEE Commun. Surv. Tutorials **17**(3), 1473–1499 (2015)
25. Llorca, J., Tulino, A.M., Guan, K., et al.: Dynamic in-network caching for energy efficient content delivery. In: 2013 Proceedings IEEE INFOCOM, Turin, Italy, pp. 14–19 (2013)
26. Jiang, A., Bruck, J.: Optimal content placement for en-route web caching. In: IEEE International Symposium on Network Computing and Applications, pp. 9–16 (2003)
27. Qing, W.G., Tao, H., Jiang, L., et al.: In-network caching for energy efficiency in content-centric networking. J. China Univ. Posts Telecommun. **21**(4), 25–31 (2014)
28. Bianchi, G., Detti, A., Caponi, A., et al.: Check before storing: what is the performance price of content integrity verification in LRU caching. ACM SIGCOMM Comput. Comm. Rev. **43**(3), 59–67 (2013)
29. Zhang, G.Q., Chen, X., Lu, Y.X.: Enabling tunneling in CCN. IEEE Commun. Lett. **20**(1), 149–152 (2016)
30. Liu, W.X., Yu, S.Z., Gao, Y., et al.: Caching efficiency of information-centric networking. IET Netw. **2**(2), 53–62 (2013)
31. Applegate, D., Archer, A., Gopalakrishnan, V., et al.: Optimal content placement for a large-scale VoD system. IEEE/ACM Trans. Networking **24**(4), 2114–2127 (2016)
32. Xi, L., Ying, A., Xin, W.J., et al.: Energy-efficiency aware probabilistic caching scheme for content-centric networks. J. Electron. Inf. Technol. **38**(8), 1843–1849 (2016)
33. Hosseini-Khayat, S.: Replacement algorithms for object caching. In: ACM Symposium on Applied Computing, pp. 90–97 (1998)
34. Chao, F., Richard, Y.F., Tao, H., et al.: A game theoretic approach for energy-efficient in-network caching in content centric networks. China Commun. **11**(11), 135–145 (2014)
35. Choi, J., Han, J., Cho, E., et al.: A survey on content-oriented networking for efficient content delivery. IEEE Commun. Mag. **49**(3), 121–127 (2011)
36. Nakamura, R., Ohsaki, H.: On the effect of scale-free structure of network topology on performance of content-centric networking. In: IEEE Computer Software and Applications Conference, pp. 686–689 (2017)
37. Li, J., Wu, H., Liu, B., et al.: Popularity-driven coordinated caching in named data networking. In: ACM/IEEE NETWORKING and Communications Systems, pp. 15–26 (2012)

Blockchain-Based Software-Defined Vehicular Networks for Intelligent Transportation System Beyond 5G

Yash Modi, Mihir Panchal, Jitendra Bhatia, and Sudeep Tanwar

Abstract Vehicular networks are the key cornerstone of Intelligent Transportation Systems (ITS). With the growth of the technical revolution in 5G the next-generation networks, it is expected to meet various future communication requirements of Intelligent ITS. However, no single technology of 5G can efficiently accommodate the broad range of requirements of vehicular networks. However, the aggregation of various communication technologies with the 5G vehicular network could help to achieve ambitious goals like efficient network management, centralized view, and computation. The need for provisioning efficient network control, management, and high resource utilization in vehicular networks motivates the hierarchical 5G Next-generation vehicular networks. Blockchain has emerged as a viable solution to offer decentralization, transparency, and immutability among the stakeholders and manage trust in networking platforms. Motivated by the preceding discussion, in this chapter, we discuss the integration of software-defined networking (SDN), cloud computing over 5G mobile communication in vehicular networks, which is the key requirement of new hierarchical 5G-SDVNs. Then, programmable, efficient, and controllable network architecture is introduced for 5G-SDVN to achieve sustainable network development. Moreover, the centralization and flexibility provided by SDN and 5G communication technologies help to meet customer demand for various ITSs and vehicular network applications like bandwidth, high speed, and ubiquity, that can be employed in next-generation ITS. Finally, we present the SDN and cloud-enabled hierarchical future 5G-vehicular network architecture, its applications, issues, and challenges based on recent advances in technology and research.

Keywords Blockchain · Software defined network · 5G · Intelligent transport system

Y. Modi · M. Panchal · J. Bhatia
Computer Department, Vishwakarma Government Engineering College, Gujarat Technological University, Ahmedabad, India

S. Tanwar (✉)
Department of Computer Science and Engineering, Institute of Technology, Nirma University, Ahmedabad, India
e-mail: sudeep.tanwar@nirmauni.ac.in

© The Author(s), under exclusive license to Springer Nature Switzerland AG 2022
P. Nicopolitidis et al. (eds.), *Advances in Computing, Informatics, Networking and Cybersecurity*, Lecture Notes in Networks and Systems 289,
https://doi.org/10.1007/978-3-030-87049-2_17

1 Introduction

Vehicular Ad-hoc NETwork (VANET) enables wireless communication capacities among vehicles in a specific range. As shown in Fig. 1, it can be classified in three ways: (i) Communication between Infrastructure to Infrastructure (I2I), (ii)Vehicle to Vehicle (V2V), and (iii) Vehicle to Infrastructure (V2I). VANET has been an exceptionally dynamic research theme in the current years due to the unusually positive effect of their execution in vehicular wellbeing, movement administration, and infotainment applications. The various diverse network components are needed to set up a vehicular ad-hoc network, for instance, vehicles outfitted with On-Board Unit (OBU), Road Side Unit (RSU), Sensor Devices, Global Positioning System (GPS), and other components. Envelop these innovations cooperatively, VANET grows to ITS. The federal communications commission (FCC), figuring out the issue of activity fatalities in the US, devoted 75 MHz of the recurrence range in the range 5.850–5.925 GHz to be utilized for a vehicle to vehicle and vehicle to roadside correspondence, known as Dedicated Short-Range Communications (DSRC) [1, 2]. Even though the leading role of DSRC is to empower car security applications, the standard considers solace applications like web access from vehicles, office-on-wheels, mixed media applications, portable web amusements, versatile shopping, downloading records, perusing email while progressing, visiting inside interpersonal organizations, and so forth.

V2I security message communication enables the wireless transmission of basic security and operational data with vehicles and other frameworks. V2I grants a wide scope of security, portability, and ecological advantages. V2I has both safety and mobility benefits, enabling the infrastructure to point out hazards lying ahead as well as optimizing the flow of traffic. Which are expected to mitigate vehicle accidents, yet besides empowering a wide scope of other security, portability, and ecological advantages [3]. Under I2V communication, the framework enables various message

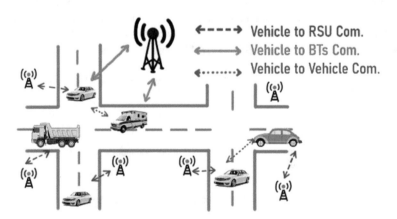

Fig. 1 Conceptual view point of VANET architecture

communication to mobile vehicles by viewing road conditions as a traffic data and RSU's are utilized as the network framework [4].

The next-generation networking is expected to acquire all the existing cellular networks like 3G, 3GPP, LTE, which is required to be heterogeneous in terms of network topologies, and various communication resources. The LTE technology provides the benefits of extensive coverage, high throughput, and minimal latency [5]. Although the high vehicular mobility results in a frequent change in network topology, it is difficult to provide satisfied ITS services using LTE technology. Also, the hand-offs occur frequently in wireless access-based infrastructures compared to the traditional wireless network. There are enormous demand and necessity to develop some unified approaches to deal with management and control problems that are rising in heterogenous VANETs. It also needs seamless services with frequent topology changes and varying QoS demands in VANETs. Despite all the efforts made in the field of heterogeneous VANETs, there is still a gap between the practical needs of ITS services and what can be given by current heterogeneous VANETs. A careful study shows that new challenges in VANETs are mainly because of the highly mobile nodes, the randomness of driver behavior, and the variable traffic density from time to time. Also, due to a lack of flexibility, programmability, and adaptability, there is a need for a new architecture that can tackle these issues. One solution is to introduce the SDN in the existing VANET architecture to resolve the problem of flexibility, programmability, and adaptability. The above issues motivate to rethink the redesign of the existing network architecture of VANETs. Consequently, the research and development for 5G (Fifth generation) systems also have been already been started which can help to overcome these issues [6–14].

1.1 Importance of SDN in VANET

SDN gives new experiences and viable solutions for the vehicular networks to handle some common challenges [15]. It can be adopted for supporting the dynamic idea of VANETs and ITS applications, by encouraging the adaptable organization of the network and streamlining on a huge scope with unified abstraction [16]. It is a revolution in computer networking that enables the provision of flexibility, programmability, and adaptability in the existing network as the data plane is decoupled from the control plane. This offers efficient control and flexibility to a system through the programming of the SDN controller at the control plane. It provides an abstraction for VANET applications to the underlying networking infrastructure, centralized logical network state, and intelligence. The adaptability of SDN makes it an alluring approach that can fulfill the requirements of VANET scenarios. The integration of SDN in VANET provides wireless resource optimization such as channel allocation, interference avoidance, packet routing in multi-hop multi-path scenarios, efficient mobility, and network heterogeneity management. Applying SDN standards to VANETs will leverage the flexibility and programmability but still, that will be inadequate in the present distributed wireless substrate for streamlining the network management and empowering new V2V and V2I administrations. Also, some of the existing decentralized VANET architecture protocol may not be able to exploit the SDN benefits and needs to redesign the protocols to become adaptable with the SDN entities [17].

1.2 Role of Cloud in VANET

A vehicular ad-hoc network (VANET) collects data of moving or fixed vehicles associated with a wireless network. To store this large amount of data using a traditional service is not efficient. Deploying the physical devices at each level of architecture will be cumbersome. So to overcome this, integrating VANET with the cloud will be more advantageous as it provides high computation speed at the lowest cost. VANET includes transaction and trade of data with different vehicles and foundations in its region utilizing wireless standards like Dedicated Short Range Communication (DSRC) and ZIGBEE. The advantage of integrating cloud over VANET will allow running global safety applications for monitoring geographical areas under the RSU coverage. It provides backup of the data on the cloud with the least cost, and also allows different web services over the internet for both client and administrator. VANET cloud will help in forming a large and flexible network that will serve many end users. It will reduce the load on the devices for performing security and privacy, data aggregation, energy efficiency, interoperability, and resource management. To design a cloud-based VANET architecture to cope with the increasing demands of emerging ITS services is a prime concern.

A combination of VANET architecture and convention cloud known as VANET Cloud (VC) is shown in Fig. 2. Here, VANET uses cloud services, in which RSUs act as gateways and provides a virtualization layer. RSUs can act as a high-speed communication gateway for the cloud service deployed for VANET applications. The next subsection discusses the involvement of various communication technologies in vehicular networks.

1.3 Need for Advance Communication Technologies

The evolution of technology in the automation area provides few advantages to the thrust areas of ITS such as security, clog control, safety, optimal resource utilization. Various communication technologies like long-term evolution (LTE), IEEE 802.11p, LTE-Advanced (LTE-A), and Wi-Fi have just been used in vehicular networks (VNs) [18]. Nevertheless, because of the limitation of the range with the complicated surrounding environment, these advancements introduce difficulties concerning data rates, dormancy, link quality, and connectivity as well. Moreover, due to expansion in the flexibility of connected vehicles and autonomous vehicles, it appearances definite difficulties such as high data transmission necessities, excessive portability in

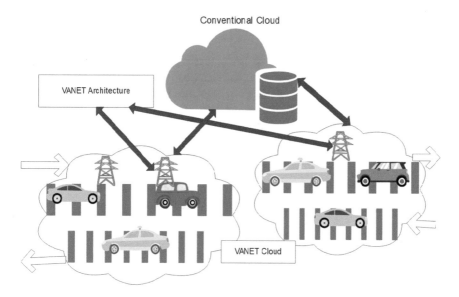

Fig. 2 VANET cloud

vehicles, adaptability of frameworks, and dependability of correspondences. To sustain the tough necessities of VNs, advancement in Fifth generation (5G) networks has been intended. Because of its energy efficiency and range rationality, it's required for increment by multiple times to the framework limit, 10–100 times the data rate, multiple times to life of the battery while keeping inactivity multiple times lower concerning 4G. Hence, 5G-based VNs can be a promising technology for addressing the limitations and difficulties related to the monstrous solicitations and data stream emerging from the associated vehicles. The emerging technologies like SDN, Cloud Computing, Edge Computing, Fog Computing are relied upon to be future competitor advances for 5G VANETs, which are necessities of future ITSs. Some underlying examinations have additionally been done to coordinate both of these innovations into vehicular communication networks [16, 19–21]. Moreover, the quality of services provided by SDN innovation gets restricted in RSUs, when the number of vehicles associated with RSU increases [20]. The frequent handover issue in the dense traffic situation in VANETs decreases the performance of SDN at RSUs [22, 23]. Nonetheless, it is understood that the versatility of wireless distributed networks (WDNs) improved by utilizing strategies like; clustering, multi-channel directing, and drafting [24, 25]. These days, cloud radio access networks (C-RAN) have been broadly acknowledged to be the promising answer for heterogeneous organizations for enabling large-scale deployment, collaborative radio support, optimal resource allocation, and network virtualization [19]. SDN can be adopted to implement wireless virtualization in C-RAN for diminishing the operational costs [26].

1.4   Lacuna in VANET with 5G Communication

Despite having many benefits, 5G networks face many problems to provide global, and dependable associations among the vehicles. So to leverage intelligence, strength, and stream programmability into 5G VNs, SDN, has been adopted due to its several benefits [27, 28]. It strengthens the progressive nature of VNs while upgrading the abilities of 5G networks. Every forwarding device (FDs) such as OpenFlow (OF) switches, routers, and gateways are available at the data plane, while the control plane is liable for making choices identified with data sending and resource allocation. The control plane works as per the guidelines given by the SDN controller and is likewise meant for conveying data identified with mobility, identity authentication, and security [29]. Regardless of current security mechanisms, there are still so many research gaps.

1.5   Research Contributions

The following are the research contribution covered in this chapter and are discussed in detail.

- A SDN enabled vehicular network is explored to overcome the lacuna in vehicular network with various communication technologies involved.
- A decentralized architecture for blockchain and different stages of amalgamation of blockchain and VANET is discussed to meet the security requirement needed in vehicular networks.
- A 5G and SDN based (5G-SDVN) architecture is explored along with its security requirement, limitations, and research challenges.

1.6   Organization

This chapter is organized as follows. Section 2 gives us a brief idea about blockchain and VANET and how they can be integrated. Section 3 presents the architecture of blockchain-based decentralized SDVN with its needs and benefits. Section 4, we highlighted the future perspective of 5G enabled blockchain-based SDVN. Finally, Sect. refsec5, concludes the chapter with some future research direction.

2   Software Defined Vehicular Network

Recently, SDN-enabled vehicular networks (SDVN) have witnessed substantial upliftments from technical and architectural aspects. The different possible components falling in traditional SDN enabled architecture are shown in Fig. 3 and their respective functions are discussed in detail.

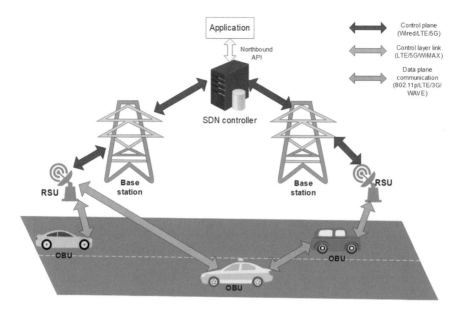

Fig. 3 The traditional architecture for SDN enabled VANET

RSU: It's a WAVE device that consists of an antenna, processor, and read/write memory and is mostly fixed alongside the roadside or some assigned site such as road junction or at off-street. RSU uses both wireless and wired interface in which wireless is used to communicate with OBU mounted on vehicles and for other RSU and internet, the wired interface is practiced. RSU comes with an equipped network device that works for a dedicated short-range communication-based technology i.e., IEEE802.11p, and also can be equipped with other network devices for communication used within an infrastructure [30, 31].

Base station (BS): Base stations are mostly located on the roadside and are used to communicate with the vehicles for sending data and receiving the data. This communication with base stations is used to get a different kind of information for further analysis for the purpose like post-accident, accident investigation, or traffic jams.

SDN Controller: It act as a central controller in which various policies can be programmed. It has a global view of entire network topology. RSU communicates with various applications deployed over SDN controller through OpenFlow standard based North bound APIs (Fig. 4).

VANET architecture can be split up into different forms based on a different perspective. The architecture of VANET falls into the main two categories as mention below:

1. Inter vehicle communication:
   The data collected from sensors is shared with nearby vehicles and can easily com-

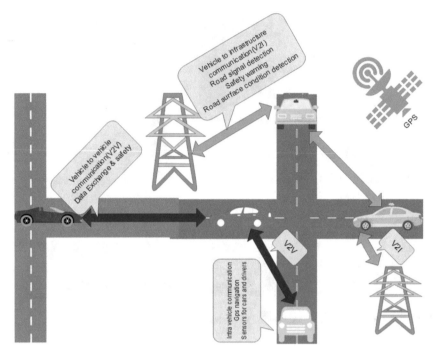

Fig. 4 Categories of VANET architecture

municate with each other without any infrastructure substructure. So it is called
vehicle-to-vehicle (V2V) communication also known as inter-vehicle communication. In the V2V approach, each SDN compatible vehicle monitors the vehicle
attributes, i.e., id, speed, direction, etc.

2. Vehicle to Infrastructure communication:
   Vehicle-to-Infrastructure (V2I) communication is the wireless exchange of data
   between vehicles and road infrastructure. Under this vehicular communication
   environment, vehicles are directly communicating with the deployed RSUs. Vehicles are communicating with these RSUs bidirectionally, without having concern
   about the information of other vehicles.

2.1 Communication Technologies Advancements

The technological advancements utilized for vehicular networks like remote correspondence, in-vehicle detecting module, and GPS helps to increase road security, traffic management, etc. To accomplish this normalization, at each layer of the
systems administration, the convention stack must be accomplished for the remote
environment with vehicles which is known as wireless access in vehicular environ-

ment (WAVE). The IEEE 802.11p's decentralized idea implements the control on the assurance of the standard chiefly inherited from the ad hoc medium access system, network overhead due to beaconing recurrence, and higher vehicle mobility. It has been outlined to address the requirements of connecting wireless services in a rapidly changing networking environment. It is challenging to obtain firm latency requirements over the cellular connection even by the LTE technology which has distinctive features are reliability, scalability, and mobility support. LTE technology-assisted cluster-head selection and management protocols for the IEEE 802.11p-based VANETs have been proposed considering in-vehicle OBUs with different interfaces. The IEEE 802.11p is the most important standard adapted for the vehicular network environment that uses the WAVE protocol stack. The physical layer with the MAC layer jointly permits ad hoc communication among On-Board Units and between On-Board Units and the Road Side Units. The physical layer lessens 20 MHz bandwidth offered by IEEE 802.11a to 10 MHz bandwidth, which halves the data rate to 3–27 Mbps from 6 to 54 Mbps. The IEEE 802.11p MAC is IEEE 802.11e Enhanced Distributed Channel Access (EDCA) protocol granting Quality of Service (QoS) support. In terms of overall network performances, it faces significant difficulties in wireless communication with high-speed vehicles over the varying channel condition. It also suffers because of the lack of an infrastructure-assisted channel access system.

The LTE has gained interest in itself because of its standard specification, like providing a truly broadband experience that allows mobile network operators (MNOs) to provide sophisticated services productively. Either with the use of the LTE-enabled OBU or using smartphones with LTE connectivity, it has been presumed to mistreat the existing LTE infrastructure to assist VANET applications. Although it has been a challenging task to transcend the time constraint data over the 4G/LTE connection and also efficiently share with existing LTE clients.

2.2 Limitations

With the severe obstruction because of higher beaconing frequency and firm vehicle mobility, one of the principal concerns is expandability as performance keeps on debasing with the rise in the number of participants. On the other side, LTE stringent latency requirements restrict the increase in VANET users. Unbounded Latency and guaranteed QoS needs are not satisfied by the DSRC based on IEEE 802.11p.

In vehicular networks, the most of the application observes a continuous delay, which is hindrance with the rise in cellular traffic load. The IEEE 802.11p offers adequate performance, but it is more sensitive to higher vehicle densities, traffic load, and high mobility. The absence of coordinated channel access and distributed congestion management mechanisms based on in-network conditions is the fundamental factor for performance degradation of IEEE 802.11p network [32]. A mobilized network for an intelligent transport system is required, which raises the limitation on the use of IEEE 802.11p. So, there is a need for the ultimate distributed system architecture of new 5G cellular networks, which help in a real-time scenario. For securing the data transfer and transfer of control and decision-making from a centralized entity (individual, organization, or group thereof) to a distributed network introducing blockchain migh be more fruitful.

2.3 Introduction to Blockchain

Blockchain is a distributed computing paradigm and decentralized infrastructure that uses cryptic chained block structures to store and validate the stored data, smart algorithms to create, update, and manipulate the data. Fundamentally, blockchain is a distributed database of records. Traditional databases are centralized, which is overseen by a solitary organization, whereas blockchain to the contrary, is decentralized, which is overseen by a variety of participants with each storing data and giving nodes on the chain. In the blockchain, every block of data is comprised of block header and body in which header encompasses current version variety, preceding block address with a timestamp. The time-stamped chain of data gives simple confirmation and recognizability [33]. Blockchain is classified into three categories, first as Public (Permission-less), then Private (Permission), and lastly as Consortium. For a public blockchain, consent from any node will be part of the network and partake in any operations. A private blockchain works in a closed limited network inside an organization, where only chosen can take an interest in the organization and are utilized when information protection rules and administrative issues are much needed. Whereas, consortium type blockchain is overseen by more than one organization and also, banks and network operators generally use this.

The features of blockchain are as follows.

- Shared database:
  It is an append-only database that keeps track of every exchange and is bestowed to each elaborate member.
- Consensus:
  For favoring the transaction, consensus alludes to an arrangement between the interested nodes of the P2P network. Once certain conditions are met, a bit of executable machine code that is preserved on the blockchain is carried out.
- Trustless:
  Using a consensus algorithm ensures trust and protects the network. It does not depend on the outsider for guaranteeing validness and insurance of data.
- Transparency:
  All network nodes know about every exchange so that they can freely check every exchange.
- Provenance:
  Data's possession has been changing over the long run and all the exchanges in the organization can be followed.

- Immutability:
  For tampering the data put away it requires control of most of the nodes in the blockchain network. Any user on the network won't be able to edit, delete or update the stored data.
- Smart contract:
  Shrewd contract is a practicable code written using a particular machine-oriented language. Smart agreements encourage decentralized automation by permitting each member's hubs in the blockchain organization to perform exchanges without utilizing an outsider substance. It can run simultaneously in-network or grew effectively. Ethereum, which supports smart contracts, is the most known blockchain platform which conveys a scripting language known as Solidity to execute the savvy contracts.

2.4 Security Aspects of Blockchain in Existing Solutions

In the VANET environment, the ground reality of data of the vehicles and their security is the key viewpoint that must be accomplished. Based on their secret information, they churn out pseudonyms for the validation of recipient for the vehicles and servers need to be confided in power just one chance to get secret data [34]. Additionally, the blockchain strategy to shield the standing from being changed is referenced in [35] which backhanded correspondence based security structure to direct the conduct of the locally available units in the VANET and diminish the assailants. As to maintain the principles for VANET frameworks, it requires a single authority, and the size of the organization expands every day—around above 750 million nodes, which raises the issue of standardization. However, conventional VANETs faces a few security issues.

2.5 Amalgamation of Blockchain and VANET

The blockchain framework is adopted for security in the SDN-enabled 5G-VANETs. These nodes form overlay P2P networks to keep up this blockchain framework. Real-time message interactions and video report service of traffic permits in the vehicular system. By victimizing options of blockchain, we tend to believe that the security and potency of this system are increasing considerably (Fig. 5).

Blockchain works in six phases to enable security in vanet scenarios. The six phases are System Initialization, System Authentication, Message Rating Generation, Trust Value Offset Calculation, Miner Election and Block Generation, Distributed Consensus. The initial phase starts with the step of initializing the system. Here phase of validating the key of the nodes and giving a declaration to the nodes after they join the network is being done. The System authentication is a subsequent phase that is responsible as a security layer to manifest nodes before nodes

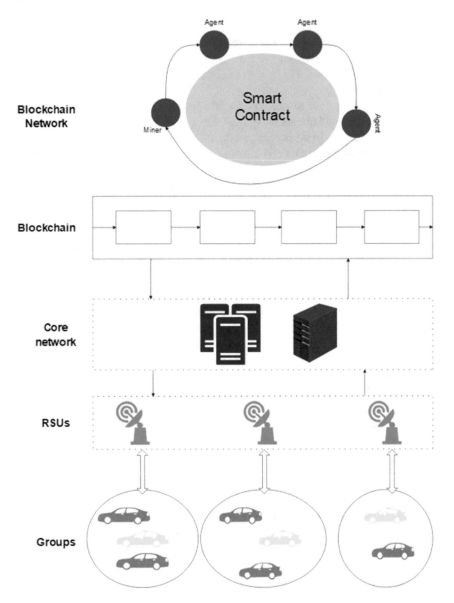

Fig. 5 Amalgamation of blockchain and VANET

begin interacting with one another after joining the architecture. The message rating phase is liable for giving a rating on a message delivered by the interacting nodes for trustiness. Succeeding phase value offset calculation is doing the calculation of the trustworthiness of each node in the framework. After this phase, phase of miner election and block generation implements the blockchain technology for an efficiency-based tracking of nodes in the system. The last phase in this system is the distribution of consensus, which works as a ledger unfolds all around the whole network [36].

In this section, we explored the traditional architecture of VANET along with the communication technologies integrated with it and also blockchain as a security framework for VANET. In the next subsection, VANET with 5G technology and SDN is explored.

3   5G-SDVN

To fulfill a high computing power requirement due to the use of blockchain with SDVN, the platform of cloud computing has been seen as an elegant solution to these services [37]. The current cloud computing technology is shown in Fig. 6. Because of requirements like ultra-low latency and high computing performance, computational intensive services with more powered computing capabilities have limitations and can't use the cloud service efficiently. The cloud-based architecture enables sharing resources by virtualizing physical foundations that may be equipped in a dynamic form to assist demands in applications, stages, and heterogeneous computing infrastructures. The network servers, storage, applications, and services are the resources that are going to be share and the physical assets that are shareable include network, server, application, storage, and services [38]. This architecture

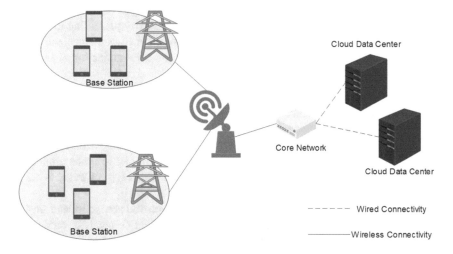

Fig. 6 Hierarchical cloud-based architecture

comprises infrastructure providers (InPs) and service providers (SPs). The role of the SPs is to sublet the resources to produce the services to the end-user and InPs are accountable to drive and directs the physical infrastructure. The same concept is in 5G networks to form virtualized wireless networks in which the operator is responsible for leasing the resource from InPs. The application of the cloud-based architecture to RAN architecture ends up in improved flexibility, measurability, and optimizing the utilization of problems. A C-RAN (Centralized RAN) is designed to supply scalable RANs to fix the capacity and coverage issue. Within the C-RAN architecture, the analog unit (RRHs) gets a separate baseband processing unit of the BS. The data collected by RAN is moved to the cloud. RRH performs digital processing, digital to analog conversion, analog to digital conversion, and power amplification of signals which is used for transmission and reception purposes of signals. C-RAN accompanied with other devices present in architecture plays a vital role in the cloud to raise network capacity, strengthen scalability, and outreach the coverage of future 5G systems.

3.1 Need of Security

The problem of security in SDN-Based 5G networks is that SDN architecture may enhance quantifiability of standard 5G network, an equivalent characteristic of unified management and scheduling related to SDN program that leads to network security problems. An important role has been played by the control plane in an SDN architecture; thus, we must protect the control plane to reduce the risk of security being compromised. It is much important to provide authentication of the access in the control plane. The coalition of progression of data with the SDN controller and switches is vulnerable. The attacker may bargain the network by simulating as an SDN controller and conjointly absence of standard rules for software development could result in a big amount of risks in security. As an example, data leakage risk by providing access to a third party that can help this third party permitting to modify and control the network and its rules without the consent of SDN controllers.

There are some reasons listed below that shows the communication channel of SDN is endangered:

1. Application might be accessed maliciously due to inappropriate authorization.
2. Southbound correspondence plagued through condemning on stream runs chiefly inside an In-band deployment.
3. Man-In-The-Middle and spoofing attacks is done due to lack of confidence and frail validation at this level, which drives the attacker simpler to snoop around on streams to envision what streams are being used and which type of traffic is being permissible over this network.
4. Indecorous authorization that leads to spawning unapproved access. Indeed, ensuring the client authenticity and environment allows client demand for an administration that prompts the monitor to form a path and undergo packets to travel through this path.

5. Due to lack of confidence and frail validation the application and the controller will undergo spoofing attack attacked by spoofing northbound API messages.

These are several justifications that advocate the need for security in this 5G-SDN system. In a basic concept, a secret key is generated by an encoding algorithm that has been maintained by all of the 5G devices. These keys are maintained in a huge database system. Throughout this process, verification of the secret key of devices is checked with the record, which is being checked by the back-end system. Since this authentication key won't be available before attackers, its demand won't pass the authentication process.

3.2 Blockchain-Based Decentralized SDVN

Few million entries approve the certificates of the various SDVN devices which are potentially stored in every BC node. It has to be stored in such a place where it can be easily accessible when needed. The enormous volume of data must not affect the short-time response. So, there's a need for a proper decentralized SDVN architecture. As some data linked to devices (certificates, revocation, and many more) are barely needed only for the nearby BC nodes located within the nearby geographical area, so there is no need to store the data in one central database and also to provide vehicles' certificates to all BC nodes is unnecessary.

We suggest separating the global BC network into totally distinguished BC sub-networks. Hence, the BC architecture no more comprises one global BC network, directing every SDVN device. However, a collection of BC sub-networks controls specific geographical areas and a limited number of devices. A country can be considered as an instance for standard Public Key Infrastructure for pilot projects in vehicular networks. In fact, to avoid legislative issues existing in different country considering a country as one geographical area is more feasible. However, cities or small areas might be considered as a small geographical area in countries having a large number of vehicles.

As shown in Fig. 7, each BC sub-network (SN1, SN2, SN3) controls a specific geographical region corresponding to some batch of SDN controllers, roadside equipment, and vehicles which are the clients right now situated in this geographical region. Each BC node can validate/abrogate/direct the access of the devices settled at intervals in its BC sub-network. Reviewing prevailing BC nodes in each BC sub-network and its activity status, a global BC network (Fig. 7: GN) accredit the sub-layers to exchange the information.

3.3 Deployment and Privacy Protection

It is always threatened by the centralized network architecture by assaults and falls in the absence of privacy. Decentralized feature of the blockchain-based technology for SDN based VANET system is adjudged as exhaustive technique concerning privacy protection.

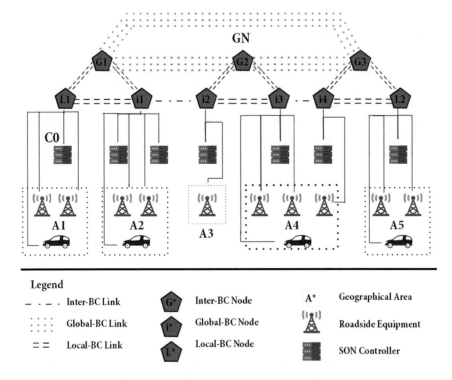

Fig. 7 BC SDVN architecture

- Transmission of data: Blocks of data are currently transmitted by distributed node trading, which has been controlled through distributed P2P network wherever a single node is identified employing a public key (PK).
- Communication between inter-nodes: Transactions are encoded by public keys, which are being broadcast with the whole network. Every node will uphold a transaction by conforming to their Public key(PK) value of the source, which certifies the suspicious agreement of blockchain.
- P2P network: An overlay P2P network to sustain a blockchain is formed by every functional node, counting vehicles, Road Side Units, and OBUs (5G base stations).
- Immutability and accountability: Accountability and immutability are the key points to make trust on blockchain. The data is encrypted during its transmission, which implies solely the source vehicle, and this data is collected by a cloud server. After that, a real-time report is generated in the vehicular system, which checks for message interaction among near vehicles.

3.4 Benefits and Challenges

In this subsection, we have highlighted some of the open issues and challenges related to this technology.

- Energy Efficiency—5G network technology will build energy utilization efficiently because of network design and expanded network gear. Endeavors to underline green correspondence, blockchain will be a major obstacle because of the computationally serious mining measure. A large portion of the power in the blockchain network is utilized because of the consensus algorithm. There have been endeavors to create energy-proficient consensus algorithms in blockchains such as Proof-of-Space and mini-blockchain. Nonetheless, it actually should be tried on a huge scale, but it requires a ton of room for additional improvisation.
- Security—blockchain is broadly acknowledged as an answer to comprehend different security issues in the 5G organizations, yet at the same time, there are a few security issues identified with blockchain itself.
- Scalability and Performance: The transactions per second (TPS) are the rate at which they endorse transactions in blockchain, and the poor transaction speed of blockchain is a significant test for a few enterprises embracing the arrangement, particularly with 5G. Most of the current blockchain arrangements experience a low transaction rate, which is unsuitable for 5G networks.
- Data Management: Same data might be shared by a few participating nodes in the network in blockchain. Perhaps on account of 5G, they might trade the data with an edge, core, and mobile clients over blockchain. Thus, the requirement for a normalized technique for ingesting, adjusting, and improving the data from various gadgets using diverse data designs into a typical data model with well-understood semantics.
- Resource Constraints and Allocation—Transaction computation is required for blockchain before it is acknowledged or dismissed. The consensus algorithms needed for this reason can be computationally serious. It isn't attainable for every node in the network to help in the transaction validation process. So, arranging a proficient consensus algorithm for resource allocation in this current network is a critical research task.
- Incentive Mechanism: 5G network will be exceptionally different in an environment where a few assets must be traded with network administrators and authorized third parties. Network computing resources from the other users may need to rent by the administrator. It presents a novel plan of action, income opportunity, and ideal use of assets in the network. So, a reasonable incentive system needs to be designed which supports cooperation in this type of resource sharing method.
- Network deployment and Interoperability—A proper network is yet not designed for 5G or beyond 5G network which includes different advancements, like, MEC, SDN, and NFV to deploy blockchain and to make it efficiently work in the design network.
- Standardization Requirement: Blockchain has recently started picking its attraction by the network operator due to its advancement in technology. Likewise, for getting

a wide acknowledgment, the network administrators require meeting up and pursue normalization of blockchain combination in the 5G network and beyond.

- Minimizing service latency—The utilization of SDN empowers the ideal usage and administration of fog computing services at network edge routers, which substantially lessens the service latency for delay-sensitive applications.
- Fast and flexible network configuration: By parting of control and logic plane in SDVNs offer help to the fast and adaptable network configurations. It will help to attune with the changes in network topology brought about by vehicle mobility.
- Heterogeneous network integration: Enabling uphold for the joining in heterogeneous networks, the controller gives the engrossment allying VANET applications and networking infrastructure in SDVN.

4   Research Challenges and Future Perspectives

Various problems starting with security, automation, optimal utilization of resources in a dispersed and decentralized manner can be efficiently puzzled out by the blockchain-SDN integration. Nevertheless, for enhancing network operation and security into the 5G network, several issues need to be resolved before integrating with blockchain-SDN. The use of 5G technology in SDN-VANET architecture leads to challenges to many problems mention here.

- Expandability and Effectiveness: Transaction per second (TPS) is a major concern in using any technology. To overcome such a problem, we need some method to increase its effectiveness and expansion in the public domain.
- Secured Framework for data management: A well-understood semantics using a shared data model with distinct data format is essential in a secured framework for consuming, disposing, and enhancing the data.
- Compatibility: It is predominant to design a suitable mechanism that can be used by all the resources used in the network for yielding opportunity and ideal utilization.
- Standardization: For the future perspective and gaining traction of network operators, some standardization of blockchain unification with 5G-SDVN network architecture is a must. It needs cooperation over industry and analysts.
- Constrained Resource Allocation: As all nodes don't need every resource that can be used and always all these resources are not available. To master these provocations, it might need the finest and ideal placement of firm validating resources in the network that demands to be inquired.
- Maintaining Ecosystem: Actionable insights and decisions have to be taken so that it ensures the ability to exchange information in the heterogeneous network and its embrace across the ecosystem.
- End-to-End security: A newly formed ITS system is anticipated to look out for the substantial challenges with the coalescence of SDN and 5G cellular connectivity under existing operating techniques based on communication overhead. In scenarios like this, the supreme significance is given to an end-to-end security solution for VNs to prevent it from unforeseen repercussions (Fig. 8).

Fig. 8 Research challenges and future perspectives

5 Conclusion

In this advancing digitized world, different technology needs to be integrated for solving technical, network, and security problems of ITS. SDN enabled 5G-VANET model with decentralized blockchain-based security framework has been introduced in this chapter. The problems associated with cooperation among network operators to perform more than one task which can be overcomed by the decentralized blockchain architecture are discussed. Maintaining and managing all sublevel nodes is a challenging task that can be addressed as a future work.

References

1. Bilgin, B.E., Gungor, V.C.: Performance comparison of IEEE 802.11 p and IEEE 802.11 b for vehicle-to-vehicle communications in highway, rural, and urban areas. Int. J. Veh. Technol. (2013)
2. Jiang, T., Alfadhl, Y., Chai, K.K.: Efficient dynamic scheduling scheme between vehicles and roadside units based on IEEE 802.11 p/wave communication standard. In: 2011 11th International Conference on ITS Telecommunications (ITST), pp. 120–125. IEEE (2011)
3. Schagrin, M.: Vehicle-to-vehicle (v2v) communications for safety (2013)

4. Engoulou, R.G., Bellaïche, M., Pierre, S., Quintero, A.: Vanet security surveys. Comput. Commun. **44**, 1–13 (2014)
5. Araniti, G., Campolo, C., Condoluci, M., Iera, A., Molinaro, A.: LTE for vehicular networking: a survey. IEEE Commun. Mag. **51**(5), 148–157 (2013)
6. Agiwal, M., Roy, A., Saxena, N.: Next generation 5g wireless networks: a comprehensive survey. IEEE Commun. Surv. Tutor. **18**(3), 1617–1655 (2016)
7. Gohil, A., Modi, H., Patel, S.K.: 5g technology of mobile communication: a survey. In: 2013 International Conference on Intelligent Systems and Signal Processing (ISSP), pp. 288–292. IEEE (2013)
8. Sun, S., Kadoch, M., Gong, L., Rong, B.: Integrating network function virtualization with SDR and SDN for 4g/5g networks. IEEE Netw. **29**(3), 54–59 (2015)
9. Ma, Z., Zhang, Z.Q., Ding, Z.G., Fan, P.Z., Li, H.C.: Key techniques for 5g wireless communications: network architecture, physical layer, and MAC layer perspectives. Sci. China Inf. Sci. **58**(4), 1–20 (2015)
10. Chen, S., Qin, F., Bo, H., Li, X., Chen, Z.: User-centric ultra-dense networks for 5g: challenges, methodologies, and directions. IEEE Wirel. Commun. **23**(2), 78–85 (2016)
11. Gong, M.X., Stacey, R., Akhmetov, D., Mao, S.: A directional CSMA/CA protocol for mmwave wireless pans. In: 2010 IEEE Wireless Communication and Networking Conference, pp. 1–6. IEEE (2010)
12. El-Keyi, A., ElBatt, T., Bai, F., Saraydar, C.: Mimo vanets: research challenges and opportunities. In: 2012 International Conference on Computing, Networking and Communications (ICNC), pp. 670–676. IEEE (2012)
13. Mitra, R.N., Agrawal, D.P.: 5g mobile technology: a survey. ICT Exp. **1**(3):132–137 (2015)
14. Zolanvari, M.: SDN for 5G. In: línea (2016). Available: http://www.cse.wustl.edu/~jain/cse570-15/ftp/sdnfor5g.pdf. Último acceso: 30 Julio 2016
15. Bhatia, J., Govani, R., Bhavsar, M.: Software defined networking: from theory to practice. In: 2018 Fifth International Conference on Parallel, Distributed and Grid Computing (PDGC), pp. 789–794, Dec 2018
16. Chen, J., Zhou, H., Ning, Z., Peng, Y., Lin, G., Sherman, S.X.: Software defined internet of vehicles: architecture, challenges and solutions. J. Commun. Inf. Netw. **1** (2016)
17. Bhatia, J., Modi, Y., Tanwar, S., Bhavsar, M.: Software defined vehicular networks: a comprehensive review. Int. J. Commun. Syst. **32**(12), e4005 (2019)
18. Eiza, M.H., Ni, Q., Shi, Q.: Secure and privacy-aware cloud-assisted video reporting service in 5g-enabled vehicular networks. IEEE Trans. Veh. Technol. **65**(10), 7868–7881 (2016)
19. Zheng, K., Hou, L., Meng, H., Zheng, Q., Ning, L., Lei, L.: Soft-defined heterogeneous vehicular network: architecture and challenges. IEEE Netw. **30**(4), 72–80 (2016)
20. Liu, K., Ng, J.K.Y., Lee, V.C.S., Son, S.H., Stojmenovic, I.: Cooperative data scheduling in hybrid vehicular ad hoc networks: Vanet as a software defined network. IEEE/ACM Trans. Netw. **24**(3), 1759–1773 (2015)
21. Zhang, N., Zhang, S., Yang, P., Alhussein, O., Zhuang, W., Shen, X.S.: Software defined space-air-ground integrated vehicular networks: challenges and solutions. IEEE Commun. Mag. **55**(7), 101–109 (2017)
22. Taleb, T., Letaief, K.B.: A cooperative diversity based handoff management scheme. IEEE Trans. Wirel. Commun. **9**(4), 1462–1471 (2010)
23. Bhatia, J., Dave, R., Bhayani, H., Tanwar, S., Nayyar, A.: SDN-based real-time urban traffic analysis in VANET environment. Comput. Commun. **149**, 162–175 (2020)
24. Li, X., Djukic, P., Zhang, H.: Zoning for hierarchical network optimization in software defined networks. In: 2014 IEEE Network Operations and Management Symposium (NOMS), pp. 1–8. IEEE (2014)
25. Abolhasan, M., Wysocki, T., Dutkiewicz, E.: A review of routing protocols for mobile ad hoc networks. Ad Hoc Netw. **2**(1), 1–22 (2004)
26. Khan, A.A., Abolhasan, M., Ni, W.: 5G next generation VANETs using SDN and fog computing framework. In: 2018 15th IEEE Annual Consumer Communications and Networking Conference (CCNC), pp. 1–6. IEEE (2018)

27. Yousaf, F.Z., Bredel, M., Schaller, S., Schneider, F.: NFV and SDN—key technology enablers for 5G networks. IEEE J. Sel. Areas Commun. **35**(11), 2468–2478 (2017)
28. Chauhan, K., Jani, S., Thakkar, D., Dave, R., Bhatia, J., Tanwar, S., Obaidat, M.S.: Automated machine learning: the new wave of machine learning. In: 2020 2nd International Conference on Innovative Mechanisms for Industry Applications (ICIMIA), pp. 205–212 (2020)
29. Bhatia, J., Kakadia, P., Bhavsar, M., Tanwar, S.: SDN-enabled network coding based secure data dissemination in VANET environment. IEEE Internet Things J. 1 (2019)
30. Budhiraja, I., Tyagi, S., Tanwar, S., Kumar, N., Rodrigues, J.J.P.C.: Tactile internet for smart communities in 5g: an insight for NOMA-based solutions. IEEE Trans. Ind. Inf. **15**(5), 3104–3112 (2019)
31. Tanwar, S., Tyagi, S., Kumar, S.: The role of internet of things and smart grid for the development of a smart city. In: Intelligent Communication and Computational Technologies, pp. 23–33. Springer (2018)
32. Gräfling, S., Mähönen, P., Riihijärvi, R.: Performance evaluation of IEEE 1609 wave and IEEE 802.11 p for vehicular communications. In: 2010 Second International Conference on Ubiquitous and Future Networks (ICUFN), pp. 344–348. IEEE (2010)
33. Vora, J., Nayyar, A., Tanwar, S., Tyagi, S., Kumar, N., Obaidat, M.S., Rodrigues, J.J.P.C.: Bheem: a blockchain-based framework for securing electronic health records. In: 2018 IEEE Globecom Workshops (GC Wkshps), pp. 1–6, Dec 2018
34. Xie, L., Ding, Y., Yang, H., Wang, X.: Blockchain-based secure and trustworthy internet of things in SDN-enabled 5G-VANETs. IEEE Access, 1 (2019)
35. Pu, Q., Wang, J., Zhao, R.: Strong authentication scheme for telecare medicine information systems. J. Med. Syst. **36**, 2609–2619 (2011)
36. Khan, A.S., Balan, K., Javed, Y., Tarmizi, S., Abdullah, J.: Secure trust-based blockchain architecture to prevent attacks in VANET. Sensors **19**(22), 4954 (2019)
37. Bhatia, J.: A dynamic model for load balancing in cloud infrastructure. Nirma Univ. J. Eng. Technol. (NUJET) **4**(1), 15 (2015)
38. Bhatia, J., Mehta, R., Bhavsar, M.: Variants of software defined network (SDN) based load balancing in cloud computing: a quick review. In: Future Internet Technologies and Trends, pp. 164–173. Springer International Publishing, Cham (2018)

Application Layer Protocols for Internet of Things

Abdon Serianni and Floriano De Rango

Abstract In this chapter an overview of Internet of Things (IoT) application layer protocols will be presented. We will briefly analyze first a layered IoT architecture and the most used communication protocols and technologies in IoT. Then, a description on the light-way protocols such as MQQT and CoAP to transmit IoT data to cloud is led out. A performance evaluation of the analyzed IoT application layer protocol will be provided with a focus on the control overhead and end-to-end delay.

Keywords IoT · M2M · MQTT · COAP · IoT communication protocols · 802.15.4 · BLE · WiFi

1 Introduction

The recent term Internet of Things was first coined by Kevin Ashton in 1999 to name real objects connected to the Internet. The IoT is gaining consensus and represents an opportunity for development especially in modern wireless telecommunications. The increase in connected devices supports the use of innovative solutions, with real objects and places that are now able to communicate with each other, collaborate with other systems, and transfer data and information. There will be an increasing number of connected objects in places and in everyday life, some examples can be:

- Thermostats
- Heating, Ventilation and Air Conditioning (HVAC) control systems
- Cameras
- Smartphones

A. Serianni · F. De Rango (✉)
DIMES (Dipartimento di Ingegneria Informatica, Modellistica, Elettronica e Sistemistica),
Università della Calabria, via Pietro Bucci, 87036 Arcavacata di Rende, CS, Italy
e-mail: f.derango@dimes.unical.it

A. Serianni
e-mail: a.serianni@dimes.unical.it
URL: https://www.dimes.unical.it/

P. Nicopolitidis et al. (eds.), *Advances in Computing, Informatics, Networking and Cybersecurity*, Lecture Notes in Networks and Systems 289,
https://doi.org/10.1007/978-3-030-87049-2_18

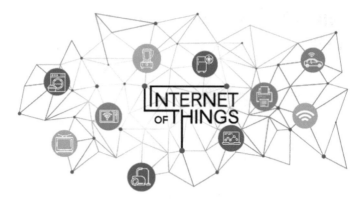

Fig. 1 Internet of Things

- Wearable objects
- Environmental sensors.

The term IoT can indicate a set of technologies that allow us to connect any type of device to the Internet [1] (Fig. 1).

The IoT can be considered as a network of physical elements composed by:

- *Sensors* used to collect data
- *Identifiers* used to identify the source of data
- *Software* used to analyze data
- *Internet connection* used to communicate and receive notifications.

An alternative re-definition of IoT could be *"IoT is the network of things, with clear element identification, embedded with software intelligence, sensors and ubiquitous connectivity to the Internet"*.

In its simplest form, the main goal of IoT is to physically connect anything/ everything through the Internet for monitoring/control functionality (Fig. 2).

Examples of IoT devices can be the refrigerator, the clock, the traffic light even if all objects can be considered examples of IoT devices.

The important thing is that the objects are connected to the network and that they can to transmit and receive data [2].

Applications of the Internet of Things can be identified in:

- Home Automation
- Robotics
- Automotive industry
- Biomedical Industry.

There are also many areas of application for IoT such as:

- Smart City
- Smart Building and Smart Home
- Smart Mobility

Fig. 2 Internet of Things
simplest form

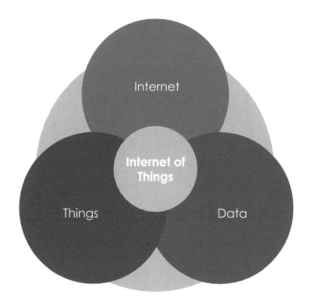

- Smart Manufacturing and Industry 4.0
- Smart Agriculture.

In these contexts is possible to provide a more complete definition of IoT as the Internet of Everything (IoE) with four main components:

- *People*: connecting people through the Internet
- *Data*: leveraging data for decision making
- *Process*: delivering the right information to the right person/machine at the right time
- *Things*: physical devices and object connected to the Internet.

The IoE combine people, things, process and data to form a network over the Internet with distributed intelligence, sensing and acting capabilities (Fig. 3).

2   IoT Architecture

It is possible to split IoT solutions into a layered architecture. There are several numbers of proposed architectures in literature [3, 4]. The basic IoT model consists of a three-layer architecture consisting of the Application, Network and Perception Layers. Some other IoT models are composed by five-layer model: Business, Application, Service Management, Object abstraction and Objects Layers (Fig. 4).

Fig. 3 Internet of
Everything

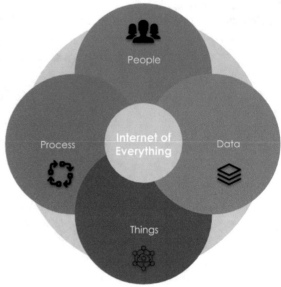

Fig. 4 The IoT architecture

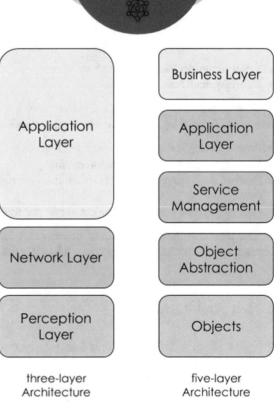

A brief description of the five-layers architecture is provided below.

- *Objects Layer*: heterogeneous physical devices (sensors and actuators) of the IoT that collect and process information.
- *Object Abstraction Layer*: transfers data produced by Object layer to the Service Management layer using various technologies such as RFID, WiFi, Bluetooth Low Energy, ZigBee, etc.
- *Service Management Layer*: process received data and enable IoT application to work with heterogeneous objects acting as middleware layer.
- *Application Layer*: provides the smart-services requested by the users.
- *Business Layer*: support decision-making processes based on Data analysis, manage system activities and services.

The Application Layer acts as an interface between end-users and smart devices. Also this layer provide an interface to the Business Layer used to high-level analysis. The management of these functionalities typically require many computational resources and exploit resources made available by Cloud computing and Fog/Edge computing. Some application domain where application layer protocols have been considered in the health-care and video-surveillance contexts are described in [20–22].

3 IoT Protocols and Technologies

IoT communication technologies connect heterogeneous devices to provide specific smart services. There are several network communication standards and protocols that allow IoT nodes to communicate typically using low power and in the presence of lossy and noisy communication links. Examples of communication protocols used in IoT solutions are IEEE 802.11 i.e. WiFi, Bluetooth Low Energy (BLE), IEEE 802.15.4 i.e. ZigBee, Long Term Evolution (LTE), Z-Wave, IPv6 over Low-Power Wireless Personal Area Networks (6LoWPAN) [5]. Other standards and protocols used in the application IoT context are Message Queuing Telemetry Transport (MQTT), Constrained Application Protocol (CoAP), REpresentational State Transfer (REST) and Extensible Messaging and Presence Protocol (XMPP) [6, 7].

A brief introduction of most used communication protocols and technologies are presented below with a separation of *IoT network communication protocols* and *IoT application layer protocols*.

3.1 IoT Network Communication Protocols

In this section a description of prominent network communication protocols in IoT context are provided.

3.1.1   IEEE 802.11

IEEE 802.11 (Wi-Fi) standards group are the most commonly used wireless standards in common networking. The IEEE 802.11 give wireless connectivity that requires quick installation inside a Wireless Local Area Network (WLAN). Wi-Fi standards have been generally adopted for digital devices, including laptops, smartphones, tablets and various smart devices. It defines the MAC methods for accessing the physical medium. Mobility is handled at the MAC layer, so handoff between adjacent cells is transparent to layers built on top of an IEEE 802.11 device [8].

A Wi-Fi WLAN is based on a cellular architecture. Each cell is called a basic service set (BSS). A BSS is a set of mobile or fixed Wi-Fi stations. Access to the transmission medium is controlled using a set of rules called a coordination function. Wi-Fi defines a Distributed Coordination Function (DCF) and a Point Coordination Function (PCF).

IEEE 802.11ah is the version of IEEE 802.11 standards which is lightweight to satisfy IoT needs and low power consumption for devices [9].

IEEE 802.11ah MAC layer features include:

- *Synchronization Frame*: Only valid stations with valid channel information can transmit by reserving the channel medium.
- *Efficient Bidirectional Packet Exchange*: with this feature, the sensors will go to sleep as soon as at the end of the communication and reduces power consumption.
- *Short MAC Frame*: IEEE 802.11ah reduces frame size from 30 bytes in traditional IEEE 802.11 to 12 bytes.
- *Null Data Packet*: Traditional IEEE 802.11 standards had acknowledgement (ACK) frames of 14 bytes. IEEE 802.11ah uses a preamble in place of ACKs and is much less in size.
- *Increased Sleep Time*: this standard is designed for power-constrained devices and it allows a long sleep period and waking up occasionally to exchange data only.

3.1.2   IEEE 802.15.4

The IEEE 802.15.4 protocol is the base of the ZigBee protocol. It is used for low-rate wireless private area networks with reliable communications based on low-power consumption, low data rate, low costs and a large number of nodes. Devices based on this protocol are used in IoT and Machine-to-Machine (M2M) communications with low data-rate services on power-constrained devices. IEEE 802.15.4 utilizes three different frequency channel bands (2.4 GHz, 915 MHz and 868 MHz) with different data-rate, distance coverage, throughput and latency. There are two different types of nodes in an IEEE 802.15.4 network:

- Full Function Devices (FFD) with capabilities of creation, control and management of network and functions of Personal Area Network (PAN) node.
- Reduced Function Devices (RFD) with reduced resources.

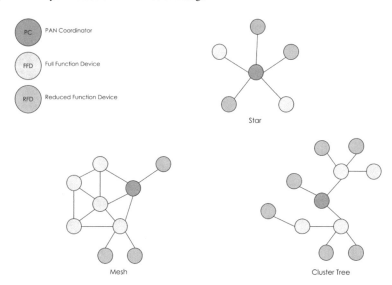

Fig. 5 IEEE 802.15.4 network topologies

The network topologies shown in Fig. 5 and used in IEEE 802.15.4 are:

- Star topology: all communications in the network are managed by the PAN coordinator.
- Mesh topology: any network nodes can communicate with any other devices with ad-hoc network management.
- Cluster Tree topology: a special case of a mesh network with a large number of FFD nodes and the RFD nodes that represents a leaf of the tree structure.

3.1.3 Bluetooth

Bluetooth is a standard for wireless communications based on a radio system designed for short-range low-cost communications devices. The devices can be used for single communications between portable devices, act as bridges between other networks, or serve as nodes of ad hoc networks. This range of applications is known as a Wireless Personal Area Network (WPAN) [8]. Bluetooth defines a full communication stack that enables the devices to find each other and advertise their offered services.

A Bluetooth device may operate in both master or slave mode; a maximum of eight devices working together.

Bluetooth devices use the 2.4 GHz band. The channels are accessed using an FHSS technique, using Gaussian-shaped Frequency Shift Keying (GFSK) modulation. Frequency hopping consists of accessing the various radio channels according to a long pseudo-random sequence generated from the address and clock of the master node of the network.

3.1.4 Bluetooth Low Energy

BLE technology is used in IoT context for short-range radio communications with very small power consumption. BLE modules are widely used in smartphones because it allows low-energy communications with other devices such as wearables, sensors and actuators. BLE as a classic Bluetooth uses adaptive frequency hopping spread spectrum to access the shared channel.

A star network topology is used by BLE devices that can operate as masters or slave. A master device can control multiple connections at the same time, but a slave can only be connected to a single master at a time.

In a BLE communication slaves use a discovery mechanism in witch sends advertisement over dedicated advertisement channels. These channels are scanned by the master periodically. To save energy a BLE device remains in sleep mode and exits this mode only to communicate with other devices [10].

3.1.5 Z-Wave

Z-Wave is a low-power wireless communication protocol designed for home automation and used IoT applications like smart homes [11].

Z-Wave communications covers about 30 m point-to-point and uses small messages. It uses a master/slave architecture with controller and slave nodes. Controllers controls the whole network topology and it manages the slaves by sending commands to them. Z-Wave devices operate around 900 MHz band with a 40 kbps transmission rate. This protocol is specified for applications like smart home control, smart energy management, wearable health care control and fire detection.

3.2 IoT Application Protocols

A description of most used IoT application protocols is provided with detailed operating mechanisms and characteristics.

3.2.1 MQTT

MQTT protocol was developed for lightweight machine-to-machine communications by Andy Stanford-Clark (IBM) and Arlen Nipper (Eurotech) in 1999 and was standardized in 2013. The intent was to have a bandwidth-efficient protocol characterized by low energy consumption [12, 13]. The protocol uses the publish/subscribe communication pattern and not the classic HTTP request/response paradigm. The publish/subscribe paradigm is event-driven and allows messages to be sent to clients. The data flow utilized by MQTT protocol is depicted in Fig. 6.

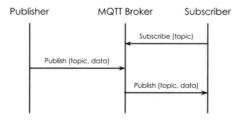

Fig. 6 MQTT publish/subscribe pattern

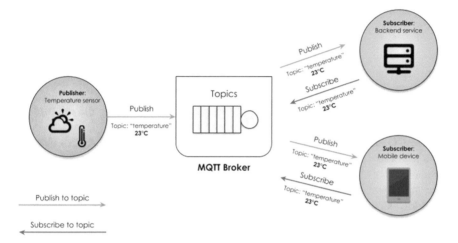

Fig. 7 MQTT architecture

MQTT consists of three components: *broker*, *publisher* and *subscriber*. The central node of communication is the MQTT broker that dispatch messages between senders and receivers. Each client who publishes a message on the broker includes a topic in the message. The topic is information that the broker uses to forward messages. Each client who wants to receive messages subscribes to a specific topic of interest and the broker forwards the messages relating to the topics of interest to the clients. This architecture allows you to create highly scalable solutions without a direct dependence between data producers and data consumers (Fig. 7).

The publish/subscribe approach used with the MQTT protocol separates the clients who send a message (publisher) and the clients who receive messages (subscribers). An MQTT client can typically present publisher and subscriber functionality and use libraries that allow connection to an MQTT broker.

MQTT connections always take place between a client and the broker, there is never a direct connection between two clients. The connection between client and broker starts with a CONNECT message from the client to the broker that replies with a CONNACK message and a status code, once the connection with the broker

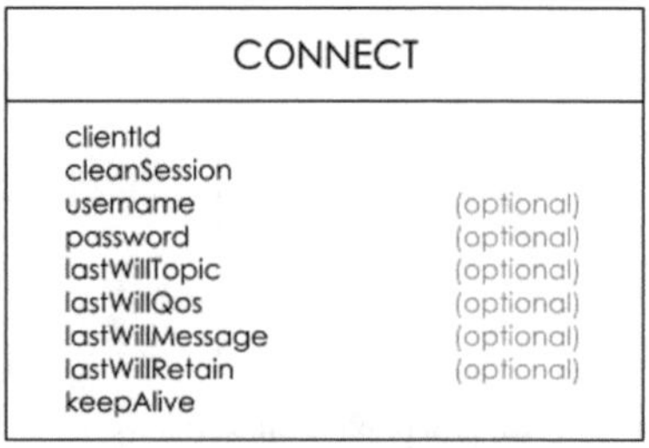

Fig. 8 MQTT CONNECT
packet structure

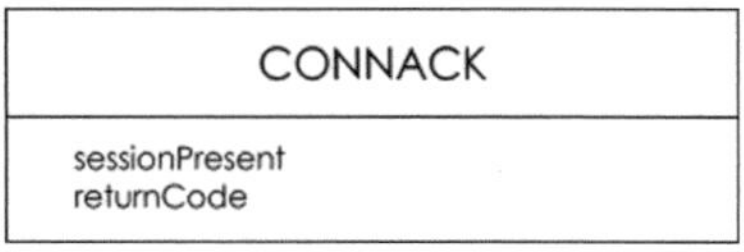

CONNECT	
clientId	
cleanSession	
username	(optional)
password	(optional)
lastWillTopic	(optional)
lastWillQos	(optional)
lastWillMessage	(optional)
lastWillRetain	(optional)
keepAlive	

Fig. 9 MQTT CONNACK
packet structure

CONNACK
sessionPresent
returnCode

is established, a client can decide to close the connection using the appropriate commands to disconnect.

The CONNECT message is sent by the client to the broker to establish the start of a new connection. If the message is malformed and does not comply with the specifications of the MQTT standard, the broker closes the connection, this allows you to prevent any attacks from "malicious" clients that can cause overloads and slowdowns and malfunctions of the broker. The typical structure of a CONNECT packet is shown in Fig. 8.

The fields that form up a CONNECT package are:

- *ClientId* represents the unique identifier assigned to each MQTT Client connected to an MQTT broker.
- *Clean Session* is a flag indicating whether the client wants to establish a persistent session or not. Using a persistent session the broker stores all the subscribe operations and the client's lost messages, in a non-persistent session no information about the client is stored by the broker.
- *Username and Password* are the client's authentication and authorization parameters.
- *Will Messages* are parameters that allow the sending of any notification messages to other clients when certain events arise, such as the disconnection of a client.
- *Keep Alive* is the time interval between the PINGs that clients and brokers carry out to check if both are online.

Upon receipt of a CONNECT packet the broker replies with a CONNACK packet which contains two fields: Session Present flag is a flag indicating whether there is already a persistent session on the broker of the client that sent the CONNECT message. Connect acknowledge flag is a signaling flag that specifies whether the connection between client and broker has been successful and possibly what connection problems have arisen. The format of a CONNACK package is shown in Fig. 9.

Table 1 shows the status codes and their respective brief description.

Table 1 MQTT CONNACK status code

Status code	Description
0	Connection accepted
1	Connection refused, unacceptable protocol version
2	Connection refused, identifier rejected
3	Connection refused, server unavailable
4	Connection refused, bad username or password
5	Connection refused, not authorized

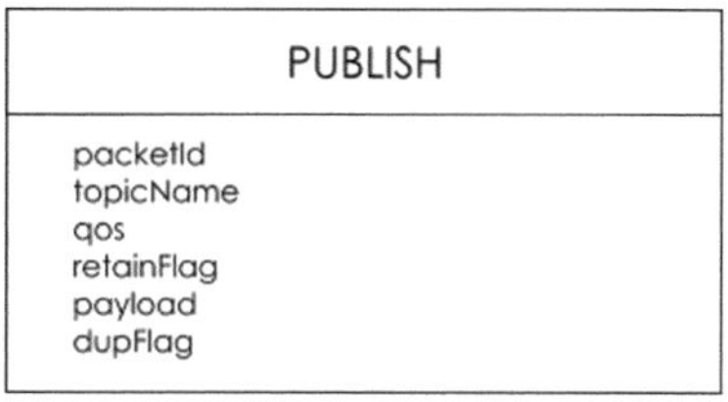

Fig. 10 MQTT PUBLISH packet structure

Once connected to the broker, an MQTT client can publish a message. The MQTT protocol is based on topic-based filtering operations of messages on the broker. Each message must contain a topic that is used by the broker to send messages to the clients "interested" in the topic. In addition to the topic, each message typically has a payload which contains the data to be transmitted in byte format. The structure of a PUBLISH package is shown in Fig. 10.

The fields that make up a PUBLISH package are:

- Topic Name is a string, typically hierarchical.
- Quality of Service (QoS) represents the Quality of Service level of the message, it can take values 0, 1 and 2.
- Retain-Flag is a flag that determines whether the message can be saved by the topic for the specific topic in order to provide the latest messages of the topic of interest to new clients who subscribe to the topic.
- Payload is the content of the MQTT message.
- Packet Identifier is the unique identifier used by clients and brokers to identify the message.
- DUP-flag represents the duplicate flag and indicates whether the message has been re-sent.

The client who wants to publish information on a specific topic, only deals with sending the PUBLISH package to the broker, it is then the responsibility of the broker to forward the information to the clients who have subscribed to the correct topic. After a publish operation, the client does not receive any feedback and has no information on how many or which clients received the information. The management of the exchange of information of the MQTT protocol provides, in addition to

Fig. 11 MQTT
SUBSCRIBE packet
structure

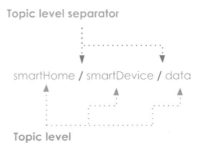

SUBSCRIBE
packetId qos1 topic1   topic + qos qos2 topic2   topic + qos ...

Fig. 12 MQTT
UNSUBSCRIBE packet
structure

UNSUBSCRIBE
packetId topic1 topic2   list of topics ...

Fig. 13 MQTT topic
structure

Topic level separator

smartHome / smartDevice / data

Topic level

sending PUBLISH messages, the consideration during the reception of messages, SUBSCRIBE messages or packets that have the following structure (Fig. 11).

The fields of a SUBSCRIBE package are:

- Packet Identifier is the unique identifier used by clients and brokers to identify the package.
- List of Subscriptions may contain an arbitrary number of topic/QoS Level pairs representing the client's topics of interest and the respective required QoS levels.

The message that can be defined as the opposite of that of SUBSCRIBE is the message of UNSUBSCRIBE which is used to remove a client's subscribe to a topic of interest. The fields of the UNSUBSCRIBE message are completely equivalent to the SUBSCRIBE message and the package format is shown in Fig. 12.

The concept that links MQTT's publish and subscribe operations is that of topic, a topic is a UTF-8 string that is used by the broker to filter the messages of the various connected clients. A topic consists of one or more topic levels, each topic level is separated by a topic level separator. A typical example of a topic is shown in Fig. 13.

A client does not need to create a topic before performing publish and/or subscribe operations, in fact the broker accepts every valid topic, without the need for initialization. A valid topic contains at least one character, can contain spaces and is case-sensitive, for example the topics *smathome/smartDevice* and *smartHome/smartDevice* will be two separate topics.

3.2.2 XMPP

The XMPP is a protocol designed for chats and messages exchange applications. It is based on XML language and was standardized by IETF more than a decade ago. XMPP is extensible and allows the specification of XMPP Extension Protocols (XEP) that increase its functionality [14].

XMPP operates over Transmission Control Protocol (TCP) and provides publish/subscribe and also request/response messaging communication patterns. It is designed for near real-time communications and thus, efficiently supports low-latency small messages.

XMPP has TLS/SSL security built in the core of the specification. However, it does not provide any QoS options that make it impractical for M2M communications. XMPP uses XML messages (eXtensible Markup Language) that create additional overhead due to unnecessary tags and require XML parsing that needs additional computational ability which increases power consumption.

3.2.3 RESTful API over HTTP

The use of web-based services based on REST architecture has now become commonplace. The REST architecture, denied in 2000 in Roy Fielding's doctoral thesis [15], collects a set of principles on how network architecture should be composed, the main characteristics of which are:

- *Client-Server*: network architecture to be used in which an entity (client) uses a series of services made available to an entity (server).
- *Stateless*: requirement of the REST architecture is that the communication is of the stateless type (CSS—Client Stateless Server) or every request made by the client to the server must contain all the information necessary for the server to understand the request, without referring to any further stored server-side information.
- *Cache*: clients have cached a response obtained following a request must be marked as cacheable or non-cacheable. If a response is cacheable, a client can take advantage of that cached response for subsequent equivalent requests. This allows you to improve scalability, decrease network traffic and have less latency.
- *Resources*: resources that servers make available to clients are indicated by a resource identifier; through it, you can access the resource and/or change its status.

Fig. 14 CoAP stack

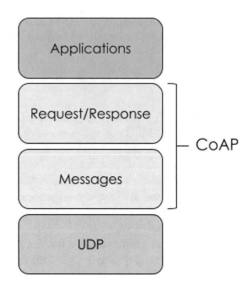

3.2.4 CoAP

CoAP is a network-oriented protocol for IoT applications that uses features similar to the HTTP protocol with low overhead [16, 17]. Unlike HTTP-based protocols, CoAP is based on the User Datagram Protocol (UDP) protocol and does not use the classic congestion control techniques provided by the TCP protocol. CoAP provides Uniform Resource Identifiers (URIs) and REST methods such as GET, POST, PUT and DELETE, allows group communication for the IoT using multicast. Improvements on the reliability of the UDP protocol are given by mechanisms for retransmission and discovery of the resources made available by the CoAP protocol.

The interaction model of the CoAP protocol is comparable to the client/server model on which the HTTP protocol is based. As illustrated in Fig. 14 CoAP can be logically divided into two distinct sub-layers:

- *Messages*: designed to manage the lower UDP layer and to provide reliable communication with message duplication detection and recovery mechanism
- *Request/Response*: manages REST communications and the exchange of information between requests and responses.

The exchange of messages in the CoAP protocol is asynchronous and the Messages level supports 4 different types of messages:

- *Confirmable*: messages that requires an acknowledgement (ACK)
- *Non-confirmable*: messages that do not require the sending of a response (ACK)
- *Acknowledgement*: response messages to confirmable messages
- *Reset*: message sent after a request that the server could not be process.

Reliability of CoAP communication is achieved by a mix of confirmable and non-confirmable messages.

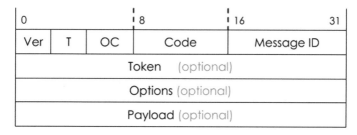

Fig. 15 CoAP packet structure

CoAP is based on the exchange of compact messages, the structure of a CoAP package is shown in Fig. 15 and includes a 4-byte header followed by a variable-length part which can include the Token, Options and Payload fields. A Token field is used to correlating requests and responses.

The header of the CoAP package consists of the following fields:

- Version (Ver): version of CoAP used
- Type (T): type of message

  0. confirmable (request)
  1. non-confirmable (request)
  2. acknowledgement (response)
  3. reset (response)

- Token Length (OC): 4 bits that indicate the length in bytes of the token field that can take values from 0 to 8 bytes.
- Code: 8 bits in which the first 3 bits indicate the class (c) and the following 5 bits indicate the detail (d) forming a code field of type *class.detail* like HTTP status codes.
- Message-ID: 16 bits that identify the messages to discriminate any duplicates and associate ACK/Reset response messages with Confirmable/Non-confirmable request messages.

3.3 Evaluation of IoT Communication Protocols

In this subsection, a performance evaluation of different IoT communication protocols is provided.

In [18], studies on the performance comparison of MQTT and CoAP are introduced using a common middleware that supports MQTT and CoAP. Experiments are used to analyse the performance of MQTT and CoAP in terms of end-to-end delay and bandwidth consumption.

In [18], the authors present and analyze the efficiency, usage, and requirements of MQTT and CoAP using a Raspberry-Pi and a temperature sensor.

Fig. 16 MQTT
publish/subscribe timing

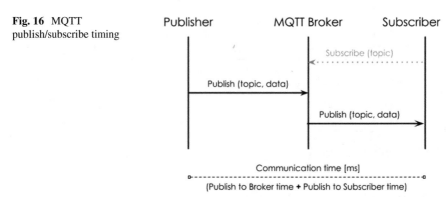

Fig. 17 REST
request/response timing

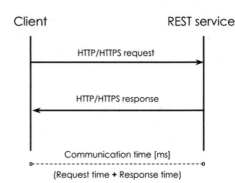

The analysis of the communication times for MQTT protocol is linked to the event-driven publish/subscribe pattern. In our tests, the communication time is given by *Publish to Broker time + Publish to Subscriber time* in milliseconds (Fig. 16).

In classic HTTP request/response paradigm, used in REST/CoAP approach, the communication time is given by *Request time + Response time* in milliseconds (Fig. 17).

For all communication tests we have used the same JSON message that contains:

- *MessageID*
- *Timestamp*
- *Payload* of variable length (10, 50 or 100 bytes).

The communication time analysis of MQTT protocol was carried out for each MQTT QoS level. The QoS level defines the guarantee of delivery for MQTT messages:

- At most once (0)—best effort delivery without guarantee of delivery
- At least once (1)—a message is delivered at least one time to the receivers
- Exactly once (2)—a message is delivered exactly once to the receivers.

Different QoS levels produce different communication times as shown in Fig. 18.

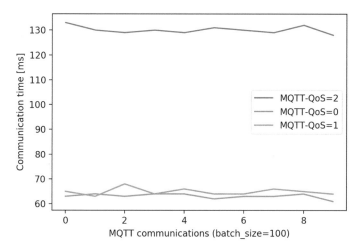

Fig. 18 MQTT communication time for different QoS

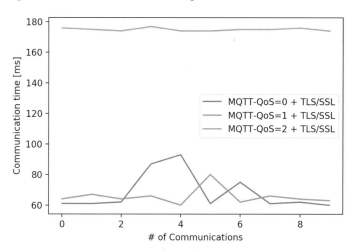

Fig. 19 MQTT communication time for different QoS with TLS security

1000 MQTT communications were used for this test for each level of QoS. The data were grouped into 10-batches and each batch represents the average communication time of 100 MQTT delivered messages.

For QoS level 0 and 1, there are similar communication times about 63 ms, for the QoS 2 level, the average communications time is about 130 ms.

The previous results are different applying QoS if also the Transport Layer Security (TLS) is applied to provide security features. IoT security is an important issue to address such as referred in [19, 23]. In this chapter we evaluated the performance of MQTT under TLS that is a cryptography protocol which enables a secure and encrypted communication at the transport layer between client and server.

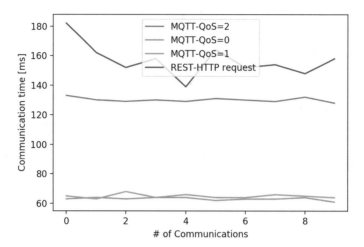

Fig. 20 MQTT versus REST communication time

For QoS level 0 and 1, there are similar communication times about 67 ms, for the QoS 2 level, the average communications time is about 175 ms.

The results of the communication time of MQTT with TLS security are higher than the results shown in Fig. 18. This is due to the construction of an encrypted channel with the exchange of keys and security certificates between client and server that allows increasing the security levels of the MQTT protocol (Fig. 19).

The result is influenced by the caching of server responses that use the same established TLS connection. The results of MQTT communication times are compared with REST request in Figs. 20 and 21.

In particular, tests have been done with 100 REST HTTP requests and 100 REST HTTPS requests grouped into 10-batches of 10 rest requests each one.

As depicted in Figs. 20 and 21 the communication times of REST requests are higher than the MQTT communication times. In particular, the average communication time of REST HTTP requests are about 156 ms and the average of REST HTTPS requests is about 295 ms.

These results have shown that MQTT usage is better than rest usage in term of communication time with or without TLS security for all QoS level. The test environment was the same for all tests, MQTT Broker and REST services were hosted into the same Cloud Virtual Machine (VM) and clients (MQTT publisher, MQTT subscriber and REST client) runs over the same test workstation.

The analysis of the messages exchanged in the communications for different used IoT protocols is presented in terms of data exchanged and protocol overhead.

As depicted in Fig. 22, the message exchange analysis for MQTT protocol was given out for each QoS level.

In MQTT communications with QoS = 0, a best-effort delivery is guaranteed, the communication can be divided into:

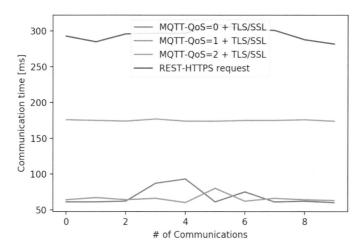

Fig. 21 MQTT versus REST communication time with security

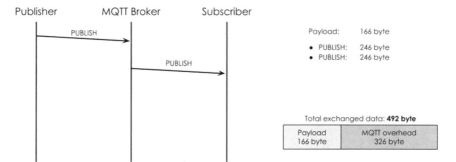

Fig. 22 MQTT with QoS = 0 data exchange

- *Publisher* sends the message to *MQTT Broker* with a PUBLISH
- *MQTT Broker* sends the message to *Subscriber* with a PUBLISH

For a message of 166 bytes of payload, the total exchanged data was 492 bytes but there is no guaranteed of delivery.

For MQTT communications with QoS = 1, a message is delivered at least one time to the receiver. As shown in Fig. 23 for every published message an ack is generated, if a sender not receives this acknowledgement re-send the message, this can generate multiple delivered messages.

The communication with QoS = 1 can be divided into:

- *Publisher* sends the message to *MQTT Broker* with a PUBLISH
- *MQTT Broker* sends the ACK message to the *Publisher*
- *MQTT Broker* sends the message to *Subscriber* with a PUBLISH
- *Subscriber* sends the ACK message to the *MQTT Broker*.

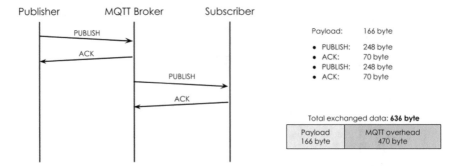

Fig. 23 MQTT with QoS = 1 data exchange

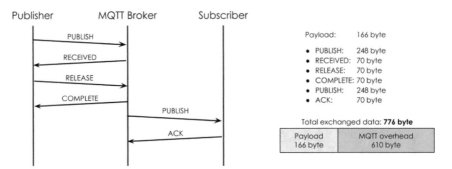

Fig. 24 MQTT with QoS = 2 data exchange

The 166 bytes of payload lead 636 bytes of total exchanged data (without multiple delivered messages).

MQTT communications with QoS = 2 is the safest and slowest QoS level with a handshake of four messages between the sender and the receiver that confirm that the message has been sent and that the acknowledgement has been received. Each message is received only once by the receivers (Fig. 24).

The communication with QoS = 2 can be divided into:

- *Publisher* sends the message to *MQTT Broker* with a PUBLISH
- *MQTT Broker* sends the RECEIVED message to the *Publisher*
- *Publisher* sends the RELEASE message to *MQTT Broker*
- *MQTT Broker* sends the COMPLETE message to the *Publisher*
- *MQTT Broker* sends the message to *Subscriber* with a PUBLISH
- *Subscriber* sends the ACK message to the *MQTT Broker*.

For a message with 166 bytes of payload, the total exchanged data was 776 bytes. The data-exchange analysis for classic HTTP/HTTPS request is presented below. The request/response paradigm of an HTTP request with a payload of 166 bytes can be divided into:

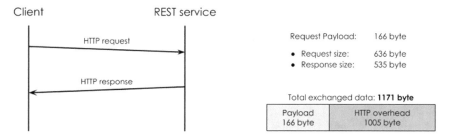

Fig. 25 HTTP request data exchange

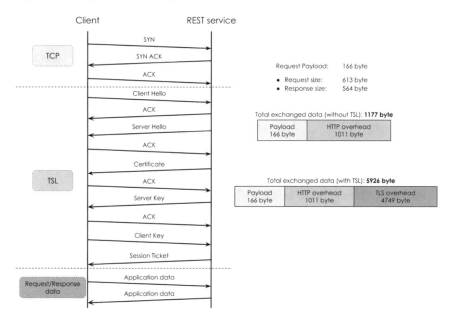

Fig. 26 HTTPS request data exchange

- HTTP request size equal to 636 bytes
- HTTP response size equal to 535 bytes.

For a request/response with an exchange of 166 bytes of payload data, there are 1171 bytes of total exchanged data with a protocol overhead of 1005 bytes (Fig. 25).

Considering an HTTPS request, there are 1011 bytes of protocol overhead to which the data exchanged for the TLS handshake must be added.

During a TLS handshake, both communicating sides exchange messages to establish encrypted communication with asymmetric encryption and use of a pair of a public and private key. The analyzed TLS overhead is equal to 4749 bytes, in Fig. 26 is depicted the complete HTTPS request data exchange.

The data-exchange analysis for CoAP request/response is presented following for *Confirmable* (CON) and *Non-Confirmable* (NON) CoAP message.

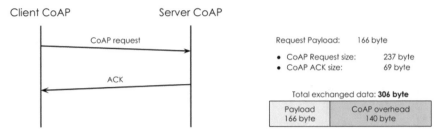

Fig. 27 CoAP CON request data exchange

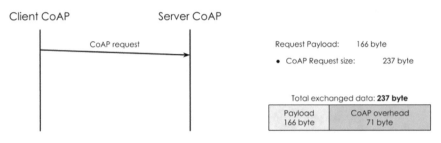

Fig. 28 CoAP NON request data exchange

In Confirmable messages an acknowledgement (ACK) is required, for a request of 166 bytes there are 306 bytes of total exchanged data with a CoAP protocol overhead of 140 bytes. The CoAP request size is equal to 237 bytes and the CoAP ACK size is equal to 69 bytes and contains the response or the error code (Fig. 27).

No acknowledgement is required in Non-Confirmable messages, for a request of 166 bytes there are 237 bytes of total exchanged data (Fig. 28).

The main differences between MQTT, CoAP and REST are:

- MQTT uses a *publish/subscriber* paradigm while CoAP and REST uses a *request/ response* paradigm
- MQTT is a many-to-many protocol that uses a central broker to dispatch messages coming from the publisher to the subscribers instead in CoAP and REST there is a one-to-one client-server communication
- MQTT is an event-oriented protocol while CoAP and REST are more suitable for state transfer
- CoAP and REST require updates on status changes with periodic requests
- CoAP runs on top of the UDP while MQTT and REST run on top of TCP.

The IoT communication protocols analyzed use different approaches and are characterized by different performances and ways of use.

MQTT is the protocol with the best performances in terms of communication times, this has been clearly observed in the previous evaluations.

In terms of ease of use and client availability, the REST approach is the most used, MQTT is the easiest protocol to use with the availability of clients for various environments (Java, Android, Python, C, C ++, JavaScript).

In terms of protocol overhead, CoAP is the analyzed protocol with the best performance. CoAP needs the use of an HTTP-CoAP proxy for remote integration and this was a drawback for the protocol.

References

1. Rayes, A., Salam, S.: Internet of Things from Hype to Reality: The Road to Digitization. Springer (2019)
2. Evans, D.: The internet of things: how the next evolution of the internet is changing everything. CISCO White Paper **1**(2011), 1–11 (2011)
3. Al-Fuqaha, A., Guizani, M., Mohammadi, M., Aledhari, M., Ayyash, M.: Internet of things: a survey on enabling technologies, protocols, and applications. IEEE Commun. Surv. Tutor. **17**(4), 2347–2376 (2015)
4. Vashi, S., Ram, J., Modi, J., Verma, S., Prakash, C.: Internet of things (IoT): a vision, architectural elements, and security issues. In: 2017 International Conference on I-SMAC (IoT in Social, Mobile, Analytics and Cloud) (I-SMAC), pp. 492–496. IEEE (2017)
5. Salman, T., Jain, R.: Networking protocols and standards for internet of things. Internet of Things and Data Analytics Handbook, vol. 2015, pp. 215–238 (2015)
6. Karagiannis, V., Chatzimisios, P., Vazquez-Gallego, F., Alonso-Zarate, J.: A survey on application layer protocols for the internet of things. Trans. IoT Cloud Comput. **3**(1), 11–17 (2015)
7. Salman, T., Jain, R.: A survey of protocols and standards for internet of things. arXiv preprint arXiv:1903.11549 (2019)
8. Ferro, E., Potorti, F.: Bluetooth and Wi-Fi wireless protocols: a survey and a comparison. IEEE Wirel. Commun. **12**(1), 12–26 (2005)
9. Park, M.: IEEE 802.11 ah: sub-1-GHz license-exempt operation for the internet of things. IEEE Commun. Mag. **53**(9), 145–151 (2015)
10. Siekkinen, M., Hiienkari, M., Nurminen, J.K., Nieminen, J.: How low energy is bluetooth low energy? Comparative measurements with ZigBee, 802.15.4. In: 2012 IEEE Wireless Communications and Networking Conference Workshops (WCNCW), pp. 232–237. IEEE (2012)
11. Gomez, C., Paradells, J.: Wireless home automation networks: a survey of architectures and technologies. IEEE Commun. Mag. **48**(6), 92–101 (2010)
12. Stanford-Clark, A., Truong, H.L.: MQTT for Sensor Networks (MQTT-SN) Protocol Specification, vol. 1, p. 2. International Business Machines (IBM) Corporation Version (2013)
13. Locke, D.: MQ telemetry transport (MQTT) V3.1 protocol specification. IBM Developer Works Technical Library, vol. 15 (2010)
14. Saint-Andre, P.: Extensible messaging and presence protocol (XMPP): Core (2004)
15. Fielding, R.T., Taylor, R.N.: Architectural Styles and the Design of Network-Based Software Architectures, vol. 7. University of California, Irvine (2000)
16. Chen, X.: Constrained Application Protocol for Internet of Things. https://www.cse.wustl.edu/jain/cse574-14/ftp/coap (2014)
17. Bormann, C., Castellani, A.P., Shelby, Z.: CoAP: an application protocol for billions of tiny internet nodes. IEEE Internet Comput. **16**(2), 62–67 (2012)
18. Thangavel, D., Ma, X., Valera, A., Tan, H.-X., Tan, C.K.-Y.: Performance evaluation of MQTT and CoAP via a common middleware. In: 2014 IEEE Ninth International Conference on Intelligent Sensors, Sensor Networks and Information Processing (ISSNIP), pp. 1–6. IEEE (2014)
19. De Rango, F., Potrino, G., Tropea, M., Fazio, P.: Energy-aware dynamic Internet of Things security system based on Elliptic Curve Cryptography and Message Queue Telemetry Transport protocol for mitigating Replay attacks. In: Pervasive and Mobile Computing, vol. 61. Elsevier (2020)
20. Santamaria, A.F., Raimondo, P., Tropea, M., De Rango, F., Aiello, C.: An IoT surveillance system based on a decentralised architecture. Sensors **19**(6), 1469 (2019)

21. Santamaria, A.F., De Rango, F., Serianni, A., Raimondo, P.: A real IoT device deployment for e-Health applications under lightweight communication protocols, activity classifier and edge data filtering. Comput. Commun. **128**, 60–73 (2018)
22. Santamaria, A.F., Raimondo, P., De Rango, F., Serianni, A.: A two stages fuzzy logic approach for Internet of Things (IoT) wearable devices. In: 2016 IEEE 27th Annual International Symposium on Personal, Indoor, and Mobile Radio Communications (PIMRC), pp. 1–6. IEEE (2016)
23. De Rango, F., Tropea, M., Fazio, P.: Mitigating DoS attacks in IoT EDGE Layer to preserve QoS topics and nodes' energy. In: IEEE INFOCOM 2020—IEEE Conference on Computer Communications Workshops (INFOCOM WKSHPS), pp. 842–847. IEEE (2020)

Resource Allocation in Satellite Networks—From Physical to Virtualized Network Functions

Franco Davoli and Mario Marchese

Abstract The integration of satellite communications (SatCom) and networking into fifth generation wireless networks (5G) and their foreseen role in the forthcoming 6th generation (6G) has been gaining relevance, especially with the advent of nano-satellites. However, full SatCom integration into 5G and beyond requires carefully revisiting radio resource allocation techniques that have been developed and successfully applied in the satellite context to fit in the highly virtualized environment that characterizes these networks. This entails adaptation of both architectural and algorithmic aspects, in order to maintain end-to-end performance requirements in a highly heterogeneous networking framework. We consider in particular bandwidth allocation schemes and suggest modeling and architectural paradigms that appear as more promising in this respect.

Keywords 5G · 6G · Satellite communications · Resource allocation · Network management and control

1 Introduction

Fifth generation (5G) mobile networks (and beyond) will allow a much higher data rate per user with respect to the current one, a greater number of connected devices implying much higher traffic volume, a significantly reduced latency allowing a huge number of new services. The requirements of 5G networks will translate in Key Performance Indicators (KPIs), such as: bandwidth, spectrum efficiency, data

F. Davoli · M. Marchese (✉)
DITEN–University of Genoa, Genoa, Italy
e-mail: mario.marchese@unige.it

F. Davoli
CNIT National Laboratory of Smart and Secure Networks (S2N), Genoa, Italy

M. Marchese
CNIT Research Unit of the University of Genoa, Genoa, Italy

rate, latency, device density, area traffic capacity, reliability, availability, energy efficiency, coverage, mobility, positioning accuracy, security metrics, among others. 5G scenarios allow envisaging applications belonging to science fiction up to few years ago, in particular in the context of Smart Cities, Smart Industries, and Smart Farms, where a huge amount of "things" connected over a high-speed network will allow: to access medical devices remotely, to control ambulances, to have municipal command and control services, smart grids, home energy management, hospital optimization, transport optimization, automated car systems, but also factory process control and optimized agriculture. The pervasiveness of the smart devices together with a ubiquitous coverage will be employed to monitor both metropolitan and hazardous isolated areas by sensors, microphones, and videos. Special attention may be deserved to protect critical infrastructures which are complex systems and assets that provide services considered vital for the society as an integrated entity: Chemical, Commercial Facilities, Communications, Critical Manufacturing, Dams, Defense Industrial Base, Emergency Services, Energy, Financial Services, Food and Agriculture, Government Facilities, Healthcare and Public Health, Information Technology, Nuclear Reactors, Materials and Waste, Transportation Systems, Water and Wastewater Systems.

So, on one hand, 5G can make concrete the intuition of Mark Weiser [1] and the paradigm of pervasive computing depicting a world where a wide set of physical quantities (vibrations, heat, light, pressure, magnetic fields, …) are acquired through sensors and transmitted through suitable seamless communication networks for information, decision, and control aim [2]; on the other hand the introduction of new components, the increased data exchange and interaction, the presence of services that, in the past, were operated over separated networks unconnected to public communication infrastructures, but now have turned to exploit the services and data provided by telecommunications networks, increase the need of managing and understanding this huge amount of data. In this challenging environment the importance of Quality of Service (QoS, identified in 5G through the concept of KPIs) is evident and each application requires a specific level of assurance from the network that is characterized by a great heterogeneity: portions managed by different Service Providers; different transmission means such as cables, satellites, and radios; different implemented solutions from the protocol viewpoint, as well as different users who can require different services and may have different availability and willingness to pay for them. In practice, we can bring out again the old concept of "network of networks" originally used for the Internet.

In this challenging environment satellites can play multiple roles as portions of a heterogeneous end-to-end network thanks to their intrinsic ubiquity and broadcasting capabilities. They can act as a main single backhaul segment for rural areas, aircrafts, vessels, and trains, as additional backhauls to provide enhanced connectivity and, to improve service continuity, as pure transport networks. They can be applied in the edge computing scenario to exploit the unicast/multicast/broadcast geographical distribution of video, audio, and application software binaries simultaneously. Associated immediate outcomes are in the field of Smart Cities, Smart Industries, and Smart Farms.

The problem of end-to-end QoS has been deeply discussed in [3]: essentially QoS requests should traverse the overall network from the source to the destination through portions that implement different technologies and different protocols; QoS requests should be received and understood by each specific portion where QoS may have different meaning and interpretation, which depend on used protocols and network features; QoS requests should be managed by control mechanisms suited for the aim; each single QoS solution is composed of layered architectures and each layer must have a specific role in QoS provision. The overall problem of QoS interworking may be structured into two different actions: Vertical and Horizontal QoS Mapping. The former is based on the idea that a telecommunication network is composed of functional layers and that each single layer must have a role for end-to-end QoS provision. Consequently, it is necessary to define an interface between adjacent layers through which to offer a specific QoS service. The latter, even if much linked to the previous concept when implemented in the field, is represented by the need to transfer QoS requirements among network portions implementing their own technologies and protocols. Special tools called QoS gateways can take charge of that, by isolating specific network portions that deserve special attention and control actions such as traffic shaping, scheduling schemes, call admission control (CAC), QoS routing, and Resource Reservation, often declined as Bandwidth Reservation.

In this context, this contribution will focus on QoS Gateways that "open the door to satellite portions", as sketched in Fig. 1, from the point of view of bandwidth allocation, by analyzing the evolution in these last years. The problem is mainly linked to the concept of Vertical QoS Mapping and service provision between adjacent layers, clarified in the next Section, in particular between higher layers (IP and upper layers) and lower layers (data link and physical ones) implemented on satellite network cards. At the same time, we will also consider the integration of the satellite

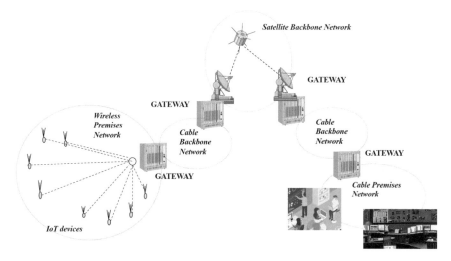

Fig. 1 End-to-end QoS service through satellite backhaul

segment in the upcoming 5G and Beyond 5G (B5G) scenario, and the evolution of the architectural solutions that were developed for vertical QoS mapping toward the virtualized networking environment that is a fundamental characteristic of this scenario.

2 Vertical QoS Mapping

An example of formal relation among layers and a clear example of vertical QoS mapping is represented by the protocol architecture proposed by ETSI [4] for the access points to a Broadband Satellite Multimedia (BSM) network portion. The architecture is reported in Fig. 2. The reference stack for the upper layers is the TCP/IP suite: application layer and transport layer implemented through either TCP or UDP or other protocols adapted for satellite communications and acting, for example, within TCP PEPs (Performance Enhancing Proxies). Either IPv4 or IPv6 act at the network layer and represent the Satellite Independent (SI) layers, i.e., those whose action is totally independent of the implementation details of the lower layers acting over the satellite link, even if IPv4/IPv6 must receive a QoS-service from the lower layers. Satellite physical and data link layers strictly dependent on the satellite features and on the specific data link (structured into Satellite MAC and Link Control sublayers, strictly satellite dependent—SD) are isolated from the rest by a Satellite Independent Service Access Point (SI-SAP), which should offer specific QoS services to the upper layers. The SD layers receive a service request from the SI layers through the SI-SAP. Satellite Dependent Layers are decoupled from Satellite Independent Layers by the SI-SAP interface. In the following, we will also use the more general terms Technology Independent (TI) and Technology Dependent (TD) to identify our vertical partitioning, when not referring explicitly to the satellite context.

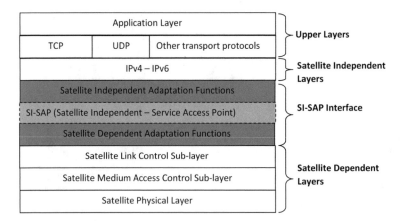

Fig. 2 ETSI BSM protocol architecture

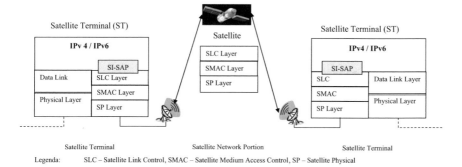

Fig. 3 SI-SAP location within a satellite network portion

Figure 3 identifies exactly the SI-SAP action point within satellite QoS Gateways, here corresponding to Satellite Terminals (STs) that give access to the satellite portion. IPv4/IPv6 has been used as relay layer because SI-SAP (even if it has a general meaning) has been formally defined only for IP-based networks, as shown in Fig. 3.

The ETSI definition implies the involvement of management, control and user planes within the framework of QoS Gateways. As said, a key point, in particular in satellite communications, is represented by resource allocation schemes. A set of queues is used to define QoS mapping operations performed at the SI-SAP interface, as similarly used in reference [5] and [6] (for DVB at the SD layer).

Along the lines drawn in [7] and [8], the control plane is composed of the following control blocks.

- The TI layer Resource Management Entity making resources available for the IP layer.
- The TD layer Resource Management Entity, which physically manages the necessary resources at the lower layers and should have feedback information from the physical state of the links. It may act together with a Network Control Centre (NCC), widely used in the case of radio and satellite networks where the bandwidth to be allocated is shared among different stations. The bandwidth to be allocated to each single station to match vertical QoS requirements may temporarily overcome the maximum limit capacity of the shared network.
- The QoS Mapping Management Entity (QoS Mapping Manager) that receives the resource allocation requests from the TI Resource Management Entity. A request may concern resource reservation, release, and modification. The communication between QoS Mapping Management and Resource Management Entity should be established through a proper interface. A proposal specific for satellite interfaces (SI-SAPs) comes from [7]. It refers to the composition of a set of primitives to establish the mentioned communication and is based on the creation of a set of abstract queues. Figure 4 contains the mentioned control modules and shows the decoupling between user and control plane, as well as the presence of the mentioned NCC.

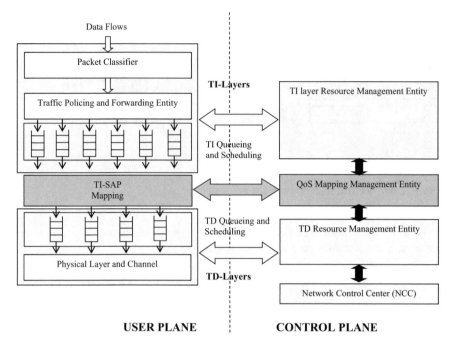

Fig. 4 Vertical QoS mapping: information forwarding and control module interaction

There are three problems arising from the action of two layers in cascade. The first two may be generically applied: change of information unit and aggregation of heterogeneous traffic. The last one is related to time varying channels. In the authors' experience it is related to satellite and radio communications. Concerning the change of information unit, at each layer, the information coming from the upper layer is encapsulated within a new frame composed of header and, possibly, trailer. It means that the TI layer packet accesses the TD queue after being encapsulated in a new frame. It is intuitive that the service rate at the TD layer must consider the additional bits of the header/trailer to keep a fixed level of service. As regards the aggregation of heterogenous traffic and referring to the satellite environment, as outlined in [9] concerning BSM systems, "it is accepted in the BSM industry that at the IP level (above the SI-SAP interface) between 4 and 16 queues are manageable for different IP classes. Below the SI-SAP these classes can further be mapped into the satellite dependent priorities within the BSM which can be from 2 to 4 generally". The due association of IP QoS classes to Satellite Dependent (SD) transfer capabilities is also limited by hardware implementation constraints. The bandwidth assigned to the queues acting at the lower layer must assure the QoS requirements to the TI queues, even if the traffic has been aggregated in a lower number of queues. It is important to note that bandwidth allocation at TD layer may be hardly standardized to avoid the violation of implementers' freedom, but it is still a very interesting technological and scientific problem that this chapter would like to treat, at least giving

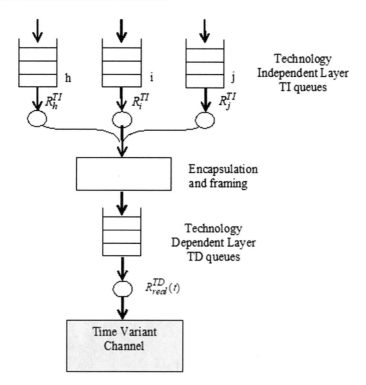

h   i   j

R_h^{TI} R_i^{TI} R_j^{TI}

Technology
Independent Layer
TI queues

Encapsulation
and framing

Technology
Dependent Layer
TD queues

$R_{real}^{TD}(t)$

Time Variant
Channel

Fig. 5 Joint model of vertical QoS mapping

some examples and theoretical indications. Finally, yet importantly, many transmission environments, such as satellite links, need to tackle time varying channel conditions due to fading. The overall model is reported in Fig. 5, by showing a set of queues identified by an index, let us say k, and served by a given bandwidth R_k^{TI}.[1] Traffic aggregation, together with the overhead imposed by encapsulation and framing, will require a specific bandwidth R^{TD} at the TD layer. From the mathematical viewpoint, the fading effect, of extreme importance for our purpose, may be modelled as a reduction of the bandwidth actually "seen" by the SD buffer, at least in satellite communications [10]. The reduction is represented by a stochastic process $\phi(t) \in [0, 1]$. At time t, the "real" service rate $R_{real}^{TD}(t)$ (available for data transfer) is $R_{real}^{TD}(t) = R^{TD} \cdot \phi(t)$, where time dependency is explicitly indicated to enforce the concept of varying channel conditions, as well as to consider traffic fluctuations.

---

[1] As is sometimes done in networking, we may refer to "bandwidth" either in terms of Hz or of bit rate. Obviously, these are not identical quantities; however, once fixed the channel characteristics, the type of modulation and the signal-to-noise ratio, a one-to-one correspondence can be established between the two, which also provides the spectral efficiency of the method in use [bits/s/Hz].

It is obvious that also other solutions based on off-line measures and on a reasonable overprovision can give good results. The problem is stimulating from the scientific viewpoint. A possible idea to solve the mentioned problems, limiting the information exchange between adjacent layers and without using off-line information will be presented in the next Section.

3   QoS Gateways for Satellite and Radio Communication

We will now focus the entire attention on a satellite network portion (to fix ideas, and without loss of generality, we consider a geostationary (GEO) satellite). The creation of the bandwidth pipe derives, in particular, from the action of bandwidth allocation, developed by the block identified as Network Control Centre (NCC). The problem is now how to allocate bandwidth to the satellite portion of the network and, in more detail, to the different earth stations. The reference general architecture is shown in Fig. 6.

To formally introduce the problem of bandwidth allocation, it is important to establish models to describe the satellite network portion, the bandwidth allocation scheme, and the channel behavior.

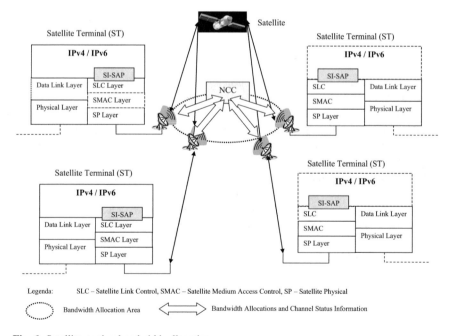

Fig. 6 Satellite portion bandwidth allocation

Fig. 7 Action of the
bandwidth allocation

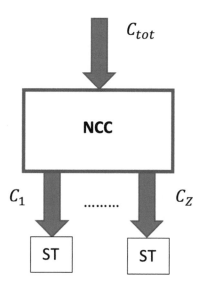

3.1 Network Topology and Bandwidth Allocation Scheme

The network portion considered is composed of Z Satellite Terminals (STs) ($Z = 4$, in
Fig. 6), modelled as nodes gathering traffic from the sources and connected through
a satellite connection. The control architecture is centralized: the NCC manages
the resources and provides the Satellite Terminals with portions C_Z of the overall
bandwidth C_{tot} (TDMA slots, for example, or any other kind of resource units), which
provide the maximum service rate of the satellite link. The NCC may be a single
physical device or a virtual function, envisaging application in the 5G environment.
The practical aim of the allocation is the guarantee of specific QoS requirements
(the Service Level Specification—SLS—offered by SD to SI layer) if the overall
available bandwidth allows it.

The general scheme of bandwidth allocation is shown in Fig. 7.

3.2 A Simple Channel Model

In satellite systems the degradation of the channel quality represents the cause of
detriment of communications services. In practice, satellite channels are affected
by the same problems of typical terrestrial wireless channels, but additionally they
are also significantly corrupted, if the transmission frequencies are high, by meteorological precipitations. In other words, satellite channels are typically affected by
rain fading, which is predominant at higher frequencies, especially above 10 GHz. It
has a negative impact on QoS. The meteorological state over an ST determines the
real availability of the channel bandwidth for transmissions. Fading effects may be

compensated (besides power allocation or diversity techniques) by a range of Forward Error Correction (FEC) coding schemes, useful at providing efficient broadband services working under different attenuation conditions. Depending on the fading conditions, the number of bits dedicated to FEC may be increased (or decreased), so improving (or not) the protection power against channel errors. In consequence of the change of the quantity of bits dedicated to protection (or of those dedicated to user information) per channel use, the size of the transported information changes. In dependence on meteorological conditions, Satellite Terminals may provide low-rate services involving powerful FEC coding schemes, when fading is severe, up to high-rate services when channel conditions are good.

So, the problem related to the usage of powerful coding schemes is the bandwidth reduction that determines the effective bandwidth availability. Due to the redundancy bits introduced at the physical layer, the capacity really "seen" by higher layers of the network is reduced, so creating possible bottlenecks. This condition may affect significantly the overall performance. The bandwidth allocation is used to compensate the reduction and to obtain simultaneously physical channel reliability.

The bandwidth reduction due to FEC may be modelled as a multiplicative factor of the overall bandwidth assigned to STs, coherently with reference [10]. The model has been presented in the previous section and it is reported again here by slightly changing the notation, so as to adapt it to bandwidth allocation and to the introduced network topology. The stress is now on the assignment of bandwidth and not on buffer service rate, even if, from the operative viewpoint, there is no difference at all: the assigned bandwidth is just the service rate "given" to the corresponding ST at a given time instant t. Mathematically, it means that the real bandwidth $C_z^{real}(t)$ available for the z-th ST is its nominal bandwidth $C_z(t)$ reduced by a factor $\phi_z(t)$, which is, in general, a variable parameter contained in the real numbers interval [0, 1]:

$$C_z^{real}(t) = C_z(t)\phi_z(t), \, \phi_z(t) \in [0, 1] \subset \mathbb{R}, z = 1, \ldots, Z \qquad (1)$$

A specific value $\phi_z(t)$ corresponds to a fixed attenuation level "seen" by the z-th ST at the instant t. The channel model maps the link layer corruption problem to a congestion problem. In this view, the capacity reduction has been considered through the factor $\phi_z(t)$ and the consequent assumption is that all the packet losses happening during communication may be supposed due to congestion events. On the other hand, in the conditions described, with the use of FEC, the loss due to link layer corruption may be supposed tending to zero and the simple proposed channel model seems to be a reasonable approximation of the satellite radio channel behavior. A numerical example of the mapping between the Carrier Power to One-Sided Noise Spectral Density Ratio P^C/N_0, $\phi_z(t)$, and used FEC is contained in Table 1 [10, 11], where the overall bit rate available for a carrier is fixed to 8.192 [Mbps]. The values reported in the table allow limiting the bit error probability below 10^{-7}.

In the scenario we consider, each fading level is supposed to happen with an associated probability and associated with a particular FEC. The interpretation of the reduction factor allows making important assumptions:

Table 1 Signal to noise ratio and related $\phi_z(t)$ level at a fixed instant t

$P^C/N_0[\text{dB}]$	$\phi_z(t)$	Code rate and bit rate per carrier
> 77.13	1	4/5, 8.192
74.63–77.13	0.8333	2/3, 8.192
72.63–74.63	0.625	1/2, 8.192
69.63–72.63	0.3125	1/2, 4.096
66.63–69.63	0.15625	1/2, 2048
< 66.63	–	Outage

- in satellite environments, the most common type of link corruption due to noise occurs in the above fashion and packet loss is due mainly to it;
- nevertheless, if FEC schemes are used, the link corruption may be considered as a congestion event.

In practice, increasing FEC bits, the errors due to a faded link may be neglected but, at the same time, the available bandwidth for information is reduced, so creating possible bottlenecks and consequent loss. The considered simple model actually neglects the instantaneous dynamics of fading and, in this view, it should be substituted with a more precise channel model. Nevertheless, within the described framework, the assumption is not too severe and seems a reasonable approximation of the satellite channel behavior, at least concerning bandwidth control algorithms [10–12].

3.3 Bandwidth Allocation Algorithms

The computation of individual $C_z(t)$ allocations at instant t depends on the total amount of resources $C_{tot}(t)$ available at the NCC and on the amounts requested by the STs at the specific decision time. The latter may stem from the vertical mapping mechanism that has to decide how to translate the service rates R_k^{TI}, $k = 1, \ldots, K$, for the Technology Independent service queues (determined by the Service Level Agreement (SLA) negotiated with the specific traffic flow each one is serving) into the aggregate service rate $R_{real}^{TD}(t)$ (considering, without loss of generality, mapping to a single TD queue), which takes also into account the channel conditions, in order to maintain the performance requirements (e.g., in terms of packet loss rate and average delay per packet) as close as possible to the values required by the SLAs.

Before providing a hint to a specific technique to perform this operation, whose treatment is reported in [3] and [13], it is worth making a general remark pertaining to this setting. Indeed, the specific architecture we have described lends itself almost naturally to a hierarchical control structure, where local decisions are taken at the earth stations or at the satellite gateways, in correspondence of a certain bandwidth assignment, and a global optimization is performed by the NCC, which plays the role of a coordinator in deciding the bandwidth allocation, which may be periodically updated if channel or traffic conditions change significantly. This vision has been followed by many papers that appeared in the first decade of the years 2000; some of them are surveyed, among others, in [14] and [15]. Regarding the local optimization, two alternative possible approaches are the aforementioned vertical QoS mapping (a different formulation of which has been also considered in [16, 17]), where differentiated queues are present for QoS categories, or a rougher division among real-time and best-effort traffic, as taken for instance in [10, 11, 18].

Whereas the approaches taken in the above-mentioned references are based on analytical queueing models, the mechanism adopted in [3] and [13] is based on Infinitesimal Perturbation Analysis (IPA), in the mathematical framework of Stochastic Fluid Models [19, 20]. Under a fluid dynamic representation of the queueing system in terms of inflow and outflow rates, suitable cost functions to be minimized are defined for the SD layer. They reflect the quadratic deviation of loss and workload, at the SD layer, from the same quantities measured at the SI layer, used as thresholds, in order to try to follow (chase) the performance at the SI layer. Following IPA, estimations of the gradient of the cost functions are obtained in real time on the basis of traffic samples acquired during the system evolution and descent steps are taken along a realization of the underlying stochastic processes.

Regarding instead the centralized NCC problem in the presence of analytical traffic and system models, it can be solved either through ordinary numerical minimization, by considering the bandwidth portions to be assigned as continuous variables, or through discrete optimization, by applying a computationally efficient dynamic programming algorithm [18], or even by posing it as a multi-objective optimization problem [12]. However, fluid models and IPA can be applied also in this setting, with the advantage of not requiring an analytical model or knowledge of the underlying stochastic processes [21, 22]. In this perspective, we outline a discussion in the next section that will lead us to consider Machine Learning (ML) implementations, which, as we will see later on, fit with the current architectural evolution towards network virtualization and with the increasing complexity of the problems at hand, caused by the growth in the number of devices, users and user-generated traffic volumes.

4   From Analytical Modeling to ML Techniques

As we have seen in the bandwidth allocation problems above, traditional approaches to network management and control often rely on analytical models. These entail some knowledge of the physical phenomenon under study, which may happen in a Cyber-Physical System (CPS) or in the elements of the network itself, like buffers, links, gateways, routers and switches, network processors, etc., along with a characterization of the traffic flows through such elements, possibly with different granularities (typically, flow- or packet-level).

Typical (stochastic) dynamic models in these settings may be in the form of Markov processes or also non-Markovian queueing systems. Depending on the time scales of the phenomenon of interest, it may be necessary to represent a non-stationary or stationary behavior (i.e., where the probability distributions do not change in time, at least over a certain time horizon). Then, if one wants to formulate an optimal control problem for the system under consideration, different performance (or cost) functions can be constructed that reflect a property or KPI of interest (often in the form of expected values). In the dynamic non-stationary case, the control law (i.e., the function mapping the system state or observations thereof to control actions) will be influenced by the system dynamics. Typically, this will be the case of Markov Decision Processes (MDPs) and, depending on the complexity of the problem, the form of the control law may be left general and unknown (giving rise to a *functional* optimization problem that may be solved analytically in very few cases), or it may be fixed a priori, limiting the optimization to control strategies of a specific form (e.g., an on–off—also called "bang-bang"—control law, a threshold function, a linear or affine one, ...), which may be represented by a finite number of parameters, transforming the optimization problem into a *parametric* one to be solved by numerical methods. Most of the analytical approaches to bandwidth allocation we mentioned in the previous section fall under this category. In all these cases, where the performance indexes may be expressed by averages computed over a finite or infinite time horizon, various techniques are available to decompose or simplify the problem (Dynamic Programming, Open-Loop-Feedback or "Receding Horizon" controls, etc.; see e.g. [23]). A situation of parametric optimization is found also in the presence of stationary queueing models, if the control can be expressed in terms of parameters (representing, e.g., load balancing coefficients, bandwidth partitions, processors' energy states, ...), as in some of the works we have cited above.

Somehow intermediate to the two approaches described (MDP with functional optimization or parametric optimization), are other ones that employ different techniques that, though possibly still based on models, require less knowledge on the stochastic phenomena underlying the system dynamics. One of these is that of Neely [24], based on Lyapunov optimization, which has actually been applied in a Low Earth Orbit (LEO) satellite context, where energy efficiency was accounted for as one of the KPIs [25] (see, e.g., [26, 27] for energy efficiency / performance trade-offs in satellite communications and networking). A concept that somehow captures "... a summary of the statistical characteristics of sources over different time and space scales" [28] (even in the case of poorly characterized traffic, as noted by the author) is that of effective (or equivalent) bandwidth. Though not in the satellite environment, another analytical approach to rate control that does not require a statistical characterization of traffic is that of [29].

Such analytical approaches have been at the basis of numerous successful achievements in network management and control, under the general context of *Teletraffic Theory*. However, the network softwarization phenomenon and the ensuing introduction of virtualization concepts in networking, along with new traffic characteristics, have rendered the complexity of network management and control problems much higher than it used to be in legacy networks with special-purpose hardware. Networks have become much more akin to computer systems, and the boundary between computation and communication tasks has turned out to be less well-defined. On the other hand, while tasks in computer systems are under the control of a central unit, the nature of computational processes occurring in networks is much more distributed. A chain of virtual network functions performing a network service is composed of multiple virtual elements that may be physically separated and, moreover, reside on shared hardware, where access to the physical resources is mediated by a hypervisor; performance requirements and QoS are often expressed and measured in end-to-end terms. For instance, regarding energy consumption KPIs, it becomes much less straightforward to attribute part of the energy consumed by the hardware to a specific virtual entity, in order to trade-off energy with performance indicators on the basis of clear analytical models. Moreover, with the growing complexity and distribution of network functions and services, the need arises for autonomous mechanisms that can perform network management and control operations autonomically and with minimal human intervention.

As was noted in [22], among others, quite a few optimization problems for resource allocation in networking have a discrete stochastic programming nature. Without consideration of the system's dynamics, which may add further complexity, a general characterization of these problems is in terms of a vector θ of decision variables, often in the form of nonnegative integers, which must be modified along time as outputs of the control policies, in order to optimize the system performance (e.g., our previously considered bandwidth allocation variables). In very broad terms, one looks for the vector θ^* such that

$$\theta^* = \underset{\theta \in \Theta}{\operatorname{argmin}} \, \underset{\omega}{E} \{L(\theta, \omega)\} \tag{2}$$

where ω represents a vector of stochastic variables, E denotes statistical expectation (over all possible realizations of the stochastic variables), $L(\cdot, \cdot)$ is a function representing a metric of interest (e.g., average delay per packet, packet loss probability, or, in the case of flows, blocking probability) and Θ is a constraint set. Still in many cases, the random variables may have a poor statistical characterization, the cost function may not be expressed analytically or, even if it is, the expectation may be very difficult to compute, requiring the use of gradient approximation techniques (with suitable relaxations of the problem in the case of integer decision variables), where a descent step is performed along a *realization* of the stochastic variables. Note that, in general, the vector θ may represent a set of parameters, but also the output of decision strategies mapping observations to control actions (which would render the problem one of *functional* optimization). Another drawback in such problems, in the presence of system dynamics, may be the difficulty of obtaining closed-form functional costs that can lead to closed-loop control strategies, limiting the search to parameter-adaptive *certainty equivalent* control [23].

We have seen that in this context, IPA provides sensitivity algorithms that allow estimation of the gradient of the cost function, on the basis of observations of the sample paths followed by the underlying stochastic process. In the case of discrete variables, suitable relaxations can be constructed (*online surrogate optimization methodologies*) that allow the use of such gradient descent techniques [30, 31]. However, such pure gradient descent optimization techniques may perform well in parametric optimization problems under a stationary behavior of the stochastic processes but may be less efficient in facing control problems where the decision variables are functions of information acquired in successive instants and the system evolution is described by dynamic equations. In [22] the IPA-based online surrogate optimization problem in a satellite environment has been compared to a solution obtained by using neural networks, whereas the first technique has been adopted in a similar setting in [21].

Though the above-mentioned IPA techniques are not ordinarily classified as ML mechanisms, nevertheless they are representative of control approaches that deal with difficulties in obtaining an analytic system and cost representation. ML approaches represent a class of more powerful techniques, and also have the advantage of not necessarily requiring an underlying model, as they are by definition based on numerical approximations. In many cases, what is required is the parametric approximation of an unknown function depending on multiple variables, by means of linear combinations of *Fixed Structure Parametrized* (FSP) [32] functions (often represented by nonlinear functions containing parameters to be tuned inside such basis functions). Among them, Neural Networks (NNs) and, more recently, Deep Neural Networks [33, 34] have become the most popular ML mechanisms based on FSP functions. Two recent works that survey this field in relation to wireless networks are [35] and [36].

Regarding NN-based ML architectures, however, a clear distinction and a thorough analysis is performed in [32] between the use of FSP functions to approximate a function of several variables (which may be known only pointwise), with a certain approximation error with respect to a given norm (which is one of the most common applications of NNs, e.g., in the recognition of features or objects), and the use of FSP functions in the approximation of multi-variable functions that constitute the solution to an Infinite Dimensional Optimization (IDO) problem, like the *functional* optimization problems arising in closed-loop optimal control. The *Extended Ritz Method* (ERIM) introduced in [32] reveals to be in many cases a powerful mechanism for this type of problems, many of which appear also in networking. Basically, by means of ERIM, the (functional) solution of many complex control and decision problems can be approximated by families of FSP functions, by expressing the approximation as a linear combination of fixed structure FSP basis functions, where parameters also appear within the basis functions themselves (e.g., One Hidden Layer (OHL) or Multi Hidden Layer (MHL) neural networks). A potential advantage here is that in quite a few cases of IDO control problems, a family of approximating functions can be found that allows avoiding the so-called "curse of dimensionality": namely, the growth in the dimension of the parametrization with increasing number of variables the function to be approximated depends on. It is worth noting that the nonlinear optimization problems stemming from this approach to the solution of IDO control problems can be solved "off-line", either at the edge itself in micro-datacenters or in the datacenter of a remote cloud. However, once computed the parameters of the approximating functions (if the iterations of the nonlinear program converge within the limits of a fixed approximation error), the approximated control law is available and, for each realization of variables upon which it performs its mapping, the calculation of the corresponding control actions is straightforward. As in all parametrized problems, decision strategies implemented in this way can be applied as long as the underlying system structure or the statistics of the random variables characterizing the input traffic do not change significantly (within certain tolerance margins: a significant change in these quantities would imply the necessity of recalculating the parameters of the approximating functions). Another potential power of the method is that it can be applied also in informationally decentralized control problems with multiple decision makers (DMs), which are usually very difficult to solve analytically, especially in the team theory context and in the presence of constraints, where the DMs have a common goal but different online information.

Still in the control setting, other widely adopted uses of neural FSP functions are in *Neuro-Dynamic Programming* [37] (which has similarities with *Reinforcement Learning*—a name mostly used in the Artificial Intelligence (AI) literature, or *Q-Learning*; see [38] for a survey), and *Approximate Dynamic Programming* for optimal control problems over finite or infinite time horizons [32]. Though differing in various respects, all these techniques construct an approximation of the *cost-to-go* of Dynamic Programming (DP). A brief description may be in order here (to fix ideas, we refer to the minimization of a cost function). In a DP decomposition, the basic step is to decompose an optimization problem into stages, and perform a minimization at each stage, by accounting for the effect of the control strategy so computed on the future stages of the problem. If the application of the control at a current state value and the state itself are associated a cost that depends only on them and the total cost is a sum of such terms, the decomposition can be applied by proceeding backwards, and by minimizing at each step t the sum of the stage cost and of the optimal cost corresponding to all forward stages from the current one, which is a function of the state value at stage $t + 1$. This second term in the minimization is exactly what is termed *cost-to-go*. In different ways, all the techniques we have mentioned in this last paragraph of the Section construct functional approximations of this term, which may be very difficult, if not impossible, to compute analytically in the presence of nonlinearities in the state equations, non-quadratic cost and non-Gaussian noise, besides the possible discretization of the state variables when their components range over a subset of the reals. Depending on the technique adopted, the system dynamics may be known or even partially or completely unknown.

5 Toward a Functional Architecture for 5G-Integrated Satellite Networking

The evolution in algorithms and techniques that we have outlined in the previous Section is in line with the architectural paradigms that accompanied the development of 5G and the growing integration between the mobile wireless and fixed network segments, and that have been spreading at a fast pace. Network Functions Virtualization (NFV) [39], Software Defined Networking (SDN) [40] and Mobile Edge Computing (MEC) [41] are among the main concepts behind such paradigms, which have spread from the cloud to embrace the networking domain and have brought forth the abstraction of network slices [42]. In this framework, as we noticed in [43], *"[t]he business/operational support systems (BSSs/OSSs) of upcoming 5G network platforms are meant to expose "customized" and isolated virtual projections of the mobile network (i.e., network slices) to vertical industries and OTT players, so as to enable them to run their applications and services on top of these network slices. To this end, a network slice is composed of a number of logical subnetworks that can have different roles and configurations. Such subnetworks can be instantiated as "private" network projections inside the slice, or shared among multiple slices (e.g., to attach multiple slices to the same radio access network)."*

The flexible and programmable networking environment stemming from this scenario lends itself, on one hand, to the full integration of the satellite segment and, on the other, to the adoption of the novel ML techniques for resource allocation that we have mentioned. With regard to the first aspect, quite a few works, besides [43], have addressed the issue (see, e.g., [44–46]). However, in the evolutionary perspective we have undertaken, it is worth outlining here the conceptual connection between the BSM architectural view we have started from and the novel architectural concepts that allow the service orchestration and logical separation (at both application and network levels) at the basis of the new paradigms.

Orchestration is needed to handle the complexity of application services that are designed and created as chains of micro-services at the application level and as chains of Virtual Network Functions (VNFs) at the network level (where a relevant framework is provided in particular by ETSI MANO—Management and Orchestration [47]). An important aspect, however, not always properly evidenced, is that of the separation of concerns between applications' and network functions' orchestrators, which has been stressed specifically by the recently concluded H2020 5G PPP European Project MATILDA [48, 49]. More specifically, the domain of vertical cloud-native applications [50], empowered with the service mesh concept [51, 52] and with suitable *sidecar proxies* [53] that allow the application developer to extend the micro-services capabilities with the specification of their communications needs, should be the concern of a Vertical Application Orchestrator (VAO). The VAO should allow application providers and application developers to operate with the mechanisms of the cloud environment they are used to; however, at the same time, it should enable them to fully exploit the advanced communication capabilities offered by 5G, by abstracting the physical network with the slice concept, transparently with respect to the heterogeneous underlying physical infrastructure (including the satellite segment) and providing the means to convey their communication needs and constraints to the Telecommunication Service Provider (TSP), and to constantly maintain this interaction during the lifecycle of their applications. Through the mediation of the OSS the TSP can receive the specifications that characterize a particular vertical application via a *slice intent*, and has the task to configure, deploy and manage the needed resources for the creation of the slice (by means of the Virtual Network Functions

Orchestrator—NFVO—provided in the MANO framework), which is then exposed to the VAO through a well-defined Northbound interface, and can be monitored and reconfigured, if necessary, to maintain QoS requirements.

We advocate that a similar separation concept is intrinsic to the BSM architecture we have briefly described in Sect. 2, entailing a clear separation between SI and SD (or, more generally, TI and TD) layers, with the interaction granted through the SI-SAP (or, more generally, TI-SAP). As such, this architectural perspective can play a similar role within the network, by separating what is closer to the physical infrastructure, where functionalities may be implemented by means of a mix of VNFs and Physical Network Functions (PNFs), orchestrated by a Satellite Dependent Orchestrator (SDO), from what pertains to the VNFs in the Technology-Independent layer. We believe that this separation concept could greatly foster the integration of the satellite segment within 5G and beyond, along the lines we have sketched in [43].

An additional remark is in order here to stress the relevance of a specific type of satellite communications in this integration. So far, to fix ideas, we have almost implicitly referred to GEO satellites. However, the growing importance of LEO small- and nano-scale satellite constellations [54] is hard denying. They are part of what is termed the "Internet of Space Things" in [55] and will constitute an essential complement to the terrestrial wireless part. In that framework, characterized by the presence of multiple constellations, intermittent connectivity, intersatellite links and, in general, more distributed data and control plane functionalities, the SI-SD separation acquires even more momentum.

We report in Fig. 8 an abstract view of the architectural layout we have described, referring to a satellite constellation with the presence of inter-satellite links, and highlighting the domains spanned and the presence of the various orchestration levels, along with the mediation points between them (OSS and NCC, as regards the VAO-NFVO and NFVO-SDO interactions, respectively). The creation of a slice in this case would start from the request of the VAO, passed along to the NFVO through the slice intent; in the case that satellite resources were needed, the NFVO would request the appropriate configuration to the SDO via the NCC.

Finally, we have already mentioned some of the reasons that would push towards the adoption of the ML techniques that were briefly surveyed in Sect. 4 in flexible and programmable networks. There is actually a mutual influence between the "softwarized" networking and application environment that was described above and the adoption of optimization techniques for resource allocation, management and control based on powerful functional approximation methodologies, especially in the Edge, where computational intelligence is being moved. In this respect, reference [36] contains an interesting discussion about ML for Communications (MLC) and Communications for ML (CML), and the vision presented in [56, Chap. 4], where "AI-as-a-Service" (AIaaS) is envisaged, is in line with it.

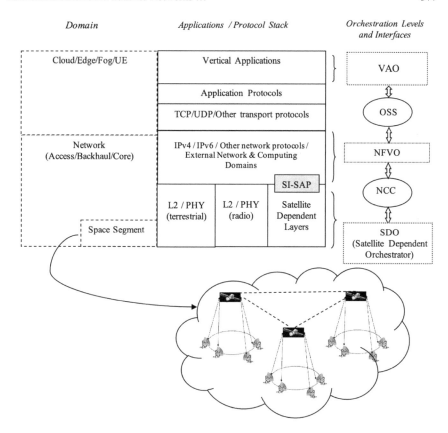

Fig. 8 Architectural layout with the inclusion of the satellite segment and its orchestration

6 Conclusions

We have attempted in this chapter to merge the vision of flexible and programmable networks that already permeates the networking scenario in the evolution toward the Next Generation Internet, which entails a growing integration between the mobile wireless and the fixed network (as already happens in 5G and is bound to strengthen in 6G), and the satellite segment. In doing so, we have taken a historical perspective and we have mixed architectural and algorithmic considerations.

Starting from the ETSI BSM vision of the early 2000s, we have outlined its relevance in the analytical and numerical approaches that marked the development of resource allocation (mainly, in terms of bandwidth) prior to the advent of network softwarization. We have then examined some of the reasons behind the adoption of ML techniques to perform the same operations in a more powerful fashion and, finally, we have tried to match these techniques to the new architectural paradigms stemming from network softwarization.

We believe that there are plenty of opportunities for research in this field to revisit some of the problems that were posed in the earlier context and to devise the new solutions and approaches that are made possible by the joint action of architectural and computational advances.

References

1. Weiser, M.: The computer for the 21st century. Sci. Am. **265**(3), 94–104 (1991)
2. Marchese, M.: Interplanetary and pervasive communications. IEEE Aerosp. Electr. Syst. Mag. **26**(2), 12–18 (2011)
3. Marchese, M.: QoS over Heterogeneous Networks. Wiley, Chichester, UK (2007)
4. ETSI: Satellite Earth Stations and Systems (SES). Broadband Satellite Multimedia, IP over Satellite. ETSI Technical Report, TR 101 985, V.1.1.2 (2002)
5. ETSI: Network Aspects (NA); General Aspects of Quality of Service (QoS) and Network Performance (NP). ETSI Technical Report, ETR 003, 2nd Ed (1994)
6. Combes, S., Goegebeur, L., Sanier, N., Fitch, M., Hernandez, G., Iouras, A., Pirio, S.: Integrated resources and QoS management in DVB-RCS networks. In Proc. AIAA Intern. Commun. Sat. Syst. Conf. (ISSCS04), Monterey, CA, USA (2004)
7. ETSI: Satellite Earth Stations and Systems (SES), Broadband Satellite Multimedia (BSM) Services and Architectures, Interworking with RSVP-based QoS (IntServ). ETSI Technical Specification, TS 102 463, V.0.4.2 (2006)
8. ETSI: Satellite Earth Stations and Systems (SES), Broadband Satellite Multimedia (BSM) Services and Architectures, Interworking with DiffServ QoS. ETSI Technical Specification, TS 102 464, V0.4.1 (2006)
9. ETSI: Satellite Earth Stations and Systems (SES). Broadband Satellite Multimedia (BSM) Services and Architectures; BSM Traffic Classes. ETSI Technical Specification, TS 102 295 V1.1.1 (2004)
10. Celandroni, N., Davoli, F., Ferro, E.: Static and dynamic resource allocation in a multiservice satellite network with fading. Int. J. Sat. Commun. **21**(4–5), 469–488 (2003)
11. Bolla, R., Davoli, F., Marchese, M.: Adaptive bandwidth allocation methods in the satellite environment. In Proc. IEEE Int. Conf. Commun. (ICC), Helsinki, Finland, pp. 3183–3190 (2001)
12. Bisio, I., Marchese, M.: E-CAP-ABASC versus CAP-ABASC: comparison of two resource allocation strategies in satellite environment. Int. J. Space Commun. **19**(3–4), 171–182 (2005)
13. Marchese, M., Mongelli, M.: On-line bandwidth control for quality of service mapping over satellite independent service access points. Comp. Netw. **50**(12), 2089–2011 (2006)
14. Giambene, G. (ed.): Resource Management in Satellite Networks: Optimization and Cross-Layer Design. Springer, New York, NY (2007)
15. Barsocchi, P., Celandroni, N., Ferro, E., Gotta, A., Davoli, F., Giambene, G., González Castaño, F.J., Moreno, J.I., Todorova, P.: Radio resource management across multiple protocol layers in satellite networks: a tutorial overview. Int. J. Sat. Commun. Netw. **23**(5), 265–305 (2005)
16. Le-Ngoc, T., Leung, V., Takats, P., Garland, P.: Interactive multimedia satellite access communications. IEEE Commun. Mag. **41**(7), 78–85 (2003)
17. Iouras, N., Le-Ngoc, T.: Dynamic capacity allocation for quality-of-service support in IP-based satellite networks. IEEE Wir. Commun. Mag. **12**(5), 14–20 (2005)
18. Celandroni, N., Davoli, F., Ferro, E., Gotta, A.: Adaptive cross-layer bandwidth allocation in a rain-faded satellite environment. Int. J. Commun. Syst. **19**(5), 509–530 (2006)
19. Cassandras, C.G., Sun, G., Panayiotou, C.G., Wardi, Y.: Perturbation analysis and control of two-class stochastic fluid models for communication networks. IEEE Trans. Automat. Contr. **48**(5), 23–32 (2003)

20. Wardi, Y., Melamed, B., Cassandras, C.G., Panayiotou, C.G.: Online IPA gradient estimators in stochastic continuous fluid models. J. Optim. Theory Appl. **115**(2), 369–405 (2002)
21. Davoli, F., Marchese, M., Mongelli, M.: Discrete stochastic programming by infinitesimal perturbation analysis: the case of resource allocation in satellite networks with fading. IEEE Trans. Wir. Commun. **5**(9), 2312–2316 (2006)
22. Baglietto, M., Davoli, F., Marchese, M., Mongelli, M.: Neural approximation of open loop–feedback rate control in satellite networks. IEEE Trans. Neur. Netw. **16**(5), 1195–1211 (2005)
23. Bertsekas, D.P.: Dynamic Programming and Optimal Control, vol. I, 4th edn. Athena Scientific, Nashua, NH, USA (2017)
24. Neely, M.J.: Stochastic Network Optimization with Application to Communication and Queueing Systems. Morgan & Claypool, San Rafael, CA, USA (2010)
25. An, Y., Li, J., Fang, W., Wang, B., Guo, Q., Li, J., Li, X., Du, X.: EESE: energy-efficient communication between satellite swarms and earth stations. In: Proc. 16th IEEE Int. Conf. Adv. Commun. Technol. (ICACT 2014), PyeongChang, Korea, pp 845–850 (2014).
26. Alagöz, F., Gür, G.: Energy efficiency and satellite networking: a holistic overview. Proc. IEEE **99**(11), 1954–1979 (2011)
27. Davoli, F.: Satellite networking in the context of green, flexible and programmable networks. In: Bisio, I. (ed.) LNICST 148, pp. 1–11. Springer (2016)
28. Kelly, F.P.: Notes on effective bandwidth. In: Kelly, F.P., Zachary, S., Ziedins, I. (eds.) Stochastic Networks: Theory and Applications, pp. 141–168. Oxford University Press, Oxford, UK (1996)
29. Kelly, F.P., Maulloo, A.K., Tan, D.K.H.: Rate control for communication networks: shadow prices, proportional fairness and stability. J. Op. Res. Soc. **49**, 237–252 (1998)
30. Gokbayrak, K., Cassandras, C.G.: Online surrogate problem methodology for stochastic discrete resource allocation problems. J. Opt. Theory Appl. **108**(2), 349–376 (2001)
31. Gokbayrak, K., Cassandras, G.C.: Generalized surrogate problem methodology for online stochastic discrete optimization. J. Opt. Theory Appl. **114**(1), 97–132 (2002)
32. Zoppoli, R., Sanguineti, M., Gnecco, G., Parisini, T.: Neural Approximations for Optimal Control and Decision. Springer Nature, Cham, Switzerland (2019)
33. Goodfellow, I., Bengio, Y., Courville, A.: Deep Learning. MIT Press, Cambridge, MA, USA (2016)
34. Liu, W., Wang, Z., Liu, X., Zeng, N., Liu, Y., Alsaadi, F.: A Survey of deep neural network architectures and their applications. Neurocomp. **234**, 11–26 (2017)
35. Chen, M., Challita, U., Saad, W., Yin, C., Debbah, M.: Artificial neural networks-based machine learning for wireless networks: a tutorial. IEEE Commun. Surv. Tut. **21**(4), 3039–3071 (2019)
36. Park, J., Samarakoon, S., Bennis, M., Debbah, M.: Wireless network intelligence at the edge. Proc. IEEE **107**(1), 2204–2239 (2019)
37. Bertsekas, D.P., Tsitsiklis, J.N.: Neuro-Dynamic Programming. Athena Scientific, Belmont, MA, USA (1996)
38. Kiumarsi, B., Vamvoudakis, K.G., Modares, H., Lewis, F.L.: Optimal and autonomous control using reinforcement learning: a survey. IEEE Trans. Neur. Netw. Learn. Syst. **29**, 2042–2062 (2018)
39. Mijumbi, R., Serrat, J., Gorricho, J.-L., Bouten, N., De Turck, F., Boutaba, R.: Network function virtualization: state-of-the-art and research challenges. IEEE Commun. Surv. Tut. **18**(1), 236–262 (2016)
40. Kreutz, D., Ramos, F.M.V., Veríssimo, P.E., Rothenberg, C.E., Azodolmolky, S., Uhlig, S.: Software-defined networking: a comprehensive survey. Proc. IEEE **103**(1), 14–76 (2015)
41. ETSI GS MEC 003: Mobile Edge Computing (MEC); Framework and Reference Architecture, version 1.1.1 (2016)
42. Ordóñez Lucena, J., Ameigeiras, P., Lopez, D., Ramos-Muñoz, J.J., Lorca, J., Folgueira, J.: Network slicing for 5G with SDN/NFV: concepts, architectures, and challenges. IEEE Commun. Mag. **55**(5), 80–87 (2017)
43. Boero, L., Bruschi, R., Davoli, F., Marchese, M., Patrone, F.: Satellite networking integration in the 5G ecosystem: research trends and open challenges. IEEE Network **32**(5), 12–18 (2018)

44. Liu, J., Shi, Y., Zhao, L., Cao, Y., Sun, W., Kato, N.: Joint placement of controllers and gateways in SDN-enabled 5G-satellite integrated network. IEEE J. Select. Areas Commun. **36**(2), 221–232 (2018)
45. Ahmed, T., Dubois, E., Dupé, J.B., Ferrùs, R., Gélard, P., Kuhn, N.: Software-defined satellite cloud RAN. Int. J. Sat. Commun. Netw. **36**(1), 108–133 (2018)
46. Gardikis, G., Koumaras, H., Sakkas, C., Koumaras, V.: Towards SDN/NFV-enabled satellite networks. Telecommun. Syst. **66**(4), 615–628 (2017)
47. ETSI GS NFV-IFA 031 V3.4.1: Network Functions Virtualisation (NFV) Release 3; Management and Orchestration; Requirements and interfaces specification for management of NFV-MANO (2020)
48. Bolla, R., Bruschi, R., Davoli, F., Fotopoulou, E., Gouvas, P., Tsiolis, G., Vassilakis, C., Zafeiropoulos, A.: Design, development and orchestration of 5G-ready applications over sliced programmable infrastructure. In: Proceedings of 1st International Workshop on Software Infrastructure for 5G and Fog Comp. (Soft5 2017), in conjunction with 29th International Teletraffic Congress (ITC 29), Genoa, Italy, pp 13–18 (2017)
49. https://www.matilda-5g.eu/. Accessed 28 Nov 2020
50. GitHub (CNCF) (2018): CNCF Cloud Native Definition v1.0. https://github.com/cncf/toc/blob/master/DEFINITION.md. Accessed 28 Nov 2020
51. Li, W., Lemieux, Y., Gao, G., Zhao, Z., Han, Y.: Service mesh: challenges, state of the art, and future research opportunities. In: Proceedings 2019 IEEE International Conference on Service-Oriented System Engineering (SOSE), San Francisco East Bay, CA, USA (2019)
52. Calçado, P.: Pattern: Service Mesh (2017). http://philcalcado.com/2017/08/03/pattern_service_mesh.html Accessed 28 Nov 2020
53. The MATILDA Consortium: Chainable application component and 5Gready application graph metamodel. Deliverable D1.2 (2018). https://private.matilda-5g.eu/documents/PublicDownload/192. Accessed 28 Nov 2020
54. Davoli, F., Kourogiorgas, C., Marchese, M., Panagopoulos, A., Patrone, F.: Small satellites and cubesats: survey of structures, architectures, and protocols. Int. J. Sat. Commun. Netw. **37**(4), 343–359 (2019)
55. Akyildiz, I.F., Kak, A., Nie, S.: 6G and beyond: the future of wireless communications systems. IEEE Access **8**, 133995–134030 (2020)
56. European Technology Platform NetWorld2020: Smart networks in the context of NGI. Strategic Research and Innovation Agenda 2021–27 (2020)

Cybersecurity

Secure D2D in 5G Cellular Networks: Architecture, Requirements and Solutions

Man Chun Chow and Maode Ma

Abstract Allowing nearby mobile devices to communicate directly without relaying the data through the conventional cellular network, device-to-device (D2D) communication is expected to improve the overall network efficiency of the 5G network and enable futuristic applications. In recent years, 3GPP has also put D2D communication under the umbrella of Proximity-based Services (ProSe) to accelerate its development. However, since D2D networks could introduce additional security issues, it is unprecedentedly important to study and mitigate the risk carefully. In this chapter, we review the security architecture, security requirements and existing solutions for the 5G D2D networks. First, we give an overview of the security architecture of the 3GPP ProSe Service. Then, we classify various security challenges and state the requirements of a secure 5G D2D network. Subsequently, we classify the major research work according to their application scenarios and present the main ideas of some of the works with security analysis. Finally, we discuss some open research issues to inspire future researchers to build more secure and robust 5G D2D solutions.

Keywords 5G · ProSe · D2D · Authentication · Network security

1 Introduction

In recent years, mobile internet traffic has soared tremendously due to the rapidly increasing number of mobile subscribers, the popularity of multimedia content and the deployment of smart Internet of Things (IoT) devices. As the conventional mobile network becomes heavily congested, the next generation mobile network with faster speed, larger capacity and more efficient resource allocation is developing actively. In the fifth generation (5G) cellular network, it is expected that many new applications such as enhanced mobile broadband (eMBB), ultra-reliable low latency network (URLLC) and massive Machine Type Communication (mMTC) can be supported

M. C. Chow · M. Ma (✉)
School of Electrical and Electronic Engineering, Nanyang Technological University, Singapore, Singapore
e-mail: EMDMa@ntu.edu.sg

© The Author(s), under exclusive license to Springer Nature Switzerland AG 2022
P. Nicopolitidis et al. (eds.), *Advances in Computing, Informatics, Networking and Cybersecurity*, Lecture Notes in Networks and Systems 289,
https://doi.org/10.1007/978-3-030-87049-2_20

in the same network. Also, it should be able to support enhanced location-based services, social applications and so on. Hence, many new strategies, including device-to-device (D2D) communication, are employed in the 5G network to satisfy all these requirements.

Being one of the promising features in the fifth generation (5G) cellular network, D2D communication is one of the key strategies for achieving the aforementioned goals. By definition, allowing adjacent mobile devices to communicate directly without relaying user data over base stations or access points, D2D communication is the technology that offloads network traffic from the conventional centralized network to peer-to-peer network. In fact, D2D communication provides numerous benefits. For example, by offloading network traffic to peer-to-peer network, it alleviates the burden of the core network to allow more concurrent users [1]. Also, with proper radio resource allocation scheme and power control scheme, it is believed that the 5G D2D network can further improve spectral efficiency thus the overall throughput of the cellular network [2, 3]. Moreover, using the D2D network as a relay to conventional base station signals, the coverage of cellular networks can be expanded and the network performance at the edge of cellular can be improved [4]. What is more, D2D communication in the 5G network also enables futuristic applications like public safety network, distributed content sharing, adjacent file sharing and so on [4].

As D2D communication in the 5G network has a huge potential, it has gained more attention from both academia and industry. For example, there is an increasing amount of research carried out to use D2D for efficient multi-hop cellular networks, content distribution service, P2P gaming and so on [5]. Apart from academia, the Third Generation Partnership Project (3GPP) has put D2D communication under the umbrella of Proximity Service (ProSe) [6]. 3GPP started a feasibility study and standardization of D2D communication since 3GPP Rel. 12 as ProSe (Proximity-based Service). It covered some basic requirements such as D2D communication via the optimized path, device discovery, one-to-one direct communication, one-to-many group communication and public safety applications [7]. Since then, many innovative services such as multi-hop relay network, push-to-talk, emergency services and location services are also included in the latest specification [1].

Despite having numerous benefits, D2D communication in the 5G network is facing more security challenges than the conventional cellular network. For example, since 5G D2D works under wireless channels, it is prone to the wireless attacks like eavesdropping and service disruption by jamming due to its nature. Also, as 5G D2D has a unique structure of connecting both ad-hoc networks and a centralized network, it could be prone to some new attacks such as impersonation attacks, free-riding behaviors, battery depletion attacks and so on. Therefore, security issues in the 5G D2D network should be carefully studied. In this chapter, we carry out an extensive literature survey on the security in the D2D communication in the 5G network. Specifically, the contribution of this chapter are as follows:

- We review the existing 5G D2D system architecture and application scenarios standardized by 3GPP.

- We specify major security requirements of 5G D2D by analyzing its security threats.
- We categorize the latest major existing work in 5G D2D and discuss their advantages and shortcomings.
- We summarize existing research gaps to suggest open research topics.

The rest of the chapter is organized as follows: Sect. 2 gives the overview of the 5G D2D system architecture. Section 3 focuses on listing the major security requirements of 5D D2D. Section 4 comprehensively reviews the latest major research work. Section 5 discusses the open research topics. Finally, a conclusion is drawn in Sect. 6 to provide a summary of the paper.

2 Overview of 5G-D2D System Architecture

In this section, we define three different network topologies of D2D communication in cellular network and explain how they are related to the 5G-D2D network. Subsequently, we discuss the formal specification of 5G-D2D proposed by 3GPP to define 5G-D2D formally. Also, the latest system architecture of 5G Core network is also demonstrated to show that there are some discrepancies between the ideal 5G-D2D network to the latest 5G Core network.

2.1 5G-D2D Network Topology

Depending on the presence of cellular network, D2D communication can be broadly classified into the following three different modes [8, 9]. Figure 1 is the graphical illustration of different topology.

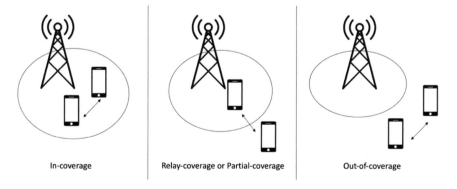

| In-coverage | Relay-coverage or Partial-coverage | Out-of-coverage |

Fig. 1 Coverage scenarios

In-coverage mode: All D2D UEs are under the coverage of base station signal, and all network resources are managed by the cellular network. Therefore, D2D connection tasks such as device identification, discovery, connection establishment and key management can be delegated to network operators. This is also the normal case of network assisted D2D network specified by 3GPP.

Relay-coverage/Partial-coverage mode: Some D2D UEs are out of coverage of base station signals. In this scenario, UEs under coverage will become network relays to allow all other UEs to access the core network. Similar to in-coverage mode, D2D connection tasks can also be delegated to network operators because all UEs can still access the core network. In 3GPP, this mode is indicated as the UE-to-Network Relay for Public Safety [6].

Out-of-coverage mode: When all devices are out of the coverage of base station signals, D2D connection tasks could be handled by UEs independently. In fact, most of the conventional D2D network without cellular network, such as Bluetooth, Wi-Fi Direct and Zigbee, can be classified into this category. This standalone connection mode could be useful for public safety applications such as walkie-talkie, because during the emergency, base stations may be heavily congested or even damaged so that UEs have to manage the connection themselves. However, 3GPP did not mention many details about the standalone D2D in the specifications, majorly due to its autonomy without relying on 5G network.

In fact, unlike the conventional cellular network which all communication is built on licensed bands, D2D communication can be implemented on both licensed bands or unlicensed bands [5]. For instance, proposed and implemented by Qualcomm, FlashLinQ is a cellular D2D protocol which provides peer discovery, link management and power management over a specially licensed band (2.586 GHz) [10]. On the other hand, Wi-Fi Direct, another dominant D2D protocol introduced by the Wi-Fi Alliance, is using the unlicensed Wi-Fi and NFC spectrum to provide D2D device discovery and D2D communication [11].

2.2 3GPP Proximity-Based Service System Architecture

3GPP proposed Proximity-based Service (ProSe) architecture models [6] based on 4G LTE Evolved Packet Core (EPC) framework, and it is expected that the overall architecture would remain similar for 5G Core Network (5GC). The proposal included various connection scenarios. For example, two UE are in the same carrier (i.e. non-roaming scenario), two UEs are connected to different carriers (i.e. inter-Public Land Mobile Network (PLMN) scenario) and more than one UEs are roaming (i.e. roaming scenario). To illustrate all these connection modes graphically, Fig. 2 shows the simplified system architecture of ProSe in core network which focuses only on interaction between UEs, ProSe Functions and ProSe App Server. In the figure, ProSe Function and ProSe App Server are located within the core network. The ProSe Function entity is responsible for UE provision, and the ProSe App Server is responsible for ProSe Direct Discovery and all other D2D related services. ProSe

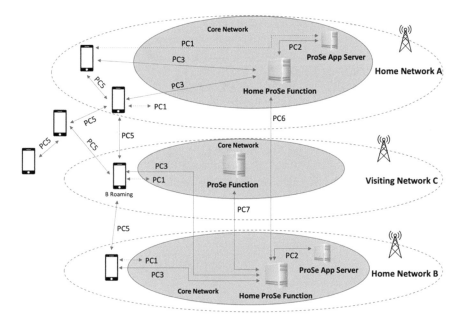

Fig. 2 3GPP ProSe system and security architecture

Function can communicate with other ProSe Function entities with path PC6 and PC7. It also communicates with UEs through path PC3 and cooperates with ProSe App Server through PC2. PC1 is the application-level connection which allows ProSe related applications in UEs to access ProSe App Server directly. Finally, the D2D communication path between UEs is labelled as PC5 [2]: this path can be either a one-to-one direct connection or a one-to-many group communication. Also, under the public safety circumstances, this path should also support relay traffic such that all UEs can still access the core network even if some UEs are outside of coverage.

Apart from the system architecture, 3GPP has also specified some ProSe use cases in the security framework [12]. Specifically, the framework has standardized the following four scenarios:

Device discovery: it includes the direct discovery that allows UEs to broadcast and scan each other autonomously, and the cellular-level discovery that allows UE to find each other with the aid of the core network. For direct discovery, open discovery means everyone can read the broadcast messages sent from other D2D UEs, and the restricted discovery means only authorized D2D UEs can read the messages. Although direct discovery generally consumes more power and introduces more security risks due to its active broadcasting, it should relieve the burden of the core network better and reduce the risk of single point of failure because of the reduced traffic.

One-to-one direct communication: when two UEs establish a peer-to-peer connection after device discovery, this is a one-to-one direct communication. Authentication and data encryption should be deployed in this mode to protect the integrity and secrecy of transferred data.

One-to-many Group Communication Establishment: when more than two UEs establish a group connection after device discovery, this is a one-to-many group communication. Authentication and data forward/backward secrecy should be guaranteed when the members are entering or leaving the group, so that all left members cannot access the new data, and the old messages cannot be recovered by the new members.

Public Safety Relay: in case of emergency, some UEs connected to the core network can act as UE-to-network relays to serve adjacent UEs with poor or no cellular coverage. In this scenario, data secrecy between both UE to UE and UE to the core network have to be carefully considered to guarantee data confidentiality of all D2D users.

2.3   3GPP 5G Core System Architecture

Although 3GPP has standardized the framework for D2D communication (a.k.a. ProSe) aforementioned, the latest specification of 5GC has omitted all ProSe related functions. Figure 3 shows the simplified architecture of the latest 3GPP 5G Core network. According to the latest 3GPP 5G Security Architecture Release 15.7 in

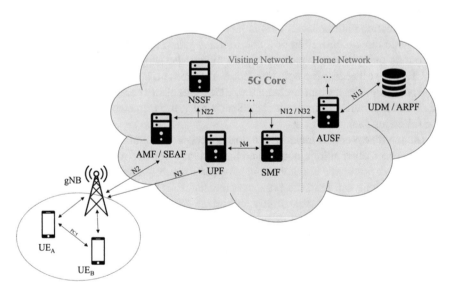

Fig. 3 3GPP 5G core system and security architecture

2019 [13], the overall system architecture involves many functional entities including the Next Generation Node B (gNB), Access and Mobility Function (AMF), Security Anchor Function (SEAF), Authentication Server Function (AUSF), Authentication credential Repository and Processing Function (ARPF), Network Slice Selection Function (NSSF), Session Management Function (SMF), Unified Data Management (UDM), User Plane Function (UPF) and so on. However, it does not contain the ProSe Function and the ProSe Application Server mentioned in ProSe system architecture specification TS 23.303 [6]. Since 3GPP did not finalize how D2D be supported in 5G network, there are many research employing different approaches: some proposed putting ProSe related functionalities into the existing functional entities, while others proposed adding back the ProSe related entities into the core network. Since D2D communication is a promising technology and its importance is rising, we hope this discrepancy would be solved in the future.

3 Security in 5G-D2D Communication

3.1 Security Threats

After understanding the system architecture and network topology, we can analyze the possible attacks systematically by making some assumptions. In this section, we assume that all ProSe Function entities providing D2D related features are located within the 5GC networks. Also, these ProSe Function entities among different operators should be able to communicate with each other using the existing inter-PLMN connection. Consequently, all security threats and solutions can be classified into the following three domains:

- **Security within 5GC network**: it refers to the security of the connection between UE and the D2D related functional entities inside the core network. It includes the connection path between ProSe Function entity and core networks (PC4) and the connection paths between ProSe Function entities and UE devices (PC1, PC3). These paths are classified into the same group because these connections should be protected and trusted under the pre-existing 5G core network security architecture.
- **Security between 5GC networks**: it refers to the security of the connection which allows D2D services to work even if two D2D UEs are using different PLMNs (i.e., carriers) or during roaming. An example would be the inter-PLMN communication between different ProSe Function entities (PC6-PC7). Similar to the case of the security within 5GC network, these paths should also be protected and trusted using the pre-existing 5G security architecture for all functional entities, just like the pre-existing PSTN (Public Switched Telephone Network) connection.
- **Security specific to D2D connection**: it refers to the security of the wireless peer-to-peer connection between two UEs, or group communication among many UEs. For example, it could be related to the device discovery procedure, connection

establishment mechanism, authentication, and the peer-to-peer communication between different UE devices (PC5). Since D2D UEs may work in under unlicensed bands or with limited cellular coverage, and it is not covered by pre-existing 5G security architecture, this category is the most relevant part of 5G-D2D security. In fact, most security research related to D2D security could also be classified into class, and they will be discussed in the Sect. 4.

Mobile devices nowadays contain much sensitive information such as personal identity, financial and health data, thus, network security has been unprecedentedly important to avoid any data leakage or misuse. However, as 5G-D2D has the unique network topology which connects to both ad-hoc network (PC5) and centralized network (PC3) wirelessly as aforementioned, it is vulnerable to many security threats pre-existed in both ad-hoc network and cellular network. According to [3, 8, 9, 14, 15], we could categorize all these attacks into two categories, namely active attacks and passive attacks. Passive attacks mean that attackers can only tap into the wireless channels to collect all messages, while active attacks mean that the attackers can not only read but also actively respond to victims with any fabricated messages:

Active Attacks

- **Data Fabrication**: an active attacker can create fake data to manipulate other D2D users.
- **Data Modification**: an active attacker can modify the data sent from a specific user to falsify other users.
- **Free riding**: some selfish users only receive contents from others but refuse to help others relay data to save battery. Availability and service quality of D2D network may be severely affected.
- **Impersonation**: an attacker can pretend to be another UE device to gain access to the network.
- **Jamming and Denial of Service (DoS) Attack**: attackers can deliberately disrupt or block the connection of normal users in a network.

Passive Attacks

- **Eavesdropping**: a passive attacker can intercept the communication channels to read user data between two UE devices, or the control messages between UE and the core network.
- **Location Spoofing**: an active attacker may send out malicious location information to confuse existing D2D users or even disrupt D2D group forming.
- **Privacy Violation**: D2D services such as device discovery may contain identity data, so an attacker can trace the device by continuously sniffing D2D broadcasts.

3.2 Security Requirements

To resolve the security concerns aforementioned, research related to D2D security requirements have been conducted extensively. To the best of knowledge, academicians have summarized the following major security requirements for a secure D2D communication system in cellular network [8, 9, 15, 16]:

- **Data confidentiality**: be able to prevent data leakage or eavesdropping.
- **Data Integrity**: be able to detect and prevent data from modification.
- **Authentication**: be able to verify the identity of UEs and prevent from impersonation attack.
- **Fine-grained access control**: be able to create data access rules or policies in small granularity such that specific unauthorized users can be barred from accessing private data.
- **Revocability**: be able to deprive privileges of D2D UEs if malicious behavior is detected.
- **Availability and Dependability**: be able to service even under jamming, DoS or DDoS attacks.
- **Non-Repudiation**: be able to prevent UEs from denying the authorship or reception of a message.
- **Privacy or Anonymity**: be able to hide identity or location information of UEs.
- **Traceability**: be able to identify and the source of a false message.

4 Existing Solutions

Many different solutions are suggested to mitigate security issues in 5G-D2D in recent years. In this section, we primarily focus on the solutions specifically related to establishments of 5G-D2D connection, which refers to the schemes securing the point-to-point connection between two UEs (i.e., the "security specific to D2D connection" aforementioned in Sect. 3.1). In fact, many different cryptographic functions are employed and combined to improve the security of 5G D2D communication in various scenarios. To differentiate them systematically, we categorize the solutions according to their application scenarios, including device discovery, one-to-one communication, group-based communication, continuous authentication, and content sharing.

4.1 Privacy Preserving Device Discovery

Abd-Elrahman et al. [17] have proposed a D2D device discovery and D2D communication encryption scheme leveraging Identity-based Encryption (IBE) and Elliptic Curve Cryptography (ECC). The proposed system model requires all DUEs to be authenticated and under the coverage of the core network, and multiple private key generators (PKG) are built across the network. The protocol has mainly three phases: the first phase is the system setup stage which allows PKGs to generate parameters for the ECC and send it to all DUEs. Next, in the restricted discovery state, all DUEs will generate one pseudo-identity and sign the discovery message with the private key derived from PKG using the elliptic curve digital signature algorithm (ECDSA) algorithm. After that, this message will be broadcasted to other DUEs and let other DUEs to ask for public keys derived from PKGs to verify the discovery message. In case that two DUEs are not in the same domain (e.g., two DUEs are using different network operators), DUE1 will make another request to its local PKG (PKG1), and the PKG1 will access the PKG at DUE2's domain (PKG2) to get the key parameters. Therefore, DUE1 can sign and broadcast the message that is verifiable to DUE2. The final stage is direct communication. After verification of the discovery message, DUE1 will establish a connection with DUE2 using Elliptic Curve Diffie-Hellman (ECDH) key exchange. Security analysis shows that with ECDSA used in the broadcast message, this protocol can guarantee non-repudiation of the signaling messages. Also, this protocol ensures device privacy, data confidentiality, perfect forward secrecy, and backward secrecy because the pseudo-identity is updated for each session and ECDH key exchange is used during connection establishment. Subsequently, another enhancement work based on this protocol has also been proposed in [18]. It provides both one-to-one direct communication and group communication by combining the concept of PKG and Group Key Management (GKM) mechanism that aggregates the session keys of small sub-groups in key server to form a larger group key.

Sun et al. [19] have proposed a privacy-preserving device discovery and authentication mechanism for 5G D2D communications. In the paper, identity-based prefix encryption is adopted to perform identification proof and encryption. This encryption mechanism is a simplified functionality of the hierarchical identity-based encryption: given a ciphertext, it can then be decrypted with a key in which the binary string correlated to that key is the prefix of the ciphertext. During the authentication process, DUEs need to perform full access authentication with both visiting network and home cellular network first, then they can use identity-based prefix encryption to encrypt then broadcast device discovery messages. If any DUE wants to connect to the broadcasting DUE, it has to decrypt the broadcast message using its prefix encryption secret keys, encrypt the session establishment message with Elliptic Curve Diffie-Hellman Key Exchange (ECDH) and send it back to the broadcaster. Finally, the broadcasting DUE can decrypt the received message and perform the remaining handshaking with the desired DUE. This scheme is proven to secure in D2D device discovery, mutual authentication and session key agreement without losing privacy of both DUEs, because their identity information is never broadcasted to the public without encryption, and the broadcast message cannot be decrypted by unintended users. Also, with ECDH and different random nonce used, eavesdropping and man-in-the-middle attacks can be prevented. This scheme can only be applied to in-coverage cases since the first step in this scheme is to perform authentication to the visiting network and the home cellular network. Also, the computational overhead of this scheme is relatively high because of the bilinear pairing in the hierarchical identity-based encryption.

4.2 One-to-One D2D: Authentication and Key Agreement (AKA)

4.2.1 Diffie-Hellman Key Exchange

Three similar lightweight D2D protocols have been suggested in [20]. Unlike most other schemes that involve many infrastructures in the core network during authentication and connection establishment, the three D2D protocols only use base stations to assist D2D network establishment. The first proposed protocol uses Diffie-Hellman Key Exchange (DHKE) to establish a shared secret over the public channel and authenticate the DHKE public keys using message authentication code (MAC) sent through the cellular channel. The second protocol is similar to the first protocol, but the base stations act as a middleman to compare the MAC code with both parties. Both two DUEs will send the encrypted message and wait for base stations to send back accept or reject acknowledgment. The third protocol is the most complicated protocol among them: first, two DUEs will exchange their symmetric key using DHKE in a public channel and create one secret channel. Then, base stations will send another key to the DUE pair through the cellular channel, and two DUEs need to encrypt and send authentication messages with keyed-hash message authentication code (HMAC). Finally, they will both use the base station as a middleman to send the accept or reject message to indicate if the authentication is successful. Performance evaluation shows that all three protocols are having less communication overhead than [21], and the security evaluation shows that they are resilient to man-in-the-middle attacks. However, since these three protocols do not use pseudo-identity or privacy-preserving device discovery, they could not protect device anonymity of DUEs.

Schmittner et al. [22] have proposed a Secure Multi-hop D2D communication (SEMUD) scheme which focuses on the secure connection and message forwarding. In the proposal, it is assumed that the network follows the ProSe architecture which consists of ProSe Application, ProSe Function and ProSe Application Server. For the connection part, SEMUD is comprised of 5 phases. The first phase is the subscription phase which registers the UE and takes the public key from the ProSe Application Server. The second phase is the discovery and authentication phase that follows the standard ProSe discovery and authentication framework. The third phase is the key distribution phase which two DUEs exchange public keys for mutual authentication and a DHKE symmetric session key for all subsequent communication. The fourth phase is the communication phase which messages will be traversed in single or multiple hops. The final phase is the termination phase which deregisters the DUE from the network. In fact, SEMUD focuses on message forwarding—during the communication phase, all messages will be appended with Blake2b hashing and then encrypted using symmetric stream cipher algorithm XSalsa20/20 and parameters from Merkle Tree. Then, the message will be forwarded to other hops based on the per-neighbor reliability metrics computed in DUEs. When the receiver DUE receives the message, an acknowledgment message will also be returned to the sender to ensure the reliability of the transmission. Security analysis shows that the proposed scheme can survive under blackhole attacks, also the message delay can remain steady even when the number of attackers is increasing.

Combining social domain knowledge and cryptographic techniques, Suraci et al. [23] have proposed a mechanism that determines the trustworthiness of adjacent D2D nodes with the Social IoT paradigm and distributes public keys with the DHKE. Social IoT parameters refer to the activity of the device and some objective features such as manufacturing data, location, and owner of the device. In the paper, the activity of the device is employed to evaluate the trustworthiness of adjacent devices. Firstly, it has assumed that all DUEs want to share the information distributed from a Service Provider (SP) located at the core network. According to the proposed scheme, all base stations (eNodeB) are the trusted parties that stored a list of blocked nodes. These blocked nodes will never be chosen as D2D transmitters because their credibility is too low. For other nodes, DHKE will be used to exchange a symmetric key between two DUEs, and all data relayed between DUEs are pre-signed by SP. To create the blocked list, during the system initialization process, each new D2D node needs to send a real identity to the core network, then the network will return a pseudo-identity and a trust value to the DUE. For freshly connected DUEs, the trust value will always be 0.5 (half trustworthy). After that, during daily network traffic, two credibility counters called "Good Behavior Amount (GBA)" and "Malicious Behavior Amount (MBA)" will be counted by the SP according to their behaviors. As a result, a value called "Absolute Player Trust Value" will be derived from all the parameters above. Only the devices that with the trust value higher than a defined threshold can be treated as trustworthy devices, otherwise they will be put into the blocked node list. Therefore, all encrypted network traffic will only traverse in those socially and behaviorally credible nodes. Simulation results show that malicious D2D nodes can be avoided with the proposed scheme, therefore the overall data secrecy can be improved, and free-riding attacks can be prevented.

Subsequently, Suraci et al. [24] also proposed a Secure and Trust D2D (SeT-D2D) mechanism to securely offload the multicast traffic in the 5G cellular network to some properly selected D2D relay nodes. In this way, when some mobile users are out of the coverage area of multicast signal, data can relay through D2D communication. In the proposal, all data sent from the gNB using multicast are hashed and then signed using the private keys of gNBs. Then, there are some relay nodes establishing connections to users with poor signal reception using D2D communication. All these connections will be secured using DHKE and signature. To improve the reliability and performance of the network, trustworthy relay nodes are selected based on a novel trustworthiness model. As a result, it can reduce malicious attacks in the network effectively. The simulation shows that the proposed mechanism achieved a much higher throughput than having no trustworthiness system and DHKE because of less wastage and fewer attacks from malicious nodes.

M. Wang et al. [25] have proposed a universal D2D AKA protocol (UAKA-D2D) which works on both roaming/non-roaming and intra/inter-operator D2D communication in 5G cellular networks. In the proposal, it is assumed that the core network is 4G LTE, and all DUEs can connect to their own HSS. The protocol has 4 stages. The first stage is the system setup stage which initializes all parameters for cryptographic algorithms. The second stage is the UE registration. During this stage, all UEs need to create a fixed or random pseudo-identity, then send this identity information with real identity to VN (for roaming case) or Authentication Centre (AuC, for non-roaming case). Then, VN or AuC will start normal EPS-AKA (the authentication and key agreement protocol used in 4G) to authenticate and register the UE. The third stage is the device discovery and random nonce exchange stage. The proposal follows the suggested device discovery procedure for open discovery in ProSe architecture, and a random nonce should be exchanged in any means such as face-to-face or social networking. The final stage is the D2D session generation. After having all the parameters such as pseudo-identity and the random nonce of each other, DUE1 need to initiate "key agreement request" to HSS/VN to verify DUE2, and then exchange their session key using a DHKE with validation check provided by HMAC. Finally, two DUEs can authenticate each other using the XOR result of the broadcasted random nonce and the common HMAC key sent from the core network. Also, one common session key can be generated using the result of DHKE common secret. Since only HMAC and DHKE are used, the computational cost of the protocol remains practically lower than most IBE or ABE-based schemes. Security analysis also shows that this protocol can provide mutual authentication, data confidentiality and device privacy. Formal analysis using AVISPA has also been carried out to verify the protocol. However, to use the protocol securely, it is suggested to make the privacy-preserving device discovery be compulsory, such that the real identities of the devices will not be leaked.

Seok et al. [26] have proposed a lightweight D2D protocol that uses tokens obtained from gNB to perform identity verification. Combining the DHKE for key exchange with the Authentication Encryption with Associated Data (AEAD) ciphers for symmetric key encryption, the protocol is said to be lightweight but effective to provide data confidentiality and authentication. The protocol has 4 steps. The first step is D2D token generation which all DUEs need to authenticate with the cellular network using 5G-AKA, and then the gNB will distribute one unique "D2D token" for every device. This token is generated using the 5G-SUCI (Subscription Conceived Identifier) of DUEs with the ECDSA algorithm. The second step is the device discovery stage which every DUEs will broadcast their own 5G-SUCI and D2D Tokens to adjacent devices. After receiving broadcast messages from the target DUE, the third step is to send the received 5G-SUCI and D2D Token to gNB for verification. If the verification is valid, two devices will enter the final step of the negotiation of the symmetric key using ECDH. Finally, all subsequent messages will be encrypted using any AEAD cipher algorithms with the common symmetric key. Security analysis shows that the protocol could provide authentication, data integrity and data confidentiality. It should also be able to mitigate various security attacks such as impersonation and eavesdropping. Moreover, the anonymous feature of 5G-SUCI also ensures the anonymity of DUEs. Nevertheless, since the proposal does not include any formal verification or logic rules derivation, loopholes of this protocol could be undiscovered.

4.2.2 Basic Public Key Infrastructure (PKI)

In-Coverage Compatible Schemes

Melki et al. [27] have proposed a lightweight authentication and key exchange protocol based on public–private key pairs, Public Key Authority (PKA) and the physical layer characteristic of the shared channel. In the proposal, all DUEs need to generate their own public–private key pair using ECC and register themselves to PKA. The PKAs are built using base stations (gNodeB), they are responsible for issuing certificates to bind public keys to the device identity. The paper also assumed that DUE1 and DUE2 are close to each other, therefore they can scan and receive similar physical characteristics in the shared channel. In case that DUE1 wants to connect to DUE2, DUE1 will first scan the shared channel to generate a nonce indicating the environment. Then, DUE1 will send a connection request message containing the hash of nonce XOR a session key, an encrypted session key, and a random token R. This connection request message will be encrypted with the public key of DUE2 and then send to DUE2. When DUE2 receives the message, DUE2 can decrypt the message using its private key, and decrypt the session key using the public key of DUE1. Since two devices are close to each other, DUE2 will also generate the same physical nonce, which can also be used to verify the hashed value. Finally, it will use the session key to encrypt the incremented value of R and send back to DUE1 as a confirmation response. Security analysis shows that the device anonymity can be protected, and various attacks such as impersonation and man-in-the-middle (MITM) attacks can be avoided. However, since the proposed algorithm uses the public key to encrypt the session key, data forward secrecy may not be guaranteed if the PKAs are under attack and some keys are leaked in the future. Also, since the physical characteristics of the wireless channel are constantly changing, and different physical devices may measure slightly different results, the generation of physical nonce could also be a challenging problem.

Baskaran et al. [28] have presented a Lightweight Incognito Key Exchange (LIKE) which uses a public key distribution entity called "Security Anchor (SA)" to assist authentication of D2D pairs. In the proposal, it is assumed that all DUEs are under the in-coverage scenario, and a new entity SA is located at every base station. This SA will be responsible for generating public–private key pairs for all DUEs, and the ProSe Function and ProSe Application Server within the core network will help find and joint D2D peers within the network. The scheme has 3 main phases. The first phase is the system setup phase which initializes all security parameters. The second phase is UE registration at SA which all UEs need to register itself to SA with its real identity. Each DUE will be assigned a pseudo-identity and a public–private key pair afterward. The third phase is LIKE protocol communication. For this protocol to work, DUE1 firstly needs to send a "D2D attach request", then the SA in base station will form a list called "Service Candidate Set (SCS)" which contains all connectable D2D nodes. After that, the ProSe Application Function will take this SCS list and prune it based on the physical distance of DUEs to choose the best D2D pair. Next, ProSe Application Server will also receive the request and start to assist pairing by sending "D2D attach response" to DUE1 and D2D initiation request to the chosen DUE2. Finally, DUE1 and DUE2 will find each other, use their assigned public–private key pair to authenticate each other and perform session key exchange using DHKE. Security analysis shows that this protocol can prevent free-riding attacks (since the core network can reject the device request if that device always rejects D2D initiation request), MITM attacks and replay attacks. Also, non-repudiation and device privacy can be guaranteed since the pseudo-identity is used for all data transmission in the unencrypted free channel.

Chow et al. [29] proposed a lightweight D2D authentication scheme that prevents free-riding attacks in the 5G network. The proposed model has three stages. In the first stage of initialization, all UEs need to register to the core network using the conventional mechanism such as 5G-AKA and retrieve one ECDSA private key and a global unique temporary identifier (GUTI). Then, in the second stage, all D2D UEs can broadcast themselves by signing their GUTI. To establish a D2D connection, two devices need to ask the core network for a shared token, and the core network will also check their history before issuing it. After that, two devices can mutually authenticate using the shared token and create a unique session key using DHKE. Security evaluation shows that the proposal guarantees mutual authentication, device anonymity, session key perfect forward secrecy, and the prevention of free-riding attacks. Performance evaluation shows that although it achieved slightly more computational overhead than Seok's proposal [26], it has achieved a higher security level, and the overhead has remained practically low.

In-Coverage and Relay-Coverage Compatible Schemes

Javed et al. [30] have proposed a D2D relay authentication scheme which uses Elliptic Curve Cryptography (ECC), ElGamal (a variant of public-key cryptography) and Public Key Infrastructure (PKI) to perform signature verification and encryption. In the proposed system model, different from the 3GPP standard 5G network, they proposed a D2D network which is constituted by many small intelligent D2D relays, and all DUEs need to connect to these small relays to join the cellular D2D network. Also, some larger relays called "Multihop Relay Base Station (MRBS)" will connect to these small relays to manage them. To establish this large D2D relay network securely, the proposed scheme requires all relays to obtain a valid certificate from CA first. Then, during authentication state, all handshaking messages will be signed and encrypted with two algorithms called "EEoP Scheme for Signature Verification" and "EEoP Scheme for Encryption" which focus on signature verification and message encryption (for the signature verification, SHA3 V3 hashing and PKI are jointly employed to generate the signature, and for the message encryption, ECC and ElGamal are used). Compared with other authentication protocols, since this scheme uses computationally lightweight ECC to generate asymmetric keys, and it has only two authentication messages (one request and one response), therefore it is suitable for deployment in IoT and mobile devices which requires high speed and low complexity. Security analysis shows that the proposed scheme can prevent various attacks such as brute force attack, man-in-the-middle attack, and interleaving attack. This scheme assumed all DUEs should connect to D2D relays or base stations, so it works for in-coverage or relay-coverage connection scenarios. Recently, A. Abro [31] has also proposed an enhanced authentication scheme (FHEEP) based on EEoP in [30] with improved hashing and encryption process using both ElGamal, ECC and Public Key Infrastructure. The proposed system model assumes all nodes in the D2D network have been registered into CA. Also, all DUEs are connected to some non-transparent relays (NTR), and these relays are connected to a large and computationally powerful MRBS relay to provide connectivity to all connected nodes. For each session, NTR needs to recompute the public/private key after a periodic interval, and SHA384 is used as the hash function during signature. Security analysis shows that since the encryption algorithm is session-based, even though if adversaries can hack into one of the sessions, only that session will be compromised and not the remaining sessions. Also, if the 192-bit or longer elliptic curve is used in this scheme, it can effectively prevent adversaries from solving the discrete logarithm problem (DLP) using the "Baby Step, Giant Step" method.

In-Coverage, Relay-Coverage, and Out-of-Coverage Compatible Schemes

Dao et al. [32] have suggested the use of the prefetched certificate of authority (CA) of D2D devices. In the proposal, all DUEs are certified by a common CA, and each DUE would save their private key, CA signed own public key (with expiry time) and CA public key locally. Then, to perform an authentication request, DUE1 will first send his CA signed own public key to DUE2, and DUE2 will verify the signature, and then send back the response message. The response should contain DUE2's CA-signed public key and a DUE1's public key encrypted random nonce (Nonce A). After that, DUE1 will decrypt the random nonce and send another response which contains the decrypted random nonce, another new nonce (Nonce B) and some other variables. This response will be signed with DUE1's private key and encrypted with DUE2's public key. Next, DUE2 will decrypt the message, verify the signature and send back the DUE1's public key encrypted Nonce B to DUE1. Finally, they can derive a common secret by XOR the Nonce A and Nonce B. To further against brute force attack, the proposed mechanism has added an expiration timestamp for the secret key, therefore a new secret key is generated before attackers can find the key. This model has demonstrated a commonly used authentication technique that promised authentication and data confidentiality. Also, it can be implemented in all connection scenarios given that the D2D certificates are downloaded in advance.

However, the paper did not mention the location and capacity of the CA server. Also, without using DHKE, there is a potential risk of data leakage if the private keys of DUEs are compromised and the channel is being eavesdropped.

Boubakri et al. [33] have proposed a PKI and Zero-Knowledge Proof (ZKP) based access control and authentication scheme specifically designed for 5G networks. In the proposal, hierarchical PKI architecture is built to delegate the issuance of certificates from root certification authority (Root CA) to terminal D2D devices. Specifically, network operators are the root CA which issue and sign "network certificates" to all its access nodes in the network. Then, these access nodes will also become intermediate CAs that issue and sign "access certificates" to all connected devices such as CUEs and other deployed facilities. For some connected devices or CUEs that can work as a virtual access node, they can also become the next intermediate CA which issue and sign "D2D certificates" for all relay-coverage DUEs. When any relay-coverage DUEs are attempting to establish a connection, ZKP based authentication and registration will start: First the DUE and the virtual access node (VAN) should have a pre-shared key. Then, DUE will issue an M2M communication request to the VAN. Next, VAN will then generate parameters for DUE to generate a public–private key pair. After that, DUE will send back the generated public keys with a secret message related to the pre-shared key. Finally, VAN will verify the secret message with ZKP to determine if the DUE is legitimate. If the DUE is a legitimate device, VAN will sign the authorization certificate and send back to the DUE. To further provide seamless handover experience, VAN (the D2D relay node) will also actively predict the next node and send a handover request which contains the certificate to target VANs. As a result, DUEs can simply reuse the existing certificate to communicate with the next targeted VANs. This scheme can be used for in-coverage and relay-coverage connection scenarios, and performance analysis showed that the proposed scheme can achieve higher performance than conventional LTE handover. This scheme can also satisfy the security requirements of data confidentiality, authentication, and device privacy because no device-related information nor messages are sent through the unencrypted channel.

4.2.3 Identity-Based Cryptography (IBE / IBC)

Hamoud et al. [34, 35] have suggested a certificateless public key cryptography (CL-PKC) based D2D authentication and key management system. This system consists of a new entity called "Regulatory Authority" which generates parameters for key centers, and two Key Generation Centers (KGCs) which generate master keys and public/private key pairs. Assuming all DUEs can connect to the cellular network (including in coverage and partial coverage), this scheme has suggested DUEs should register themselves with two KGCs which located in cellular operators and ProSe Service Provider (ProSe SP). During the setup stage, it allows two KGCs to have different master keys, but the same sub-public key which represents the specific DUE, so totally two different public/private key pairs will be generated. Then, during registration, these two pairs of keys and a pseudo identifier generated by DUE will be combined into a single public–private key pair. After registration, DUEs can now pair with others under any D2D coverage scenario (including out-of-coverage) because it should have all the parameters stored locally. To pair two DUEs, a triplet containing an identifier, a public key and a random number will be exchanged between two DUEs. Finally, after some validity checks, a new symmetric key can be calculated and used for ongoing data confidentiality protection. Since multiple KGCs are used in this scheme, it is believed that the system is resistant to "Malicious but Active KGC Attack" and "Malicious but Passive KGC Attack". Furthermore, this scheme provides device privacy using a pseudo-identity but also remains traceability of devices due to the use of KGCs, so network operators or SP can still track and detect the misbehaving DUEs.

As a variant of IBC, Ciphertext Policy Attribute-Based Encryption (CP-ABE) is an encryption method that encrypts the message according to some access control policies, such that only devices satisfying the same policies and attributes can decrypt the message. As an example, Kwon et al. [36] have proposed an authentication scheme using CP-ABE and an out-of-band personal identification number (PIN). Similar to Bluetooth pairing procedures, this scheme used out-of-band shared PIN code to prevent MITM attack. However, while connecting to a large D2D group, the pin may not be easily shared across many user devices. Hence, in this proposal, it has presented two different protocols. The first protocol is called "Device to Device Authentication 1 (D2DA1)". Assuming D2D discovery is finished, DUE1 needs to enter a PIN code first, then it will send the pin code encrypted with the attribute of DUE2 to DUE2. Then, DUE1 will also send a random nonce and a hashed secret to DUE2. Since DUE2 has the attribute of itself, it can decrypt the PIN, and further ask for the PIN from the user to double confirm. DUE2 will also check if the hashed secret is correct. Finally, both devices will calculate the common secret key using the aforementioned nonce and PIN. For its second protocol D2DA2, it assumed that DUE is joining a D2D group, and there is one group manager (GM) UE. To create group D2D connection, DUE1 will first enter a PIN, the PIN will be encrypted with the attribute of the group, and it will be sent to GM DUE. GM DUE will relay the message to the group, and group members will decrypt the PIN and continue all the authentication procedure similar to D2DA1. For security analysis, although this scheme can prevent MITM attack, it has not mention about how can DUE1 get the attribute information of DUE2 or the group securely. If the adversary can also easily get the attribute information through an unsecured D2D discovery procedure, the adversary could decrypt the message and get the PIN. Then, it could potentially be more dangerous than the traditional Bluetooth pairing procedure which never sends PIN over the air. Also, since the DUE2 should be able to decrypt the PIN, PIN verification from the user becomes a procedure to ensure mutual authentication. It should be possible to further polish the scheme such that it can run safely without user intervention.

A D2D access control scheme that uses the CP-ABE algorithm to encrypt the symmetric keys has been suggested in [37]. To serve different connection scenarios, two levels of trust, namely "General Trust (GT)" for in-coverage and "Local Trust (LT)" for relay-coverage/out-of-coverage are proposed. The paper has assumed that the core network should have both ProSe Function Server and ProSe App Server, which record and evaluate the GT trust levels of all DUEs. For the GT only case, DUE needs to register with the core network first, then it will be assigned with parameters such as pseudo-identity, GT trust levels and attribute keys. After that, a symmetric key will be encrypted by the aforementioned attribute keys and all D2D communication data will be encrypted by that symmetric key. If the targeted DUE also have the same attribute value, it should be able to decrypt the symmetric key and use it to decrypt the D2D communication data. Finally, DUEs can send LT values and feedback votes to the core network for future trust value evaluation. For the LT only case, attribute keys and local trust values are controlled by DUEs instead of the core network, and the remaining steps are similar to the GT case. In fact, when the core network is available, GT and LT can also be used together to cater to both in-coverage and relay-coverage scenarios. Security analysis shows that the proposed scheme can provide fine-grained access control (i.e., only a specific trust level can access the data), data confidentiality and device privacy. However, since this protocol has introduced two levels of trust and mixed with CP-ABE cryptography, the overall complexity of this protocol remains high.

Wang et al. [38] have proposed an anonymous authentication and key agreement protocol (AAKA-D2D) based on the assumption of the 3GPP EPS system model, and it is also suitable for 5G cellular network. Using pseudo-identity, identity-based signature (IBS) and DHKE, the proposed protocol works in the in-coverage scenario to provide device anonymity, mutual authentication, and session key agreement. There are totally four steps in the proposal, including system initialization, user registration, D2D discovery and D2D session establishment. For the initialization and user registration stages, CN initializes the elliptic curve parameters for computing bilinear pairings and distribute pseudo-identities to all DUEs. Then, in the D2D discovery process, DUEs broadcast their own pseudo-identity over-the-air for other discoverers to find them. For a discoverer to initiate a connection, it has to send one request including the received DUE's pseudo-identity to CN. Then, CN distributes a common intermediate key and a session ID to both two DUEs. Subsequently, DHKE messages between two devices are encrypted using the intermediate key, and it is digitally signed using IBS such that the message recipient can verify the signature without having to request public keys from CN. Finally, a session key can be obtained using DHKE messages between two devices. For security analysis, this scheme can provide mutual authentication, perfect forward secrecy, device privacy protection and it should prevent message fabrication attacks. Also, since the pseudo-identity is appended with an expiry timestamp, it provides time-based key revocation to prevent key reuse if the public key is leaked to the adversary. However, with no formal verification proof provided in the proposal, the security of this protocol is yet to be assured. A simulation shows that the performance of the scheme is 20% faster than other IBE-based D2D schemes, which makes it more practical to be deployed in mobile devices than other IBE-based schemes (Table 1).

Table 1 Schemes for securing one-to-one D2D communication

Scheme	Technology used	Addressed security properties								
		DC	I	A	F	R	AD	NR	PA	T
[17]	IBE, ECDSA and ECDH	Y	Y	Y				Y	Y	
[19]	Identity-based prefix encryption and ECDH	Y	Y	Y				Y	Y	
[20]	DHKE, MAC and HMAC	Y	Y	Y			Y	Y	Y	
[22]	DHKE, Blake2b hashing and XSalsa20/20	Y	Y	Y						
[23]	Social parameters, DHKE, two credibility counters	Y	Y	Y		Y			Y	Y
[24]	Social parameters, DHKE, signature	Y	Y	Y						
[25]	DHKE, XOR and HMAC	Y	Y	Y					Y	
[26]	5G-SUCI, ECDSA and ECDH	Y	Y	Y					Y	Y
[27]	PKI and PLS	Y	Y	Y					Y	
[28]	PKI, ECDH, ECDSA, SHA	Y	Y	Y				Y	Y	Y
[29]	PKI, ECDSA, ECDH	Y	Y	Y				Y	Y	Y
[30]	PKI, ECC, ElGamal, SHA	Y	Y	Y						
[31]	PKI, ECC, ElGamal, SHA	Y	Y	Y						
[32]	Prefetched PKI	Y		Y						
[33]	PKI, ZKP			Y	Y					
[34, 35]	Multiple KGC	Y	Y	Y		Y			Y	Y
[36]	CP-ABE, PIN	Y		Y	Y				Y	
[37]	CP-ABE	Y		Y	Y				Y	
[38]	IBS, ECDH	Y	Y	Y				Y	Y	

4.3 One-to-Many D2D: Group-Based Authentication and Key Agreement (Group AKA)

4.3.1 Basic Public Key Infrastructure (PKI)

Hsu et al. [39] have proposed two PKI based D2D authenticated key exchange protocols, namely "Group Anonymity for D2D communication with CN Assistance (CN-GD2C)" for the in-coverage scenario and "Group Anonymity for Network-absent D2D communication (NA-GD2C)" for the out-of-coverage scenario in 5G network. For CN-GD2C protocol, it is assumed that both DUEs have registered themselves to the core network with their own unique key, and all DUEs have installed the public/private key of home subscriber service (HSS). First, DUE1 will encrypt the D2D request message containing the Diffie-Hellman variable using the public key of HSS and send it to DUE2. Then, DUE2 will also encrypt its DH variable and concatenate it with the received request message from DUE1 and send it to HSS. As HSS and ProSe Server have all group and device information, if DUE1 and DUE2 are found in the same group, after several times of validation, DUE1 and DUE2 will finally authenticate each other and have their shared session key generated by DHKE key exchange for session communication. For NA-GD2C, it is assumed that in a network, there are many D2D groups, and each group will have its own certificate issued by the CA. For every DUE, it has stored certificates of different groups in advance, but each DUE only belongs to one specific group. During D2D authentication of two DUEs, since the cellular network is absent, these certificates will be used instead of the public key of HSS to encrypt request messages. To make sure that eavesdroppers will not be able to know their group, DUE1 will send a list of certificates to DUE2, and DUE2 will respond with another list of certificates and a ciphertext if DUE2 is also within the listed groups. After various message exchanges, they will be able to confirm if both devices belong to the same group. Security analysis of these two protocols shows that they can provide secure mutual authentication, group anonymity, data confidentiality, device anonymity and traceability. Simulation on Android devices also shows that the proposed algorithm is computationally implementable on modern mobile devices. Subsequently, an enhancement based on this protocol has also been proposed in [40].

Wang et al. [41] have proposed a constant-round D2D group communication protocol designed for both in-coverage, partial coverage and out-of-coverage scenario. This paper focuses on authentication, key exchange, and group key update policy. During the protocol setup, one sponsor will be chosen in the D2D network. In case that no cellular network is available, this sponsor DUE in the D2D network will be chosen to generate a timestamp and related parameters. After D2D device discovery is completed, the sponsor will assign all DUEs with their own public–private key pair and pseudo-identity for joining the D2D network. These keys and identity will be used to ensure proper authentication and privacy of DUEs. Then, DUEs will enter a two-round process which makes use of DHKE to generate session keys and establish sessions with other DUEs. While encountering scenarios that some group members are joining or leaving the D2D group, the group session key of that D2D network will be updated with the two-round process accordingly, such that forward and backward data secrecy of that group can be protected. The simulation shows that with elliptic curve cryptography (ECC) used, all the keys can be generated with a low computational burden of DUEs. It is also believed that the protocol is resilient to denial-of-service attacks because it can work even without the cellular network. However, since the paper has not focus on D2D device discovery before group formation and data encryption scheme after group formation, privacy and data integrity can only be ensured if proper privacy-preserving discovery processes and data encryption schemes are also employed.

Shang et al. [42, 43] have proposed a certificateless authentication and key agreement protocol for D2D group communication in the 5G cellular network (CAKA-D2D). For a certificateless system, the complete set of public–private key pairs consists of two parts, namely the "first part" generated by a trusted entity (i.e., the CA), and the "second part" generated by the users registered into the trusted entity. It guarantees that even if the keys in the trusted entity (i.e., the "first part") are leaked, the data is still secure without the private keys from the "second part". In the proposal, all DUEs are connected to a 3GPP 5G cellular network, and there are totally 5 stages for establishing secure group authentication. The first stage is system setup and user registration. The AMF in 5GC generates a master key and publicizes the elliptic curve parameters to all DUEs, and every DUE register themselves to AMF to get their pseudo-identities and the first part of public–private key pair. The DUE will also generate the second part of the public–private key pair locally. The second stage is D2D discovery, all DUEs broadcast their pseudo-identities and their complete set of public keys to other DUEs. The third stage is the group session request. After receiving broadcasts from other intended group member DUEs, every participant needs to send the aggregated pseudo-identities of other group members to AMF. AMF will return a new session identity of that group. The fourth stage is the key agreement. Each DUE generates a random key hint and sends the digitally signed encrypted key hint to their group members. Each DUE also needs to verify the key hints from other group members and aggregate them as one final group session key. Finally, in the fifth stage of group activation, each DUE sends the hashed group session key to AMF for verification. If AMF finds all the received hash keys are identical, it will claim the D2D group as activated, and all group members can now start to communicate using that common group session key. For security analysis, the proposed scheme is tested using the AVISPA tool to guarantee that it prevents common active and passive attacks including eavesdropping, MITM and desynchronization attacks. To further protect grouped forward and backward secrecy, the proposal has also mentioned about the key update procedures of different connection scenarios. Also, the ECDSA signature used in the scheme guarantees the prevention of impersonation attacks, and the pseudo-identity makes sure device anonymity. The use of a certificateless system has also prevented the key escrow problem which allows the single point of failure in AMF. The simulation shows that the computational cost of the scheme is lower than most IBE-based schemes, but more analysis is yet to be carried out to understand its performance compared with other group AKA schemes.

Paula et al. [44] proposed a D2D group authentication protocol based on asymmetric cryptography and aggregated signature. Aggregated signature is a mechanism to combine all signatures from different participants in a group into a single signature, such that the group leader only needs to send one combined signature to the verifier to reduce bandwidth consumption. In the proposal, all D2D UEs get a pseudo-identity and other encryption parameters from an authentication center. The center server will also elect one D2D group leader among all surrounding users based on its battery life, computational power, and other factors. To start a group D2D communication, each group member UEs need to generate one signature based on its pseudo-identity, current timestamp, and a random number. When the group leader UE received all signatures from its members, it aggregates them and sends the result to the authentication center. If the validation succeeded, then all incoming signatures are legitimate, and the session key generation procedure with DHKE continues. Finally, the authentication center verifies the session keys from all users in the D2D group and returns success indications to them one-by-one. Security evaluation with the AVISPA tool shows that the proposal provides mutual authentication, replay attack prevention, redirection attack prevention, and group perfect forward and backward secrecy. Performance evaluation shows that it achieved similar computational overheads to PPAKA-HMAC [45]. However, it could also introduce more computational overheads to the authentication servers: it requires doing much extra work such as forming D2D groups dynamically, picking group leaders, verifying the aggregated signatures from group leaders, and sending session key success events to all D2D users one-by-one.

4.3.2 Identity-Based Cryptography (IBE / IBC)

As an enhancement of [17], Abd-Elrahman et al. have proposed the concept of combining Group Key Management (GKM) mechanism with Identity-based Encryption (IBE) to secure D2D multicast group communication [18]. In the paper, it assumes that the connection between UE and core network is pre-secured, and multiples private key generators (PKG) are built in the core network. Additionally, each D2D key generator will only distribute keys to its corresponding D2D user group, and all D2D communication is based on a smaller subgroup that only has two members to share one symmetric key. After the D2D device discovery and discovery message verification phase which is similar to [17], each D2D UEs will be assigned into some groups based on Logical Key Hierarchy (LKH) protocol. For example, in the 4-ary key tree structure, 4 D2D communication pairs (subgroups) will be assigned into one larger group, and 4 groups will be managed by one KGC. After that, each subgroup will generate their own symmetric subgroup key, and KGC will generate another symmetric group key for multicast messages encryption. If there is anyone joining or leaving the multicast group, subgroup and KGC will do re-keying together to achieve perfect forward and backward secrecy. This protocol inherits all the security characteristics of [17], also it does not suffer from the key escrow problem. However, since coordination of core network and computationally expensive group formation in PKG are required in this protocol, it can be applied only at the in-coverage connection scenario and it may not benefit the public safety applications mentioned in the paper.

Wang et al. [45] have proposed two authentication and key agreement protocols (PPAKA-HMAC and PPAKA-IBS) for D2D group communication. These two algorithms work in similar ways, while PPAKA-IBS uses the identity-based signature rather than the HMAC to provide stronger resistance to internal attacks by malicious users in the group. In the proposal, it is assumed that all DUEs are under the coverage and registered to the cellular network (i.e., in coverage scenario). Both protocols have 6 stages. The first stage is the system setup, the serving network will initialize all cryptographic parameters for group formation. The second stage is user registration, the serving network will distribute pseudo identities to all DUEs, and for PPAKA-IBS, an extra public–private key pair will also be distributed. The third stage is the D2D discovery, DUEs will discover each other with their pseudo identities. The fourth stage is the session request, DUEs will ask the serving network for D2D group formation and the serving network will reply with a list of pseudo identities of the same group. For PPAKA-HMAC, an extra HMAC key will also be sent to all group members for future message signing. The fifth stage is the session establishment. For PPAKA-HMAC, there will be a two-round algorithm in which the first round uses HMAC to authenticate all users and the second-round computes session keys. For PPAKA-IBS, the two-round algorithm is more complicated since it involves verification of many identity-based signatures sent from group member DUEs. The final stage is group session activation, the common session key computed by UEs will be hashed as a seed, and it will be sent to the serving network for verification. If all seeds sent from UEs in the same group are the same, the serving network will store the seed into a management table, and the group formation is completed. For scenarios such as the expiration of the session key or a new user is joining or leaving, a new session key will be calculated and updated within the group immediately to make sure group backward and forward secrecy of data. Security analysis also showed that the proposed protocols can fulfill the requirements of data confidentiality, authentication, and device privacy.

Improving the two protocols (CN-GD2C and NA-GD2C) mentioned in [39] with Boneh and Franklin Identity-Based Encryption (BF-IBE), k-anonymous secret handshake, DHKE and zero-knowledge proof, Hsu et al. [40] have proposed two group D2D communication and key exchange protocols, namely "Group Anonymous and Accountable D2D Communication in Mobile Network (GRAAD)". BF-IBE scheme has a total of four algorithms: "Setup" is to initialize a master key and its parameters. "Extract" is to return the private key given the identity ID. "Encrypt" is to encrypt a message with given encryption parameters and identity. "Decrypt" is to decrypt the ciphertext using the private key. For CN-GD2C (cellular network presence) scenario, using BF-IBE, two DUEs firstly choose a random session identity, then DUE1 will send an encrypted D2D request message containing Diffie-Hellman variable and other parameters to DUE2. Then, DUE2 will also send similar information concatenated with DUE1's request message and send it to HSS and ProSe Server. Since HSS (the KGC of BF-IBE) and ProSe Server have all device information and group belonging information, after several times of validation, if DUE1 and DUE2 are in the same group, they will finally be able to authenticate each other. For NA-GD2C (cellular network absence) case, although the procedure is similar to, it is a more complicated mechanism because four new functions including gSelect, uSelect, gSelectVer and uSelectVer are introduced to generate encryption parameters to perform IBE-based encryption locally instead of exchanging a list of certificates. Security analysis shows that the two schemes can provide D2D group anonymity, data confidentiality (with IND-CCA and forward secrecy proven), device privacy while preserving device traceability and revocability.

As an advancement of [19], Sun et al. [46] have proposed an Efficient Anonymity Proximity Device Discovery and Batch Authentication Mechanism (EAP-DDBA) for one-to-many D2D connection establishment in 3GPP 5G network. The proposed scheme combines the privacy-preserving discovery procedure in [19] that encrypts the broadcast message with identity-based prefix encryption. Therefore, only authorized D2D users can read the content of the broadcast message and initiate connection requests using the ECDH parameters in the encrypted broadcast message. Subsequently, certificateless batch signature (CLBS) is employed such that the broadcasting UE can verify multiple signatures from different connection initiator DUEs in a single batch. After that, to respond to the connection initiators that all incoming connections are approved in just one single message, the broadcasting UE combines all hashed identities of connection initiator DUEs into one single number using the Chinese Remainder Theorem (CRT). Finally, these connection initiator DUEs can calculate the remainder using their parameters and the received number to verify if the broadcasting UE has approved the connection. The scheme has three phases, including the system parameter generation phase that generates all encryption parameters, initial access authentication phases that uses 5G-AKA or EAP-AKA' to build secure channels between the home network and UEs, and the device discovery and authentication phase that does not rely on the home network to establish a new connection. Security analysis shows that this scheme can provide mutual authentication, device anonymity and perfect forward secrecy. It can also resist many attacks including stolen private key attacks, replay attacks, desynchronization attacks and battery depletion attacks. Performance analysis shows that the scheme performs faster than [25, 40, 45]. However, unlike most other schemes that share the same session key among all group users, this scheme is designed for one-to-many D2D communication. As a result, users within the same group use their session keys to connect to a single point, and star topology is adopted instead of mesh topology for subsequent communications (Table 2).

Table 2 Schemes for securing one-to-many group D2D communication

Scheme	Technology	Security properties								
		DC	I	A	F	R	AD	NR	PA	T
[18]	IBE, ECDSA and ECDH	Y	Y	Y				Y	Y	
[39]	PKI, DHKE	Y	Y	Y					Y	Y
[40]	IBE, DHKE	Y	Y	Y	Y				Y	Y
[41]	PKI, Gap-DH group signature, XOR	Y	Y	Y			Y		Y	
[42, 43]	Certificateless, XOR	Y	Y	Y						
[44]	Aggregated signature, DHKE	Y	Y	Y		Y		Y	Y	Y
[45]	PPAKA-HMAC: HMAC, DHKE PPAKA-IBS: IBS, DHKE	Y	Y	Y					Y	
[46]	ECDH, CLBS, CRT	Y	Y	Y	Y				Y	

4.4 Continuous Authentication

Some schemes focused on authenticating the DUE continuously after successful initial authentication and key agreement are also proposed to secure D2D communications in 5G network.

Abualhaol et al. [47] have proposed the idea of using legitimacy patterns to continuously authenticate the D2D connections. Legitimacy pattern refers to a secret pattern agreed on both D2D senders and receivers. During usual data transmission, this pattern will be inserted at the end of one packet, while the location of that packet can be anywhere. If this pattern is found altered or disappeared in the D2D receiver side, the D2D communication is likely under any physical layer attacks including unauthorized attack or data alternation. To evaluate the security of this protocol, instead of revisiting the whole standard, a mathematical representation called security score (SeS) of the protocol is also proposed to evaluate how secure the system is. By definition, SeS is the probability that given the physical attack has happened, the chance that the system can detect this attack successfully. Security analysis showed that although this protocol may not easily against eavesdropping attacks, some physical attacks such as jamming, restricting access and packet injection can be detected. The simulation also shows that by increasing the percentage of legitimate patterns in the message, the SeS can also be improved. However, with longer legitimate patterns, it is expected that attackers may also find and replicate the pattern. Also, the overall throughput of the network could decrease. As a result, the proposed scheme still needs to be improved for actual deployment in the 5G network.

Tan et al. [48] have proposed a D2D authentication scheme that uses certificateless authentication for initial connection setup and smartphone sensors data for continuous authentication. In the proposal, it is assumed that all DUEs should be in the in-coverage or relay coverage scenario, and all DUEs are smartphones equipped with both accelerometer and gyroscope. For connection establishment, the scheme has 3 main phases. The first phase is "Offline Registration Phase" which registers the DUE to the serving network (SN) with user identification information. After that, a unique license ID will be assigned to the DUE and saved securely for the usage in the next phase. The second phase is the "Authentication Phase". During this phase, decisional bilinear Diffie-Hellman (DBDH) assumptions will be used to authenticate the DUE. The final stage is "Group Key Distribution State". If the DUE can be verified by SN, DUE will join a D2D group, and SN will construct a new group key and distribute it to all group members in this stage. Unlike most authentication schemes that only focus on the key agreement, this scheme also introduced the concept of continuous authentication. By making use of the activity data collected from accelerometer and gyroscope, DUE will also use machine learning techniques including SVM (Support Vector Machine) and Radial Basis Function (RBF) to classify the behavior of the user and send to SN. Finally, SN will collect all data within the group and compare the list of activities done by many users to determine if the DUE is trustworthy or it needs reauthentication. Security analysis shows that by using certificateless authentication, key escrow problems in IBE can be solved and it also provided device privacy since license ID is used. However, collecting behavioral data of users and sending it to SN will introduce new privacy concerns, and all users within the group may not behave the same. This could potentially create redundant reauthentication work.

4.5 Content Sharing Models

Unlike the traditional way of building a secured communication channel with symmetric key exchange schemes, some other proposals use different strategies for encrypting messages in 5G D2D systems. These schemes aim to provide more efficient content sharing and data relaying. It could also provide new features such as message aggregation and manipulations.

Zhang et al. [21] have proposed a secure D2D data sharing model (SeDS) which is based on the cooperation of both gateway (GW), base stations (such as eNB), the service provider (SP) and DUEs in a cellular network. This model uses different types of cryptographic techniques such as public-key based signature and symmetric encryption to ensure data confidentiality, authentication, and device privacy. Its non-repudiation characteristics can also help to prevent free-riding attacks. For any authenticated DUEs in the same core network, base stations will assign one pseudo-identity for each DUEs, and they will exchange public–private key pairs with DUEs through the secured channel. Then, GW will actively search for potential D2D pairs and assist DUEs to pair each other securely without revealing the real identity of another DUE automatically. After that, these DUEs can share the information provided by SP using D2D communication. To secure the integrity of data and ensure non-repudiation, all data shared by SP are pre-signed and all messages sent through DUEs will be re-signed with either public-key based signature or HMAC. These data would also be encrypted using symmetric key encryption to provide data confidentiality. Finally, to decrypt and verify the D2D messages, recipient DUE needs to send a key hint request message to base stations. Base stations will decide to either assist data decryption or punish the recipient DUE if it behaves maliciously. This scheme is particularly useful for P2P content sharing during the in-coverage connection scenario such as video streaming in a highly dense area. However, since it involves the support of many infrastructures in the cellular network, the availability and dependability of the D2D data transmission are heavily dependent on the performance of these infrastructures. Also, since this proposed model focuses only on data sharing which downloads data from a single source (i.e., the SP), it should be further enhanced to fit other D2D use cases.

A D2D-assist data transmission protocol for M-Health systems has been proposed by Zhang et al. [49] and has been revised by Zhou [50]. This protocol consists of two parts: the first part is an algorithm called "Certificateless Generalized Signcryption (CLGSC)" which performs both encryption and signature at the same time. In short, CLGSC is the algorithm which combines key generation centers, public/private key encryption, and hash functions to provide both key generation and data protection. The second part is a protocol called "Lightweight and Robust Security-Aware (LRSA)" D2D-assist data transmission protocol which transmits the data through the data source, relays, and destinations securely. To start data transmission from the health device to the physician, health device needs to undergo 4 phases including system initialization, data formulation, data transmission and data receiving and processing. For the system initialization stage, a network manager (NM) will generate all encryption parameters such as public–private key pairs and secure hash functions. Then all health devices with pseudo-identities and all physicians need to register to the NM to get their keys. After that, the health device and the physician will start the initial session key agreement to generate a symmetric key using DHKE. The second phase is data formation that health devices would encrypt the content using CLGSC with the physician's public key. The third phase is data transmission. Since data may pass through many relays, the signature of the data will be verified in each relay and signed with the relay's public key using CLGSC. After the message is relayed to NM, NM will verify all the signatures and decrypt the relayed content. At this moment, the content encrypted with the physician's public key will be stored. The final stage is data receiving and processing. When the physician receives the data from NM, he will verify the signature and decrypt the content using his private keys. Security analysis shows that the scheme provided data confidentiality, data integrity, mutual authentication, and device privacy. Although the papers assumed that data senders are healthcare devices and data receiver is a server for physicians, this scheme can be applied to any D2D client–server model which NM or KGC exists.

Homomorphic encryption can also be used for building D2D content sharing protocols. For instance, Wang et al. [51] have proposed a secure content sharing protocol (SCSP) which does the D2D matching using social network and shares common hot topics automatically. In the paper, it is assumed that both DUEs are within the same core network, and both are visiting the same social network server (SNS). The proposed protocol has 3 main phases. The first phase is system initialization which allows trust authority to generate encryption parameters, register SNS with a public/private key pair and register all DUEs. The second phase is user profile matching which every DUEs will submit their public key encrypted interest index table (a table indicating preferred topics or content of a user) to SNS. After that, SNS will calculate their common interest score and credibility score using homomorphic multiplication, decryptions, and other operations. The final phase is the secure content sharing phase that DUE can get the common interest topics by requesting base stations with a request message. This message is encrypted, and its integrity is protected by the HMAC code. Finally, the base station will assist connection and secure message sharing between two adjacent DUEs using ECDH key exchange. Although the proposed scheme can protect data confidentiality, data integrity, device privacy, non-repudiation and provide mutual authentication, it only does homomorphic multiplication in SNS, but SNS owned the private key of the encrypted content. This could be further improved if the SNS does not have a private key, and all credibility is calculated locally on DUEs.

5 Open Research Issues

After studying the existing literature extensively, we have found some unresolved issues in the current research. In this section, we analyze these issues and propose some open research directions:

A universal 5G-D2D protocol: most of the existing D2D protocols are not designed following the system model of 3GPP 5G security architecture. Even though some of them can be converted to the 3GPP 5G system model due to their similarities, these protocols can only satisfy a few security requirements and therefore may not be secure enough for deployment. Moreover, most of the existing protocols cannot provide all functional requirements for the 5G-D2D network including open device discovery, restricted device discovery, public safety relay, mutual authentication, and key agreement for both one-to-one and one-to-many connection scenarios. Therefore, a universal 5G-D2D protocol with full functionalities and a high level of security should be designed in the future.

Post-quantum cryptography: as the development of quantum computers is accelerating and the Shor's algorithm in quantum computers is proven to be implementable, most existing cryptographic functions based on modular arithmetic and computational Diffie-Hellman, such as RSA and ECC mentioned in the literature, are likely to be vulnerable soon. While physical layer security remains impractical and it is still vulnerable to many other attacks, some post-quantum cryptography such as lattice-based cryptography, code-based cryptography and symmetric encryption should be considered to further provide quantum attack resistivity in 5G D2D.

Lightweight protocol for 5G IoT devices: while most of the literature uses modular arithmetic, identity-based cryptography, and attribute-based cryptography to provide mutual authentication and message signature, these cryptographic functions are computationally expensive at key generation, encryption, and decryption. However, in the 5G network, it is expected that there will be a tremendous amount of 5G IoT devices, connected sensors and cyber-physical systems that have very limited battery and computational resources. To enable all these resource-constrained devices to enjoy the benefits of D2D communication, lightweight protocols that use efficient algorithms should be carefully designed such that it can provide an efficient 5G D2D network without sacrificing major security requirements.

Fine-grained access control: most literature follows the 3GPP ProSe security framework which only provides either open discovery that everyone can discover nearby DUEs or the restricted discovery that all registered users in the 5GC can discover DUEs. However, with new technology such as social relationship-based security, homomorphic encryption and attribute-based encryption introduced, it is expected that fine-grained device discovery such as friends only or common interests only discovery procedures can also be facilitated. The same concept can also be extended to provide fine-grained access control to the message content in a group D2D communication.

Decentralized authentication: most literature follows the assumption that the AUSF in 5GC is the single certificate authority for all devices in the 3GPP 5G network. However, with the rapid development of blockchain applications, multi-party computation (MPC) and decentralized network, it is possible that the 5G D2D communication in the future could build its trust model based on a decentralized blockchain with the concept of proof of work (PoW), Merkle tree and digital signatures. This could further improve system reliability by preventing the single point of failure in one authority and potentially improve the overall security of the network.

6 Conclusion

5G D2D communication is one of the promising features in 5G network that brings various benefits to the network, but there are also new security challenges yet to be solved. In this book chapter, we provided a comprehensive overview to the security issues in 5G D2D network based on the security architecture of 3GPP ProSe Service and the system architecture of 3GPP 5G Core Network. We firstly discussed various 5G D2D application scenarios such as discovery, one-to-one communication, one-to-many communication, and then we classified some security challenges and suggested security requirements for them. Based on the above requirements, we also reviewed the existing work related to 5G D2D security extensively and proposed some open research issues that are yet to be explored in the future. We hope that the discussions in this chapter could be served as a pointer for future researchers to design a more secure and robust 5G D2D solution.

Acknowledgements This work is supported by the MOE AcRF Tier 1 funding for the project of RG 26/18 by Ministry of Education, Singapore.

References

1. Ansari, R.I., Chrysostomou, C., Hassan, S.A., Guizani, M., Mumtaz, S., Rodriguez, J., Rodrigues, J.J.P.C.: 5G D2D networks: techniques, challenges, and future prospects. IEEE Syst. J. **12**(4), 3970–3984 (2018). https://doi.org/10.1109/JSYST.2017.2773633
2. Feng, D., Lu, L., Yi, Y.W., Li, G.Y., Feng, G., Li, S.: Device-to-device communications underlaying cellular networks. IEEE Trans. Commun. **61**(8), 3541–3551 (2013). https://doi.org/10.1109/TCOMM.2013.071013.120787
3. Jameel, F., Hamid, Z., Jabeen, F., Zeadally, S., Javed, M.A.: A survey of device-to-device communications: Research issues and challenges. IEEE Commun. Surv. Tutor. **20**(3), 2133–2168 (2018). https://doi.org/10.1109/COMST.2018.2828120
4. Zhang, A., Lin, X.: Security-aware and privacy-preserving D2D communications in 5G. IEEE Network **31**(4), 70–77 (2017). https://doi.org/10.1109/MNET.2017.1600290
5. Asadi, A., Wang, Q., Mancuso, V.: A survey on device-to-device communication in cellular networks. IEEE Commun. Surv. Tutor. **16**(4), 1801–1819 (2014). https://doi.org/10.1109/COMST.2014.2319555
6. 3GPP: Technical Specification Group Services and System Aspects; Proximity-based services (ProSe) (Release 15); TS23.303 (2018). [Online]. Available: https://www.3gpp.org/DynaReport/23303.htm
7. 3GPP: Technical Specification Group Services and System Aspects; Feasibility study for Proximity Services (ProSe) (Release 12); TS22.803 (2013). [Online]. Available: https://portal.3gpp.org/desktopmodules/Specifications/SpecificationDetails.aspx?specificationId=653
8. Wang, M., Yan, Z.: A survey on security in D2D communications. Mobile Netw. Appl. **22**(2), 195–208 (2017). https://doi.org/10.1007/s11036-016-0741-5
9. Nait Hamoud, O., Kenaza, T., Challal, Y.: Security in device-to-device communications: a survey. IET Networks **7**(1), 14–22 (2018). https://doi.org/10.1049/iet-net.2017.0119
10. Wu, X., Tavildar, S., Shakkottai, S., Richardson, T., Li, J., Laroia, R., Jovicic, A.: FlashLinQ: a synchronous distributed scheduler for peer-to-peer ad hoc networks. IEEE/ACM Trans. Networking **21**(4), 1215–1228 (2013). https://doi.org/10.1109/TNET.2013.2264633
11. Wi-Fi P2P Technical Specification v1.7. Wi-Fi Alliance, Dec 2016
12. 3GPP: Universal Mobile Telecommunications System (UMTS); LTE; Proximity-based Services (ProSe); Security aspects (3GPP TS 33.303 version 15.0.0 Release 15) (2018)
13. 3GPP: System architecture for the 5G System (5GS) (Release 15.7.0); TS 23.501. p. 353 (2019). [Online]. Available: https://portal.etsi.org/TB/ETSIDeliverableStatus.aspx
14. Haus, M., Waqas, M., Ding, A.Y., Li, Y., Tarkoma, S., Ott, J.: Security and privacy in device-to-device (D2D) communication: a review. IEEE Commun. Surv. Tutor. **19**(2), 1054–1079 (2017). https://doi.org/10.1109/COMST.2017.2649687
15. Zhang, S., Wang, Y., Zhou, W.: Towards secure 5G networks: a survey. Comput. Netw. **162**, 106871 (2019). https://doi.org/10.1016/j.comnet.2019.106871
16. Wang, M., Yan, Z.: Security in D2D communications: a review. In: Proceedings—14th IEEE International Conference on Trust, Security and Privacy in Computing and Communications, TrustCom 2015, Aug. 2015, vol. 1, pp. 1199–1204. https://doi.org/10.1109/Trustcom.2015.505
17. Abd-Elrahman, E., Ibn-Khedher, H., Afifi, H., Toukabri, T.: Fast group discovery and non-repudiation in D2D communications using IBE. In: IWCMC 2015—11th International Wireless Communications and Mobile Computing Conference, Aug. 2015, pp. 616–621. https://doi.org/10.1109/IWCMC.2015.7289154

18. Abd-Elrahman, E., Ibn-Khedher, H., Afifi, H.: D2D group communications security. In: International Conference on Protocol Engineering, ICPE 2015 and International Conference on New Technologies of Distributed Systems, NTDS 2015—Proceedings, Jul. 2015, pp. 1–6. https://doi.org/10.1109/NOTERE.2015.7293504
19. Sun, Y., Cao, J., Ma, M., Li, H., Niu, B., Li, F.: Privacy-preserving device discovery and authentication scheme for D2D communication in 3GPP 5G HetNet. In: 2019 International Conference on Computing, Networking and Communications, ICNC 2019, Feb. 2019, pp. 425–431. https://doi.org/10.1109/ICCNC.2019.8685499
20. Sedidi, R., Kumar, A.: Key exchange protocols for secure Device-to-Device (D2D) communication in 5G. In: IFIP Wireless Days, Mar. 2016, vol. 2016-April, pp. 1–6. https://doi.org/10.1109/WD.2016.7461477
21. Zhang, A., Chen, J., Hu, R.Q., Qian, Y.: SeDS: Secure data sharing strategy for d2d communication in LTE-advanced networks. IEEE Trans. Veh. Technol. 65(4), 2659–2672 (2016). https://doi.org/10.1109/TVT.2015.2416002
22. Schmittner, M., Asadi, A., Hollick, M.: SEMUD: secure multi-hop device-to-device communication for 5G public safety networks. In: 2017 IFIP Networking Conference, IFIP Networking 2017 and Workshops, Jun. 2017, vol. 2018-Janua, pp. 1–9. https://doi.org/10.23919/IFIPNetworking.2017.8264846
23. Suraci, C., Pizzi, S., Iera, A., Araniti, G.: Enhance the protection of transmitted data in 5G D2D communications through the Social Internet of Things. In: IEEE International Symposium on Personal, Indoor and Mobile Radio Communications, PIMRC, Sep. 2018, vol. 2018-Septe, pp. 376–380. https://doi.org/10.1109/PIMRC.2018.8580860
24. Suraci, C.,Pizzi, S., Garompolo, D., Araniti, G., Molinaro, A., Iera, A.: Trusted and secured D2D-aided communications in 5G networks. Ad Hoc Netw. 114(November 2019), 102403 (2021). https://doi.org/10.1016/j.adhoc.2020.102403
25. Wang, M., Yan, Z., Niemi, V.: UAKA-D2D: universal authentication and key agreement protocol in D2D communications. Mobile Netw. Appl. 22(3), 510–525 (2017). https://doi.org/10.1007/s11036-017-0870-5
26. Seok, B., Sicato, J.C.S., Erzhena, T.,. Xuan, C., Pan, Y., Park, J.H.: Secure D2D communication for 5G IoT network based on lightweight cryptography. Appl. Sci. (Switzerland) 10(1) (2020). https://doi.org/10.3390/app10010217
27. Melki, R., Noura, H.N., Chehab, A.: Lightweight and Secure D2D authentication & key management based on PLS. In: IEEE Vehicular Technology Conference, Sep. 2019, vol. 2019-Septe, pp. 1–7. https://doi.org/10.1109/VTCFall.2019.8891531
28. Baskaran, S.B.M., Raja, G.: A lightweight incognito key exchange mechanism for LTE-A assisted D2D communication. In: 2017 9th International Conference on Advanced Computing, ICoAC 2017, pp. 301–307 (2018). https://doi.org/10.1109/ICoAC.2017.8441370
29. Chow, M.C., Ma, M.: A Lightweight D2D authentication scheme against free-riding attacks in 5G cellular network. In: Proceedings of the 2020 2nd International Electronics Communication Conference, Jul. 2020, pp. 143–149. https://doi.org/10.1145/3409934.3409952
30. Javed, Y., Khan, A.S., Qahar, A., Abdullah, J,: EEoP: a lightweight security scheme over PKI in D2D cellular networks. J. Telecommun. Electron. Comput. Eng. 9(3–11), 99–105 (2017). [Online]. Available: http://journal.utem.edu.my/index.php/jtec/article/view/3191
31. Abro, A., Deng, Z., Memon, K.A.: A lightweight elliptic-elgamal-based authentication scheme for secure device-to-device communication. Future Internet 11(5), 108 (2019). https://doi.org/10.3390/fi11050108
32. Dao, N.N., Na, W., Lee, Y., Vu, D.N., Cho, S.: Prefetched asymmetric authentication for infrastructureless D2D communications: feasibility study and analysis. In: 9th International Conference on Information and Communication Technology Convergence: ICT Convergence Powered by Smart Intelligence, ICTC 2018, Oct. 2018, pp. 1053–1054. https://doi.org/10.1109/ICTC.2018.8539475
33. Boubakri, W., Abdallah, W., Boudriga, N.: Access control in 5G communication networks using simple PKI certificates. In: 2017 13th International Wireless Communications and Mobile Computing Conference, IWCMC 2017, Jun. 2017, pp. 2092–2097. https://doi.org/10.1109/IWCMC.2017.7986606

34. Hamoud, O.N., Kenaza, T., Challal, Y.: Towards using multiple KGC for CL-PKC to secure D2D communications. In: 2018 International Conference on Smart Communications in Network Technologies, SaCoNeT 2018, Oct. 2018, pp. 283–287. https://doi.org/10.1109/SaCoNeT.2018.8585671

35. Hamoud, O.N., Kenaza, T., Challal, Y.: A new certificateless system construction for multiple key generator centers to secure device-to-device communications. In: ICETE 2019—Proceedings of the 16th International Joint Conference on e-Business and Telecommunications, vol. 2, pp. 84–95 (2019). https://doi.org/10.5220/0007841500840095

36. Kwon, H., Kim, D., Hahn, C., Hur, J.: Secure authentication using ciphertext policy attribute-based encryption in mobile multi-hop networks. Multimed. Tools Appl. **76**(19), 19507–19521 (2017). https://doi.org/10.1007/s11042-015-3187-z

37. Yan, Z., Xie, H., Zhang, P., Gupta, B.B.: Flexible data access control in D2D communications. Futur. Gener. Comput. Syst. **82**, 738–751 (2018). https://doi.org/10.1016/j.future.2017.08.052

38. Wang, M., Yan, Z., Song, B., Atiquzzaman, M.: AAKA-D2D : anonymous authentication and key agreement protocol in D2D communications. In: 2019 IEEE SmartWorld, Ubiquitous Intelligence & Computing, Advanced & Trusted Computing, Scalable Computing & Communications, Cloud & Big Data Computing, Internet of People and Smart City Innovation (2019). https://doi.org/10.1109/SmartWorld-UIC-ATC-SCALCOM-IOP-SCI.2019.00248

39. Hsu, R.H., Lee, J.: Group anonymous D2D communication with end-to-end security in LTE-A. In: 2015 IEEE Conference on Communications and NetworkSecurity, CNS 2015, Sep. 2015, pp. 451–459. https://doi.org/10.1109/CNS.2015.7346857

40. Hsu, R.H., Lee, J., Quek, T.Q.S., Chen, J.C.: GRAAD: group anonymous and accountable D2D communication in mobile networks. IEEE Trans. Inf. Forensics Secur. **13**(2), 449–464 (2018). https://doi.org/10.1109/TIFS.2017.2756567

41. Wang, L., Tian, Y., Zhang, D., Lu, Y.: Constant-round authenticated and dynamic group key agreement protocol for D2D group communications. Inf. Sci. **503**, 61–71 (2019). https://doi.org/10.1016/j.ins.2019.06.067

42. Shang, Z., Ma, M., Li, X.: A certificateless authentication protocol for D2D group communications in 5G cellular networks. In: 2019 IEEE Global Communications Conference, GLOBECOM 2019—Proceedings, pp. 1–7 (2019). https://doi.org/10.1109/GLOBECOM38437.2019.9014047

43. Shang, Z., Ma, M., Li, X.: A secure group-oriented device-to-device authentication protocol for 5G wireless networks. IEEE Trans. Wirel. Commun. **300350**(c), 1–1 (2020). https://doi.org/10.1109/TWC.2020.3007702

44. Paula, A., Lopes, G., Gondim, P.R.L.: Group authentication protocol based on aggregated signatures for D2D communication. Comput. Netw., 107192 (2020). https://doi.org/10.1016/j.comnet.2020.107192

45. Wang, M., Yan, Z.: Privacy-preserving authentication and key agreement protocols for D2D group communications. IEEE Trans. Industr. Inf. **14**(8), 3637–3647 (2018). https://doi.org/10.1109/TII.2017.2778090

46. Sun, Y., Cao, J., Ma, M., Zhang, Y., Li, H., Niu, B.: EAP-DDBA: efficient anonymity proximity device discovery and batch authentication mechanism for massive d2d communication devices in 3GPP 5G HetNet. IEEE Trans. Dependable Secure Comput. **14**(8), 1–1 (2020). https://doi.org/10.1109/tdsc.2020.2989784

47. Abualhaol, I., Muegge, S.: Securing D2D wireless links by continuous authenticity with legitimacy patterns. In: Proceedings of the Annual Hawaii International Conference on System Sciences, Jan. 2016, vol. 2016-March, pp. 5763–5771. https://doi.org/10.1109/HICSS.2016.713

48. Tan, H., Song, Y., Xuan, S., Pan, S., Chung, I.: Secure D2D group authentication employing smartphone sensor behavior analysis. Symmetry **11**(8), 969 (2019). https://doi.org/10.3390/sym11080969

49. Zhang, A., Wang, L., Ye, X., Lin, X.: Light-Weight and robust security-aware D2D-assist data transmission protocol for mobile-health systems. IEEE Trans. Inf. Forensics Secur. **12**(3), 662–675 (2017). https://doi.org/10.1109/TIFS.2016.2631950

50. Zhou, C.: An improved lightweight certificateless generalized signcryption scheme for mobile-health system. Int. J. Distrib. Sens. Netw. **15**(1), 1550147718824465 (2019). https://doi.org/10.1177/1550147718824465
51. Wang, L., Li, Z., Chen, M., Zhang, A., Cui, J., Zheng, B.: Secure content sharing protocol for D2D users based on profile matching in social networks. In: 2017 9th International Conference on Wireless Communications and Signal Processing, WCSP 2017—Proceedings, Oct 2017, vol. 2017-Janua, pp. 1–5. https://doi.org/10.1109/WCSP.2017.8171117

New Waves of Cyber Attacks in the Time of COVID19

Izzat Alsmadi and Lo'ai Tawalbeh

Abstract For hackers and many actors in the cyber world, crises can make opportunities. With the continuous spread of Covid-19 pandemics, new waves of cyberattacks are witnessed. Attackers are taking advantage of possibly increasing vulnerabilities due to lack of awareness of best security practices when using online resources by many new users. While the data and network security specialist do their best to protect the cyber users, and Information Technology companies strive to look for threats countermeasures, in this chapter, we are presenting, revising, and analyzing different types of cyberattacks which might exist before, but evolved at the COVID-19 era. Moreover, we are addressing the new cyberattacks that flourished during COVID-19 pandemic from different prospective including their economical and social impact. Finally, we are discussing possible countermeasures to detect and prevent these attacks.

Keywords Cyberattacks · Vulnerabilities · Data security · Privacy

1 Introduction

Over the time, the humanity passed through many disasters and crisis that had huge impact on the humans life. Besides the wars, there is the earthquakes, volcanos, floods, hurricanes, contagious diseases, and many others. These crises have impact on many aspects of our lives, including: health, social, psychological, and financial.

One of the recent pandemics is the COVID-19 which has an impact on almost every aspect of our life. It changed the way we live, meet, socialize, communicate, and even the way we work [1]. With the general closure of many cities and countries all over the world, people can't see or meet each other, so they have no choice but

I. Alsmadi (✉) · L. Tawalbeh
Department of Computing and Cyber Security, Texas A&M University, San Antonio, TX 78224, USA
e-mail: ialsmadi@tamusa.edu

L. Tawalbeh
e-mail: Ltawalbeh@tamusa.edu

P. Nicopolitidis et al. (eds.), *Advances in Computing, Informatics, Networking and Cybersecurity*, Lecture Notes in Networks and Systems 289,
https://doi.org/10.1007/978-3-030-87049-2_21

to use online applications to socialize. Adding to that, the schools and university are closed as well, so the students are being taught from home using video conferencing applications including: Webex, Zoom, Google Meets, Microsoft Teams, etc. [2].

At the business level, a new working environment took place, which is virtual through online applications mentioned above. So, work meetings, discussions, and even business deals are conducted online and mostly from our homes. In summary, we can say that almost the whole world moved to the internet. In other words, there is a huge number of new users to the online applications and from different ages from Pre-K to high school students and college students to senior citizens.

Moreover, not all the Internet users have the same technical awareness about the best security practices of using online applications. This fact made the Internet and these applications users a target for cyber-attacks. For example, there is an increasing attacks on video conferencing applications that expose their vulnerabilities such as Zoom attack 2020 [3].

People increased their usage of the Internet and social networks as they are spending more time isolated in their homes. So, while Covid-19 created lots of business challenges, it also creates some business opportunities. For example, it creates new opportunities for online businesses, especially large online retailers such as Amazon and eBay. It also creates new business opportunities that were less common such as online orders for food and meals.

We can say that COVID-19 made the environment easier for the cyber hackers and attackers, to evolve and develop cyber attacks worldwide. For example, the phishing attacks did exist before COVID-19, but it is made more powerful and more damaging on the users all over the world. Another attack that is flourished during this time is using social engineering. Hackers utilized the new environment that has plenty of new users who lack the needed training on using certain online applications, and the lack of enough awareness of appropriate measures and practices when being online which resulted in exploiting human vulnerabilities and launching cyber attacks [4].

In this chapter, we are analyzing how COVID-19 enabled such cyber attacks to be more sever and powerful by discussing related factors including: the new business models that depend on the wide usage of online communication tools, and the lack of authentic details related to Covid-19 possible treatments, origin, etc., and related fake news. The third factor that people need to socialize and communicate, and with all this closures all over the world (no travel, no shopping, no picnics, no schools, no family visits, …, etc.), people have no other option but to spend more time online using social media platforms such as twitter and many others in order to communicate. In the next section, we will focus on those three factors.

2   False Information and Covid19

Covid-19 pandemic large scale impacts on all humans around the world and the lack of reliable and clear response from most countries caused waves of conspiracies around the virus, how it started, and the best ways for humans to protect themselves

True ⓘ	Mostly True ⓘ	Half True ⓘ	Mostly False ⓘ	False ⓘ	Pants on Fire ⓘ
4%	7%	10%	11%	47%	17%
20 Checks	30 Checks	42 Checks	48 Checks	195 Checks	72 Checks

Fig. 1 Covid19 fact-checking from politifact.com

against its infection. Several facts checking websites list large volumes of fake news related to Covid-19. Following are a few examples of those websites:

- The CoronaVirusFacts/DatosCoronaVirus Alliance Database: https://www.poynter.org/ifcn-covid-19-misinformation/
- Covid-19 fact-checking from factcheck.com: https://www.factcheck.org/issue/coronavirus/
- Covid-19 fact-checking from politifact.com https://www.politifact.com/coronavirus/. Figure 1 from politifact.com shows that majority of "rumors" around Covid-19 are not true
- Coronavirus Disease, 2019: Myth versus Fact: https://www.hopkinsmedicine.org/health/conditions-and-diseases/coronavirus/2019-novel-coronavirus-myth-versus-fact
- Debunking COVID-19 (coronavirus) myths: https://www.mayoclinic.org/diseases-conditions/coronavirus/in-depth/coronavirus-myths/art-20485720
- Coronavirus myths explored: https://www.medicalnewstoday.com/articles/coronavirus-myths-explored
- Covid-19: Myths and facts: https://www.avert.org/coronavirus/covid-19-myths-and-facts.

Snopes.com, one of the popular fact-checking websites, lists several categories of claims or miss-information associated with Covid-19, Fig. 2.

The main two aspects of inaccurate information related to Covid19 are related to three categories:

1. The origin of Covid19: The main story of the birth of Covid-19 is related to patients in Wuhan with a background of working in the wholesale animal or wet market [5]. Other stories that the virus was manufactured by a government agency or lab (e.g., China) are still circulating. There have been suggestions that the virus may have come through the biological warfare laboratories.
2. Hoax circulates on also how Covid19 came to humans in the first place and through which animal. Though the origin is still not cleared well, most studies indicate that Covid-19 originates from a strain found in bats [6]. From Wuhan, then China, Covid-19 spreads now to almost every single country in the world. Some people try to differentiate between 'first recorded outbreak' and 'the origin.' Many believe that while the first recorded outbreaks were in Wuhan,

Fig. 2 Covid-19
Miss-information categories
in Snopes.com

- **<u>Origins and Spread</u>**
- **<u>Prevention & Treatments</u>**
- **<u>Prevention & Treatments II</u>**
- **<u>International Response</u>**
- **<u>U.S. Government Response</u>**
- **<u>Trump & the Pandemic</u>**
- **<u>Trump & the Pandemic II</u>**
- **<u>Conspiracy Theories</u>**
- **<u>Prophecies & Predictions</u>**
- **<u>Memes & Misinformation</u>**
- **<u>Memes & Misinformation II</u>**
- **<u>Viral Videos</u>**
- **<u>Business & Industry</u>**
- **<u>Entertainment & Media</u>**

the origin of the virus is yet unknown. This is in comparison to H1N1 Flue in 1918 that was incorrectly correlated to Spain as origin [7].

How is that relevant to cybersecurity and attacks? Recent years witnessed large scale spreading of fake news and rumors by government agencies. Russian Troll Tweets (RTT) dataset and the attempts to influence the US 2016 election was one significant example. Between the US and China, in particular, Covid-19 extended already strained relations. Rumors spread across the two countries where each one is trying to connect the roots of Covid-19 to the other country. Through OSNs in particular, many accounts belong to robots, social bots, or trolls who claim to be humans. Different results showed increased activities of social bots in the time of Covid19 [8, 9].

3. The statistics: Different countries around the world reported their Covid-19 statistics in terms of reported cases, new, recovered, and death cases. Nonetheless, the accuracy of those numbers was questioned. Different factors can be behind why reported numbers are not accurate:

   a. To avoid public response, some governments may not report actual infected cases. They may not be associated with all death cases with Covid-19 for the same reason.

   b. Availability and accurate of Covid-19 test facilities. Many countries may not have the resources to establish testing facilities. Additionally, as a new disease, several reports indicated accuracy problems of test results and that in a certain period of the disease or for certain humans, symptoms may not be visible.

4. Possible treatments: As of the time of this paper, no confirmed treatment is announced for Covid-19. Nonetheless, there are many popular candidates such as Chloroquine and hydroxychloroquine, Lopinavir and ritonavir, Nafamostat and camostat, Famotidine, Umifenovir, Nitazoxanide, Ivermectin, Corticosteroids, Tocilizumab and sarilumab, Bevacizumab, and Fluvoxamine.

People around the world started trying different types of treatments, including chemicals in response to the information they received or learned through TV, the Internet, or OSNs. Some people took advantage of humans' fear and eagerness to find treatment and start making their treatments and sell them through online stores. Some of those fake treatments use popular candidate treatments such as Chloroquine and hydroxychloroquine.

Beyond the candidate drugs mentioned earlier, some of the widespread rumors that spread about possible treatments include possible food types, vitamins [10], and even chemicals: internal usage of disinfections (internal injection). Snopes website lists the following rumors about Covid-19 treatments that include claims related to some food (e.g., Banana, Garlic, Lemons, etc.)

2.1 The Transition to Online Business Models and Security Issues

Before the pandemic, working online options for businesses was optional. Limited number of employees worked from home before COVID-19. In early 2020 things changed and new business models were forced. Majority of the companies world wide had to move their business from their offices to online virtual offices. The employees are staying home and not coming to offices, companies headquarters are empty and all the meetings are done via video conferences applications.

Moreover, Universities and schools had to adopt virtual teaching and virtual class rooms. The ability to deal with such a transition can be different from one institution or school district to another based on their previous experience with online options and based on existing support resources.

Definitely, this new business models imposed extra overhead on the employees in all general and on IT employees in particular. Majority of the companies were not ready for such transition given the short period of time. They had to upgrade/build new IT infrastructure to support the new online business model. A lot of effort was needed to install new hardware and new applications. Adding to that, there is a continuous need to secure users' and organizations data being online against cyberattacks. From security prospective, it is also important to configure the applications used to communicate online correctly to avoid any vulnerabilities that might lead to data breach which will cost the organizations not only money, but also their reputation specially if an incident resulted in exposing customers private information [11, 12].

2.2 Cyber-Attacks During COVID-19 Crisis

Many of the employees worked from home and in many cases they had to use their personal computers. Some of these computers might not had the security measures and patches required to protect against many cyberattacks. Moreover, even of these devices were up to date, many of the online users are new to use the online applications. Lack of appropriate settings and configurations, and lack of user's experience might but the employees data and their employers data at risk. An example of such an incident occurred in Mongolia, where the attackers targeted the public sector employees. The phishing attack involved an email sent on the prevention of Coronavirus [13].

Adding to that, big number of the online users are students who easily can be attracted to access video games or entertainment websites. Such websites, might compromise the device's security and even put's the user's privacy in danger which is a more serious issue when those users are children. The parents are doing their best on monitoring their children, but they can't be always watching their computers screens and what they access.

Besides the mentioned vulnerabilities mentioned above, and possible threats they might cause, one big security risk comes from the state-backed hackers [14]. These are people who work on behalf of the governments and use the illegal methods of gaining information from other institutions or governments. As a result of different approaches in handling the pandemic, there is an increase in such attacks to gain information from other countries.

There is similarity between the cyber attacks and the natural disasters in terms of the infrastructure, social, and economical damage they cause. As an example, Hurricane Florence, which hit North Carolina did not only affect the people in terms of destruction of property and loss of lives [15], but also all the people there and the infrastructure too were prone to attacks by cybercriminals. This is because, at the time, the people are more concerned about helping others overcome the problem as opposed to enhancing the security of the information systems. At the same time, the corporate and the business interests may face more significant risks because of the compromised systems, and the IT personnel may focus on rebuilding the destroyed infrastructure as opposed to the monitoring.

According to Crick and Crick, during disasters, the victims are dealing with government agencies, non-governmental organizations as well as the donors who the cybercriminals may want to impersonate [16]. He easiest way is to use phishing attacks where they send emails that impersonate the organizations offering aid. It is a technique that makes it easier for people to believe that they are legitimate. The main aim in such instances is to get the personal information, credit card, and social security numbers from the victims. At the same time, the cybercriminals may pose as charity organizations which are soliciting funds for the victims and hence obtain the bank information.

The cyber attacks did exist before, but during COVID-19 era, many of these attacks have increased and made more powerful in terms of the impact they have on people, infrastructure, institutions, and even on governments:

1. **Phishing**
   Phishing: It is the technique of fraudulently acquiring users information by mimicking use familiar websites and lure then to enter confidential data. There are several anti-phishing solutions implemented by corporate to prevent these attacks [17]. Phishing emails are just one common way which the hackers can use to gain access of any network. Another form of phishing is though advertisements where the users receive ads that tell them they have discounts coupons or they won prizes to claim.

2. **Info-Stealing Trojans**
   The main aim of most attackers is to steal the information which they can use to ask for ransom. The cybercriminals depend on social engineering and the complexity of events to harvest the passwords and other sensitive information of the victims. During this pandemic, the Trojan AZORult is used, and it is invading the computers in the guise of an interactive COVID-19 map application [17]. However, once in the device, it executes the stealer component, which is a second-stage payload.

3. **Denial of Service (DoS)**
   DoS is a technique used by attackers to make certain network resources unavailable issuing a large number of requests more than the system is designed to handle. There are several DoS detection systems available like, firewalls, intrusion detection have been developed to circumvent these situations. Related from of attack that help launching DoS attack are the Botnets. Botnets are a malware infected computers which can be controlled by the attacker using integrated commands and can be grouped them into a network. The cyber attacker then uses these botnets to launch the many requests simultaneously (TCP flooding for example) to the targeted serves resulting in bringing it down so it will not be able to fulfill legitimate request (Denial of Service).

4. **Teleworkers**
   During this time of COVID-19, teleworkers are used to pass the right information to the general public. At the same time, the remote workforce has become a norm for many businesses. This means that people are holding virtual meetings to interact with. VPNs are used to ensure that the remote workers connection and the enterprise data assets remain secure. However, the use of videoconferencing software and the VPN tools has led to the increase in the use of black hats to identify the areas of vulnerabilities within a system [17]. Exploiting such vulnerabilities might lead to stealing corporate information and secrets.

5. **Pharma Span Splashes**
   The users are always trying to stay connected to have the updates on the coronavirus treatment. However, cybercriminals are using the fake pharmacies, which are a trick they were using in the past. Hackers are using this online marketplace to promote fake products. They are using the bots and the scripts that

deluge the popular websites and provide comments which have sketchy links [17]. This attack aims at deceiving the user to access the given website link. And in majority of situations, the returned results are a drugs shop presenting harmful medical products at very low prices to encourage people to buy.

To summarize, we can say that the increasing usage of the Internet, web applications, and technological devices has also led to the increased cases of cyber attacks during the COVID pandemic compared to the time before. This problem has been there before in any forms of online information exchange. There are many types of malware which are; network-based which include spyware, adware, and sniffers or the ordinary malware which consists of viruses, worms, and logic bombs. Malwares disrupt the normal functioning of systems, loss of information, corruption of files, and they can be used by hackers to access information or even commit cyber-attacks [18]. It is for these reasons that organizations have invested in malware detection and prevention systems in order to keep their data and information safe [19].

In the same context, many information security firms have reported that there is a growing trend in the number of cyber-attacks during 2020. Figure 3 shows the significant increase in the inbound attacks and attacks on routers and devices. This increase is due to that the cyber-criminals want to take advantage of the inadvertent security gaps and vulnerabilities resulted from the tremendous usage of the online services and applications [20].

On the other hand, and among the common tactics used by the cybercriminals at the time of COVID-19 consists of phishing attacks that try to make the user believe that certain products are medication to the virus. Motivated by the curiosity, online users may access these phishing links and be victims of the cyber attack.

Adding to that, the health care sector is being targeted by the cyber attackers. During the pandemic these cyber attacks did increase since this sector involves huge amount of important data (patient records, medications, possible vaccinations, and statistics about the cases and many other important information. This valuable

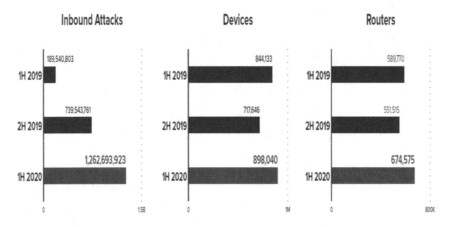

Fig. 3 There was a significant increase in the inbound attacks and attacks on routers and devices

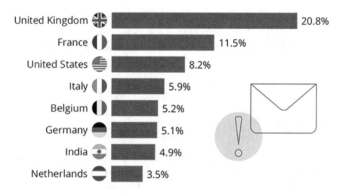

Fig. 4 Percentages of cyberattacks. *Source* Trend Micro [22]

information if obtained by hackers, they might use it to ask for a ransom, otherwise such data breach might destroy the reputation of these medical institutions. The attacks against the Brno University Hospital and the Champaign-Urbana Public Health District are indications of how the cybercriminals are working and targeting the healthcare systems [21].

Someone might ask what is the motivation to launch these cyberattacks and cause all this damage? Usually the attackers are motivated by the money that might be paid by the organization being under attack to retrieve back it's stolen data and to protect it's reputation. These threats imposes extra overhead on the IT engineers and professionals in these organizations to secure there infrastructures and to employ best security practices in their corporations.

Not only health institutions are targeted severely by the hackers during the pandemic time, but also the government agencies dealing with vaccinations are considered to have proprietary information on tests and vaccines, which is valuable, especially now that businesses are competing to find a cure. Therefore, these institutions are a target of the attacks. For this reason, most of the targeted countries by the cyber-attacks during this pandemic are the developed ones [22]. Figure 4 shows the countries that have become a major target for cybercriminals.

3 Impacts Cyber-Attacks During COVID-19 Pandemic

3.1 Disruption of Businesses

The ability of the cybercriminals to compromise the devices used by the remote workers has put the organizations at a greater risk. This is because once it is accessed, it can function as a conduit to the core network of the organization. The results are a massive impact on the business, especially where a ransom for taking the business offline is the target [23] With the growth of the pandemic, the domains using the word Corona or COVID-19 have increased and hence making it easier for the people to fall, victims, as they search for information. Successful cyber-attacks have several impacts on businesses.

3.2 Economic Costs

A successful cyber-attack can cause substantial financial losses in the organization. This is because it can include loss of corporate information such as the patents and the new designs of the products. The loss of such information threatens the financial capacity of the organization. At the same time, these attacks can cause loss of financial information [24]. This includes the bank details and payment card details. Such information can be used by the cyber-criminals to make successful transfers from the banks. Apart from this, it may result in disruption of business where successful transactions are not able to be carried out—such a disruption results in loss of money. Businesses that have suffered from successful attacks also incur the cost of repairing the affected system as well as devices and the networks.

3.3 Reputational Damage

The organizations perform better when they can maintain positive relationships with the customers. Cyber-attacks can damage the reputation of the organization as the customers feel that their information is not secure with the company's security systems. It erodes the trust that the customers have and hence can damage the business. The results of a damaged reputation include loss of customers [24]. When they are unable to trust the organizational ability to protect their information, they seek the services of the competing firms. This is coupled with the loss of sales because as the number of customers declines, the sales also reduce. Organizations, therefore, suffer from declined profits as a result of the successful attacks.

3.4 Legal Consequences

Data protection and privacy laws require that the organizations manage the security of all the personal data that is held [24]. This includes information about the customers and the staff. When the security measures are deliberately or accidentally breached due to failure of the organization to deploy the necessary security measures, then they face fines and regulatory sanctions.

4 Preventing Cyber-Attacks During COVID-19

The organizations, institutions and corporates are trying their best to secure their IT systems, infrastructures, and data against possible cyber attacks that increased during the pandemic. They employ different countermeasures to prevent these cyber attacks in the first place and to detect it in case it happened. Securing the IT infrastructure and building a robust systems are main keys to protect against the cyber attacks. Now, within short amount of time, everything was moved to online including the meetings, customer service, product marketing, and even the offices were moved to homes. This new working environment, imposed extra overhead and huge pressure on the IT engineers and professionals in these organizations to secure there infrastructures and to employ best security practices in their corporations.

Moreover, to protect the corporate, employees, and customers data, the organizations and institutions should force efficient security policies to prevent against possible data breaches.

In the same context, there should be a well prepared incident response plan in case any incident happened. This way, it is easier to effectively recover from an attack as well as regaining the lost information. A well written incident plan will identify the tasks of every individual in the team, during the different phases starting from the response to the incident till the recovery phase and going back to normal operation.

There are many countermeasures to protect against cyberattacks, we identify some of them in the subsections below:

4.1 Limited Data Access by Employees

As we described earlier, many of the employees of the institutions and companies all over the world found themselves using online resources heavily and as never before. Many of those employees didn't have the required training and enough awareness about the possible cyber threats. This situation created an environment suitable for the hackers to perform malicious activities on the user's data and on the corporates information as well which might be very critical in many situations such as financial data and health records. In order to prevent such attacks on crucial data, the organizations should limit the employees access to their data. There are many data access models that can be adopted. For example, the employees should be given access to the data that they need to know to accomplish their assigned tasks [25]. No extra access to other data records should be granted to avoid cyber attacks. And at the worst case scenario, if an attack occurred on that employee data, the impact will be very limited and will not affect other user's information.

4.2   Following Best Practices in Configuring Network Devices and Updating Software

The hardware and software components of any Information Technology (IT) system should be maintained regularly. Software components including the operating systems, applications, anti-viruses, and intrusion detection systems should be kept up to date for any security patches and updates for the routers, switches, and hardware networking components, the network administrators should follow best security practices when configuring these devices. For example, avoid using common passwords and default settings.

4.3   Using Software and Hardware Firewalls

The main purpose of the firewalls is to thwart any malicious hackers and to stop the employees from accessing inappropriate websites [25]. Therefore, having up-to-date firewalls plays an important role in enhancing the security of a network.

5   Conclusion

In conclusion, a disaster is utilized by cybercriminals to increase their cybercriminal acts. The fact that many people focus on mitigating the disaster gives them a good environment to perform their malicious activities. There is a noticeable increase in the cyber-attacks during the time of COVID-19. Given the new business model and working from home environment, Attackers are taking advantage of possibly increasing vulnerabilities due to lack of awareness of best security practices when using online resources by many new users. This means that the security systems of the organizations are affected, and hence they can be easily penetrated. At the same time, there is a huge chance that the employees will not adhere to the information security policies determined by their institutions in their new working environment which is their homes. The cyber attacks targeted almost every valuable data transaction and any organization that might have such data including banks, schools, corporates, factories, and many others. Attackers paid more focus towards health institutes and hospitals since they deal with the information directly related to vaccinations and the management of the pandemic. Many of the cyber attacks are related to phishing in forms of spam emails asking providing false information about the pandemic and asking the users to click provided links. Motivated by the curiosity, many of the users were deceived and fell as victims for those cybercriminals. It makes it easier for individuals to follow the provided links and hence fall to the trap of the cybercriminals. The main aim is to disrupt business or to gather the information that is considered sensitive. Therefore, as the governments and organizations work to achieve a vaccination and a remedy to Covid-19 pandemic, cybercriminals are at work to maximize the number of successful attacks carried out.

Moreover, the following recommendations can be considered:

- Organizations limit access to the information system by ensuring that the employees access the data that is necessary to complete their tasks.
- The employees need to be educated on the best ways of ensuring that they maintain high-security standards even when working from home. At the same time, they need to understand some of the signs which might be used to depict an attack. The employees should also learn that it is insecure to follow unknown links.

Organizations need to have a proper incidence response plan, which helps in ensuring that in case of a successful attack, the right measures and procedures are used to return the system to the same state it was before the attack within a short time.

References

1. Brohi, S.N., Jhanjhi, N.Z., Brohi, N.N., Brohi, M.N.: Key Applications of State-of-the-Art Technologies to Mitigate and Eliminate COVID-19 (2020)
2. Perez, S.: Videoconferencing apps saw a record 62M downloads during one week in October (2020). [Online]. Available: https://techcrunch.com/2020/03/30/video-conferencing-apps-saw-arecord-62m-downloads-during-one-week-in-march/. Accessed: 5 May 2020
3. Vigliarolo, B.: Who has banned Zoom? Google, NASA, and more (2020). [Online]. Available: https://www.techrepublic.com/article/who-has-banned-zoomgoogle-nasa-and-more/ Accessed: July 2020
4. Humayun, M., Niazi, M., Jhanjhi, N.Z., Alshayeb, M., Mahmood, S.: Cyber security threats and vulnerabilities: a systematic mapping study. Arab. J. Sci. Eng., 1–19 (2020)
5. Huang, C., Wang, Y., Li, X., Ren, L., Zhao, J., Hu, Y., Zhang, L., et al.: Clinical features of patients infected with 2019 novel coronavirus in Wuhan, China. The Lancet **395**(10223), 497–506 (2020)
6. Li, T., Wei, C., Li, W., Hongwei, F., Shi, J.: Beijing Union Medical College Hospital on "pneumonia of novel coronavirus infection" diagnosis and treatment proposal (V2.0). Med. J. Peking Union Med. Coll. Hosp. (2020). http://kns.cnki.net/kcms/detail/11.5882.r.20200130.3991430.002.html. Accessed: Aug 2020
7. Trilla, A., Trilla, G., Daer, C.: The 1918 "Spanish Flu" in Spain. Clin. Infect. Dis. **47**(5), 668–673 (2008). https://doi.org/10.1086/590567
8. Gallotti, R., Valle, F., Castaldo, N., Sacco, P., De Domenico, M.: Assessing the risks of "infodemics" in response to COVID-19 epidemics. arXiv preprint arXiv:2004.03997 (2020)
9. Alsmadi, I., O'Brien, M.J.: How many bots in Russian troll tweets? Inf. Process. Manage. **57**(6), 102303 (2020)
10. Bae, M., Kim, H.: The role of vitamin C, vitamin D, and selenium in immune system against COVID-19. Molecules **25**(22), 5346 (2020)
11. St. John, A.: It's not just Zoom. Google Meet, Microsoft Teams, and webex have privacy 406 issues, too (2020). [Online]. Available: https://www.consumerreports.org/video-conferencing services/videoconferencing-privacy-issues-google-microsoftwebex/. Accessed: 21 May 2020
12. Warren, T.: Zoom faces a privacy and security backlash as it surges in popularity (2020). [Online]. Available: https://www.theverge.com/2020/4/1/21202584/zoom-securityprivacy-iss ues-video-conferencing-software-coronavirus-demandresponse. Accessed: 21 May 2020

13. Chamola, V., Hassija, V., Gupta, V., Guizani, M.: A comprehensive review of the COVID-19 Pandemic and the role of IoT, drones, AI, blockchain, and 5G in managing its impact. IEEE Access **8**, 90225–90265 (2020)
14. Chapman, P.: Are your IT staff ready for the pandemic-driven insider threat? Netw. Secur. **2020**(4), 8–11 (2020)
15. Thakur, K., Kopecky, S., Nuseir, M., Ali, M.L., Qiu, M.: An analysis of information security event managers. In: 2016 IEEE 3rd International Conference on Cyber Security and Cloud Computing (CSCloud), pp. 210–215. IEEE (2016)
16. Crick, J.M., Crick, D.: Coopetition and COVID-19: collaborative business-to-business marketing strategies in a pandemic crisis. Ind. Market. Manag. (2020)
17. Alsmadi, I., Easttom, C., Tawalbeh, L.: The NICE Cyber Security Framework: Cyber Security Management. Springer Nature (2020). ISBN 978-3-030-41987-5
18. Maleh, Y., Baddi, Y., Alazab, M., Tawalbeh, L., Romdhani, I. (eds.): Artificial Intelligence and Blockchain for Future Cybersecurity Applications. Springer (2021). ISBN 978-3-030-74575-2. https://www.springer.com/us/book/9783030745745
19. Elavarasan, R.M., Pugazhendhi, R.: Restructured society and environment: a review on potential technological strategies to control the COVID-19 pandemic. Sci. Total Environ., 138858
20. 1H 2020 cyber security defined by Covid-19 pandemic. Online https://www.trendmicro.com/en_us/research/20/i/1h-2020-cyber-security-defined-by-covid-19-pandemic.html. Last accessed 20 Sept 2021
21. Smith, W.R., Atala, A.J., Terlecki, R.P., Kelly, E. E., Matthews, C.A.: Implementation guide for rapid integration of an outpatient telemedicine program during the COVID-19 pandemic. J. Am. Coll. Surg. (2020)
22. Trend Micro. Available online: https://www.trendmicro.com/en_us/research/21/i/midyear-2021-cybersecurity-landscape-review.html. Last accessed 5 Oct 2021
23. Beaunoyer, E., Dupéré, S., Guitton, M.J.: COVID-19 and digital inequalities: reciprocal impacts and mitigation strategies. Comput. Human Behav. **111**, 106424 (2020)
24. Kott, A., Ludwig, J., Lange, M.: Assessing mission impact of cyberattacks: toward a model-driven paradigm. IEEE Secur. Priv. **15**(5), 65–74 (2017)
25. Sharifi, A.Z., Zaheer, H., Azizi, M.F., Faizi, J.: Detection and prevention of distributed denial of service attacks in SMEs: the case of CloudPlus. In: 2019 Sixteenth International Conference on Wireless and Optical Communication Networks (WOCN), pp. 1–4, December 2019. IEEE (2019)

A Comparison of Performance of Rough Set Theory with Machine Learning Techniques in Detecting Phishing Attack

Arpit Singh and Subhas C. Misra

Abstract Phishing is a deceptive social engineering trick that lures online users to disclose personal and confidential information to the fake websites disguising as legitimate ones in electronic communication. Several machine learning methods have been proposed to detect phishing websites. The major limitation of assumption of a peculiar distribution of the data plagues analysis with machine learning (ML) especially when the number of data points are limited in number. This paper applies classical rough sets analysis (CRSA) for detecting phishing websites from a collection of legitimate and fake websites sourced from an open UCI Repository. The work focuses primarily on comparing the classification performance of CRSA with the traditional ML tools. CRSA does not require any distributional assumption for analysis and does not need any user inputs for the parameters in the analysis unlike the traditional ML tools. The comparison revealed no significant difference in the classification performances of the two sets of algorithms. The precision of CRSA was 95.6% which is comparable and in some cases better than the other ML tools. Similarly, high ROC and f-measure places CRSA at par with the statistical ML tools. The relaxation in considering the statistical properties of data clearly gives CRSA edge over other data analysis tools. The CRSA based techniques provide a robust method of detecting phishing websites. The generalization of the analysis makes CRSA a quick-to-use framework for detection of phishing websites without consideration of statistical properties of the data.

Keywords Phishing · Cyber-security · Rough sets · Machine learning · Classification

A. Singh
O.P. Jindal Global University, Jindal Global Business School, Sonipat, Haryana, India

S. C. Misra (✉)
Kanpur, India
e-mail: subhasm@iitk.ac.in

© The Author(s), under exclusive license to Springer Nature Switzerland AG 2022
P. Nicopolitidis et al. (eds.), *Advances in Computing, Informatics, Networking and Cybersecurity*, Lecture Notes in Networks and Systems 289,
https://doi.org/10.1007/978-3-030-87049-2_22

1 Introduction

Phishing is categorized as a cyber-attack that uses disguised e-mail as a weapon. The intention is to trick the users into believing that the e-mail is something they want or need, such as a request from the bank, or a note from an acquaintance requiring them to click on a link or download an attachment [11]. The distinguishing feature of phishing emails is the form it takes: attackers normally deceive users to gather the personal or confidential information by pretending to be a trusted entity of some kind, often a real person or a company the victim might be doing some business with [25]. The phishing e-mails and webpages possess a very high degree of visual and content similarity with the legitimate e-mails and webpages it tries to mimic. Such phishers often attach at least one login form that gathers personal and confidential information such as the bank account number, etc. [45]. This is also accompanied by persuading users to click on malicious links leading to the installation of malware, freezing of the system, and ransomware attack, or revealing classified information. Broadly, phishing is a term that indicates a socially engineered attack that indicates the stealing of personal data, fake offers of goods and services, impersonation of authorities, and fraudulent medical offers, gifts, and donations [20].

The basic tool of phishing is link manipulation. The user is connected to webpages, e-mails, and other online portals through superficially reliable links that lead to a fake destination other than the expected one. It is easy for the users to be tricked into clicking on such malicious links as it is easily created in a simple HTML code. The title of the link can be easily altered at will that leads the users to land in webpages that are visually identical to the authentic pages but belong to malicious users' servers [14]. The problem of tracing phishing mails and webpages becomes even more pronounced on occasions where the phishers resort to untraceable means. Particularly, when phishers dupe users by International domain names (IDN) spoofing, a phenomenon where identical URLs lead to different webpages [24]. Phishing attacks can have irreparable damage to the individuals that include unauthorized purchases, identity thefts, and financial losses. For large organizations, phishing attacks can have a larger adverse impact such as advanced persistent threat (APT). Under APT, employees manipulate software to bypass security perimeters and the distribution of malware inside a closed environment [27]. The large scale organizations that are under the attack of phishers suffer high financial losses in addition to the loss of market share, reputation, and consumer trust [36].

A real-life case of the hacking of the webpage of government universities of India, namely, Delhi University, Indian Institute of Technology, Delhi, Aligarh Muslim University, and Indian Institute of Technology, BHU took the nation by shock where a group of hackers from Pakistan with the name PHC posted pictures of alleged "brutalities" by Indian forces in "Kashmir" [19].

Online hackers and criminals rely on deception to create a sense of exigency in users to attain success with their manipulative crimes. Such activities gain a stronger ground during the time of crises. People are in a vulnerable state during the time of a national crisis. They are constantly looking for instructions and directions from the government and authorities to be updated with the information. The defenseless situation of people arising from these crises allows the online attackers to lure people in their phishing trap. Corona-virus pandemic has accelerated the incidents of phishing attacks and online scams in the last few months. Scammers are taking advantage of the COVID-19 panic among people by disguising their scams and phish emails as legitimate mails and messages about the virus.

According to Anti Phishing Working Group (APWG) Phishing Activity Trend Report that analyzes the phishing attacks and other identity theft techniques reported about 86,276 unique phishing web sites in the third quarter of 2019, approximately 42,273 unique phishing e-mail reports, and about 425 brands that succumbed to phishing attacks [4].

Various techniques are employed on an individual basis to circumvent the issue of identifying phishing emails and webpages. For instance, educating users on identifying key errors in the fraudulent emails and notices such as spelling errors, erroneous punctuation marks, and mistakes in spellings of names and organizations.

A wide range of techniques has been proposed in detecting phishing mails and other related scams. These techniques are broadly classified into list-based, heuristics, and machine learning approaches. The list-based and heuristics methods of detecting and predicting scams and spoofed emails have been of little use due to the limited capacity in collecting and analyzing a large database of malicious links and spams. Subsequently, based on a selected set of features that are attributable to the set of fraud and bogus emails and webpages, an algorithm is developed that predicts the scam emails and webpages. Since these techniques rely heavily on a documented collection of phish emails and websites, it is practically impossible to create a database that collects the exponentially increasing number of scam websites and mail domains daily [38]. Also, the detection capabilities of these heuristics can be easily bypassed by a knowledgeable hacker who can ensure that the phishing websites do not contain the detected features.

The machine learning approach also identifies key attributes that aid in detecting fraudulent emails and webpages. These attributes are found by the mining repository of legitimate websites for training the algorithm. The collection of a rich and diverse set of genuine websites and webpages for training fraud-detection algorithms in the machine learning approach is a challenging task [3].

Despite the limitation of the machine learning approach for detecting fraudulent emails and websites, it is widely used for classifying and predicting phishing emails and webpages. Several machine learning algorithms are employed to extract features and use them to predict the online scams and bogus mails [2, 5, 21]. The quality of classification and prediction of malicious e-mails and websites depends largely on the set of features chosen from the database. Traditionally, machine learning algorithms mine the features characteristic of fraudulent and bogus e-mails and websites.

However, there are some limitations of machine learning and statistical algorithms that make these techniques conceptually difficult to use for classification purposes. One prominent limitation being the distributional assumption of the data. Almost, every statistical procedure and some machine learning algorithms assume a normal distribution of the data before analysis. This assumption can prove to be restrictive in situations where the data are limited in number. Some machine learning algorithms also lack generalizability due to strict constraints imposed on the parameters and other metrics needed to carry out the classification. In the case of detecting phishing e-mails and websites through these algorithms, effective feature selection is crucial to achieving generalizability. However, this is not an easy task especially because of the difference in the nature of the websites these scammers try to imitate thus limiting the generalizability of the results. The solutions to the problems mentioned above coupled with many additional advantages, Rough set theory provides a convenient method of identification of phishing messages and e-mails efficiently [46].

This paper attempts to extract the most effective attributes to detect phishing websites using Rough set theory. Subsequently, the performance of classification is compared with other well-established classification machine learning algorithms. The dataset employed for the study is taken from an online repository that comprises examples of legitimate and phishing websites [44]. The rest of the paper is organized as follows.

2   Literature Review

Machine learning-based approaches to build a phish-detection system for websites is an active research field that employs a wide range of supervised algorithms to detect phishing websites through a set of characteristic attributes. To improve the classification ability of the classifiers, a novel classification algorithm was proposed to identify fraud websites based on heuristic features extracted from URL, source code, and third party services. Eight different machine learning techniques were applied for the classification purpose. Principal component analysis Random forest (PCA-RF) demonstrated the best performance in classification with an accuracy level of 99.55% [38]. Combinations of different learning algorithms improve the classification performance and yield an effective set of features that aid in detecting phishing websites. The principal component analysis combined with random forest, neural network, bagging, support vector machines, and naïve Bayes improved the classification accuracy of the framework [48].

To detect phishing attacks using the hyperlinks found in the HTML source code of the websites, several machine learning algorithms were used in classification including support vector machines, logistic classifier, naïve bayes, ada-boost, neural network, and C4.5. Logistic classifier showed the highest classification accuracy of about 98.42% [21]. Often when a parameter or a threshold is altered, the classification performance of the classifiers increases exponentially. In one such study, when a logistic regression algorithm was applied in the logistic classifier, it outperformed other machine learning algorithms including support vector machines and naïve Bayes [1]. In a similar study of classifying genuine websites from the phishing ones based on URLs, 1353 real-world URLs consisting of suspicious, legitimate, and phishing websites were used. Four classifiers, namely, decision tree, naïve Bayes, support vector machine, and neural network were used to classify each website into the corresponding category. The classification was done correctly 90% of the time [23]. The quality of classification of phishing and legitimate websites using the machine learning approach depends largely on the selection of appropriate feature sets. In a study of detection of phishing e-mails, seven machine learning algorithms and natural language processing (NLP) based classifier was used. Random Forest algorithm combined with NLP based classifier yielded the best performance with an accuracy rate of 97.98% for detecting phishing URLs [10]. Researchers are actively engaged in developing intelligent web-phishing detection and analysis systems where various machine learning algorithms are applied either standalone or combined with other algorithms [12].

The efficiency of machine learning algorithms in the classification of phishing e-mails and websites lean heavily on the parameters selected by the users. Approximate string matching algorithms were devised to determine the relationship between the content and the URL of the page to detect phishing attacks in internet banking [28]. The performance of anti-phishing machine learning approaches is influenced by the selection of an effective feature vector set. This was highlighted in the experiment consisting of 2541 phishing instances and 2500 benign instances, where a feature selection module was proposed using support vector machine and Naïve Bayes algorithms [31]. Deep learning methods such as artificial neural networks have been employed to detect phishing activities online [8]. In all the studies undertaken to detect phishing e-mails and websites involving machine learning algorithms, user-specified parameters are employed. This cannot be avoided as the functioning of the algorithms is parameter dependent. This issue affects the classification performance of the classifiers as users select the parameters of the machine learning algorithms heuristically. It greatly limits the generalizability of the results, especially when dealing with out-of-sample training datasets. Besides, prior distributional assumption of the data inhibits the appropriate usage of the machine learning algorithms.

In this paper, we propose using Rough set theory as the feature selection tool. The benefits of Rough set theory outweighs that of the other machine learning algorithms for classification purposes. It does not assume any prior distribution of the data and hence, applies effectively in almost all kinds of data available. This property is helpful particularly because we have to deal with real-life data that cannot be expected to display a characteristic distribution. Secondly, it does not need any parameter specification from the users. It operates on the given data and deciphers patterns that are hidden in the dataset. We demonstrate the classification of phishing websites using Rough set theory from the datasets taken from the online repository. Additionally, we will be conducting comparisons of classification performance between the Rough set theory and other machine learning algorithms.

3   Rough Set Theory in Selection of Features

3.1   Classical Rough Set Approach (CRSA)

CRSA is a method of identifying patterns in the data by grouping similar objects based on some specific characterisitic [32, 33]. The grouping of objects is made on the basis of similarity between the objects. The indiscernibility relation between the objects is the mathematical foundation of CRSA. Further details about CRSA is presented in the following sections.

3.2   Mathematical Preliminaries of CRSA

The data is arranged as an information matrix given by (U, A, V, f), where 'U' represents the non-empty finite set of entities, 'A' indicates the attribute set which is given as $A = C \cup Dec$, where C is the set of conditions features and Dec is the set of decision variables. 'V' represents the values that an object assumes for each of the condition and decision attributes $a \epsilon A$. 'func' is the function represented as $func : U \times A \rightarrow V$ that maps the conditions A to its value. The indiscernibility relation forms the backbone of the CRSA. Essentially, similar objects on the basis of certain attributes are grouped together. Thus, for a given condition attribute $b \epsilon A$ two objects x_i and x_j are said to be indiscernible or similar if $f(x_i, b) = f(x_j, b)$. Put mathematically, the indiscernible relation is given by

$$R(B) = (x_i, x_j) \epsilon U : \forall b \epsilon A, f(x_i, b) = f(x_j, b) \text{ [18].}$$

The classes that are equivalent are specified by the object x_i with respect to R(B) is denoted as $[x_i]B$. This way it forms the basic granules of knowledge that is indicative of the fact that there is always data by means of information associated with every entity in the universe of discourse. Keeping these similar groups as the foundation, CRSA proceeds with the mathematics where similar sets are approximated by lower and upper approximations that shall be explained in the following section. CRSA operates on the dataset and treats the inconsistencies present in the dataset [47]. It is this inconsistency in the data that interferes with the correct inferences from the data. The decision about whether the data is consistent or inconsistent depends on the decision class or set Y. If the objects can, for certainly be attributed as belonging to the decision class Y we say that the information is consistent and thus can render crisp decision rules. However, if there is any ambiguity about the belonging of an element to decision class Y then inconsistencies creeps in the data which is dealt by CRSA by way of approximating the sets of such objects [9].

The group of entities that can be said to be included in a particular set with utmost certainity are said to be in lower approximation whereas entities that are possibly included in a set are said to be a part of upper approximation which, in terms of set notation can be written as

$R_{Blow}(Y) = x \epsilon U : [x]_B \subseteq Y$ and
$R_{Bupp}(Y) = x \epsilon U : [x]_B \cap Y \neq \phi$ respectively [18].

There are also sets of objects that can neither be grouped in definitely belonging to set Y nor as not belonging to the set Y and such sets are referred to as the boundary regions given in set notation as follows

$$R_{Bupp}(Y) - R_{Blow}(Y) = BND_B(Y) \text{ [33]}.$$

In other words, these are those elements that possibly belong to set Y but not certainly.

3.3 Brief Overview of Machine Learning Tools

3.3.1 $C_{4.5}$

$C_{4.5}$ follows Id3 algorithm for classification. Conceptually, it is based on the information entropy. The node that gets split possesses the maximum information at the time of splitting [40]. The event with the highest probability of occurrence is said to have the highest entropy. The information associated with such an event is the least. Thus it is not very useful to consider the branching of the event with the highest entropy. Rather the splitting yields better and granular results when the event with the highest information gain is selected. $C_{4.5}$ is implemented using the J48 classifier in WEKA 3.8.3 following the configuration:

(weka.classifiers.trees.J48 -C 0.25 -M 2)

where C 0.25 is the confidence factor of 0.25 which is used for pruning purposes and M 2 represents the minimum number of instances per leaf which is 2 in this case.

3.3.2 Random Forest

Random forests builds an average model of many decision classes for the classification problems. It addresses the issue of overfitting of data [26]. The basic premise of a Random Forest algorithm is based on an ensemble of decision trees. The two rules that are needed for the classification using Random Forest classifier is that the feature set selected should be able to classify the instances better than random guessing and secondly the predictions (hence errors) between the candidate models should have least correlations among themselves. Random forest is accomplished using the following configuration in WEKA:

(weka.classifiers.trees.RandomForest -P 100 -I 100)

where P 100 represents the size of each bag as a percentage of training set size, I 100 is the number of iterations performed, num-slots 1 are the number of execution slots or threads to use for constructing the ensemble, K sets the number of features to consider where the value less than 1 implies Log M + 1 inputs where M is the number of inputs, S is the random number seed that is to be used and V is the minimum variance for split that is it sets the minimum numeric class proportion of train variance for split (default is 0.001).

3.3.3 Naïve Baiyes

Naive Bayes randomly draws a set of labels from a finite feature set. Subsequently, assignment of labels to different instances takes place. It is important to note that Naïve Bayes is not a single algorithm but a collection of algorithms [39]. This classifier is based on the Bayes theorem which determines the posterior probability of an event from the prior probability. The theorem is given as follows

$$p(y|X) = \frac{p(X|y)p(y)}{p(X)} \tag{1}$$

where "y" is the predicted variable and "X" represents the set of predictor variables. The key assumption made in this classification process is the set of X are mutually independent and the predictor variables exert equal reaction on the predicted variable.

3.3.4 Support Vector Machines (SVM)

Support vector machines (SVMs) accomplish the regression and classification objectives. SVM establishes the classification task by creating a hyper-plane that distinguishes between the classes of objects as distinctly as possible. The data points close to the hyper-plane are called support vectors. The method of optimizing the algorithm is to minimize the margin between the support vectors of different classes. The shape of the hyper-plane depends on the number of features. For a two feature set the hyper-plane is just a line. For a two dimensional feature set the hyper-plane assumes the form of a two-dimensional plane.

SVM is implemented using John Platt's sequential minimal optimization algorithm [43] for training a support vector classifier using the following configuration:

(-S 0 -K 2 -D 3 -G 0.001 -R 0.0 -N 0.5 -M 40.0 -C 250.0 -E 0.001 -P 0.1 -model)

where P 0.1 represents the epsilon for round-off error, M represents the fitting of logistic models to SVM outputs, S is the random number seed for cross validation, K is the Kernel to use, D is the number of decimal after the outputs in the model, R sets the ridge in the log-likelihood, G is the gamma for RBF Kernel, N represents whether to normalize, standardize or neither of the two, M fits calibration models to SVM outputs, E is the exponent value, C is the cache size for the poly-kernel and model is the SVM chosen for the classification.

3.3.5 Logistic Regression

Logistic regression is a member of the nonlinear family of regressions where the regressand is dichotomous or binary in nature and regressors can be of any type (i.e. continuous or discrete). There are other forms of logistic regressions available, depending upon whether the dependent variable is nominal, ordinal, or categorical [22]. In WEKA it is implemented using the configuration:

(weka.classifiers.functions.Logistic -R 1.0E-8 -M -1 -num)

where R 1.0E−8 is the ridge value in the log-likelihood, M represents the maximum number of iterations to be performed and num-decimal-places 4 represents the number of decimal places for the output numbers.

3.4 Performance Evaluation Metrics

Some of the important metrics are described below that are useful in the assessment of the classification performance of the classifiers. These metrics are applicable for all the classifiers irrespective of the algorithms employed for classification [34].

Confusion Matrix

Confusion matrix is a matrix representation of the correctly and incorrectly classified instances from the data [30]. The following terms are essential to understand various components of a confusion matrix.

TrPo: true positives: instances shown positive that are actually positive
FaPo: false positives: instances shown positive that are actually positive
TrNe: true negatives: instances shown negative that are actually negative
FaNe: false negatives: instances shown negative that are actually positive.

So, confusion matrix is defined as in Table 1 where Predicted is given by Pred, Actual is given by Act, Positive by POS and Negative by NEG.

Sensitivity

It is also known as the True Positive Rate (TPR) or the recall value. It is indicative of the fact of how good a classifier is in not missing out a positive case [7]. Mathematically, it is represented as

Table 1 Matrix representing positive and negative cases

Pred	Act	
	POS	NEG
POS	TrPo	FaPo
NEG	FaNe	TrPo

$$Sensitivity = \frac{TrPo}{TrPo + FaNe} \tag{2}$$

Precision

When the classifier results in less number of false positives it is considered to be more precise and the precision [13] is given mathematically by

$$Precision = \frac{TrPo}{TrPo + FaPo} \tag{3}$$

f-measure

The precision of classifiers is indicated with the help of f-measure. Essentially, it is an estimate of the robustness of classification algorithms. The precision of classifiers is the number of instances classified correctly and robustness implies the number of instances left unconsidered in the analysis. f-measure is a metric that provides equal weight-age to precision and recall and thus is a proper metric to assess the precision and robustness of the classifiers simultaneously.

Mathematically, it is given as $\frac{2}{1/p+1/r}$, where p and r are the precision and recall estimates respectively [35, 41].

Receiver Operating Characteristics (ROC)

The classifiers' potential in discriminating the entities contained in different decision classes is measured in the form of Receiver Operating Characteristic (ROC) Curve or as an approximation of the Area Under the Curve (AUC) where the characteristics are a plot of sensitivity versus 1—specificity in the range 0–1, where the straight line from (0, 0) to (1, 1) represents no discriminatory power and a line from (0, 0) to (0, 1) to (1, 1) represents perfect discrimination [16].

The cross validation procedure is utilized to verify the accuracy of the classifiers. In particular, 10-fold method is used for estimating the generalized error of the prediction model.

4 Experiments

The paper attempts to apply and compare the classification ability of Rough sets methods with the traditional machine learning algorithms using the two benchmark data-sets for phishing attacks. Table 2 describes the data used in the study.

Table 2 Description of data-sets

Dataset	Sample size	# features	# phishing websites
UCI1	11,055	30	3793
Mendeley	10,000	48	5000

Table 3 Features of the data

Feature	Description
Abnormal URL	Mismatch in the actual hostname and claimed identity
Abnormal DNS record	Absence of domain record in WHOIS database
Abnormal anchors	Anchors pointing to different domains unlike in legitimate websites
Server form handler	Processing of information takes place in different domains
Abnormal cookie	Cookies conflicts with website identity
Abnormal certificate in SSL	Distinguished names within the certificates interfere with the claimed identities

The first data is UCI1[1] that consists of a total of 11,055 phishing and legitimate websites. The second data is Mendeley[2] that comprises a total of 10,000 websites with 5000 websites classified as phishing websites. As per [29], the features of the above mentioned data-set are classified into the following categories: Abnormal URL, Abnormal DNS (Domain Name System) record, Abnormal anchors, Server form handler, Abnormal cookie, and Abnormal certificate in SSL (Secure Sockets Layer). The description of the features is provided in Table 3.

The sample sizes for the two data-sets are remarkably different as is noted in Table 2. UCI1 data-set comprises 3793 phishing websites with a total of 11,055 websites including the legitimate ones. On the other hand, Mendeley consists of 5000 phishing websites with a total of 10,000 websites inclusive of legitimate websites. The difference in the sample sizes and the number of phishing websites might indicate some differences in the characteristics been investigated. But, since we are dealing with machine learning and non-invasive data mining techniques such as Rough sets, the issue of sample size is not too overwhelming.

We will be comparing the classification performance of Rough sets based methods with the traditional machine learning methods namely, C4.5, logistic regression, Naive Bayes, Random Forest, and SVM. The parameters for the machine learning algorithms are kept as per [6, 37]. The algorithms were applied to the two data-sets separately. It was done in order to assess the generalizability of the classification performance of the algorithms. Table 4 presents the number of correctly and incorrectly classified instances for the case of UCI1 data-set.

Table 5 lists the number of correctly and incorrectly classified instances for Mendeley data-set.

The performance metrics of various classifiers for UCI1 data-set are tabulated in Table 6.

Table 7 lists the performance metrics for Mendeley data-set.

---

[1] https://archive.ics.uci.edu/ml/datasets/phishing+websites.
[2] https://archive.ics.uci.edu/ml/datasets/Phishing+Websites.

Table 4 List of correctly and incorrectly classified instances for UCI1

	Correctly classified instances (% of total)	Incorrectly classified instances (% of total)
Classical Rough sets	10,566 (95.57)	489 (4.43)
C4.5	10,559 (95.87)	456 (4.12)
Logistic regression	10,391 (93.99)	664 (6.01)
Naïve Bayes	10,279 (92.98)	776 (7.02)
Random forest	10,752 (97.26)	303 (2.75)
SVM	10,370 (93.80)	685 (6.20)

Table 5 List of correctly and incorrectly classified instances for Mendeley

	Correctly classified instances (% of total)	Incorrectly classified instances (% of total)
Classical Rough sets	9604 (96.04)	396 (3.96)
C4.5	9731 (97.31)	269 (2.69)
Logistic regression	9449 (94.49)	551 (5.51)
Naïve Bayes	8515 (85.15)	1,485 (14.85)
Random forest	9837 (98.37)	163 (1.63)
SVM	9387 (93.87)	613 (6.13)

Table 6 Performance metrics for UCI1

	Precision	Recall (sensitivity)	ROC	F-measure
Classical Rough set	0.956	0.956	0.987	0.956
C4.5	0.959	0.959	0.984	0.959
Logistic regression	0.940	0.940	0.987	0.940
Naïve Bayes	0.930	0.930	0.981	0.930
Random forest	0.973	0.973	0.996	0.973
SVM	0.938	0.938	0.936	0.938

Table 7 Performance metrics for mendeley

	Precision	Recall (sensitivity)	ROC	F-measure
Classical Rough set	0.961	0.960	0.984	0.960
C4.5	0.973	0.973	0.976	0.973
Logistic regression	0.945	0.945	0.984	0.945
Naïve Bayes	0.864	0.852	0.949	0.850
Random forest	0.984	0.984	0.999	0.984
SVM	0.939	0.939	0.939	0.939

Reflecting on the number of correctly and incorrectly classified instances in Tables 4 and 5, it is observed that the performance of classical rough sets is comparable or only marginally better than the other classifiers. This is true for both the data-sets thus affirming the generalizability of the results. However, this observation should not eclipse the advantages that classical rough sets offer as compared to the other classifiers. As was mentioned in Sect. 2, there were no distributional assumption of the data for rough sets classification procedure. In addition, there were no user-defined parameters to tune the claasifier which was done for the other classifiers. With relaxations in the above adjustments, the marginal improvement in the classification performance of classical rough sets is significantly better than the other traditional classifiers.

The performance metrics in Tables 6 and 7 also suggest a similar fact. Classical rough sets share similar metrics with the other classifiers. In fact, rough sets show higher performance metrics than the majority of the other classifiers. It has significantly higher ROC than C4.5, naive bayes, and SVM.

The robustness check of the classifiers is assessed by the significance testing of the classification performance of each classifier [41]. The statistical test for evaluating the performance is the *paired sample T-test*. The test statistic used is *"Accuracy"* and *"Area under ROC"*. The decision for choosing the aforementioned metrics for calibrating the performance is the direct estimate of the classification ability given by these metrics. The statistical hypotheses to be tested are given as follows:

- H_{0a}: There is no difference in the accuracy of the pair of the classifiers
- H_{1a}: There is some difference in the accuracy of the pair of the classifiers
- H_{0b}: There is no difference in the area under the ROC curve for the pair of the classifiers
- H_{1b}: There is some difference in the area under the ROC curve for the pair of the classifiers.

Table 8 Significance testing of classifiers performance for UCI1

Parameters	Logistic	CRSA	Random forest	Naïve Bayes	C4.5	SVM
Accuracy	0.94	0.96^V	0.95^V	0.93*	0.95^V	0.93
Area under ROC	0.99	0.98*	0.98	0.98*	1^V	0.94*

V Significantly higher at 5% level
* Significantly lower at 5% level

Table 9 Significance testing of classifiers performance for Mendeley

Parameters	Logistic	CRSA	Random forest	Naïve Bayes	C4.5	SVM
Accuracy	0.94	0.96^V	0.97^V	0.98^V	0.85*	0.94*
Area under ROC	0.98	0.98	0.98*	1^V	0.95*	0.94*

V Significantly higher at 5% level
* Significantly lower at 5% level

The significance level is set at 0.05 (5%). The classifiers are compared pairwise with the logistic classifier. The test statistic is given as follows

$$t = \frac{\hat{diff}}{\sigma_{diff}^{\hat{}}/\sqrt{n}} \tag{4}$$

where $\hat{diff}$ is the difference in the mean accuracy of the two classifiers, n is the number of trials and $\hat{\sigma}_{diff}$ is the sample standard deviation of the mean difference. The outputs of the *paired sample T-test* for the two data-sets are shown in Tables 8 and 9.

The experiment was conducted using 10-fold cross validation method. As is shown in Tables 8 and 9, classical rough sets have statistically significantly higher accuracy than the reference classifiers for both the data-sets. Other classifiers show variability in statistical significance of accuracy for both data-sets. For instance, Naive Bayes is statistically significantly higher accurate than the reference category for UCI1 data-set when compared to Mendeley. This illustrates the limited generalizability of other machine learning algorithms. Similarly, classical rough sets show comparable area under ROC with the reference category with 98% which is similar for both the data-sets.

Figures 1 and 2 give the graphical representation of the ROC curves of the classifiers for UCI1 and Mendeley data-sets respectively.

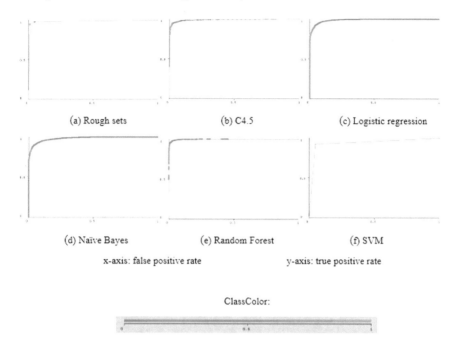

(a) Rough sets (b) C4.5 (c) Logistic regression

(d) Naïve Bayes (e) Random Forest (f) SVM

x-axis: false positive rate y-axis: true positive rate

ClassColor:

Fig. 1 ROC curves for UCI1 data-set

The ROC curve is a representation of the trade-off between sensitivity also called *true positive rate* and specificity also known as 1-*false positive rate*. Classifiers that give the ROC curve towards the top-left corner indicate a better classification performance. In this regard, the ROC curve lying exactly on the diagonal indicates worst classification performance. The ROC curves for all the classifiers clearly lie in the top left corner indicating good classification performance. Classical rough sets display comparable classification performance with the other classifiers for both the data-sets as shown in Figs. 1 and 2. To compare the classifiers, the one important metric derived from ROC curves is the area under the ROC curve (AUC). AUC is the probability that a randomly chosen positive instance is ranked higher than a randomly chosen negative instance. A higher value of AUC indicate a better classification ability of the classifiers. As is shown in Table 7, classical rough sets show the highest AUC, next to Random forest. Thus, rough sets display a significantly better class separation capability than the other traditional classifiers.

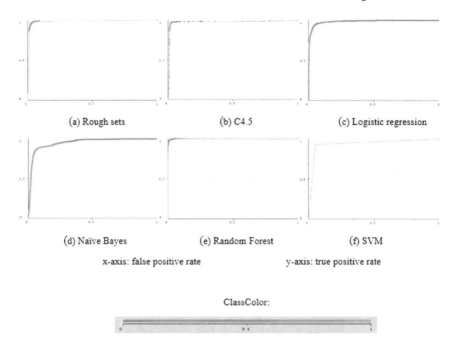

(a) Rough sets   (b) C4.5   (c) Logistic regression

(d) Naïve Bayes   (e) Random Forest   (f) SVM

x-axis: false positive rate   y-axis: true positive rate

ClassColor:

Fig. 2 ROC curves for Mendeley data-set

5 Conclusions

In this work, we attempt comparison of the classification ability of Rough sets methods with other traditional machine learning algorithms in detecting phishing websites. We evaluated the classification performance of classical rough sets with five other machine learning tools based on two benchmark phishing data-sets. The classification models used more than 20,000 websites inclusive of legitimate and phishing websites, thereby avoiding the possibility of over-fitting. Comparison with the traditional machine learning tools, classical rough sets displayed a reasonably comparable and based on some metrics, better classification performance. With an overall precision of 96.1%, ROC measure of 98.4%, and f-measure of 96%, classical rough sets can be considered as a robust mechanism to classify phishing websites from the legitimate ones with a high level of accuracy. Since ther assumption of statistical properties of the data is relaxed in the analysis in CRSA, it makes it a readily available plug-and-play software for quick detection of phishing websites.

6 Appendix: Description of Features in the Data-Set

Attribute	Description	Scale[a]
having IP address	Using IP address of the hostname in the URL part	B
URL length	Length of URL	T
Shortening service	A shortened URL or not	B
Having at symbol	@ existing in the URL	B
Double slash redirecting	URL path starts with a double slash	B
Prefix suffix	Adding prefix and suffix to the URL	B
Having sub domain	Adding sub-domains to the URL	T
Domain registration length	Domain registration length greater than a threshold	B
Favicon	Favicon is from a different domain	B
HTT ps_token	whether a token is sent to the attacker or not	B
Request URL	The ratio of the objects loaded from a different domain	B
URL of anchor	Links within the webpage might refer to a different domain name typed on the URL	T
SFH	Processing of the submitted information in the same domain where the webpage is loaded	T
Submitting to email	HTML source code contains mailto:	B
Abnormal URL	Mismatch of website identity with the record in WHOIS database	B
Redirect	The number of times user is redirected to a new link within the same page	B
On mouseover	HTML source code contains onmouseover	B
Rightclick	Right click of the webpage is disabled or not	B
Popupwidnow	Website asking users to submit credentials through a pop-up window	B
Iframe	HTML source code contains frame/iframe	B
Age of domain	Age of the domain greater than 2 years, or between 2 and 1 year	B
DNS record	DNS record is empty or not	B
Web traffic	Total visits to a webpage	T
Page rank	Page rank assigned by google above a particular threshold	B
Google index	Website is indexed in google	B
Result	Phishing site or a legitimate site	B

[a] *B* binary; *T* ternary

References

1. Abbas, A.R., Singh, S., Kau, M.:. Detection of phishing websites using machine learning. In: Inventive Communication and Computational Technologies, pp. 1307–1314. Springer, Singapore (2020)
2. Ahmed, K., Naaz, S.: Detection of phishing websites using machine learning approach. In: Proceedings of International Conference on Sustainable Computing in Science, Technology and Management (SUSCOM). Amity University Rajasthan, Jaipur, India, Feb 2019
3. Alloghani, M., Al-Jumeily, D., Hussain, A., Mustafina, J., Baker, T., Ijaaf, A.J.: Implementation of machine learning and data mining to improve cybersecurity and limit vulnerabilities to cyber attacks. In: Nature-Inspired Computation in Data Mining and Machine Learning, pp. 47–76. Springer, Cham (2020)
4. APWG. Accessed on 2 May 2020. https://docs.apwg.org/reports/apwg_trends_report_q3_2019.pdf
5. Aydin, M., Butun, I., Bicakci, K., Baykal, N.: Using attribute-based feature selection approaches and machine learning algorithms for detecting fraudulent website URLs. In: 2020 10th Annual Computing and Communication Workshop and Conference (CCWC), pp. 0774–0779. IEEE (2020)
6. Babagoli, M., Aghababa, M.P., Solouk, V.: Heuristic nonlinear regression strategy for detecting phishing websites. Soft Comput. 23(12), 4315–4327 (2019)
7. Bujang, M.A., Adnan, T.H.: Requirements for minimum sample size for sensitivity and specificity analysis. J. Clin. Diagn. Res. JCDR 10(10), YE01 (2016)
8. Chavan, S., Inamdar, A., Dorle, A., Kulkarni, S., Wu, X.W.: Phishing detection: malicious and benign websites classification using machine learning techniques. In: Proceeding of International Conference on Computational Science and Applications, pp. 437–446. Springer, Singapore (2020)
9. Cekik, R., Telceken, S.: A new classification method based on rough sets theory. Soft Comput. 22(6), 1881–1889 (2018)
10. Chiew, K.L., Tan, C.L., Wong, K., Yong, K.S., Tiong, W.K.: A new hybrid ensemble feature selection framework for machine learning-based phishing detection system. Inf. Sci. 484, 153–166 (2019)
11. CSO. Accessed on 1 May 2020. https://www.csoonline.com/article/2117843/what-is-phishing-how-this-cyber-attack-works-and-how-to-prevent-it.html
12. Cuzzocrea, A., Martinelli, F., Mercaldo, F.: A machine-learning framework for supporting intelligent web-phishing detection and analysis. In: Proceedings of the 23rd International Database Applications and Engineering Symposium, pp. 1–3, June 2019
13. Davis, J., Goadrich, M.: The relationship between precision-recall and ROC curves. In: Proceedings of the 23rd International Conference on Machine Learning, pp. 233–240, June 2006
14. Demertzis, K., Iliadis, L.: Cognitive web application firewall to critical infrastructures protection from phishing attacks. J. Comput. Modell. 9(2), 1–26 (2019)
15. Dziak, J.J., Coffman, D.L., Lanza, S.T., Li, R., Jermiin, L.S.: Sensitivity and specificity of information criteria. Brief. Bioinform. 21(2), 553–565 (2020)
16. Eralda, H., Vaes, B., Van Pottelbergh, G., Matheï, C., Verbakel, J., Degryse, J.M.: Predictive accuracy of frailty tools for adverse outcomes in a cohort of adults 80 years and older: a decision curve analysis. J. Am. Med. Dir. Assoc. (2019)
17. Greco, S., Matarazzo, B., Slowinski, R., Stefanowski, J.: Variable consistency model of dominance-based rough sets approach. In: International Conference on Rough Sets and Current Trends in Computing, pp. 170–181. Springer, Berlin, Heidelberg (2000)
18. Greco, S., Matarazzo, B., Slowinski, R.: Rough sets theory for multicriteria decision analysis. Eur. J. Oper. Res. 129(1), 1–47 (2001)
19. Hindustan times. Accessed on 1 May 2020. https://www.hindustantimes.com/delhi/pakistan-group-hacks-iit-delhi-du-websites-posts-about-kashmir-violence/story-lsQ8Q08ksLfsRMoA0gcaEO.html

20. Imperva. Accessed on 1 May 2020. https://www.imperva.com/learn/application-security/phishing-attack-scam/
21. Jain, A.K., Gupta, B.B.: A machine learning based approach for phishing detection using hyperlinks information. J. Am. Intell. Human. Comput. **10**(5), 2015–2028 (2019)
22. Kleinbaum, D.G., Dietz, K., Gail, M., Klein, M., Klein, M.: Logistic Regression. Springer, New York (2002)
23. Kulkarni, A.: Phishing Websites Detection Using Machine Learning (2019)
24. Le Pochat, V., Van Goethem, T., Joosen, W.:. Funny accents: exploring genuine interest in internationalized domain names. In: International Conference on Passive and Active Network Measurement, pp. 178–194. Springer, Cham (2019)
25. Li, Y., Xiong, K., Li, X.: Understanding user behaviors when phishing attacks occur. In: 2019 IEEE International Conference on Intelligence and Security Informatics (ISI), p. 222. IEEE (2019)
26. Liaw, A., Wiener, M.: Classification and regression by RandomForest. R News **2**(3), 18–22 (2002)
27. Lv, K., Chen, Y., Hu, C.: Dynamic defense strategy against advanced persistent threat under heterogeneous networks. Inform. Fusion **49**, 216–226 (2019)
28. Moghimi, M., Varjani, A.Y.: New rule-based phishing detection method. Exp. Syst. Appl. **53**, 231–242 (2016)
29. Mohammad, R.M., Thabtah, F., McCluskey, L.: An assessment of features related to phishing websites using an automated technique. In: 2012 International Conference for Internet Technology and Secured Transactions, pp. 492–497. IEEE (2012)
30. Ohsaki, M., Wang, P., Matsuda, K., Katagiri, S., Watanabe, H., Ralescu, A.: Confusion-matrix-based kernel logistic regression for imbalanced data classification. IEEE Trans. Knowl. Data Eng. **29**(9), 1806–1819 (2017)
31. Orunsolu, A.A., Sodiya, A.S., Akinwale, A.T.: A predictive model for phishing detection. J. King Saud Univ. Comput. Inf. Sci. (2019)
32. Pawlak, Z.: Vagueness a rough set view In: Mycielski, J., Rozenberg, G., Salomaa, A. (eds.) Structures in Logic and Computer Science, vol. 1261, pp. 106–117. Springer, Berlin, Heidelberg (1997). https://doi.org/10.1007/3-540-63246-8_7
33. Pawlak, Z., Skowron, A.: Rudiments of rough sets. Inf. Sci. **177**(1), 3–27 (2007)
34. Portugal, I., Alencar, P., Cowan, D.: The use of machine learning algorithms in recommender systems: a systematic review. Exp. Syst. Appl. **97**, 205–227 (2018)
35. Powers, D.M.: Evaluation: from precision, recall and F-measure to ROC, informedness, markedness and correlation (2011)
36. Prasad, R., Rohokale, V.: Phishing. In: Cyber Security: The Lifeline of Information and Communication Technology, pp. 33–42. Springer, Cham (2020)
37. Rajab, M.: An anti-phishing method based on feature analysis. In: Proceedings of the 2nd International Conference on Machine Learning and Soft Computing, pp. 133–139, Feb 2018
38. Rao, R.S., Pais, A.R.: Detection of phishing websites using an efficient feature-based machine learning framework. Neural Comput. Appl. **31**(8), 3851–3873 (2019)
39. Rish, I.: An empirical study of the naive Bayes classifier. In: IJCAI 2001 Workshop on Empirical Methods in Artificial Intelligence, vol. 3(22), pp. 41–46 (2001)
40. Siahaan, H., Mawengkang, H., Efendi, S., Wanto, A., Windarto, A.P.: Application of classification method C4.5 on selection of exemplary teachers. J. Phys. Conf. Ser. **1235**(1), 012005 (2019)
41. Singh, A., Misra, S.C.: A dominance based Rough set analysis for investigating employee perception of safety at workplace and safety compliance. Saf. Sci. **127**, 104702 (2020)
42. Słowiński, R., Greco, S., Matarazzo, B.: Rough set analysis of preference-ordered data. In: International Conference on Rough Sets and Current Trends in Computing, pp. 44–59. Springer, Berlin, Heidelberg (2002)
43. Suykens, J.A., Vandewalle, J.: Least squares support vector machine classifiers. Neural Process. Lett. **9**(3), 293–300 (1999)

44. UCI repository. Accessed on 3 May 2020. https://archive.ics.uci.edu/ml/machine-learning-databases/00327/
45. Verkijika, S.F.: "If you know what to do, will you take action to avoid mobile phishing attacks": self-efficacy, anticipated regret, and gender. Comput. Hum. Behav. **101**, 286–296 (2019)
46. Vluymans, S., Mac Parthalain, N., Cornelis, C., Saeys, Y.: Weight selection strategies for ordered weighted average based fuzzy rough sets. Inform. Sci. **501**, 155–171 (2019)
47. Wei, W., Miao, D., Li, Y.: A bibliometric profile of research on rough sets. In: International Joint Conference on Rough Sets, pp. 534–548. Springer, Cham (2019)
48. Zamir, A., Khan, H.U., Iqbal, T., Yousaf, N., Aslam, F., Anjum, A., Hamdani, M.: Phishing web site detection using diverse machine learning algorithms. Electron. Libr. (2020)

Towards Owner-Controlled Data Sharing

Sabrina De Capitani di Vimercati, Sara Foresti, Giovanni Livraga, and Pierangela Samarati

Abstract We discuss some of the main problems related to allowing data owners to share their data with interested consumers in a controlled way in the context of digital data markets. Since resorting to the cloud for data storage reduces the burden at the owner's side, we first address the problem of supporting owners in selecting suitable cloud plans for storing their data collections. To this end, we illustrate some recent proposals for the specification and enforcement, in a friendly and flexible way, of the owners' requirements and preferences that should guide the selection process. We also address the problem of ensuring that data owners remain in control of their data and that they receive rewards for making their data available to others, and illustrate recent proposals addressing it.

Keywords Digital data markets · Data privacy · Data ownership · User empowerment · Controlled data sharing · Cloud plan selection

1 Introduction

It goes without saying that data is the oil fueling a constantly growing number of activities and businesses of our society. Data are increasingly used to extract useful information and to create knowledge and predictions, generating an immense profit for the parties using them. Often, this happens without the involved users (i.e., the subjects *behind* the data themselves) being aware of their data being collected and

S. De Capitani di Vimercati · S. Foresti · G. Livraga · P. Samarati (✉)
Università degli Studi di Milano, 20133 Milano, Italy
e-mail: pierangela.samarati@unimi.it

S. De Capitani di Vimercati
e-mail: sabrina.decapitani@unimi.it

S. Foresti
e-mail: sara.foresti@unimi.it

G. Livraga
e-mail: giovanni.livraga@unimi.it

© The Author(s), under exclusive license to Springer Nature Switzerland AG 2022
P. Nicopolitidis et al. (eds.), *Advances in Computing, Informatics, Networking and Cybersecurity*, Lecture Notes in Networks and Systems 289,
https://doi.org/10.1007/978-3-030-87049-2_23

used, and even more frequently without them being benefiting from the profits that directly derive from the usage of their data [1]. This has fostered a vision towards a fairer scenario, where individuals are first-class citizens and active participants of the data sharing ecosystem. An interesting direction concerns the development of *digital data markets*, where datasets can be traded between their owners and interested consumers (i.e., subjects wishing to access data) to generate knowledge and profits and where: (i) sharing is controlled by the data owners (who can decide whether to share a piece of their data, and with whom); and (ii) the owners get a reward for sharing their data with consumers (e.g., [2]). The awareness of the importance of these aspects is becoming more and more common among individuals. It is then easy to envision the possibility of such data markets being realized and possibly used.

In this context, the first aspect that owners should address concerns reasoning on how and where their data collections should be stored, for being easily (and selectively) available to interested data consumers. A possible approach for limiting the overhead at the owners' side consists in delegating data storage to external third parties. A natural possibility is then represented by outsourcing data collections to the cloud, leveraging one or more cloud plans available from the rich and diverse cloud market characterizing today's society. In this way, the overhead of storing and managing data is pushed from their owners to the chosen cloud providers, reducing owners' burden while, at the same time, enjoying the benefits of the cloud for making data available 24/7 from everywhere in the world. In this context, selecting the most suitable plan among the multitude available in the market is a critical problem, since different plans can exhibit different features and characteristics that can make them suitable to different application scenarios. Seemingly a non-critical and easy-to-solve problem, it is unfortunately far from being trivial. As a matter of fact, it requires a careful analysis of the features that can make the difference among plans, and a way to evaluate those features against possible requirements and preferences of users. It can also possibly require some technical skills or training, since the scenario is clearly characterized by cutting-edge ICT solutions. Selecting the right cloud plan hence represents a key factor for realizing an attractive digital data market: it is intuitive that if the market is built on cloud plans showing, for example, frequent downtimes or high latency, this would negatively impact the experience of all involved actors. In this chapter, we address such selection problem and illustrate recent approaches that can be adopted for cloud plan selection.

Once data have been moved to the cloud, they can be easily searched for and accessed by interested consumers. To ensure that owners maintain control over the sharing of their data, two aspects need to be considered. The first aspect, common to any scenario where a data owner resorts to external (and hence possibly not fully trusted) platforms for storing and managing data, consists in guaranteeing that each piece of data is shared with a subject (e.g., a consumer or the cloud provider) only with the will of its owner. This also requires to ensure that owners know, at any time and with high confidence, which consumer has had access to which portion of their data. A second aspect to be addressed, more specific to market scenarios, concerns the management of rewards to data owners for contributing their data. This is a critical aspect, especially when the interacting parties (the data owner, the cloud provider, and the interested data consumers) do not fully trust each other. In this chapter, we discuss these two aspects and illustrate recent approaches for solving them.

The reminder of this chapter is organized as follows. Section 2 addresses the problem of specifying requirements and selecting cloud plans for realizing a digital data market. It illustrates recent proposals based on a flexible specification language for formulating requirements, possibly using natural language expressions. Section 3 focuses on the issue of ensuring owner-controlled sharing of data. It illustrates a recent proposal building on selective owner-side encryption to ensure that owners remain in control of who can access which portions of their data collections. Section 4 addresses the problem of the management of rewards to data owners. It illustrates a possible solution building on blockchain and smart contracts to ensure that owners receive an agreed reward whenever an interested consumer gets access to some of their data. Finally, Sect. 5 concludes this chapter.

2 Cloud Plan Selection

A first problem to be addressed when moving data collections to the cloud concerns the selection of the most suitable set of cloud plans for data storage. This demands the evaluation of the features of the different candidate cloud plans with respect to possible requirements and preferences the owner could have, in turn demanding for approaches supporting owners in formulating such requirements. In this section, we first discuss some basic concepts and approaches addressing this problem (Sect. 2.1). We then illustrate recent solutions that permit to specify and enforce arbitrary requirements and preferences (Sect. 2.2), possibly leveraging natural language expressions (Sect. 2.3).

2.1 Basic Concepts

The first aspect to address in our scenario concerns determining the features, among those characterizing the different plans, which are relevant to the needs of the owner, and which can be used to drive the selection process. As an example, intuitive features can encompass the performance or the availability characterizing the different plans [3, 4]. Traditional proposals for cloud plan selection are based on Quality-of-Service (QoS) attributes that are guaranteed in Service Level Agreements (SLAs) by the providers for their services (e.g., [5–7]). In this regard, the scientific community has proposed several solutions for selection. These include the evaluation of low-level characteristics (such as CPU and network throughput) and cost of available plans (e.g., [8]), the definition of standardized bodies of more complex QoS attributes (e.g., [9]), the combination of QoS evaluation and other criteria such as subjective assessment and personal/past experience (e.g., [10–13]). It is interesting to note that some approaches have suggested to evaluate the *user-side* QoS rather than the *provider-side* QoS, meaning the values of QoS attributes measured at the user side (i.e., at the owner) rather than those declared by providers (e.g., [14]). Also, specific approaches have put forward the idea of considering security-related attributes in the selection (e.g., [15–18]).

Recently, the scientific community has started to investigate the possibility of supporting owners in arbitrarily selecting the features and characteristics over which their requirements should be defined (e.g., [19, 20]). This paradigm shift is typically based on the existence of a broker [19], that is, an entity in charge of collecting and understanding such arbitrary requirements and assessing their satisfaction by the candidate plans. By permitting the definition of arbitrary requirements on arbitrary features (clearly, representing characteristics that can be evaluated), such proposals allow for greater flexibility and user-friendliness than traditional approaches that are limited to a pre-defined set of features. In the remainder of this section, we illustrate two recent proposals that pursue this direction by allowing owners to leverage a friendly specification language [19] and natural language expressions [21] to formulate arbitrary requirements and preferences.

2.2 Crisp Requirements and Preferences

In this section, we illustrate how it is possible for data owners to easily formulate requirements to guide the selection of the cloud plan that best fits their needs. We will focus on the framework in [19], since it proposes a flexible and user-friendly language for supporting the specification of both *requirements* and *preferences* (i.e., hard and soft constraints, respectively) that can then be used to identify those plans that can be considered acceptable (requirements) and, among the acceptable ones, the ones that are preferable (preferences). As mentioned in Sect. 2.1, the proposal in [19] permits to formulate requirements and properties over arbitrary *attributes* of interest (configuration parameters) such as those appearing in the SLAs of the different plans, as well as metadata associated with them (e.g., the provider of a plan, or the nationality of the provider). To this end, plans are represented as vectors with one element for each attribute of interest reporting the value that such attribute assumes for the plan (or the special value '—' if the value of such attribute is unknown/unspecified for the plan). Figure 1 presents an example of such vectors for four plans P_1, P_2, P_3, and P_4, defined over seven different attributes. For instance, $P_1[\text{prov}] = \text{provA}$ means that plan P_1 is offered by provider provA. $P_1[\text{aud}] = -$ means that the frequency of security auditing for P_1 is not specified/unknown.

2.2.1 Requirements and Preferences Specification

The proposal in [19] builds on the idea that requirements and preferences should identify the values of the attributes that are considered mandatory or preferable for a plan to satisfy the needs of the data owner. The building block for the definition

	P_1	P_2	P_3	P_4	
prov	provA	provA	provA	provB	*cloud provider*
loc	locB	locA	locB	locB	*geographical location of servers*
encr	AES	AES	AES	3DES	*adopted encryption*
band	25	25	20	15	*bandwidth (Gb/s)*
test	med	top	top	low	*penetration test authority*
cert	certC	certA	certC	certB	*security certification*
aud	—	—	—	1Y	*security auditing frequency*

Fig. 1 Abstract representation of cloud plans

Complex requirement	**Semantics**
ANY$(b_1, \ldots, b_n)$	alternatives among base requirements
ALL$(b_1, \ldots, b_n)$	sets of base requirements to be jointly satisfied
IF ALL$(b_1, \ldots, b_k)$ THEN ANY$(b_{k+1}, \ldots, b_n)$	conditional requirements
FORBIDDEN$(b_1, \ldots, b_n)$	forbidden configurations
AT_LEAST$(m, (b_1, \ldots, b_n))$	at least m base requirements $(m < n)$
AT_MOST$(m, (b_1, \ldots, b_n))$	at most m base requirements $(m < n)$

(a)

b_1 : prov(provA, provB, provC)
b_2 : ¬band(0.2, 5)
c_1 : ALL({loc(locA, locB), ¬encr(DES)})
c_2 : ANY({test(top, med), cert(certA, certB)})
c_3 : ANY({loc(locA), cert(certC)})
c_4 : IF ALL({loc(locB), encr(3DES)) THEN ANY(aud(3M, 6M), cert(certA))
c_5 : IF ALL(test(−)) THEN ANY(cert(certA))
c_6 : FORBIDDEN({¬loc(locA), test(low)})
c_7 : AT_MOST(2, {prov(provB), band(15), encr(3DES)})
c_8 : AT_LEAST(2, {loc(locA), encr(AES), prov(provA, provB)})

(b)

Fig. 2 Complex requirements supported by the language in [19] (**a**) and an example of a set of requirements over the plans in Fig. 1 (**b**)

of requirements is the concept of *base requirement*, which is denoted by b. Given a set $\{v_1, \ldots, v_n\}$ of values in the domain of an attribute attr, a base requirement on attr imposes that attr can assume $(\text{attr}(v_1, \ldots, v_n))$ or cannot assume $(\neg \text{attr}(v_1, \ldots, v_n))$ such a set of values. For instance, a base requirement of the form prov(provA, provB, provC) states that, to be acceptable, a plan must be offered by provider provA, provB, or provC.

Starting from base requirements, the specification language in [19] permits to express a variety of *complex requirements,* summarized in Fig. 2a. An ANY requirement models alternatives among base requirements, and demands that at least one base requirement in the set $\{b_1, \ldots, b_n\}$ be satisfied. Similarly, an ALL requirement demands that all the listed base requirements be satisfied. A conditional IF-THEN requirement demands that if all the requirements appearing in the IF part are satisfied, then at least one of those in the THEN part must be satisfied. A FORBIDDEN requirement demands that the involved requirements must not be satisfied together in a plan, since they represent a forbidden configuration. An AT_LEAST (AT_MOST, respectively) requirement demands that at least (at most, respectively) m out of the n related base requirements be satisfied. Figure 2b illustrates an example of a set of base (b_1 and b_2) and complex ($c_1, \ldots, c_8$) requirements specified over the plans in Fig. 1. For instance, base requirements b_1 and b_2, respectively, demand that an acceptable plan must be offered by provider provA, provB, or provC (b_1), and must not have a bandwidth equal to 0.2 Gb/s or 5 Gb/s (b_2). Complex ALL requirement c_1 states that a plan can be considered acceptable only if its servers are geographically located in locA or locB, and if the adopted encryption is different from DES. Complex FORBIDDEN requirement c_6 states that a plan whose servers are not located in locA and for which the penetration test is enforced by the authority named 'low' cannot be considered acceptable.

As for the preferences, the framework in [19] permits two levels of specification, on the attribute values and on the attributes themselves. In particular, preferences on attribute values model the fact that, for a certain attribute, some values are preferred over other ones. A natural and intuitive interpretation can then model such preferences as a total order relationship among sets of values that the different attributes can assume. In other words, specifying preferences on attribute values can be done by first partitioning the values that an attribute can assume (indeed, those that are acceptable according to the requirements) in sets of equivalently-preferred values, and then specifying an order relationship among these value sets. Such kind of preference can be graphically represented as a hierarchy where, for example, preferred elements appear higher than less preferred ones. Figure 3 illustrates an example of preferences over attribute values for the plans in Fig. 1. Here, for example, value locA is preferred over value locB for attribute loc. The figure also reports, for each value, its *score* (used in the enforcement phase) given by its relative position in the induced ranking. Given the number h of partitions in which the values are grouped, the least preferred value(s) has score of $1/h$. Going up in the hierarchy, scores increase of $1/h$ at each step. The most preferred value(s) has then a score equal to 1. To illustrate, consider attribute prov in Fig. 3: its values have been partitioned in three sets ($h =$

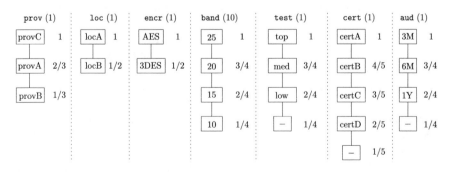

Fig. 3 Preferences for the plans in Fig. 1

3), and hence value provB (the least preferred) has score of $1/h = 1/3$. Value provA, immediately better than provB, has score $1/3$ higher than that of provB, and hence equal to $1/3 + 1/3 = 2/3$. Value provC, immediately better than provA, has score $1/3$ higher than that of provA, and hence equal to $2/3 + 1/3 = 1$.

Preferences on attributes, on the other hand, can be used to specify the relative importance that the data owner assigns to the attributes. The proposal in [19] models them through a weight function assigning higher weights to more important attributes. Figure 3 illustrates also an example of weights assigned to the attributes of our running example, reported as the number appearing in brackets close to the name of the attributes. In this example, all attributes have weight 1, except attribute band, which has weight 10.

2.2.2 Requirements and Preferences Evaluation

As mentioned above, the framework in [19] leverages requirements to determine *acceptable plans*, which are then ranked according to how much they respond to preferences.

A plan can be considered acceptable if and only if it satisfies *all* the stated requirements. The approach in [19] for determining the acceptable plans is based on a Boolean interpretation of the requirements, which also offers the possibility of checking whether the overall set of requirements is satisfiable or there are con-flicting requirements (e.g., a requirement demands a certain value for an attribute, and another requirement excludes that value from those acceptable). With reference to the plans in Fig. 1 and the requirements in Fig. 2b, it is easy to see that plans P_1, P_2, and P_3 are acceptable, while P_4 is not (since it does not satisfy requirements c_3, c_4, c_6, c_7, c_8).

As for preferences, the solution in [19] proposes different approaches for ranking plans. The first, and more intuitive, is a Pareto-based ranking, where a plan P_a is ranked better then a plan P_b iff P_a shows values that are equally or more preferred than those showed by P_b for all attributes and, for at least one attribute, P_a shows a value that is preferred than that showed by P_b. For instance, as illustrated in Fig. 4a, plan P_2 Pareto-dominates (and is hence preferable to) P_1 and P_3, which are instead not comparable between them. To overcome the possibility of returning incomparable solutions, a distance-based approach can be used instead of the Pareto-based ranking. The distance-based approach is based on the notion of *ideal plan*, a plan that shows the most preferred values for all attributes. The 'closer' a plan is to such an ideal plan, the better it satisfies the stated preferences. Plans are then modeled as points in an m-dimensional space, with m the number of attributes characterizing plans. Given a plan, its position in the space is given by a set of coordinates being the scores assigned to its attribute values based on the ranking induced by the preferences. For instance, plan P_2 has coordinates $[2/3, 1, 1, 1, 1, 1, 1/4]$ given by the scores of its attribute values in the ranking in Fig. 3 (e.g., $2/3$ is the score associated with value $P_2[\text{prov}] = \text{provA}$). Clearly, the values of the coordinates of the ideal plan will be, by definition, 1 for all dimensions. Leveraging such interpretation of plans it is then immediate, for

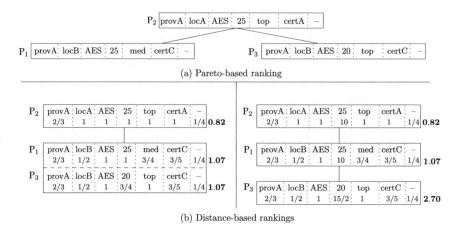

(a) Pareto-based ranking

(b) Distance-based rankings

Fig. 4 Rankings of plans P_1, P_2, and P_3 in Fig. 1 according to the preferences in Fig. 3

example through a simple computation of the Euclidean distance, to evaluate how close the plans are to the ideal plan. Such distance-based approach also permits to consider the preferences on attributes, by simply weighting the different elements in the coordinates of a plan by the weight given to the corresponding attribute. Figure 4b graphically illustrates the distance-based rankings over the acceptable plans of our running example (where the figure on the right-hand side considers attribute weights). In the figure, the distance of each plan to the ideal plan is reported in boldface on the right-hand side of each plan in the ranking.

2.3 Requirements with Natural Language

The approach illustrated in Sect. 2.2 provides data owners with a flexible and user-friendly language for specifying arbitrary requirements and preferences. However, such approach can still show the complication, intrinsic to the definition of crisp and precise requirements, of demanding domain-specific technical knowledge to fully understand the attributes characterizing cloud plans and the meaning of their values. As a matter of fact, requirement evaluation crosses out plans simply based on acceptable/unacceptable values. Whenever the owner formulating requirements is not technically skilled or trained, this may represent a barrier to requirement specification, ultimately leading to non-optimal solutions. Intuitively, it would be easier for non-skilled owners to specify their requirements using linguistic labels (such as 'high' or 'low' or 'important') instead of crisp values, and possibly on high-level properties (such as 'security', or 'performance') instead of on attributes representing low-level (configuration) parameters such as those in SLAs. In this section, we present how it is possible to support owners in an easy and intuitive requirement specification, leveraging natural language and high-level properties [21].

2.3.1 Requirements Specification

The proposal in [21] builds on the definition of *abstract parameters* and *abstract concepts*. Abstract parameters permit owners to specify their requirements over low-level configuration parameters (e.g., attribute `bandwidth` used in the example for the framework in Sect. 2.2) through natural language expressions. For instance, an owner could state a requirement for 'high' `bandwidth`, without specifying a precise threshold and using instead the linguistic label 'high'. Abstract concepts complement abstract parameters by representing higher-level abstractions over parameters. For instance, an owner could state a requirement over `performance`, which intuitively represents an abstraction over lower-level parameters `throughput` and `bandwidth`. Owners (and especially non-skilled or non-trained ones) can then operate on abstract parameters and concepts to specify in a friendly and flexible manner their desiderata.

Operating with abstract parameters and concepts requires the existence of a set of *linguistic labels* that can be associated with them as values (in contrast to domain-specific values). Such sets can be arbitrarily defined in such a way to 'quantify' a parameter or a concept. As an example, abstract parameter `bandwidth` could be associated with a set {small, large} of linguistic labels. Similarly, abstract concept `performance` could be associated with a set {low, med, high} of linguistic labels.

As already mentioned, abstract concepts and parameters are clearly strictly connected, since concepts represent abstractions over parameters, more easily accessible by non-skilled owners. For instance, with reference to the example above, a high `throughput` and a large `bandwidth` can result in a high `performance` value. Following this intuition, the relationship between concepts and the involved parameters is modeled through a set of *implication rules*, specifying conditions on which combination of values (linguistic labels) for parameters imply a given value (linguistic label) for concepts. To avoid ambiguities, each label associated with a concept should be regulated by an implication rule. For instance, since concept `performance` is associated (in the examples above) with three linguistic labels, it is governed by three inference rules:

$$\langle \text{throughput} = \text{high} \rangle \vee \langle \text{bandwidth} = \text{large} \rangle \Longrightarrow \langle \text{performance} = \text{high} \rangle;$$
$$\langle \text{throughput} = \text{med} \rangle \Longrightarrow \langle \text{performance} = \text{med} \rangle;$$
$$\langle \text{throughput} = \text{low} \rangle \vee \langle \text{bandwidth} = \text{small} \rangle \Longrightarrow \langle \text{performance} = \text{low} \rangle.$$

Equipped with a vocabulary for reasoning over parameters and concepts with linguistic labels, it is then possible to formulate requirements. Intuitively, in line with the 'quantitative' approach permitted for parameters and concepts, requirements represent how much a certain combination of values (linguistic labels) for the abstract parameters and concepts satisfies the owner formulating them. The rationale is to support the specification of rules that say that a certain combination of characteristics is, for example, *highly satisfactory*, while another combination is *not satisfactory*. Requirements are then modeled as the implication rules above-mentioned: a combination of linguistic expressions over a set of parameters and concepts implies a certain linguistic expression (e.g., 'low' or 'high') of ad-hoc variable `satisfaction`, modeling the overall satisfaction for a plan. For instance, "⟨performance = high⟩ ⟹ ⟨satisfaction = high⟩" is an example of a requirement, defined over concept performance, stating that a high level of performance highly satisfies the owner.

2.3.2 Requirements Evaluation

To evaluate the specified requirements, it is necessary to establish a mapping between the elements included in the requirements (i.e., expressions using natural language and possibly referred to abstract concepts) to the actual (crisp) characteristics of the different plans. This requires to map: (i) concepts to actual configuration parameters; and (ii) linguistic labels to crisp (domain-specific) values. An intuitive approach for this second problem could involve associating with a certain linguistic label a set of crisp values. For instance, given the set of crisp values that can be assumed by parameter bandwidth (e.g., [0.2 Gb/s, 25 Gb/s]), one may partition such set in disjoint intervals (e.g., [0.2 Gb/s, 10 Gb/s), [10 Gb/s, 25 Gb/s]), and associate each interval with a linguistic label (e.g., [0.2 Gb/s, 10 Gb/s) with small and [10 Gb/s, 25 Gb/s] with large). This solution, while indeed viable, carries the drawback of creating sharp boundaries among the sets of values that correspond to the different linguistic labels. With reference to the example above, a sharp boundary is created around value 10 Gb/s: a value of 9.99 Gb/s would be considered small, while the slightly larger value 10 Gb/s would be considered large. A more flexible approach, which enjoys the advantage of also providing with a means to manage the correspondence between abstract concepts and actual parameters, can be based on the adoption of fuzzy logic [21]. Abstract parameters and concepts can be interpreted as *fuzzy variables*, and the adopted linguistic labels as *fuzzy sets*. Fuzzy variables are variables that can assume crisp values and linguistic labels, while fuzzy sets are sets whose elements have a degree of membership, expressed as a value in the continuous interval [0, 1]. Given an element, in classical set theory it either belongs or does not belong to a set. In fuzzy set theory, an element belongs to a fuzzy set with a certain *degree of membership* (and of course can belong to different fuzzy sets, possibly with different degrees of membership). The degree of membership μ of elements to a fuzzy set is governed by the definition of *membership functions*, which can assume different shapes and permit a gradual assessment of the membership of elements to sets. Figure 5a illustrates an example of membership functions that can be associated with parameter bandwidth, one for each linguistic label that can be associated with it (i.e., small and large).

As it can be seen in the figure, the functions operate over the domain of crisp values that can be assumed by the parameter, and define how much a certain value belongs to the fuzzy set represented by the linguistic labels, meaning how much a certain value is 'representative' of each linguistic label. For instance, consider a value x in the domain [0.2, 25] of bandwidth and the membership function regulating label large: it is immediate to see that the more value x grows, the more it belongs (i.e., the higher its degree μ of membership) to the fuzzy set large. Similarly, considering

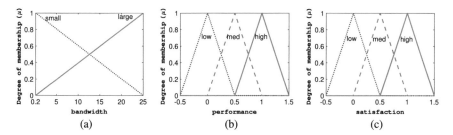

Fig. 5 An example of membership functions for parameter bandwidth (**a**), concept performance (**b**) and the ad-hoc variable satisfaction (**c**)

the membership function regulating label small, the more value x grows, the less it belongs (i.e., the lower its degree μ of membership) to the fuzzy set small.

Abstract parameters have then a natural interpretation in terms of fuzzy logic, establishing a correspondence (through membership functions) between the linguistic labels adopted in the requirements and the original crisp values assumed by a configuration parameter. A similar approach can be used also for managing abstract concepts and the variable satisfaction used in the requirements with the note that, being abstractions, they do not have a natural domain of crisp values. An intuitive approach is then to arbitrarily define their domains, for example in the continuous interval [0, 1], and then again establish a mapping through membership functions. Figure 5b and c illustrates examples of membership functions for the abstract concept performance and for the ad-hoc variable satisfaction.

To quantify concepts and the satisfaction of the owner as well as to reason about and evaluate the requirements, the proposal in [21] uses *fuzzy logic* and, more precisely, *fuzzy inferences*. In a nutshell, a fuzzy inference process takes as input a (set of) crisp value(s), interprets it (them) with a fuzzy modeling as illustrated before, evaluates a set of if-then rules based on such fuzzy modeling obtaining a (fuzzy) result, and returns such result after having transformed it again into a crisp value. Fuzzy inference can then easily provide the framework in which evaluating requirements: a first inference process leverages the implication rules linking concepts and parameters as the if-then rules of the process. It takes as input the crisp parameter values characterizing the plans under analysis and quantifies the concepts used in the requirements. A second inference process leverages instead the requirements themselves as if-then rules. It takes as input the quantification of the concepts done in the first inference (and, if present in the requirements, also the crisp values for parameters) and quantifies the level of satisfaction.

To illustrate the working of the fuzzy inference process, consider requirement "⟨performance = high⟩ ⟹ ⟨satisfaction = high⟩", taken from a set of requirements, and suppose that concept performance depends on low-level parameters throughput and bandwidth. Clearly, candidate plans exhibit values (e.g., in their SLAs) for throughput and bandwidth but not for performance. The framework in [21] would then apply a first inference process as follows: (i) the

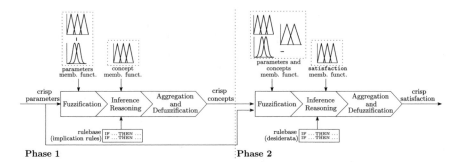

Fig. 6 Fuzzy inference processes for evaluating requirements (©2019 IEEE. Reprinted, with permission, from [21])

crisp values for `throughput` and `bandwidth` are taken as input; (ii) the inference rules governing concept `performance` are used as if-then rules and applied to the (fuzzified) input values; (iii) depending on the input values and on the evaluated rules, a quantification of `performance` is then produced. With such value for `performance`, a second inference process would then operate to assess owner's satisfaction. The second inference process operates exactly as the first one, with the difference that it takes as input the quantification of `performance` and operates on it with the rules expressed in the requirements. Figure 6 graphically illustrates this two-steps approach. We close this section with the note that the inference processes require the definition of different domain-specific operations, such as the fuzzification of crisp values and the defuzzification of fuzzy values, which can be performed adopting different existing approaches.

3 Controlled Sharing

While moving data to the cloud reduces the overhead left at the owners' side, it also causes the owners to lose direct control over their data and on who can access them [22]. To ensure a fair data market, it is crucial to empower the owners with control over the sharing of their data, and to guarantee that they receive rewards for making their data available to others (which is addressed in Sect. 4). In this section, after briefly recalling some basic concepts (Sect. 3.1), we illustrate a recent approach enabling controlled sharing. This approach allows owners to share their data (to which, for the sake of generality, to which we refer with the term 'resources') to interested consumers ensuring that owners remain in control of who can access which resources (Sect. 3.2).

3.1 Building Blocks

The solution in [23] ensures controlled data sharing by leveraging two main building blocks: (i) selective owner-side encryption and (ii) key derivation.

Selective owner-side encryption. Selective owner-side encryption protects the confidentiality of the resources to which it is applied through encryption. It is particularly appealing in scenarios of selective sharing as it consists in encrypting, at the owner-side, different resources with different keys. Keys are then distributed to consumers in such a way that each consumer can decrypt all and only the resources she is authorized to access. Since the encryption layer is applied at the owner side, resources self-enforce the access restrictions defined over them. Also, owner-side encrypted resources are protected against the cloud provider storing them as, without encryption keys, it cannot perform decryption. This represents a convenient feature in the considered scenario, since the owner can enjoy the benefits of resorting to the cloud for data storage while resting assured that only consumers knowing the encryption keys will be able to decrypt resources. A straightforward solution to enforce access restrictions through selective encryption consists in encrypting each resource with a different key, and in distributing to each consumer the keys of the resources she can access. However, this practice would imply a considerable key management burden for owners and consumers. To mitigate such overhead, the proposal in [23] leverages key derivation.

Key derivation. Key derivation permits to derive an encryption key k_y from the knowledge of another encryption key k_x and of a public label l_y (i.e., a piece of information) associated with k_y [24, 25]. The derivation of k_y from k_x is enabled by a public token $t_{x,y}$ computed as $k_y \oplus \mathcal{H}(k_x, l_y)$, with $\oplus$ the bitwise xor operator, and $\mathcal{H}$ a cryptographic hash function. The derivation of k_y from k_x can be *direct*, leveraging a single token $t_{x,y}$, or *indirect*, through a chain $\langle t_{x,z_1}, \ldots, t_{z_n,y} \rangle$ of tokens. Key derivation structures can be graphically represented as directed acyclic graphs, where vertices represent encryption keys (and their labels), and edges represent tokens among them. Since tokens do not need to be kept private, they can be physically stored in a public catalog $\mathcal{T}$. Figure 7 illustrates an example of derivation among three keys k_α, k_β, and k_γ (Fig. 7a) and the corresponding token catalog $\mathcal{T}$ (Fig. 7b). For simplicity, in our examples, we use x to denote the label of key k_x and, in the figures, we use the label x of key k_x to denote the corresponding vertex v_x (e.g., vertex

Fig. 7 An example of key derivation structure (**a**) and token catalog (**b**)

$\mathcal{T}$		
From	**To**	**Value**
α	γ	$k_\gamma \oplus \mathcal{H}(k_\alpha, l_\gamma)$
β	γ	$k_\gamma \oplus \mathcal{H}(k_\beta, l_\gamma)$

(a)   (b)

v_α, denoted α, in Fig. 7 represents key k_α and its label α). In the following, when clear from the context, we will use the terms keys and vertices (tokens and edges, respectively) interchangeably.

3.2 Ensuring Controlled Sharing

We now illustrate how selective owner-side encryption and key derivation can be used to effectively enforce the authorization policy specified by the data owner, hence providing for owner-controlled sharing in the data market.

In principle, the authorization policy can be represented in different ways, including access control lists (reporting, for each resource, the list of consumers authorized for access) as well as capability lists (reporting, for each consumer, the list of authorized resources). Given the dynamic scenario considered, characterized by consumers leveraging the market to purchase sets of resources, it is natural to think in terms of sets of resources and hence to represent authorizations as capability lists. Given a consumer c, $\mathsf{cap}(c)$ represents the set of resources for which c has purchased access (see Sect. 4). The capability $\mathsf{cap}(c)$ is then updated whenever c purchases access to a new resource r and is then authorized for r (i.e., $\mathsf{cap}(c) := \mathsf{cap}(c) \cup \{r\}$).

Ensuring controlled sharing requires that the content of the resources remains protected and only authorized consumers can access it. Moreover, despite resorting to the cloud for storage and hence losing direct control over their resources, the data owners must be aware, at all times, of which consumers have access to which resources. Selective owner-side encryption (Sect. 3.1) represents a promising solution that enjoys the advantage of maintaining resources confidential also to the cloud provider. We will now illustrate how sharing of resources can be controlled.

With selective owner-side encryption, the owner needs to agree a key k_c with every consumer c with which she wants to share resources. To grant c access to the resources in her capability list $\mathsf{cap}(c)$, the owner publishes a set of tokens enabling the derivation of the keys used to encrypt the resources in $\mathsf{cap}(c)$ starting from k_c. A (basic) key derivation structure includes a vertex v_c for each consumer c (representing k_c, which is known to c), a vertex v_r for each resource r (representing k_r, which is used to encrypt r), and a set of edges (representing tokens for derivation) connecting, for each consumer c, vertex v_c to vertex v_r for each $r \in \mathsf{cap}(c)$. This means that it is possible to derive, from key k_c, the encryption key k_r for each resource r in $\mathsf{cap}(c)$. To keep the size of the token catalog under control, the key derivation structure can be enriched with additional vertices representing keys used for derivation purposes only. In line with the authorization policy being represented as capability lists, such additional vertices represent sets of resources [23]. A possible approach for correctly enforcing an authorization policy in a market consists in connecting vertices through edges (i.e., keys through tokens) in such a way that [23]:

- for each consumer c, vertex v_c is connected to vertex $v_{\mathsf{cap}(c)}$ representing the set of resources in her capability list;

Consumer c	cap(c)
w	α, β, γ
x	$\gamma, \delta, \epsilon, \zeta$
y	α, β, γ
z	β, γ

(a)

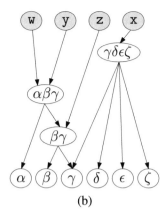

(b)

Fig. 8 An example of an authorization policy for four consumers and six resources (**a**), and of a key derivation structure that enforces it (**b**)

- each vertex v_{R_i}, representing a set R_i of resources, is directly connected to other vertices according to the subset containment relationship (i.e., for each edge (v_{R_i}, v_{R_j}), $R_i \supset R_j$) and in such a way that set R_i is fully covered by the resources represented by these other vertices;
- for each consumer c, there is a path connecting vertex v_c to all vertices v_r such that $r \in$ cap(c).

Figure 8 illustrates an example of an authorization policy regulating access to a set $\{\alpha, \beta, \gamma, \delta, \epsilon, \zeta\}$ of six resources for four consumers w, x, y, and z (Fig. 8a), and of a key derivation structure that enforces it (Fig. 8b). The structure contains a vertex for each consumer (gray vertices), representing the key agreed between the owner and the consumer (and known to the consumer herself), and a vertex for each resource. It also contains three additional vertices $\beta\gamma$, $\alpha\beta\gamma$, and $\gamma\delta\epsilon\zeta$ representing the capability lists of the consumers. Vertices of consumers are directly connected to the vertices of their capability lists, which are in turn connected to other vertices respecting the subset containment relationship and such that they are covered (e.g., vertex $\alpha\beta\gamma$ is connected to vertices α and $\beta\gamma$, in turn connected to vertices β and γ). The derivation structure correctly enforces the authorization policy in the figure, since there is a path linking each consumer vertex to all and only the vertices representing the resources in her capability list. For instance, consider consumer w with cap(w) $= \{\alpha, \beta, \gamma\}$. From her key k_w, she can use token $t_{w,\alpha\beta\gamma}$ to derive $k_{\alpha\beta\gamma}$ (note that this key is used for derivation purposes only and does not encrypt any resource). From $k_{\alpha\beta\gamma}$, she can then use token $t_{\alpha\beta\gamma,\alpha}$ to derive key k_α, and token $t_{\alpha\beta\gamma,\beta\gamma}$ to derive key $k_{\beta\gamma}$, from which she can then use tokens $t_{\beta\gamma,\beta}$ and $t_{\beta\gamma,\gamma}$ to derive k_β and k_γ, respectively. Hence, starting from the knowledge of a single key k_w, w can derive the keys of all the resources shared with her.

Having illustrated how the key derivation structure can be built, we now illustrate how it can be updated to reflect changes to the authorization policy. The structure must be updated when: (i) new resources are placed in the market; and (ii) new accesses are requested by (and granted to) consumers. We note that authorization revocation is not in line with the fact that access is provided upon payment [23].

The insertion of a new resource r into the market is reflected in the key derivation structure by simply adding a vertex v_r. This requires the owner to generate an encryption key k_r and a label l_r, encrypt r with k_r, and place the encrypted version of r in the market (i.e., outsource r to the chosen cloud platform). For instance, with reference to the structure in Fig. 8b and assuming that the owner publishes all six resources before any consumer request, the derivation structure includes only the leaves of the structure.

The purchase of a set R of resources by consumer c is reflected in the key derivation structure by ensuring that there is a path from vertex v_c of the consumer c to the vertices of all the resources in the updated capability list $\mathsf{cap}(c) = \mathsf{cap}(c) \cup R$. It is first necessary to create, if not already in the structure, vertices v_c for c and $v_{\mathsf{cap}(c)}$ for the updated capability list of c (creating also the corresponding encryption keys and labels). Vertex v_c is then connected to vertex $v_{\mathsf{cap}(c)}$ that, as illustrated previously, is connected to the other vertices in the hierarchy following the subset containment relationship and coverage principle. Each added edge corresponds to a token, which is inserted into the catalog. Figure 8b illustrates a possible key derivation structure after a series of granted requests, allowing the consumers to access resources according to the policy in Fig. 8a.

4 Rewards to Owners

In this section, we illustrate a possible approach for ensuring that owners receive a reward every time they grant access to some of their resources to interested consumers. This approach nicely complements the selective owner-side encryption approach illustrated in Sect. 3.

Ensuring rewards is a complex aspect in the addressed scenario, for two main reasons: (i) consumers and owners might not completely trust each other, and (ii) both parties could in principle misbehave to obtain illicit benefits. The main misbehaviors that may happen are related to the payment of a reward (i.e., a malicious consumer does not pay the reward for a resource she accessed, or a malicious owner falsely claims that a reward has not been paid, demanding a new payment) as well as to the access to a resource (where a malicious owner does not grant access despite having received a payment, or a malicious consumer falsely claims that access has not been granted despite the payment, requesting her money back). All misbehaviors could hamper the adoption of markets to trade (personal) data, disincentivizing owners to contribute with their data. In this section, after a discussion of some basic concepts (Sect. 4.1), we illustrate how the solution in [23] ensures rewards to owners, while preventing possible misbehaviors (Sect. 4.2).

4.1   Building Blocks

The solution in [23] ensures rewards to owners leveraging two main building blocks: (i) blockchain and (ii) smart contracts.

Blockchain. A blockchain is a shared and public ledger of transactions organized as a list of blocks, linked in chronological order [26]. Each block contains a certain number of transaction records as well as a cryptographic hash of the previous block. The blockchain is then maintained in a distributed way by a decentralized network of peers. While the single peers might not trust each other, the content and status of the blockchain is continuously agreed upon since each transaction is validated by the network of peers, and is then included in a block through a consensus protocol. A peculiarity of blockchains, which provide trust even if the peers singularly taken are not fully trusted, is that everyone can inspect a blockchain but no single peer can tamper with it, since modifications to the content of a blockchain requires mutual agreement among peers. Consequently, nobody can modify a committed block, and possible updates are reflected in a new block containing the updated information. This permits to trust the content and the status of a blockchain, while not trusting the single peers.

Smart contracts. Smart contracts can be used to establish an agreement among multiple, possibly distrusting, parties through a blockchain. A smart contract is a piece of software deployed on a blockchain, and is typically composed of a set of rules on which the interacting parties have to agree. The rules are of the form 'if-then' and define events and subsequent actions. Such rules formalize the clauses of the contract to be agreed upon (and virtually signed) by the parties. By leveraging the underlying blockchain consensus protocol, the execution of a smart contract can be trusted for correctness, meaning that all the conditions of the agreement have been met (as validated by the network). However, smart contracts and their execution do not provide confidentiality and privacy guarantees, as open visibility over the content of a contract and over the data it manipulates is a necessary condition for validation [27].

4.2   Ensuring Rewards to Owners

The possibility of consumers not paying for a granted resource (and conversely the possibility of a malicious owner claiming that she did not receive a payment for a resource, while she actually has) can be easily prevented adopting blockchain and smart contracts. The basic idea consists in ensuring that access to a resource be granted *upon* a monetary transaction (i.e., the payment of the reward) occurring between the owner and the consumer purchasing access. In this way, the money transfer can be inspected and validated by the blockchain network. Considering that access is granted through the possibility of decrypting resources, two straightforward approaches could be envisioned. A first approach could directly trade encryption keys through a smart contract. A second approach could trigger the updates to the derivation hierarchy (see Sect. 3.2) to ensure access directly through the smart contract, computing and communicating tokens within the smart contract. However, neither of these approaches is feasible since, as mentioned above, smart contracts are public and hence their simple observation would disclose the traded secrets (i.e., the encryption keys in both approaches, since also updating the derivation structure requires knowledge of the keys used in the system).

A possible solution to such issues leverages blockchain and smart contracts only to finalize the monetary transaction and, at the same time, obtain a public and verifiable commitment of the owner to grant the paid access. The smart contract can then be arbitrarily formulated in such a way to securely keep track of the purchase, according to the following logic: "upon receiving price from c for the set R of resources, $\mathsf{cap}(c) := \mathsf{cap}(c) \cup R$, and the token catalog is updated to grant c access to $\mathsf{cap}(c)$" [23], where price is the agreed price to be paid by consumer c to the owner to get access to R. The overall interplay between the owner and consumer c, including the deployment of the smart contract on the blockchain, is then regulated by the interaction protocol in Fig. 9, which works as follows.

1. Consumer c communicates off-chain the set R of resources that she wants to purchase to the owner.
2. If necessary (i.e., in the first interaction, where c has not yet received the encryption key k_c from the owner), the owner generates k_c and sends it off-chain to c, who signs it with her own private key and sends back to the owner the signed key, denoted in Fig. 9 as $[k_c]_{priv_c}$ (the signature is needed for the audit process described next).
3. The owner deploys the smart contract following the logic illustrated above.
4. Consumer c accesses, executes and signs the smart contract, triggering the money transfer to the owner.
5. The owner updates the key derivation structure as needed to grant c the agreed accesses.
6. The owner stores the updated token catalog on the blockchain.

It is interesting to note that such a slim interaction protocol prevents the possibility for a malicious consumer of not paying for a resource (the key derivation structure is updated after the execution of the smart contract) as well as for a malicious owner of claiming that she has not received the agreed money (thanks to the public nature of blockchains). Also, keys are safe since the derivation structure is updated locally by the owner. However, the interaction protocol cannot prevent malicious owners from refusing to update the derivation structure after receiving a payment, nor malicious consumers from claiming that access has not been provided despite the payment. Unfortunately, such misbehaviors cannot be prevented since the key derivation structure is updated locally. However, they can be easily detected and exposed, through a very simple *audit process* that can be executed whenever they are suspected. The audit protocol leverages the fact that the token catalog and the

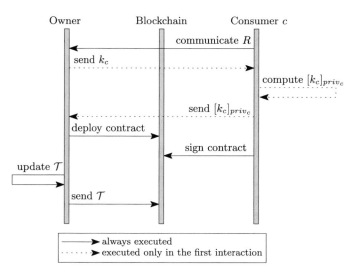

Fig. 9 Interaction protocol for token catalog update in [23]

capability lists are stored on-chain (and hence are tamper-proof), and that the keys agreed with consumers are signed by them.

The audit process can be invoked by both owners and consumers, and simply requires a designated trusted auditor to explicitly check whether, starting from the key k_c of the involved consumer c, the set of tokens stored on-chain actually permits to correctly derive the keys k_r for all $r \in \mathsf{cap}(c)$. This is simply done by querying the token catalog and performing the key derivation that is allowed by tokens. If the entire derivation succeeds, then the owner has behaved correctly and hence the malicious behavior is at c's side, falsely claiming of not having being granted an access that actually is enabled. On the contrary, if the derivation fails, the owner has not done what she promised in the smart contracts, preventing a legitimate access. The availability of an audit process detecting and exposing not only misbehaviors, but also the misbehaving party, counteracts the possibility of misbehaviors and incentivizes all parties to behave correctly not to reduce their credibility and reputation that (like in any real-world transaction) are key factors for having subjects engaging in negotiations and transactions.

We close this section with some notes on the audit process. The reliability of its results is based on the correctness and freshness of the tokens and labels, of the capability lists, and of the starting key k_c. Correctness and freshness of tokens, labels, and capability lists are guaranteed by the fact that they are stored on-chain. For keys, these properties are guaranteed by the fact that they are signed by the respective owners (the owner cannot forge and c cannot repudiate her signature). This is the reason why keys are signed in the interaction protocol (and signed keys are of course used in the computation and update of tokens in the key derivation structure). It is also interesting to note that the combined adoption of on-chain storage, interaction protocol, and audit process permits to arbitrarily check the correctness of the derivation structure at any point in time, since all communications and relevant data (i.e., tokens and capability lists) are stored on-chain and hence tamper-proof.

5 Conclusions

We have addressed the problem of allowing data owners to share data with interested consumers in the context of digital data markets. First, we have investigated the specification and enforcement of requirements and preferences that can be used by owners to select the most suitable cloud plans for storing their data collections. Then, we have investigated the problem of controlled data sharing in the market, ensuring that owners remain in control of who accesses which portions of their data, and that they receive incentives for trading and sharing their data with others. For both issues, we have illustrated recent solutions that can be adopted. These are two key aspects, out of many, to ensure that owners remain in control in digital data markets. Other relevant aspects include, but are not limited to, the specification and enforcement of complex access and usage policies, possibly based on purpose and secondary use, and the enforcement of privacy-aware retrieval and analytics.

Acknowledgements This work was supported in part by the EC within the H2020 Program under projects MOSAICrOWN and MARSAL, by the Italian Ministry of Research within the PRIN program under project HOPE, and by JPMorgan Chase & Co under project "k-anonymity for AR/VR and IoT/5G".

References

1. Zuboff, S.: The Age of Surveillance Capitalism: The Fight for a Human Future at the New Frontier of Power. PublicAffairs, New York (2019)
2. De Capitani di Vimercati, S., Foresti, S., Livraga, G., Samarati, P.: Towards owners' control in digital data markets. IEEE Syst. J. (ISJ) (2020, in press)
3. Jhawar, R., Piuri, V., Santambrogio, M.: Fault tolerance management in cloud computing: a system-level perspective. IEEE Syst. J. (ISJ) **7**(2), 288–297 (2013)
4. Jhawar, R., Piuri, V.: Fault tolerance and resilience in cloud computing environments. In: Vacca, J. (ed.) Computer and Information Security Handbook, 2nd edn. Morgan Kaufmann, pp. 125–142 (2013)
5. Guo, Y., Mi, Z., Yang, Y., Ma, H., Obaidat, M.S.: Efficient network resource preallocation on demand in multitenant cloud systems. IEEE Syst. J. (ISJ) **13**(4), 4027–4038 (2019)
6. De Capitani di Vimercati, S., Foresti, S., Livraga, G., Piuri, V., Samarati, P.: Supporting users in cloud plan selection. In: Samarati, P., Ray, I., Ray, I. (eds.) From Database to Cyber Security: Essays Dedicated to Sushil Jajodia on the Occasion of his 70th Birthday. Springer (2018)
7. Xie, Y., Guo, Y., Mi, Z., Yang, Y., Obaidat, M.S.: Loosely coupled cloud robotic framework for QoS-driven resource allocation-based Web service composition. IEEE Syst. J. (ISJ) **14**(1), 1245–1256 (2020)

8. Li, A., Yang, X., Kandula, S., Zhang, M.: CloudCmp: comparing public cloud providers. In: Proceedings of the 10th ACM Internet Measurement Conference (ACM IMC), Melbourne, Australia, Nov 2010
9. Garg, S.K., Versteeg, S., Buyya, R.: A framework for ranking of cloud computing services. Future Gener. Comput. Syst. (FGCS) **29**(4), 1012–1023 (2013)
10. Ding, S., Wang, Z., Wu, D., Olson, D.L.: Utilizing customer satisfaction in ranking prediction for personalized cloud service selection. Decis. Support Syst. **93**, 1–10 (2017)
11. Ghosh, N., Ghosh, S.K., Das, S.K.: SelCSP: a framework to facilitate selection of cloud service providers. IEEE Trans. Comput. (TCC) **3**(1), 66–79 (2015)
12. Qu, L., Wang, Y., Orgun, M.A., Liu, L., Liu, H., Bouguettaya, A.: CCCloud: context-aware and credible cloud service selection based on subjective assessment and objective assessment. IEEE Trans. Serv. Comput. (TSC) **8**(3), 369–383 (2015)
13. Tang, M., Dai, X., Liu, J., Chen, J.: Towards a trust evaluation middleware for cloud service selection. Future Gener. Comput. Syst. (FGCS) **74**, 302–312 (2017)
14. Zheng, Z., Wu, X., Zhang, Y., Lyu, M.R., Wang, J.: QoS ranking prediction for cloud services. IEEE Trans. Parallel Distrib. Syst. (TPDS) **24**(6), 1213–1222 (2013)
15. Casola, V., De Benedictis, A., Eraşcu, M., Modic, J., Rak, M.: Automatically enforcing security SLAs in the cloud. IEEE Trans. Serv. Comput. (TSC) **10**(5), 741–755 (2017)
16. Luna, J., Suri, N., Iorga, M., Karmel, A.: Leveraging the potential of cloud security service-level agreements through standards. IEEE Cloud Comput. **2**(3), 32–40 (2015)
17. Cloud Security Alliance: Cloud Control Matrix v3.0.1. https://cloudsecurityalliance.org/research/ccm/
18. Jhawar, R., Piuri, V., Samarati, P.: Supporting security requirements for resource management in cloud computing. In: Proceedings of the 15th IEEE International Conference on Computational Science and Engineering (IEEE CSE), Paphos, Cyprus, Dec 2012
19. De Capitani di Vimercati, S., Foresti, S., Livraga, G., Piuri, V., Samarati, P.: Supporting user requirements and preferences in cloud plan selection. IEEE Trans. Serv. Comput. (TSC), Nov 2017 (in press)
20. Sundareswaran, S., Squicciarini, A., Lin, D.: A brokerage-based approach for cloud service selection. In: Proceedings of the 5th IEEE International Conference on Cloud Computing (IEEE CLOUD), Honolulu, HI, USA, June 2012
21. De Capitani, S., di Vimercati, S., Foresti, G., Livraga, Piuri, V., Samarati, P.: A fuzzy-based brokering service for cloud plan selection. IEEE Syst. J. (ISJ) **13**(4), 4101–4109 (2019)
22. Samarati, P., De Capitani di Vimercati, S.: Cloud security: issues and concerns. In: Murugesan, S., Bojanova, I. (eds.) Encyclopedia on Cloud Computing. Wiley (2016)
23. De Capitani di Vimercati, S., Foresti, S., Livraga, G., Samarati, P.: Empowering owners with control in digital data markets. In: Proceedings of the 12th IEEE International Conference on Cloud Computing (IEEE CLOUD), Milan, Italy, July 2019
24. Atallah, M., Blanton, M., Fazio, N., Frikken, K.: Dynamic and efficient key management for access hierarchies. ACM Trans. Inf. Syst. Secur. (TISSEC) **12**(3), 18:1–18:43 (2009)
25. De Capitani di Vimercati, S., Foresti, S., Jajodia, S., Paraboschi, S., Samarati, P.: Encryption policies for regulating access to outsourced data. ACM Trans. Database Syst. (TODS) **35**(2), 12:1–12:46 (2010)
26. Zheng, Z., Xie, S., Dai, H., Chen, X., Wang, H.: An overview of blockchain technology: architecture, consensus, and future trends. In: Proceedings of the 2017 IEEE International Conference on Big Data (IEEE BigData), Boston, MA, USA, Dec 2017
27. Cheng, R., Zhang, F., Kos, J., He, W., Hynes, N., Johnson, N., Juels, A., Miller, A., Song, D.: Ekiden: a platform for confidentiality-preserving, trustworthy, and performant smart contracts. In: Proceedings of the 4th IEEE European Symposium on Security and Privacy (IEEE EuroS&P), Stockholm, Sweden, June 2019

Emerging Role of Block Chain Technology in Maintaining the Privacy of Covid-19 Public Health Record

Jana Shafi, Amtul Waheed, and P. Venkata Krishna

Abstract Amid covid-19 outbreak contact tracing becomes an essential tool for protection and connecting to the most possible connections with the infectious patients which leads to addition of high number of individuals data to be recorded in the location wise public health system. The public health records surface complications concerning security, reliability and supervision. Here the novel technology Blockchain is on the track to take over a part to battle the Covid-19 and overthrow the pandemic. This technology affords a solution for the adjustment between public health and private record. Now as blockchain technology has entered the area of possible units against covid-19.The blockchain technology protected anonymized private data in the search for record generated by the pandemic. The utmost valued defence against the Covid-19 is information which includes one's contacts and connections as well path tracing. However, it derives at a great cost of private data. A blockchain-based covid-19 infected-patient data record would guarantee untampered neutral data with cryptographically secured decentralized data record protection. Moreover, the blockchain acquired knowledge confirms that data records are only used for its planned purpose, as it authenticates the applied algorithms. Additionally, block chain and secure hardware together can eliminate hacks and annoying data-selling. Hence, block chain can efficiently provide data privacy. This paper in the context of block chain introduces Gravity: a distributed protocol for handling the amount of all potential data and the usage of distributed data provisions such as IPFS is inevitable in this charge. Medical-Chain platform for protecting the identities of Covid infected and traced patients and their history. One must be sure of data records to be following HIPPA compliance to use anonymous nature and proper encryption.

Keywords Covid-19 · Blockchain · IPFS · Off-chain · Gravity · Public health

J. Shafi (✉) · A. Waheed
Prince Sattam Bin Abdul Aziz University, Wadi ad-Dawasir, Saudi Arabia

P. Venkata Krishna
Sri Padmavati Mahila Visvavidyalayam, Tirupati, India

© The Author(s), under exclusive license to Springer Nature Switzerland AG 2022
P. Nicopolitidis et al. (eds.), *Advances in Computing, Informatics, Networking and Cybersecurity*, Lecture Notes in Networks and Systems 289,
https://doi.org/10.1007/978-3-030-87049-2_24

1 Introduction: Blockchain in Pandemic

Blockchain is a technology of distributed ledger that assists users to share information in a reliable and certified decentralized way. Together with the help of cryptography and security technology, blockchain able to guard the user's privacy who add information while sharing the data source, improving reliance. This technology offers an effective safe way to correct documents, preserve, pile, and transport health proceedings to financial connections. Also with the help of blockchain, the public can engage directly with others to accept facilities other common everyday tasks. With the gush of blockchain technology in the field of health-care, the prospects are endless, emerging potential of blockchain technology is in healthcare settings of protected health information (PHI). PHI is from any demographic data that can be used to find a healthcare patient identity information. In a model of blockchain healthcare patients have possession of their health records in the form of records data packets. Patients could give allowance to doctors to use their records by switching their decryption key. Here new technology of blockchain healthcare can overlap with the HIPAA compliance of healthcare. The Health Insurance Portability and Accountability Act (HIPAA) established the standard for sensitive persistent document guards. Providing cure, expense, also operations in healthcare and corporate contacts that have access to patient information and provides support in treatment, payment, or actions must encounter HIPAA Compliance. As it possesses tight privacy as well as security standards for the practice and confession of PHI, blockchain healthcare implementation can stance encounters to the HIPAA rules. The real-world use of blockchain in healthcare is ahead of the proper supervision and the directive, from the Department of Health and Human Services (HHS) also from the Office for Civil Rights (OCR) as trends continue to progress. Under recent prototypes for storage of PHI in EHR, platforms have a high risk of a data breach. If the stage is bare to malware or hacking the entire PHI held by the provider can be a threat because the data is all stored in a single central warehouse. Block-chain healthcare is impersonated to defend PHI in extraordinary new techniques as of its decentralized nature [1]. It's an underestimation that the world is changed due to the pandemic. The field of Healthcare is, almost unquestionably, set the revolution in the upcoming years. Preserving privacy agreement is also more challenging. Increasing risk issues of private health information contain: The online visit to healthcare health-care provider has risen steeply due to Covid-19. Visitor's data records protection over the Internet is problematic if proper provisions are ignored. Bigger Patient numbers after lockdown at workplaces when combined with physical distancing guidelines of WHO, are often petite on work when schedules are pass with flying colors out. This condition generates a chance for HIPAA agreement errors. Patients often see several physicians. Principal care doctors getting updates from various challenging laboratories, patients, or hospitals means records of data are moving from one place to another at a speedier pace while dealing with prospective virus cases.

Management of private data is the foremost space of blockchain applications. The abrupt arrival of Covid-19 swift and abandoned dissemination everywhere in the world has exposed us not only the disappointment of current healthcare investigation systems in punctually handling the public health crisis but also the marked absence of unconventional projecting structures based on the large scale distribution of clinical records. The blockchain-based system steadily accomplishes health data, confirming interoperability without cooperating safety and patient confidentiality. Entire personal records are encrypted and kept in cloud servers based upon blockchain technology even though the businesses unable to repossess the original data and also it will eradicate all the public data after the pandemic [2]. The EHR management with block-chain technology may decrease clinical unfairness, thus refining the global healthcare results [3]. The problem of interoperability between different EHR systems may be solved using distinct block-chain systems that would work as a link to confirm cross-communication: Block-chain is a chance to guarantee securing data interactions cryptographically among two or more users: lately, this occasion has caught the attention within the scientific group, which targets mostly to simplify collaboration between dissimilar protected links; this will make sure a reliable decentralization of events similar as an asset and message conversation [3–5]. Block-chain permits for rival groups to share patient's data in an auditable mode, while preserving their competitive freedom and privacy worries. It is these vital abilities that enable block-chain to arise as a feasible result for several dire healthcare functions whose reputation has full-fledged during this pandemic, for example, contact tracing and patient data-sharing. Contact tracing in Covid-19: To follow the possible diffusion of the novel Covid-19, infected persons are inquired to enlist the people they've come into contact with over a certain defined period. Here decentralization of data aids expedites contact tracing since the method is needful on using rough, delicate data to notify public health officers of are subject to the risk of contact based on their schedules. In contract tracing, upholding people's confidentiality is critical. Before this year, block-chain dais Nodle hurled a Covid contact tracing app namely "Coalition," which highlighted user confidentiality. One more valued block-chain use case is the aggregation of Covid-19 patient accounts during a crisis to form an electronic health records structure, which distinct sets of sources can use to share patient data while treating unacquainted patients throughout the pandemic or other emergencies. This kind of platform will permit sources to work with patients who are deprived of access to their common provider, but able to receive a variety of required facilities and treatments. The key notion of the explanation is that patient's HER shadow them anywhere they go and they can obtain essential health facilities. This patient data can be sent via the blockchain wallet, thus providing access, safety, and reliability of data [6].

Chapter Organization: *Sect.* 1 *Introduction*, *Sect.* 2 *Background work*, *Sect.* 3 *discussing Gravity and Medical-chain Platform.* *Section* 4 *Conclusion.*

2 Background

The World Health Organization (WHO) has endorsed that worldwide nations pull up a "COVID-19 Pandemic Plan", because of the amplified likelihood of risk. A Pandemic Strategy is normally established conferring to the pandemic stages professed by the WHO and targets to attain strong outcomes in dealing with pandemics from the primary phases [7]. In the field of healthcare, different approaches may be recognized in formulating a crisis which is categorized by different parts such as modification, preparation, reaction, and repossession [8]. An appropriate method could afford a common outline and a psychological model replicating the seamless setting for an upcoming resolution [8]. On the day of 30th January 2020, the World Health Organization (WHO) declared a public health crisis due to the diffusion of a novel coronavirus named SARS-CoV2, related with the COVID-19 infection, followed by on 11th March 2020 the epidemic converted a pandemic [9]. The COVID-19 outburst is representing the weakness of the global public to new and extremely transmissible biotic means. In this setting, some nations have measured the strengthening of policies of "risk administration" as urgent. The key fear of sensible risk supervision is interrelated to data sharing amongst clinicians and mass broadcasting, as usually, they are responsible for creating havoc in the general inhabitants. Though, in some nations, healthcare establishments are used to stop or suspend the distribution of vital data to prevent comprehending the exposed risk to the public and also to accurately limit the transmission of unsafe pathologies. In the latest spans, the advancement of diagnostic skills also biomedical tools have assisted the industrial based nations in supporting and regulating risk supervision in the field of healthcare [10]. The SARS-CoV2 contagion has complicated in all continents, challenging risk administration in all the foremost universal institutions [11]: in this framework, Wendelboe et al. have apprehended and planned a precise tabletop use for academies and firms to recommend consistent goals and thorough orders to avoid and cope COVID-19 contagion: (i) Investigation of COVID-19 cases with identified travel-related disclosure, (ii) Study of COVID-19 cases with no identified (i.e., public) exposure, (iii) COVID-19 outburst in the native region, (iv) COVID-19 rescue from (phases 2 and 3) [12]. A consistent strategy to escort and formulate communication criteria among organizations also in health staff is planned for broadcasting the information that the public requires, as individuals must be effectively ready for the crisis, and qualified to recover their aids and planning [13]. The healthcare strategy in Italy of accommodating Covid-19 department which expanded in only five days was amended, also to convert the ground floor into emergency rooms for safe treatment of symbiotic patients is highly valued, appreciations to the pains of many clinicians, fosters, managerial staff as well hospital board, over a precise time [14]. In February 2020, the Netherlands also turn to be involved in the pandemic. The team of Dutch national epidemic management (DNEM) come across in March to converse restrictions and comprehend the pandemic range in the whole nation. The Netherlands approved a swift two-day investigation of 9 clinics to notice the working of health professionals in these country areas, notifying native experts when they exhibited mild respirational symptoms [15]. Infirmaries were probed to offer the transmission check to workers and this course signified to investigate a representative trial. The provincial establishments decided to use preventive size to edge the contagion to a large fragment of the inhabitants due to this data [16]. Suitable controlled actions, balanced to the development of the epidemiological state, based on the pandemic plans drawn up according to WHO directives, have prevented irregular health risk and have matched nationwide reaction to the medical crisis [17]. Management of Healthcare can use numerous planned tools to be active in sharing data and mining data, machine learning models, artificial intelligence techniques, and block-chain are the most impactful policies [18]. Recently, the block-chain technology has been gradually smeared to healthcare, to support effective procedures, and to construct the proper base for a well-organized and actual evidence-based decisional course. Block-chain plays a planned part in securely data sharing among groups of people, reliability independent also validating these groups.

3 Concern: Huge Sensitive Covid-19 Related Data Records

As the result of contact tracing a huge amount of data records collected which requires tight privacy from the hackers and other third partys. Data driven if secured with the help of blockchain ensures to the maximum extent of safety is discuses in the following sections.

- Gravity Protocol-Block-chain to store the bulk of Covid-19 associated patients and tracers data in blocks which are known to be limited.
- Medical-chain Maintaining all Patients history including tracing records which are highly sensitive in a core protective manner.

3.1 Gravity Protocol: Managing Huge Data in Covid-19 Proposed Model

Covid-19 cases data is quite huge in number as it includes infectious records as well list of traced contact including all personal information and limited size is one of the main concerns in the scalability of blockchain which makes it unmanageable to save complex records such as any data transaction account history, registry records, and hashes in a block. So, the public block-chains to be common have not till now exceeded the transactional stage of growth. Here Gravity based protocol is the best method which tries to turn into not merely transactions of block-chain, but beyond all, circulated conventions for handling the amount of all potential data cases records, and the stocks circulated cases data usage such as IPFS which is inescapable in this charge. IPFS arranges circulated all cases records room on charges' computers. This is the same concept blocks are alike to the distribution in the Kademlia DHT. These registries handle only with peer-to-peer connections devoid of bandwidth, latency, or network accessibility, creating IPFS a multipurpose and decentralized extensible case system. Thus, subsequently IPFs integration, it turns out to be one of the probable stores only for the hash data on-chain. The case data records themselves will be kept off-chain, in IPFS. The capability to stockpile just the on-chain hash data of cases pointedly drops the data size to be put in storage, which reduces down the swiftness of chain bloat, therefore declining the required resources for the nodes of blockchain, also as the phase, it takes to resync and repeat, which is a giant problem on current blockchains that attempt to stock needless data records on their chains.

Working: The file's hash completely agrees to the loaded data cases into IPFS which is the address of records. If the folder is altered, its hash key will be amended and the file will be unavailable at the address of the content. This is how no third parties can make changes in uploaded data of cases and hereby also making room for various opportunities for practical applications in the areas of storage and transmission of consistent personal data records. Folder Multi-versioning (such as git) is maintained to let the holder amend the content. Large file processing and storage as reliable data records is the utmost important element for the procedure of blockchain technologies as well as the execution of decentralized solutions zones for example Internet-of-Things, Big Data, AI Deep Learning, Smart House, etc.

Data hash has a relationship with smart contracts, which establish the data records transaction cycle. This is an extra vital feature of block-chain IPFS combination where authorized users can control access to the records themselves, counting on a possible waged basis, without any intercessors. The hash key is encrypted with the recipient's private key. This is an assurance that only individuals who have funded or permitted will be able to get data access. This attribute permits broad assimilation of block-chain skills in the covid data records to work of permitted access control, for instant video and audio files, training ways related to covid cases data records, also copyrighted items access to control.

Gravity is a protocol that executes a tool for reassigning the download rights of case records such as files from media within a blockchain transaction, with the help of IPFS encryption on storage blocks. An asymmetric encryption algorithm with a public key is used to create the signature that has the hash address of the data and the password used to decrypt the blocks of data (see Fig. 1). The data blocks on their own are encrypted using the AES-256 symmetric algorithm, as the asymmetric encryption algorithm is not effective when it comes to functioning with big sums of data in the relation to performance and source necessities when matched to symmetric. The procedure of reassigning the data download rights is executed by in-built transaction protocol which is imitated in all blockchain transaction records. As such, in the case of data recipients (doctors), processes are implemented in the opposite order (see Fig. 2).

The files automatically are set on the computers and servers of the participant's network. All network members will get the chance to stock Gravity IPFS data on their servers and plans and will obtain expense for this. The Gravity wallet interface will be used to trigger IPFS storage. As a result, the loaded data records into Gravity IPFS

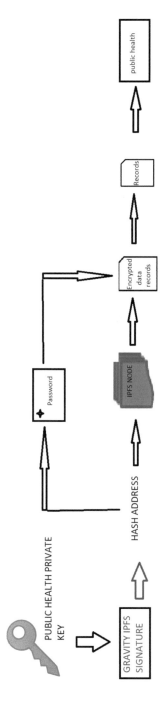

Fig. 1 Asymmetric encryption algorithm

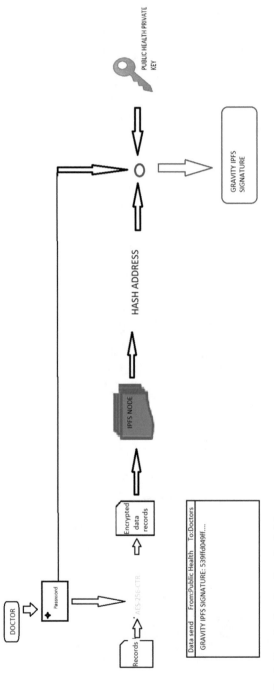

Fig. 2 Reassigning data download rights

will have maximum storage security. With this in mind, the IPFS-storage system of the Gravity network preferably needs to be self-reliant (*refer Algorithm* 1).

Algorithm 1: Algorithm of Data Records and Reassigning Data Download Rights
1 Initialization
    2 Inputs: Data Files (R), AES-256 symmetric, IPFS encryption, Gravity Trigger
    3 Outputs: IPFS Loaded data records Secured
    4 for Gravity wallet, where data on IPFS storage
    5 if R already loaded on IPFS then
    6 Gravity Triggers for R loaded on IPFS
    7 Grant maximum security
    8 else
    9 if R not loaded then
    10 R pileon participant network
    11 then stock up R on IPFS Goto step 6 and 7
    12 R get encrypted using symmetric AES-256
    13 Transcation Protocol initiated
    14 Reassigned R download rights
    15 End

3.2 Protecting Covid-19 Records Medical-Chain Platform

Medical-chain is a block-chain for securing electronic health records which can be used in protecting identities of novel coronavirus patients as well list of individuals acquired in contact tracing. Medical-chain will permit various healthcare managers such as doctors, hospitals, laboratories, pharmacists, and insurers to request consent to gain access and check patients or people's medical records. Each interaction made is auditable, clear, safe, and will be logged as a transaction on a distributed ledger [19]. It is constructed on the consent-based Hyperledger Fabric construction which permits changing access steps; patients can control who can view their accounts, how much they see as well time duration. Medical-chain targets to deal with the following core aids to patients and Covid-19 assisted providers:

- The patient's private key can access data only; although if the file is got third party access, the data of the patient will be all encrypted.
- Patients have complete control over accessing their healthcare data.
- Immediate assignment of medical data in the distributed network of the Coronavirus block-chain would have identical data for the patient which leads to lessen the risk of inaccuracies also improved patient care.

Features: Medical-chain offers all of the following features:

- Improved Confidentiality and Control Access where the managers can setup approved access and allow providers to compose information to their block-chain.
- Real-time and Secure Physician To Patient Interaction where Patients are allowed to share their records.

The flow procedure used in a Medical-chain Covid-19 records transaction and data entry is given in Fig. 3.

- Covid-19 patients and tracers data is generated where even the patient's mobile app could generate data, a specialist could inscribe a note to the block-chain, or a chemist could distribute the drug.
- The data get encrypted and provided an identity that is kept on the patient's block-chain which is then sent to store on the cloud.
- The ID on the block-chain is used to fetch the encrypted data when the data is requested.
- Lastly, the requested data is decrypted and then presented on the related means or application.

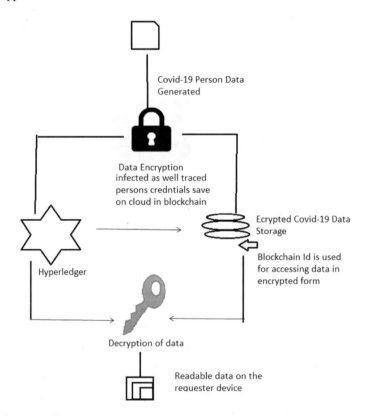

Fig. 3 Procedure flow

4 Conclusion

The block-chain technology is progressively implemented in this pandemic to form the proper center for a capable and active proof-based transparent decisional procedure. This is an effective and valid method to securely sharing data among groups, freely irrespective of reliability and the verifying of these groups every time with the help of the Medical Chain Platform. As the data generated is in huge form collected from tracers and infectious person blockchain solved the problem by using Gravity protocol. Block-chain can be used in a new workflow or improved protocols with particular attention to risk management. Based on this, we can say that block-chain plays a strategic role in COVID19-safe clinical practice in protecting sensitive information.

Acknowledgements Authors are thankful to Deanship of Scientific Research, Prince Sattam bin Abdul Aziz University for their sustenance in this work.

References

1. https://compliancy-group.com/hipaa-compliant-blockchain-healthcare/
2. https://towardsdatascience.com/blockchain-technology-ensuring-data-security-immutability-7150d309352c
3. Angeles, R.: Blockchain-based healthcare: three successful proof-of-concept pilots worth considering. J. Inf. Technol. Manag. **27**, 4 (2018)
4. Tatullo, M., Marrelli, M., Amantea, M., Paduano, F., Santacroce, L., Gentile, S., Scacco, S.: Bioimpedance detection of oral lichen planus used as preneoplastic model. J. Cancer **6**, 976–983 (2015)
5. Ballini, A., Cantore, S., Scacco, S., Coletti, D., Tatullo, M.: Mesenchymal stem cells as promoters, enhancers, and playmakers of the translational regenerative medicine. Stem Cells Int. **2018**, 6927401 (2018) [PubMed]
6. https://hitconsultant.net/2020/11/17/covid-19-healthcare-blockchain-use-cases/#.X7zEBW gzY2w
7. Madhav, N., Oppenheim, B., Gallivan, M., Mulembakani, P., Rubin, E., Wolfe, N.: Pandemics: risks, impacts, and mitigation. In: Jamison, D.T., Gelband, H., Horton, S., Jha, P., Laxminarayan, R., Mock, C.N., Nugent, R. (eds.) Disease Control Priorities: Improving Health and Reducing Poverty, 3rd ed. The World Bank Group, Washington (DC),WA, USA (2018)
8. Public Health Emergency: Emergency Management and the Incident Command System. US Department of Health and Human Services, Washington, DC, USA (2012)
9. World Health Organization (WHO): Naming the Coronavirus Disease (COVID-19) and the Virus That Causes It. World Health Organization (WHO), Geneva, Switzerland (2020)
10. Palmer, M.J.: Learning to deal with dual-use. Science **367**, 1057 (2020)
11. He warned of coronavirus. Here's what he told us before he died. New York Times. Available online: https://www.nytimes.com/2020/02/07/world/asia/Li-Wenliang-china-corona virus.html. Accessed on 7 Feb 2020
12. World Health Organization: Coronavirus disease (COVID-19) training: simulation exercise. In: Programme HE. World Health Organization, Geneva, Switzerland (2020)
13. Wendelboe, A.M., Miller, A., Drevets, D.: Tabletop exercise to prepare institutions of higher education for an outbreak of COVID-19. J. Emerg. Manag. **18**, 1–20 (2020) [PubMed]

14. Asperges, E., Novati, S., Muzzi, A.: Rapid response to COVID-19 outbreak in Northern Italy: How to convert a classic infectious disease ward into a COVID-19 response center. J. Hosp. Infect. **105**, 477–479 (2020)
15. Reusken, C.B., Buiting, A., Bleeker-Rovers, C.: Rapid assessment of regional SARS-CoV-2 community transmission through a convenience sample of healthcare workers, the Netherlands, March 2020. EuroSurveill **25**, 2000334 (2020)
16. Rijksinstituut voorVolksgezondheidenMilieu(RIVM).COVID-19: Nieuwe aanwijzing voor inwonersNoord-Brabant [COVID-19: Advice for residents of Noord-Brabant]. RIVM, Bilthoven, The Netherland (2020)
17. Kubo, T., Yanasan, A., Herbosa, T., Buddh, N., Fernando, F., Kayano, R.: Health data collection before, during and after emergencies and disasters—the result of the Kobe expert meeting. Int. J. Environ. Res. Public Health **16**, 893 (2019) [PubMed]
18. Davenport, T., Kalakota, R.: The potential for artificial intelligence in healthcare. Future Healthc. J. **6**, 94–98 (2019)
19. https://medium.com/crypt-bytes-tech/medicalchain-a-blockchain-for-electronic-health-records-eef181ed14c2

Malware Forensics: Legacy Solutions, Recent Advances, and Future Challenges

Farid Naït-Abdesselam, Asim Darwaish, and Chafiq Titouna

Abstract Malware applications are continuing to grow across all computing and mobile platforms. Since the last decade, every day, half a million malware applications emerge as a real threat and hamper both mobile and computer ecosystems. With the rapid evolution and expansion of the smartphone market, malware detection and prevention for handheld devices need a paramount attention and effective solutions. Enrichment and diversification of smart applications, tools, sensors, and various services with underlying sensitive information always allure malware writers to exploit vulnerabilities. As Android is open-source and the smartphone's market leader, it is therefore more vulnerable and has contributed to the boom of a variety of android malware applications. Malware is the payload created by an attacker to compromise the system's integrity, availability, and confidentiality. Mostly, the interest is to steal confidential information, financial data, or crippling critical infrastructure and servers. Malware leads sometimes to severe damages and huge financial losses to businesses, institutions, and individuals. Malware writers keep coining novice methods and sophisticated routes for creating malware for mobile devices, computers, and servers. Therefore, it is very essential to devise a system that can detect and prevent legacy and new malware with as high accuracy as possible under different settings. In this chapter, we will cover the background of malware across different platforms, the existing solutions and techniques for malware analysis, prevention, and detection, as well as the recent advances in malware research domain employing cutting edge technologies.

Keywords Malware · Mobile applications · Detection systems · Machine learning

F. Naït-Abdesselam (✉)
University of Missouri Kansas City, Kansas City, USA
e-mail: naf@umkc.edu

A. Darwaish · C. Titouna
University of Paris, Paris, France
e-mail: asim.darwaish@etu.u-paris.fr

C. Titouna
e-mail: chafiq.titouna@u-paris.fr

© The Author(s), under exclusive license to Springer Nature Switzerland AG 2022
P. Nicopolitidis et al. (eds.), *Advances in Computing, Informatics, Networking and Cybersecurity*, Lecture Notes in Networks and Systems 289,
https://doi.org/10.1007/978-3-030-87049-2_25

1   Introduction

Medical campaigns and awareness programs are launched every year before the flu season to save people's lives by getting flu shots. This is a proactive measure before starting any flu season. The recent outbreak of COVID-19 pandemic also took millions of lives and now special measures are taken to stop its spread along with the discovery of a vaccine. Contrary to this, there is no special predictable season for computers, handheld devices, smartphones, servers, and sensor devices to get infected with a virus. There is a continuous threat to computing machines and it's always a flu season.

Malware infections come at us like a torrent of arrows on the battlefield, each with its poison and method of attack with stealth, deceitful and subtle. Malware detection and isolation studies are preventative immunization against infection and wounds. We enlighten a short study on malware, what it is, how to deal with this, its symptoms, how it can impact, and how to prevent it, the legacy and obsolete solutions to deal with and their shortcomings, recent advances and future direction in research and challenges.

1.1   History of Malware

Given the daily generation of massive malware variants for different platforms, it is impossible to enumerate the complete history of malware here. However, the malware trends in recent decades are convenient to manage as circumvented below.

The foundation of viruses "automatically self-reproducing" was encountered in pre-personal computer platforms in the 1970s. The history of malware and viruses starts in 1982 which impacted the Apple II systems by a program called Elk Cloner [40]. The virus is scattered via an infected floppy disk and spread to all the disks connected to a system. The virus itself was not harmful but it is considered as the first massive outbreak in computer security history and it was before the era of Windows PC malware. Since the Windows PC had emerged in the 1990s, viruses, worms, and malware start the journey of widespread. The malware spread starts with its flexible macros of window's application and provided the foundations to malware authors to write the malicious code using macros for MS Word and various other programs. These macro based infected codes impacted the documents and templates rather than portable executable (PE) program.

From early 2002 to 2007 the famous instant messaging worm outbreak happened which spread the self-replicating malicious code via an instant messaging (IM) network by leveraging the network loopholes. It impacted the famous Yahoo Messenger and MSN Messenger and various other corporate IM systems. Similarly, from 2005 to 2009 a famous malware proliferated to various devices named Adware. It displays the advertisements in the form of new windows and pop-ups and deliberately disabled the user to turn them off. This adware piggybacked on a few famous companies for its spread but sooner the companies started legally sue them for the fraud. As a result, these adware companies got shut down.

With the boom of social networks, scammers targeted the famous MySpace in 2007 for delivering scoundrel advertisements, offering fake security tools and anti-viruses. They deceive the users via social engineering cheats. Later on, Facebook and Tweeter became the main target as MySpace's popularity vanished. The common methods for social networks were links to phishing pages and advertising Facebook applications with malicious extensions. A new variant of malware named ransomware was introduced using the name of CryptoLocker and targeted the Windows-based PCs from September 2013 to May 2014. The CryptoLocker attack has gathered 27 million dollars from victims. Similarly, a different variant of ransomware remained effective till 2015 and gathered millions of dollars. In 2017, ransomware become the king of malware by affecting every domain and business by encrypting the victims' data and asking for payments to release it. From 2018 till cryptocurrency is on hype, which gave birth to a new malware named Cryptojacking which operates in a stealth mode on a victim's device and uses his resources to mine for cryptocurrency.

1.2 Definition of Malware

Malware is an umbrella term that means malicious software intents to infect the computing system. In the computer security domain, "the program aims to breach the system's security policy by compromising confidentiality, integrity, or availability". Numerous other definition exists in literature which loosely defines the malware as "Annoying software program", "Running a code without the user's consent". Malware infiltrates to damage, disable, make the system unresponsive, and impact the PCs, desktops, smartphones, tablets, and networks by taking complete or partial control over target devices. The financial goal is to make more money illegally.

1.3 Symptoms of Malware Infection

The symptoms of malware infection reveal through various anomalous behaviors. The following are a few pointers which help you to alarm against malware infection:

- Slowing down your computer, tablet, or smartphone devices with or without internet connectivity.
- Bombardment of annoying ads on your screen is a typical sign of adware and especially you cannot close them. Moreover, they allure the users to click and offer various fake prizes, etc.
- Recurrent system crashes for example a blue screen on Windows PC.
- Mysterious loss of storage space is another hint for hidden malware in your system.

- Sudden the resources of your system reached the peak and the cooling fan starts at full swing is a clear indication of malware consuming the resources in the background.
- Frequent and automatic change of your home browser without permission, Similarly, New plugins install and appear on browser without your information.
- Anti-virus scanner or security tools installed on your system become unresponsive and cannot be updated
- Loss of your important data from your system in a stealth mode by the famous ransomware and then demands you to pay to recover it back.
- Few other malware resides in your system and collect your personal data like bank cards and passwords etc. for illegitimate activities.

1.4 Types of Malware

There are various kind of malware and million of their families, but the most common and threatening and scoundrel's group of malware discussed below:

- Adware is a kind of malware which throws numerous kind of advertisements to your system via the web browser. Normally, it infiltrates into the system by impersonating itself as legitimate software or piggyback on other software.
- Spyware is deceitful and resides in the target system in a stealth mode and report the activities to malware's author without the victim's permission and knowledge.
- A virus is a kind of malware usually appended to other software and executed unintentionally by the end-user. It starts replicating itself and modify other installed programs by infecting its piece of code.
- Worms are similar to viruses and destroy the system's file and user's data.
- Trojan also known as Trojan horse, is a special type of malware that portrays himself as necessary and useful to deceive the user. Once it's entered into the system, it also gave birth to other malware like viruses, ransomware, and more importantly, it is designed to steal financial information.
- Ransomware is the most dangerous malware and considered a cyber criminal's weapon which encrypts and steals your important files and data and then asks for a ransom to make them available for you.
- Rootkit also operates in a stealthy fashion on the system from OS and other software and it grants the root permission to the malware author.
- A keylogger is a spy on keystrokes of touchpad and keyboard, it stores and sends the gathered information to the attacker to manipulate the sensitive information like usernames, passwords, credit card, and financial information.
- Cryptojacking is a specialized malware that steals the victim's resources to mine cryptocurrencies such as Bitcoin or Monero and many other coins.

2 Malware for Handheld Devices

Malware authors love to penetrate the smartphone market because mobiles are sophisticated handheld computers. Smartphones carrying multiple sensors endowed with important and sensitive information which allures malware criminals to make more money. There are a number of ways for malware from Trojan, adware, spyware, ransomware, and many others to capture your smartphone and start exploiting the malicious activities. One of the sources presented the statistics about smartphones and reported almost 3.5 billion smartphone users in 2020 and predicted 4.4 billion users for the year 2022.

Mobile users are a relatively easy target for malware authors as compared to conventional desktop users. Most smartphone users do not protect their phones by installing security applications and do not follow security measures as they do for their computers. Due to this negligence, they are more vulnerable to even primitive malware. Moreover, there are various cheap smartphones available in the market with pre-installed malware and leverage open source android operating system. In the smartphone market, iOS and Android are the two big giants, Android capturing the 80% market and iPhone holding 15%. Due to the biggest market capture and open-source platform, Android is the hive for malware authors. The subsequent subsection shreds the lights on both Android and iPhone separately.

2.1 Symptoms of Android's Infection

Following are the most common indications if your Android device is infected with malware

- Sudden appearance of pop-ups with annoying advertisement and allure you to click the link which may take you to sketchy websites. This is a clear indication adware is hiding in your device and prompting you to click on the advertisement.
- surprisingly increases in your data usage. Malware consumes your data plan by displaying numerous ads and sending information outworld without your permission and knowledge.
- The malware sends SMS and call premium numbers in s stealth mode resulting in unknown and false charges on the bill.
- The sudden and frequent drainage of battery charge. The malware uses your resources and sucks the battery's juice very fast as compared to normal usage. Trojan can wipe out your battery by doing stealth processing which can dead the battery and Phone both.
- obnoxious and strange calls and SMS received by your contacts from your phone.
- The stealth installation of unknown apps on your phone by piggybacking on other famous apps.
- Automatically turning on the WiFi, internet, and Hotspot on your phone.

2.2   Symptoms of iPhone's Infection

Apple iPhone is more secure and studies revealed that they are not easy to invade by malware authors but it doesn't mean that malware doesn't exist on iPhone. Malware can target the iPhone in two scenarios which are rare. The first way is the state-level adversary authorized by the government by targeting the obscure security loophole in iOS. Apple has done a great job of making the iOS and app store secure and has a rigorous scanning mechanism on the device and App store. The second route is the jail-breaking of the phone by the user and removes all restrictions imposed by Apple. Using the iPhone and iOS doesn't mean the user is safe from malware. If the user taps a link received in a message from an unknown source it can lead to the website where login details and other personal information is requested by alluring the user.

2.3   General Guidance to Prevent Malware

Always stay vigilant and avoid domain names ending with odd letters. Pay close attention to all indications and symptoms as discussed in earlier sections to prevent malware infection. Do not click ads and pop-ups and stay away from unknown email attachments. Always use up to date OS, browsers, and software and plugins. For smartphone users, only download and install the applications from a trusted source such as the Google play store for Android and App Store for iOS and block third-party sources for app download and installation. The business and corporate sector should install strict security policies and their compliance to prevent malware. Finally, always install the updated and trusty anti-virus or malware detection and prevention systems.

3   Legacy Solutions

The techniques used for detecting malware can be categorized into two main categories: Static and Dynamic analysis. Figure 1 presents a brief taxonomy of malware detection approaches.

3.1   Static Analysis Techniques

In static analysis-based approaches, the detection of malicious software is performed without its execution or installation on the target system. The properties and the characteristics of the internal structure are used for detecting malicious codes. Hence, the program executable should be unpacked using a disassembler or debugger before

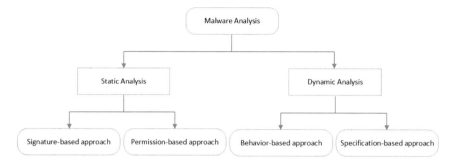

Fig. 1 Taxonomy of malware detection approaches

analyzing any codes. To do that, reverse engineering methods are used in order to analyze deeply the structure of the software. A lot of tools are employed to make the work easy, like IDA Pro [36], OllyDbg [37], XXD [38], and Hexdump [39]. IDA Pro and OllyDbg are considered as disassembly programs that are able to create maps of software execution in order to illustrate the binary instructions as an assembly program. The method of disassembly helps the software experts to check the structure of programs and detect the malicious ones. On the other hand, XXD is a simple Linux command that produces hexadecimal or binary dump of a file in several different formats. It can also do the reverse, converting from its hex dump format back into the original data. Hexdump is a tool that shows the contents of binary files in hexadecimal, decimal, octal, or ASCII. Hence, XXD and Hexdump are very useful to inspect the details of the functionalities of any program.

The above static techniques can be categorized into a signature-based approach and a permission-based approach.

3.1.1 Signature-Based Approach

The signature-based approach is widely employed for detecting viruses and malware. A signature is a known pattern or simply a sequence of bytes that can be used to identify specific malware. In this type of approach, multiple patterns matching schemes are used to scan for signatures using a database where malware signatures are stored. Updating this database is very recommended in order to deal with the new types of malware. So, detecting malware begins by creating the appropriate signatures for each software that should be analyzed and comparing them with the signatures in its databases. If a match is found, this software is considered as malware where any further actions are blocked.

Sung et al. [6] proposed a robust signature-based malware detection technique in order to detect obfuscated and mutated malware. The proposal technique called SAVE for Analysis for Vicious Executable. The authors supposed that all variants of malware use the same core signature which is presented as a set of features of the code. So, detecting one of the variants leads to identify all the rest of the variants by examining the signature of the detected malware. The SAVE algorithm decompresses the Microsoft Windows Portable Executable (PE) binary code in order to extract the Windows API calling sequences. The latter are represented by a 32-bit integer id number where the 16-bit most significant represents the Win32 module and the other 16-bit represents a specific API in this module. SAVE compares the extracted API calling sequence to a set of signatures that are already stored in a database using a similarity measure module. This module uses Euclidean distance and a set of well-known functions like Cosine, extended Jaccard, and the Pearson correlation measures. The authors experimented their work for detecting malware like MyDoom, Bika, Beagle, and Blaster worm. The results of their comparison with eight antivirus and scanners demonstrated that SAVE outperformed the compared detectors. Similar to the previous work, Christodorescu et al. [7] presented a tool called SAFE (Static Analyzer For Executables) to detect malicious patterns in x86 executables. The SAFE tool is resilient to common obfuscation transformations. Later, the authors [8] proposed a program to evaluate the resilience of malware detectors against obfuscation transformations usually employed by hackers to mask the malware. In fact, the developers of malware try to implement new methods in order to escape the signature-based detection. They developed metamorphic-based techniques that morph the code after each infection, i.e., the structure of the malware code will be changed but the functionality of malicious is always present [13]. Several works focused on detecting metamorphic malware [10–12], where the authors used opcode-graph [14] or statistical analysis [15]. Shanmugam et al. [9] proposed a similarity-based technique in which a simple substitution cipher cryptanalysis is used to measure the distance between executables. The signature-based technique is also used for android malware detection [16, 17]. The authors in [16] proposed an approach called DroidAnalytics which is a signature-based analytic system to automatically collect, manage, analyze, and extract android malware. Gao et al. [17] proposed an approach based on the function call graphs which are extracted from android App packages. The authors claimed that their approach can identify malware and detect its variants.

Signature-based detection approaches are mostly used by antivirus and malware vendors for their flexibility, simplicity, and efficiency against various categories of malware. A drawback of these approaches is that: (a) it requires an updated repository of malware signature; (b) the detection always happens after the malware harms systems; (c) it fails to identify new malware that utilizes polymorphic methods.

3.1.2 Permission-Based Approach

Permission-based approaches have attracted researchers since the development of smart devices that used Android, iOS, etc. In such systems, threats have increased by developing a new kind of malware based on permissions. In fact, the operating systems deployed on smart devices (Android) operate on permissions in order to authorize new applications to access to stored data and the device itself. So, the system forces applications to declare the permissions required for them to function properly. If permissions are not granted, the application will not be able to access to this service, otherwise, it can use it without limit. Permission-based approaches are considered as a faster detector to identify malware and application with suspicious permission requests.

Wu et al. [18] proposed a static technique called DroidMat in order to detect Android malware using a set of information such as API calls, Intent messages passing, deployment of components, and permissions. All these information are extracted from the manifest file and regards components of the application in order to track the API calls. Then, the K-means algorithm is applied to enhance the malware modeling capability. Finally, DroidMat uses the k-Nearest Neighbors algorithm (kNN) to classify applications as malware or benign. Similar to the previous work, Peiravian and Zhu [19] presented a permission-based approach using a set of classifiers. The advantage of this technique is that it does not require the dynamical tracing of the API calls used by the system. In [20], the authors proposed almost the same previous work using a feature set containing permissions and API calls for Android malware. They conducted a series of experiments using various machine learning algorithms such as Multilayer Perceptron (MLP), Support Vector Machine (SVM), Naive Bayes, and Radial Basis Function (RBF). Liang and Du [21] developed a security scheme called Droid Detective, in which they combined permissions in sets. Droid Detective obtains permission combinations that generally required by malware applications, then, it generates a set of rules to detect malware. The results showed that the proposal has a detection rate over than 96%. Recently, Nivaashini et al. [22] proposed an approach in which detection of permission is performed automatically. Then, applying a classification model in order to classify applications. The authors claim that achieving 97.88% as detection rate. Kapoor et al. [23] classify applications on malware into benign using six machine learning algorithms. Logistic Regression Algorithm provides high accuracy rate equal to 99.34%. In [24], the authors employed a nonparametric technique called Kernel Density Estimation to estimate the probability distribution of data. Then, in order to simplify the analysis, the authors categorized permission into a signature, dangerous, and normal.

Despite the performance of the permission-based approach, it still presents several disadvantages. In fact, this type of approach generates a high false positive rate since a lot of benign applications require permissions to function correctly, which are considered as dangerous by the malware detector. In addition, malware can perform without requesting any permissions.

3.2 Dynamic Analysis Techniques

In dynamic-based approaches, the detection of malware is performed after the installation of applications. In fact, this type of technique analyzes the behavior of apps during their execution. Dynamic analysis checks which API calls are used by apps, instruction traces and memory writes, etc. The advantage of dynamic based approaches is the ability of detecting new, unknown, and obfuscated polymorphic malware. The main categories of these approaches are behavior-based approaches and specification-based approaches.

3.2.1 Behavior-Based Approach

The behavior-based malware detection approach examines the execution of the application in order to ascertain if it works as malicious or benign software. When an application is launched in the system, the behavior-based approach analyzes its activities such as the system calls used by the application. Generally, in such approaches, the execution is done within a sandboxed situation and a set of run-time tests are verified. The main advantage is that allowing comprehension of all malware behaviors.

Kolbitsch et al. [25] proposed an approach based on graphs for detecting malware. A behavior model is developed after analyzing a malware application in a controlled environment called Anubis. This behavior model is represented as a graph where the edges are the call's transition while the nodes are the system calls. Then, the obtained model is compared to known graphs in order to detect malware programs. The authors claim that the proposed approach presents effective detection with an acceptable performance overhead.

Burguera et al. [26] proposed an efficient tool that can realize a deep analysis to detect Android malware dynamically. In fact, the proposed approach is based on observing the behavior of applications to identify malware. Thus, the tool employs the concept of crowdsourcing to collect traces from real users, and then, performs an analysis on a remote server. The latter is used as a collector of information from its clients (Crowdroid). All data are analyzed and a behavior database is generated. This database is used to detect malware by applying the k-means algorithm. The authors achieved a detection rate of 100% when using self-written malicious apps and 85% when using the Monkey Jump 2 app.

Mohaisen et al. [27] presented a behavior-based malware system called AMAL. This proposed system comprises two sub-systems named AutoMal and MaLabel. AutoMal ensures the collection of the low granularity behavioral artifacts that indicate the utilization of the memory, registry, and file system. Besides that, the MaLabel uses artifacts to build and train classifiers that will be employed to classify and detect malware programs. The result of the evaluation showed that MaLabel realizes a 99.5% of precision and 99.6% of recall.

Similar to the approach presented in [25], Polenakis and Nikolopoulos [28] proposed a graph-based technique where they developed a model in which exploiting the relationships that exist between system calls. In fact, the proposed solution uses the system call dependency graphs that already generated. The authors employed three metrics: Δ-similarity, SaMe-similarity, and NP-similarity metrics for the process of detection and classification.

Saracino et al. [29] presented a system called MADAM, a host-based malware detection system designed for Android equipment which can perform analysis on four levels: package, application, user, and kernel. MADAM can classify malware basing on their behavioral classes. MADAM detects and blocks at least 96% of malware.

Choi et al. [30] proposed a new technique where they used the FPGrowth algorithm and Markov Logic Networks algorithm. The FPGrowth algorithm is used to find associations that exist between patterns in order to create malware behavior patterns. And then, the Markov algorithm is applied as an inference step to measure the correlation between patterns.

Behavior-based detection is considered an efficient and effective method to deal with new variants of malware. But these approaches consume more execution time to determine behaviors, extract features, and then detect abnormal behaviors. In addition, some malware programs are very complex which makes building behavior patterns a big challenge.

3.2.2 Specification-Based detection

Specification-based detection is a derivative of a behavior-based approach where the execution of the program is monitored. In fact, in this type of approach, if the behavior of an application deviates from the behavior predefined in the specification, the application is considered as a malware program. In most cases, specification-based approaches involve using a training phase in which the detector learns the normal behavior of systems that will be checked.

Chaugule et al. [32] developed a tool called SBIDF for specification Based Intrusion Detection Framework. The proposed approach is based on the utilization of keypad and touchscreen interrupts to detect unauthorized malicious behaviors. The authors used TLCK (Temporal Logic of Causal Knowledge), an application that allows describing the normal behavior patterns. The results showed that SBIDF can detect malware that tries to use sensitive services without the authorization of the users.

Song et al. [31] developed a tool called PoMMaDe for Push down Model-checking Malware Detector. The latter detector models the binary program as a pushdown system (PDS) in order to track the stack of the program. The authors claim that PoMMaDe can detect 600 real malware where 200 of them are new variants. The same authors in [33] proposed an approach where they model the Android apps as a PDS. This technique permits the detection of private data leaking working at Smali code level.

Battista et al. [34] used model checking to detect Android malware families (the DroidKungFu and the Opfake families). The proposed approach analyzes and verifies the java bytecode generated after a source code compilation phase. The authors affirm that the preliminary results seem significant and the detection of the malicious payloads can be achieved with a very high accuracy rate and acceptable time.

A similar approach has been proposed by Canfora et al. [35]. The authors developed a tool called LEILA (formaL tool for idEntifying mobIle maLicious behAviour), which is based on a model checking where they employed an algebras process named CCS (Calculus of Communicating Systems). The results of conducted experiences show that LEILA can detect malicious behavior efficiently with an accurate rate over than 0.97%.

The main disadvantage of the specification-based approach in detecting malware is that the complete and accurate specification of normal behaviors is considered a very complex task. Even for moderately complex apps, designing a complete specification of its all correct behaviors cannot be achieved. In addition, if we can specify all behaviors of an application with natural language, we cannot ensure the correct translation into the language machine.

4 Recent Advances and Challenges

The detection and isolation of malware is a challenge due to variant nature, diversity, code obfuscation, and other sophisticated ways of malware creation. It is imperative to establish a robust defense line against malware apps for mobile and IoT devices. With an exceptional growth of mobile apps, it is notably challenging to examine each application manually for malicious behavior. Traditional machine learning-based approaches are using handcrafted features and need extensive source code analysis for signature detection.

Advances in the field of Artificial Intelligence AI is the paradigm shift in the technology landscape and advancing enormously in various domains like computer vision, Natural language processing, self-driving vehicles, data analytic, security diagnostic, e-health diagnosis and prognosis, and many more. For example, machine learning-based data-driven analytic has revolutionized healthcare, finance, and other business models. Detection, monitoring, and classification in the security domain extracting valuable information from text and images from a plethora of social media data and other means are possible due to these emerging technologies. The taxonomy of recent advances for malware detection is divided into Machine learning-based and deep learning-based techniques, and further in Deep Learning, it circumvents computer vision specific image-based and Natural language processing related text-based techniques.

4.1 Machine Learning Based Malware Detection

To protect computers, servers and handheld devices, anti-virus software provided by anti-malware companies set up the first line of defense. Traditionally, signature and heuristic-based malware detection methods are used, which are usually the short sequence of unique byte-code assigned to each known malware by a domain expert.

Since the last decade, the research community has incurred tremendous efforts for malware detection using machine learning and data mining techniques. The machine learning approaches for malware detection are further grouped into two main types *Supervised/Classification* techniques and *unsupervised/Clustering* techniques. These techniques are widely adopted for the signature detection of previously unseen malware and/or malware family. All the machine learning techniques are heavily dependent on feature extraction using either the static, dynamic, or hybrid analysis of a file. The first step of all ML-based approaches is to perform feature engineering to capture the characteristics of a malware sample file, for instance, API calls, permissions, intent, services, protected string, opcode sequences, etc. The second step is input representation and training the classifier for signature detection and assigning the family label using classification or clustering technique. The trained classifier is then deployed locally or on a cloud for the detection of unknown sample. The basic accuracy metrics of a malware classifier is usually determined through True Positive, False Positive, True Negative, and the False Negative Rate. However, for clustering-based techniques, Macro and Micro F-1 measure is used to evaluate the system. In the subsequent subsections, we have further enlightened the details of these metrics.

4.1.1 Supervised Learning for Malware Detection

Supervised learning is the mechanism in which samples are provided with correct labels to a statistical model in the preliminary phase called training. The model can correctly classify the unseen sample based on the distribution learned from training data. An example is a listing of one thousand android APKs (Android Application Packaging kit) with their malware and benign signature. The supervised classification task is completed in two sequential steps, firstly, classifier construction the second is model deployment and usage. For the abovementioned example, how a supervised machine learning model will work?

In the first step, one thousand android APKs with correct malware and benign labels, are available to the system. The system processes each APK file to extract the underlying characteristics, converts it into a feature vector, and assigns the provided label from the training set. The converted feature vector and class label of each APK file from the training set is used to train the classifier, for example, K-Nearest neighbor (KNN), Support Vector Machine (SVM), Artificial Neural Network (ANN), Decision Tree, Bayesian Belief Network, etc. In the second step of model usage, the constructed model is deployed as a part of an application and unknown APK file, either malware or benign are exposed to model. The system performs the feature extraction of a new sample (similar to the training set) and gives a verdict of malware or benign based on learning from the training set. Researchers have presented various supervised learning techniques for malware detection; a few commonly used are discussed below.

- **Decision Tree (DT)**: A decision tree is a conventional classifier used for categorical data. The Decision trees are built using if-then-else rule-based approach, presenting the application features or variables in a tree structure. A heuristic-based greedy mechanism is used to construct the tree, which involves the information gain and entropy score. Leaf node represents the class labels, while the root/decision node represents the features/variables and has two or more branches and are calculated using information gain and entropy. The dataset is split into smaller subsets and entropy measures the homogeneity/heterogeneity of the data using Eq. 1. If the number of samples in a dataset is equally distributed, then entropy is 1, if all the samples are homogeneous, then entropy is zero. For example, a dataset having ten samples and all are benign, then entropy becomes zero if five are benign and five are malware, then entropy is one.

$$Entropy = -\sum p(x) * \log_2 p(x) \tag{1}$$

where p(x) represents the number of examples in a given class. The decision trees are computationally less expensive and easy to understand as the complete path from the root node to the leaf node makes one rule and shows the decision. On the other hand, DT are more prone to over-fitting and noisy data leads to wrong classification, the problem of over-fitting can be solved by pruning the decision tree. Numerous studies are available in literature to solve the malware detection issue but lack robustness and need heavy feature engineering.

- **Random Forest (RF)** Random forest is the advanced and improved version of decision tree. More specifically, it is the collection of decision trees and always yields better results as compared to the decision tree by choosing an independent dataset for each tree in the forest using random or bootstrap data sampling. These sampling techniques ensure that decision trees in the forest have diversity and are dissimilar. For constructing each tree, the same approach is used, as discussed above for DT. The algorithm makes voting on each decision tree's result in the forest and performs the final classification. A few popular studies like [3, 4] used random forest for android malware detection and yielded state of the art results.

- **K-Nearest Neighbour (KNN)** KNN is the oldest and simplistic algorithm used for classification and regression by measuring the similarity/distance metric between samples or variables. An unknown query instance is classified into the same class as its nearest neighbor belonging or depend on the wining class of its neighbor samples. K represents the size of the neighbors involved in the decision process for classification. Usually, K is an odd number and typical values lie between 3 and 8. Different distance metrics used to measure the similarity of variables of an unknown sample to a training set sample, for example, Manhattan distance, Hamming distance, Minkowski distance, but the most popular one is Euclidean distance. KNN is not an optimal choice for malware detection because practically, malware datasets are highly imbalanced with a high number of benign samples and fewer malware samples. Therefore the majority of neighbor belongs to the dominant class, which results in high False Positive and False Negative Rate.

- **Naive Bayes Classifier (NBC)** Naive Bayes is a supervised learning technique applicable to binary and multi-class classification problem using Bayes Theorem as given in Eq. 2. NBC evaluates the probability of each attribute independently and calculates the class probability and conditional probability to make the prediction. An advanced version of Naive Bayes is the Bayesian Belief Network, which shows the graphical structure representing the dependencies of variables but not conditionally independent. Bayesian Belief Network comprises of two components. The first is directed acyclic graphs depicting all random variables as nodes and the edges between nodes represent the probabilistic probabilities between these variables. Second is the conditional probability table for variables. Several studies are available in literature using NBC and Bayesian Belief Network for static and dynamic malware detection using opcode sequences, API calls and permission analysis.

$$P(A|B) = \frac{P(B|A)P(A)}{P(B)} \tag{2}$$

- **Support Vector Machine (SVM)** Support Vector Machine is a machine learning-based classification method mostly adopted for binary classification problems and the objective of this classifier to search for a hyperplane that maximizes the distance between two given classes. Term "support Vector" refers to data points lie close to hyperplane and the distance between support vector and hyperplane is referred to as margin. The objective function of SVM is to maximize these margins. SVM with multiple hyperplanes can be used for multi-class classification. SVM has been widely used in malware detection; the famous Drebin paper [5] uses SVM for Android malware detection after applying feature engineering to extract relevant features that contribute to benign and malware.
- **Artificial Neural Network (ANN)** Artificial Neural network, as the name depicts, is inspired by the human brain and mimics the functionality of the human biological nervous system. As a layman term, Neural Networks are a collection of input and output neurons, each associated by weight except the final decision-making neuron. During the training phase weights are randomly assigned initially and then network adjust using backpropagation to correctly predict the true class. There are various variant of Neural Networks for Malware detection using images, binary and text data.
- **Deep Neural Network (DNN)** also known as Deep Learning (DL) is a focal concern for academia and industry for different applications, for instance, computer vision, Natural Language Processing (NLP), speech recognition, and cognitive learning, etc. DNN with multi-layer architecture allows superiority in feature learning as compared to ANN and overcome various challenges like layer wise pre-training etc. DNN involves more than one hidden layer in contrast to simple ANN and conducive to learn multiple abstractions, advanced feature extraction, and higher-level concepts. However, DNNs are compute-intensive with GPU requirement and time consuming to learn the underlying model. There are various variants of DNN which are extensively used for malware detection such as Convolutional Neural Networks (CNN), Recurrent Neural Networks (RNN), Deep Belief Networks (DBN), and Deep Autoencoders. The details of these malware detection systems using DL is discussed in Sect. 4.2.

4.1.2   Un-Supervised Learning for Malware Detection

In machine learning, all the classification algorithms require labeled inputs for predicting the correct outcome. However, semi-supervised learning does not require a fully labeled training datasets. The semi-supervised techniques are handy in scenarios where the scarcity of labeled inputs or partially labeled inputs are provided. The research community is actively working in automatic malware detection and categorization using unsupervised learning where no labels are available. Clustering is a famous unsupervised learning technique for segregating the data into groups called clusters using different predefined distance measures like Euclidean distance, Manhattan distance, etc. The clustering aims to maximize the inter-cluster distance while minimizing the intra-cluster distance. K-mean clustering and Hierarchical clustering are two famous techniques heavily used for unsupervised learning. Hierarchical clustering is suitable for malware datasets which exhibits irregular cluster structures while K-means clustering is more effective in globular representation.

In malware detection, different malware families represent different clusters. A file sample shares the same trait and features can be assigned to a respective cluster using the clustering technique by measuring the distance measure. Similarly, clustering can be used in malware detection in both static and dynamic analysis. In dynamic analysis, the grouping is based on profiles behavior and runtime execution and API calling structures. In static analysis, Malware can be identified and grouped using the same permissions, features, and similarity analysis. Researchers have presented various unsupervised approaches for automated malware detection using structural information (function call graph analysis). After the pre-processing, the fine-grained feature extraction step is performed using call graphs which determines the file sample belong to which malware cluster by applying discriminate distance metric to marginalize the inter-cluster distance. The choice of clustering technique in malware detection is dependent on the dataset, extracted features, platform OS, Windows or Android, and their underlying feature distribution. The malware expansion is not limited to a single platform but hitting all platforms with the same force. It is imperative for the research community to set up an apt combat mechanism across all platforms.

4.1.3   Cross Platform Malware Detection

Malware is a kind of software developed with the intention to deliberately harm the target system. The exponential growth of malware is targetting every platform including Linux, macOS, Windows, Android, etc. due to recent internet advancement. The concept of malware exists since the inception of computing in 1949 and numerous malware attacks have posed serious threats to legitimate users and evolving expansion of malware is not limited to one platform. According to the report of the AV-Test Institute each month, half a million android malware is created. Similarly, the number is near to one million for Windows, and in total 350,000 malware are created daily for different platforms. With the emergence and popularity of IoT devices, malware attackers have broader target markets to penetrate. Malware is generally not only categorized into various families but also platform-specific and may have various variants of one family. Malware can infiltrate into the system via multiple routes (a) vulnerable services over the network allow automatic entry (b) Download from the internet and studies depicted that about 80% of the malware came from famous websites (c) malware authors allure the individuals to execute the malicious code on their machines, for instance, asking them to install a codec to watch a movie or a tv program online or clicking the adds. It has been noted in some cases the intention of the malware author is to use as a spy for collecting sensitive information and sometimes malware affects the system performance and creates overload processes.

It is a universal truth that non of the platform is 100% fully bulletproof against malware and spyware. There are different pros and cons associated with each platform regardless of how well engineered an operating system is. Due to the never-ending race of defenders and attackers, there is always room for malware injection. Studies have revealed that the Windows platform is more prone to malware attacks. The vulnerability is not due to any inherent flaw of the operating system but due to its high market share and sheer size of users. Hackers and malware authors create numerous attacks against the window as compared to Linux or macOS due to low effort and high return ratio. Therefor various studies and research community targets mostly Windows (PE) for malware detection and presented unsurpassed approaches. Linux is less prone to malware due to its very little market share of 3%. Therefore, attackers do not surf energies to create attacks for Linux OS and Linux also does not grant admin access by default which limits the attacks by clicking different Adds and Malware. Similarly, macOS and iOS also have less market share in contrast to Windows and Android. Therefore the attack pool is marginally less to compromise and infiltrate. Moreover, macOS has apple system integration protection (SIP) which doesn't allow to access and modify the core files and processes. On the other hand, Android is open source and capturing the 85% smartphone market and is a hive for malware authors. In addition to the Google Play store, there are various third-party app stores that allow the end-user to install various applications. The reaming part of the chapter focuses on android malware detection techniques.

4.1.4 Android Malware Detection

The smartphone is the biggest revolution of the 21st century and is proven as a game-changer in everyday life, including social interaction, finance management, entertainment, education, and many more. Smartphones with more than 4 billion apps including iOS and Android are transforming human interaction. Android occupying the largest market share of 85% and being an open-source platform is the most vulnerable for hackers and malware authors. According to AV-Test anti-virus institute, half a million android malware crafted each month to invade unsuspecting users. The expansion of such malware posed higher security threats, for instance, stealing credentials, leaking sensitive information, financial transactions, calling premium numbers, sending SMS messages, etc., and using their private data for other means without individual knowledge and permission. Therefore, detection of android malware is of utmost concern for both the antivirus industry and researchers to protect legitimate users from these threats. In order to understand the preliminaries and basic structure of android app hierarchy, let's deep dive into the structure of an Android application called APK (Android application packaging Kit). Unlike desktop-based Portable executable PE files, an android application is packaged and wrapped in APK zip format used for the distribution and installation of an android app on a smartphone. Java is the language choice for all android applications and compiled into Dalvik bytecode wrapped in a .dex file. The main structure of APK include manifest file. dex file, assets and resources. The main components are discussed as follow

- **Manifest.xml file**: Android OS uses the component-based framework for mobile application development, deployment, installation, and maintenance. Manifest.xml file act as a configuration file for the android application which manages the access control, authorization, versioning control, permissions and app components. App components have four main modules namely *Activities, Services, Receivers, and Providers*. Activities enable Graphical User Interface (GUI) interaction while services are responsible for background communication and processing of inter and intra app components. Providers act as a database management system for the app and Receivers are background process liable to respond to messages broadcast by the system. The android app declares all its components in the android manfiest.xml file to manage the structural information. Before initiating any component the system reads the manifest file and check the app components, permissions, and filter intents. Further, the actions of each component are governed by filtered intents which specifies the type of intents an activity, provider, services or receivers can respond to. for instance, an activity can initiate a phone call via filtered intent or the Receiver can monitor SMS. The manifest file governs the application functionality and acts as a road map of the android application. The manifest also enumerates all the permissions requested by the app to perform certain functionalities (e.g. wifi access, internet access, SMS send permission, etc.). Hence, permissions, intents, and app components are used to handle the interaction between app to app and with OS. All the android malware detection techniques using machine learning or deep learning on the track of static analysis, dynamic analysis, or hybrid extract these ingredients from manifest file to represent the feature set and then train a learning algorithm to detect the malware. Different studies follow their respective methodologies and use all app components and permission or selective app components and permissions for malware detection. Fewer studies only relying on permissions with different accuracy metrics, and reasoning.
- **Classes.dex**: Dex is an executable file in android containing the Dalvik bytecode compiled from Java source code and interpreted by a Dalvik Virtual Machine. Android applications are always written in java and compiled into Dalvik bytecode. Dex file is the file format containing this compiled code which includes all the user-implemented classes, functions, and objects. Dex file also contains different API calls to be used by the android app to access the different operating system activities, creating new instances and resources. In short, API calls, new instances,

and different opcode sequences for a dex file can be used to determine the behavior of an android app. It is imperative to mention that the dex file is unreadable without any reverse engineering tool. Mostly android malware detection studies use different reverse engineering tools to decompile the dex file into a smali file which is the intermediate between java and Dalvik bytecode and then parse the smali file to extract features like ope code sequences, system calls graph, API calls, etc. to detect malicious behavior.

- **Feature Representation** The research community has used different approaches to represent the android application, most of the studies use a reverse engineering tool to extract the features from the manifest file and dex file and represent the android app as a binary vector space. After the preprocessing and feature representation different machine learning and deep learning-based algorithms are trained to detect legacy and new malware. The feature extraction and representation for android application is illustrated in Table 1

A more generic way to express the feature vector $1, …, M$, including all permission, API calls, system calls, and intents is to represent in a binary vector $X \epsilon \{0, 1\}^M$, where X_i represents whether the application has particular feature i, i.e. $X_i = 1$, or not, i.e. $X_i = 0$. Due to different functionalities and purposes, numerous applications are available in the market, and pertinent to these functionalities, M can be massive, while one application might contain a few features, and this results in a very sparse feature vector. ach Android app can be represented as a binary vector as depicted in Eq. 3 in the illustration to Table 1.

$$
x_i = \begin{pmatrix} 1 \\ 0 \\ \cdots \\ 0 \\ 1 \\ \cdots \\ 1 \\ 1 \\ \cdots \\ 1 \\ 0 \\ \cdots \end{pmatrix} \left. \begin{cases} \text{ACCESS\_NETWORK\_STATE} \\ \text{CALL\_PHONE} \\ \cdots \\ \text{Splash\_View} \\ \text{lib.activity.SmsPatternForm} \\ \cdots \\ \text{intent.action.GTALK\_CONNECTED} \\ \text{BROADCAST\_CHANNEL} \\ \cdots \\ \text{Landroid/telephony/TelephonyManager:getDeviceId} \\ \text{Landroid/app/AcitivityManager:killBackgroudProcess} \\ \cdots \end{cases} \right\} \begin{array}{l} \text{Permissions} \\ \\ \\ \text{Activities} \\ \\ \\ \\ \text{Filtered Intents} \\ \\ \text{API Calls} \end{array} \quad (3)
$$

After the binary vector representation of an android application different machine learning techniques are applied to train the malware detection system. Apart from binary representation, numerous studies are presented representing the app into the two-dimension array, pixel values (images) and word sequences, etc. and then on top of these representations, learning-based classifiers are trained to detect the malware and classify the corresponding malware's family.

Table 1 Illustration of feature representation of an Android app

File type	Feature	Sample
Manifest	Permissions	android.permission.ACCESS_NETWORK_STATE
		android.permission.CALL_PHONE
	Activities	Activity_Splash_View
		com.haunsoft.moneyapp_lib.activity.SmsPatternForm
	Services	com.android.providers.update.OperateService
		com.android.providers.sms.SMSService
	Filtered Intents	android.intent.action.GTALK_CONNECTED
		android.intent.category.BROADCAST_CHANNEL
Dex	API Call	Landroid/telephony/TelephonyManager:getDeviceId
		Landroid/app/AcitivityManager:killBackgroudProcess
	Opcode	invoke-super(parameter),method_to_call

4.1.5 Challenges for Machine Learning Based Android Malware Detection Systems

With the proliferation of handheld android-based devices, a plethora of malware apps have emerged as a real threat to the end-users. To defend against these malware apps, various machine learning approaches were designed for their detection and isolation. Since the last decade, ML-based frameworks have achieved unsurpassed versatility and attraction to defend various malware by leveraging different feature representations includes Application Programming Interface (API), permissions, App components, dynamic behaviors, system calls graph, etc. These approaches use various ML algorithms like Random Forest, K-nearest neighbors (KNN) Support Vector Machine (SVM), and different variants of Deep Neural Networks on top of extracted features for malware classification.

The detection of malware is challenging due to the polymorphic nature, obfuscation, and numerous sophisticated routes of creating malware. Moreover, with the proliferation of mobile apps, it is remarkably challenging to analyze each application manually for malicious activities. However, traditional machine learning-based approaches heavily dependent on fine-grained feature extraction and extensive source code analysis for malware detection. Most of the machine learning-based approaches are analyzing the malicious behavior of android APK files, either using static application behavior or dynamic application behavior. The research community is a bit inclined toward the static behavioral analysis of malicious apps using machine learning, which involves small run-time overhead, scalable, considered more efficient, and features are extractable without the execution of a program. However, the common issues pertinent to these approaches are high false-positive rate, required heavy feature engineering, and usual failure on dynamic code load and obfuscation. While approaches rely on dynamic behavior analysis, extract features based on execution/emulation by incorporating run-time code loading, obfuscation, API, and system calls but need proper inputs to determine execution paths.

Malware is continuously emerging, and new variants are being introduced daily, traditional approaches like signature-based, handcrafted feature selection have limitations to cope with this challenge and struggles to detect new malware and corresponding families. Rule-based and various machine learning techniques have been in practice for a decade to detect malware in both static and dynamic environments but these are not robust to detect new malware families and struggling on the automatic extraction of fine-grained features. Moreover, recent studies revealed that Machine Learning is vulnerable to adversarial attacks and the trained models can be compromised in a production-ready environment. This imposes a new challenge and malware authors launch different adversarial attacks on machine learning-based approaches to compromise their integrity and availability. An adversary aims to mislead the trained classifier to produce the least True positives by classifying malware samples as Benign.

4.2 Deep Learning Based Android Malware Detection Systems

Advances in the field of Artificial Intelligence AI is the paradigm shift in the technology landscape. Due to the latest research and innovation since 2011, they are widely adopted in every domain of computer science and other fields, including Medical, Engineering, social sciences, business processes, and management. For example, machine learning-based data-driven analytic has revolutionized healthcare, finance, and other business models. Detection, monitoring, and classification in the security domain plus extracting valuable information from text and images from a plethora of social media data and other means are possible due to these emerging technologies. Progression in Machine Learning and Deep Learning opened gateways to solve complex systems and often make the software industry beyond human capabilities.

Deep learning is the subfield of machine learning focusing more on various levels of representations analogous to feature ranking and importance. The higher level of features are inferred from the lower level and at times lower level features help in determining the higher-level features. Prior to deep learning, techniques are applied to draw a separation of feature levels from malware samples to perform classification, which lacks in mirroring the overarching behaviors of malware. Moreover, classification formulated on varying and diverse attributes leads to doubt on dimensions, computational resources, and training time. Deep learning-based malware classification offers scalable models to handle a huge amount of data, without deep feature analysis and without expanding the resources like memory, etc. Deep learning classifies the malware by observing the general pattern, and able to distinguish a variety of malware attacks and their variants. The studies revealed that deep learning-based solutions offer profound classification, robust malware detectors with improved accuracy, generalized due to relying on multiple levels of feature orchestration, and learning as compared to traditional machine learning methods.

As discussed in Sect. 4.1.1 DNN, RNN, CNN, DBN, RBM, and deep auto-encoders are mostly used in android malware detection systems by researchers. The selection of an apt classification algorithm for detection is made in consideration of its impact on detection accuracy and performance. Few studies in the literature for android malware detection systems trained the deep learning method DBN (Deep belief Network) by leveraging the combo of dangerous API calls and risky permissions as a feature set to automatically recognize the malware from benign samples. Few approaches use continuous space and convert the feature set into an image and train a CNN for malicious behavior, for instance, one of the studies converts the requested permissions into an image file and trained a CNN classifier for malware detection. Few studies follow the NLP route and designed a model which deals with system call sequences as texts and considers the malware detection function as theme extraction and trained a Recurrent Neural Network (RNN) to identify the malicious sample. Another NLP base approach treated the dex file as a text and automatically learns the features from raw data and then apply multiple CNN layers for classification.

4.2.1 Open Issues with Deep Learning Based Malware Detection

This section elaborates on a few well known open issues relating to deep learning in android malware detection.

- The first and the most severe issue pertinent to all machine learning and deep learning approaches are misleading the classifier, all the classifier can be deceived.
- Malware analysts are always interested in the decision and need to understand the reasoning of the decision made but deep learning lacks visibility and transparency to explain the interpretation of detection decision formulated by its method.
- Deep learning-based android malware detection systems are not fully autonomous and need retraining if totally out of distribution malware sample coined, It needs continual retraining and parameter tuning.
- Human experts are required for feature engineering and continual model updates are mandatory and there is no guarantee that deep learning-based trained models are equally effective and efficient against the new samples that are not related to previously trained distribution.
- Deep learning analyzes the complex correlations of input and output feature sets with no inherent explanation of causality.
- Deep learning model hosted on the cloud for malware detection send data to a remote server for analysis, this mechanism needs data privacy protection which is not yet addressed fully.

4.2.2 Adversarial Android Malware Detection

Since the last decade, machine learning and deep learning-based solutions provided unparalleled flexibility and versatility to address android malware detection but unfortunately, these approaches are self vulnerable to adversarial attacks. In computer security, defenders and attackers always engage in never-ending arms. They always examine the system and assess strengths, opportunities, and vulnerabilities to safe or exploit the system as per their interest and always try to beat each other methodologies. Competition leads to considerably versatile and secure solutions versus sophisticated and polymorphic routes to exploit vulnerabilities. For instance, to make the traditional signature-based malware detection techniques ineffective by repackaging and code obfuscation. Attackers create polymorphic and dynamic malware by defeating attempts to analyze the inner mechanism. Researchers had successfully evaded various malware detection systems in various domains like PDF malware detection, Windows-based malware detection, Cloud-based anti-virus engines, and android malware detection systems.

An adversary can generate adversary samples to launch attacks on the target model by capitalizing the knowledge and parameter settings. Various methods have been proposed in the literature to attack the deep learning model by carefully crafted adversarial samples to make the deep learning model fool. Adversarial examples are closer to real distribution with a slight perturbation that is imperceptible to a human eye and mislead the model to predict the targeted or incorrect class. The malware authors are getting attracted by this vulnerability of machine learning and misleading the learning-based android malware detection systems to produce minimum True positive. The metric used to measure the security of machine learning and deep learning models depends on the goals and capabilities of an adversary. The adversarial goals are mainly related to compromise the output integrity of learning models and can be categorized into two main types, i.e., Confidence Reduction and Misclassification. However, misclassification can be further categorized into three categories, i.e., random misclassification, targeted misclassification, and source/target misclassification as enumerated below:

- **Confidence Reduction** Mostly the output of the machine learning model is the predicted class label and associated confidence score, the goal of the adversary is to reduce the confidence score for the target model. For instance, predicting a network packet as malicious with a low confidence score.
- **Random Misclassification** The adversary intends to change the output of a target model to a random class different from the actual class. For example, the class of a network packet changed to any other class except the malicious.
- **Targeted Misclassification** An Adversary attempts to change the output of a class into the specific target class. For example, Input fed to a malware classifier is classified as benign.
- **Source/Target Misclassification** The goal of an adversary is to change the output of a specific class into a specific target class. For example, the Benign class is always classified as Malicious.

The adversarial capabilities are directly proportional to the access and amount of information known to an adversary, An adversary is stronger if the information available to him is more as compared to others. At a higher level, the adversarial capabilities are grouped as training phase capabilities and testing phase capabilities. In the training phase, the adversary tries to corrupt the target model by altering the training dataset with partial or full access to training data. The scenario where the adversary has complete access is known as WhiteBox attack settings, and where no access is referred to as BlackBox attacks and in partial access scenario, it is termed as GrayBox attacks. Researchers have launched different adversarial attacks in Whitebox and BlackBox settings to compromise the integrity and availability of deep learning-based android malware detection systems. It's a hotspot area for the research community and anti-virus industry, few approaches are presented to cope with these challenges but still, substantial efforts are required to combat these attacks. Hardening and making deep learning models robust against adversarial attacks need paramount attention and the biggest challenge for deep learning-based detection models.

5 Conclusion

In this chapter, we have presented a summary of legacy malware analysis techniques, recent approaches, and challenges. A classification of different approaches is also given where we reviewed the well-known papers that exist in the literature. For the classical proposed solutions, the approaches can be classified into two main categories: static and dynamic. The static analysis consists of two known approaches, namely, the signature-based approaches and the permission-based approaches. In these techniques, the detection of malware is performed without its execution or installation of the application. The properties and characteristics of the internal structure are investigated in order to detect malicious codes. On the other hand, dynamic analysis is represented by two main categories: behavior-based approaches and specification-based approaches. Not like static analysis, this type of techniques authorizes the installation of apps and then analyzes their behavior during execution. In most cases, dynamic analysis checks which API calls are used by apps and instruction traces. However, the previous approaches can not identify all new variants and sophisticated malware. Recently, advances in the field of artificial intelligence have led to the emergence of a new philosophy for detecting malware. The recent malware detection techniques can be categorized into machine learning-based approaches and deep learning-based approaches. The first category consists of a supervised/unsupervised learning-based approach in which extraction of features, classifier training, and classification are the main malware detection adopted phases. The second category uses deep learning algorithms such as RNN and CNN to have more accurate detection rates. We have also presented in this paper challenges and open issues for machine and deep learning-based approaches for malware detection.

References

1. Ye, Y., Li, T., Zhu, S., Zhuang, W., Tas, E., Gupta, U., Abdulhayoglu, M. (2011). Combining file content and file relations for cloud based malware detection, pp. 222–230. https://doi.org/10.1145/2020408.2020448
2. Hardy, W., Chen, L., Hou, S., Ye, Y., Li, X.: DL4MD: a deep learning framework for intelligent malware detection. In: Proceedings of the International Conference on Data Mining (DMIN), p. 61. The Steering Committee of the World Congress in Computer Science, Computer Engineering and Applied Computing (WorldComp) (2016)
3. Alam, M.S., Vuong, S.T.: Random forest classification for detecting android malware. In: 2013 IEEE International Conference on Green Computing and Communications and IEEE Internet of Things and IEEE Cyber, Physical and Social Computing, pp. 663–669. IEEE (2013)
4. Zhu, H.J., Jiang, T.H., Ma, B., You, Z.H., Shi, W.L., Cheng, L.: HEMD: a highly efficient random forest-based malware detection framework for Android. Neural Comput. Appl. **30**(11), 3353–3361 (2018)
5. Arp, D., Spreitzenbarth, M., Hubner, M., Gascon, H., Rieck, K., Siemens, C.E.R.T.: Drebin: effective and explainable detection of android malware in your pocket. In: NDSS, vol. 14, pp. 23–26 (2014)
6. Sung, A., Xu, J., Chavez, P., Mukkamala, S.: Static analyzer of vicious executables (SAVE). In: Proceedings—Annual Computer Security Applications Conference, ACSAC, pp. 326–334 (2005). https://doi.org/10.1109/CSAC.2004.37
7. Christodorescu, M., Jha, S.: Static analysis of executables to detect malicious patterns. In: Proceedings of the 12th Conference on USENIX Security Symposium—Volume 12 (SSYM'03). USENIX Association, USA (2003)
8. Christodorescu, M., Jha, S.: Testing malware detectors. ACM SIGSOFT Softw. Eng. Notes **29**, 34–44 (2004). https://doi.org/10.1145/1007512.1007518
9. Shanmugam, G., Low, R., Stamp, M.: Simple substitution distance and metamorphic detection. J. Comput. Virol. Hacking Tech. **9** (2013). https://doi.org/10.1007/s11416-013-0184-5
10. Wong, W., Stamp, M.: Hunting for metamorphic engines. J. Comput. Virol. **2**, 211–229 (2006). https://doi.org/10.1007/s11416-006-0028-7
11. Sorokin, I.: Comparing files using structural entropy. J. Comput. Virol. **7**, 259–265 (2011). https://doi.org/10.1007/s11416-011-0153-9
12. Baysa, D., Low, R., Stamp, M.: Structural entropy and metamorphic malware. J. Comput. Virol. Hacking Tech. **9** (2013). https://doi.org/10.1007/s11416-013-0185-4
13. Ször, P., Ferrie, P.: Hunting for metamorphic. In: Virus Bulletin Conference (2001)
14. Runwal, N., Low, R., Stamp, M.: OpCode graph similarity and metamorphic detection. J. Comput. Virol. **8** (2012). https://doi.org/10.1007/s11416-012-0160-5
15. Stamp, M., Toderici, A.: Chi-squared distance and metamorphic virus detection. J. Comput. Virol. Hacking Tech. **9**, 1–14 (2013). https://doi.org/10.1007/s11416-012-0171-2
16. Zheng, M., Sun, M., Lui, J.C.S.: Droid analytics: a signature based analytic system to collect, extract, analyze and associate Android malware. In: 2013 12th IEEE International Conference on Trust, Security and Privacy in Computing and Communications, Melbourne, VIC, 2013, pp. 163–171. https://doi.org/10.1109/TrustCom.2013.25
17. Gao, T., Peng, W., Sisodia, D., Saha, T.K., Li, F., Al Hasan, M.: Android malware detection via graphlet sampling. IEEE Trans. Mob. Comput. **18**(12), 2754–2767 (2019). https://doi.org/10.1109/TMC.2018.2880731
18. Wu, D., Mao, C., Wei, T., Lee, H., Wu, K.: DroidMat: Android malware detection through manifest and API calls tracing. In: 2012 Seventh Asia Joint Conference on Information Security, Tokyo, 2012, pp. 62–69. https://doi.org/10.1109/AsiaJCIS.2012.18
19. Peiravian, N., Zhu, X.: Machine learning for android malware detection using permission and API calls. In: 2013 IEEE 25th International Conference on Tools with Artificial Intelligence, Herndon, VA, 2013, pp. 300–305. https://doi.org/10.1109/ICTAI.2013.53

20. Chan, P.P.K., Song, W.-K.: Static detection of Android malware by using permissions and API calls. In: 2014 International Conference on Machine Learning and Cybernetics, Lanzhou, 2014, pp. 82–87. https://doi.org/10.1109/ICMLC.2014.7009096
21. Liang, S., Du, X.: Permission-combination-based scheme for Android mobile malware detection. In: 2014 IEEE International Conference on Communications (ICC), Sydney, NSW, 2014, pp. 2301–2306. https://doi.org/10.1109/ICC.2014.6883666
22. Wang, Z., Li, K., Hu, Y., Fukuda, A., Kong, W.: Multilevel permission extraction in Android applications for malware detection. In: 2019 International Conference on Computer, Information and Telecommunication Systems (CITS), Beijing, China, 2019, pp. 1–5. https://doi.org/10.1109/CITS.2019.8862060
23. Kapoor, A., Kushwaha, H., Gandotra, E.: Permission based Android malicious application detection using machine learning. In: 2019 International Conference on Signal Processing and Communication (ICSC), Noida, India, 2019, pp. 103–108. https://doi.org/10.1109/ICSC45622.2019.8938236
24. Saleem, M.S., Mišić, J., šić, V.B.: Examining permission patterns in Android apps using Kernel density estimation. In: 2020 International Conference on Computing, Networking and Communications (ICNC), Big Island, HI, USA, 2020, pp. 719–724. https://doi.org/10.1109/ICNC47757.2020.9049820
25. Kolbitsch, C., Comparetti, P., Kruegel, C., Kirda, E., Zhou, X.-y., Wang, X.: Effective and efficient malware detection at the end host. In: USENIX Security Symposium, pp. 351–366 (2009)
26. Burguera, I., Zurutuza, U., Nadjm-Tehrani, S.: Crowdroid: behavior-based malware detection system for Android. In: SPSM '11, pp. 15–26 (2011). https://doi.org/10.1145/2046614.2046619
27. Mohaisen, A., Alrawi, O., Mohaisen, M.: AMAL: high-fidelity, behavior-based automated malware analysis and classification. Comput. Secur. (2015). https://doi.org/10.1016/j.cose.2015.04.001
28. Nikolopoulos, S., Polenakis, I.: A graph-based model for malware detection and classification using system-call groups. J. Comput. Virol. Hacking Tech. **13** (2016). https://doi.org/10.1007/s11416-016-0267-1
29. Saracino, A., Sgandurra, D., Dini, G., Martinelli, F.: MADAM: effective and efficient behavior-based Android malware detection and prevention. IEEE Trans. Depend. Secure Comput. **15**(1), 83–97 (2018). https://doi.org/10.1109/TDSC.2016.2536605
30. Choi, C., Esposito, C., Lee, M., Choi, J.: Metamorphic malicious code behavior detection using probabilistic inference methods. Cognit. Syst. Res. **56**, 142–150 (2019)
31. Song, F., Touili, T.: PoMMaDe: pushdown model-checking for malware detection, pp. 607–610 (2013). https://doi.org/10.1145/2491411.2494599
32. Chaugule, A., Xu, Z., Zhu, S.: A Specification Based Intrusion Detection Framework for Mobile Phones, pp. 19–37 (2011). https://doi.org/10.1007/978-3-642-21554-4_2
33. Song, F., Touili, T.: Model-Checking for Android Malware Detection, pp. 216–235 (2014). https://doi.org/10.1007/978-3-319-12736-1_12
34. Battista, P., Mercaldo, F., Nardone, V., Santone, A., Visaggio, C.A.: Identification of Android malware families with model checking. https://doi.org/10.5220/0005809205420547
35. Canfora, G., Martinelli, F., Mercaldo, F., Nardone, V., Santone, A., Visaggio, C.A.: LEILA: formal tool for identifying mobile malicious behaviour. IEEE Trans. Softw. Eng. **45**(12), 1230–1252 (2019). https://doi.org/10.1109/TSE.2018.2834344
36. IDApro: Available at: https://www.hex-rays.com/. Accessed 28 Nov 2020
37. Debugger, O.: Available at: http://www.ollydbg.de/. Accessed 28 Nov 2020
38. XXD. Available at: https://linux.die.net/man/1/xxd. Accessed 28 Nov 2020
39. Miller, P., Hexdump (2000). Available at: https://man7.org/linux/man-pages/man1/hexdump.1.html. Accessed 28 Nov 2020
40. The first computer virus was designed for an apple computer, by a 15 year old. https://blogs.quickheal.com/the-first-pc-virus-was-designed-for-an-apple-computer-by-a-15-year-old/

Security of Cyber-Physical-Social Systems: Impact of Simulation-Based Systems Engineering, Artificial Intelligence, Human Involvement, and Ethics

Tuncer Ören

> *Efficiency is doing things right; effectiveness is doing the right thing.*
> Peter Drucker (1909–2005)

Abstract Almost every aspect of life is becoming connected. The types of the connectedness of the cyber-physical-social systems are outlined. They span from Internet of Things to Internet of senses (including Internet of eyes and ears) and Internet of brains. Some sources of failures in Cyber-Physical-Social systems as well as the fact that with increased connectedness, the systems become more vulnerable are outlined. Some possibilities to get prepared for unexpected conditions are discussed. They include experience and experimentation aspects of simulation, systems engineering, simulation-based systems engineering, as well as artificial intelligence. The possible dangers of AI are pointed out. Often solutions to technical problems are sought in technical realms. In the article, impact of human involvement is elaborated as sources of additional and severe problems.

Keywords Security of cyber-physical-social systems · Connected world · Internet of things · Internet of senses · Internet of brains · Internet of intangibles · Failure avoidance · Human biases · Dysrationalia · Cognitive biases · Cultural biases · Human malevolent behavior · Hackers · Machiavellianism · Dark triad · Truth decay · Ethics

T. Ören (✉)
School of Electrical Engineering and Computer Science, University of Ottawa, Ottawa, ON, Canada
e-mail: oren.tuncer@sympatico.ca; oren@eecs.uOttawa.ca

P. Nicopolitidis et al. (eds.), *Advances in Computing, Informatics, Networking and Cybersecurity*, Lecture Notes in Networks and Systems 289,
https://doi.org/10.1007/978-3-030-87049-2_26

711

1 Introduction

It is a privilege to be invited to contribute a chapter to a book dedicated to my long-time dear friend and internationally acclaimed scientist Prof. Mohammad Obaidat. Recently, Prof. Obaidat was the General Chair of the 2020 IEEE International Conference on Communications, Computing, Cybersecurity, organized virtually at the University of Sharjah, in UAE. He continued to honor me also in this conference by inviting me as a keynote speaker and a member of the closing panel. The theme I selected for this chapter is closely related with the topics of my contribution as a keynote speaker (Grand Challenges in Modelling and Simulation: What M&S can do and what we should do for M&S?) [62] and as a member of the closing panel he organized on: "Recent Trends and Challenges in Communications, Computing, Cybersecurity and Informatics" My contribution was focussed on: Some Desirable Features: Communications, Computing, Cybersecurity and Informatics and Quality in a World Connected by Computers [63].

Cyber-physical-social systems are becoming pervasive. With this, their security becomes a major issue. Unexpected emerging conditions arise in all aspects of life and often necessitate fast and effective decisions. Lack of appropriate interventions by policy makers, by fact-based rational and timely decisions, often hinder proper functioning of the affected systems and can be detrimental even mortal.

Both experience and experimentation aspects of simulation can be very pertinent for the preparation of decision makers and as decision tools in real time. These aspects of simulation are: (1) providing three types of experience even under extreme conditions and (2) the ability to perform experiments which can also be used to prepare dynamic predictive displays to facilitate fact-based rational decision support. In purely technical fields, both aspects of simulation have been used successfully for the preparation of decision makers and as decision tools in real time.

Non-myopic decision making requires not only fact-based rational decisions but also a systemic approach to consider all aspects of problems for which one needs to intervene as well as consideration of the future implications of the current decisions or lack of them. Hence systems approach is paramount for the proper conception and management of complex cyber-physical-social problems. In addition to this, systems engineering has proven advantages for design, analysis, and control of complex systems as well as their simulations. Simulation-based systems engineering enhances already powerful systems engineering with experimentation and experience aspects of simulation.

Decision makers for cyber-physical-social systems can benefit from the power of simulation-based systems engineering. Furthermore, the reliance on AI (Artificial Intelligence) needs to be counterbalanced by an emphasis on IA (Intelligence Amplification). However, ethical commitments by humans as well as by computationally intelligent systems is paramount for the effectiveness of the activities.

Scientific, engineering, and technological advancements have been phenomenal. However, to understand the success or failure of systems, including cyber-physical-social systems, the advantages as well as disadvantages of human involvement need also to be scrutinized.

In Sect. 2, cyber-physical-social systems are discussed, including their ubiquity, importance, and severity of their security, as well as several types of sources of their failures.

In Sect. 3, Some possibilities to get prepared for unexpected conditions are discussed. These possibilities include simulation, simulation-based systems engineering, intelligence for fact-based rational decision-making, as well as ethics. In this section some possible dangers of AI are also pointed out.

Section 4 is dedicated to the importance of human involvement especially with negative (sometimes detrimental) implications and some remedial actions.

Last section is for the conclusions of the chapter which is followed by the list of references.

2 Cyber-Physical-Social Systems—Their Increasing Omnipresence and Importance

Cybernetics, founded in 1948 by Norbert Wiener and originally defined as: *"the scientific study of control and communication in the animal and the machine"* [98] is a transdisciplinary approach for exploring all aspects of regulatory systems.

> The early contributions of cybernetics were mainly technological and gave rise to feedback control devices, communication technology, automation of production processes and computers. Interest moved soon to numerous sciences involving man, applying cybernetics to processes of cognition, to such practical pursuits such as psychiatry, family therapy, the development of information and decision systems, management, government, and to efforts to understand complex forms of social organization including communication and computer networks. The full potential of cybernetics has not yet been realized in these applications. Finally, cybernetics is making inroads into philosophy. This started by providing a non-metaphysical teleology and continues by challenging epistemology and ethics with new ideas about limiting processes of the mind, responsibility and aesthetics. (Krippendorff) [74]

2.1 Importance and Severity of the Security of Cyber-Physical-Social Systems

"From a systemic perspective, we realize that connectedness exists even in the universe where objects are connected, to attract or to repel, by several types of forces such as gravitational forces, weak and strong forces, and electromagnetic forces" [72]. The term *"Cyber-Physical Systems denotes interconnected word we live in where this interconnectedness is increasing rapidly"* [60, 72]. In Cyber-Physical

Systems, the basic element which assures connection is knowledge. Characteristics of the connected world, some examples of the connected entities as well as evolution of the connected world were elaborated in a previous article [72]. *"The internet of everything (IoE), connecting people, organizations and smart things, promises to fundamentally change how we live, work and interact, and it may redefine a wide range of industry sectors"* [39]. *Though observation and perception, including introspection, are important in interconnected world, another possibility for these entities is to understand and be aware of what is going on in their neighbourhood and be aware of their own status to be able to decide for needed control functions"* [60].

Cyber-Physical Systems (CPS) consists of three types of components:

(1)   *Smart physical systems* which have, in addition to their physical abilities, abilities to process information.
(2)   *Cyber systems* as information processing components to assure communication and control; and
(3)   *Network infrastructures* which can be internet-based or non-internet-based, to provide the underlying structure for the communication and control to occur.

In **Cyber-Physical-Social systems** (CPSS), in addition to the components of CPS, human elements exist as:

(1)   *Passive recipients* of the positive and/or negative outcomes of the CPS.
(2)   *Active actors* to act as human-in-the-loop elements to provide inputs to mobile quasi-autonomous components of CPS; or
(3)   *High-level strategic decision makers*: (i) to activate or allow functioning of CPS or (ii) to unplug CPS, unless it is too late for Artificial Superintelligence (ASI) to take control from the humans.

Advancements on every aspects of CPS continues [50, 79, 92, 97].

Connected world may be limited in scope, such as connection of the appliances in a home, or connection of a frail person to a central station for monitoring health reasons, such as reporting vital signals and/or any non-recovered fall. It might restrict the capacity of a vehicle when it is well above the allowed speed limit and/or a log may be maintained for all infringement of proper driving. Connection can have a larger scope as it is the case in smart cities.

Military robots, including robot soldiers can be autonomous (i.e., human out of the loop) or semi-autonomous (i.e., human in the loop) remote-controlled mobile entities. Military army would consist of robotic soldiers and other robotic equipment. Their communications and several types of control are also example of cyber-physical systems [82].

Internet of Things (IoT) can be considered as Internet of Tangibles [1]. Two recent applications of Internet of Everything can be considered as Internet of Intangibles (IoI): A recent extension of the connected world is Internet of Behaviors (IoB) as listed at the top of the people-centric strategic trends for 2021 at Gartner's IT Symposium/Xpo conference in 2020 [21].

The following is part of the reported knowledge on Internet of Behaviors (IoB):

The IoB is about using data to change behaviors. ... According to Gartner, with an increase in technologies that gather what the analysts call 'digital dust' of daily life --data that spans the digital and physical worlds-- that information can be used to influence behaviors through feedback loops. ... According to Gartner, the Internet of Behaviors can gather, combine, and process data from many sources including:

- Commercial customer data
- Citizen data processed by public-sector and government agencies
- Social media
- Public domain deployments of facial recognition
- Location tracking
- The increasing sophistication of the technology that processes this data has enabled the IoB trend to grow.

Indeed, IoB does have ethical and societal implications depending on the goals and outcomes of individual uses. As it happens with every other technology available, the Internet of Behaviors can be used for good or for bad. With employees under the magnifier glass much of their privacy can be lost [21].

Internet of Senses (IoS) is yet another recent type of connected world [16, 22]. As anticipated even within next ten years, with the advancements of Brain-Machine Interfaces (BMI), thought-controlled communication may open new vistas in connected world. Digital communication of thoughts is part of the foreseen Internet of senses [22]. Full two-way communication over BMI which may also allow uploading to brains may be the beginning of Internet of Brains.

The scope of interconnected world of everything is still widening, A recent term, "digital ecosystems" is coined to represent this concept:

In the world of the Internet of Things (IoT), the rapid growth and exponential use of digital components leads to the emergence of intelligent environments namely "digital ecosystems" connected to the web and composed of multiple and independent entities such as individuals, organizations, services, software and applications sharing one or several missions and focusing on the interactions and inter-relationships among them [43].

Another extension of Internet of Things is the Internet of Eyes (IoEyes): "*Similar to the Internet of Things, the IoEyes is a network of cameras and visual sensors connected via the internet enabling the collection and exchange of visual data on a scale unimaginable before*" [52]. Another development of internet of senses is Internet of Ears [10].

With the increasing scope and functionalities of cyber-physical-social systems, their importance as well as our reliance on them are growing. This reliance that makes the relevant systems dangerously fragile necessitates to take proper countermeasures.

2.2 Sources of Failures in Cyber-Physical-Social Systems

Systems can fail unless countermeasures are carefully considered and taken in the life cycles of systems. Classical remedy is reliability engineering, a sub-discipline of systems engineering. The aim of reliability engineering is to build reliable systems with unreliable components. Reliability has the following two meanings:

(1)   *The quality of being trustworthy or of performing consistently well.*
(2)   *The degree to which the result of a measurement, calculation, or specification can be depended on to be accurate* [55].

A through analysis of what can go wrong and taking necessary precautions may help to prevent reliability disasters. Failure Avoidance was proposed to avoid failures in modeling and simulation studies [70]. Failure Avoidance covers errors in modelling, simulation software, behavior generation by simulation, rule-based systems, autonomous systems, agents with personality, emotions, and cultural background, externally or internally generated inputs, and in systems engineering.

Security of cyberspace is of paramount importance [27, 45, 46]. A recent book and some of its chapters cover reliability issues of CPS [28, 37, 38, 63, 78].

Loper elaborates on the *"Dimensions of Trust in Cyber Physical Systems."* She gives a detailed list of applications IoT in cities and asserts that the *"Cities may be the first to benefit from the IoT, but being surrounded by billions of sensors, devices, and machines has profound implications for security, trust, and privacy. The more technology a city uses, the more vulnerable it is, so the smartest cities face the highest risks"* [38, p. 410]. Increasing connectivity in urban environments will end up by having a large number of sensors and other equipment which will be used for autonomic decisions, hence their trust issues need to be elaborated and well understood. About the Internet of simulation things Loper asserts that *"Since simulation models can be used to control or give commands to sensors and actuators or provide faster-than-real-time prediction to systems, we need to enhance trust relationships when the simulation is part of the IoT system"* [38, pp. 420–422]. In the section on Internet of trusted simulations Loper points out a *"NIST report which identifies 17 technical concerns that negatively affect the ability to trust IoT products and services* [95]" [38, pp. 422–424].

Haque et al. elaborate on cloud-based simulation platform for quantifying cyber-physical systems resilience. In a section on CPS security threats, they elaborate on external threats, internal threats, technology threats, ICS (Industrial Control System) and ITS (Information Technology System) integration threats, and physical infrastructure security threats [28, pp. 358–360]. Under cloud security issues, they cover multi-tenancy (i.e., "the resource sharing characteristics of the cloud"), data security, software security, virtualization issues, access control, and malicious inside [28, pp. 360–362].

Lazarova-Molnar and Mohamed elaborate on the following aspects of the reliability of CPS: hardware reliability, software reliability, reliability related to human interaction, combined reliability of CPS, and CPS reliability approaches [37].

Ören gives a list of categories of reliability issues and their definitions. They include error, failure, malfunction, accident, defect, fault, flaw, glitch, blunder, bug, mistake, shortcoming, inaccuracy, misconception, misinterpretation, misunderstanding, and falsehood [64, p. 432]. A list of categories of fallacies that may create reliability issues as well as their definitions include fallacy, including formal fallacy, material fallacy, sophism, and verbal fallacy (amphibology, fallacy of composition, fallacy of division) [64, p. 433]. Reliability and failure avoidance in computation and simulation studies are covered in another section [64, pp. 433–434]. References for failure avoidance in three types of Artificial Intelligence [64, pp. 435–436] as well as failure avoidance in or due to simulation studies are also given [64, p. 436]. Finally, eight categories of aspects of sources of failures are listed and clarified: (1) ethics and value systems, (2) decision-making biases, (3) improper use of information: misinformation, disinformation, and mal-information, (4) attacks, (5) flaws, (6) accidents, and (7) natural disasters [64, pp. 436–440].

2.3 Some Additional References About Cyber Security

The Body of knowledge of Cyber Security is being developed [12]. The following statement gives an idea about the severity of cyber attacks: *"An increasing political, societal and economic concern, cyber attacks cost an estimated $400 (according to Lloyds) to global economies"* (CyBOK, about). Lists of MSC programs in cybersecurity are available. For example, 75 Master of Science degree programs in cyber security is given in Cybersecurity [11] and 94 MSc Programs in Cybersecurity are listed in Keystone [36]. Dictionaries covering cybersecurity terms are by Ayala [3], Global Knowledge [26], NICCS [51], NIST-CSD [53] and Radely [76]. Morgan and Morgan [48] provide a rich list of sources of cybersecurity glossaries. A recent document covers cybersecurity in the quantum era [96].

3 How to Get Prepared for Unexpected Conditions

Ignorance of threat cannot diminish the severity of the consequences. Hence, there is a need to be prepared. In the sequel, some possibilities offered by simulation, systems engineering, and Artificial Intelligence are elaborated.

3.1 Simulation

Experience and experimentation aspects of simulation can be useful to get prepared for unexpected conditions.

Simulated experience

As it is widely known as virtual, constructive, and live simulations, simulation provides three types of experience opportunities to develop/enhance three types of skills:

(1)   Motor skills by using virtual equipment (Virtual simulation, use of simulators)
(2)   Decision making skills by constructive simulation (all types of serious games, business games, war games, and peace games).
(3)   Operational skills by live simulations.

Simulated experiences have several advantages over real-life experiences.

Simulated experiments

Simulated experiment can be done independent of the real systems or simultaneously with the real-system's operation. Integrated use of simulation can be done to generate predictive displays or for on-line diagnosis abilities.

Predictive displays. During the operation of a real system, a simulator of the system gets the time-varying information about the environment of the system, directly through sensors and A/D convertors, and gets the values of the control variables, from the system through transducers, and displays the predicted state (trajectory) of the system. By using a predictive display, decision maker/operator can base his/her decision(s) on system characteristics (as represented in the model) and facts as generated by the simulator instead of using an undocumented mental model.

On-line diagnosis abilities can be provided by comparing the outputs of the real system and the simulator working under same conditions. A discrepancy may indicate a misfunction of the system.

Synergies of simulation with several disciplines offer powerful paradigms [61, 71]. Furthermore, synergy of simulation and nature inspired modelling and nature-inspired computation [65] as well as synergy of simulation and computationally aware systems [33], [60] open powerful new vistas.

3.2   Systems Engineering and Simulation-Based Systems Engineering

The Guide to the Systems Engineering Body of Knowledge clarifies the fact that *"The definitions of systems engineering has evolved over time."* Hence the Guide includes three definitions of Systems Engineering as follows:

(1)   Interdisciplinary approach governing the total technical and managerial effort required to transform a set of customer needs, expectations, and constraints into a solution and to support that solution throughout its life. (ISO/IEC/IEEE 2010)

(2)   An interdisciplinary approach and means to enable the realization of successful systems. It focuses on defining customer needs and required functionality early in the development cycle, documenting requirements, then proceeding with design synthesis and system validation while considering the complete problem:

  • Operations
  • Performance
  • Test

- Manufacturing
- Cost & Schedule
- Training & Support
- Disposal

Systems engineering integrates all the disciplines and specialty groups into a team effort forming a structured development process that proceeds from concept to production to operation. Systems engineering considers both the business and the technical needs of all customers with the goal of providing a quality product that meets the user needs. (INCOSE 2012)

(3) A transdisiplinary and integrative approach to enable the successful realization, use, and retirement of engineered systems, using systems principles and concepts, and scientific, technological, and management methods.

We use the terms "engineering" and "engineered" in their widest sense: "the action of working artfully to bring something about". "Engineered systems" may be composed of any or all of people, products, services, information, processes, and natural elements. (INCOSE Fellows 2019) [80].

As it can be seen from its definitions, Systems Engineering is based on mathematical system theories and is the most comprehensible approach for engineered systems. Furthermore, systems engineering is used successfully for a large group of areas. A recent article, adopted from [71], gives a list of 64 types of systems engineering by application areas and a list of domain-independent 20 types of systems engineering [65, pp. 159–161]. The first table also includes cyber-physical systems engineering, cyber-social-physical systems engineering, and cyber-social systems engineering. *"Modeling and Simulation Framework for Systems Engineering"* is well established [13, 25, 90, 104].

Due to mutually beneficial synergies of simulation with many disciplines, benefits of simulation-based disciplines are well established [47, 66, 69]. Simulation-based systems engineering provides experience and experimentation possibilities of simulation to systems engineering. Hence, simulation-based systems engineering is most comprehensive engineering discipline to tackle cyber-physical-social problems.

3.3 Artificial Intelligence and Fact-Based Rational Decision-Making

Artificial intelligence (AI), the ability of a digital computer or computer-controlled robot to perform tasks commonly associated with intelligent beings. The term is frequently applied to the project of developing systems endowed with the intellectual processes characteristic of humans, such as the ability to reason, discover meaning, generalize, or learn from past experience. Since the development of the digital computer in the 1940s, it has been demonstrated that computers can be programmed to carry out very complex tasks—as, for example, discovering proofs for mathematical theorems or playing chess—with great proficiency. Still, despite continuing advances in computer processing speed and memory capacity, there are as yet no programs that can match human flexibility over wider domains or in tasks requiring much everyday knowledge. On the other hand, some programs have attained the performance levels of human experts and professionals in performing certain specific tasks, so that artificial intelligence in this limited sense is found in applications as diverse as medical diagnosis, computer search engines, and voice or handwriting recognition [9].

Some of the abilities of computers, such as speed, replicability, and memory of learned knowledge, are far superior compared to human abilities. For example, human learning is individual and is very slow; however, the product of computational learning can be shared among a large number of computational systems. As a futuristic application, when a robot soldier or any robot with cognitive abilities becomes inoperative, its memory can be transferred to other robot (soldiers); hence its accumulated knowledge based on lived and learned experience may be shared and may be recuperated. **A book on** Implications of Artificial Intelligence for Cybersecurity was **recently** published by [49]. As part of advanced Artificial Intelligence, several types of computational understanding systems are being developed [67, 68]. Computationally aware systems are becoming important research and advanced application areas in general [33, 60] as well as for cyber-physical systems [60].

Especially fact-based rational decision making as opposed to human decision making based on truth value questionable beliefs or opinions is one of the advantages of Artificial Intelligence. Other advantages of Artificial Intelligence include self-improving software such as Amazon's affinity analysis [4, p. 73]. Ethical behavior is also a prerequisite for Artificial Intelligence systems [14, 17, 102].

3.3.1 Possible Dangers of Advance Artificial Intelligence

We should not fear from intelligence alone; but from entities (not only machines, but especially humans) without ethical behavior when they hold power and exercise their power selfishly. In the case of machines, potential dangers may come from systems self-aware and self-improving autonomic systems which have several additional self-abilities. Potential dangers of advanced Artificial Intelligence are foreseen from autonomic systems which may change their initial goals. For example, Omohundro [4, 5, 83], and Bostrom and Yudkowsky [6] elaborate on this issue. Autonomic systems are advanced autonomous systems. Among other characteristics, autonomic systems are self-aware and self-improving AI systems. Additional features of autonomic systems are clarified by [87]. It is even argued that Computer-Based Systems Should be Autonomic [88]. A fundamental question is how to assure that the original goal of an AI system will not be changed?

In studies pointing out dangers of advanced AI one point has not yet been raised. Definitions of AI often stress the fact that AGI (Artificial General Intelligence) *"is an emerging field aiming at the building of 'thinking machines', that is, general-purpose systems <u>with intelligence comparable to that of the human mind</u>(emphasis added)"* [34]. However, it is a fact that some humans lie, don't act ethically and/or have many biases. In developing AGI systems, it would be highly desirable to assure that they will not have human-like flaws, namely, they will not lie, will act ethically and/or will not exhibit human biases.

4   Impact of Human Involvement

Most of the technological developments advanced the quality of civilized life. Due to our reverence to advanced technology, often solutions to technical problems are also sought in technical realms. However, without neglecting this possibility, it might be helpful to realize that some problems may also be caused by human involvements. In cybersecurity, the organizers of a recent conference titled "Understanding the Human Side of Cybersecurity" asserted:

> The use of personal and Internet of Things (IoT) devices, wearable technologies, cloud computing and wireless networks all create opportunities for malicious actors to gain access to and exploit our personal and professional information. The Government of Canada is investing more than $1 billion over 10 years to defend Canadians against cyber threats, but the greatest vector for cyber attack isn't a piece of technology—it's us [23].

Indeed, as part of **human fallibility** [94], several types of human biases exist to hinder fact-based rational decision making. In the sequel, problems due to human involvements are outlined in two categories, namely problems due to human biases and problems due to human malevolent behavior.

4.1   Human Biases

4.1.1   Dysrationalia

"*The inability to think and behave rationally despite having adequate intelligence*" is an important human handicap and should be taken into account to avoid disasters in systems [86].

4.1.2   Cognitive Biases

"*Cognitive biases describe the irrational errors of human decision making and they are a crucial part of understanding behavioral economics*" [32]. The following 49 important and interesting types of cognitive biases listed and explained in [32] are worth perusing: Affect heuristics, anchoring, availability heuristics, bounded rationality, certainty effect, choice overload, cognitive dissonance, commitment, confirmation bias, decision fatigue, decoy effect, Dunning-Kruger effect, time discounting/present bias, diversification bias, ego depletion, elimination-by-aspects, hot–cold empathy gap, endowment effect, fear of missing out (FOMO), framing effect, gambler's fallacy (Monte Carlo fallacy), habit, halo effect, hedonic adaptation, herd behavior, hindsight bias (knew-it-all-along effect), IKEA effect, less-is-better effect, licencing effect, loss aversion, mental accounting, naïve diversification, optimism bias, overconfidence effect, over justification effect, pain of paying, partitioning, peak-end rule, priming, procrastination, projection bias, ratio bias, reciprocity, regret aversion, representativeness heuristic, scarcity, social proof, sunk cost fallacy, and zero price effect.

4.1.3 Cultural Biases

Cultures differ and the differences have consequences. Cultural differences and biases across nations are well documented by [30]. Geert Hofstede's website has additional valuable information [31]. Hofstede's cultural dimensions theory is clarified at [101].

4.2 Human Malevolent Behavior

Human malevolent behavior encompasses actions of hackers, Machiavellianism, and truth decay.

4.2.1 Hackers

Hackers are humans (at the current state of sophistication of the computational intelligence), and they can create profoundly serious problems to most sophisticated technical systems including Cyber-Physical-Social systems. In Sect. 3.2 it was mentioned that *"cyber attacks cost an estimated $400 billion (according to Lloyds) to global economies"* (CyBOK, about). Even *"Microsoft Confirms Hackers Accessed Its Source Code"* [42].

4.2.2 Machiavellianism

There is another well known aspect of human involvement that may create societal problems. It is based on the book titled "The Prince" written in early sixteenth century where Machiavelli analyses power of head of states [81]. Even though the foundation was developed for the head of states, Machiavellianism is also applicable to workplace and other human relationships. Machiavellianism is defined in the Dictionary of Psychology of American Psychological Association as follows:

> n. a personality trait marked by a calculating attitude toward human relationships and a belief that ends justify means, however ruthless. A **Machiavellian** is one who views other people more or less as objects to be manipulated in pursuit of his or her goals, if necessary, through deliberate deception. [Niccolò **Machiavelli**, who argued that an effective ruler must be prepared to act in this way] [2].

Combined with narcissism and psychopathy, Machiavellianism is one of the traits of dark triad personality [91]. A recent article documents *"The Dark Triad and Insider*

Threats in Cyber Security" [40]. The authors had already elaborated on "*The Dark Side of the Insider: Detecting the Insider Threat through Examination of Dark Triad Personality Traits*" [41]. The insider threat is one of the serious threats to cyber security.

4.2.3 Truth Decay

In an in-dept study of RAND corporation, truth decay is clarified as follows: "*a central component of Truth Decay is people's inability or unwillingness to distinguish between and assign different values to different types of information (e.g., facts versus opinion, disinformation versus anecdote)*" [35]. Four trends that characterize Truth decay are reported as:

1. Increasing disagreement about facts and analytical interpretations of facts and data
2. A blurring of the line between opinion and fact
3. The increasing relative volume and resulting influence of opinion and personal experience over fact
4. Declining trust in formerly respected sources of facts [35].

Four terms with negative connotations to "information" exist. They are, "misinformation," "disinformation," "mal-information," and "myth-information." Their definitions follow:

Information ("first used in fourteenth century in the quoted meaning") "*knowledge obtained from investigation, study, or instruction; Intelligence, News; Facts, data*" (More clarification at Merriam-Webster-information) [44].

Misinformation ("first used in 1605") "*incorrect or misleading information*" [44].

Disinformation ("first used in 1939") "*false information deliberately and often covertly spread (as by the planting of rumors) in order to influence public opinion or obscure the truth*" [44].

Mal-information "*information, that is based on reality, used to inflict harm on a person, social group, organisation or country*" [93]

Myth-information (1960s) [54]: "*North American-Information which is widely held to be true, but which is in fact flawed or unsubstantiated; common knowledge based on hearsay rather than fact.*"

In addition to above, early twenty-first century saw additional terms such as fake news, alternative fact and post truth.

"Alternative facts" is a very recent addition to the literature [99].

Post Truth was selected as the word of the year in 2016 by Oxford Dictionary [73].

4.3   Remedial Actions

Durant and Durant define civilization as follows:

> Civilization is social order promoting cultural creation. Four elements constitute it: economic provision, political organization, <u>moral traditions</u> (emphasis added), and the pursuit of knowledge and the arts. It begins where chaos and insecurity end. For when fear is overcome, curiosity and constructiveness are free, and man passes by natural impulse towards the understanding and embellishment of life [15, p. 1].

Survival of humanity requires a concerted effort. Luckily, the aim of some institutions is or includes survival of human civilizations.

4.3.1   UN Secretary General's

High-level Panel on Digital Cooperation is very significant. The forward of their report includes: *"No one knows how technology will evolve, but we do know that our path forward must be built through cooperation and illuminated by shared human values. We hope this Report will contribute to improved understanding of the opportunities and challenges ahead, so that together we can shape a more inclusive and sustainable future for all."* [92]. An interpretation of the report includes: *"The panel's work comes as the world collectively wakes up to the pressing need for a new approach to how we handle the urgent questions that technologies present each day. The very real economic, political, and social costs of inaction are all too clear. The past two years have witnessed a proliferation of efforts to develop new rules, principles, and cooperation mechanisms around new technologies. But we face a moment where, while appreciating the risk of inaction, a multitude of uncoordinated actions still fail to address the most pressing risks"* [79].

4.3.2   FHI (Future of Humanity Institute)

Future of Humanity Institute (FHI)'s *"stated objective is to focus research where it can make the greatest positive difference for humanity in the long term"* [19, 20, 100]. **GCRI** (Global Catastrophic Risk Institute) *"is a think tank that analyzes risks to the survival of human civilization. Their mission is to develop the best ways to confront humanity's gravest threats"* [24].

4.3.3   RAND Corporation's View

RAND Corporation's efforts to countering Truth Decay is an important positive development [35]. *"The four main drivers of truth Decay are reported as:*

(1)   *cognitive processing and cognitive biases;*

(2)   *changes in the information system, including the rise of social media and a transformation of the media industry;*

(3)   *competing demands on the U.S. educational system that have prevented school curricula from keeping pace with the challenges of the new information system; and*

(4)   *polarization, both political and sociodemographic"* [35, p. 79].

After outlining main drives of Truth Decay, their assertion is: *"RAND's research agenda addresses these issues and much more. But research and analysis alone cannot solve the complex problem of Truth Decay. Policymakers, media companies, and individuals must also act on the basis of this research"* [77].

4.3.4 A Personal View

To solve problems, we must first change our attitude (that created the problems at the first place). If even super intelligence would be developed to solve human problems, some human leaders may simply "unplug" them. Hence, human development is paramount. As pointed out in Chap. 14 on Discovery of Ignorance of his book Sapiens—A Brief History of Humankind, Harari [29, p. 247] points out that those Homo sapiens who were clever enough to realize that they did not know, i.e., they were ignoramus, started to learn and discover. Indeed, inventions started from the ancient times [18] and continued [103] even with an accelerated rate [56, 84, 85, 89]. However, the phenomenal rate of scientific and technological advancements could not be matched by human development. Comparison of the current *truth decay* in some "advanced" countries [35] and *post truth* [73] with the Age of Enlightenment [7] and The Age of Reason of humanity [15] during seventeenth and eighteenth centuries may be a testimony that human values have not been advancing.

To assure sustainment of high-level civilisations, civility and ethical behavior need to be promoted. It might be beneficial to create a culture of empathy and ethical behavior among humans. For example, it might be useful to promote success metrics which are not limited to benefits only; especially success metrics limited to financial benefits only may be detrimental to civilized behavior. If in a society, the only success metric is benefit (especially financial benefit), then risk of "buying" persons or their votes may be extremely high. Hence, success metrics should have multi-dimensional values but not dominated by benefits only. This may take a long time but may decrease polarizations in human societies. Furthermore, in societies where empathy and ethical behavior are the norms, decision makers having different (especially Machiavellian) norms may not be followed by large masses and may loose their powers.

In advanced communication between humans to—computer controlled—devices or machines, a possibility is brain-machine interfaces (BMI). BMI can be used for communication with and control of personal equipment such as prosthetic devices as well as thought-controlled activation of many types of devices. Any flaw in BMI—whether original or hacker-induced—may cause malfunctions and need to be under control.

Ethics and Virtue

Intelligence—natural or computational—can help solving problems. However, may also create problems and even disasters, if devoid of ethics. This is similar to the fact that it is impossible to find the correct direction with a faulty compass. Both natural and artificial types of intelligence need to be guided by ethics. Ethical behavior for humans and machines is highly desirable for sustainable civilized societies [59]. Ethics as a branch of philosophy is paramount for many disciplines and for everyday activities. In the rationale for a code of ethics for simulationists [58]. it was argued that the respect for the rights of others is the essence of ethical conduct [59]. In spite of "truth decay," publication on virtue ethics and professional journalism is very comforting [75].

Virtue is *"conformity to a standard of right as well as a particular moral excellence"* [44]. Center for Ethical Leadership of Notre Dame Deloitte's view is enlightening: *"What Most Leaders Miss About the Value of Virtue - New research suggests that to achieve the highest levels of performance, leaders need both character and competence"* [8].

Our activities should serve a worthwhile **goal**. However, even (and especially) our goals need to be scrutinized [57]. Goal-directed adoption of proper goals and their evaluations would be helpful. Peter Drucker's (1909–2005) aphorism *"Efficiency is doing things right; effectiveness is doing the right things"* is a useful guide. We may aim to be efficient while being effective. Efficiency without taking into account effectiveness may be most often counterproductive.

5 Conclusion

Based on the immediacy of the need, two types of solutions exist for problems. They are short-term and long-term solutions. For example, in an operating room, a medical doctor cannot wait for long-term development of advanced and most effective devices or medicines. She or he must use the available means. However, one should not forget that the available knowledge, devices, and medicines have been there based on the accumulation of efforts, including research and development, some of which might have been initiated long time ago.

Similarly, some societal problems may require immediate interventions, including at extreme cases intervention of law enforcement agencies. The following example may elucidate the benefit of long-term solutions. In Canada part of the municipal tax income goes to education of youth. A person complained and claimed that since she does not have any school-age children, her municipal tax should be reduced. The reply was: "Would you prefer that this amount should be paid to police force to protect you from delinquent youth?".

Scientific and technological developments have been phenomenal and most of them helped to raise quality of life. Hope we can also nurture empathy and respect to the rights of others, as the essence of ethical behavior. Respectful treatment of others, animate, including humans and animals, and inanimate, including environment may also be desirable and very effective.

References

1. Angelini, L., et al.: Internet of tangible things (IoTT): challenges and opportunities for tangible interaction with IoT. Informatics **5**(7), 1–34 (2018)
2. APA DoP−Machiavellianism: American Psychological Association, Dictionary of Psychology−Machiavellianism. https://dictionary.apa.org/machiavellianism. Accessed 11 Dec 2020
3. Ayala, L.: Cybersecurity Lexicon. Apress, Springer Science, New York, NY (2016)
4. Barrat, J.: Artificial Intelligence and the End of the Human Era—Our Final Invention. Thomas Dunne Books, St. Martin's Press, New York, NY (2013)
5. Bostrom, N.: Superintelligence—Paths, Dangers, Strategies. Oxford University Press, Oxford, UK (2014)
6. Bostrom, N., Yudkowsky, E.: The ethics of artificial intelligence. In: Frankish, K., Ramsey, W. (eds.) Chapter 15 of Cambridge Handbook of Artificial Intelligence, pp. 316–334. Cambridge University Press, New York. https://intelligence.org/files/EthicsofAI.pdf. Accessed 19 Dec 2020
7. Bristow, W.: Enlightenment, Stanford Encyclopedia of Philosophy (2017). https://plato.stanford.edu/entries/enlightenment/. Accessed 15 Dec 2020
8. Center for Ethical Leadership: What Most Leaders Miss About the Value of Virtue—New research suggests that to achieve the highest levels of performance, leaders need both character and competence. University of Notre Dame. https://ethicalleadership.nd.edu/news/what-most-leaders-miss-about-the-value-of-virtue/. Accessed 19 Dec 2020
9. Copeland, B.J.: Artificial Intelligence, Britannica (2020). https://www.britannica.com/technology/artificial-intelligence/The-Turing-test. Accessed 28 Dec 2020
10. CWU: What's next for smart homes: An 'Internet of Ears?' Case western University (2018). https://techxplore.com/news/2018-11-smart-homes-internet-ears.html?utm_source=TrendMD&utm_medium=cpc&utm_campaign=TechXplore.com_TrendMD_1. Accessed 24 Dec 2020
11. Cybersecurity Guide: A planning guide for a master's degree in cybersecurity (2020). https://cybersecurityguide.org/programs/masters-in-cybersecurity/. Accessed 10 Dec 2020
12. CyBOK: The Cyber Security Body of Knowledge (2019). https://www.cybok.org/. Accessed 10 Dec 2020
13. Diallo, S., Tolk, A., Gore, R., Padilla, J.: Modeling and simulation framework for systems engineering. In: Gianni., D., D'Ambrogio, A., Tolk, A. (eds.) Modeling and Simulation-Based Systems Engineering Handbook, pp. 377–400. CRC Press, Taylor and Francis Group, Boca Raton, FL (2015)
14. Dignum, V.: Ethics in artificial intelligence: introduction to the special issue. Ethics Inf. Technol. **20**, 1–3 (2018). https://link.springer.com/article/10.1007%2Fs10676-018-9450-z. Accessed 19 Dec 2020
15. Durant, W., Durant, A.: The Story of Civilization VII—The Age of Reason Begins (A History of European Civilization in the Period of Shakespeare, Bacon, Montaign, Rembrant, Galileo and Descartes: 1558–1648). Simon and Schuster, New York, NY (1961)
16. Ericsson CL: Ericsson ConsumerLab: Ten Hot Consumer Trends 2030—The Internet of Senses. 10 Hot Consumer Trends 2030 (2020). https://www.ericsson.com/4ae16a/ass

ets/local/reports-papers/consumerlab/infographs/10hct_2030_infographic.pdf. Accessed on 9 Dec 2020

17. Evenstad, L.: Ethics Key to AI Development, says Lords Committee—House of Lords Select Committee calls for government to draw up an ethical code of conduct, which organisations developing AI can sign up to. ComputerWeekly.com (2018). https://www.computerweekly.com/news/252438994/Ethics-key-to-AI-development-says-Lords-committee. Accessed 19 Dec 2020

18. Fagan, B.M. (ed.): The Seventy Great Inventions of the Ancient World. Thames & Hudson, London, England (2004)

19. FHI—Future of Humanity Institute, University of Oxford. https://www.fhi.ox.ac.uk/. Accessed 22 Dec 2020

20. FHI Research Areas. https://www.fhi.ox.ac.uk/research/research-areas/. Accessed 22 Dec 2020

21. Fourtané, S.: Top Strategic Technology Trends for 2021, Interesting Engineering, 19 Nov 2020

22. Fourtané, S.: Emerging Consumer Trends Evolving Toward 2030: The Internet of Senses, Interesting Engineering, 9 Dec 2020

23. GC: Understanding Human Side of Cybersecurity, 8–9 Oct 2020, Virtual Conference, Government of Canada (2020). https://www.csps-efpc.gc.ca/events/human-cybersecurity/index-eng.aspx. Accessed 19 Dec 2020

24. GCRI—Global Catastrophic Risk Institute. http://gcrinstitute.org/. Accessed 19 Dec 2020

25. Gianni, D., D'Ambrogio, A., Tolk, A.: Modeling and Simulation-Based Systems Engineering Handbook. CRC Press, Taylor and Francis Group, Boca Raton, Florida (2015)

26. Global Knowledge—Cybersecurity Glossary of Terms. https://www.globalknowledge.com/us-en/topics/cybersecurity/glossary-of-terms/. Accessed 10 Dec 2020

27. Goodman, S.E., Lin, H.S.: Toward a Safer and More Secure Cyberspace. NAP (The National Academies Press), Washington, DC (2007). https://www.nap.edu/catalog/11925/toward-a-safer-and-more-secure-cyberspace. Accessed 19 Dec 2020

28. Haque, Md.A., Gochhayat, S.P., Shetty, S., Krishnappa, B.: Cloud-based simulation platform for quantifying cyber-physical systems resilience. In: Risco-Martin, J.-L., S. Mittal, T. Ören (eds.) Simulation for Cyber-Physical Systems Engineering: A Cloud-based Context, pp. 349–384. Springer (2020)

29. Harari, Y.N.:. Sapiens—A Brief History of Humankind. Signal—Penguin Random House Canada, Toronto, Ontario, Canada (2014)

30. Hofstede, G.: Culture's Consequences: Comparing Values, Behaviors. Sage Publications, Thousand Oaks, CA, Institutions and Organizations Across Nations (2001)

31. Hofstede-website: https://geerthofstede.com/landing-page/. Accessed 12 Dec 2020

32. HumanHow: The Ultimate List of Cognitive Biases: Why Humans Make Irrational Decisions (2020). http://humanhow.com/list-of-cognitive-biases-with-examples/. Accessed 12 Dec 2020

33. Irani, R., Monfard, P.S., Kazemifard, M., Ören, T.: Computationally aware systems: recommended features through a review of 188 definitions and its relevance to advanced simulation. Int. J. Model. Simul. Sci. Comput. (2021—In Press—Invited contribution)

34. JAGI – The Journal of AGI Society. https://content.sciendo.com/view/journals/jagi/jagi-overview.xml (Accessed 2020–12–31).

35. Kavanagh, J., Rich, M.D.: Truth Decay—An Initial Exploration of the Diminishing Role of Facts and Analysis in American Public Life. RAND Corporation, Santa Monica, CA (2018). https://www.rand.org/pubs/research_reports/RR2314.html. Accessed 14 Dec 2020

36. Keystone-Master: 94 MSc Programs in Cybersecurity. https://www.masterstudies.com/MSc/Cyber-Security/. Accessed 10 Dec 2020

37. Lazarova-Molnar, S., Mohamed, N.: Reliability analysis of cyber-physical systems. In: Risco-Martin, J.-L., S. Mittal, T. Ören (eds.) Simulation for Cyber-Physical Systems Engineering: A Cloud-based Context, pp. 385–405. Springer (2020)

38. Loper, M.L.: Dimensions of trust in cyber physical systems. In: Risco-Martin, J.-L., S. Mittal, T. Ören (eds.) Simulation for Cyber-Physical Systems Engineering: A Cloud-based Context, pp. 407–427. Springer (2020)

39. Langley, D.J., et al.: The internet of everything: smart things and their impact on business models. J. Bus. Res. **122**(January 2021), 853–863 (2021). https://www.sciencedirect.com/science/article/pii/S014829631930801X (Accessed 2020–12–31).

40. Maasberg, M., Van Slyke, C., Ellis, S., Beebe, N.: The dark triad and insider threats in cyber security. Commun. ACM **63**(12), 64–70 (2020). https://cacm.acm.org/magazines/2020/12/248799-the-dark-triad-and-insider-threats-in-cyber-security/fulltext. Accessed 31 Dec 2020

41. Maasberg, M., Warrn, J., Beebe, N.:. The dark side of the insider: detecting the insider threat through examination of dark triad personality traits. In: 2015 48th Hawaii International Conference on System Sciences, pp. 3518–3526 (2015). https://www.computer.org/csdl/proceedings-article/hicss/2015/7367d518/12OmNz2TCGs. Accessed 1 Jan 2021

42. McFadden, C.: Microsoft confirms hackers accessed its source code. Interesting Engineering, 1 Jan 2021. https://interestingengineering.com/microsoft-confirms-hackers-accessed-its-source-code?_source=newsletter&_campaign=Bz1yoKMzEY54v&_uid=4oeERrJWa0&_h=919ba87c882eb99a495edecc59557c1837f70ce6&utm_source=newsletter&utm_medium=mailing&utm_campaign=Newsletter-01-01-2021. Accessed 1 Jan 2020

43. MEDES'21—The 13th International ACM Conference on Management of Digital EcoSystems, 1–3 Nov 2021, Hammamet, Tunisia. http://medes.sigappfr.org/21/call-for-papers. Accessed 11 Dec 2020

44. Merriam-Webster: https://www.merriam-webster.com/dictionary/. Accessed 13 Dec 2020

45. Miller, M.: Senate passes bill to secure internet-connected devices against cyber vulnerabilities. The Hill (2020). https://thehill.com/policy/cybersecurity/526605-senate-passes-bill-to-secure-internet-connected-devices-against-cyber. Accessed 19 Dec 2020

46. Miller, M.J.: Gartner's Top Strategic Technology Trends for 2021 (2020). https://www.pcmag.com/news/gartners-top-strategic-technology-trends-for-2021

47. Mittal, S., Durak, U., Ören, T. (eds.): Guide to Simulation-Based Disciplines: Advancing our Computational Future. Springer (2017)

48. Morgan, S., Morgan, C.: The motherlist glossary of cybersecurity and cybercrime definitions. Cybercrime Magazine (2020). https://cybersecurityventures.com/cybersecurity-glossary/. Accessed 10 Dec 2020

49. NAP: National Academies of Sciences, Engineering, and Medicine. Implications of Artificial Intelligence for Cybersecurity: Proceedings of a Workshop. The National Academies Press, Washington, DC (2019). https://www.nap.edu/download/25488. Accessed 28 Dec 2020

50. NASEM: Information Technology Innovation: Resurgence, Confluence, and Continuing Impact. NAP (The national Academies press), Washington, DC (2020)

51. NICCS (National Initiative for Cybersecurity Careers and Studies): Cybersecurity Glossary (2020). https://niccs.cisa.gov/about-niccs/cybersecurity-glossary. Accessed 10 Dec 2020

52. Nisselson, E.: It's coming! The Internet of Eyes will allow objects to see. TNW (2016). https://thenextweb.com/insider/2016/05/13/the-internet-of-eyes-and-the-personification-of-everything-around-you/. Accessed 24 Dec 2020

53. NIST-CSD (Information Technology Laboratory—Computer Security Resource Center): Glossary. https://csrc.nist.gov/glossary. Accessed 10 Dec 2020

54. OED-myth-information: https://www.lexico.com/definition/myth-information. Accessed 13 Dec 2020

55. OED-reliability: https://www.lexico.com/definition/reliability. Accessed 27 Nov 2020

56. Olson, S.: The Endless Frontier—The Next 75 Years in Science. The National Academies Press, Washington, DC (2020) (Rapporteur). https://www.nap.edu/catalog/25990/the-endless-frontier-the-next-75-years-in-science. Accessed 22 Dec 2020

57. Ören, T.: Concepts and criteria to assess acceptability of simulation studies: a frame of reference. CACM **24**(4), 180–189 (1981)

58. Ören, T.I.: Rationale for a code of professional ethics for simulationists. In: Proceedings of the 2002 Summer Computer Simulation Conference, pp. 428–433 (2002)

59. Ören, T.I.: Ethics as a basis for sustainable civilized behavior for humans and software agents. Acta Systemica **2**(1), 1–5. Also published in the Proceedings of the InterSymp 2002—The 14th International Conference on Systems Research, Informatics and Cybernetics of the IIAS, July 29–August 3, Baden-Baden, Germany (2002—Invited Plenary Paper). https://www.researchgate.net/publication/249853648_Ethics_as_a_Basis_for_Sustainable_Civilized_Behavior_for_Humans_and_Software_Agents. Accessed 19 Dec 2020
60. Ören, T.: Simulation of cyber-physical systems of systems: some research areas—computational understanding, awareness and wisdom. J. Chin. Simul. Syst., issue 2 (2018)
61. Ören, T.: Powerful higher-order synergies of cybernetics, systems thinking, and agent-directed simulation for cyber-physical systems. Acta Syst. **18**(2) (July 2018), 1–5 (2018)
62. Ören, T.: Grand challenges in modeling and simulation: what M&S can do and what we should do for M&S? In: Keynote at CCCI 2020—IEEE International Conference on Communications, Computing, Cybersecurity, and Informatics, Virtual Conference, 3–5 Nov 2020. www.site.uottawa.ca/~oren/pres/2020-CCCI.ppsx. Accessed 19 Dec 2020
63. Ören, T.: Statement at the panel recent trends and challenges in communications, computing, cybersecurity and informatics. In: CCCI 2020—IEEE International Conference on Communications, Computing, Cybersecurity, and Informatics, Virtual Conference, 3–5 Nov 2020. www.site.uottawa.ca/~oren/pres/CCCI-panel.ppsx . Accessed 19 Dec 2020
64. Ören, T.: Ethical and other highly desirable requirements for cyber-physical systems engineering. In: Risco-Martin, J.-L., S. Mittal, T. Ören (eds.) Simulation for Cyber-Physical Systems Engineering: A Cloud-Based Context, pp. 429–445. Springer (2020)
65. Ören, T.: Agent-directed simulation and nature-inspired modeling for cyber-physical systems engineering. In: Risco-Martin, J.-L., S. Mittal, T. Ören (eds.) Simulation for Cyber-Physical Systems Engineering: A Cloud-based Context, pp. 143–166. Springer (2020)
66. Ören, T.: Shift of Paradigm from Model-based to Simulation-based. In: Obaidat, M.S., T. Ören, H. Szczerbicka, (eds.) Simulation and Modeling Methodologies, Technologies and Applications—9th International Conference, SIMULTECH 2019, Prague, Czech Republic, July 29–31, 2019, Revised Selected Papers. Springer (In the Series: Advances in Intelligent Systems and Computing), pp. 29–45 (2020—Invited Chapter)
67. Ören, T., Ghasem-Aghaee, N., Yilmaz, L.: An ontology-based dictionary of understanding as a basis for software agents with understanding abilities. In: Proceedings of the Spring Simulation Multiconference (SpringSim'07), March 25–29, pp. 19–27, Norfolk, VA (2007)
68. Ören, T., Kazemifard, M., Noori, F.: (2015-Feature article). Agents with four categories of understanding abilities and their role in two-stage (deep) emotional intelligence simulation. Int. J. Model. Simul. Sci. Comput. (IJMSSC) **6**(3) (September), 1–16 (2015)
69. Ören, T., Mittal, S., Durak, U.: A shift from model-based to simulation-based paradigm: timeliness and usefulness for many disciplines. Int. J. Comput. Softw. Eng. **3**(1) (2018—Invited Paper)
70. Ören, T., Yilmaz, L.: Failure avoidance in agent-directed simulation: beyond conventional V&V and QA. In: Yilmaz, L., Ören, T. (eds.) Agent-Directed Simulation and Systems Engineering. Systems Engineering Series, pp. 189–217. Wiley-Berlin, Germany (2009)
71. Ören, T., Yilmaz, L.: Synergies of simulation, agents, and systems engineering. Expert Syst. Appl. **39**(1), 81–88 (2012)
72. Ören, T., Yilmaz, L.: The Age of the Connected World of Intelligent Computational Entities: Reliability Issues including Ethics. Autonomy and Cooperation of Agents (Invited ebook chapter) Faria Nassiri Mofakham. Frontiers in Artificial Intelligence—Intelligent Computational Systems, Bentham Science Publishers **2017**, 184–213 (2017)
73. Oxford Dictionary—Word of the year 2016: Post-truth. https://languages.oup.com/word-of-the-year/2016/. Accessed 14 Dec 2020
74. Principia_Cybernetica_Web: Cybernetics. Web Dictionary of Cybernetics and Systems. http://pespmc1.vub.ac.be/ASC/CYBERNETICS.html. Accessed 21 Dec 2020
75. Quinn, A.: Virtue Ethics and Professional Journalism. Springer, Cham, Switzerland (2018)
76. Radley, A.: Cybersecurity: Lexicon (2020). https://scienceofcybersecurity.com/lexicon/. Accessed 10 Dec 2020

77. RAND: Countering Truth Decay—A RAND Initiative to Restore the Role of Facts and Analysis in Public Life. https://www.rand.org/research/projects/truth-decay.html. Accessed 13 Dec 2020
78. Risco-Martin, J.-L., Mittal, S., Ören, T. (eds.): Simulation for Cyber-Physical Systems Engineering: A Cloud-based Context. Springer (2020). https://www.springer.com/gp/book/978 3030519087. Accessed 27 Dec 2020
79. Roberts, M.: Welcome to the Age of Digital Interdependence. United Nations Foundation (2019). https://unfoundation.org/blog/post/welcome-to-the-age-of-digital-interdepende nce/. Accessed 24 Dec 2020
80. SEBoK: The Guide to the Systems Engineering Body of Knowledge, INCOSE-International Council on Systems Engineering (2013). https://www.sebokwiki.org/wiki/Systems_Enginee ring_(glossary). Accessed 27 Dec 2020
81. SEP-Machiavelli: Stanford Encyclopedia of Philosophy—Machiavelli. https://plato.stanford. edu/entries/machiavelli/1. Accessed1 Dec 2020
82. Sharre, P.: Army of None: Autonomous Weapons and the Future of War. W.W. Norton & Company, New York | London (2018)
83. Shulman, C.: Omohundro's "Basic AI Drives" and Catastrophic Risks. The Singularity Institute, San Francisco, CA (2010). https://intelligence.org/files/BasicAIDrives.pdf. Accessed 30 Dec 2020
84. Smithsonian: Timelines of Science—The Ultimate Visual Guide to the Discoveries that Shaped the World. DK Publishing, New York, NY (2013)
85. Smithsonian: Science—The Definitive Visual Guide. Revised and, 2nd edn. DK Publishing, New York, NY (2016)
86. Stanovich, K.F. (2015). Rational and Irrational Thought: The Thinking That IQ Tests Miss—Why smart people sometimes do dumb things. Scientific American Mind, 1 Jan 2015. https://www.scientificamerican.com/article/rational-and-irrational-thought-the-thi nking-that-iq-tests-miss/. Accessed 12 Dec 2020
87. Sterritt, R.:. Apoptotic computing: programmed death by default for computer-based systems. IEEE Computer, January 2011, pp. 37–43 (2011). https://pure.ulster.ac.uk/ws/portalfiles/por tal/11822513/r1ster4.pdf. Accessed 26 Dec 2020
88. Sterritt, R., Hinchey, M.G.: Why computer-based systems should be autonomic. In: Unknown Host Publication, pp. 406–412. IEEE Computer Society (2005)
89. TIME: 100 New Scientific Discoveries—Fascinating, Momentous and Mind-Expanding Breakthroughs). Special TIME edition (2017). https://www.amazon.ca/TIME-Scientific-Dis coveries-Mind-Expanding-Breakthroughs/dp/1683307356. Accessed 22 Dec 2020
90. Tolk, A., Glazner, C.G., Pitsko, R.: Simulation-based systems engineering. In: S. Mittal, U. Durak, T. Ören (Eds.), Guide to Simulation-Based Disciplines. Simulation Foundations, Methods and Applications, pp. 75–102. Springer, Cham, Switzerland (2017)
91. Travers, M.: How to Recognize Dark Triad Traits. Psychology Today, 9 Aug 2020 (2020). https://www.psychologytoday.com/ca/blog/social-instincts/202008/how-recognize-dark-triad-traits. Accessed 11 Dec 2020
92. UN-HLP DC.: The Age of Digital Interdependence. Report of the UN Secretary-General's High-level Panel on Digital Cooperation (2019). https://www.un.org/en/pdfs/DigitalCooperat ion-report-for%20web.pdf. Accessed 24 Dec 2020
93. UNESCO: Journalism, 'Fake News' & Disinformation: Handbook for Journalism Education and Training (2018). https://unesdoc.unesco.org/ark:/48223/pf0000265552. Accessed 14 Dec 2020
94. Veldkamp, C.L.S.: The human fallibility of scientists: dealing with error and bias in academic research. Ph.D. Thesis, Tilburg University, Tilburg, The Netherlands (2017)
95. Voas, J., Kuhn, R., Laplante, P., Applebaum, S.: Internet of things (IoT) trust concerns. NIST cybersecurity white paper, 17th October (2018). https://csrc.nist.gov/CSRC/media/Publicati ons/white-paper/2018/10/17/iot-trust-concerns/draft/documents/iot-trust-concerns-draft.pdf. Accessed 22 Dec 2020

96. Wallden, P., Kashefi, E.: Cyber security in the quantum era. CACM **62**(4), 120 (2019). https://cacm.acm.org/magazines/2019/4/235578-cyber-security-in-the-quantum-era/fulltext. Accessed 24 Dec 2020
97. WEF: World Economic Forum—Why we urgently need a Digital Geneva Convention (2017). https://www.weforum.org/agenda/2017/12/why-we-urgently-need-a-digital-geneva-convention. Accessed 24 Dec 2020
98. Wiener, N.: Cybernetics or Control and Communication in the Animal and the Machine, 2nd edn. The MIT Press, Cambridge, MA, USA (March 15 1965) (1948)
99. Wikipedia-AF: Alternative facts (2020). https://en.wikipedia.org/wiki/Alternative_facts. Accessed 19 Dec 2020
100. Wikipedia-FHI: Future of Humanity Institute (2020). https://en.wikipedia.org/wiki/Future_of_Humanity_Institute. Accessed 23 Dec 2020
101. Wikipedia-HCDT: Hofstede's cultural dimensions theory (2020). https://en.wikipedia.org/wiki/Hofstede%27s_cultural_dimensions_theory. Accessed 12 Dec 2020
102. Wikipedia-ME: Machine Ethics (2020). https://en.wikipedia.org/wiki/Machine_ethics. Accessed 19 Dec 2020
103. Williams, T.I.: (Updated and Revised by William E. Schaaf, J. and Arianne E. Burnette) (1999). A History of Invention from Stone Axes to Silicon Chips. Little, Brown Company, UK (1987)
104. Yilmaz, L., Ören, T.I. (eds.): (All Chapters by Invited Contributors). Agent-Directed Simulation and Systems Engineering. Wiley Series in Systems Engineering and Management, Wiley-Berlin, Germany, 520 p (2009)

Machine Learning Methods for Enhanced Cyber Security Intrusion Detection System

M Satheesh Kumar, Jalel Ben-Othman, K G Srinivasagan, and P Umarani

Abstract In the ever-changing world of information security, networks had expanded in scale and complexity that integrates wide range of business functions, intrusion threats have increased in occurrence and intelligence. Network administrators and vendors are now moving beyond conventional Intrusion-Detection Systems (IDS), that only identify problems after they have occurred, to a novel, constructive approach termed Artificial Intelligence (AI) based intrusion detection system. Conventional network Intrusion Detection Systems and firewalls are usually preconfigured to spot malicious network attacks. Now-a-days attackers have become profounder and can try evading common detection rules. There are a few targeted areas where Artificial Intelligence will distribute the extreme evolution for Cybersecurity. To design a proactive defence mechanism, the system has to understand the intelligence of threats that are currently targeting the organization. The implementation of Machine Learning (ML) and threat intelligent-based solutions into blend can revolutionize the landscape in cyber security industry against any kinds of network attacks. Machine Learning is an application of AI that uses a system which is capable of learning from experience. Even in the era of extremely large amount of data and cybersecurity skill shortage, ML can aid in solving the most common tasks including regression, prediction, and classification. In this chapter, the origin and evolution of IDS has been described, followed by the classification of IDS. This

M. Satheesh Kumar (✉)
Department of Electronics and Communication Engineering, National Engineering College, Kovilpatti, Tamil Nadu 628503, India

J. Ben-Othman
L2S Lab CentraleSupélec, Université de Paris Sud et Université de Paris 13, Rue Juliot Curie, 91190 Gif sur Yvette, France
e-mail: jalel.benothman@l2s.centralesupelec.fr

K. G. Srinivasagan
Department of Information Technology, National Engineering College, Kovilpatti, Tamil Nadu 628503, India

P. Umarani
School of Information Technology and Engineering, Vellore Institute of Technology, Katpadi, Tamil Nadu 632014, India

© The Author(s), under exclusive license to Springer Nature Switzerland AG 2022
P. Nicopolitidis et al. (eds.), *Advances in Computing, Informatics, Networking and Cybersecurity*, Lecture Notes in Networks and Systems 289,
https://doi.org/10.1007/978-3-030-87049-2_27

733

chapter will provide a truly interactive learning experience to help and prepare the researchers for the challenges in traditional IDS and the contributions of ML in IDS. This comprehensive review briefs the prominent current works, and an outline of the datasets frequently used for evaluation purpose. Moreover, this chapter will also describe the Collaborative Intrusion Detection that enhances the Big Data Security. Finally, it presents the IDS research issues and challenges; and the skills that need to survive and thrive in today's threat-ridden and target-rich cyber environment.

Keywords Intrusion detection system · Machine learning · Network traffic · Dataset · Attributes

1 Introduction

Simple antivirus packages and firewall are not sufficient in protecting an organisation's network from cybercrimes. A well knowledge and idea of what is happening on the networks and systems is required to proactively control it [1]. A security system called Intrusion Detection System (IDS) is expected to serve as a layer of security for the infrastructure. The Intrusion Detection System (IDS) can be implemented as a software application or may be a hardware [2]. This IDS constantly monitor the network traffic or system activity [3] if any malicious activity or policy violations happens. IDS detect the activity and then sends warnings to supervisors who will take action afterwards. A network intrusion detection system is crucial for securing an organization's network [4]. IDS make it easy to detect and respond to malicious traffic and also alert the organization. The primary aim of the intrusion detection system is to confirm that the Information Technology (IT) personnel gets warned when an outbreak or network intrusion takes place. In the next section, the various categories of IDS are briefed.

1.1 Types of Intrusion Detection System (IDS)

IDS can be classified into different categories based on the operation, actions and technologies applied [5].

1.1.1 IDSs Based on the Operations

IDS can be categorized based on the procedures it performs. This has been grouped into the following types.

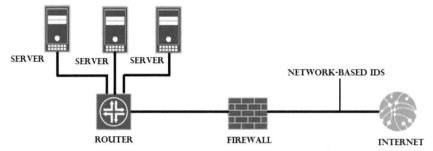

Fig. 1 General overview of implementing IDS

Network-Based IDS (NIDS)

NIDS is used to detect any malicious activity and monitor the traffic which is heading into a firewall. The main purpose of NIDS is to detect the possible attempts when it breaches the firewall security [6]. The NIDS can be placed along with a network segment or at the boundary of the network. As a means to track all network activity, it may be strategically located at single or multiple locations as illustrated in Fig. 1.

Host-Based IDS (HIDS)

The host Intrusion detection systems (HIDS) is used to monitor only the end point on which the software is installed, and it could not monitor the entire network. The most important role of a HIDS is to take snapshots of internals computing system files and analyse them for any changes. If any inappropriate activity is detected, then it has been notified to the administrator as an alert. A host-based IDS can continuously monitor system logs, important files and detect suspicious modifications that originate by insider attacker. It operates on all network computers that are connected to the organization's Internet/intranet. It can track malicious traffic coming from inside the organization., For e.g., when a malware [7] is attempting to propagate from a server to other hosts in the enterprise to other networks.

1.1.2 IDSs Based on the Actions

Based on their actions accomplished, IDS can also be categorized. Often known as an intrusion detection and prevention system, Active IDS generates warnings and records entries along with commands to change the network security configuration [8]. Passive IDS is another one that detects malicious operations and produces an alert or records, but still doesn't make any decisions.

1.1.3   IDSs Based on the Approaches

Based on the methodologies, the IDS utilize, it has been categorized into the following types.

Signature-Based IDS

This approach utilizes the existing set of known network threats or signature for comparing with incoming network traffic. In signature- based IDS [9], a set of pre-defined rules describing malicious activities has been incorporated earlier. It detects possible threats by looking for pre-define rules and makes alert when the pattern match. Signature-based IDS can easily detect known attack but if a new sort of attack is happened it is impossible to detect, for which no patterns will be available.

Anomaly-Based IDS

It is like signature-based method but used to detect unknown or new malware/suspicious activities. This approach [10] uses machine learning techniques for creating a model that simulates normal activity and then matches the behavior change with pre-existing model. In this detection method, machine learning is used to create a trustworthy activity in which what the normal network traffic looks like. Then it has been compared with new behaviour against this trust model. When the monitored data does not match with trust model then it enables alert notification.

1.1.4   Limitations of Conventional NIDS

1.   When there is a high false-alarm rate generated by noise, there is a possibility of real threats to go unnoticed.
2.   It requires regular updates in signature-based IDS to continue the detection against new malwares and threat actors.
3.   Protocol-based attacks will cause the IDS to malfunction while detection [11].
4.   Network based IDS is limited to detect only network-based anomalies, it is also a bottleneck for all incoming and outgoing traffic.
5.   It there is attack based on audit log modification, Host based IDS which works on audit log may suffer and lose integrity.

1.1.5   Transformation from Conventional IDS to Machine Learning IDS

Conventional IDS as a function remains vital to current enterprises and also it is not as a standalone solution, Network security is one of the crucial parts in information security of a Cyber Physical System (CPS) as illustrated in Fig. 2. NIDS were rule

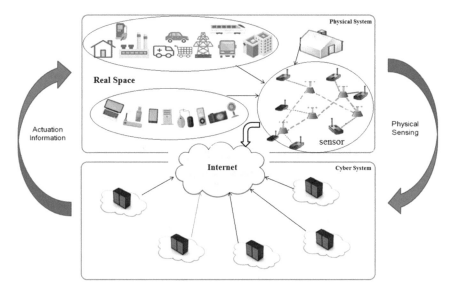

Fig. 2 Cyber Physical System

based method which can only detect specific events in the network. However, the nature of attacks keeps on changing which requires intelligent and adaptive systems to detect various types of attacks. Machine learning based technique have shown promise and is very successful in accomplishing tasks related to intrusion detection. Machine learning based IDS techniques [12] exhibit good accuracy and they perform extremely well against novel attack and also against security issues introduced by the attackers to disrupt the network operations. Thus, machine learning based IDS technique performs smartly and its need is very essential in the current era of zero-day attacks [13] and will help replacing the blacklist database that has to be updated regularly.

Thus, this section gives a brief introduction about the categorization of IDS. The next section describes the related works and research gaps. The section includes different types of machine learning algorithms employed by different researchers. The third section incorporates the evaluation of machine learning algorithms for intrusion detection systems. In this section, datasets have been reviewed elaborately and KDD cup dataset has been dissected and examined the attributes in detail. In Sect. 4, different evaluation metrics have been discussed with results comparison of different machine learning algorithms. In Sect. 5, closing of the manuscript has been deliberated.

2   Related Works and Research Gaps

In the previous decades, quantity of research was processed in IDS and were
tested against various dataset. Comparative studies [14] and reviews were conferred
on IDS and concerning of machine learning techniques were also discussed.
Many researchers presented their IDS techniques for different application which
is illustrated in Table 1.

Intrusion detection is conducted on a per packet basis also called as stateless or per
flow basis also called as stateful in which both signature-based and anomaly-based
detection involves. Most recent IDSs are stateful, as "context" is provided by the flow,
while this context is not provided by packet analysis. The analyst is responsible for
this to determine which technique is most suitable for their use. Based on the training
framework used, anomaly-based IDS can be divided into subcategories, which can

Table 1 IDS techniques and applications

Intrusion Techniques	Author	Source of data	Utilized for application	Dataset
Signature based IDS	Mohan et al. [15]	Host IDS and Network IDS	Cloud computing	Real Time-Generated Dataset
	Toumi et al. [16]	Network IDS	Cloud computing	
	Guerroumi et al. [17]	Host IDS	Internet of Things (IoT)	
	Modi et al. [18]	Host IDS	Cloud computing	
Anomaly based IDS	Harsh et al. [19]	Network IDS	Cloud Ccomputing	Real Time-Generated Dataset
	Ali Alheeti et al. [20]	Network IDS	Transport	
	Hong et al. [21]	Network IDS	Substation	
	Satam et al. [22]	Network IDS	Telecommunication	
	Haider et al. [23]	Host IDS	Cyber space	KDD98 and UNM
	Dipika et al. [24]	Network IDS	Information systems	KDD99 Cup
	Aissa et al. [25]	Network IDS	Computer systems	KDD99
Hybrid based IDS	Vasudeo et al. [26]	Host IDS and Network IDS	Computer systems	Real Time-Generated Dataset
	Banerjee et al. [27]	Network IDS	Computer systems	

be mathematical, knowledge-based or Machine Learning based [28]. Design-based blends the power to form a composite type with both signature and anomaly-based.

The approaches to validating networking models that refer to IDS are outlined [29] into four categories: mathematical models, simulation, emulation and actual experiments and as discussed each of these methods has its own pros and cons.

2.1 Machine Learning Based IDS

Machine learning based IDS is better than the conventional model IDS because signature-based detection is designed to detect the known attacks by using the signature of those attack alone. Machine Learning Based Intrusion Detection system [30] is software defined networks in which a flow-based IDS model is built to provide modular threat detection in this design. Machine Learning is implemented in flow-based anomaly detection to overcome the limitation of signature-based IDS. Machine Learning is a system of data processing that learns from data to draw solution and makes a decision based on the gathered knowledge. Traditional approaches in machine learning require a significant to understand and interpret data to contend with big data. IDS can overcome several problems, such as speed and computing time, and create an effective IDS, using big data and deep learning techniques. In order to minimize processing time and ensure efficient classification, the big data strategies [31] are used nowadays that deal with big data in IDS.

2.2 Machine Learning Techniques for NIDS

Machine Learning technique plays an important role in identifying the malicious intrusions by applying the classification technique. It is a process of giving computers with data or observation to learn and make patterns out of data. Through this data can be processed and make better predictions based on the example provided and primarily these things should happen without any human assistance. There are three types of major techniques under machine learning [32], they are supervised, unsupervised and semi-supervised classifier.

2.2.1 Supervised Machine Learning Algorithm

In this machine learning process, past datasets have been taught to forecast future ones using labeled instances. The machine will then earn decisions about the outcomes after adequate training. Supervised learning is based on valuable knowledge in labelled data and the most prominent role in supervised learning is classification which is often found in IDS also. Manually labelling results, however, is costly and time consuming. As a consequence, the absence of adequate classified data constitutes the primary bottleneck for supervised learning.

2.2.2 Unsupervised Machine Learning Algorithm

In this machine learning algorithm, the data for training is not labelled or classified. The role of the machine is to aggregate unlabeled data without any previous training data according to patterns, variations, and similarities. It derives useful attribute information from unlabelled data in this unsupervised learning, making it much simpler to collect the training data. The identification efficiency of unsupervised learning methods, however, is typically inferior to that of supervised methods of learning.

2.2.3 Semi-supervised Machine Learning Algorithms

A mixture of supervised and unsupervised methods of machine learning is semi-supervised machine learning. The machine computes from a dataset consisting of both labeled and unlabeled information. Here labelled data is blended with a lot of unlabeled data for training. Based on the labeled data, it easily generates models and assigns them to the unlabeled data, and thereafter utilizes such dataset to build more models. This significantly cuts the period of time it would take to hand-label a dataset by an analyst or data scientist, providing a raise to efficacy and efficiency.

2.3   The Classification Algorithms

A dataset is defined in supervised learning [29] as a finite set of real vectors with m characteristics,

$$A = \{\overline{a}_1, \overline{a}_2, \ldots, \overline{a}_m\}, \text{ where } \overline{a}_i \in \mathbb{R}^m, \tag{1}$$

$$B = \{b_1, b_2, \ldots, b_m\}, \text{ where } b_m \in (0, 1) \text{ and } b_i \in \mathbb{R}^+ \tag{2}$$

It attempts to extract the features of a relation or function $X: A \rightarrow B$ from a training set (A, B). As compared to other classifier, supervised learning is more accurate and reliable.

It has an input set A with x-length vectors in an unsupervised learning process, and the clustering function with m target clusters can be defined as

$$k_t = C(\overline{x,\theta}), \text{ where } k_t \in (0, 1, 2, \ldots, m). \tag{3}$$

2.3.1 Bayesian Classifier

Naive Bayes are the most intuitive and reliable classifiers that decide the possibility of an outcome for a given series of conditions using Bayes' theorem. The Bayes' theorem interprets two X and Y probabilistic cases. $P(X)$ and $P(Y)$ are the marginal probabilities that can be correlated with $P(X|Y)$ and $P(Y|X)$ conditional probabilities using the product law:

$$P(X \cap Y) = P(X|Y)P(Y) \tag{4}$$

$$P(Y \cap X) = P(Y|X)P(X) \tag{5}$$

The Bayes theorem can be derived as

$$P(X|Y) = \frac{P(Y|X)P(X)}{P(Y)} \tag{6}$$

Consider a dataset:

$$A = \{\overline{a}_1, \overline{a}_2, \ldots, \overline{a}_n\} \quad \text{where } \overline{a}_i \in \mathbb{R}^m \tag{7}$$

where, every other b can adhere to one of the various classes of P.

$$B = \{b_1, b_2, \ldots, b_n\} \quad \text{where } b_n \in \{0, 1, 2, \ldots P\} \tag{8}$$

Under conditional autonomy, considering Bayes' theorem, it can be written like:

$$P(b|a_1, a_2, \ldots, a_m) = \alpha P(b) \prod_i P(a_i|b) \tag{9}$$

Naive Bayes variants based on different probabilistic distributions: Bernoulli (binary distribution), multinomial (discrete distribution), and Gaussian (continuous distribution) can be described by its mean and variance.

Gaussian Naïve Bayes

When dealing with categorical features for which the probabilities could be superimposed using a Gaussian distribution,

$$P(a) = \frac{1}{\sqrt{(2\pi\sigma^2)}} e^{-\frac{(a-\mu)^2}{2\sigma^2}} \tag{10}$$

Conditional probabilities $P(a_i|c)$ are also Gaussian distributed; thus, the mean and variance can be calculated using the maximum likelihood method. It deems the Gaussian's property,

$$L(\mu; \sigma^2; a_i|b) = \log \prod_k P(a_i^{(k)}|b) = \sum_k \log P(a_i^{(k)}|b) \tag{11}$$

the k index indicates to the trials in our dataset and $P(a_i|b)$ is Gaussian. By means of reducing the inverse of this Eq. (8), the mean and variance for each Gaussian associated with $P(a_i|b)$, is generated and the model is hence trained.

Bernoulli Naïve Bayes

Only two values are assumed, if X is random variable and is Bernoulli-distributed. The two values can be 0 and 1 and their probability is:

$$P(X) = \begin{cases} a, X = 1 \\ b, X = 0 \end{cases} \text{where } b = 1 - a \text{ and } 0 < a < 1 \tag{12}$$

Multinomial Naïve Bayes

Here the multinomial distribution is beneficial to model the feature vectors where each value represents an attribute, for instance, the sum of events at a period of time or its proportional rate of recurrence. If the features extracted have m components and every one of them will presume with probability pk for different values of k, then:

$$P(A_1 = a_1 \cap A_2 = a_2 \cap .... \cap A_k) = \frac{m!}{\prod_i a_i!} \prod_i p_i^{a_i} \tag{13}$$

The conditional probabilities $P(a_i|b)$ are computed with a frequency count.

2.3.2 Support Vector Machines (SVM)

SVM is very powerful and can be used to classify data points by find a plane in n-number of features. It internally uses some kernel tricks which is smart transformations that used to resolve hard cases. For convenience, it can be taken it as a binary grouping for all other situations, the one-versus-all approach should be used automatically, and class labels can be set as -1 and 1:

$$B = \{b_1, b_2, \ldots, b_n\} \quad \text{where } b_n \in \{-1, 1\} \tag{14}$$

To find the separating hyperplane, this eqn. is used

$$\overline{w}^T \overline{a} + b = 0 \ where \ \overline{w} = \begin{pmatrix} w_1 \\ .. \\ w_m \end{pmatrix} and \ \overline{x} = \begin{pmatrix} a_1 \\ .. \\ a_m \end{pmatrix} \tag{15}$$

The classifier can be written as,

$$\tilde{b} = f(\overline{a}) = \mathrm{sgn}(\overline{w}^T \overline{a} + x) \tag{16}$$

The two classes are separated by two boundaries, that elements are "support vectors". It lies between two hyperplanes. To reduce the probability of misclassification, the distance between hyperplanes is maximized.

$$\begin{cases} \overline{w}^T \overline{a} + x = -1 \\ \overline{w}^T \overline{a} + x = 1 \end{cases} \tag{17}$$

SVM's high dimension problem was overcome by Kernel tricks. It can project in very large number of dimensions.

$$K(\overline{a}_i, \overline{a}_j) = \emptyset(\overline{a}_i)^T \emptyset(\overline{a}_j) \tag{18}$$

Radial Basis Function kernel is $K(\overline{a}_i, \overline{a}_j) = e^{-\gamma |\overline{a}_i - \overline{a}_j|^2}$ where γ determines the amplitude of the function.
Polynomial kernel function is

$$K(\overline{a}_i, \overline{a}_j) = (\gamma \overline{a}_i^T \cdot \overline{a}_j + r)^c \tag{19}$$

where the degree of the parameter determines the exponent c and the constant term r is coef0.
Sigmoid kernel function is

$$K(\overline{a}_i, \overline{a}_j) = \frac{1 - e^{-2(\gamma \overline{a}_i^T \cdot \overline{a}_j + r)}}{1 + e^{-2(\gamma \overline{a}_i^T \cdot \overline{a}_j + r)}} \tag{20}$$

in which the constant term r is defined using the coef0 parameter.

2.3.3 K-Nearest Neighbor (KNN)

K-Nearest Neighbor (KNN) is a classifier technique that widely used, where it uses the nonparametric to classify data. In this algorithm, the data are classified by voting from the nearest neighbors. Based on this voting, the objects in the image will be assigned to relevant classes. For malicious intrusion classification, the KNN algorithm is used to get better classification result to determine if the intrusions are normal or abnormal.

2.3.4 Decision Tree

Decision tree (DT) is the best possible model with little dataset preparation requirement. It is used to minimize the impurity of each node. Each node is associated with a label. A node is impure if it has misclassified data points. In a decision tree model, there are two ways to measure the impurity. One is the Gini index, and the other is the Entropy. DT is highly prone to overfitting.

Let us consider an input dataset as described in Eq. (1) where $a_n \in \{-1, 1\}$. Every vector is made up of n features, so each of them can create a node based on the feature, threshold called tuple.

$$\sigma = \prec i, t_n \succ \tag{21}$$

where i is the index of the feature, whereas the threshold that defines left and right divisions is the tn. The performance of the tree is determined by the best threshold. The aim is to minimize the impurity in the minimum average of differences so that the sample data and the classification outcome have a very short decision path. Total impurity measure can be defined by considering the two branches:

$$I(D, \sigma) = \frac{N_{left}}{N_D} I\left(D_{left}\right) + \frac{N_{right}}{N_D} I\left(D_{right}\right) \tag{22}$$

where D_{left} and D_{right} are the resulting subsets for the whole dataset D at the selected node, and the I are impurity measures.

The Gini impurity index can be defined as

$$I_{Gini}(b) = \sum_i p(a|b)(1 - p(a|b)) \tag{23}$$

and the cross-entropy measure is given by

$$I_{cross-entropy}(b) = -\sum_i p(a|b) \log p(a|b) \tag{24}$$

2.3.5 Random Forest

Random Forest is a powerful ML technique with ensemble learning. Overfitting problem is overcome by this algorithm. RF is basically a collection of DT models.

Each decision tree has random set of data. The average over all trees in the forest is computed as follows,

$$Importance\,(a_i) = \frac{1}{N_{Trees}} \sum_t \sum_k \frac{N_k}{N} \Delta I_{a_i} \tag{25}$$

2.3.6 Artificial Neural Network (ANN)

A neural network is composed of groups of neurons interconnected though synapses in biology. An artificial neural network (ANN) [33] is a guided framework which integrates an input vector with an output vector. All these operations are typically distinctive, and the cumulative vector function can easily be written as:

$$\overline{b} = f(\overline{a}) \tag{26}$$

where $\overline{a} = (a_1, a_2, ..., a_n)$ and $\overline{b} = (b_1, b_2, ..., b_m)$.

There are three layers: the input layer receives the input vectors; and the output one is responsible for producing the output; in-between input and output layers that conduct nonlinear transformations of the inputs inserted into the network are hidden layers. All the neurons belonging to the next layer are bound to each neuron and have two weight matrices, $W = (w_{ij})$ and $H = (h_{jk})$. Using this convention to refer the first index to the previous layer and the second to the current layer and the net input to each hidden neuron and the resulting output can be derived as,

$$\begin{cases} c_j^{Input} = w_{0j}a_0 + w_{1j}a_1 + \cdots + w_{nj}a_n = \sum_i w_{ij}a_i \\ c_j^{Output} = f_p^{Hidden}\left(r_j^{Input} + b_j^{Hidden}\right) \end{cases} \tag{27}$$

The network output is computed by,

$$\begin{cases} b_k^{Input} = h_{0k}c_0^{Output} + h_{1k}c_1^{Output} + \cdots + h_{xk}c_x^{Output} = \sum_j h_{jk}c_j^{Output} \\ c_k^{Output} = f_p^{Output}\left(b_k^{Input} + q_k^{Output}\right) \end{cases} \tag{28}$$

3 Evaluating ML Algorithms for IDS

Machine learning is a vast area of science that inherits concepts from many different areas, such as artificial intelligence. The focus of the field is learning, that is, acquiring skills or knowledge from experience. This most often involves synthesizing useful

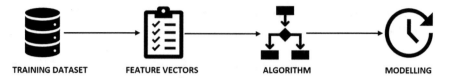

Fig. 3 Learning phase in Machine learning techniques

ideas from historical knowledge. The first thing is to build a collection of training data and the variables will be used in this training set. The algorithm predicts and defines on the basis of this training dataset as illustrated in the Fig. 3 of learning phase of Machine learning techniques.

3.1 IDSs and Datasets Review

Prominent datasets are summed up in this segment, and their shortcomings are illustrated. In addition, recent IDSs are checked, describing the algorithms used and the datasets to which the IDSs is analyzed. To measure the outcomes, researchers focused on benchmark datasets. Networks are continuously evolving, so the performance of IDS does not help, relying entirely on old datasets. This relentless reality of transition should be taken into account in the process of creating new datasets. The available datasets are summarized and classified in Table 2 based on the realm of which they belong, and attacks contained in each are also shown.

3.2 KDD Dataset Summary

In this section, the database of KDD Cup is investigated and reviewed. With reference to this, it will give an implied structure of any dataset to be processed for the researcher while implementing a machine learning technique. Data preprocessing is a method of filtering up the raw data by extracting from the dataset [1] the basic information and calculating the data consistency. The cleaned data set type is then fed into the model for machine learning. For machine learning implementation, the essential factor is to begin with getting reliable data. For example, for the dataset of KDD Cup 1999. This database has a standard dataset of 4,898,431 interventions with 41 characteristics that provide a wide spectrum of virtual activities in a military network system. Each record of a link consists of around 100 bytes.

Attacks fell predominantly into four key categories,

1. DOS—Denial of Service.
2. R2L—unauthorized remote access.
3. U2R—unauthorized root privileges' access
4. Probing—reconnaissance and other inquiries.

Table 2 Dataset, algorithm and attacks

Author	Dataset used	Algorithms used	Attacks detected
[34]	Real Time-Generated Dataset	Isolation Forest and Local Outlier Factor	Port Scanning, SYN Flooding, HTTP and SSH Brute Force
[35]	KDD-99	Naïve Bayes, Random Forest, Decision Tree and Ontology	Probing, R2L, U2R and DoS
[36]	NSL-KDD	Deep Learning and Neural Network	Probing, R2L, U2R and DoS
[37]	KDD-99, NSL-KDD, WSN-DS, CICIDS2017, UNSW-NB15 and Kyoto2006+	Support Vector Machine, Logistic Regression, Naïve Bayes, Random Forest, Decision Tree, K Nearest Neighbor	Probing, R2L, U2R DoS, Blackhole, Grayhole, Backdoors, Exploits, DDoS, Web-based, Brute force, and Botnet Scan
[38]	NSL-KDD, ISCX 2012 and Kyoto2006+	Support Vector Machine, Instance Based Learning, Multi-Layer Perceptron, Information Gain, and Principal Component Analysis	Probing, R2L, U2R and DoS
[39]	Real Time-Generated dataset using httperf web server performance tool	Mapping	SQL Injection and XSS
[40]	gureKddcup6percent	Support Vector Machine	Probing and R2L
[41]	Banking and Credit Card Dataset	Decision Tree	Scam
[42]	DARPA 2000	Markov Chain and K-means Clustering	Distributed Denial of Service
[43]	1998 DARPA	Support Vector Data Description	Probing and U2R
[44]	1999 DARPA	Radial Basis Function and Neural Network	Probing, R2L, U2R and DoS
[45]	Banking Transaction Dataset	Fuzzy Association Rules	Credit Card Scam
[46]	PIERS, Emergency Department Dataset and KDD-99	Bayesian Network Likelihood, Conditional Anomaly Detection and Anomaly Pattern Detection	Probing, R2L, U2R and DoS

The attributes grouped in the datasets which is collected using any connections implemented based on TCP/IP are described in Table 3.

These data have to be processed for machine learning algorithm. Some information can be easy to find but others have to found by experimenting and running tests [47]. Using all these features of dataset may cause computation cost and error rate of

Table 3 Dataset and protocol attributes

Attribute name	Narrative	Type
Period	The total duration of the link in seconds	Constant
ProtocolType	Protocol form, Example: UDP or TCP	Distinct
NetworkService	The destination network service, Example: telnet or http	Distinct
SourceBytes	Number of bytes of data from sender to receiver	Constant
DestinationBytes	Number of bytes of data from receiver to sender	Constant
FlagStatus	The connection's status, either normal or error	Distinct
Landing	'1' if the connection is from/to the identical host/port '0' if the connection is from/to the different host/port	Distinct
WrongFragment	Number of fragments that are 'wrong'	Constant
UrgencyStatus	Number of packets that are 'urgent'	Constant

Table 4 Dataset and Attack Attributes

Attribute name	Narrative	Type
HotIndicator	Number of indicators for 'hot'	Constant
FailedLogins	The number of unsuccessful login attempts	Constant
LoggedIn	'1' if signed in successfully; 0 otherwise	Distinct
Compromised	Number of conditions compromised	Constant
RootShell	'1' if the root shell has been obtained; 0 if not	Distinct
SUAttempted	'1' if attended by the 'Super User root' command; 0 otherwise	Distinct
RootAccess	Number of accesses to 'root'	Constant
FileCreations	Number of Operations for File Formation	Constant
Shells	Number of prompts for shells	Constant
AccesFiles	Number of operations on files for access control	Constant
OutboundCommands	In an ftp session, the number of outbound commands	Constant
HotLogin	'1' if the username belongs to the list 'hot'; 0 if not	Distinct
GuestLogin	'1' if the login is for 'guest'; 0 if not	Distinct

system can be increased. The introduction of expert suggested attributes which help to understand attacks [48] and their behaviors and described in Table 4.

3.3 Semi Supervised Machine Learning Based IDS

Machine learning approaches can be a supervised or unsupervised learning, and the use of these learning depends on the factors related to the composition and scale of the dataset and the use cases of the issue. Semi-supervised learning can be implemented to detect intrusion in network, since huge number of input data are usually generated in

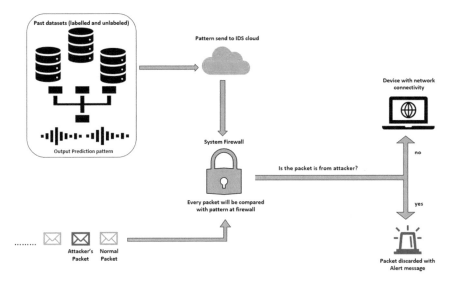

Fig. 4 Semi Supervised Machine Learning based IDS with labelled Dataset

the network. Some data can be labelled, but much of it is unlabeled, and it is necessary to use a combination of supervised and unsupervised techniques. If, unsupervised machine learning algorithms is used, the IDS will grasp the patterns of network and can account suspicious acts devoid of any labelled data but be able to also cause or trigger more false positive alarms. To reduce this, a supervised machine learning approach [9] is used with certain labelled datasets can be employed as described in Fig. 4. With this approach, difference among benign and malicious packets in the network can be identified, recognize attack variations, and can also handles the unknown attacks.

4   Evaluation Metrics

The basic performance metrics to assess the effectiveness of any intrusion detection system [48] are true positive, false positive, true negative, false negative rates and accuracy. Let the concerns be

M	Total number of malicious instances,
L	Total number of legitimate instances,
$L \to L$	Number of the legitimate instances classified as legitimate,
$L \to M$	Number of the legitimate instances misclassified as malicious,
$M \to M$	Number of the malicious instances classified as malicious,
$M \to L$	Number of the malicious instances misclassified as legitimate.

True Positive Rate (TPR) is the rate of malicious instances classified as malicious out of the total malicious instances. It can be termed as 'Detection Rate' and defined as follows,

$$TPR = \frac{M \to M}{M} \times 100 \tag{26}$$

False Positive Rate (FPR) is the rate of legitimate instances classified as malicious out of the total legitimate instances.

$$FPR = \frac{L \to M}{L} \times 100 \tag{27}$$

False Negative Rate (FNR) is the rate of malicious instances classified as legitimate out of the total malicious instances.

$$FNR = \frac{M \to L}{M} \times 100 \tag{28}$$

True Negative Rate (TNR) is the rate of legitimate instances classified as legitimate out of the total legitimate instances.

$$TNR = \frac{L \to L}{L} \times 100 \tag{29}$$

Accuracy (A) is the rate of malicious and legitimate instances which are identified correctly with respect to all the instances.

$$Accuracy = \frac{L \to L + M \to M}{L + M} \times 100 \tag{30}$$

The metrics are compared for different machine learning algorithms as portrayed in Table 5. These algorithms present excellent results as illustrated in Figure, it is also explained that these results are also owns the means of production of datasets, coverage of attacks and taxonomy of updated threats. Real-life modelling

Table 5 Result Comparison

Machine learning classifiers	Total malicious instances	Correctly classified instances	Incorrectly classified instances	Accuracy (%)
Random Forest	71,000	66,255	4745	92.94
Decision Tree	71,000	65,855	5145	93.11
SVM	71,000	65,452	5548	92.44
Naïve Bayes	71,000	64,750	6250	92.07
KNN	71,000	64,339	6661	91.41

and expandable databases, zero-day attacks treatment, using detailed measurement metrics, using network monitoring to produce real traffic and apply models to introduce practical attacks are suggested for the researchers for the future work proceedings (Figs. 5 and 6).

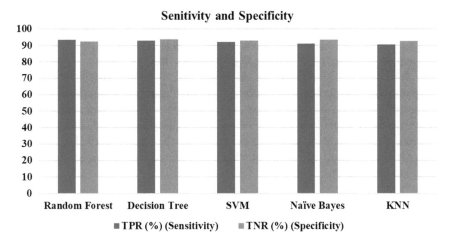

Fig. 5 TPR and TNR evaluation

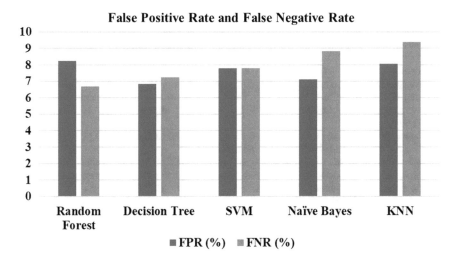

Fig. 6 FPR and FNR evaluation

5 Conclusion

To address ever-increasing cyber threats to the networks, all present IDS are migrating towards machine learning techniques. This mostly focuses on automating the intrusion detection process with high precision. These outcomes can be used by companies to combat against the root of threats, block zero-day attacks and refine their network. Another vital benefit is that firms will not have to charge for individual patented signatures to defend themselves against new threats. IDS using many Machine Learning Techniques were discussed in this chapter. Various datasets for assessing different algorithms were reviewed and detection against various attacks were also discussed. The performance of IDS depends on the methods utilized for detection with different metrics, and based on the specifications, the most appropriate solutions can be offered by the IDS. In the near future, if any enterprise struggling to implement these methods may risks in compromising their data or even malevolent their servers. So, the adoption of ML in IDS will effectively provide a best solution for detecting against new emerging cyber threats.

References

1. Mukherjee, B., Heberlein, L.T., Levitt, K.N.: Network intrusion detection. IEEE Netw. **8**(3), 26–41 (2002). https://doi.org/10.1109/65.283931
2. Chaturvedi, P.: A Systematic Literature Survey on IDS, pp. 671–676
3. Cheng, T.H., Lin, Y.D., Lai, Y.C., Lin, P.C.: Evasion techniques: sneaking through your intrusion detection/prevention systems. IEEE Commun. Surv. Tutorials **14**(4), 1011–1020 (2012). https://doi.org/10.1109/SURV.2011.092311.00082
4. Sommestad, T., Hunstad, A.: Intrusion detection and the role of the system administrator. Inf. Manag. Comput. Secur. **21**(1), 30–40 (2013). https://doi.org/10.1108/09685221311314400
5. Liao, H.J., Richard Lin, C.H., Lin, Y.C., Tung, K.Y.: Intrusion detection system: a comprehensive review. J. Netw. Comput. Appl. **36**(1), 16–24 (2013). https://doi.org/10.1016/j.jnca.2012.09.004
6. Estevez-Tapiador, J.M.: Book review: practical intrusion prevention. IEEE Distrib. Syst. Online **7**(6), 5–5 (2006). https://doi.org/10.1109/mdso.2006.39
7. Satheesh Kumar, M., Ben-Othman, J., Srinivasagan, K.G.: An investigation on wannacry ransomware and its detection. In: Proceedings—IEEE Symposium on Computers and Communications, vol. 2018-June (2018). https://doi.org/10.1109/ISCC.2018.8538354
8. Sheldon, F.T., Weber, J.M., Yoo, S.M., Pan, W.D.: The insecurity of wireless networks. IEEE Secur. Priv. **10**(4), 54–61 (2012). https://doi.org/10.1109/MSP.2012.60
9. Shafi, K., Abbass, H.A.: Evaluation of an adaptive genetic-based signature extraction system for network intrusion detection. Pattern Anal. Appl. **16**(4), 549–566 (2013). https://doi.org/10.1007/s10044-011-0255-5
10. Xiong, W., et al.: Anomaly secure detection methods by analyzing dynamic characteristics of the network traffic in cloud communications. Inf. Sci. (Ny) **258**(60773192), 403–415 (2014). https://doi.org/10.1016/j.ins.2013.04.009
11. Atighetchi, M., Pal, P., Webber, F., Schantz, R., Jones, C., Loyall, J.: For Survival and,\" no. December, pp. 25–33 (2004)
12. KishorWagh, S., Pachghare, V.K., Kolhe, S.R.: Survey on intrusion detection system using machine learning techniques. Int. J. Comput. Appl. **78**(16), 30–37 (2013). https://doi.org/10.5120/13608-1412

13. Borkar, A., Donode, A., Kumari, A.: A survey on intrusion detection system (IDS) and internal intrusion detection and protection system (IIDPS). In: Proceedings of International Conference on Inventive Computing and Informatics (ICICI 2017), no. Icici, pp. 949–953 (2018). https://doi.org/10.1109/ICICI.2017.8365277

14. Hindy, H., et al.: A taxonomy of network threats and the effect of current datasets on intrusion detection systems. IEEE Access **8**, 104650–104675 (2020). https://doi.org/10.1109/ACCESS.2020.3000179

15. Danda, J.M.R., Hota, C.: Attack identification framework for IoT devices. In: Advances in Intelligent Systems and Computing, vol. 434 (2016). https://doi.org/10.1007/978-81-322-2752-6_49

16. Toumi, H., Talea, M., Sabiri, K., Eddaoui, A.: Toward a trusted framework for cloud computing (2015). https://doi.org/10.1109/CloudTech.2015.7337013

17. Guerroumi, M., Derhab, A., Saleem, K.: Intrusion Detection System against Sink Hole Attack in Wireless Sensor Networks with Mobile Sink (2015). https://doi.org/10.1109/ITNG.2015.56

18. Modi, C., Patel, D.: A feasible approach to intrusion detection in virtual network layer of Cloud computing. Sadhana—Acad. Proc. Eng. Sci. **43**(7) (2018). https://doi.org/10.1007/s12046-018-0910-2

19. Vaid, C., Verma, H.K.: Anomaly-based IDS implementation in cloud environment using BOAT algorithm (2015). https://doi.org/10.1109/ICRITO.2014.7014762

20. Alheeti, K.M.A., Gruebler, A., McDonald-Maier, K.D.: An intrusion detection system against malicious attacks on the communication network of driverless cars (2015). https://doi.org/10.1109/CCNC.2015.7158098

21. Hong, J., Liu, C.C., Govindarasu, M.: Detection of cyber intrusions using network-based multicast messages for substation automation (2014). https://doi.org/10.1109/ISGT.2014.6816375

22. Satam, P.: Cross layer anomaly based intrusion detection system (2015). https://doi.org/10.1109/SASOW.2015.31

23. Haider, W., Hu, J., Yu, X., Xie, Y.: Integer Data Zero-Watermark Assisted System Calls Abstraction and Normalization for Host Based Anomaly Detection Systems (2016). https://doi.org/10.1109/CSCloud.2015.11

24. Narsingyani, D., Kale, O.: Optimizing false positive in anomaly based intrusion detection using Genetic algorithm (2016). https://doi.org/10.1109/MITE.2015.7375291

25. Aissa, N.B., Guerroumi, M.: A genetic clustering technique for Anomaly-based Intrusion Detection Systems (2015). https://doi.org/10.1109/SNPD.2015.7176182

26. Vasudeo, S.H., Patil, P., Kumar, R.V.: IMMIX-intrusion detection and prevention system (2015). https://doi.org/10.1109/ICSTM.2015.7225396

27. Banerjee, S., Nandi, R., Dey, R., Saha, H.N.: A review on different Intrusion Detection Systems for MANET and its vulnerabilities (2015). https://doi.org/10.1109/IEMCON.2015.7344466

28. Liu, H., Lang, B.: Machine learning and deep learning methods for intrusion detection systems: a survey. Appl. Sci. **9**(20) (2019). https://doi.org/10.3390/app9204396

29. Buczak, A.L., Guven, E.: A survey of data mining and machine learning methods for cyber security intrusion detection. IEEE Commun. Surv. Tutorials **18**(2), 1153–1176 (2016). https://doi.org/10.1109/COMST.2015.2494502

30. Hodo, E., Bellekens, X., Hamilton, A., Tachtatzis, C., Atkinson, R.: Shallow and deep networks intrusion detection system: a taxonomy and survey, arXiv, pp. 1–43 (2017)

31. Tan, Z., et al.: Enhancing big data security with collaborative intrusion detection. IEEE Cloud Comput. **1**(3), 27–33 (2014). https://doi.org/10.1109/MCC.2014.53

32. Aburomman, A.A., Reaz, M.B.I.: Survey of learning methods in intrusion detection system. In: 2016 International Conference on Advances in Electrical, Electronic and Systems Engineering, ICAEES 2016, no. Ml, pp. 362–365 (2017). https://doi.org/10.1109/ICAEES.2016.7888070

33. Shah, B., Trivedi, B.H.: Artificial neural network based intrusion detection system: a survey. Int. J. Comput. Appl. **39**(6), 13–18 (2012). https://doi.org/10.5120/4823-7074

34. Eskandari, M., Janjua, Z.H., Vecchio, M., Antonelli, F.: Passban IDS: an intelligent anomaly-based intrusion detection system for IoT edge devices. IEEE Internet Things J. **7**(8), 6882–6897 (2020). https://doi.org/10.1109/JIOT.2020.2970501

35. Sarnovsky, M., Paralic, J.: SS symmetry Learning and Knowledge Model, pp. 1–14 (2020)
36. Liu, Z., Ghulam, M.-U.-D., Zhu, Y., Yan, X., Wang, L., Jiang, Z., Luo, J.: Deep Learning Approach for IDS (2020), pp. 471–479. https://doi.org/10.1007/978-981-15-0637-6_40
37. Vinayakumar, R., Alazab, M., Soman, K.P., Poornachandran, P., Al-Nemrat, A., Venkatraman, S.: Deep learning approach for intelligent intrusion detection system. IEEE Access 7(c), 41525–41550 (2019). https://doi.org/10.1109/ACCESS.2019.2895334
38. Salo, F., Nassif, A.B., Essex, A.: Dimensionality reduction with IG-PCA and ensemble classifier for network intrusion detection. Comput. Networks 148, 164–175 (2019). https://doi.org/10.1016/j.comnet.2018.11.010
39. Sonewar, P.A., Thosar, S.D.: Detection of SQL injection and XSS attacks in three tier web applications. In: Proceedings—2nd International Conference on Computing Communication Control and automation (ICCUBEA). ICCUBEA 2016 (2017). https://doi.org/10.1109/ICC UBEA.2016.7860069
40. Masduki, B.W., Ramli, K., Saputra, F.A., Sugiarto, D.: Study on implementation of machine learning methods combination for improving attacks detection accuracy on Intrusion Detection System (IDS). In: 14th Int. Conf. QiR (Quality Res. QiR 2015—conjunction with 4th Asian Symp. Mater. Process. ASMP 2015 Int. Conf. Sav. Energy Refrig. Air Cond. ICSERA 2015, pp. 56–64 (2016). https://doi.org/10.1109/QiR.2015.7374895
41. Sahin, Y., Bulkan, S., Duman, E.: A cost-sensitive decision tree approach for fraud detection. Expert Syst. Appl. 40(15), 5916–5923 (2013). https://doi.org/10.1016/j.eswa.2013.05.021
42. Shin, S., Lee, S., Kim, H., Kim, S.: Advanced probabilistic approach for network intrusion forecasting and detection. Expert Syst. Appl. 40(1), 315–322 (2013). https://doi.org/10.1016/j.eswa.2012.07.057
43. Kang, I., Jeong, M.K., Kong, D.: A differentiated one-class classification method with applications to intrusion detection. Expert Syst. Appl. 39(4), 3899–3905 (2012). https://doi.org/10.1016/j.eswa.2011.06.033
44. Tong, X., Wang, Z., Yu, H.: A research using hybrid RBF/Elman neural networks for intrusion detection system secure model. Comput. Phys. Commun. 180(10), 1795–1801 (2009). https://doi.org/10.1016/j.cpc.2009.05.004
45. Sánchez, D., Vila, M.A., Cerda, L., Serrano, J.M.: Association rules applied to credit card fraud detection. Expert Syst. Appl. 36(2 PART 2), 3630–3640 (2009). https://doi.org/10.1016/j.eswa.2008.02.001
46. Das, K., Schneider, J., Neill, D.B.: Anomaly pattern detection in categorical datasets. In: Proceedings of ACM SIGKDD Conference on Knowledge Discovery and Data Mining, pp. 169–176 (2008). https://doi.org/10.1145/1401890.1401915
47. Kompella, R.R., Singh, S., Varghese, G.: On scalable attack detection in the network. IEEE/ACM Trans. Netw. 15(1), 14–25 (2007). https://doi.org/10.1109/TNET.2006.890115
48. Satheesh Kumar, M., Srinivasagan, K.G. Ben-Othman, J.: Sniff-Phish: A novel framework for resource intensive computation in cloud to detect email scam. Trans. Emerg. Telecommun. Technol. 30(6) (2019). https://doi.org/10.1002/ett.3590

5G Network Slicing Security

Tin-Yu Wu and Tey Fu Jie

Abstract 5G technologies are envisaged supporting novel use-cases for vertical industries. All the verticals, however, originate numerous use scenarios with different needs that 5G networks have to sustain effectively in the future. At present, "network slicing" is the key research direction that has attracted a lot of attention and we therefore focus on the 5G network slicing security in the following. First, we introduce the technical background and architecture of 5G, explain the architecture of SDN/NFV and the threats to information security, and discuss the privacy of 5G. Besides spotlighting "Network Slicing Security," including inter-slice and intra-slice security threats and harmonized resource between inter-domain slice segments, the final part also mentions security risks related to 5G network slicing and corresponding solutions identified by 3GPP.

Keywords 5G mobile communication · Network slicing · SDN/NFV · 5G security · Privacy · Authentication · Cryptography

1 5G Introduction

5G, the fifth generation technology, is a new mobile broadband standard that is going to replace or augment today's 4G LTE connection eventually. In 4G, the smart device coupled with 4G connectivity, creates a new mobile online community experience that we are now usually using for shopping, socializing, financial services and more. We can say that each new generation of mobile communication networks brings us a better mobile Internet experience and makes life much more convenient. However, with the arrival of 5G, everything will radically change. According the International Telecommunications Union (ITU)'s International Mobile Telecommunications-2020

T.-Y. Wu (✉)
1, Shuefu Road, Neipu, Pingtung 91201, Taiwan
e-mail: tyw@mail.npust.edu.tw

T. F. Jie
No. 43, Keelung Rd., Sec. 4, Da'an District, Taipei City 106335, Taiwan

© The Author(s), under exclusive license to Springer Nature Switzerland AG 2022   755
P. Nicopolitidis et al. (eds.), *Advances in Computing, Informatics, Networking and Cybersecurity*, Lecture Notes in Networks and Systems 289,
https://doi.org/10.1007/978-3-030-87049-2_28

Standard (IMT-2020 Standard), three main usage scenarios include: enhanced mobile broadband (eMBB), ultra-reliable and low-latency communications (URLLC), and massive machine type communications (mMTC) [1–4].

- **Enhanced Mobile Broadband (eMBB)**

Mobile Broadband provides improved access to multimedia content, services and data for human-centric applications. When there is an increasing need for mobile broadband services, further enhancements must be made in mobile broadband services. Besides existing mobile broadband applications, the eMBB scenario will stimulate novel applications and demands for enhanced efficiency, and offer an increasingly sophisticated user experience. This use case includes a great variety of situations, like hotspots and wide-ranging coverage, each of which has distinct needs. As for hotspots, i.e., areas with a high density of users, network traffic capacity is required to be very high, and user data rates, for lower mobility requirements, must be higher than that for wide-ranging coverage. As for wide-ranging coverage that seamless coverage and medium to high mobility are especially required, compared with existing data rates, user data rate significantly increases. But, compared with hotspots, data rate requirements for wide-ranging coverage can be reduced.

- **Ultra-reliable and low latency communications (URLLC)**

The URLLC scenario has rigorous performance requirements for throughput, latency, and reliability. Several instances for URLLC include industrial automation in manufacturing or production, remote medical diagnosis and surgery, smart-grid distribution automation, and public safety.

- **Massive machine type communications (mMTC)**

The mMTC scenario features numerous linked apparatus that normally transmit small amounts of non-latency sensitive data. It is also necessary to have low-cost devices with a very long battery life.

Figure 1 shows several examples of envisaged use cases for IMT-2020 and beyond.

1.1 Capabilities of IMT-2020

MT-2020 has an extensive range of capabilities deeply associated with purposed use case scenarios and applications that will give rise to more extensive requirements along with present and future trends. Designed to be more flexible and versatile, the IMT-2020 capabilities will be able to serve many different use scenarios and provide diverse services. Considerations must be given to not only network energy consumption but also spectrum resource limitations.

Crucial IMT-2020 capabilities include the following eight parameters:

- **Peak data rate**: The maximum data rate (Gbit/s) that each user/device can achieve in optimal conditions.

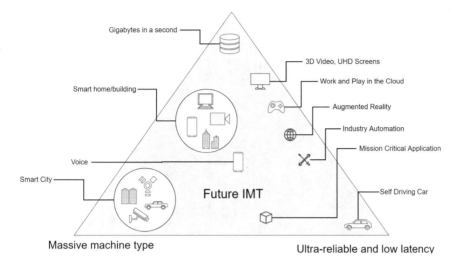

Fig. 1 IMT-2020 use case scenarios

- **User experienced data rate**: Ubiquitous and attainable data rates (in Mbit/s or Gbit/s) for mobile users/devices throughout the coverage area.
- **Latency**: The radio network's contribution (in ms) to the time it takes to send a package from the data source to the desired destination.
- **Mobility**: The maximum speed (km/h) to achieve the specified QoS as well as seamless transmission between radio nodes that may be parts of different layers and/or radio access technologies (multilayer/RAT).
- **Connection density**: Total number of linked and/or available devices every unit area (per km^2)
- **Energy efficiency**: There are two sides of energy efficiency:

  1. Network-side energy efficiency is defined as the number of bits of information that is delivered or obtained by the radio access network (RAN) to the user every unit of energy consumption (bit/Joule).
  2. Device-side energy efficiency is defined as the amount of information bits each energy consumed unit by the communication module (bit/Joule).

- **Spectrum efficiency**: The average data throughput e each spectrum resource unit and every cell[3] (bit/s/Hz).
- **Area traffic capacity**: Whole traffic throughput served every geographic range (Mbit/s/m^2).

As mentioned above, when all crucial capabilities are important to some extent in the great majority of usage scenarios, the connections of certain crucial capabilities could greatly vary according to the use cases/scenarios. Figure 2 shows how crucial capabilities are important for different usage scenarios, such as low latency communication, ultra-reliable, enhanced Mobile Broadband and massive machine-type communication. Three levels, high, medium and low, are used as the indicative scale.

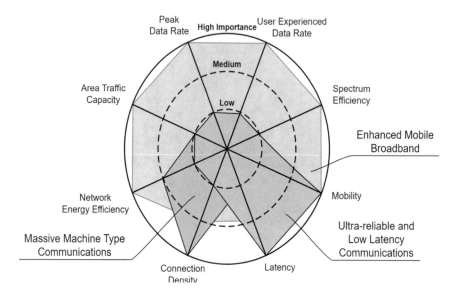

Fig. 2 Crucial capabilities for usage scenarios

In the eMBB scenario, capabilities of high importance are data rate for user experienced, area traffic capacity, rate, mobility, energy efficiency, peak data and spectrum efficiency, but in some cases, mobility and the data rate for user experienced may not be equally important. For example, compared with the wide-ranging coverage, hotspots require higher user experienced data rate but lower mobility.

In the URLLC scenarios, low latency is the most important capability to enable safety–critical applications, for instance. In some high mobility cases, this capability is also required, as high data rates may be not so important in terms of public safety, for example.

In the mMTC scenario, to sustain numerous devices in the network, high connection density is definitely needed since it may only occasionally be transmitted at low bit rates and zero or very low mobility. For this usage scenario, it is necessary to have a low cost device and long operating lifetime.

2 Network Slicing [5, 6]

5G Network Slicing allows multiple virtualized networks to be created on the self-same concrete network infrastructure [7]. Every slice of the network is an independent E2E network to meet various needs required by a specific application.

This technology therefore serves a principal role in supporting 5G networks, which are presented to effectively support a large number of services with various service level requirements (SLRs). In the fulfillment of this service oriented view of the

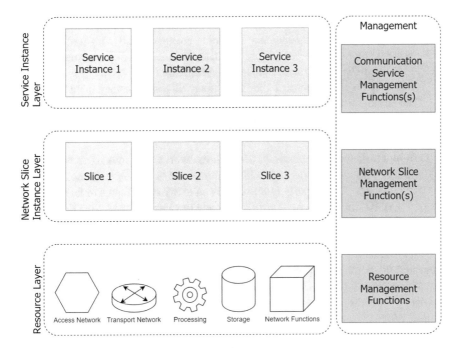

Fig. 3 Network slicing architecture

network, the purposes of software-defined networking (SDN) and network functions virtualization (NFV) are revealed: adaptable and expandable network slices can be enabled on top of a common network infrastructure.

Each layer of the network slice has its own management function [8], and the whole structure is divided into three layers [9], as shown in Fig. 3.

1. Resource Layer

    The base layer includes network resources and network management functions, and the purpose of these components is to serve the requesting end-user. Network resources include storage, processing, and transmission nodes. Another network management functions include switching and routing functions. These components can be logical/virtual or physical. Network resources and network management functions can be provided individually for one or more network slices.

2. Network Slice Instance Layer

    The central layer is made of slices, which provide the network capabilities required for service instances. A slice can also act like a resource layer to provide functionality for one or more services or to run on another slice. Each slice may run on different architectures or run on the same physical one. Depending on the architecture, the network slice instance layer may or may not be able to share resource layer's resources and network functions

3. Service Instance Layer

The upper layer describes the services provided to customers. For service instances, resource management functions are associated with resources and network functions, and each function can be assigned to a different management field. The network segment management function is responsible for managing the life cycle of the segment and interacting with other management functions. The communication service management function is responsible for managing the life cycle of the service and interacts with the network slice management function.

Different from former generations of mobile networks, 5G provides a new business role model to allow third parties to have more control and system capabilities at these three layers. A third party in this case is an entity other than a mobile network operator (MNO) that wants to be able to manage its own resources, functions, slices and services. The service instance layer can split the management control of each layer among MNOs or third parties.

After the concept of network slicing was introduced, various works in the past have presented the potential approaches, use scenarios, architectures and the advantages that network slicing brings so that the vertical application needs of 5G networks can be met.

First proposed and introduced by the Next Generation Mobile Network (NGMN), the 5G network slice, as defined by NGMN, is an E2E logical network/cloud operating on a common underlying architecture, isolated from each other, with individual control and management that can be generated on demand. Network slice may include cross-domain segments from various fields of the same or distinct administrations, or of segments suitable to access network, transport network, edge networks, and core network. As a result, network slices can be independent, manageable, mutually isolated, and programmable to sustain multi service and tenancy. Figure 4 displays an example of several 5G slices running simultaneously on the same architecture.

ITU envisions network slicing as the basic idea of network softening to promote the realization of logically isolated network partition (LINP), which consists of multiple virtual resources, isolated and equipped with data planes and programmable control devices. Network slicing is a technology defined by 3GPP that "enables operators to create customized networks to provide optimized solutions for different market scenarios that require different requirements (such as functionality, performance, and isolation)." From a business perspective From a point of view, a shard contains all the relevant network resources, functions and asset combinations required to complete a specific business case or service, as well as BSS, OSS and DevOps processes. Therefore, slices can be divided into two types:

1. Internal slices—Understand that partitions are used for the provider's internal services, maintaining full control and management of them.
2. External slices—Those partitions that host customer service are seen by the customer as a private network, cloud or data center.

Network slicing can provide radio, cloud and network resources to application providers or different vertical market components that do not have a physical network

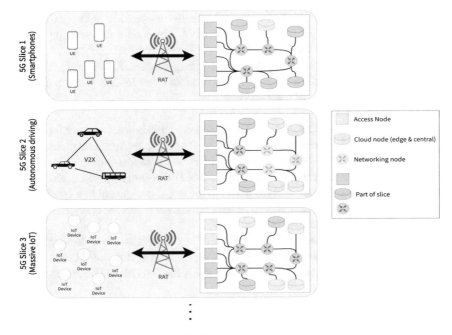

Fig. 4 The 5G infrastructure of network slices implemented

infrastructure. In this way, it can customize its network operations to meet the requirements of its customers based on their type of business, thus differentiating its services. The basic principles covering network slicing on 5G softwarized networks and their associated operations are the following [10]:

- **Network Operation Automation**
  Automation enables dynamic lifecycle management of network slices, like deploying, changing, deleting, optimization of network resources, as auto-scaling, migration, and dynamic interaction between management plane and data plane.
- **High-Reliability, Scalability and Isolation**
  These are key characteristics of 5G network slicing, which uses instant error detection methods for services with various performance requirements to ensure performance assurance and security for each tenant.
- **Programmability**
  Programmability can provide services, network manageability, and integration and operational challenges, focus on supporting communication services. For example, programmability allows third parties to control resources of an assigned slice (e.g. use open APIs that expose network functionality to control network and cloud resources). In 5G softwarized and virtualized networks can facilitate on-demand service-oriented customization and resource resiliency.
- **Hierarchical Abstraction**

Network slicing can create logically or physically independent network resource groups and NFs configuration for virtual, thereby introducing abstract supplementary layers. This abstraction can facilitate the provision of services from network slicing services on top of previous network slicing services. For example, network operators and ISPs use network slicing to enable other industrial companies to use the network as part of their services.

- **Slice Customization**
  Slice customization is implemented at all layers of the abstract network topology, using SDN to decouple the data and control planes. On the data plane, NFV capabilities provide service customization NFs and data forwarding mechanisms where artificial intelligence can be used to enable value-added services. Moreover, customization can ensure that the network resources designated to a specific 5G customer are utilized efficiently to fit the demands of a specific service.

- **Network Resources Elasticity**
  Network resources elasticity can be achieved via an efficient, non-destructive re-provisioning mechanism to scale up and down the allocated resources. As a result, elasticity can ensure that the required SLA/ELA can be achieved regardless of the user's location.

3 SDN/NFV Technology

This chapter will provide an introduction to Software-defined networking (SDN)/Network Functions Virtualization (NFV), the components of network slicing technology implementation, including its technical principles.

3.1 Software-Defined Networking (SDN)

To improve the network performance and monitoring, Software-defined Networking (SDN) technology makes network configuration dynamic and programmatically efficient, more like cloud computing instead of traditional network management.

Traditional networks face some challenges, like complex network management, inflexibility, troubleshooting problem, making it difficult to update either the hardware or the protocols (e.g., from IPv4 to IPv6) without affecting the normal use of the users, let alone a major project, such as a complete replacement of the hardware or the protocols.

To separate the forwarding process of network packets (data plane) from the routing process (control layer), SDN centralizes the network in one network component [11]. In such an environment, when any new network application (e.g., firewall, intrusion detection system, etc.) or functionality (e.g., bypass mechanism) changes, network administrators will have to change the entire system architecture directly at

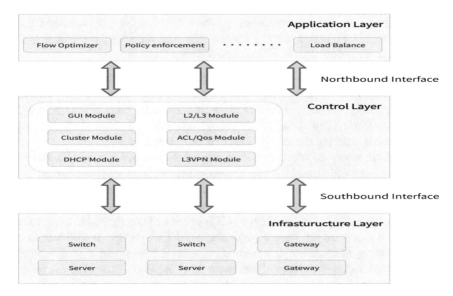

Fig. 5 SDN architecture

the infrastructure level. SDN is an attempt to solve the costly task of time and labor, while at the same time facilitating the evolution of network systems.

As Fig. 5 shows, SDN architecture includes three distinguishable layers: the application layer, the control layer and the infrastructure layer that are linked via northbound and southbound APIs.

- **Application Layer**

    The application layer uses specific network applications and/or functions. These functions include intrusion detection systems, load balancing, and firewalls. When traditional networks use specific devices such as load balancers and firewalls, the software-defined network will replace the device with an application that uses a controller to control the activities of the data plane.

- **Control Layer**

    The control layer symbolizes the consolidated SDN controller software that operates as the SDN brain. This controller dwells in a server to manage policies and the traffic flows via the network.

- **Infrastructure Layer**

    The infrastructure layer is assembled by the physical switches in the network.

- **Principle**

    The three layers use individual southbound and northbound application programming interfaces (APIs) to communicate with each layer. For instance, applications communicate with the controller via the northbound interface, as the controller and switches communicate through the southbound interfaces, like OpenFlow or other existing protocols. In the SDN world, as time goes by, Openflow has been accepted by the majority of internet service providers (ISP) and communities as the communication protocol between control plane and data plane. An SDN solution with OpenFlow therefore requires the protocol to be implemented in both controller and network elements.

SDN includes technologies such as separation of functions, network virtualization, and automation through programmability. These SDN technologies initially focused only on separating the network control plane from the data plane. When the control plane is considering how data packets should flow through the network, the data plane is actually moving the data packets to their destination.

The switch, also recognized as a data plane device, requests the controller for guidance as it is needed, and the switch also offers the message about the traffic it deals with to the controller. The switch delivers all packets to the same target through the same route and treats every packet exactly in the same way. SDN uses an adaptive or dynamic operation mode, in which a switch issues a route required to a controller for a packet that does not have a particular path. This process is different from adaptive routing, because adaptive routing is based on routers and algorithms built on the network topology instead of sending routing requests through the controller.

The virtualization aspect of SDN is achieved through a virtual overlay, which is a logically independent network above the physical network. In order to abstract the basic network and segment network traffic, users implement end-to-end coverage. This fine classification has many benefits for service providers and operators with multi-tenant cloud environments and cloud services, because they can provide each customer with a single virtual network with a specific strategy.

- **SDN Features and Advantages**

  - **Simplified Operations**
    Decouple the control and data planes to reduce complexity and provide highly secure and scalable automation.
  - **Faster Deployment**
    Leverage open APIs to deploy applications and services faster so that SDN can easily merge third-party products.
  - **Programmatic Network**
    This can be removed from manual configuration. Arrange and manage data centers, campuses and extensive networks.
  - **Centralized Management**
    This one platform will include management, configuration, monitoring, control, service delivery and cloud automation.

3.2 Network Functions Virtualization (NFV)

Network Functions Virtualization (NFV) abstracts network functions so that functions are allowed to be established, controlled, and created by software operating on standardized compute nodes. NFV drives fast advancement of new network services with flexible scale and automation by amalgamating cloud and virtualization technologies. NFV technologies frequently cooperate with software-defined networking (SDN), but they connect to different domains so some NFV components could complement the SDN functions effectively.

Though relying on traditional server-virtualization techniques, NFV is different from those that are used in enterprise IT. A VNF (virtualized network function) can include one or more virtual machines or containers operating different software and processes, these standard high-capacity servers, switches, and storage devices are all built on the cloud computing structure, instead of having customized hardware devices for each network function.

Presented by the European Telecommunications Standards Institute (ETSI), the NFV architecture helps to define the standards for NFV implementation [12]. Every segment of the infrastructure is rooted in those standards to optimize stability as well as interoperability.

Figure 6 reveals that the NFV architecture includes:

- Virtualized network functions (VNFs)

  - This software application that provides network functions includes directory services, file sharing and IP configuration.

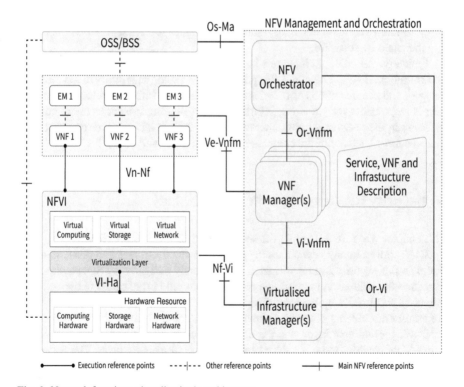

Fig. 6 Network functions virtualization's architecture

- Network functions virtualization infrastructure (NFVi)

  - The infrastructure components, like compute, storage, and networking on a platform to support software. For example, to run network apps, it is necessary to have a container management platform or a hypervisor like KVM.

- Management, automation and network orchestration (MANO)

  - This mechanism is used to manage the NFV infrastructure and the framework for supplying new VNFs.

NFV offers the flexibility to providers to operate VNFs on distinct servers and move them around when there is a change in demand. Such flexibility allows service providers deliver applications and services more quickly.

3.3 NFV Versus SDN

NFV stands for the virtualization of network components, while SDN stands for a network structure that will insert automation and programmability into the network by separating network control and forwarding functions. It can be seen that the main job of NFV is to virtualize network infrastructure, while SDN centralizes network control. Using a combination of SDN and NFV helps to start a network built, operated and managed by software.

Typically, the SDN architecture includes an SDN controller, northbound API, and southbound API. The controller permits network administrators to examine the network and decide actions or policies to the underlying infrastructure. Southbound API is responsible for gathering information and sending it back to the controller, to maintain the network smooth operation. Applications and services communicate their resource needs with the controller through northbound API [13].

4   Security Challenge in SDN and NFV [14]

This chapter will further explore the security challenges of SDN/NFV.

SDN, similar to any new technology, has its own advantages and disadvantages. For example, security can be mitigated or completely mitigated by SDN technology to reduce the risks or vulnerabilities commonly found in traditional networks. But, the addition of SDN technology also adds new vulnerabilities and threats, which are common problems of new architectures. As a matter of fact, the division of the control layer and data layer, and the centralization of all network logic, allows the entire SDN network to be paralyzed by the failure of a single layer. The following section describes the most obvious attacks and dangers that have been identified in the layers and interfaces of the SDN architecture, and next discusses the problems, implications, and consequences of exploiting the attacks and dangers that exist in the SDN architecture.

Network security attacks are usually divided by the primary target of the attack. For example, to eavesdrop on a control interface is one type of attacks, while others such as tampering with private personal data exchanged between devices are broadly considered unauthorized information leaks. The next section explains the general problems and threats in SDN architectures, including discovered attacks, security vulnerabilities, and abnormal behavior [15].

4.1 Overview of Attacks on SDN

- **Unauthorized Access**

  Attackers can exploit invalid or weak access mechanisms to launch brute-force attacks on the REST APIs of managed devices to expose login sessions, or exploit vulnerabilities in network components to install compromised devices or bind remote communications to give themselves unsupervised access to SDN components.

- **Unauthorized Information Disclosure**

  There are several attack methods that can be used to access sensitive network information. For example, an attacker can assume that network forwarding has occurred by broadcasting probe packets to the target. If an attacker successfully exploits a vulnerable network application with an attack, it can access the network policy database and other internal network data storage. In addition, the vulnerable channel can easily capture snooping and eavesdropping of data packets. Finally, the attacker can also use the device to simulate an attack to receive network messages intended to be sent to the attacker.

- **Unauthorized Modification of Network Information**

  Vulnerable protocols and APIs that lack validation and authentication mechanisms can allow attackers to override existing traffic rules by publishing conflicting traffic rules using malicious applications. In addition, unpermitted access to internal storage offers the possibility of attacking this to introduce contradictory network policies or to modify existing policies. On the other side, attackers can use protocol misconfiguration, device imitation and counterfeit packet injection attacks to make changes in network topology.

- **Network Information Destruction**

  The most obvious cases of this type of attack are inflicted flow rule flushing in switches, malicious applications dropping them and then exploiting control packets from another application or service chain, and to delete the network policy by not allowing access to the management or internal network database.

- **Service Disruption**

  Three major sources of service disruption include:

- Malformed packet injection attack: Using the malformed packet to cause the connection disconnected or target device collapse.
- Flooding attacks: Control packet flooding and flow rule tables flooding.
- Topology poisoning attacks: Network controller spoofing resulting in disconnection of the target device.

- **Misconfigurations**
  Misconfiguration of protocols, API interfaces and systems can lead to new vulnerabilities, conflicting network policies and traffic rules.
- **Poorly Configured Authentication, Trust and Verification Mechanisms**
  The most common attacks because of plain-text channels and weak authentication mechanisms include packet injection, eavesdropping, network information poisoning, traffic interception, and identity hijacking.

4.2 Attacks to the SDN Architecture

Certain specific attacks can compromise network components on that layer as well as those on other layers [13]. The following list describes the most common attacks on various layers of the SDN architecture.

- **Application Layer**

  - Application termination by abusing fixed privileges and authority
    Third party or control applications that have no permissions on the network system are adopted to disconnect/shut down some sensitive network APIs whose permissions cover the execution of system commands
  - Service disruptions
    Malicious applications that have been successfully installed back on the controller can be disconnected from service by manipulating the packet in the following ways:
    1. Dropping the control packet prevents the control packet from being delivered to the intended application
    2. Adjusting the order of application access control packets
    3. Interrupting control packet that needs to be forwarding
    4. Sniffing sensitive network information and using it to execute specific attacks
  - Attacks to vulnerable northbound API
    An attacker can issue a system command to terminate the target application or expose the controller and the target application in exchange for a packet via misconfiguration and vulnerability in northbound API.

- **Control Layer**

  - Dynamic flow rule tunneling

Malicious applications that run conflicting traffic rules may be able to bypass firewalls by attackers who run malicious applications with conflicting traffic rules, taking advantage of the controller's inability to distinguish between conflicting traffic rules and binding these rules to different network and control strategies.

– Controller Poisoning

Weak network protocols and malevolent applications are used to damage the controller's messages and topology, which also facilitates attacks on the data layer.

– Network Operating System Abuse

Corrupted applications and malicious data plane devices can use controller vulnerabilities and misconfigurations to achieve different goals. For example, the disclosure of sensitive information can reside in any instance of internal network data storage.

– Packet-in Flooding

An attacker broadcasts a large number of malformed network packets to a host or switch and the network infrastructure enters its packets into a controller to occurrence high probability of switch table misses. The controller responds to a large number of incoming packets and wastes all computing resources.

– Controller's Switch Table Flooding

Without authentication and verification mechanisms of incoming packet sender identity in the controller's incoming control packet, possible entry points for attacks may take place. Switch table flooding is the result of an absence of a proper mechanism, resulting in a continuous fake message. This situation allows an attacker to add false switches to the switch table of the target controller. Performing such an attack continuously will degrade the performance f the controller.

– Forced Switch Disconnection

An attacker can initiate any of the following attacks and the controller will be disconnected from the legitimate exchange

1. Legitimate Exchange Identity Hijacking

    An attacker will attempt to build a connection with the controller by impersonating a legitimate switch using a unique DPID (datapath identifier). The fake switch will successfully connect, take control and disconnect a legitimate switch to make itself the master of the real DPID. The target is then allowed to regain connection, and the period between connection and disconnection is now entirely under the attacker's control.

2. Spanning Tree Poisoning

    To deceive the controller, an attacker may use an untrue redundant link usually between the target host and the switch and force the controller to disconnect the untrue redundant link.

3. Malware

    Deployment of malware in the network operating system that is capable of corrupting the switch table information of the controller.

- **Infrastructure Layer**

  - Denial of Service Leveraging ARP Poisoning Attack

    An attacker can use a fake controller to achieve switch isolation. An attacker uses ARP poisoning attack to hijack the controller's identity, forcing the target switch to disconnect from the actual controller and link to the fake one instead. This terminates the disconnection of the switch from the network.
  - Flow-rule Modification/Flushing

    Attackers can modify, overwrite or refresh the information in the flow-tables that exist on switches. Usually an attacker can launch such an attack via a compromised application or network controller.
  - Flow-rule Flooding

    Attackers use auxiliary channel attack techniques to infer that the switch table is full, or use packets that miss the table to force the switch to send a request to the controller to install a new flow rule. In this case, an attacker can launch a table flooding attack, forcing the switch to continuously request new rules to fill the flow table, thereby reducing the performance and stability of the switch.
  - Malformed Control Packet Injection

    This attack makes use of manual control packets with malformed or misused header messages that are smartly staged to allow present vulnerabilities or invalid inputs to cause undesirable conditions on the switch.
  - Side-channel Attacks

    Also known as a reconnaissance attack, a side-channel attack usually uses the response of a particular device to a particular network situation to infer certain hidden messages that an attacker can use to launch an effective attack. For example, an attacker records the Round-Trip time of certain packets sent to certain switches, which can be used to determine which switch to launch a flow table flooding attack against.

Figure 7 summarizes the attacks that can be launched from various SDN architectures

4.3 Overview of Attacks on NFV

The VNF runs on virtual machines, so the security threat to the VNF would be theoretically related to a technically virtualization or physical network [16]. In the following, we will discuss NFVI-related attack scenarios and potential security risks [12].

- **Isolation Failure Risk**

  Also known as VM escape attack, a successful attack of this type would bring a big threat. In this attack scenario, we consider that the attacker has successfully exploited some VNFs on the hypervisor via an intrusion manager, so the attacker

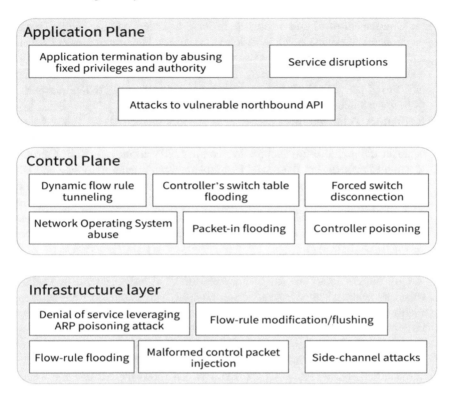

Fig. 7 Attacks to the SDN architecture

would first take control of it by accessing the operating system of one of the VNFs. Using tools or VNF network connections to cloud management networks, an attacker can gain access to the hypervisor API and take advantage of the opportunity to hack into the hypervisor, causing a big impact. This attack can occur if the hypervisor is not properly isolated from the VNF. For example, an application launches a handcrafted network packet that invades the heap overflow of a virtualization process and causes the hypervisor to execute arbitrary code, thereby gaining control of the host machine.

Another scenario is that one VNF may orchestrate the other VNFs so that an attacker can be given access to the VNF API to access the virtualization infrastructure and create a new VNF through its API and gain full access to the infrastructure resources to achieve the intrusion.

- **Network Topology Validation and Implementation Failure**

    With NFV, it is simple to create virtual network components such as virtual routers or virtual networks. Fast and dynamic service decisions may lead to perceived errors if no firewall is used when creating virtual routers to be used in virtual network interconnections. The dynamic virtual network devices and its connectivity can lead to incorrect division of networks and subnets, unlike in the deployment of physical network devices. Attackers can use the previously mentioned VM escape attack intrusion virtual firewall feature to restrict firewalls and gain sufficient privileges to carry out attacks. In reality, attackers may also take advantage of the elastic nature of NFVI and the lower security protection of NFVI to trigger VNF instantiation or migration to gain access to more resources that are available on the network infrastructure.

- **Regulatory Policy Failure**

  Attacks that place or relocate workloads outside the legal boundaries do not occur in traditional infrastructure. We can use NFV to move a VNF from one location to another, but there may be problems with regulatory policy and law. Violation of local regulatory policies and laws may result in a complete ban of the service or the imposition of financial penalties, allowing attackers to achieve their goal of harming the service provider. Another scenario is that attackers use insecure VNF APIs to violate user privacy, for example by dumping users' personal data information.

- **Denial of Service Protection Failure**

  DoS attacks focus on virtual networks or VNF public interfaces to drain network resources and affect service accessibility. The exploited VNF will send a large amount of traffic in parallel to other VNFs, which may be the same hypervisor or different hypervisors. Similarly, some VNFs consume high resources of CPU, hard disk, and memory, thus consuming the resources of the hypervisor. For example, the NFVI infrastructure sets up a virtual DNS service as a segment of the virtual evolved Packet Core (vEPC), which allows the orchestrator to automatically arrange additional virtual DNS services when traffic loads increase. So, when there is a large volume of DNS lookup traffic, the orchestrator instantiates new virtual machines in order to handle the traffic and the attacker continuing their attack will eventually exhaust their resources until interrupted or unavailable.

- **Security Logs Troubleshooting Failure**

  This type of security attack uses the compromised VNF to generate a large number of logs in the hypervisor and analyzing other VNF logs therefore becomes difficult. The risk exists when infrastructure logs are disclosed, as this could allow an attacker to see how one VNF is related to another VNF in a virtual network through the associated logs, and thus extracting relevant sensitive information.

- **Malicious Insider**

  This is usually a malicious behavior caused by internal administrators and is an internal security risk. One attack scenario involves a malicious administrator accessing a memory dump of a user's virtual machine. Since a malicious administrator can access the system management program, it can search for user IDs, passwords, and SSH keys through memory dumps, thus violating user privacy and data confidentiality. Another situation is that malicious administrators may directly retrieve user data from hard disk volumes managed by cloud storage devices. The attacker will first create a backup copy of the virtual machine drive and then use open source tools (such as kpartx or vgscan) to obtain sensitive data from it.

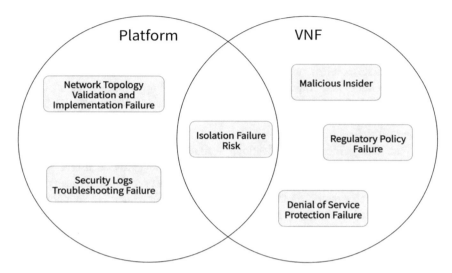

Fig. 8 NFV safety-related targets

Figure 8 provides an overview of the security risks related to NFV as mentioned above and the attack targets for these risks.

5 5G Security—Privacy [17–19]

The security challenges of each component of the network slice were described in the previous section, and this section will discuss the important security challenges of 5G-privacy.

Privacy generally refers to the protection of personal information, as this information should leak or hint at details of a particular user's personal information or activities. If such private and personal information is not protected properly, any attacker can use it to observe the user's daily activities and even do harm. For example, some users may wish to provide certain information about themselves to other users or to authorize a certain level of privacy rights. This also shows the control the user has over the privacy of the data. However, complete access to information is not possible in the electronic world, but protection of privacy can be achieved under certain conditions.

Privacy is of key importance in 5G networks, as it will transform daily life applications as well as access to digital services. Compared to older mobile networks, 5G has a lot to improve on in its service requirements, so privacy policies and regulations need to be addressed and possibly resolve privacy issues for 5G to be accepted and adopted by the general public [20]. In 5G mobile networks, user privacy can be divided into three main kinds:

- **Data Privacy**

  Data privacy stands for confidentiality and privacy of stored data. With the rapid development of mobile networks, consumers are more likely to prefer mobile networks. 5G mobile networks will have higher data rates and lower latency which results in large data volumes. For instance, critical applications like healthcare and financial related applications contain sensitive user data, which must be well protected.

- **Location Privacy**

  With the plethora of applications available today, location-based services (LBS) are becoming increasingly influential as they provide useful services according to the user's location. Apps on mobile devices, for example, can provide users with information on the location of the nearest restaurant, shopping mall or ATM based on the user's location, while social networking sites such as Facebook also have a 'check-in' location sharing function that allows users to let their nearest friends/relatives know where they are. However, with the introduction of such services, it also means that the application can continuously track the location of the user, which can lead to privacy issues.

- **Identity Privacy**

  Identity privacy signifies the protection of the user and the device or the identity information of the user's device. With the development of 5G and IoT, a substantial number of devices will be linked by the network. In this trend, each user or device will be categorized by a certain identity to facilitate the provision of relevant services. For example, medical applications will obtain patient information via identity, and online purchases will be paid for via credit cards, which also include identity information. Likewise, the privacy of individuals is affected if this identity information is revealed.

- **End-to-End (E2E) Data Confidentiality**

  In the 5G ecosystem, different services will be provided according to the needs of the service providers. However, these heterogeneous services may store and use users' personal data without users' authorization. The data may pass through different services during the use of the service, so a mechanism to ensure end-to-end confidentiality is necessary.

- **Responsibility Ambiguity**

  There are different roles involved in the 5G network, such as network operators, cloud service providers, or application developers. However, the lack of a clear statement on the responsibilities of each role can lead to legal or commercial disagreements. Therefore, each role should be clearly defined by the weight of responsibility, service agreement and appropriate control of user data.

- **Location of Legal Issues**

  The location of the victim, the location of the offender or the location of the service provider may have different applicable jurisdictional rules depending on the country in which the user's data is hosted.

- **Shared Environment**

  It is possible for different network service users to use virtualized network resources, such as mobile virtual network operators (MVNOs) or to use the same infrastructure shared with a competitor. In this environment, vulnerabilities in the hypervisor or distributed denial of service (DDoS) attacks can be exploited to affect the execution of other virtual machines if not properly secured. This could lead to unauthorized access to user data, which could compromise the privacy of the user [21, 22].

- **Different Objectives for Trust**

  Roles related to physical network infrastructure, such as network infrastructure providers, mobile network operators or mobile virtual network enablers (MVNEs) can work together, but may have different security objectives in mind.

- **Loss of Control**

  In the case of mobile network operators, moving part of the network to the cloud also means transferring control to Communications Service Providers (CSPs). This transfer requires the network operator to work with the CSPs to discuss the scope of their responsibilities and avoid losing control of the network management. In previous generations of networks, the operator controls some system components directly. But now, due to the move to the cloud, some of the information and control responsibilities will be transferred to the CSP, so the operator must work with the CSP.

- **Visibility**

  CSPs, in common cases, are not willing to share security and privacy measures with mobile operators. Therefore, mobile operators may not know the CSP's security measures or have a privacy management plan in place, and mobile operators will not be able to see the whole network.

- **Trans-border Data Flow**

  As global connections become more and more tightly linked, it is important to develop ways of storing and processing data and transferring it outside the country. However, data protection mechanisms could vary in different countries and the law enforcement authorities in one country require that data interception is also acceptable in another country. In general, the value of personal data differs from one country to another. For example, sexual orientation or religious beliefs may not be a sensitive issue in some countries, but may be very sensitive in others. With current Internet routing protocols, for maximum redundancy and flexibility, there are no restrictions on how to get to the destination after the destination IP has been defined, so it is possible for packets transmitted domestically to cross into other countries.

- **Hacking**

  The addition of an open IP-based open architecture to mobile networks makes them vulnerable to IP-based and network-based attacks. In addition, 5G's heavy reliance on cloud technology may further increase the likelihood that hackers will be able to attack, ultimately leading to serious user privacy issues.

- **Providing Information for Third Party**

  5G third-party application development provides a new interface for developers to share or sell personal information using privileged access to 5G systems. Moreover, data based on cloud systems may raise many privacy issues due to the sharing principle, leaving data open to unpredictable privacy damage in the future.

In addition, to increase the dynamism of the 5G network, network programmability and connectivity support for various vertical industries have been added. This is why security configurations and privacy policies are regularly verified to prevent the formation of privacy-threatening vulnerabilities.

- **IoT Privacy**

5G technologies enhance the use of the Internet of Things (IoT), but lots of security issues with IoT devices have not been incorporated into the design of IoT devices, including not considering security issues or not encrypting communications. The main reason for these problems is cost.

In addition, typical IoT devices such as webcams, thermostats and refrigerators can be used to launch DDoS. Therefore, if these devices are controlled, the information they share can also be at risk. These data can involve personal data, and this data is sensitive information that can be used by criminals to provide useful information.

6 Network Slicing Security for 5G [23, 24]

Network slicing for service-oriented networks is one of the important technologies of 5G. It can provide differentiated and tailored features, functions, and performances to suit the tenants' needs. Like every new technology, consideration must be given to security, including security issues and mitigations in each domain. Network slicing has raised many unprecedented security challenges, including inter-slice and intra-slice security threats and the issues of resource harmonization between inter-domain slice segments [25]. Moreover, many security risks related to 5G network slicing have been identified by 3GPP. Since network slices are E2E logical networks, the security should be discussed holistically. Next, we will use brief passages to explain the life-cycle security, inter-slice security and intra-slice security of network slicing [9, 26–28].

6.1 Life-Cycle Security of Network Slicing

There are four phases in the life cycle of a slice: (1) Preparation, (2) Installation, Configuration, and Activation, (3) Run-time, and (4) Decommissioning. In the first Preparation phase, the network slice template must be prepared. However, due to poor template design, tampering, lack of the latest security patches or incorrect process procedures, the risks may occur in the preparation phase, even leading to exposure of sensitive information. To solve the problems, the template must be prevented from being detected, the correctness of the template must be confirmed, and the security of the network during the template transmission process must be confirmed to ensure the integrity and authenticity of tshe template.

In the next Installation, Configuration, and Activation phase, API attacks may happen to generate fake slices, or change the configuration of slices. In order to ensure the API security, it is necessary to confirm the API operation authority and the tenant's identity verification. In addition, there must be strict specifications for API audit and monitoring, and secure communications between slices are also important at this stage.

During the Run-time phase, API and slice services remain the entry points for threats: DoS attacks, data breaches and privacy leaks may happen. Some of these problems are caused due to convenient management issues and the way to solve them is to take corrective actions. To prevent fake slices, network slices require verification and slice isolation technology are adopted to mitigate DDoS attacks.

The final part of the life-cycle is the Decommissioning phase. In this phase, slices may not be properly deactivated, leading to a DoS attack caused by leakage of sensitive data or improper release of resources. Therefore, in order to avoid the problem, sensitive data must be deleted when it is deactivated, and resources must be reallocated to ensure that the resources can be released. It is very important to manage and protect the logs that are responsible for the recording process in the life-cycle. Because the logs will not be deleted when the slice is deactivated, a dedicated and isolated security area to manage and save the logs is necessary to avoid any potential leakage.

6.2   Intra-slice Security of Network Slicing

When a slice is established and the 5G customer devices access it using non-3GPP networks, the risks will increase since unauthorized access may impact the resource consumption and increase the chance of DoS attacks. These interconnections will cause identification of slices, which can become the possible attack entry point related to the permanent identifiers of customers' devices. These security risks may attack the slice's service and damage the slice and other services running on it. In addition, when the slice is defined as a chain of several sub-slices, the weakest sub-slice will become the security gap for the entire security in a chain of sub-slices. To deal with these risks in end-to-end logical networks, first consideration should be given to end-to-end security. Therefore, the correct isolation must be achieved between services and between fragments and services and no correlation between identifiers should be leaked. Appropriate levels and mechanisms should be used to ensure the desired security level; minimum requirements should include confidentiality, integrity, data authenticity, and mutual authentication between peers. Moreover, network slicing performance and fault monitoring are recommended by the 3GPP standard in different customers' environments. Using both primary and secondary authentication, tenant devices can be authenticated strongly and intra-slice security can be maintained.

6.3 Inter-slice Security of Network Slicing

5G customer devices may increase the risk due to different access technologies (3GPP and non-3GPP, for example.) A security risk occurs if a customer's device permitted to access one slice try to get access to another unpermitted slice. The device may be allowed to connect to various slices at the same time for diversified service access. However, it is possible that the device reveals sensitive data from a slice with a higher security level to a slice with a lower security level.

With the belief that services operating on various slices are independent and communication is not necessary, most slice managers and system designers define it a low security risk. However, an attacker may damage other services running on other chips by attacking certain services. To solve these problems, it is necessary to strengthen the isolation between slices and the communication between slices must be controlled and protected to prevent one slice from being damaged and affecting other slices. How to avoid leakage should be previously referred to for communication between services. Each slice will generate new and independent keys using a key derivation function, and key management therefore should be further investigated in the future. The allocation of resources between slices is also a risk factor. A minimal level of resource availability should be ensured for every slice.

7 Summary

When 5G network is continually evolving, Software Defined Networking (SDN), Network Function Virtualization (NFV) and other virtualization key technologies have substantially improved network performance. However, still more and more security threats occur since network slicing is a very new technology in virtualization. To achieve network slicing security in 5G network environment, several security mechanisms must be implemented to reach a secure network slicing system. Since network slicing is an end-to-end logical network, its end-to-end isolation is a prerequisite for end-to-end security. Considerations must be given to different aspects of isolation: isolation between network slices, isolation between network functions, isolation between users, and so on.

Currently, using a cloud-based architecture environment for slice isolation may also be difficult because of the wide variety of technologies used in the network. Therefore, isolation should be performed in different levels and heterogeneous environments (such as OS kernel, firmware, upper-level software systems, etc.).

At present, using cloud-based architecture environment for slice isolation might also be challenging because there are a great diversity of technologies used in the network. For this reason, isolation should be performed at different levels and within heterogeneous environments, such as OS kernel, firmware, upper-level software systems etc.

In addition, to achieve end-to-end security, communication security issues between intra-slices and inter-slices must be considered, and a trustworthy model must be established. Current secure communication system needs to frequently exchange keys and security parameter updates with these Network slice core security elements, making it one of the most important threats to the current network slice security. So in response to the 5G communication environment has a variety of scenarios that support 3GPP and Non-3GPP of multi-domain environments, and several layers of imbricated tenants, which can play different roles and have different rights.

References

1. 3GPP: Release 16. https://www.3gpp.org/release-16 (2019)
2. 3GPP: Release 17. https://www.3gpp.org/release-17 (2020)
3. Shi, Y., Han, Q., Shen, W., Zhang, H.: Potential applications of 5G communication technologies in collaborative intelligent manufacturing. IET Collaborative Intell. Manuf. 1(4) (2019). http://doi.org/10.1049/iet-cim.2019.0007
4. Fifth Generation of Mobile Telecommunications Networks (5G). ENISA, Heraklion, Greece (2019)
5. The Evolution of Security in 5G: A, 'Slice' of Mobile Threats. 5G Americas, Bellevue, WA, USA (2019)
6. Network Slicing for 5G Networks& Services. 5G Americas, Bellevue, WA, USA (2016)
7. Ji, X., et al.: Overview of 5G security technology. Sci. China Inf. Sci. 61(8) (2018). http://doi.org/10.1007/s11432-017-9426-4
8. Study on Security Aspects of 5G Network Slicing Management (Release 15) V15.0.0, document 3GPP TR33.811 (2018)
9. Olimid, R.F., Nencioni, G.: 5G network slicing: a security overview. IEEE Access 8, 99999–100009 (2020)
10. Jinsong, M., Yamin, M.: 5G network and security (2020). http://doi.org/10.23919/indiacom49435.2020.9083731
11. Chica, J.C.C., Imbachi, J.C., Vega, J.F.B.: Security in SDN: a comprehensive survey. J. Netw. Comput. Appl. 159, 102595 (2020). http://doi.org/10.1016/j.jnca.2020.102595
12. Lal, S., Taleb, T., Dutta, A.: NFV: security threats and best practices. IEEE Commun. Mag. 55(8) (2017). http://doi.org/10.1109/mcom.2017.1600899
13. Ahmad, I., Kumar, T., Liyanage, M., Okwuibe, J., Ylianttila, M., Gurtov, A.: Overview of 5G security challenges and solutions. IEEE Commun. Stan. Mag. 2(1) (2018). http://doi.org/10.1109/mcomstd.2018.1700063
14. Murillo, A.F., Rueda, S.J., Morales, L.V., Cárdenas, Á.A.: SDN and NFV security: challenges for integrated solutions. In: Computer Communications and Networks, pp. 75–101. Springer International Publishing, Berlin (2017)
15. Ordonez-Lucena, J., Ameigeiras, P., Lopez, D., Ramos-Munoz, J.J., Lorca, J., Folgueira, J.: Network slicing for 5G with SDN/NFV: concepts, architectures, and challenges. IEEE Commun. Mag. 55(5), 8087 (2017)
16. Blanc, G., Kheir, N., Ayed, D., Lefebvre, V., de Oca, E.M., Bisson, P.: Towards a 5G security architecture (2018). http://doi.org/10.1145/3230833.3233251
17. Huawei: Huawei: 5G Security Architecture White Paper. Huawei, Shenzhen, China (2017)
18. Ting, T.-H., Lin, T.-N., Shen, S.-H., Chang, Y.-W.: Guidelines for 5G end to end architecture and security issues (2019). http://arxiv.org/abs/1912.10318

19. Schinianakis, D., Trapero, R., Michalopoulos, D.S., Crespo, B.G.-N.: Security considerations in 5G networks: a slice aware trust zone approach. In: Proceedings of IEEE Wireless Communication Networks Conference (WCNC), p. 18, Apr 2019
20. Liyanage, M., Salo, J., Braeken, A., Kumar, T., Seneviratne, S., Ylianttila, M.: 5G privacy: scenarios and solutions (2018). http://doi.org/10.1109/5gwf.2018.8516981
21. Ahmad, I., Kumar, T., Liyanage, M., Okwuibe, J., Ylianttila, M., Gurtov, A.: 5G security: analysis of threats and solutions (2017). http://doi.org/10.1109/cscn.2017.8088621
22. Arfaoui, G., Bisson, P., Blom, R., Borgaonkar, R., Englund, H., Félix, E., Klaedtke, F., Nakarmi, P.K., Näslund, M., O'Hanlon, P., Papay, J., Suomalainen, J., Surridge, M., Wary, J.-P., Zahariev, A.: A security architecture for 5G networks. IEEE Access **6**, 22466–22479 (2018)
23. Study on Security Aspects of Enhanced Network Slicing (Release 16) V0.8.0, document 3GPP TR33.813 (2019)
24. Security Architecture and Procedures for 5G System (Release 16) V16.2.0, document 3GPP TS33.501 (2020)
25. NMNG Alliance: 5G Security Recommendations, Package #2: Network Slicing, NMGN, Frankfurt, Germany (2016). https://www.ngmn.org/wp-content/uploads/Publications/2016/160429_NGMN_5G_Security_Network_Slicing_v1_0.pdf
26. 5G Security White Paper-Security Makes 5G Go Further. ZTE, Shenzhen, China (2019)
27. Cao, J., Ma, M., Li, H., Ma, R., Sun, Y., Yu, P., Xiong, L.: A survey on security aspects for 3GPP 5G networks. IEEE Commun. Surveys Tuts. **22**(1), 170195 (2020)
28. Cunha, V.A., da Silva, E., de Carvalho, M.B., Corujo, D., Barraca, J.P., Gomes, D., Granville, L.Z., Aguiar, R.L.: Network slicing security: challenges and directions. Internet Technol. Lett. **2**(5), e125 (2019). http://doi.org/10.1002/itl2.125

A Security-Driven Scheduling Model for Delay-Sensitive Tasks in Fog Networks

Surendra Singh and Sachin Tripathi

Abstract Nowadays, the uses of delay-sensitive applications are rapidly increasing due to their performance, QoS, and enrich the user experience. Therefore, security and scheduling aspects for delay-sensitive tasks have become more critical issues that are not incorporated in the current existing algorithms. To solve these intricate issues, "A security-driven scheduling model for delay-sensitive tasks in Fog networks" has been introduced and abbreviated as "SDSM". The proposed method is the integration of binary integer programming, Min_Heap algorithm, and modified Earliest Deadline First policy (m-EDF). The binary integer programming is used to evaluate optimal average security value for delay-sensitive tasks from basic security service categories, i.e., confidentiality, integrity, and authentication whereas, only one security service can be selected from a category, however, each category has some distinct security services along with their normalized performance value. The Min_Heap algorithm is used to find an optimal node in Fog networks, in which system load and load threshold values are used as the key parameters which are based on the weighted sum of the square method. And the m-EDF scheduling policy is used for the delay-sensitive tasks. The contribution of the proposed method is two-fold: first, is to provide the robust security service to the delay-sensitive tasks, and second, is to enhance the performance of the system (in terms of success ratio) without violating the scheduling constraints of delay-sensitive tasks. The novelty of the proposed method is proven in terms of success ratio, average security value, and overall performance through extensive experimental result analysis and compared to some existing baseline algorithms. The Network Simulator (NS-3) with python scriptwriting and a synthetic data set is used to obtain the experimental results.

Keywords Delay-sensitive tasks · Security-driven · Scheduling · Fog networks

S. Singh (✉)
National Institute of Technology, Uttarakhand, India
e-mail: surendra@nituk.ac.in

S. Tripathi
Indina Institute of Technology (IMS) Dhanbad, Dhanbad, India
e-mail: sachin2781@iitism.ac.in

© The Author(s), under exclusive license to Springer Nature Switzerland AG 2022
P. Nicopolitidis et al. (eds.), *Advances in Computing, Informatics, Networking and Cybersecurity*, Lecture Notes in Networks and Systems 289,
https://doi.org/10.1007/978-3-030-87049-2_29

1 Introduction

Nowadays, secure real-time systems for delay-sensitive tasks are becoming more practical because of the integration of enhanced security services and optimal load balancing mechanisms. The application of such type of platforms are flight control systems [1], rail traffic control systems [2], cyber-physical systems [3], radar systems [4], on-line transactions [5], and defense monitoring systems. These all kinds of systems are needed to finish its task/process within the defined time limit (no delay) because failing which, can produce logical or physical disasters. Task security and optimal load balancing mechanisms become more crucial aspects. Therefore, A Security-Driven Scheduling Model for Delay-Sensitive Tasks in Fog Networks has been introduced in this study. In which, the mechanism for providing optimal security and enhanced load balancing algorithm are combined together to fulfillment of requirements.

The basic security services are divided into various categories to make a suitable combination of these as per the requirements. These categories are as follows: (i) Confidentiality, which protects against the disclosure of information in transit, (ii) Data integrity, which assures against the data modification in transit, and (iii) Authentication, which is responsible for the user authentication process [6, 7]. The key aspect of this study is the on-demand combination of the above security categories as per the task requirements. The binary integer programming approach is used to make an on-demand combination of security categories, which produced higher security value as compared to the needed one (refer motivational Example 1.1). The initialization of evaluated security value may increase the initial execution (processing) time, which is known as security overhead. Further, an enhanced load balancing mechanism is required to uniformly distribute the load among the nodes without violating the scheduling constraints.

Now, the secure task with a new processing time needs to be scheduled to complete its process within the deadline. The proposed modified earliest deadline first (m-EDF) scheduling policy is used to assign a task at a node for its execution [8–11]. The scheduling process becomes very easy if a system has a large number of resources and a suitable load distribution approach. Hence, Fog computing along with the proposed load distribution algorithm (Min_Heap) has used to uniformly distribute the load among the various nodes in a network [12–15]. The dynamic load distribution mechanism is used in Fog or connected Fog networks [16–19]. In which, Fog network called FN is used to select an optimal node at ant instance of time (arrival of a task). The optimal node is a node that has the least computing load (utilization value) at the time of arrival of a task against all the nodes in a Fog. Further, FN holds a utilization table that has the entire details of each node and Fog in a network (refer Sect. 3.3). For more clarity, a motivational example has been considered to calculate optimal security value, security overhead, and the optimal node through the min-heap algorithm.

1.1 Motivational Example

It has assumed that each category has various distinct security algorithms along with their category weights for providing enhanced security to the tasks [20] subject to the sum of all category weights equals to one. The normalization process of security values has been used to simplify the mathematical calculation only, and for more details of the normalization process, refer [21]. Now, it is assumed that the normalized security values are varied between 0.1 and 1.0, where 0.1 indicates the week but fastest security mechanism and 1.0 shows the strongest but the slowest security mechanism in a category. The sum of each category weights are equals to one, i.e., $w_1 + w_2 + w_3 = 1$ (refer to Table 1).

Let us assume that a category is targeting for confidentiality security service then it is responsible only to provide the confidentiality security services, and similarly, if other categories are aiming for integrity and/or authentication security service(s) then these are responsible to offer only aforementioned security service. For instance, the Seal, RC4, Blowfish, Khafre, IDEA can be used for task/message authentication and MAC-MD5, HMAC-SHA-1, CBC-MAC-AES for authentication security service, and similarly for integrations which can be used in collaboration with authentication service [21].

Let us assume that if a category is targeting for providing confidentiality security service, then that is responsible for offering the confidentiality security services only. And similarly, if the remaining categories are aiming to offers other kinds of security services, then they are responsible only for respective ones, not others. For example, security services like Seal, RC4, Khafre, IDEA, Blowfish are using for task authentication will provide only task authentication service only. Similarly, user authentication services like CBC-MAC-AES, MAC-MD5, HMAC-SHA-1 are responsible only for user authentication only, not for other kinds of security services, and so on. Therefore, the best suitable combination of these categories as per the task need can be made using a binary integer programming approach [21].

Assumed that there are Nc categories, and each has various kinds of security services along with their normalized value of security service (Sv) that offers the same security service but in a different type of performance. The category weight is represented through w_k, and Sr_i^{min} represents minimum security requirements of ith task.

The task set $T \in \{t_1, t_2, \ldots, t_n\}$ has the following predefined attributes, i.e., $< a_i, e_i, D_i, Sr_i^{min} >$, here a_i represents the arrival time, e_i denotes the processing time, D_i is the absolute deadline, and Sr_i^{min} is the minimum security requirement of ith

Table 1 Categorization of security services for ith task

#Categories (Nc)	Service values (Sv)										Weight (w_k)	Sr_i^{min}
Confidentiality{C}	0.1	0.2	0.3	0.4	0.5	0.6	0.7	0.8	0.9	–	0.4	
Integrity{I}		0.1	0.2	0.3	0.4	0.5	0.6	0.7	0.8	0.9	1.0	0.4	0.46
Authentication{A}	0.3	0.6	0.9	–	–	–	–	–	–	–	0.2	

Table 2 Set of 5 tasks with predefined and calculated values

Existing values

# Task	a_i	e_i	D_i	Sr_i^{min}
t_1	01	08	20	0.53
t_2	04	11	30	0.25
t_3	**06**	**15**	**40**	**0.46**
t_4	07	20	60	0.68
t_5	09	25	70	0.73

The bold value (task t_3) is used to show the calcualtions as given in the Eqs. 1, 2, and 3, respectively

task, respectively. The various calculations of the task set have been given in Table 2, where times are in milliseconds. In Table 2, Ne_i is the new processing time, Sr_i^{min} and So_i are the minimum security requirements and security overhead, respectively. A brief description of security overhead is given in Sect. 3.2.2. The mathematical evaluations of all respective parameters for task t_3 are given below.

The mathematical equation for calculation of security value (Sr_i^{Eval}) for ith task is given below.

$$Sr_i^{Eval} = \sum_{k=1}^{Nc} Sv_k \times w_k \geq Sr_i^{min} \tag{1}$$

i.e.,

$$Sr_i^{Eval} = 0.5 \times 0.4 + 0.5 \times 0.4 + 0.3 \times 0.2 = 0.46 \geq 0.46$$

Now, the security overhead for ith task (So_i) can be evaluated as follows:

$$So_i = e_i \times Sr_i^{Eval} \tag{2}$$

i.e.,

$$So_i = 15 \times 0.46 = 06.90$$

Hence, the new processing time can be calculated as follows:

$$Ne_i = e_i + So_i \tag{3}$$

i.e.,

$$Ne_i = 15 + 06.90 = 21.90$$

The calculated new processing time (Ne_i), evaluated security value (Sr_i^{Eval}), and security overhead (So_i) values based on the above equations for given task sets are as follows (Table 3):

The new processing time can be calculated using Sr_i^{Eval} or Sr_i^{min} because both are equal. Now, the tasks with new processing time are going to be assigned to the optimal nodes in Fog or networks. However, the start time of a task is measured at

Table 3 Set of 5 tasks with and calculated values

Calculated values

# Task	Ne_i	Sr_i^{Eval}	So_i
t_1	12.24	0.53	04.24
t_2	13.75	0.25	02.75
t_3	**21.90**	**0.46**	**06.90**
t_4	33.6	0.68	13.60
t_5	43.25	0.73	18.25

the assigned node (refer to Eq. 16). The concept of SDSM is unique and novel, which has not been applied in this field as per state-of-the-art. The experimental analysis for a synthetic dataset using Python scripting and NS-3 tool shows the feasibility of the proposed work relative to the state-of-the-art methods. The main contributions of this work are as follows:

1. Offer enhanced security service using a binary integer programming approach, which combines the essential security services viz., authentication, confidentiality, and data integrity.
2. The outcome of Min_Heap algorithm is an optimal node, which has maximum computing power and minimum workload.
3. The newly arrived tasks are selected based on m-EDF policy.

The outcome analysis of the proposed SDSM approach has proved the novelty. The rest of the paper is organized as follows: the literature survey has been given in Sect. 2. The proposed system architecture of SDSM is given in Sect. 3, and the proposed SDSM approach in Sect. 4. Section 5 talks about the experimental setup and results analysis. Finally, the Conclusion and future work are illustrated in Sect. 6.

2 Literature Survey

Scheduling and robust security requirements are the most crucial aspects of any real-time system. In real-time systems, several considerations have been taken to ensure the reliability of scheduling with the high-security interest. The research articles have been discussed for task scheduling with enhanced security, in which various methods have been used to assign a task at optimal computing nodes in a Fog network while balancing the utilization either statically or dynamically.

Lin et al. [22] address the issue of security performance in real-time systems and proposed "Static security optimization for real-time systems". In which a grouped based security model has discussed, where security services have been divided into various groups according to the respective type of security mechanism. However, services provide the same kind of security service within the same security group.

The group services may acquire various security performances. In this analysis, the conventional real-time scheduling algorithm, i.e., earliest deadline first (EDF) [23] has integrated with the group-based security model and also build an EDF schedulability test facility for safety-aware [24]. "An improved hierarchical load balancing algorithm for cluster environment" abbreviated as i-HLBA has proposed by Lee et al. [25]. In which, system load is used as a critical parameter to determine the threshold value of the system load. The scheduler dynamically adjusts the threshold value when system load changes.

Lee et al. [25] has proposed "Minimum energy semi-static scheduling of a periodic real-time task on dvfs enabled multi-core processor"for dynamic scheduling along with safety information and energy minimization. Some traditional scheduling algorithm does not protect real-time tasks. Therefore, Xie has designed a security algorithm that applied an EDF scheduling to increase the performance of the system. Chetto et al. [26] presented "Dynamic scheduling of periodic skippable tasks in an overloaded real-time systems" named as RLP/T, which deals with the quality of service. RLP/T is a variant of EDF, which enhances the performance in terms of the periodic task. The author has introduced the issues of periodic task scheduling, where the primary goal is to increase the QoS by executing the tasks before their deadlines.

Further, Chetto et al. [27] has presented "Dynamic scheduling of skippable tasks in a Hard real-time system", which increases the robustness of a real-time system. RLP/T algorithm uses the task workload according to their deadlines and skips the unnecessary attributes to assure better robustness. However, the skip model consists of tasks either in worst-case execution time or skip aspects that have been used to bear the missed deadlines for the task set. Earliest Deadline as soon as possible (EDS) algorithm has been used when no delay is required in between the tasks. Simultaneously, Earliest Deadline as late as possible has is used when there is an urgency not to delay in the execution of a task. Liang et al. [28] have proposed "Task scheduling with load balancing using multiple ant colonies optimization in grid computing".

In this literature, to achieve task scheduling with load balancing, a bio-inspired algorithm called as multiple ant colonies optimization (MACO) approach has been proposed. In which numerous ant colonies operate together and exchange to jointly find out the solution with a dual goal of reducing of missing execution period and the degree of software node mismatch.

Singh et al. [21] has proposed "Secured Dynamic Scheduling Algorithm for Real-Time Applications on Grid (SDSA)". This algorithm works for real-time applications in a grid environment. The unsatisfied packets have been sending to the node, which has the minimum accepted queue length. Still, this algorithm suffers from (i) it might be possible for a node with minimum accepted queue length to have a higher workload, and (ii) if two or more packets have the same deadline then the selection of packet becomes difficult. To overcome these drawbacks, Singh [29] has presented "utilization based Secured Dynamic Scheduling Algorithm for Real-Time Applications on Grid (u-SDSA)". This algorithm has provided the solution to the above problems. The major drawback of the u-SDSA algorithm is that the security of the packets waiting in the accepted queue has decreased if a packet set is not feasible at the arrived node. Hence u-SDSA might not be able to offer the optimal security service. The next version of u-SDSA has been proposed by Singh et al. [30] known as "Security Aware Dynamic Scheduling Algorithm (SADSA) for Real-Time Applications on Grid". In SADSA, the securities have been distributed according to uniform and exponential distribution. It enhances the security quality of the real-time tasks but has not to offer optimal scheduling quality, which can be further explored.

The baseline algorithms, which are taken in consideration to prove the novelty of proposed SDSM algorithm are as follows.

The "Secured Packet Scheduling Strategy (SPSS)" is proposed in [31], which has mainly focused on security service even when the packet rate is much higher. Hence, no process exists to adjust the security services according to the packet arrival rate. Therefore, SPSS suffers from a higher packet drop rate. To dynamically adjusting the security services and maintaining the high quality of services (QoS) Zhu [32] has introduced "An Improved Security Aware Packet Scheduling algorithm (ISAPS)". Zhu has used round-robin fashion for adjusting the security level of packets waiting in the accepted queue and mainly focuses on security services subject to higher packet arrival rate.

The "Enhancing Security Real-Time Applications on Grids Through Dynamic Scheduling (SAREG)" algorithm is introduced by Xie [33]. The SAREG works in a grid environment to improve the performance of systems. This algorithm sends the unsatisfied packets to the other nodes in the grid. The main drawback of the algorithm is as follows: (i) there is no pre-defined policy for transferring the unsatisfied packets from one node to other nodes, (ii) this may send the packets that have higher processing time causing new incoming packet might be incompatible with the system. That will increase the packet drop rate.

Further, Singh et al. [34] has proposed a model in a grid environment to solve the complex problems called as "SLOPE: Secure and Load Optimized Packet scheduling model in a grid environment. SLOPE is the combination of improved Hierarchical Load Balancing Model (i-HLBM), weighted Combined Security Service Model (w-CSSM), and modified Earliest Deadline First (m-EDF) algorithm. The SLOPE has achieved an enhanced security service without violating the scheduling constraint even in higher workload conditions. But still, there are some improvements required in the calculation of optimal security service and to find an optimal node for load balancing.

Hence, A Security-Driven Scheduling Model for Delay-Sensitive Tasks in Fog Networks has been introduced, which shows better performance against all the baseline algorithms and prove the novelty of SDSM.

3 Proposed System Architecture of SDSM

The primary components in the proposed system architecture of SDSM are as follows: (i) Task model, (ii) Security controller, (iii) Load balancing model, and (iv) Scheduling model. The overview of the proposed SDSM model is given in Fig. 1. The

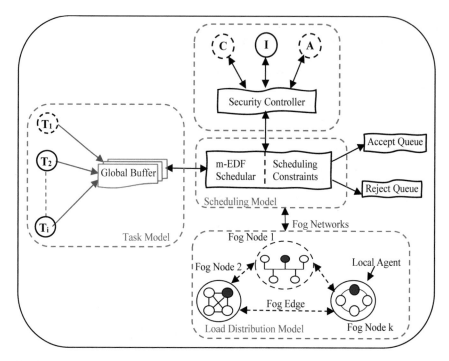

Fig. 1 Proposed system architecture of SDSM

global buffer is used to store newly arrived tasks with its predefined tuples according to their deadline[1] (refer Sect. 3.1). The scheduling model plays a vital role, which is the integration of an m-EDF scheduler and scheduling constraint. Now, m-EDF scheduler selects a task from the global buffer, and simultaneously, evaluate the security value (Sr_i^{Eval}) through the security controller (SC) and find out an optimal node in Fog Networks (FN) using load distribution model and then check the scheduling constraints (the second component of scheduling model). A task is inserted into the accepted queue of an optimal node (assigned), if and only if that task satisfies the scheduling constraints and there exist an optimal node at that instance of time, else inserted into the rejected queue of a node.[2] A brief discussion of each primary component is given below.

---

[1] The global buffer is assumed as a finite size buffer.

[2] The accepted and rejected queue are used to calculate the success ratio only.

3.1 Task Model

Assume that the task set $T \in \{t_1, t_2, t_3, \ldots, t_i\}$ has i aperiodic and independent tasks, which has to be scheduled with optimal security level at any node[3] in Fog Network. The newly arrived tasks are stored in the global buffer in the non-decreasing fashion according to their absolute deadlines. Here, i is the index of a task, and each task is sorted according to the assigned indexed number only. Now, the optimal security value through Eq. 1 is evaluated and assigned to the task. If a task selected (using m-EDF policy) is assigned to an optimal node in the Fog Networks as per the scheduling constraints, then set the binary flag value to 1 (one), else remain 0 (zero) or the initial flag value.

3.2 Security Controller

The security controller is used to integrate the traditional security services to offer enhanced security service to real-time tasks. Hence, traditional security services are divided into three categories, viz., confidentiality, data integrity, and authentication. The security controller (SC) is fully responsible for combining these security services and their respective category weights, constraint to only one security service can be selected from a category at any instance of time. Moreover, the sum of all category weight equals to 1 (refer Table 1). The security controller evaluates security value, which is known as evaluated security service value (Sr_i^{Eval}), and that must be greater than or equals to the predefined minimum required security service value (Sr_i^{min}).

For example, IDEA cryptographic algorithm can be used in combination with an HMAC-SHA-1 authentication function to enhance the security level of a real-time task (for more details, refer to [29]). However, knowledge of various security categories is highly required to integrate them, which is the responsibility of the developer. The SC uses a **binary integer programming approach** to select an optimal security value with their respective category weight from each category and refer Eq. 1 to calculate enhanced security value. The brief introduction of the binary integer programming approach is given below.

3.2.1 Binary Integer Programming

The binary integer programming approach is used to dynamically choose the best combination of security services from the predefined security categories. Assume that there are three traditional security services called confidentiality (C), authentication (A), and data integrity (I). These security services are involved to find an optimal weighted combined security service value that is also greater than the minimum required value. The main focus of the binary integer programming is to combine the best values with their weights from each category subject to that only one value can

---

[3] Here node has multi-core processor.

be selected at a time. Here, each category values are placed in a separate array (refer Table 1).[4] Our problem of selection of best combination can be formulated into a binary integer programming problem as follows:

Let us assume Y_{jk} is a variable, which represent the jth security value and $1 \leq j \leq Sc_k$ from kth category, where k represents $1 \leq k \leq N_c$ and Pv_{jk} denotes the normalized performance value of the jth security service from kth category [22]. Here, Sc_k represents the total number of security values in kth category and N_c is the total number of categories used. The only constraint is that one security value from a category can be chosen which is described as following:

$$Y_{jk} = \begin{cases} 1, & \text{if the } j\text{th security service from category } k \text{ is selected}, \\ 0, & \text{otherwise.} \end{cases}$$

The evaluated weighted security value function for ith packet is given below:

$$Sr_i^{Eval} = \frac{\sum_{k=1}^{N_c} w_k \times (\sum_{j=1}^{Sc_k} Y_{jk} \times Pv_{jk})}{\max_{k=1}^{N_c} [max\{Pv_k\}]} \geq Sr_i^{mim} \tag{4}$$

where w_k represents the kth category weight[5] and $max\{Pv_k\}$ represents the maximum normalized performance security value of kth category.

Let us assume that the three security service categories named as C, I, A (confidentiality, authentication, and integrity) are used to compute the weighted combined security service value, where each category has distinct security services along with their normalized performance value and the evaluation of minimum-security value depends on the type of communication or real-time applications (application dependent). For example, the railway signaling control systems, air traffic control systems, and radar control systems need more security as compared to the on-line chat system and so on. Therefore, the administrator will first identify the type of communication or type of application, which may require a thorough knowledge of the communication systems. However, minimum-security value is considered in this work is random and based on the existing literatures (refer Table 4) [35]. Hence, the given problem can be formulated as follows:

$$Sr_i^{Eval} = \frac{\sum_{k=1}^{3} w_k \times (\sum_{j=1}^{Sc_k} Y_{jk} \times Pv_{jk})}{\max_{k=1}^{3} [max\{Pv_k\}]} \geq Sr_i^{mim}$$

$$Sr_i^{Eval} = \Big[[\{w_1 Y_{11} Pv_{11} + w_1 Y_{12} Pv_{12} + \cdots + w_1 Y_{18} Pv_{18}\}$$
$$+ \{w_2 Y_{21} Pv_{21} + w_2 Y_{22} Pv_{22} + \cdots + w_2 Y_{27} Pv_{27}\} \tag{5}$$
$$+ \{w_3 Y_{31} Pv_{31} + w_3 Y_{32} Pv_{32} + \cdots + w_3 Y_{37} Pv_{37}\} \Big]$$
$$/ \max_{k=1}^{3} [max\{Pv_k\}] \Big] \geq Sr_i^{min}$$

---

[4] Starting index of an array is zero (0).

[5] sum of all assigned weight of each category equals to 1 [22].

Table 4 Categorization of security services

#Categories\#Services	1	2	3	4	5	6	7	8	9	10	Weight(w_k)	Sr_i^{min}
Confidentiality{C}	0.1	0.2	0.3	0.4	0.5	0.6	0.7	0.8	0.9	–	0.4	
Integrity{I}	0.1	0.2	0.3	0.4	0.5	0.6	0.7	0.8	0.9	1.0	0.2	0.64
Authentication{A}	0.3	0.6	0.9	–	–	–	–	–	–	–	0.4	

by using Table 4 values Sr_i^{Eval} can be calculated as given below:

$$Sr_i^{Eval} = \Big[[0.04Y_{11} + 0.08Y_{12} + 0.12Y_{13} + 0.16Y_{14} + 0.20Y_{15} + 0.24Y_{16} + 0.28Y_{17} + 0.32Y_{18} + 0.36Y_{19}$$
$$+ 0.02Y_{21} + 0.04Y_{22} + 0.06Y_{23} + 0.08Y_{24} + 0.10Y_{25} + 0.12Y_{26} + 0.14Y_{27} + 0.16Y_{28} + 0.18Y_{29} + 0.20Y_{210} \quad (6)$$
$$+ 0.12Y_{31} + 0.24Y_{32} + 0.36Y_{33}]/[1.0]\Big] \geq 0.64$$

subject to only one security value is selected from a category, i.e.,

$$Y_{11} + Y_{12} + Y_{13} + Y_{14} + Y_{15} + Y_{16} + Y_{17} + Y_{18} + Y_{19} = 1$$
$$Y_{21} + Y_{22} + Y_{23} + Y_{24} + Y_{25} + Y_{26} + Y_{27} + Y_{28} + Y_{29} + Y_{210} = 1 \quad (7)$$
$$Y_{31} + Y_{32} + Y_{33} = 1$$

Now solve Eqs. 6 and 7 using binary integer programming:

- $Y_{11} = 0, Y_{12} = 0, Y_{13} = 0, Y_{14} = 0, Y_{15} = 1, Y_{16} = 0, Y_{17} = 0, Y_{18} = 0, Y_{18} = 0$
- $Y_{21} = 0, Y_{22} = 0, Y_{23} = 0, Y_{24} = 1, Y_{25} = 0, Y_{26} = 0, Y_{27} = 0, Y_{28} = 0, Y_{29} = 0, Y_{210} = 0$
- $Y_{31} = 0, Y_{32} = 0, Y_{33} = 1$

$$Sr_i^{Eval} = \frac{0.5 \times 0.4 + 0.4 \times 0.2 + 0.9 \times 0.4}{1.0} = 0.64 \geq 0.64 \quad (8)$$

After solving Eqs. 6 and 7, using binary inter programming [36], we get the most optimal value of Sr_i^{Eval} is 0.64, which is close to 0.64 (minimum required value). The values of variable Y_{jk} at which the Sr_i^{Eval} value becomes most closer to 0.64 are shown in the above and the normalized performance value of selected security values are given in Table 1 using different colors.

3.2.2 Security Overhead

The primary focus of this work is to offer optimal securities to the tasks without violating the predefined scheduling constraints. It is noteworthy that the scheduling and security needs are the crucial aspects of any real-time systems. But offering enhanced security to the real-time tasks increases the original processing time. This delay in processing time is known as security overhead. Hence, security overhead and increased processing can be calculated by Eqs. 2 and 3, respectively (refer Table 2).

When the new processing time is under calculation, meanwhile, the scheduler sends a call in Fog Networks (FN) to find an optimal node using a load distribution model approach, which uses the Min_Heap algorithm (Algo. 1) to find the best optimal node.[6] The detailed description of the Min_Heap algorithm is given in the below sections.

3.3 Load Distribution Model

Load distribution is one of the crucial parameters for every real-time systems. In this study, various constraints have been applied to evenly distribute the load at each individual nodes in the Fog networks. For that Min_Heap algorithm is used to uniformly distribute the load among all the nodes in Fog networks. The Min_Heap algorithm is used to find an optimal node, which has a minimum load value (utilization value) among all other nodes in a network and also less than equals to the current threshold value of the load (λ_N^{thresh}), i.e., $U_{(1,k)}, U_{(2,k)}, \ldots, U_{(r,k)}, \ldots, U_{(m,k)} \leq \lambda_{thresh}^N$, where m is the total number of node in a Fog networks. The load distribution model is fully responsible for uniformly distributing the load, which uses the weighted sum of squares method as [37] to calculate the utilization of a node and Fog network. The load calculation of a node and Fog is given in the below subsection.

3.3.1   Load Calculation

The mathematical formulation of current and average load calculation of a node and Fog using weighted sum of squares method are given in the below equations. The calculation of current utilization of rth node in kth Fog Network using is as

$$U_{(r,k)} = \sqrt[2]{\sum_{i=1}^{I} w_i \times [U_{(pr)}]^2} \quad (\forall\ 1 \leq i \leq I : r \in k) \tag{9}$$

where, $U_{(r,k)}$ illustrates the utilization of rth node in kth Fog network, which has m number of nodes and $1 \leq r \leq m$. The w_i indicates the weight of each parameters used in the calculation of utilization, and $U_{(pr)}$ represents the utilization parameters of node r in the kth Fog network, where $w_i > 0$ and addition of all parameters weight are equals to 1, i.e., $w_1 + w_2 + w_3 + \cdots + w_i = 1$. The i denotes the number of parameters are involved in calculation of node utilization.

This work has considered four parameter for measuring the node's utilization in kth Fog network, i.e., CPU uses (U_{cpu}^r), memory uses (U_{mem}^r), disk uses (U_{disk}^r), and network uses ($U_{n/w}^r$) at rth node. So, the utilization of rth node in kth Fog network

---

[6] The scheduling mode (SM), m-EDF scheduler, FN, task model, and scheduling constraints have developed using Python scriptwriting.

can be calculated as follows:

$$U_{(r,k)} = \sqrt[2]{w_1 \times (U_{cpu}^r)^2 + w_2 \times (U_{mem}^r)^2 + w_3 \times (U_{disk}^r)^2 + w_4 \times (U_{n/w}^r)^2} \quad (10)$$

here, w_1, w_2, w_3, w_4 are the weights of respective utilization parameter. Now, the current utilization of network $U_{(Fog)}^k$, which has K number of Fogs can mathematically be defined as given below.

$$U_{(Fog)}^k = \frac{1}{m} \sum_{r=1}^{m} U_{(r,k)} \quad (\forall\, 1 \le r \le m : k \in K) \quad (11)$$

here, $U_{(Fog)}^k$ denotes the utilization of kth Fog in the network, which has K number of Fogs in a network. Similarly, the utilization value of a node in a Fog (λ_N^{thresh}) can be calculated through below given equation, which is less than or equals to utilization threshold value.

$$\lambda_N^{thresh} = U_{(Fog)}^k + \rho \quad (12)$$

here, ρ represents the standard deviation of node's utilization in kth Fog that is given below:

$$\rho = \sqrt{\frac{1}{m} \sum_{r=1}^{m} (x_r - \bar{x})^2} \quad \forall\, r \in \{1, 2, 3, \ldots, m\} \quad (13)$$

here, x_r is the node's utilization in kth Fog, and $\bar{x}$ is the average utilization value of kth Fog.

3.3.2 Process of Load Distribution Through Min_Heap Algorithm

The Min_Heap algorithm uses the above-defined equations to calculate the utilization nodes and Fogs in a network. The outcome of Min_Heap algorithm is a optimal node. The Min_Heap algorithm spread the nodes in the form of a tree, in which, the node is a pair of $< i, U >$, where i and U represents the level and utilization values of a node in the tree, respectively. The node $< i, U >$ exists at the ith level and Uth position in the tree. The position of an optimal node in the tree is shown in Fig. 2 and pseudo code of same is given in Algorithm 1.

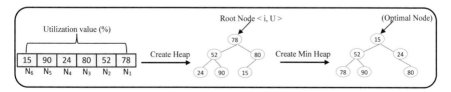

Fig. 2 Process of Min_Heap algorithm to find an optimal node

In the above figure, it is assumed that Fog has six nodes out of which 6th node (N_6) has minimum utilization value (15%). Hence, a new arrival task can be assigned to that node. Once, the new task is assigned to that node then immediately recalculate the utilization values of a node and threshold value. This process is repeated for all the tasks waiting in the global buffer.

If a task is not assigned to any one of the nodes in the Fog networks, it means, either Fog or nodes are currently overloaded in that network, then wait till FN updates its database for utilization values. Once the FN updates its utilization table then immediately recall Min_Heap algorithm (Algorithm 1) to find an optimal node in the Fog networks. The utilization tables updating process of FN are given in the below subsection and Algorithm 1 represents the Pseudo-code of the same.

3.3.3 Updation of Utilization Tables

The FN maintain a utilization table having five attributes viz. Fog number (Fog_{id}), node number (N_{id}), Fog utilization (U_{id}^{Fog}), node utilization (U_{id}^{N}), and number of nodes in a network (N_n^{Fog}). The following steps are used to create or update theses utilization tables.

1. If an entry does not exist for given Fog_{id}, or N_{id} then insert.
2. If Fog_{id}, or N_{id} already exists, then entry against corresponding *id's* has to be updated.

3.3.4 Time Complexity of Min_Heap Algorithm

The time complexity of the Min_heap algorithm is $O(|N| \log_2 |N|)$, where N is the finite number of nodes in a network.

Proof The asymptotic complexity for creating a Heap is $O(|N|)$, where $|N|$ represents the finite number of nodes (refer step 2 to 5). Now this Heap is to be converted into the Min_Heap, so the complexity will be $O(|N| \log_2 |N|)$. In which root node has minimum utilization value compared to its successor node (child nodes) from left and right (refer step 6 to 20). Moreover, the time complexity for deleting k_1 number of existing nodes from a Min_Heap is equals to $k_1 \times$ **height of tree** or $O(k_1 \times \log_2 |N|)$ in both the average and worst cases (refer step 21 to 39), but if k_1 is very very less than N, i.e., $k_1 <<< N$ then time complexity is $O(\log_2 |N|)$. Similarly, the time complexity for inserting k_2 number of nodes in a Min_Heap tree is $O(1)$ in average case and $O(k_2 \times \log_2 |N|)$ in worst case (refer step-40 to 43) but if k_2 is very very less than the N, i.e., $k_2 <<< N$ then the time complexity is $O(\log_2 |N|)$. Hence, the total time complexity of Min_Heap algorithm is $O(|N|) + O(|N| \log_2 |N|) + O(\log_2 |N|) + O(\log_2 |N|) \implies O(|N|) + O(|N| \log_2 |N|) + 2 \times O(\log_2 |N|) \implies O(|N| \log_2 |N|)$ where, $O(|N| \log_2 |N|)$ is much higher as compared to $O(|N|)$ and $2 \times O(\log_2 |N|)$. Finally, the time complexity of Min_Heap algorithm is $O(|N| \log_2 |N|)$.

Algorithm 1: Pseudo code of Min_Heap algorithm

Input : Nodes with initial utilization value.
Output: Node having optimal utilization value.
1 **BEGIN**
2 **Build_Heap (N)**   // Heap creation process from line 2 to 5
3 N.heapsize = N.length
4 **for** $i = \lfloor \frac{|N|}{2} \rfloor$ *to 1* **do**
5 $\quad \lfloor$ Min_Heapify(N, i)

6 **Min_Heapify (N, i)**   // Min_Heap creation process from line 6 to 20
7 l $\leftarrow$ Left(i) and r $\leftarrow$ Right(i)
8 **if** $l \leq |N|$ *and N[l]* < *N[i]* **then**
9 $\quad |$ min $\leftarrow$ l
10 **else**
11 $\quad \lfloor$ min $\leftarrow$ i
12 **if** $r \leq N$ *and N[r] < N[min]* **then**
13 $\quad |$ min $\leftarrow$ r
14 $\quad |$ **if** *min* $\neq$ *i* **then**
15 $\quad |$ $\quad |$ Exchange N[i] and N[min]
16 $\quad |$ **else**
17 $\quad |$ $\quad \lfloor$ break
18 **else**
19 $\quad \lfloor$ break
20 Min_Heapify(N, min)   // Min_Heap process is called for all Nodes
21 **Heap_Extract_Min(N)**   // Node deletion process from line 21 to 30
22 **if** *N.heapsize* < *1* **then**
23 $\quad |$ error "heap underflow"
24 **else**
25 $\quad \lfloor$ break
26 min =(N[1])
27 N[1]= N(N.heapsize)
28 N.heapsize = N.heapsize -1
29 Min_Heapify(N, 1)
30 return (min)
31 **Heap_Decrease_Key(N, i, key)**   // Node insertion process from line 31 to 43
32 **if** *key > N[i]* **then**
33 $\quad |$ error "new key (utilization value) is greater than current key"
34 **else**
35 $\quad \lfloor$ break
36 N[i] = key
37 **while** $i > 1$ *and N[parent(i)] > N[i]* **do**
38 $\quad \lfloor$ exchange N[i] and N[parent(i)]

39 i = parent(i)
40 **Min_Heap_Insert(N, key)**
41 N.heapsize= N.heapsize + 1
42 N[N.heapsize]= +∞
43 Heap-Decrease-Key(N, N.heapsize, key)
44 **END**

3.4 Scheduling Constraints

The newly arrived tasks are scheduled according to m-EDF scheduler and scheduling constraints based on its processing time and absolute deadlines. The m-EDF scheduler is used for scheduling the non-primitives and aperiodic tasks, which offers better scheduling results as compared to traditional EDF scheduler. Due to some constraints are used to identify whether the task is schedulable or not (refer definition). The brief descriptions of scheduling constraints are given below.

Definition 1 Scheduling constraint: A task $t_i \in T$ is said to be pass in scheduling test at rth optimal node in kth Fog, if and only if it satisfies the following conditions [21, 30, 34]:

1. The current utilization of rth optimal node for ith task should less than equals to 1.
2. The tasks whose execution order are later than ith task in a accepted queue of rth optimal node should have a utilization value less than equals to 1.

The mathematical representation of aforesaid conditions are as follows:

$$
\begin{aligned}
&F_i^r \leq D_i, \quad and \\
&\forall \ t_n \ where \ o_n > o_i \ and \ q_n = 1 : F_n^r \leq D_n
\end{aligned}
\tag{14}
$$

here, F_i^r is the finish time of ith task at rth optimal node in a Fog and are calculated using Eq. 15 given below. The o_i and o_n are the indexed number of ith and nth task, respectively, such that n = (i + 1), (i + 2), and so on. The $q_n = 1$ represents the total number of tasks waiting in the local queue of a node whose execution orders are later than the ith task. The calculation of finish time for ith task at rth node can be defined as follows:

$$
F_i^r = S_i^r + Ne_{n_i}
\tag{15}
$$

here, S_i^r is the start time of ith task at rth optimal node in a Fog and Ne_{n_i} is the new processing time for ith task. The start time is defined as follows:

$$
S_i^r = a_i^r + Re_m^r + \sum_{n=m+1; \ o_n^r < o_i^r \ and \ q_n=1}^{i-1} Ne_n^r
\tag{16}
$$

where, Re_m^r denotes the remaining processing time of current running task at rth optimal node in a Fog. The a_i^r and S_i^r are arrival and start times for ith task at rth optimal node. And the Ne_n^r is denotes the total new processing time of all the tasks that are waiting in the local queue of rth optimal node.

Definition 2 m-EDF scheduler: The task which has minimum processing time will get selected first, if two or more tasks have same deadlines.

It becomes very difficult to select one task from the global buffer, when two or more tasks have same deadlines through traditional EDF scheduler. Therefore, m-EDF scheduler has been proposed and used in this work [34]. It has also reduced the total average waiting time besides removing the ambiguity of selecting the tasks from global buffer [34].

4 Proposed SDSM Algorithm

The pseudocode of the proposed SDSM algorithm is given in Algorithm 2. The proposed SDSM algorithm is the concatenation of a security model, load distribution model, and scheduling constraints along with an m-EDF scheduler. The input tasks are taken from a global buffer, and output tasks are stored in the respective local accepted and rejected queues of the nodes.

The primary focus of SDSM is to offer enhanced security and scheduling through a load distribution mechanism for any real-time tasks. The m-EDF scheduling policy is used to sort (increasing order of their deadlines) the tasks in a global buffer. Moreover, the m-EDF scheduler picks a task from the global buffer and sent request messages to the security controller (SC) for calculating the optimal security value through binary integer programming. Simultaneously sends a request command to FN to find an optimal node in the Fog network according to Min_Heap algorithm.

As long as security controller obtained the optimal security value for ith task (refer Eq. 1), meanwhile, FN finds an optimal node r (refer Fig. 2 and Algorithm). Now the scheduling constraint test is admitted, whether ith task with applied optimal security value passed the scheduling constraint test at rth optimal node or not. If yes, means ith task passes the test then assign ith task to rth optimal node and simultaneously increase the counter value of accepted tasks in rth node's local queue. Else, drop the task, and increase the counter value of rejected tasks in the rth node's local queue. The counter values of the accepted queue and rejected queue are used to calculate the success ratio of rth node and Fogs in a network. It is mandatory that whenever a task is assigned or released from a node, the recalculation of node's utilization value, average utilization value of a Fog and a threshold value of utilization for a node happens immediately and also updates the utilization table of FN subsequently. A similar process is repeated for all the tasks stored in a global buffer.

4.1 Mathematical Example

A set of eight tasks have been considered to simplify the task assignment process in the respective Fogs, and the same is given in the first part of Table 5. Moreover, the second part of the Table shows the sorted order (according to deadlines) of tasks. A set of eleven tasks have considered simplifying the process of task assignment in the respective Fogs, and the same is shown in the first part of Table 5. The second part of

---

Algorithm 2: Pseudo code of SDSM algorithm (*Task $t_i \in T$*)

Input : Select a task from global buffer (G_b)
Output: Counter value of accepted queue at rth node : q_a^r
Output: Counter value of rejected queue at rth node : q_r^r
1 **BEGIN**
2 **Select** $t_i \in T$ *from* G_b // G_b is the global buffer
3 m-EDF scheduler call SC and FN
4 SC calculates the optimal security value (Sr_i^{Eval}) through eq. 1
5 Simultaneously, FN find rth optimal node in the kth Fog in a network using Algorithm 1
6 Calculate start time (S_i^r) of ith task at rth optimal node in the kth Fog in a network using Eq. 16
7 **Procedure** Call SDSM for task $t_i \in T$
8 **while** *($G_b \neq 0$ or (NULL))* **do**
9   **if** *($U_{(r,k)} \leq \lambda_N^{thresh}$)* **then**
10   rth optimal node is selected in kth Fog
11   **if** *($t_i^r \vdash Definition1$)* **then**
12   assign ith task to rth optimal node
13   q_a^{r++} and G_b^{--}
14   FN, immediately update its the utilization tables
15   goto **Procedure**
16   **else**
17   drop task t_i
18   q_r^{r++} and G_b^{--}
19   goto **Procedure**
20   **else**
21   current utilization of nodes in a Fog network is higher
22   wait till updating of utilization tables
23 **END**

Table 5 Set of 5 tasks with predefined and calculated values

# Task	a_i	e_i	D_i	Sr_i^{min}
t_1	01	08	20	0.53
t_2	04	11	30	0.25
t_3	**06**	**15**	**40**	**0.46**
t_4	07	20	60	0.68
t_5	09	25	70	0.73

Existing values

Table 5 shows the sorted order (according to deadlines) of tasks. Table 6 depicts the assignment of tasks on various nodes in a Fog as per the SDSM algorithm, which has also included the evaluated values of start time, finish time, node number, and Fog id values according to above said equations. However, tasks t_3 and t_8 have not satisfied the scheduling constraint test on rth optimal node. Therefore, task t_3 and t_8 are not assigned to any node in a Fog and inserted into the dropped queue.

Table 6 Task assignment according to SDSM algorithm

# Task	Existing values				Calculated values						
	a_i	e_i	d_i	Sr_i^{min}	Sr_i^{Eval}	So_i	Ne_i	S_i^r	F_i^r	N_{id}	Fog_{id}
t_2	01	08	20	0.53	0.53	4.24	12.24	1.0	13.24	2	3
t_1	04	11	30	0.25	0.25	2.75	13.75	4.0	17.75	8	1
t_3	06	15	40	0.46	0.46	6.90	21.90	6.0	28.90	3	2
t_4	07	20	60	0.68	0.68	13.6	33.60	17.75	51.35	1	1
t_5	09	25	70	0.73	0.73	18.25	43.25	28.90	72.15	Nil	Nil

Table 7 Simulation parameters

Settings	Value
# Fog networks	03
# Nodes	20–30 nodes par Fog
Arrival rate	$10–90\ s^{-1}$
Packet size	20–100kB
Deadline	50–1200ms
Bandwidth	10 Mbps
Average security value	0.1–1.0

5 Simulation Results and Analysis

The NS-3 simulation tool with *python* script is used for analytical result analysis. Further, these results are compared with existing similar kinds of traditional algorithms. The experimental results using synthetic data set are compared to the existing four baseline algorithms.[7] Hence, the proposed SDSM algorithm can be compared to any baseline algorithm, if the data set of that algorithm is available, i.e., SPSS [31], ISAPS [32], SAREG [33], and SLOPE [34], which shows the novelty of proposed SDSM algorithm.

The statistical analysis of the results has been through a statistical tool called ANOVA. The novelty of the proposed SDSM algorithm is proved in terms of (A) Success ratio, (B) Average security value, and (C) The overall performance. The simulation parameters used are as follows (refer Table 7).

---

[7] Author have selected these baseline algorithms due to having sufficient available synthetic data set.

5.1 Success Ratio

The success ratio is a direct measurement of the number of tasks that have been executed at a node and Fog, respectively. It means the success ratio is an indirect measurement of nonexecuted tasks (rejected) and a direct measurement of accepted tasks. The success ratio of a node and Fog respectively can be calculated through Eq. 17. The value of the success ratio is in percentage [21, 32].

$$Sr_{(r,k)} = \left[\frac{Number\ of\ accepted\ tasks}{Total\ number\ of\ tasks} \right] \times 100\% \tag{17}$$

here, $Sr_{(r,k)}$ denotes the success ratio of rth node in kth Fog network. Similarly, the success ratio of a Fog network can be evaluated as follows:

$$Sr_k = \frac{\sum_{r=1}^{m} Sr_{(r,k)}}{m} \% \tag{18}$$

here, Sr_k illustrates the success ratio of kth Fog network in percentage (%), where m represents the total number of nodes in a Fog network.

5.2 Average Security Value

The average security value is a fraction of the total security value applied on accepted tasks and the total number of accepted and rejected tasks at rth node. The average security value illustrates the efficiency of the proposed SDSM algorithm and can be evaluated as follows:

$$Asv_{(r,k)} = \left[\frac{Security\ value\ of\ accepted\ tasks}{Total\ number\ of\ accepted\ and\ rejected\ tasks} \right] \times 100\% \tag{19}$$

here, $Asv_{(r,k)}$ represents the average security value of rth node in kth Fog. Similarly, average security value of kth Fog network can be evaluated as follows:

$$Asv_k = \frac{\sum_{r=1}^{m} Asv_{(r,k)}}{m} \% \tag{20}$$

here, Asv_k is the average security value of kth Fog and m represents the total number of nodes in a Fog network.[8]

This study has shown that the total number of resources (nodes and Fogs) is directly proportional to the success ratio and overall performance.

---

[8] This study has considered three Fog networks for simulation.

5.3 Overall Performance

The cross production of success ratio and average security value represents the overall performance. The overall performance of a node and Fog network can be calculated by Eq. 21 and denoted as $OP_{(r,k)}$ and OP_k, respectively.

$$
\begin{aligned}
OP_{(r,k)} &= \frac{Sr_{(r,k)} \times Asv_{(r,k)}}{100} \% \\
OP_k &= \frac{Sr_k \times Asv_k}{100} \%
\end{aligned}
\tag{21}
$$

here, $OP_{(r,k)}$ illustrates the overall performance of rth node in kth Fog and OP_k is denotes the overall performance of kth Fog network.

The impact of various parameters, i.e., the deadline of tasks, task arrival rate, task size, and the number of resources on success ratio, average security value, and overall performance are addressed in the below subsections, respectively.

5.4 Impact of Deadline on Success Ratio, Average Security Value, and Overall Performance

The impact of the task's deadline on success ratio, average security value, and overall performance is shown in Fig. 3 for deadline range 50–1200 ms while keeping other parameters are fixed. The figure illustrates the comparative result analysis for SPSS, ISAPS, SAREG, SLOPE, and proposed SDSM algorithms, respectively, which proves the novelty of the SDSM algorithm. However, a detailed description of each parameter is given in the below sub-figures.

Figure 3a shows the impact of the deadline on the success ratio for the deadline range 50–1200 ms. The deadline is directly proportional to the waiting time of a task, which represents that the task with the maximum deadline and minimum execution time have more chance to pass the scheduling constraint test, which enhances the success ratio. The performance of SDSM in case of a maximum deadline, i.e., 1200 ms is 1.88% increased than SLOPE, which shows the improvements as compared to the baseline algorithms, i.e., SAREG, ISAPS, and SPSS, respectively. Similarly, the performance of SDSM is 7.14% better than SLOPE, which shows a significant improvement as compared to the baseline algorithms in case of a minimum deadline, i.e., 50 ms.

The impact of a deadline on average security value has been given in Fig. 3b for deadline range 50–1200 ms. The proposed SDSM algorithm offers enhanced security value compared to the baseline algorithms. The average security value is directly proportional to the absolute deadline, which shows that an optimal security value is assigned in case of a maximum deadline, i.e., 1200 ms and vice versa. In case of a maximum deadline, the performance of the SDSM algorithm has been 5.3% increased as compared to the SLOPE, which illustrates the improvements in the proposed SDSM algorithm.

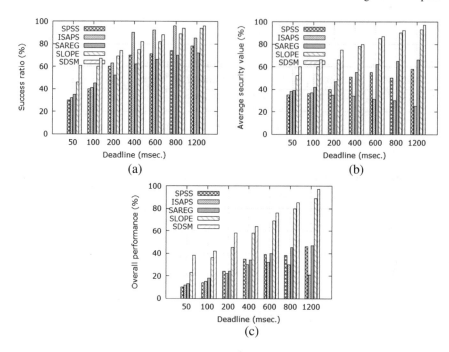

Fig. 3 Impact of deadline, when arrival rate = 45 s^{-1}, task size = 50 KB, and bandwidth = 10 Mbps

The overall performance[9] has shown in Figure 3c. The experimental results of SDSM are compared to the baseline algorithms to prove the novelty. The overall performance of SDSM has increased due to optimal (more accurate) security value as needed compared to the baseline algorithms.

The aforesaid statements have proved the novelty of the proposed SDSM algorithm against the baseline algorithms in terms of success ratio, average security value, and overall performance.

5.5 Impact of Task Arrival Rate on Guarantee Ratio, Average Security Value, and Overall Performance

The impact of task arrival rate on success ratio, average security value, and overall performance is illustrated in Fig. 4 for 10–90 tasks per second while other parameters are remain fixed. This figure depicts the comparative analysis of baseline algorithms and the proposed SDSM algorithm to proves the novelty. A brief description is given in the below sub-figures.

---

[9] Overall performance is measured in terms of Fog networks.

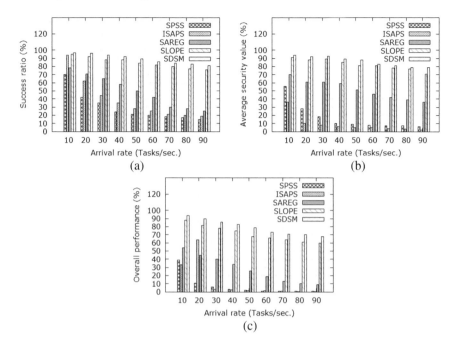

Fig. 4 Impact of task arrival rate, when deadline = 900 ms, task size = 50 KB, and bandwidth = 10 Mbps

Figure 4a represents, the impact of arrival rate on success ratio for 10–90 tasks per second. The task arrival rate is directly proportional to the success ratio, which shows that fewer tasks arrived per second have maximum chances to satisfy the scheduling constraints that enhances the success ratio. The performance of SDSM for max task arrival rate, i.e., 90 tasks per second has significant improvement compared to the baseline algorithms and vice versa.

The impact of task arrival rate on average security value is given in Fig. 4b for 10–90 tasks per second. The proposed SDSM algorithm provides enhanced security as compared to the baseline algorithms due to not have any security reduction policy in all the cases that is maximum task arrival rate or minimum task arrival rate, while others have. The average security value is directly proportional to the number of tasks that arrived per second. It means when the task arrival rate is maximum, standard security can be assigned. Similarly, robust security can be assigned in case of a minimum arrival rate, which shows that the performance of SDSM has been enhanced compared to the baseline algorithms.

Similarly, overall performance is illustrated in Fig. 4c. The experimental results of SDSM are compared to the existing baseline algorithms to prove the novelty in terms of success ratio, average security value, and overall performance. The overall performance of SDSM has enhanced due to offering optimal security in both cases.

5.6 Impact of Task Size on Success Ratio, Average Security Value, and Overall Performance

The impact of task size on success ratio, average security value, and overall performance has illustrated in Fig. 5 for 10–100 KB task size, and keeping other aspects are remained fixed. The figure shows the comparative analysis of experimental results of baseline algorithms and proposed SDSM algorithms to proves the novelty. A brief description of each aspect has been given in the below sub-figures.

Figure 5a shows the impact of task size on the success ratio for task size of 10–100 KB. More execution time is needed to initialize security for large task size, which may increase the task execution time. The delay in execution time can increase the total execution time of a task and have more chances to violate the scheduling constraints. This will reduce the success ratio of tasks and overall performance. Hence, it can be noticed that the success ratio is inversely proportional to task size. The SDSM has provided an enhanced success ratio due to having the best load balancing mechanism in Fog networks. Therefore, the success ratio and overall performance of SDSM has been increased as compared to the baseline algorithms for both max and min task sizes.

The impact of task size on average security value has given in Fig. 5b for task size of 10–100 KB. A max task size needed more execution time due to having max security value, which can generate maximum workload on a node. Therefore, the size of a task is directly proportional to the workload of a node. It has been noticed that the baseline algorithm reduced the security values for max workload conditions, which may reduce the overall security of a task. Hence, in both cases, the performance of the proposed SDSM algorithm is much higher than the baseline algorithms, which proves the novelty.

The overall performance of a node has been given in Fig. 5c. The experimental results of SDSM have been compared with the baseline algorithms to prove the novelty. The overall performance has depended on the success ratio and average security value of a task. Hence, for both the cases (higher workload and lower workload), the success ratio and average security value has been increased, which proves the novelty of SDSM.

6 Conclusion and Future Works

The main objective of this study is to provide enhanced security service without violating scheduling constraints. For this, A Security-Driven Scheduling Model for Delay-Sensitive Tasks in Fog Networks has been introduced and abbreviated as SDSM. In SDSM, each task has uniformly distributed among the nodes such that optimal utilization can be possible. But the baseline algorithms have the policy to reduce the assigned security service in case of higher workload conditions, and focus has shifted towards scheduling only. Therefore, security threats can be possible in

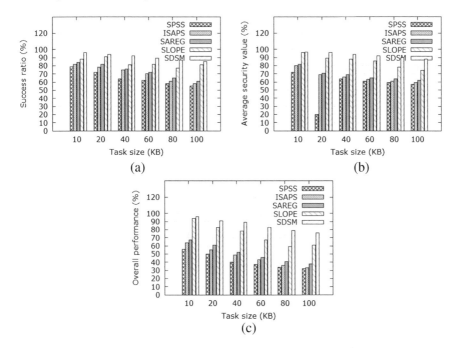

Fig. 5 Impact of task size, when deadline = 900 ms, task arrival rate = 45 s^{-1}, and bandwidth = 10 Mbps

case of a lack of security. These issues have been solved using the SDSM algorithm, which has no security reduction mechanism either in a higher workload situation because SDSM using an optimal load balancing (Min_Heap) algorithm (refer to Eq. 1). The experimental result analysis has illustrated that enhancement in success ratio and overall performance along with increased security without violating the scheduling constraints even in higher workload conditions. Hence, the novelty of the proposed SDSM algorithm has been proven in terms of success ratio, average security value, and overall performance.

In the future, this work can extend for minimizing the energy consumptions for battery-operated systems.

References

1. Faller, W.E., Schreck, S.J.: Real-time prediction of unsteady aerodynamics: application for aircraft control and manoeuvrability enhancement. IEEE Trans. Neural Netw. **6**(6), 1461–1468 (1995)
2. Gregor, T., Sergej, V.: Railway signalling & interlocking. In: International Compendium, Hamburg, p. 448. Eurail-press Publ. (2009)

3. Mo, Y., Kim, T.H., Brancik, K., Dickinson, D., Lee, H., Perrig, A., Sinopoli, B.: Cyber-physical security of a smart grid infrastructure. Proc. IEEE **100**(1), 195–209 (2012)
4. Mahafza, B., Welstead, S., Champagne, D., Manadhar, R., Worthington, T., Campbell, S.: Real-time radar signal simulation for the ground based radar for national missile defense. In: Proceedings of the 1998 IEEE Radar Conference, RADARCON'98. Challenges in Radar Systems and Solutions (Cat. No.98CH36197), pp. 62–67 (1998)
5. Nilsson, J., Dahlgren, F.: Improving performance of load-store sequences for transaction processing workloads on multiprocessors. In: Proceedings of the 1999 International Conference on Parallel Processing, pp. 246–255 (1999)
6. Son, S.H., Zimmerman, R., Hansson, J.: An adaptable security manager for real-time transactions. In: Proceedings 12th Euromicro Conference on Real-Time Systems. Euromicro RTS 2000, pp. 63–70 (2000)
7. Singh, G.: A study of encryption algorithms (RSA, DES, 3DES and AES) for information security. Int. J. Comput. Appl. **67**(19) (2013)
8. Dulana, R., Hassan, S.: Task allocation, migration and scheduling for energy-efficient real-time multiprocessor architectures. J. Syst. Archit. **98**, 17–26 (2019)
9. Liu, F., Narayanan, A., Bai, Q.: Real-time systems (2000)
10. Xie, T., Sung, A., Qin, X.: Dynamic task scheduling with security awareness in real-time systems. In: 19th IEEE International Parallel and Distributed Processing Symposium, pp. 8–14 (2005)
11. Singh, S., Ranvijay: Improve real-time packet scheduling algorithm with security constraint. In: 2014 Annual IEEE India Conference (INDICON), pp. 1–6 (2014)
12. Syed, M., Fernández, E., Ilyas, M.: A Pattern for Fog Computing, pp. 1–10 (2016)
13. Huang, R., Sun, Y., Huang, C., Zhao, G., Ma, Y.: A survey on fog computing. In: Wang, G., Feng, J., Alam Bhuiyan, M.Z., Lu, R. (eds.) Security, Privacy, and Anonymity in Computation, Communication, and Storage, pp. 160–169 (2019)
14. Yi, S., Li, C., Li, Q.: A survey of fog computing. In: Proceedings of the 2015 Workshop on Mobile Big Data—Mobidata-15. ACM Press (2015)
15. Mouradian, C., Naboulsi, D., Yangui, S., Glitho, R., Morrow, M., Polakos, P.: A comprehensive survey on fog computing: state-of-the-art and research challenges. IEEE Commun. Surv. Tutor. (2017)
16. Mehta, H., Kanungo, P., Chandwani, M.: Performance enhancement of scheduling algorithms in web server clusters using improved dynamic load balancing policies. In: 2nd National Conference, INDIACom-2008 Computing For Nation Development, pp. 651–656 (2008)
17. Ababneh, M., Hassan, S., Bani-Ahmad, S.: On static scheduling of tasks in real time multiprocessor systems: an improved GA-based approach. Int. Arab. J. Inf. Technol. **11**(6), 560–572 (2014)
18. Mittal, A., Manimaran, G., Siva Ram Murthy, C.: Integrated dynamic scheduling of hard and QoS degradable real-time tasks in multiprocessor systems. J. Syst. Archit. **46**(9), 793–807 (2000)
19. Casavant, T.L., Kuhl, J.G.: A taxonomy of scheduling in general-purpose distributed computing systems. IEEE Trans. Softw. Eng. **14**(2), 141–154 (1988)
20. Saleh, M., Dong, L.: Real-time scheduling with security enhancement for packet switched networks. IEEE Trans. Netw. Serv. Manag. **10**(3), 271–285 (2013)
21. Singh, S., Tripathi, S., Batabyal, S.: Secured dynamic scheduling algorithm for real-time applications on grid. In: Ray, I., Gaur, M.S., Conti, M., Sanghi, D., Kamakoti, V. (eds.) Information Systems Security, pp. 283–300. Springer (2016)
22. Lin, M., Xu, L., Yang, L.T., Qin, X., Zheng, N., Wu, Z., Qiu, M.: Static security optimization for real-time systems. IEEE Trans. Ind. Informatics **5**(1), 22–37 (2009)
23. Krishna, C.M.: Real-Time Systems (1999)
24. Burns, A.: Scheduling hard real-time systems: a review. Softw. Eng. J. **6**(3), 116–128 (1991)
25. Lee, Y.-H., Leu, S., Chang, R.-S.: Improving job scheduling algorithms in a grid environment. Future Gen. Comput. Syst. **27**(8), 991–998 (2011)

26. Chetto, M., Marchand, A.: Dynamic scheduling of skippable periodic tasks in weakly-hard real-time systems. In: 14th Annual IEEE International Conference and Workshops on the Engineering of Computer-Based Systems (ECBS'07), pp. 171–177 (2007)
27. Marchand, A., Chetto, M.: Dynamic scheduling of periodic skippable tasks in an overloaded real-time system. In: 2008 IEEE/ACS International Conference on Computer Systems and Applications, pp. 456–464 (2008)
28. Bai, L., Hu, Y., Lao, S.-Y., Zhang, W.: Task scheduling with load balancing using multiple ant colonies optimization in grid computing. In: 2010 Sixth International Conference on Natural Computation, vol. 5, pp. 2715–2719 (2010)
29. Singh, S., Tripathi, S., Batabyal, S.: Utilization based secured dynamic scheduling algorithm for real-time applications on grid (u-SDSA). In: 2017 IEEE 31st International Conference on Advanced Information Networking and Applications (AINA), pp. 606–613 (2017)
30. Singh, S., Batabyal, S., Tripathi, S.: Security aware dynamic scheduling algorithm (SADSA) for real-time applications on grid. Cluster Comput. 1–17 (2019)
31. Qin, X., Alghamdi, M., Nijim, M., Zong, Z., Bellam, K., Ruan, X., Manzanares, A.: Improving security of real-time wireless networks through packet scheduling [transactions letters]. IEEE Trans. Wirel. Commun. **7**(9), 3273–3279 (2008)
32. Zhu, X., Guo, H., Liang, S., Yang, X.: An improved security-aware packet scheduling algorithm in real-time wireless networks. Inf. Process. Lett. **112**(7), 282–288 (2012)
33. Xie, T., Qin, X.: Enhancing security of real-time applications on grids through dynamic scheduling. In: Feitelson, D., Frachtenberg, E., Rudolph, L., Schwiegelshohn, U. (eds.) Job Scheduling Strategies for Parallel Processing, pp. 219–237 (2005)
34. Singh, S., Tripathi, S.: SLOPE: secure and load optimized packet scheduling model in a grid environment. J. Syst. Archit. **91**, 41–52 (2018)
35. Xie, T., Qin, X., Sung, A.: SAREC: a security-aware scheduling strategy for real-time applications on clusters. In: 2005 International Conference on Parallel Processing (ICPP'05), pp. 5–12 (2005)
36. MIT. Binary Integer Programming. http://web.mit.edu/15.053/www/AMP-Chapter-09.pdf
37. Chang, R.S., Lee, Y.H., Leu, S.: Improving job scheduling algorithms in a grid environment. Future Gen. Comput. Syst. **27**(8), 991–998 (2011)

Printed in the United States
by Baker & Taylor Publisher Services